ANNUAL REPORTS ON THE PROGRESS OF CHEMISTRY

SECTION B

Annual Reports on the Progress of Chemistry Volume 76, 1979

SECTION B
Organic Chemistry

The Royal Society of Chemistry
Burlington House, London W1V 0BN

British Library Cataloguing in Publication Data

Annual reports on the progress of chemistry.
Section B: Organic chemistry.
Vol. 76: 1979.
1. Chemistry, Organic
I. Royal Society of Chemistry
547 QD251.2

ISBN 0-85186-111-3
ISSN 0069-3030

Set in Times on Linotron and printed offset by
J. W. Arrowsmith Ltd., Bristol, England

Made in Great Britain

Contributors

Senior Reporters

G. P. Moss, B.Sc., Ph.D., A.R.C.S., *Queen Mary College* (*University of London*)

J. H. P. Utley, B.Sc., Ph.D., D.Sc., C.Chem., F.R.S.C., *Queen Mary College* (*University of London*)

Reporters

R. Baker, B.Sc., Ph.D., *The University, Southampton*

R. Bolton, B.Sc., Ph.D., *Bedford College* (*University of London*)

A. J. Boulton, B.A., Ph.D., *University of East Anglia*

J. W. S. Bradshaw, B.A., Ph.D., *The University, Southampton*

R. Brettle, B.Sc., M.A., D.Phil., C.Chem., F.R.C.S., *The University, Sheffield*

D. G. Buckley, M.Sc., Ph.D., A.R.A.C.I., *Queen Mary College* (*University of London*)

R. J. Bushby, D.Phil., B.Sc., A.R.C.S., *The University, Leeds*

C. Christophersen, Cand. Scient., *University of Copenhagen, Denmark*

H. M. Colquhoun, M.A., Ph.D., *ICI Corporate Laboratory, Runcorn, Cheshire*

M. J. Cook, B.Sc., Ph.D., *University of East Anglia*

A. Cox, M.A., B.Sc., Ph.D., D.I.C., *University of Warwick*

A. G. Davies, B.Sc., Ph.D., D.Sc., *University College London*

M. J. Gait, B.Sc., Ph.D., *M.R.C. Laboratory of Molecular Biology, Cambridge*

C. D. Hall, B.Sc., Ph.D., *King's College* (*University of London*)

J. R. Hanson, M.A., B.Sc., Ph.D., D.Sc., C.Chem., F.R.S.C., *University of Sussex*

J. Holton, B.Sc., D.Phil., *ICI Corporate Laboratory, Runcorn, Cheshire*

H. R. Hudson, Ph.D., D.Sc., C.Chem., F.R.S.C., *The Polytechnic of North London*

N. Jacobsen, Ph.D., *Cheminova Ltd., Lemvig, Denmark*

M. Jarman, Ph.D., *Institute of Cancer Research, Sutton, Surrey*

D. W. Knight, B.Sc., Ph.D., *The University, Nottingham*

P. F. Leadlay, M.A., D.Phil., *University of Cambridge*

R. Lines, B.Sc., Ph.D., *Raychem Ltd., Dorcan, Wilts*

S. Neidle, B.Sc., Ph.D., A.R.C.S., D.I.C., *King's College* (*University of London*)

W. E. Paget, B.Sc., Ph.D., *University College of Swansea*
H. S. Rzepa, B.Sc., A.R.C.S., Ph.D., D.I.C., *Imperial College* (*University of London*)
J. K. M. Sanders, B.Sc., M.A., Ph.D., A.R.C.S., C.Chem., F.R.S.C., *University of Cambridge*
G. Shaw, B.Sc., Ph.D., D.Sc., A.R.C.S., D.I.C., *University of Bradford*
K. Smith, B.Sc., M.Sc., Ph.D., C.Chem., M.R.S.C., *University College of Swansea*
A. R. Tatchell, M.Sc., Ph.D., C.Chem., F.R.S.C., *Thames Polytechnic, London*
K. J. Toyne, B.Sc., Ph.D., *University of Hull*
M. V. Twigg, B.Sc., Ph.D., *ICI Agricultural Division, Billingham, Cleveland*
J. L. Wardell, B.Sc., Ph.D., *University of Aberdeen*

Preface

Up until 1967 *Annual Reports on the Progress of Chemistry* had been published as a single volume. In that year it was recognized that coverage of the major advances in the whole field of chemistry in any one year could no longer easily be encompassed within the confine of one volume, and *Annual Reports* were split into two separate volumes: Section A – General Physical, and Inorganic Chemistry; and Section B – Organic Chemistry. For the same general reason, and also for the specific reasons outlined in the Prefaces to the 1977 and 1978 *Reports*, a further subdivision has now been introduced into the 1979 *Reports*, *viz*. Section A – Inorganic Chemistry, Section C – Physical Chemistry.

Thus in Section A, for which Professor D. C. Bradley continues as Senior Reporter, more space is available for the coverage of inorganic chemistry and the text of the 1979 Reports takes up 260 pages (*cf.* 208 for 1978 and 164 for 1977). The new Section C consists of 9 chapters covering 313 pages and the general philosophy behind their choice is described in the Introduction to that Section by Professor M. C. R. Symons, whose services as Senior Reporter we were indeed fortunate to obtain. He has been successful in recruiting some distinguished physical chemists as Reporters for this first *Annual Reports* volume devoted solely to physical chemistry. The response to this volume will be observed with considerable interest; in particular, of course in this hard commercial world, the sales will need to be monitored.

Section B continues with the same broad policy for the coverage of organic chemistry as in recent years, and the Senior Reporters' Introduction outlines the areas highlighted in 1979. With these *Reports* Dr. J. H. P. Utley completes his term of office as Senior Reporter and we thank him warmly for his hard work and willing co-operation over the past four years. Dr. G. P. Moss will provide continuity in the present policy by collaborating with Dr. Utley's successor for Section B of the 1980 *Reports*.

It is doubly unfortunate this year, which sees an important development in the evolution of *Annual Reports*, that industrial problems in the printing industry in the early part of 1980 should have seriously delayed production. An effort was made to counter this, at least partially, by restricting proofing to one stage, but at the time of going to press it seems that publication of all three Sections, particularly the long Section B, will regrettably be nearer the end of the Autumn than the beginning. We trust that subscribers will be patient in the face of these delays and appreciate that every effort is being, and will be, made to overcome these difficulties.

Contents

Section A, Inorganic Chemistry, contains the following items

1 Introduction. By D. C. Bradley

2 The Typical Elements

Part (i) Groups I, IIA, and IIB. By F. A. Hart

Part (ii) Group III. By A. G. Massey

Part (iii) Groups IV and V. By P. G. Harrison

Part (iv) Groups VI, VII, and VIII. By J. H. Holloway

3 The Transition Elements

Part (i) Scandium, Yttrium, the Lanthanides, and the Actinides. By F. A. Hart

Part (ii) Groups IVA, VA, and VIA. By J. E. Newbery

Part (iii) Groups VIIA, VIIIA, and IB. By P. Thornton

4 Organometallic Chemistry

Introduction. By D. C. Bradley

Part (i) Alkyls, Aryls, Carbonyls, Cyanides, Carbenes, and Carbynes. By J. Evans

Part (ii) π-Bonded Organometallics. By A. J. Deeming

5 Radiochemistry. By D. S. Urch

Section C, Physical Chemistry, contains the following items

1 Introduction. By M. C. R. Symons

2 Catalysis. By D. A. Dowden

3 Gas-phase Molecular Spectroscopy. By M. T. Macpherson and R. F. Barrow

4 Spectroscopic Studies of Intermolecular Forces in Dense Phases. By J. Yarwood

5 Kinetics of Reactions in Solution. By J. R. Jones and J. E. Crooks

6 Molecular Structure by Diffraction Methods. By M. R. Truter

7 Neutron and *X*-Ray Diffraction Studies of Concentrated Aqueous Electrolyte Solutions. By G. W. Neilson and J. E. Enderby

8 Macromolecular Chemistry. By E. A. Boucher

9 Thermochemistry. By M. N. Jones and H. A. Skinner

10 Electron Solvation Phenomena. By B. Webster

1 Introduction

By G. P. MOSS and J. H. P. UTLEY
Department of Chemistry, Queen Mary College (University of London)
Mile End Road, London E1 4NS

Continuing our policy of selective rather than comprehensive coverage, once again readers are referred to the Society's series of *Specialist Periodical Reports* for further details of most of the topics covered in this volume. Chapters on organoboron chemistry, organophosphorus chemistry, and terpenoids and steroids are included this year to review the recent advances which have been largely confined to the appropriate SPR volumes. The first two are timely in view of the award of the 1979 Nobel Prize for Chemistry to H. C. Brown and G. Wittig.

In the biological chemistry section, it is the turn of alkaloids and insect chemistry for special coverage. A chapter on nucleic acids is re-introduced, together with a timely report on marine natural products.

The year has seen advances in the application of two-dimensional n.m.r. and solid-state n.m.r. to organic chemistry. When combined with various difference techniques, it is possible to assign completely all the signals of a complex molecule, together with all the coupling constants. Determination of the structure of a hexanucleotide fragment of DNA by *X*-ray crystallography is expected to lead to a radical review of many ideas in molecular biology.

Our attention has been drawn to the advances in the use of organometallic reagents in aliphatic chemistry, with special comment on hydroalumination and the use of dilithiopropyne. Other advances in this area include the use of h.p.l.c. and g.l.c. for resolution and asymmetric synthesis with an enantiomer excess >60%.

In enzyme chemistry, the first really detailed analysis of an enzyme-catalysed transfer of a methyl group has appeared. The synthesis of the first pure diastereomer of cyclic AMP is a landmark in the many studies of the stereochemistry of reactions at phosphorus.

The chapter on organoboron chemistry draws attention to the emergence of four-co-ordinate borate salts as key reagents for synthetically useful transformations and, more recently, the use of vinyloxy- and vinylamino-boranes as enolates with very specific reactivity.

On the mechanistic side, special attention has been paid to the intrinsic properties of polar reactions in the gas phase. In solution and in the solid state, detailed information on the nature of reaction intermediates continues to accumulate; the first high-resolution ^{13}C n.m.r. spectrum of a carbo-cation has been reported. The effects of micellar catalysis in various reactions are becoming better understood.

The substitution of H by Li and of C by Si continues to provide theoretical predictions as targets for experimentalists. By contrast, for small molecules such as methylene, theoretical predictions of their properties are now considered sufficiently reliable that remaining discrepancies between theory and experiment have been attributed to the neglect of relativistic corrections in the calculations!

Yet again we must congratulate our contributors, for meeting the deadlines for manuscripts, and the staff of The Royal Society of Chemistry, for their prompt handling of material. Neither of these operations is a rate-limiting step in the kinetic scheme which now describes the production of Annual Reports.

Finally, we draw attention to the updated table of topics recently covered in Annual Reports.

Table *Synopsis of Reports 1975—1979*

Subject	1975	1976	1977	1978	1979
Physical Methods:					
Mass spectrometry	√	×	√	×	√
X-Ray crystallography	√	×	√	×	√
N.m.r. spectroscopy	√	×	×	√	√
Theoretical chemistry	√	√	√	*i*	√
Reaction Mechanisms:					
Pericyclic	√	√	√	√	√
Polar reactions	√	√	√	√	√
Free-radical reactions	√	√	√	√	√
Arynes, carbenes, nitrenes, and related species	√	√	√	√	√
Organic electrochemistry	√	√	√	√	√
Photochemistry	√	√	√	√	√
Aliphatic compounds	√	√	√	√	√
Alicyclic compounds	√	√	√	√	√
Aromatic compounds	√	√	√	√	√
Heterocyclic compounds	√	√	√	√	√
Organometallic compounds	√	√	√	√	√
Synthetic methods	√	√	√	√	√
Biological Chemistry:					
Alkaloids	×	√	×	×	√
Nucleic acids	√	√	×	×	√
Enzymes	√	√	√	√	√
Biosynthesis	√	√	×	√	×
Peptides and proteins	×	√	×	√	×
Insect chemistry	√	×	√	×	√
Other chapters*	*a, b*	*c, d, e*	*f, g, h*	*i, j*	*k, l, m, n*

* (*a*) Laser Raman and infrared spectroscopy. (*b*) Steroids. (*c*) Optical rotatory dispersion/circular dichroism. (*d*) Gas chromatography–mass spectrometry. (*e*) Neurochemistry. (*f*) U.v. and visible spectroscopy. (*g*) Monosaccharides. (*h*) Tetrapyrroles. (*i*) Molecular mechanics. (*j*) Organic peroxides; synthetic and biological aspects. (*k*) Organoboron chemistry. (*l*) Organophosphorus chemistry. (*m*) Terpenoids and steroids. (*n*) Marine natural products.

2 Physical Methods and Techniques
Part (i) Mass Spectrometry

By M. JARMAN
Institute of Cancer Research, Block F, Clifton Avenue, Sutton, Surrey SM2 5PX

Since this subject was last reviewed,[1] the progress already made in the application of ionization methods alternative to the conventional electron-impact (EI) procedure has been consolidated and extended, and is therefore emphasized in the present review. Also, new applications have been found for other important procedures such as linked-scan techniques, a topic also highlighted here. The publication in 1978 of the Proceedings of the 7th International Mass Spectrometry Conference,[2] although representing the 'state of the art' as of August 1976, provides a comprehensive background against which recent progress can be assessed, as does the fifth Specialist Periodical Report.[3] Topics of other published conference proceedings have included quantitative analysis,[4] stable isotopes,[5,6] biomedical applications,[7] and natural products.[8] Books have appeared on the chemistry and physics of organic ions in the gas phase,[9] quantitative mass spectrometry,[10] and computers.[11] A compendium of biomedical and environmental applications of stable isotopes has been made.[12] Subjects of substantial reviews have included an extensive overview of mass spectrometry,[13] biomedical and environmental applications of field ionization and field desorption mass spectrometry,[14] and mass spectrometry in cancer research.[15]

[1] T. P. Toube, *Ann. Reports (B)*, 1977, **74**, 3.
[2] 'Advances in Mass Spectrometry', ed. N. R. Daly, Vols. 7A, 7B, Heyden, London, 1978.
[3] 'Mass Spectrometry', ed. R. A. W. Johnstone, (Specialist Periodical Reports), The Chemical Society, London, 1979, Vol. 5.
[4] 'Quantitative Mass Spectrometry in Life Sciences', ed. A. P. De Leenheer, and C. Van Peteghem, Vol. 2, Elsevier, Amsterdam, 1978.
[5] 'Stable Isotopes. Applications in Pharmacology, Toxicology and Clinical Research', ed. T. A. Baillie, MacMillan Press, London, 1978.
[6] 'Stable Isotopes', Proceedings of the Third International Conference, ed. E. R. Klein and P. D. Klein, Academic Press, New York, 1979.
[7] 'Recent Developments in Mass Spectrometry in Biochemistry and Medicine', ed. A. Frigerio, Vol. 1, Plenum, New York, 1978.
[8] 'Mass Spectrometry of Natural Products': IUPAC International Mass Spectrometry Symposium on Natural Products, ed. B. Sklarz, Rehovot, Israel, 1978.
[9] K. Levsen, 'Fundamental Aspects of Organic Mass Spectrometry'. Progress in Mass Spectrometry, Vol. 4, Verlag Chemie, Weinheim, W. Germany, 1978.
[10] B. J. Millard, 'Quantitative Mass Spectrometry', Heyden, London, 1978.
[11] J. R. Chapman, 'Computers in Mass Spectrometry', Academic Press, London, 1978.
[12] E. Klein and P. D. Klein, *Biomed. Mass Spectrometry*, 1978, **5**, 91 (Pt. I), 321 (Pt. II), 373 (Pt. III), 425 (Pt. IV); 1979, **6**, 515 (Pt. V).
[13] A. L. Burlingame, C. H. L. Shackleton, I. Howe, and O. S. Chizhov, *Analyt. Chem.*, 1978, **50**, 346R.
[14] H.-R. Schulten, *Internat. J. Mass Spectrometry Ion Physics*, 1979, **32**, 97.
[15] J. Roboz, 'Mass Spectrometry in Cancer Research', in 'Advances in Cancer Research', Vol. 27, pp. 201–267, Academic Press, New York, 1978.

The IUPAC recommendations on symbolism and nomenclature for mass spectrometry have been published[16] and summarized in an editorial review.[17]

1 Ion Structure and Fragmentation

A review[18] of the chemistry of isolated cations contains a useful classification of competing reactions of isomeric ions into three types. If the activation energy for interconversion is smaller than the energy required for decomposition of either ion, isomerization precedes decomposition, resulting in common fragmentation pathways. Conversely, if the energy barrier for interconversion is higher than for dissociation of either ion, independent fragmentation pathways result. If the energy barrier to interconversion is intermediate between those for decomposition of the isomeric ions, the consequences are more complex.[19] All three categories of metastable decomposition appear to be represented in the fragmentation pathways[20] for the various $C_4H_8O^{+\cdot}$ ions.

The dissociation of oxonium ions of the general structure $H_2C{=}\overset{+}{O}CHR^1R^2$ has been rationalized[21] in terms of the relative proton affinities of CH_2O and the potential olefin $H_2C{=}CR^1R^2$. Since ethylene has a lower proton affinity than formaldehyde, the $CH_2{=}\overset{+}{O}H$ ion is formed preferentially from the ethyl derivative ($R^1 = Me$, $R^2 = H$) and ethylene is expelled as the neutral species. Conversely, the n-propyl ($R^1 = Et$, $R^2 = H$) and isopropyl ($R^1 = R^2 = Me$) analogues give protonated propene, since the latter has a higher proton affinity than formaldehyde.

Pulsed ion cyclotron resonance (ICR) spectroscopy was used to show[22] that compounds of the structure $CH_3O(CH_2)_nSCH_3$ prefer to form sulphur-containing ($H_2C{=}\overset{+}{S}CH_3$) rather than oxygen-containing ($H_2C{=}\overset{+}{O}CH_3$) ions in gas-phase ion–molecule reactions.

The fragmentation of cyclic ethers continues to attract attention. The predicted[23] existence of the ethylene oxide ($C_2H_4O^{+\cdot}$) ion in the open form ($H_2C{=}\overset{+}{O}{-}\dot{C}H_2$) as opposed to the C–C- bonded form ($\dot{C}H_2{-}CH_2{-}\overset{+}{O}$) was confirmed by examining the ICR and normal mass spectra of [4,4,5,5-$^2H_4$]-1,3-dioxolan (1), which afforded an ion of m/z 46 (cleavage *a*) as opposed to m/z 48 (cleavage *b*). Four stable isomeric $C_4H_5O^+$ ions, the most stable being a cyclic species, are fragmentation products of various unsaturated linear and cyclic ethers.[24] A cyclic ([furan]$^+$) rather

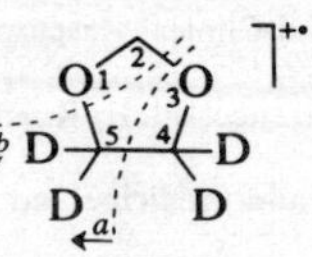

(1)

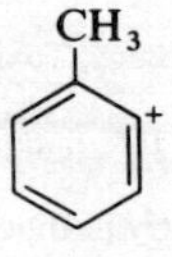

(2)

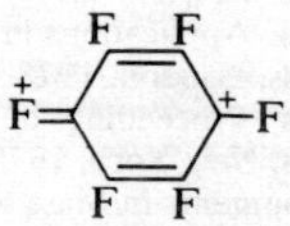

(3)

[16] *Pure Appl. Chem.*, 1978, **50**, 65.
[17] A. Maccoll, *Org. Mass Spectrometry*, 1979, **14**, 1.
[18] R. D. Bowen, D. H. Williams, and H. Schwarz, *Angew. Chem. Internat. Edn.*, 1979, **18**, 451.
[19] R. D. Bowen, D. H. Williams, G. Hvistendahl, and J. R. Kalman, *Org. Mass Spectrometry*, 1978, **13**, 721.
[20] D. J. McAdoo, C. E. Hudson, and D. N. Witiak, *Org. Mass Spectrometry*, 1979, **14**, 350.
[21] R. D. Bowen, B. J. Stapleton, and D. H. Williams, *J.C.S. Chem. Comm.*, 1978, 24.
[22] J. K. Pau, M. B. Ruggera, J. K. Kim, and M. C. Caserio, *J. Amer. Chem. Soc.*, 1978, **100**, 4242.
[23] W. Bouma, J. K. MacLeod, and L. Radom, *J.C.S. Chem. Comm.*, 1978, 724.
[24] J. K. Terlouw, P. C. Burges, and H. Hommes, *Org. Mass Spectrometry*, 1979, **14**, 387.

than an acyclic (vinylketen$^+$) structure is favoured on the basis of collision activation (CA) studies for the stable $C_4H_4O^+$ ion formed from 4-pyrone.[25]

A predictive model[26] for unimolecular reactions of organic ions in which dissociation proceeds by simple σ-cleavage in the final step has been successfully applied to the decomposition of hydrocarbon ions $[C_nH_{2n-3}]^+(n = 3—8)$. Elimination of methane from metastable molecular ions of n-butane and methylpropane is suggested to yield [cyclopropane]$^+$ as the common daughter ion.[27] using pulsed ICR spectrometry, butanes and butyl halides containing isobutyl or t-butyl groups were shown to give t-butyl cations, but n- and s-butyl derivatives gave both s- and t-butyl cations.[28]

In the area of aromatic hydrocarbon studies, interest in the various isomeric $C_7H_7^+$ ions continues. The isomerization of benzyl and tropylium cations has been investigated in 2-phenylethyl-derivatives.[29] Photon-impact measurements of appearance energies of $C_7H_7^+$ for alkyl-benzenes support a threshold tropylium structure.[30] The proposed[31] formation of pure tolyl ions, *e.g.* (2) from fluoro-toluenes, has been confirmed[32] in a CA study.

The problem of distinguishing linear from cyclic ion structures takes on a new dimension when doubly charged ions are considered. Thus the release of large amounts of kinetic energy (reflected in 'dish topped' metastables) accompanies scission into singly charged species, owing to Coulombic repulsion. Measurement of this energy release enables charge separation to be calculated, and this approach has been used to show that the fragmentation $C_6H_6^{2+} \rightarrow C_5H_3^+ + CH_3^+$ proceeds after rearrangement of the benzene ion to a linear structure $H_3C{-}\overset{+}{C}{=}C{=}C{=}\overset{+}{C}{-}CH_3$. In an elegant extension[33] of this approach to halogenated benzenes, $C_6F_6^{2+}$ was concluded to fragment (to CF^+ and $C_5F_5^+$) mainly from a cyclic transition state (3), though minor loss of CF_3^+ (acyclic state) was seen. Surprisingly, hexachlorobenzene gave some CCl_3^+; thus, despite its large size, Cl can migrate to each end of a linear structure. Bromine (hexabromobenzene) does not do so, however.

2 Chemical Ionization

A major motivation behind the continuing growth of interest in chemical ionization (CI) has been its capacity to yield abundant molecular ion species, thus enabling molecular weights to be unequivocally established, as well as minimizing interference from competing ions in quantitative studies. Consequently, the fragmentations of protonated molecules in contrast to odd-electron molecular ions have been relatively little studied. Collision-induced decomposition of mass-selected ions can give information on the fragmentation behaviour of individual molecular ion species. Thus the main fragmentation pathways of protonated ethers and ketones,[34]

[25] J. L. Holmes and J. K. Terlouw, *J. Amer. Chem. Soc.*, 1979, **101**, 4793.
[26] R. D. Bowen, B. J. Stapleton, and D. H. Williams, *Org. Mass Spectrometry*, 1978, **13**, 330.
[27] P. Wolkoff and J. L. Holmes, *J. Amer. Chem. Soc.*, 1978, **100**, 7346.
[28] D. M. Shold and P. Ausloos, *J. Amer. Chem. Soc.*, 1978, **100**, 7915.
[29] D. Kuck and H.-Fr. Grützmacher, *Org. Mass Spectrometry*, 1979, **14**, 86.
[30] R. G. McLoughlin, J. D. Morrison, and J. G. Traeger, *Org. Mass Spectrometry*, 1979, **14**, 104.
[31] H.-W. Leung, H. Ichikawa, Y.-H. Li, and A. G. Harrison, *J. Amer. Chem. Soc.*, 1978, **100**, 2479.
[32] F. W. McLafferty and F. M. Bockhoff, *Org. Mass Spectrometry*, 1979, **14**, 181.
[33] M. Medved, B. Kralj, J. Marsel, V. Kramer, T. Ast, and J. H. Beynon, *Org. Mass Spectrometry*, 1979, **14**, 307.
[34] M. L. Sigsby, R. J. Day, and R. G. Cooks, *Org. Mass Spectrometry*, 1979, **14**, 273.

the condensation products from benzyl acetate,[35] and the tautomeric equilibrium between hydroxy-pyridines, hydroxy-quinolines, and their *O*- and *N*-ethyl derivatives[36] have been investigated. The rate at which the commonly observed fragmentation $RX{\cdot}H^+ \rightarrow R^+ + HX$ occurs should be inversely proportional to the proton affinity of the neutral HX molecule.[37] Two further studies[38,39] lend support to this prediction; one,[39] on the fragmentation of protonated glycol derivatives, was partly aimed at rationalizing the markedly different CI fragmentation behaviour of protonated epimeric hexitol derivatives.[40] Other CI fragmentation studies have concerned halogeno- and amino-benzene derivatives,[41] carboxylic acids,[42] esters,[43,44] and amines.[44]

Analogy can be drawn between fragmentation behaviour under CI and EI[22] in the not unexpected preference for the formation of $C_6H_5\overset{+}{S}{=}CH_2$ from $C_6H_5O(CH_2)_nSC_6H_5$ precursors in methane CI.[45] In contrast, the *ortho*-effect observed for loss of water from *ortho*-substituted benzoic acids under CI[46] has a mechanism different from the *ortho*-effects observed[47,48] in the EI mode. Thus under EI conditions [(4) → (5)] the *ortho*-substituent must carry a proton. In the CI mode, this proton can be provided by the reagent gas, and the proposed mechanism [(6) → (8)] requires its ready transfer to the adjacent carboxyl group. Thus a 2-halogeno-group, but not a 2-methyl substituent, can promote loss of water from protonated benzoic acids.

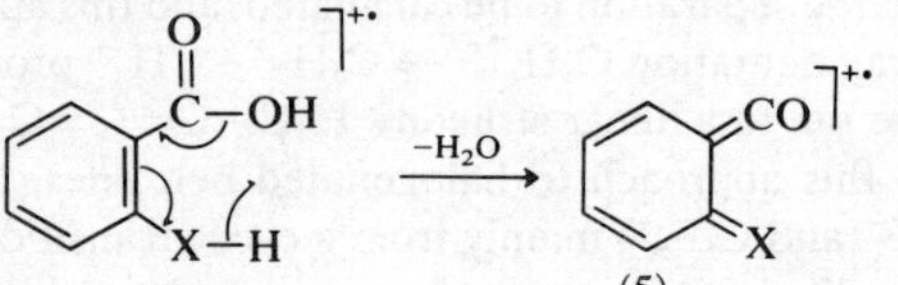

(4) (5)

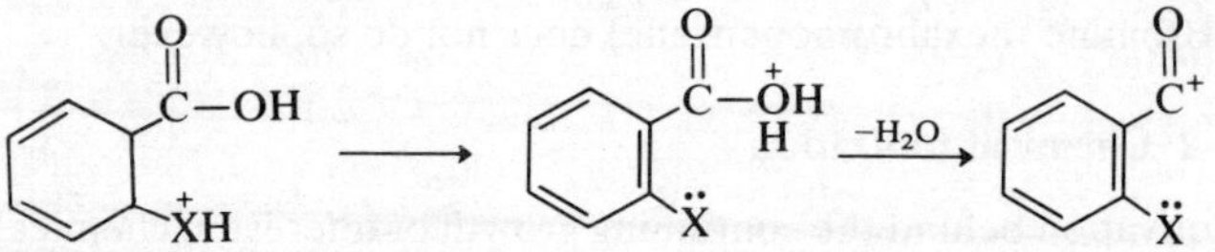

(6) (7) (8)

[35] D. Cameron and R. G. Cooks, *J. Amer. Chem. Soc.*, 1979, **101**, 3162.
[36] A. Maquestiau, Y. van Haverbeke, R. Flammang, H. Mispreuve, A. R. Katritzky, J. Ellison, J. Frank, and Z. Mészáros, *J.C.S. Chem. Comm.*, 1979, 888.
[37] F. H. Field, in 'Mass Spectrometry', ed. A. Maccoll, Ch. 5, p. 49. MTP Review of Science, Butterworths, London, 1972.
[38] A. G. Harrison and F. I. Onuska, *Org. Mass Spectrometry*, 1978, **13**, 35.
[39] R. J. Schmitt and R. H. Shapiro, *Org. Mass Spectrometry*, 1978, **13**, 715.
[40] M. NcNeil and P. Albersheim, *Carbohydrate Res.*, 1977, **56**, 239.
[41] H.-W. Leung and A. G. Harrison, *J. Amer. Chem. Soc.*, 1979, **101**, 3168.
[42] N. E. Middlemiss and A. G. Harrison, *Canad. J. Chem.*, 1979, **57**, 2827.
[43] F. M. Benoit and A. G. Harrison, *Org. Mass Spectrometry*, 1978, **13**, 128.
[44] M. L. Sigsby, R. J. Day, and R. G. Cooks, *Org. Mass Spectrometry*, 1979, **14**, 556.
[45] C. C. Van de Sande, F. Van Gaever, L. Dhaenens, and R. Myngheer, *Org. Mass Spectrometry*, 1979, **14**, 191.
[46] H. Ichikawa and A. G. Harrison, *Org. Mass Spectrometry*, 1978, **13**, 389.
[47] J. T. Bursey, M. M. Bursey, and D. G. I. Kingston, *Chem. Rev.*, 1973, **73**, 191.
[48] A. R. Butcher and C. B. Thomas, *Org. Mass Spectrometry*, 1979, **14**, 448.

In CI, the extent of fragmentation can be influenced by the choice of reagent gas. Because of its high proton affinity, coupled with low energy transfer in ion–molecule reactions, ammonia has proved a particularly versatile reagent gas, being superior to methane and isobutane in producing $[M+H]^+$ or $[M+NH_4]^+$ ions and stereochemical information on cardiac glycosides,[49] in recognizing specific functional groups,[50] in giving $[M+H]^+$ and $[M+NH_4]^+$ ions of steroidal spiro-lactones[51] and anthrone glycosides[52] and, when used in conjunction with trideuterioammonia, in locating and enumerating labile protons (*e.g.* NH, OH, SH).[53] However, methane CI was better for distinguishing 1,2- from 1,4-linked disaccharides.[54] Neither ammonia nor isobutane CI gave useful molecular weight information on trimethylsilylated glucuronides, for which purpose pyridine, giving an intense $[M+80]^+$ ion, was the reagent of choice.[55]

Among other reagent gases for CI, nitric oxide continues to be used occasionally. It was superior to isobutane for locating the triple bond in alkynes,[56] but should be used with caution. Thus it reduces aromatic nitro-compounds to amines, giving anomalous $[MH-30]^+$ ions.[57]

Although there has been little evidence of recent growth in interest in EI studies of negative ions, reports[58,59] that, under CI conditions, negative-ion currents can exceed positive-ion currents by several orders of magnitude have prompted growing interest in negative-ion CI. Naturally, comparisons between positive- and negative-ion CI have been made. Thus, in the OH^--ion CI spectra of terpenols and their esters, $[M-H]^-$ ions were more consistently seen than the corresponding $[M+H]^+$ ions in the positive-ion (isobutane) spectra.[60] The appearance of intense carboxylate anions in the negative-ion CI spectra of esters identifies the appropriate acid,[60] although it should be noted that the negative-ion EI spectra of *p*-nitrophenyl and *p*-nitrobenzyl esters also give intense anion peaks.[61] The number of hydroxyl groups in hydroxy-steroids is defined[62] in their negative-ion CI spectra by sequential losses of 18 *u* from $[M-1]^-$ for each such group present. The influence of geometrical isomerism on fragmentation pathways is illustrated by a comparison between the OH^--ion CI spectra of isomeric cyclohexanediols.[63] Thus negative ions from the *cis* isomers are stabilized (9) against fragmentation by intramolecular bonding between O^- and OH moieties. The spectra of *trans* isomers, which cannot thus be stabilized (10), contain intense $[M-H_3]^-$ ions.

49 J. Vine, L. Brown, J. Boutagy, R. Thomas, and D. Nelson, *Biomed. Mass Spectrometry*, 1979, **6**, 415.
50 Y. Y. Lin and L. L. Smith, *Biomed. Mass Spectrometry*, 1978, **5**, 604.
51 D. R. Boreham, G. C. Ford, N. J. Haskins, C. W. Vose, and R. F. Palmer, *Biomed. Mass Spectrometry*, 1978, **5**, 524.
52 F. J. Evans, M. G. Lee, and D. E. Games, *Biomed. Mass Spectrometry*, 1979, **6**, 374.
53 Y. Y. Lin and L. L. Smith, *Biomed. Mass Spectrometry*, 1979, **6**, 15.
54 E. G. de Jong, W. Heerma, and C. A. X. G. F. Sicherer, *Biomed. Mass Spectrometry*, 1979, **6**, 242.
55 L. P. Johnson, S. C. Subba Rao, and C. Fenselau, *Analyt. Chem.*, 1978, **50**, 2022.
56 E. Busker and H. Budzikiewicz, *Org. Mass Spectrometry*, 1979, **14**, 222.
57 J. J. Brophy, V. Diakiw, R. J. Goldsack, D. Nelson, and J. S. Shannon, *Org. Mass Spectrometry*, 1979, **14**, 201.
58 D. F. Hunt and F. W. Crow, *Analyt. Chem.*, 1978, **50**, 1781.
59 J. R. Hass, M. D. Friesen, D. J. Harvan, and C. E. Parker, *Analyt. Chem.*, 1978, **50**, 1474.
60 A. P. Bruins, *Analyt. Chem.*, 1979, **51**, 967.
61 Y. Hirata, T. Takeuchi, and K. Matsumoto, *Analyt. Chem.*, 1978, **50**, 1943.
62 T. A. Roy, F. H. Field, Y. Y. Lin, and L. L. Smith, *Analyt. Chem.*, 1979, **51**, 272.
63 F. J. Winkler and D. Stahl, *J. Amer. Chem. Soc.*, 1978, **100**, 6779.

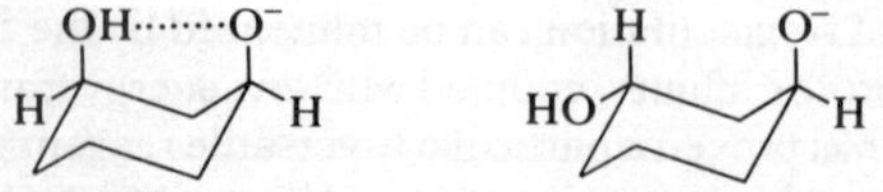

(9) (10)

The higher sensitivity of CI spectra in the negative-ion mode, as compared with the positive-ion mode, has been noted for polychlorinated dibenzo-*p*-dioxins[59] and diphenyl ethers[64] and for various classes of drugs, including dopamine and amphetamine,[58] phenothiazines,[65] and methadone and its metabolites.[66] However, the authors of the last-mentioned study re-emphasize that the greatly enhanced sensitivity of negative-ion CI is reported only when negative ions are formed by an electron-capture process.[58] Finally, negative-ion CI shows great promise in GC–MS analysis[58,67] down to attomole levels.

3 Mass Spectrometry of Non-volatile and Thermally Labile Compounds

Compounds which will not volatilize without undergoing structural alteration provide a challenge for the mass spectroscopist. One solution is to use ionization procedures which do not require the sample to vaporize prior to ionization. Field desorption (FD) remains the most widely used such technique, and many of the problems formerly associated with its use have been overcome. Thus, although the FD mode may rival EI or CI in total ion yield per sample, ion production per unit time is low and ion currents may fluctuate,[68] problems which have now been circumvented by using integrating methods.[68,69] Other developments which have aided the practical application of FD have included the use of a laser to control emitter temperature[70] and further advances in emitter preparation.[71,72] Analytical applications have included the determination of the molecular weight of large molecules, *e.g.* of corrinoid metal complexes related to vitamin B_{12},[73,74] and of di-β-hydroxyacyl-trehaloses (up to m/z 1767);[75] analysis of quaternary ammonium salts, *e.g.* quantification of berberine chloride,[76] identification of carpronium chloride and a metabolite in urine,[77] and quantification of choline and acetylcholine in rat brain;[78] and identification of herbicides in river water[79] and of organochlorine

64 K. L. Busch, Å. Norström, M. M. Bursey, J. R. Hass, and C.-A. Nilsson, *Biomed. Mass Spectrometry*, 1979, **6**, 157.
65 R. Ryhage and H. Brandenberger, *Biomed. Mass Spectrometry*, 1978, **5**, 615.
66 A. L. C. Smit and F. H. Field, *Biomed. Mass Spectrometry*, 1978, **5**, 572.
67 A. J. Lewy and S. P. Markey, *Science*, 1978, **201**, 741.
68 G. L. Peele and D. A. Brent, *Biomed. Mass Spectrometry*, 1978, **5**, 180.
69 W. D. Lehmann and H.-R. Schulten, *Biomed. Mass Spectrometry*, 1978, **5**, 208.
70 H.-R. Schulten, W. D. Lehmann, and D. Haaks, *Org. Mass Spectrometry*, 1978, **13**, 361.
71 T. L. Youngless, M. M. Bursey, and C. E. Rechsteiner, Jr., *Analyt. Chem.*, 1978, **50**, 1951.
72 T. Matsuo, H. Matsuda, and I. Katakuse, *Analyt. Chem.*, 1979, **51**, 69.
73 H. M. Schiebel and H.-R. Schulten, *Tetrahedron*, 1979, **35**, 1191.
74 R.-P. Hinze, H. M. Schiebel, H. Laas, K.-P. Heise, A. Gossauer, and H. H. Inhoffen, *Annalen*, 1979, 811.
75 G. Puzo, G. Tissie, C. Lacave, H. Aurelle, and J. C. Prome. *Biomed. Mass Spectrometry*, 1978, **5**, 699.
76 H. Miyazaki, E. Shirai, M. Ishibashi, K. Hosai, S. Shibata, and M. Iwanaga, *Biomed. Mass Spectrometry*, 1978, **5**, 559.
77 M. Sano, K. Ohya, and R. Ito, *Biomed. Mass Spectrometry*, 1979, **6**, 467.
78 W. D. Lehmann, H.-R. Schulten, and N. Schröder, *Biomed. Mass Spectrometry*, 1978, **5**, 591.
79 Y. Yamato, M. Suzuki, and T. Watanabe, *Biomed. Mass Spectrometry*, 1979, **6**, 205.

pesticides in soil.[80] The use of alkali-metal salts to induce cationization has been severally exemplified,[74,75,81] and this technique has also been used in conjunction with collision activation[82,83] to combine determination of the molecular weight with obtaining structural information from fragmentation.

Other ionization techniques applicable to involatile compounds have been compared with FD in a study of quaternary ammonium salts.[84] Californium (^{252}Cf) fission-fragment-induced desorption (FFD),[85] laser-induced desorption (LID), and secondary-ion mass spectrometry (SIMS) are techniques in which intact molecules of involatile substances desorb when energetic particles impact on the sample material, coated on a metal surface.[86] The high-temperature bursts from the Cf fission products or the sub-microsecond laser pulse are too short for the molecule to absorb energy into vibrational modes and decompose before ionizing. LID has given molecular ions from a digitalis pentaglycoside (m/z 1228) and underivatized adenylyl-3′,5′-cytidine.[86] In SIMS,[87,88] the target sample is bombarded with ions of inert gases. The metal of the target can produce cationized molecular ions, and the ion bombardment of these leads to collision-induced decomposition, so that the molecular weight is determined and fragmentation information is obtained in one technique. Finally, electrohydrodynamic ionization, in which ions are generated from an electron-doped liquid surface by an applied electric field, has been applied to nucleotides[89] and to doubly charged ionic salts (*e.g.* 2-naphthol-3,6-disulphonic acid disodium salt),[90] a class of compound which usually fails to desorb in FD.

The problem of thermal lability may sometimes be circumvented by introducing the sample into, or very near to, the ionizing beam, using a modified direct-insertion probe, in otherwise conventional EI or CI operation. Variations in the probe material, its distance from the ion beam, and the heating rate all influence the results obtained. Although not essential,[91] inert materials such as Teflon,[92] Vespel,[93] and silicone coatings on conventional probes[94] appear to disfavour surface adsorption of applied compounds, and thus to minimize thermal decomposition. Conversely, metal probes of stainless steel[49] and rhenium wire[95] probably aid rapid heating of the sample, the importance of which is stressed by several authors.[95—97] Such techniques have yielded molecular or related ions from amino-sugars in the EI mode,[91] from sucrose in EI[95] and CI,[94] and (in CI) from creatine, arginine, and aspartylglycine,[92]

[80] M. Suzuki, Y. Yamato, and M. Koga, *Biomed. Mass Spectrometry*, 1978, **5**, 518.
[81] G. Puzo and J. C. Prome, *Adv. Mass Spectrometry*, 1978, **7B**, 1596.
[82] H. J. Veith, *Org. Mass Spectrometry*, 1978, **13**, 280.
[83] F. W. Röllgen, U. Giessmann, F. Borchers, and K. Levsen, *Org. Mass Spectrometry*, 1978, **13**, 459.
[84] B. Schueler and F. R. Krueger, *Org. Mass Spectrometry*, 1979, **14**, 439.
[85] R. D. MacFarlane and D. F. Torgerson, *Science*, 1976, **191**, 920.
[86] M. A. Posthumus, P. G. Kistemaker, H. L. C. Meuzelaar, and M. C. Ten Noever de Brauw, *Analyt. Chem.*, 1978, **50**, 985.
[87] A. Benninghoven and W. K. Sichtermann, *Analyt. Chem.*, 1978, **50**, 1180.
[88] H. Grade and R. G. Cooks, *J. Amer. Chem. Soc.*, 1978, **100**, 5615.
[89] S.-T. F. Lai and C. A. Evans, Jr., *Biomed. Mass Spectrometry*, 1979, **6**, 10.
[90] S.-T. F. Lai and C. A. Evans, Jr., *Org. Mass Spectrometry*, 1978, **13**, 733.
[91] M. Ohashi, S. Yamada, H. Kudo, and N. Nakayama, *Biomed. Mass Spectrometry*, 1978, **5**, 578.
[92] G. Hansen and B. Munson, *Analyt. Chem.*, 1978, **50**, 1130.
[93] R. J. Cotter, *Analyt. Chem.*, 1979, **51**, 317.
[94] D. I. Carroll, I. Dzidic, M. G. Horning, F. E. Montgomery, J. G. Nowlin, R. N. Stillwell, J.-P. Thenot, and E. C. Horning, *Analyt. Chem.*, 1979, **51**, 1858.
[95] W. R. Anderson, Jr., W. Frick, and G. Doyle Daves, Jr., *J. Amer. Chem. Soc.*, 1978, **100**, 1974.
[96] R. J. Cotter and C. Fenselau, *Biomed. Mass Spectrometry*, 1979, **6**, 287.

and from guanosine and *p*-nitrophenylglucuronides,[96] none of which give such ions when conventional probes are used. The greatest intensity of the molecular ion is observed when the sample is on the fringe of the beam[91] or *ca.* 2.5—3 mm from it.[92,96] Presently, one drawback of these 'in beam' techniques appears to be the relatively large sample size (*ca.* 50 μg^{96}) needed to obtain maximum intensity of the molecular ion. Finally, although the controversy generated as a result of the finding that the FD emitter could also be used as a probe for the generation of molecular ion species from thermally sensitive molecules under EI[97] and CI[98] conditions was not fully resolved by a correspondence[99—101] between two of the relevant groups, the need for the field in generating many of the ions observed in FD spectra is not disputed.[100]

4 Recent Applications of Linked-Scan Techniques

Organic compounds in mixtures are conventionally separated by gas chromatography prior to introduction into the ion source of the mass spectrometer. Though 'soft' ionization procedures (*e.g.* CI[102] and FD) can sometimes be used to detect or to quantify individual components in a mixture introduced directly into the ion source, the simplification of spectra on which success of this method depends entails loss of fragmentational information. In the technique of mass-analysed ion kinetic energy spectrometry (MIKES) the mass spectrometer alone is used both to resolve and to fragment selected components in a mixture that is introduced directly into the ion source, using a spectrometer of reversed geometry. The selected ions are first isolated in the magnetic sector by appropriate adjustment of the magnetic field (B). Collision activation in the field-free region between magnet and electrostatic analyser fragments these ions, and the spectrum of the daughter ions is recorded by scanning the electrostatic voltage (E). Thus, 1-methyladenine has been detected[103] in intact salmon sperm DNA, and distinguished from isomers by its different daughter-ion spectrum. Similarly, the ion of m/z 180 was assigned to MH^+ for 5,6-benzoquinoline rather than to acridine in the MIKES analysis of coal liquid samples.[104] Cocaine and cinnamoylcocaine were detected in coca leaf by simultaneously altering the accelerating voltage V (to select the reactant ion) and E (to select the daughter ions) with B constant, to achieve multiple reaction monitoring.[105] Similarly, acetophenone and butabarbital were quantified by stable-isotope dilution. Negative-ion CI was used in the detection, by MIKES, of glucose in urine[106] and of individual carboxylic acids in a mixture.[107] Artifact peaks in MIKES spectra arise mainly from metastable transitions in the first field-free region. Methods for identifying these and other interfering ions have been reviewed.[108] MIKES analysis

[97] J. F. Holland, B. Soltmann, and C. C. Sweeley, *Biomed. Mass Spectrometry*, 1976, **3**, 340.
[98] D. F. Hunt, J. Shabanowitz, F. K. Both, and D. A. Brent, *Analyt. Chem.*, 1977, **49**, 1160.
[99] H. D. Beckey and F. W. Röllgen, *Org. Mass Spectrometry*, 1979, **14**, 188.
[100] J. F. Holland, *Org. Mass Spectrometry*, 1979, **14**, 291.
[101] H. D. Beckey, *Org. Mass Spectrometry*, 1979, **14**, 292.
[102] R. J. Weinkam and H.-S. Lin, *Analyt. Chem.*, 1979, **51**, 972.
[103] A. E. Schoen, R. G. Cooks, and J. L. Wiebers, *Science*, 1979, **203**, 1249.
[104] D. Zakett, V. M. Shaddock, and R. G. Cooks, *Analyt. Chem.*, 1979, **51**, 1849.
[105] R. W. Kondrat, G. A. McClusky, and R. G. Cooks, *Analyt. Chem.*, 1978, **50**, 2017.
[106] G. A. McClusky, R. W. Kondrat, and R. G. Cooks, *J. Amer. Chem. Soc.*, 1978, **100**, 6045.
[107] R. W. Kondrat, G. A. McClusky, and R. G. Cooks, *Analyt. Chem.*, 1978, **50**, 1222.
[108] T. Ast, M. H. Bozorgzadeh, J. L. Wiebers, J. H. Beynon, and A. G. Brenton, *Org. Mass Spectrometry*, 1979, **14**, 313.

has also been used in functional-group analysis; for example, to differentiate side-chains in steroids having similar nuclear structure.[109]

Instruments of normal geometry can also be used to obtain linked scans. Scanning with constant ratio B/E gives a spectrum of all daughter ions arising from a common precursor ion; B^2/E scans yield all precursor ions giving a common daughter ion.[110] These scans achieve good resolution of daughter ions, but poor resolution of precursor ions; in this respect, they are complementary to MIKES.[111] The B/E-scan mode has been used to distinguish isomeric androstane-3,17-diols[112] and (using GC–MS) for monitoring selected metastable peaks of 4,5α-dihydrotestosterone t-butyldimethylsilyl ether[113] and selected single metastable ions of the analogous testosterone derivative, using epitestosterone as an internal standard.[114] This quantitative application was comparable in sensitivity and selectivity to conventional selected-ion monitoring at high resolving power, and superior to the corresponding low-resolution procedure. As in MIKES spectra, artifact ions in B/E scans can result from metastable transitions of ions of higher mass (often the molecular ion) in the first field-free region.[115]

5 Conclusion

The continuing growth of interest in biomedical and biochemical applications of mass spectrometry is reflected in the quest for information on molecular weight and structure on ever more complex molecules. In addition to examples previously given, peptide sequencing continues to be an outstandingly active area.[116–119] One such study[119] shows how the analysis of thermally labile compounds has also been aided by further developments in methods for coupling the mass spectrometer with the high-performance liquid chromatograph.[120] It remains to be seen what are the ultimate physical constraints on obtaining molecular weights of large molecules, but at least the problem of calibration in the high mass range (up to m/z *ca.* 10^4) appears to have been ingeniously solved[121] by the choice of polystyrene, a polymer that has a high molecular weight and also mass-defining repeating monomer sequences and reasonable volatility. It has been suggested[122] that compounds of a molecular size such that bond cleavage, rather than vaporization, will occur on heating will probably require one of the newer (and as yet relatively unexplored) techniques, such as FFD or LID, if analysis is to be successful.

109 M. H. Bozorgzadeh, A. G. Brenton, J. L. Wiebers, and J. H. Beynon, *Biomed. Mass Spectrometry*, 1979, **6**, 340.

110 M. J. Lacey and C. G. Macdonald, *Org. Mass Spectrometry*, 1979, **14**, 465.

111 W. F. Haddon, *Analyt. Chem.*, 1979, **51**, 983.

112 S. J. Gaskell, A. W. Pike, and D. S. Millington, *Biomed. Mass Spectrometry*, 1979, **6**, 78.

113 S. J. Gaskell and D. S. Millington, *Biomed. Mass Spectrometry*, 1978, **5**, 557.

114 S. J. Gaskell, R. W. Finney and M. E. Harper, *Biomed. Mass Spectrometry*, 1979, **6**, 113.

115 J. N. Bilton, N. Kyriakidis, and E. S. Waight, *Org. Mass Spectrometry*, 1978, **13**, 489.

116 P. Roepstorff, *Biomed. Mass Spectrometry*, 1978, **5**, 362.

117 R. J. Anderegg, K. Biemann, A. Manmade, and A. C. Ghosh, *Biomed. Mass Spectrometry*, 1979, **6**, 129.

118 A. D. Auffret and D. H. Williams, *J.C.S. Chem. Comm.*, 1979, 692.

119 B. G. Dawkins, P. J. Arpino, and F. W. McLafferty, *Biomed. Mass Spectrometry*, 1978, **5**, 1.

120 G. Guiochon and P. J. Arpino, *Analyt. Chem.*, 1979, **51**, 682A.

121 T. Matsuo, H. Matsuda, and I. Katakuse, *Analyt. Chem.*, 1979, **51**, 1329.

122 J.-P. Thenot, J. Nowlin, D. I. Carroll, F. E. Montgomery, and E. C. Horning, *Analyt. Chem.*, 1979, **51**, 1101.

Regarding the future of linked-scan techniques (obviously a growth area), further development of the tandem quadrupole mass spectrometer[123] should complement the more sophisticated instrumentation used for the studies reported here, and lead to wider application of these techniques.

[123] R. A. Yost and C. G. Enke, *J. Amer. Chem. Soc.*, 1978, **100**, 2274.

2 Physical Methods and Techniques
Part (ii) *X*-Ray Crystallography

By S. NEIDLE
Department of Biophysics, Kings College, (University of London), 26—29 Drury Lane, London WC2B 5RL

The literature of organic crystal structures continues to grow at a steady pace of about 2000 newly reported structures *per annum.* As in previous years, the reader is referred to the latest extensive bibliographies from the Cambridge Crystallographic Data Centre.[1] The detailed files of the Data Centre are now available on computer installations world-wide, and are extensively used for speedy access to details of all published organic molecular structures.[2] As detailed below, this powerful facility is being utilised as a structural tool in its own right, capable of extracting new and meaningful data by means of structural correlations.

Several trends in organic crystallography have become apparent during the past several years. The power of spectroscopic methods is now such that *X*-ray crystallography only rarely needs to be called upon as a determinant of molecular structure, its traditional role in organic chemistry. Thus, crystallographers are increasingly defining new areas for their attention. Studies relating structure to theoretical calculation and prediction, either for bonding geometry or molecular conformation, are of particular interest at present, as are analyses involving solid-state reactivity *vis à vis* structure. It is also increasingly apparent that small-molecule as well as protein crystallography can provide important data on biological molecules, and such molecules serve as models for biological systems. Thus, organic crystallography is now moving rapidly into the phase of its development where the crystallographer himself defines his problems and performs his own chemistry and biochemistry, as well as solving and analysing his structures.

1 Molecular Structure and Bonding

A number of studies have been concerned with the influence of crystal structure (*i.e.* non-bonded forces, including inter-molecular hydrogen bonds) on molecular conformation. In one approach, the conformational polymorphism of *N*-(*p*-chlorobenzylidene)-*p*-chloroaniline has been examined,[3] by combining molecular orbital calculations with an analysis of the molecular packing in the two space groups found

1 'Molecular Structures and Dimensions', Crystallographic Data Centre, Cambridge University, 1978, Vol. 9; 1979, Vol. 10.

2 F. H. Allen, S. Bellard, M. D. Brice, B. A. Cartwright, A. Doubleday, H. Higgs, T. Hummelin, B. G. Hummelink-Peters, O. Kennard, W. D. S. Motherwell, J. R. Rodgers, and D. G. Watson, *Acta Cryst.*, 1979, **B35**, 2331.

3 J. Bernstein and A. T. Hagler, *J. Amer. Chem. Soc.*, 1978, **100**, 673.

for this molecule. Extensive conformational polymorphism has been noted in the three crystal forms of iminodiacetic acid, with each showing a distinct molecular conformation.[4] Intermolecular forces in hydrogen-bonded crystals of carboxylic acids and amides have been examined[5] with consistent-force-field calculations. *Ab initio* molecular orbital calculations of hydrogen bonds in carbohydrates[6] suggest that, for O—H · · · O systems, the energy minimum is not at a 180° bond angle, but at ~163°; neutron-diffraction data gives an average of 166°. The Cambridge data files have been used for an alternative approach to the problems of conformational variability in crystals, based on the examination of all structures determined, and employing standard statistical methods to analyse the data.[7] In one instance, the geometry of the β-1′-aminofuranoside ring (in nucleic acids and their constituents) was examined in some 100 structures, and statistically meaningful correlations among several geometric parameters were noted.[8] The geometry of substituent-induced deformation of bond length in cyclopropane-containing structures has recently been investigated;[9] the trends found are not large, but support the predictions of molecular orbital theory. The structure analysis of 10,10-dimethyl-3,4-dioxatricyclo[5.2.1.$0^{1,5}$]decane-2-spiro-2′-adamantane (1) has led to a comparison of C—O and O—O bond lengths in small rings as a function of ring size.[10]

The exploration of reaction pathways *via* examination of relevant crystal structures is proving to be most instructive. Dunitz and his colleagues have approached the problem by examining related compounds that all react in the same manner, but, say, with varying rates. A crystallographic study of four compounds that display aspects of the keto-acid–hydroxy-lactone isomerization (2) has shown two to be on

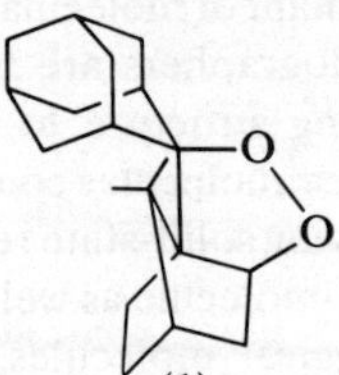

(1)

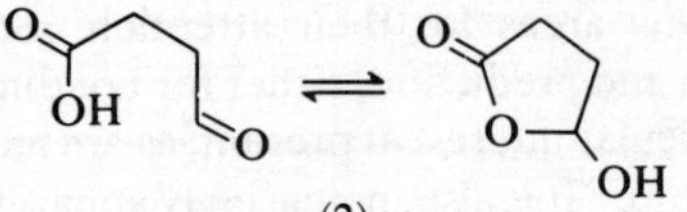

(2)

the ring-opened side, but with short O · · · C=O distances. The two ring-closed compounds have lengthened C—O bonds.[11] Explorations of solid-state reactions continue to be a fruitful area; for example, a plausible explanation in terms of differences in molecular packing has been advanced for the racemic → chiral thermal conversion of 2,6-dimethyl-4-(α,α-diphenylmethylene)cyclohexa-2,5-dien-1-one.[12] *X*-Ray data on eight derivatives of *cis*-4a,5,8,8a-tetrahydro-1,4-naphthoquinone have been correlated with their solid-state photochemistry.[13]

[4] J. Bernstein, *Acta Cryst.*, 1979, **B35**, 360.
[5] S. Lifson, A. T. Hagler, and P. Dauber, *J. Amer. Chem. Soc.*, 1979 **101**, 5111.
[6] M. D. Newton, G. A. Jeffrey, and S. Takagi, *J. Amer. Chem. Soc.*, 1979, **101**, 1997.
[7] P. Murray-Rust and W. D. S. Motherwell, *Acta Cryst.*, 1978, **B34**, 2518, 2527.
[8] P. Murray-Rust and W. D. S. Motherwell, *Acta Cryst.*, 1978, **B34**, 2534.
[9] F. H. Allen, *Acta Cryst.*, 1980, **B36**, 81.
[10] P. B. Hitchcock and I. Beheshti, *J.C.S. Perkin II*, 1979, 126.
[11] D. J. Chadwick and J. D. Dunitz, *J.C.S. Perkin II*, 1979, 276.
[12] T. W. Lewis, I. C. Paul, and D. Y. Curtin, *Acta Cryst.*, 1980, **B36**, 70.
[13] J. R. Scheffer and A. A. Dzakpasu, *J. Amer. Chem. Soc.*, 1978, **100**, 2163.

It is not uncommon for crystal structures to provide the theoretician with new phenomena to explain. This is well illustrated by, for example, [3]- and [4]-rotane, (3) and (4), both of which have abnormally short mean central C—C bond distances of 1.470 Å, suggesting that there are π-bonding contributions in these structures.[14]

(3)

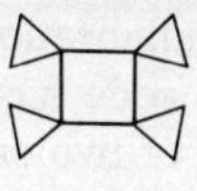
(4)

The central C—C bond length in 9,9-bitriptycyl has been found to be shorter than predicted by force-field calculations, even though these correctly predicted other aspects of the geometry.[15] An authentic example of an intramolecular O—H $\cdots$ π hydrogen bond has been reported in the crystal structure of 2,6-diphenylphenol.[16]

2 Natural Products and Biological Molecules

The natural world continues to cause surprises with the bizarre nature of the secondary metabolites of plants. Noteworthy among these crystal structures are mintsulphide (5), a sulphur-containing sesquiterpene from peppermint oil,[17] and solanscone (6), a novel sesquiterpene ketone from *Nicotiana tabacum* (as the oxime).[18] DL-Bi-(*O*-trimethyl-*cis*-brazilane) (7), a derivative of brazilin, has an

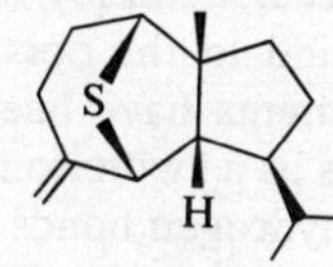

(5)

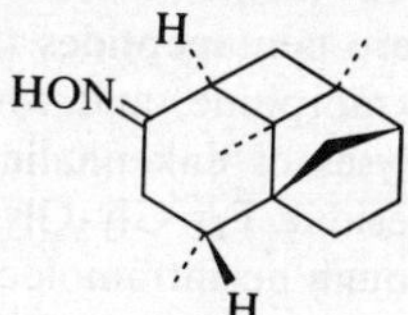

(6)

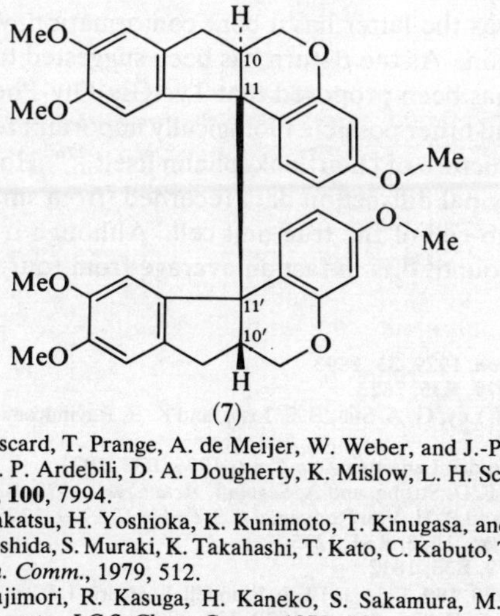

(7)

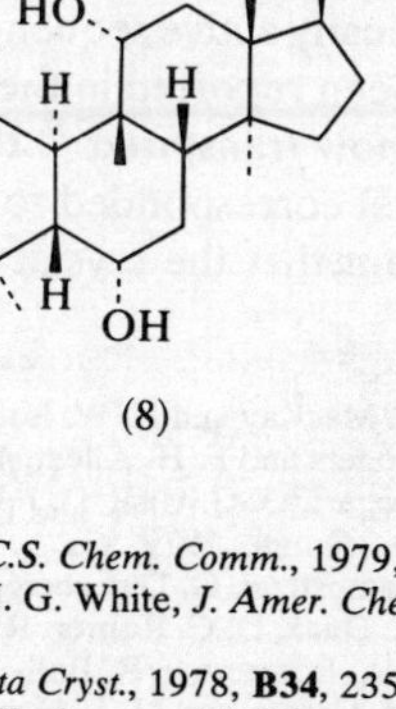

(8)

[14] C. Pascard, T. Prange, A. de Meijer, W. Weber, and J.-P. Barnier, *J.C.S. Chem. Comm.*, 1979, 425.
[15] M. H. P. Ardebili, D. A. Dougherty, K. Mislow, L. H. Schwartz, and J. G. White, *J. Amer. Chem. Soc.*, 1978, **100**, 7994.
[16] K. Nakatsu, H. Yoshioka, K. Kunimoto, T. Kinugasa, and S. Ueji, *Acta Cryst.*, 1978, **B34**, 2357.
[17] T. Yoshida, S. Muraki, K. Takahashi, T. Kato, C. Kabuto, T. Suzuki, T. Uyehara, and T. Ohnumen, *J.C.S. Chem. Comm.*, 1979, 512.
[18] T. Fujimori, R. Kasuga, H. Kaneko, S. Sakamura, M. Noguchi, A. Furusaki, N. Hashiba, and T. Matsumoto, *J.C.S. Chem. Comm.*, 1978, 563.

expected staggered conformation; however, the three central bonds C-10—C-11, C-11—C-11′, and C-11′—C-10′ are all longer than normal C(sp^3–sp^3) bonds, with the central one of length 1.61 Å.[19] This is most likely a result of the steric overcrowding in the molecule.

An important code of recommended practice and standardized procedures for the assignment of absolute configurations has been established.[20] The use of the principles laid down in this code has led to the revision of the chirality originally assigned for clerodin,[21] one of two errors found so far among the *ca.* 900 absolute configurational assignments derived by *X*-ray crystallography that are in the literature.

The analysis of steroid structures continues to be an active topic. The two independent molecules of 3β-acetoxy-16β-methylpregn-5-en-20-one have their side-chains in very different conformations;[22] energy calculations show that the minimum is a broad one. The careful analysis of the structures of four oestranes in the light of conformational data on other 1,3,5(10)-oestratriene structures has shown the role of conformational transmission, in that exocyclic non-bonded interactions are crucial factors in determining conformation.[23] The structures of three 4,4,14α-trimethyl-19(10 → 9)-*abeo*-5β,9β,10α-pregnane-6,11-diols [typified by (8)] have been determined.[24] Their conformations are as predicted by force-field methods, which have also been used to analyse puckering parameters and steric energies.

Interest in structural studies of small peptides has been revitalized by the discovery, and subsequent intensive biochemical study in many laboratories, of the enkephalins. These are pentapeptides that occur naturally in animal brains, that mimic the action of morphine, and which bind to the opiate receptor. Several crystallographic analyses of enkephalin fragments have been reported. The *N*-terminal tripeptide residue Tyr-Gly-Gly exists as a zwitterion with an α-helix-like conformation,[25] although no intramolecular hydrogen bonds were observed. The structures of two tetrapeptide enkephalin fragments, Tyr-Gly-Gly-Phe and Gly-Gly-Phe-Leu, have been determined.[26] Their conformations are quite distinct; the former has a Gly-Gly β-turn whereas the latter has a bent conformation, without intramolecular head-to-tail interaction. As the β-turn has been suggested to occur for enkephalin itself in solution, it has been proposed that Tyr-Gly-Gly-Phe is the biologically active sequence. This and other possible biologically important features have been reported in the crystal structure of [Leu^5]enkephalin itself.[27a] However, it has now transpired[27b] that the original diffraction data recorded (from small thin crystals) corresponded to a small sub-cell of the true unit cell. Although it is now apparent that the crystal structure found[27a] is in fact an average from four closely

[19] M. F. MacKay and N. W. Isaacs, *Tetrahedron*, 1979, **35**, 1893.
[20] D. Rogers and F. H. Allen, *Acta Cryst.*, 1979, **B35**, 2823.
[21] D. Rogers, G. G. Ünal, D. J. Williams, S. V. Ley, G. A. Sim, B. S. Joshi, and K. R. Ravindranath, *J.C.S. Chem. Comm.*, 1979, 97.
[22] H. Campsteyn, O. Diderberg, L. Dupont, and J. Lamotte, *Acta Cryst.*, 1979, **B35**, 2971.
[23] W. L. Duax, D. C. Rohrer, R. H. Blessing, P. D. Strong, and A. Segaloff, *Acta Cryst.*, 1979, **B35**, 2656.
[24] J. C. A. Boeyens, J. R. Bull, A. Tuinman, and P. H. Van Rooyen, *J.C.S. Perkin II*, 1979, 1279.
[25] W. M. Carson and M. L. Hackert, *Acta Cryst.*, 1978, **B34**, 1275.
[26] T. Prangé and C. Pascard, *Acta Cryst.*, 1979, **B35**, 1812.
[27] (*a*) G. D. Smith and J. F. Griffin, *Science*, 1978, **199**, 1214; (*b*) T. L. Blundell, L. Hearn, I. J. Tickle, R. A. Palmer, B. A. Morgan, G. D. Smith, and J. F. Griffin, *Science*, 1979, **205**, 220.

related molecules, it has not as yet been possible to solve the true structure, and hence unequivocally to define all the conformational features, especially those of the side-chains.

Among the several structures of cyclic peptides published during the period under review, two are especially worthy of note. The structure of the cyclic decapeptide gramicidin S (as a complex with urea) has finally been solved,[28] after many years of effort. It has many of the features predicted in an early model,[29] as well as some unexpected ones, such as an intramolecular hydrogen bond. The conformation of the cyclic decapeptide antamanide is the same as that found in several solvents as well as in several derivatives with variants in the side-chains.[30] It has thus been concluded that the conformation found is an intrinsic property of the molecule.

The crystal structures of several fragments of the nucleic acid DNA have attracted much attention. The tetranucleotide d(pApTpApT) forms Watson–Crick base-pairs in the solid state,[31] but not as a continuous fragment of double-stranded DNA. Instead, two short segments of base-pairing are formed by each tetranucleotide, by means of hydrogen-bonding to two other molecules in the crystal structure. On the other hand, the hexanucleotide d(CpGpCpGpCpG) crystallizes as a double-helical Watson–Crick base-paired duplex.[32] This structure has a number of remarkable and unexpected features, which may well have considerable biological implications; for example, the helicity is left-handed rather than right-handed, as in the classical model of DNA.

[28] S. E. Hull, R. Karlsson, P. Main, M. M. Woolfson, and E. J. Dodson, *Nature*, 1978, **275**, 206.
[29] D. C. Hodgkin and B. M. Oughton, *Biochem. J.*, 1957, **65**, 752.
[30] I. L. Karle, T. Wieland, D. Schermer, and H. C. Ottenhegm, *Proc. Nat. Acad. Sci. U.S.A.*, 1979, **76**, 1532.
[31] M. A. Viswamitra, P. Kennard, P. G. Jones, G. M. Sheldrick, S. Salisbury, L. Falvello, and Z. Shakked, *Nature*, 1978, **273**, 687.
[32] A. H.-J. Wang, G. J. Quigley, F. J. Kolpak, J. L. Crawford, J. H. van Boom, G. van der Marel, and A. Rich, *Nature*, 1979, **282**, 680.

2 Physical Methods and Techniques
Part (iii) N.M.R. Spectroscopy

By J. K. M. SANDERS
University Chemical Laboratory, University of Cambridge, Lensfield Road, Cambridge CB2 1EW

1 Introduction

Last year's Report[1] was both long and concentrated, describing major advances published during 1976–78 in a variety of areas of n.m.r. spectroscopy. This Report on the 1979 literature retains the same format as before but concentrates on the development of those ideas and techniques as tools for solving chemical problems. No attempt has been made to restrict examples to specifically organic applications: each section contains illustrations from physical, inorganic, or biological applications, chosen because they are instructive and interesting.

2 Spin–Lattice Relaxation

For organic molecules in solution the dominant relaxation mechanism for both ^1H and ^{13}C is dipole–dipole relaxation, induced by nearby protons. The relaxation time is given by equation (1), where r is the distance from neighbouring protons and τ_c is a correlation time for molecular motion.

$$T_1^{-1} \propto \tau_c \sum r^{-6} \qquad (1)$$

Thus T_1 gives information about mobility or about inter-nuclear distances and both effects have been put to good effect this year in assigning spectra. For example, the methine protons of frangulanin (1) are expected to experience similar tumbling rates, and so they can be distinguished on the basis of their different distances from neighbouring protons; the methyls, which all have the same geometry (and hence the same $\sum r^{-6}$), can be assigned through their differing mobilities.[2] Similarly, C-18 in asperdiol (2) is the most mobile methylene carbon and has a significantly longer T_1 than the ring methylenes.[3] As expected from equation (1), the ring methine carbons of (2) relax at half the rate of the methylenes.

More quantitatively, equation (1) can be used to extract detailed geometrical information, particularly when substitution by deuterium[4,5] or selective pulse experiments[6] allow the contribution of a particular proton to the relaxation of

[1] J. K. M. Sanders, *Ann. Reports*, (*B*), 1978, **75**, 3.
[2] E. Haslinger and W. Robien, *Monatsh.*, 1979, **110**, 1011.
[3] G. E. Martin, J. A. Matson, and A. J. Weinheimer, *Tetrahedron Letters*, 1979, 2195.
[4] L. D. Hall, K. F. Wong, W. E. Hull, and J. D. Stevens, *J.C.S. Chem. Comm.*, 1979, 953.
[5] L. M. Jackman and J. C. Trewella, *J. Amer. Chem. Soc.*, 1979, **101**, 6428.
[6] L. D. Hall, K. F. Wong, and H. D. W. Hill, *J.C.S. Chem. Comm.*, 1979, 951.

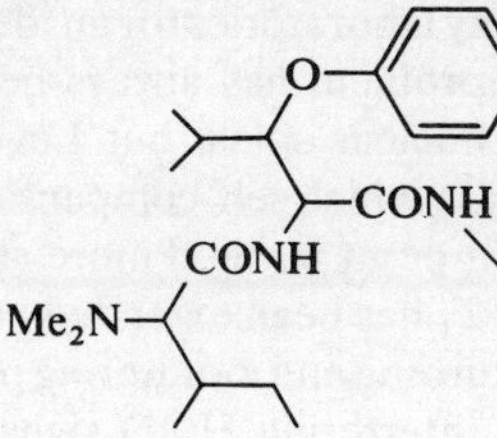

(1)

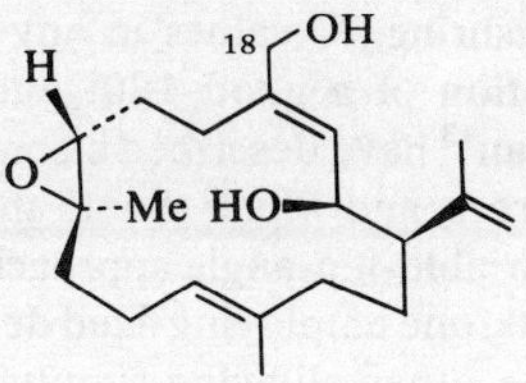

(2)

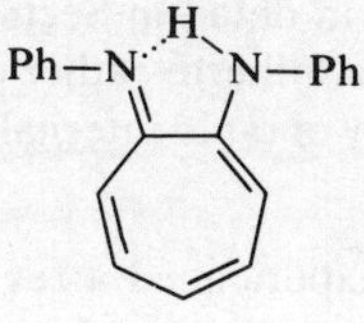

(3)

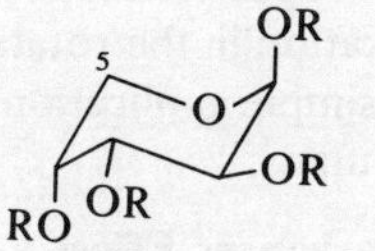

(4) R = $COCD_3$

another nucleus to be determined. In the case of (3), ^{13}C and ^{15}N relaxation rates enabled precise measurement of the hydrogen (deuterium) bond geometry and demonstrated its asymmetry rather convincingly,[5] whilst substitution by deuterium at C-5 in (4) gave values for all the inter-proton distances that were in complete agreement with those from neutron-diffraction measurements.[4] All the above examples report suitable control experiments which ensure that indeed dipole–dipole interactions are responsible for the observed effects, and all are for molecules that are essentially rigid. More difficult is the determination of mobility, given an essentially known geometry. For example, interpretation of ^{13}C relaxation data is crucially dependent on choosing an appropriate value for the C—H bond length in both small organic molecules[7] and proteins.[8] Even then, *detailed* interpretation of relaxation rates in terms of molecular motion is fraught with problems,[9] and should be attempted only by the brave.

Paramagnetics are very efficient inducers of relaxation, and therefore should be removed from solutions where T_1 measurements are contemplated. The most common approach for aqueous solutions is to add EDTA, but when this was tried with a solution containing Fe^{3+} and phosphate, the T_1 of ^{31}P actually shortened because of the formation of ternary EDTA–Fe–P_i complexes.[10] An alternative approach that is sometimes successful is extraction with 8-hydroxyquinoline,[11] but the important point is that care must be exercised. Of course paramagnetic effects on T_1 may be exploited, and one fine example of this — inadvertently omitted from last year's Report [Vol. 75, p. 3] — is the determination of the *bound* conformation of propionyl-CoA on the metal-containing enzyme transcarboxylase.[12]

7 R. K. Harris and R. H. Newman, *Mol. Phys.*, 1979, **38**, 1315.
8 K. Dill and A. Allerhand, *J. Amer. Chem. Soc.*, 1979, **101**, 4376.
9 P. Stilbs and M. E. Moseley, *J. Magn. Reson.*, 1979, **33**, 209.
10 G. A. Elgavish and J. Granot, *J. Magn. Reson.*, 1979, **36**, 147.
11 D. H. Live, H. R. Wyssbrod, A. J. Fischman, W. C. Agosta, C. H. Bradley, and D. Cowburn, *J. Amer. Chem. Soc.*, 1979, **101**, 474.
12 C. H. Fung, R. J. Feldmann, and A. S. Mildvan, *Biochemistry*, 1976, **15**, 75.

Experimental Methods.—The search continues in many laboratories for an ideal way of measuring T_1 values in any given sample. One problem has always been the generation of a good 180° pulse to invert all the nuclear spins, but Levitt and Freeman[13] have described a composite 'sandwich' pulse which self-compensates for any errors, and which should therefore make the null-point method more reliable. The variable-flip-angle approach to measurement of T_1 has been extended into two variants, one employing fixed delays between the perturbing and monitoring pulses[14] and the other allowing simultaneous nulling of an interfering HOD signal.[15] A completely different way of removing unwanted resonances is to use spin echoes in T_1 sequences.[16,17] Spin echoes are discussed in detail in Section 5 of this Report. Spin–lattice relaxation in the rotating frame, a hitherto rather obscure subject, has proved to have a simple application in the study of rapid internal motions, and it may well become popular.[18]

The Nuclear Overhauser Effect.—This has experienced a revival, in the guise of n.O.e. difference spectroscopy, and is now proving to be perhaps the most subtle and powerful manifestation of spin–lattice relaxation phenomena in organic chemistry. The 'traditional' n.O.e. experiment is very powerful, and it has been exploited skilfully to determine the structure, conformation, and absolute configuration of the polypeptide antibiotic ristocetin (mol. wt. 3000),[19] but it is limited in scope because the signal to be observed must be resolved and because the minimum credible effect using integration is *ca.* 5%. In n.O.e. difference spectroscopy a control spectrum without n.O.e. is subtracted from the spectrum with n.O.e., so that only spectral changes appear. The signal of interest need no longer be resolved in the normal spectrum, and the lower limit of observable effect is determined only by instrument stability. Thus all the protons of the steroids 1-dehydrotestosterone (5) and 11β-hydroxyprogesterone (6) have been resolved and assigned, using n.O.e.'s in the

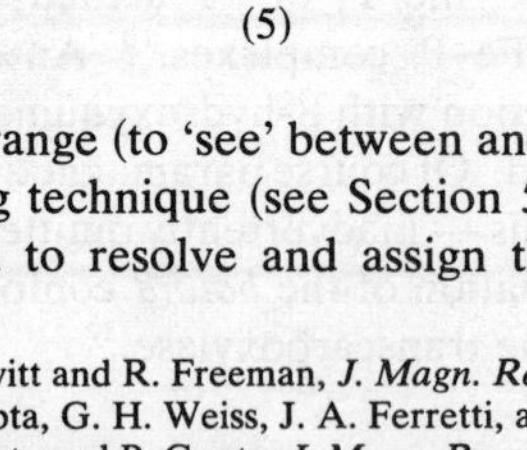

(5)

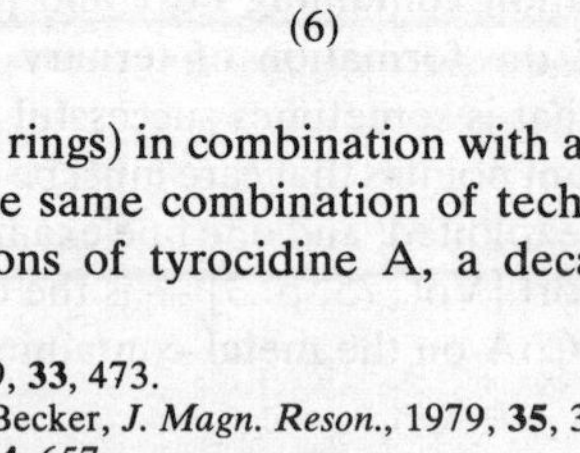

(6)

0.5—5% range (to 'see' between and across rings) in combination with a difference decoupling technique (see Section 5).[20] The same combination of techniques has been used to resolve and assign the protons of tyrocidine A, a decapeptide.[21]

[13] M. H. Levitt and R. Freeman, *J. Magn. Reson.*, 1979, **33**, 473.
[14] R. K. Gupta, G. H. Weiss, J. A. Ferretti, and E. D. Becker, *J. Magn. Reson.*, 1979, **35**, 301.
[15] R. K. Gupta and P. Gupta, *J. Magn. Reson.*, 1979, **34**, 657.
[16] J. Hochmann, R. C. Rosanske, and G. C. Levy, *J. Magn. Reson.*, 1979, **33**, 275.
[17] D. L. Rabenstein, T. Nakashima, and G. Bigam, *J. Magn. Reson.*, 1979, **34**, 669.
[18] D. M. Doddrell, M. R. Bendall, P. F. Barron, and D. T. Begg, *J.C.S. Chem. Comm.*, 1979, 77.
[19] D. H. Williams, V. Rajananda, G. Bojesen, and M. P. Williamson, *J.C.S. Chem. Comm.*, 1979, 906 and refs. therein.
[20] L. D. Hall and J. K. M. Sanders, *J.C.S. Chem. Comm.*, 1980, 368.
[21] M. Kuo and W. A. Gibbons, *J. Biol. Chem.*, 1979, **254**, 6278.

Nuclear Overhauser effect difference spectroscopy has also been used[22] to assign non-equivalent protons in the primary amide group—$CONH_2$, to study the conformation of trimethoprim bound to dihydrofolate reductase,[23] and to characterize the interaction of ADP with creatine kinase.[24]

Most n.O.e. experiments are carried out under steady-state conditions, but in large proteins, where spin diffusion eventually dilutes the useful r^{-6} dependence of the n.O.e., it is helpful to acquire 'truncated' n.O.e.'s just after the saturation has begun to build up.[25] Alternatively 'transient' n.O.e.'s arising from the selective inversion of a single resonance can be useful in geometry measurements[9] or spectral assignments[20] for organic molecules. Since the build-up of transient and truncated effects depends on the relaxation rates of both nuclei involved, the rates of build-up can be used to measure T_1 values of signals that are otherwise inaccessible because they have large negative n.O.e.'s.[26]

Finally we come to Chemically Induced n.O.e., or deceptive CIDNP,[27] an elegant technique for generating enhancements in a small part of the molecule of interest: that molecule must (for the present) contain an exchangeable OH or NH proton. To the solution is added an aromatic acceptor and a suitable tertiary amine. Irradiation with light in the probe leads to electron transfer [equation (2)] and the resulting amine radical cation exchanges its now acidic protons (**H**) with the molecule under study:

$$\text{Acceptor} + R_2NCH_2Me \xrightleftharpoons{h\nu} A^{\bar{\cdot}} \quad (R_2NC\mathbf{H}_2Me)^{\dot{+}} \qquad (2)$$

The exchanged protons carry into their 'host' CIDNP, which then imparts *negative* enhancements to any nucleus that is *J*-coupled to them and a *positive* enhancement to protons that are dipolar-coupled to them. This trick has promise both as an assignment tool and as a sensitivity-enhancement method for nuclei such as ^{15}N.

3 Other Nuclei

Deuterium, ^{2}H.—Direct observation of ^2H spectra can be a very powerful way of studying reaction mechanisms, and one of this year's finest examples is the determination of the mechanism and of the conformation of the transition state of the vitamin-D_3–previtamin-D_3 equilibrium.[28] Direct comparison of proton, deuterium, and tritium chemical shifts in intramolecular hydrogen bonds gives a detailed picture of the shape of the potential well, easily distinguishing a symmetrical hydrogen bond from two unsymmetrical rapidly interconverting bonds.[29] This approach nicely complements Jackman's,[5] which was described in Section 2. Deuterium is an almost ideal nucleus owing to the ease of incorporation into biological molecules, the

[22] A. G. Redfield and S. Waelder, *J. Amer. Chem. Soc.*, 1979, **101**, 6151.
[23] P. J. Cayley, J. P. Albrand, J. Feeney, G. C. K. Roberts, E. A. Piper, and A. S. V. Burgen, *Biochemistry*, 1979, **18**, 3886.
[24] M. Vasák, K. Nagayama, K. Wüthrich, M. L. Mertens, and J. H. R. Kägi, *Biochemistry*, 1979, **18**, 5050.
[25] G. Wagner and K. Wüthrich, *J. Magn. Reson.*, 1979, **33**, 675; A. A. Bothner-By and J. H. Noggle, *J. Amer. Chem. Soc.*, 1979, **101**, 5152.
[26] S. J. Opella, R. A. Friedman, M. C. Jarema, and P. Lu, *J. Magn. Reson.*, 1979, **36**, 81.
[27] J. Bargon and G. P. Gardini, *J. Amer. Chem. Soc.*, 1979, **101**, 7732.
[28] M. Sheves, E. Berman, Y. Mazur, and Z. V. I. Zaretskii, *J. Amer. Chem. Soc.*, 1979, **101**, 1882.
[29] L. J. Altman, D. Laungani, G. Gunnarsson, H. Wennerström, and S. Forsen, *J. Amer. Chem. Soc.*, 1978, **100**, 8264.

simplicity of the ensuing spectra, and the straightforward interpretation of its relaxation parameters. For example, the histidine of lysozyme has been labelled with deuterium by exchange and the ^{2}H nucleus then used to study aggregation of the enzyme;[30] deuterium has been incorporated, by chemical modification, into proteins and glycoproteins for n.m.r. studies.[31]

More important for the organic chemist are the isotope effects of deuterium on ^{1}H and ^{13}C chemical shifts (relaxation effects having been discussed in Section 2). Anet and Dekmezian have very carefully distinguished between equilibrium isotope effects, which are due to a change in the population of two or more equilibrating species (usually conformations), and intrinsic effects that are observed in single species.[32] As an example of the latter they describe a shift of 0.0137 p.p.m. in H* of the half-cage (7) when X is changed from H to D. Isotope effects on ^{13}C shifts of

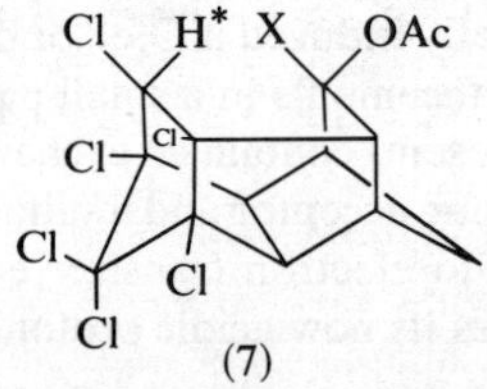

(7)

carbohydrates in H_2O *vs.* D_2O provide an elegant assignment method,[33] and small deuterium effects are strongly amplified at pH values near the pK_a, allowing extraction of all the microscopic pK_a values.[34] A similar amplification effect has long been known with lanthanide shift reagents.[35]

Nitrogen-15.—Last year's Report stated that this nucleus was probably, after ^{1}H and ^{13}C, the most interesting nucleus for many chemists and biochemists, its main drawback being low sensitivity. Section 5 describes several methods which may increase its effective sensitivity sufficiently to make ^{15}N a reasonably accessible nucleus; chemically induced n.O.e. may also help in this regard. Meanwhile, it is possible (with patience, and adequate access to spectrometers) to acquire natural-abundance spectra of, *e.g.*, reserpine and related alkaloids, and to draw up all the expected correlations of chemical shift with structure and conformation.[36]

Most ^{15}N spectra are taken with molecules that have been highly enriched synthetically [see, for example, (3)] or biosynthetically. Thus, measurements of T_1 and of the n.O.e. of labelled oxytocin yield information on its conformation,[11] and the mechanism of action of a serine protease has been thoroughly investigated, using ^{15}N-labelled histidine in the 'catalytic triad'.[37] Perhaps most remarkable of all are the ^{15}N spectra of biosynthetically labelled bacterial cell walls, which supply extraordinarily detailed pictures of the chemical and physical similarities and differences

[30] J. B. Wooten and J. S. Cohen, *Biochemistry*, 1979, **18**, 4188.
[31] M. A. Bernstein, L. D. Hall, and W. E. Hull, *J. Amer. Chem. Soc.*, 1979, **101**, 2744.
[32] F. A. L. Anet and A. H. Dekmezian, *J. Amer. Chem. Soc.*, 1979, **101**, 5449.
[33] P. E. Pfeffer, K. M. Valentine, and F. W. Parrish, *J. Amer. Chem. Soc.*, 1979, **101**, 1265, 7438.
[34] J. J. Led and S. B. Petersen, *J. Magn. Reson.*, 1979, **33**, 603.
[35] J. K. M. Sanders and D. H. Williams, *J.C.S. Chem. Comm.*, 1972, 436.
[36] S. N. Y. Fanso-Free, G. T. Furst, P. R. Srinivasan, R. L. Lichter, R. B. Nelson, J. A. Panetta, and G. W. Gribble, *J. Amer. Chem. Soc.*, 1979, **101**, 1549.
[37] W. W. Bachovchin and J. D. Roberts, *J. Amer. Chem. Soc.*, 1978, **100**, 8041.

between one species and another.[38] It is worth noting that ^{15}N seems to be far more sensitive to the presence of paramagnetics than are ^{1}H and ^{13}C.[11]

The great potential in biosynthetic investigations of detecting ^{15}N through its coupling with the common nuclei was discussed in last year's Report. It should be emphasized that the couplings looked for in such experiments must be reliably known or estimated from model experiments; the dangers are stressed in a study of labelled adenine which found that some $^{1}J_{NC}$ are actually very small indeed, and may not be resolved:[39] see also Section 5.

'Other' Other Nuclei.—Phosphorus-31 continues to be a powerful probe of intact biological entities. It has been used to monitor the internal pH of the chromaffin granules of adrenal gland and to determine the co-ordination state of the ATP within those granules,[40] and also to monitor release of ATP in blood platelets.[41] More exciting possibilities include the *simultaneous* acquisition of ^{31}P and ^{13}C spectra from a single sample, using a probe that is quadruply tuned to both these frequencies and to ^{1}H (for decoupling) and to ^{2}H (lock),[42] and obtaining n.m.r. spectra from parts of bodies, using probes which simply lie on the skin.[43]

Isotope effects of ^{18}O on ^{13}C and ^{31}P chemical shifts are proving to be very useful mechanistic tools: the former has only been used in organometallic chemistry[44] as yet, but it should be generally applicable. The latter has rather quickly become an established method in enzyme chemistry.[1,45] This year, such isotope effects have been used to show that purine nucleotide phosphorylase catalyses exchange of phosphate on α-D-ribose 1-phosphate with C—O cleavage (rather than the expected P—O cleavage)[46] and to show that in two amino-acyl tRNA synthetases the nucleotidyl-transfer step takes place with inversion at phosphorus.[47]

Oxygen-17 has much organic chemical potential also,[48] this year's most intriguing example being the direct observation of H_3O^+ (quartet, $J_{OH} = 106$ Hz) in cold, wet, acidic carbon tetrachloride.[49] Finally, we should note that it is relatively easy to observe ^{113}Cd–^{113}Cd coupling in suitably labelled samples of the protein metallothionein.[50]

4 Paramagnetics

Shift Reagents and Metal Complexes.—It is a measure of the maturity of the field that there is relatively little of novelty to report on shift reagents. Reuben[51] has developed a system for the n.m.r. resolution of chiral molecules in aqueous solutions:

38 A. Lapidot and C. S. Irving, *Biochemistry*, 1979, **18**, 704.
39 M. Kainosho, *J. Amer. Chem. Soc.*, 1979, **101**, 1031.
40 H. B. Pollard, H. Shindo, C. E. Creutz, C. J. Pazoles, and J. S. Cohen, *J Biol. Chem.*, 1979, **254**, 1170.
41 K. Ugurbil, H. Holmsen, and R. G. Shulman, *Proc. Natl. Acad. Sci. USA*, 1979, **76**, 2227.
42 P. Styles, C. Grathwohl, and F. F. Brown, *J. Magn. Reson.*, 1979, **35**, 329.
43 J. J. H. Ackerman, T. H. Grove, G. G. Wong, D. G. Gadian, and G. K. Radda, *Nature*, 1980, **283**, 167.
44 D. J. Darensbourg and B. J. Baldwin, *J. Amer. Chem. Soc.*, 1979, **101**, 6447.
45 M. C. Summers, *Ann. Reports (B)*, 1978, **75**, 391.
46 F. Jordan, J. A. Patrick, and S. Salamone, Jr., *J. Biol. Chem.*, 1979, **254**, 2384.
47 S. P. Langdon and G. Lowe, *Nature*, 1979, **281**, 320.
48 See, for example, M. Katoh, T. Sugawara, Y. Kawada, and H. Iwamura, *Bull. Chem. Soc. Japan*, 1979, **52**, 3475 and refs. therein.
49 G. D. Mateescu and G. M. Benedikt, *J. Amer. Chem. Soc.*, 1979, **101**, 3959.
50 J. D. Otvos and I. M. Armitage, *J. Amer. Chem. Soc.*, 1979, **101**, 7734.
51 J. Reuben, *J.C.S. Chem. Comm.*, 1979, 68.

α-hydroxy-carboxylic acids form 2:1 complexes with lanthanide ions, so the addition of a 1:1 mixture of, *e.g.*, D-mandelate [$PhCH(OH)CO_2^-$] and lanthanide to another, optically impure, α-hydroxy-acid will result in the formation of diastereoisomeric complexes with different stabilities and geometries. Alternatively, enantiotopic groups in a prochiral molecule (the methyls of α-hydroxybutyrate or the methylene protons of glycollate, $HOCH_2CO_2^-$) can be rendered non-equivalent.[51]

The fitting of shift data to geometrical models is perhaps becoming less risky as both the models and their statistical evaluation improve. Thus for adamantan-2-one the goodness of fit improves dramatically as the model is refined from a single metal-binding site that is collinear with the carbonyl group to a chemically sensible model with two sites along the carbonyl lone pairs, and improves again in a four-site model that minimizes steric interactions between reagent and substrate.[52] The standard Hamilton *R* test for the statistical evaluation of such geometrical models has been shown to be quite inappropriate, and simple alternatives have been proposed.[53] Indeed the apparent failure of shift reagents to distinguish diastereoisomers which was discussed in last year's Report (p. 14, ref. 125) turns out to be only a failure of the statistical method — the use of valid statistics shows that the diastereoisomers were distinguishable after all. It is a matter of taste whether one wants chemical problem-solving to be reduced to such statistics, particularly when 'significance testing [can] only test whether one *mathematical* model of a *real* situation fits the data better than another such model .... [It] can never determine whether such a model is correct.'[53]

The temperature dependence of lanthanide-induced shifts has given rise to some controversy (see last year's Report, p. 13), but McGarvey[54] has now confirmed theoretically that the dependence should indeed be *approximately*, and not exactly, T^{-2}. However, in cases where reagent–substrate binding is bi- or ter-dentate, the exchange rate becomes so low that it is directly observable, in the form of line broadening and separate free/bound spectra. In such cases it is possible not only to measure temperature dependences of the *shifts* directly but also to determine whether the exchange mechanism is associative or dissociative.[55]

Biologically oriented applications of paramagnetic effects include the determination of conformation of enzyme-bound propionyl-CoA already referred to[12] and the assignment of the five indole *N*-H resonances of hen egg-white lysozyme.[56] N.m.r. spectroscopy of haems has now developed to the point where it is detecting localized conformational flexibility in cytochromes, using a methyl group attached to the haem as reporter,[57] and is giving exquisite detail of electronic structure, co-ordination chemistry, and geometry of complexes in model porphyrins and real proteins.[58]

[52] R. J. Abraham, D. J. Chadwick, L. Griffiths, and F. Sancasson, *Tetrahedron Letters*, 1979, 4691.
[53] M. F. Richardson, S. M. Rothstein, and W.-K. Li, *J. Magn. Reson.*, 1979, **36**, 69.
[54] B. R. McGarvey, *J. Magn. Reson.*, 1979, **33**, 445.
[55] L. F. Lindoy and H. W. Louie, *J. Amer. Chem. Soc.*, 1979, **101**, 841.
[56] R. E. Lenkinski, J. L. Dallas, and J. D. Glickson, *J. Amer. Chem. Soc.*, 1979, **101**, 3071.
[57] P. D. Burns and G. N. LaMar, *J. Amer. Chem. Soc.*, 1979, **101**, 5844.
[58] D. L. Budd, G. N. LaMar, K. C. Langry, K. M. Smith, and R. Nayyir-Mazhir, *J. Amer. Chem. Soc.*, 1979, **101**, 6091 and references therein.

CIDNP.—Three contributions have captured this Reporter's imagination. Chemically Induced n.O.e. has been described in detail in Section 2 of this report. The detection of CIDNP with a time resolution of microseconds after a laser flash was achieved[59] by first saturating the equilibrium magnetization by a series of pulses, then a laser flash of several ns duration is followed by a variable delay (to allow evolution of CIDNP) and an r.f. pulse, to give the n.m.r. signal. Only pure CIDNP responses are observed, and they can be monitored from 1 μs after the flash. The potential for mechanistic chemistry is clearly very significant. Finally, a series of papers from Kaptein's laboratory, developing laser photo-CIDNP as a way of simplifying protein spectra (see last year's Report, p. 15), has culminated in selective detection of the single tryptophan in bovine pancreatic phospholipase A_2 and the use of that tryptophan to characterize a pH-dependent conformational change.[60]

5 New Experimental Methods

Last year's Report highlighted two-dimensional (2-D) n.m.r. and high-resolution solid-state n.m.r. as fundamental developments that would soon have a great impact on many areas of chemistry, and indeed the application of these techniques is already moving from physics into organic chemistry. Thus, *all* the geminal and vicinal proton coupling constants in a steroid have been resolved by 2-D *J* spectroscopy; the sensitivity of ^{15}N n.m.r. has been increased, in some cases by one hundred fold, using population-transfer tricks; solid-state chemical reactions at catalyst surfaces have been observed. This is only the beginning of a new phase in n.m.r. spectroscopy, which will develop rapidly in the next few years as the implications are digested by the organic chemist.

Two-dimensional (2-D) and Spin-echo Techniques.—As explained more fully in last year's Report, 2-D spectra display the response of the nuclear spins to two different sets of frequencies, in contrast to 1-D (traditional) spectra, which are a function of only one frequency. As a bonus, most 2-D techniques are based on spin-echo methods, which means that all effects due to inhomogeneity of the magnetic field are removed and thus that linewidths are determined only by the spin–spin relaxation time T_2.

2-D J *Spectroscopy.* A simple pulse sequence, *i.e.* 90°–delay–180°–delay–collect FID, gives spin-echo spectra in which the signal phases depend on the size and number of *J* couplings. Repetition of the sequence with a set of different delays gives, after suitable manipulation of the data,[1] spectra in which chemical shifts and coupling constants are separated *in different dimensions* (for weakly coupled systems). This leads, of course, to the resolution of signals that are otherwise hopelessly overlapping. The most spectacular use of proton–proton 2-D *J* spectroscopy known to the Reporter (in January 1980) is shown in Figure 1: the spectrum allowed the measurement of all the chemical shifts and the geminal and several long-range couplings of the steroid 1-dehydrotestosterone (5),[61] but others will

[59] G. L. Closs and R. J. Miller, *J. Amer. Chem. Soc.*, 1979, **101**, 1639.
[60] E. H. J. M. Jansen, G. J. M. van Scharrenburg, A. F. Slotbloom, G. H. de Haas, and R. Kaptein, *J. Amer. Chem. Soc.*, 1979, **101**, 7397.
[61] L. D. Hall, J. K. M. Sanders, and S. Sukumar, *J.C.S. Chem. Comm.*, 1980, 366.

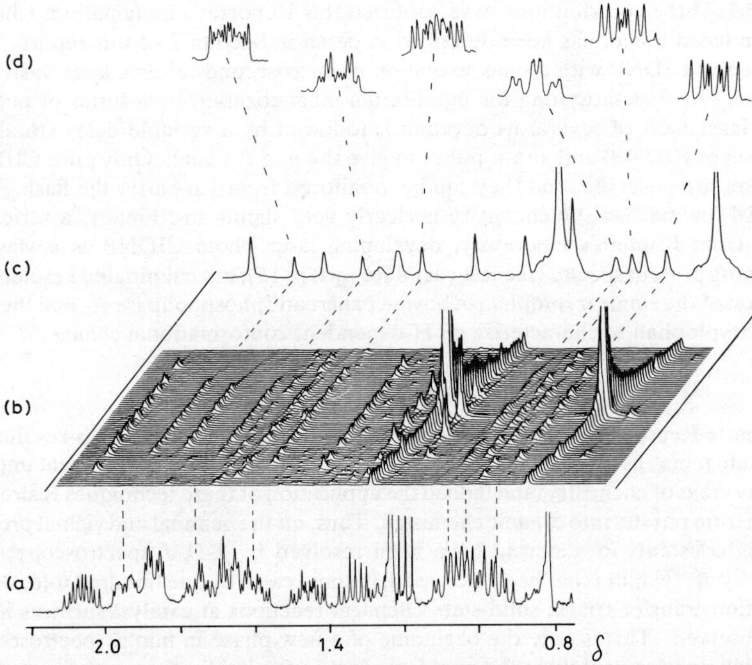

Figure 1 *Partial spectra* (400 MHz) *of* 1-*dehydrotestosterone* (5) (*data are from unpublished spectra and from ref.* 61). (*a*) *Normal one-dimensional spectrum.* (*b*) *Two-dimensional J spectrum of the same region.* (*c*) *'Proton-decoupled' proton spectrum obtained by summing all the two-dimensional cross-sections onto the chemical-shift axis.* (*d*) *Some representative cross-sections, showing the multiplet structure of individual proton resonances*

surely follow. In addition, as heteronuclear couplings behave like chemical shifts in the spin-echo experiment, it is possible to remove (without a decoupler) either all the homo- or all the hetero-nuclear couplings from a proton at will, merely by changing the angle of projection of the proton in the 2-D spectrum.[62] It is possible also to achieve ^{13}C–^{1}H 2-D *J* spectroscopy,[63] in which all the J_{CH} appear in one dimension and the ^{13}C chemical shifts in the other.*

There are several technical problems associated with 2-D *J* spectroscopy, some of which have yet to be solved satisfactorily. The most important (apart from the difficulty of second-order spectra) is that the spectra experience a 'phase twist' around each peak. In early work this was dealt with by plotting absolute value (AV) mode† spectra, but these have the disadvantage of inducing line broadening and generating long 'tails' around intense peaks. One way round the problem is to use a

* Other nuclear pairs are possible, of course.

† $AV = (real^2 + imaginary^2)^{\frac{1}{2}}$

[62] L. D. Hall and S. Sukumar, *J. Amer. Chem. Soc.*, 1979, **101**, 3120; J. R. Everett, D. W. Hughes, A. D. Bain, and R. A. Bell, *ibid.*, p. 6777.

[63] R. Freeman, S. P. Kempsell, and M. H. Levitt, *J. Magn. Reson.*, 1979, **34**, 663.

Lorentzian to Gaussian transformation for the AV spectra;[64] another is to plot the power spectrum (AV^2), which cuts off tails but also distorts multiplets, so that a 1:2:1 triplet becomes 1:4:1.[65] Better, for the $^1H-^1H$ experiment, one can phase each signal of interest, although some distortions remain,[66] and for the $^{13}C-^1H$ experiment several other possibilities are also available.[63]

It is possible to remove interfering solvent signals that have long T_1's by a standard 180° pulse followed by a delay before the spin-echo sequence,[67] or to decouple during a 2-D *J* acquisition; the latter turns out to be an experiment fraught with complications, and it looks unlikely that it will be of general applicability in organic chemistry.[68] Finally, 2-D *J* spectra contain extra spinning side-bands, but these are (fortunately) at predictable places.[69]

2-*D Correlation Spectroscopy*. In this form of 2-D spectroscopy, both frequency axes contain chemical shifts: in the heteronuclear variants, suitable pulsing of both the proton and another nucleus yields 2-D spectra in which just one signal appears in the map, at the co-ordinate for chemical shifts corresponding to the heteronucleus and its coupled proton (Figure 2a). The advantages of this technique are that the

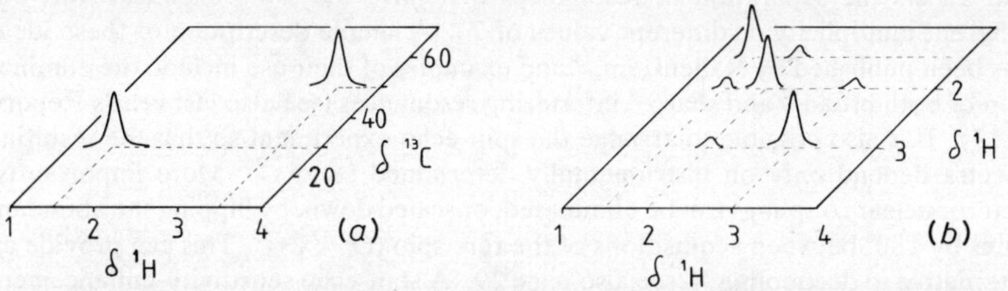

Figure 2 (*a*) *Schematic two-dimensional correlation spectrum, showing a carbon* (δ 30) *coupled to a proton at* δ 1.5 *and a carbon* (δ 50) *coupled to a proton at* δ 3.5. (*b*) *Schematic two-dimensional exchange-correlation spectrum showing a proton* (*X*) *at* δ 3 *exchanging with a proton* (*Y*) *at* δ 2, *the latter also exchanging with a proton* (*Z*) *at* δ 1.5.

heteronucleus is observed with a sensitivity approaching that of protons (by virtue of a cross-polarization effect), that the shifts of coupled nuclei are automatically correlated, abolishing many assignment problems, and that only the nuclei involved in such coupling will appear in the spectrum. Thus in the $^{31}P-^1H$ correlation experiment only 'local' protons that are coupled to ^{31}P are seen, provided that the proton spectrum is first-order.[70] Examples of $^{13}C-^1H$ correlation experiments were given in last year's Report (refs. 174–176): once the technique becomes established, it could eventually be a routine way of acquiring spectra of rare spins. The basis of these experiments is well described by Bodenhausen and Freeman, with an intuitive approach that employs a relatively simple physical picture of population levels.[71]

[64] J. C. Lindon and A. G. Ferridge, *J. Magn. Reson.*, 1979, **36**, 277.
[65] L. D. Hall, S. Sukumar, and G. R. Sullivan, *J.C.S. Chem. Comm.*, 1979, 292.
[66] M. H. Levitt and R. Freeman, *J. Magn. Reson.*, 1979, **34**, 675; L. D. Hall and S. Sukumar, *ibid.*, 1980, **38**, 555.
[67] L. D. Hall and S. Sukumar, *Carbohydrate Res.*, 1979, **74**, C1.
[68] K. Nagayama, *J. Chem. Phys.*, 1979, **71**, 4404.
[69] G. Bodenhausen, S. P. Kempsell, R. Freeman, and H. D. W. Hill, *J. Magn. Reson.*, 1979, **35**, 337.
[70] P. H. Bolton and G. Bodenhausen, *J. Amer. Chem. Soc.*, 1979, **101**, 1080.
[71] G. Bodenhausen and R. Freeman, *J. Magn. Reson.*, 1979, **36**, 221.

Homonuclear 2-D correlation methods allow the unambiguous elucidation of chemical exchange pathways[72] or assignments of coupling constants.[73] As with almost all the 2-D experiments mentioned above and several 1-D experiments that are described at the end of this section, these techniques were developed by Ernst in a series of definitive and brilliant papers. In the chemical-exchange experiment[72] the slowly exchanging peaks are 'labelled' by a 90° pulse, allowed to evolve, and then are monitored by further pulses. The resulting spectrum (Figure 2b) has two identical axes for proton chemical shift, with the main peaks appearing on a diagonal. Peaks representing exchanged magnetization are off-diagonal, and instantly show (in this example) that X exchanges with Y and Y with Z. The technique seems to this Reporter to have an appealing simplicity which will make it most valuable. In its present form at least, the closely related coupling-correlation-pulse sequence[73] gives, by contrast, spectra that are highly complex and low in effective sensitivity, thus limiting its general applicability.

Other Spin-echo Techniques. Even without resort to 2-D methods, the spin echo can be very useful, as it removes effects arising from inhomogeneity of the magnetic field and allows the separation of resonances that have the same chemical shift but different multiplicity or different values of T_2. A simple description of these ideas has been published by Rabenstein,[74] and examples of their use include the elimination of both broad[17] and sharp[75] interfering resonances (see also last year's Report, p. 17). It is also possible to arrange the spin-echo experiment so that the resulting spectra depend *only* on instrumentally determined factors.[76] More importantly, heteronuclear couplings can be eliminated, or scaled down, by flipping the abundant spins by 180° between acquisitions of the rare spin (*i.e.* ^{13}C).[77] This can provide an alternative to decoupling — see also page 29. A spin-echo sensitivity-enhancement technique is described in the following paragraph.

Sensitivity Enhancement.—A series of papers[78,79] have described methods for greatly increasing the effective sensitivity of nuclei such as ^{13}C, ^{15}N, and ^{29}Si. Morris and Freeman[78] employed spin echoes to give a cross-polarization effect between protons and the insensitive nucleus. This is the same effect that operates in the 2-D correlation experiments described above; it requires an estimate of the H–X coupling constant, and enhances all X with that (or similar) coupling. The other papers[79] employ a *selective* population transfer from one proton at a time, giving a time saving for the coupled ^{15}N of 10^2- to 10^4-fold. Both techniques allow the pulse rate to be determined by the proton T_1 rather than the slower X spin T_1, and neither is (in principle) very demanding of hardware or software.

N.M.R. of Organic Solids.—Last year's Report described how a combination of 'magic-angle' spinning, high-power (dipolar) decoupling, and cross-polarization

72 B. H. Meier and R. R. Ernst, *J. Amer. Chem. Soc.*, 1979, **101**, 6441; J. Jeener, B. H. Meier, P. Bachmann, and R. R. Ernst, *J. Chem. Phys.*, 1979, **71**, 4546.
73 K. Nagayama, K. Wüthrich, and R. R. Ernst, *Biochem. Biophys. Res. Comm.*, 1979, **90**, 305.
74 D. L. Rabenstein and T. T. Nakashima, *Analyt. Chem.*, 1979, **51**, 1465A.
75 D. L. Rabenstein and A. A. Isab, *J. Magn. Reson.*, 1979, **36**, 281.
76 A. Bax, A. F. Mehlkoff, and J. Smidt, *J. Magn. Reson.*, 1979, **35**, 167.
77 R. Freeman, S. P. Kempsell, and M. H. Levitt, *J. Magn. Reson.*, 1979, **35**, 447.
78 G. A. Morris and R. Freeman, *J. Amer. Chem. Soc.*, 1979, **101**, 762.
79 H. J. Jakobsen and W. S. Brey, *J. Amer. Chem. Soc.*, 1979, **101**, 774; *J.C.S. Chem. Comm.*, 1979, 478.

techniques rendered high-resolution ^{13}C spectra accessible from organic solids. Applications published this year include the demonstration of conformational preferences in the solid diethoxycarbonium ion $[HC(OEt)_2]^+$ by observation of non-equivalent methylene carbons,[80] the observation of solid-state organometallic fluxional processes,[81] and a study of the gelation of sickle-cell haemoglobin.[82] Significant industrial potential is clearly indicated in observations of toluene adsorbed on zeolite.[83] Styrene can be manufactured by alkylation of toluene with methanol over zeolite catalysts, and n.m.r. experiments revealed the identity of at least some of the intermediates involved.

A neat and simple modification of the standard cross-polarization experiment has allowed Opella's group to observe non-protonated carbons in solids:[84] all that is required is to turn off the proton decoupler very briefly before FID acquisition, and the protonated carbons almost instantly broaden. The technique is not only applicable to solid amino-acids, sugars, steroids, and other organics, but sometimes turns out to be quicker for solid proteins than for protein solutions, giving linewidths of *ca.* 20 Hz.

Finally, it should be noted that it is possible to carry out double cross-polarization experiments from ^1H to ^{13}C to ^{15}N (or from ^1H to ^{15}N to ^{13}C), enabling either the observation of the natural-abundance ^{13}C nuclei that are directly attached to a ^{15}N label or the direct measurement of the relative concentration of ^{15}N–^{13}C and ^{15}N–^{12}C pairs without separation or purification of the solid sample.[85a] This has opened the way for some fine biosynthetic and metabolic experiments,[85b] as the experiment monitors the history of a *bond*, rather than an atom, in an even more direct way than the ^{13}C–^2H methods discussed in last year's Report (p. 8).

Miscellaneous.—Selective excitation of a single resonance, using DANTE, was described in last year's Report, on p. 19. Extensions this year include its use to generate transient n.O.e.'s[20] and a slight modification to yield off-resonance decoupled spectra.[3] An alternative, named CASS*, has been described which does not require such modern computer control, but instead needs a rather good decoupler system.[86] Other new data-acquisition methods include coils for body surfaces,[43] the simultaneous observation of ^{13}C and ^{31}P,[42] and improved ways of noise-decoupling which do not cause overheating of the sample.[77,87]

The flow of data-processing tricks also continues. A simple resolution-enhancement routine which requires no programming has been proposed,[88] and Marshall has shown that a plot of absorption *vs.* dispersion for a broad line will give valuable information on the broadening mechanism involved.[89] Two groups have used

* A clever allusion to the senior author's history.

80 J. R. Lyerla, C. S. Yannoni, D. Bruck, and C. A. Fyfe, *J. Amer. Chem. Soc.*, 1979, **101**, 4770.
81 J. R. Lyerla, C. A. Fyfe, and C. S. Yannoni, *J. Amer. Chem. Soc.*, 1979, **101**, 1351.
82 J. W. H. Sutherland, W. Egan, A. N. Schecter, and D. A. Torchia, *Biochemistry*, 1979, **18**, 1797.
83 M. D. Sefcik, *J. Amer. Chem. Soc.*, 1979, **101**, 2164.
84 S. J. Opella, M. H. Frey, and T. A. Cross, *J. Amer. Chem. Soc.*, 1979, **101**, 5856.
85 (*a*) J. Schaefer, R. A. McKay, and E. O. Stejskal, *J. Magn. Reson.*, 1979, **34**, 443; (*b*) J. Schaefer, E. O. Stejskal, and R. A. McKay, *Biochem. Biophys. Res. Comm.*, 1979, **88**, 274.
86 G. T. Andrews, I. J. Colquhoun, B. R. Doggett, W. McFarlane, B. E. Stacey, and M. R. Taylor, *J.C.S. Chem. Comm.*, 1979, 89.
87 V. J. Basus, P. D. Ellis, H. D. W. Hill, and J. S. Waugh, *J. Magn. Reson.*, 1979, **35**, 19.
88 B. Clin, J. de Bony, P. Lalanne, J. Blais, and B. Lemanceau, *J. Magn. Reson.*, 1979, **33**, 457.
89 A. G. Marshall and D. C. Roe, *J. Magn. Reson.*, 1979, **33**, 551.

decoupling difference spectroscopy to simplify and assign spectra:[20,21] a spectrum with decoupling is subtracted from one without, leaving responses only from the decoupled nuclei. This is particularly useful when the resonances are not otherwise resolved.

'Conventional' flow n.m.r. has proved to be useful for studying reactions that are too fast for standard n.m.r. techniques, but which are slow on the time-scale of FID collection: for example, it is possible, at −40 °C, to see intermediates in the nucleophilic substitution of 2,4,6-trinitroanisole by butylamine.[90] Ernst has now developed a theory, and tested it experimentally, for stopped-flow n.m.r. when chemical evolution is occurring on the time-scale of one FID collection, *i.e.* rates of around 50—100 $l\ mol^{-1}\ s^{-1}$.[91] Characteristic line shapes, quite different from those seen in exchange equilibria, were predicted theoretically and then observed experimentally. The beautifully designed apparatus achieved a dead time of only *ca.* 2 ms by abandoning the idea of a sample tube: the receiver coil was simply wound around the mixing chamber.[91]

The determination of molecular geometry by dissolving a compound in a nematic solvent to restore some dipolar coupling interactions is a well-established technique. However, in a definitive paper from Diehl *et al.*, the structure of [1-^{13}C]benzene has been determined with a precision of *better than* 0.001 *Angstroms* for all the interatomic distances.[92] This paper discusses lucidly, and in great detail, all the problems which may arise from, for example, vibrational motion. Ernst has shown that geometrical information can be re-introduced into nematic-phase proton-decoupled ^{13}C spectra by the selective introduction of ^{2}H or ^{15}N. These added nuclei induce splittings in the ^{13}C signals that can be directly converted into distance information.[93] In the same paper, a 2-D *J* technique is also described that was used to give geometrical information.[93]

The principles of spin imaging (also called spin mapping, or Zeugmatography) were outlined in last year's Report. The technique is potentially of medical importance if it can be made sufficiently sensitive that it can map portions of the body in a short time. To this end, Ernst has carried out a detailed theoretical study of the potential sensitivity characteristics of the various different imaging methods.[94] An imaging method has been described which is based on 2-D n.m.r. methods,[95] and n.m.r. images have been obtained both from water flowing in the pores of ceramics[96] and from solids.[97]

[90] C. A. Fyfe, S. W. H. Damji, and A. Koll, *J. Amer. Chem. Soc.*, 1979, **101**, 951.
[91] R. O. Kühne, T. Schaffhauser, A. Wokaun, and R. R. Ernst, *J. Magn. Reson.*, 1979, **35**, 39.
[92] P. Diehl, H. Bösiger, and H. Zimmermann, *J. Magn. Reson.*, 1979, **33**, 113.
[93] A. Höhener, L. Müller, and R. R. Ernst, *Mol. Phys.*, 1979, **38**, 909.
[94] P. Brunner and R. R. Ernst, *J. Magn. Reson.*, 1979, **33**, 83.
[95] D. I. Hoult, *J. Magn. Reson.*, 1979, **33**, 183.
[96] R. J. Gummerson, C. Hall, W. D. Hoff, R. Hawkes, G. N. Holland, and W. S. Moore, *Nature*, 1979, **281**, 56.
[97] R. A. Wind and C. S. Yannoni, *J. Magn. Reson.*, 1979, **36**, 269.

3 Theoretical Chemistry

By H. S. RZEPA
Department of Chemistry, Imperial College of Science and Technology, (University of London), London SW7 2AY

1 Introduction

The activity in this field is illustrated by the citation, in volumes 89—91 of *Chemical Abstracts*, of about 420 references to *ab initio* calculations alone! This necessarily selective report reviews recent advances in quantum-mechanical methods and the application of these techniques to the electronic structure and geometry of molecules and to dynamic processes and reaction hypersurfaces. A new journal concerned with computational chemistry is appearing,[1] and recent advances have been summarized in excellent series on the theory and application of quantum-mechanical methods[2] and in the Specialist Periodical Reports of the Chemical Society.[3] Particularly characteristic of the past few years has been the use of standard computer programs distributed by QCPE,[4] which has greatly facilitated the direct comparison of the results of different investigations.

2 Advances in Theoretical Techniques

Geometry Optimization.—Despite misgivings by Woolley[5] concerning the Born–Oppenheimer approximation and whether or not the geometry of a molecule can be usefully defined, much effort has been expended in this area. The desirability of complete geometry optimization and the characterization of stationary points in a potential surface by inspection of the corresponding force-constant (Hessian) matrix has been amply demonstrated by, amongst others, Dewar and co-workers, who have concentrated in their studies on the use of computationally fast semi-empirical methods.[6] Efficient optimization at the *ab initio* level was not feasible until the development of analytical methods for obtaining the first derivatives of the energy with respect to the nuclear co-ordinates. Pulay has been prominent[7] in the development of these gradient methods, and recently a number of programs have

[1] *J. Computational Chem.*, Wiley–Interscience, 1980.
[2] (*a*) J. A. Pople, W. J. Hehre, L. Radom, and P. von R. Schleyer, 'Ab initio Molecular Orbital Theory. II: Assessment and Applications', Academic Press, New York, 1979; (*b*) 'Modern Theoretical Chemistry', ed. H. F. Schaefer, Plenum, New York, 1977, Vols. 3 and 4; *ibid.*, ed. G. A. Segal, 1977, Vols. 7 and 8.
[3] 'Theoretical Chemistry', ed. R. N. Dixon and C. Thompson, (Specialist Periodical Reports), The Chemical Society, London, 1978, Vol. 3.
[4] Quantum Chemistry Program Exchange, Indiana University, Bloomington, In.
[5] R. G. Woolley, *J. Amer. Chem. Soc.*, 1978, **100**, 1073.
[6] (*a*) M. J. S. Dewar, M. L. McKee, and H. S. Rzepa, *J. Amer. Chem. Soc.*, 1978, **100**, 3607; (*b*) M. J. S. Dewar and G. P. Ford, *ibid.*, 1979, **101**, 5558.
[7] (*a*) P. Pulay, in ref. 2*b*, Vol. 4, pp. 152—185; (*b*) P. Pulay, *Theor. Chim. Acta*, 1979, **50**, 299.

been reported which are capable of carrying out geometry optimization more or less automatically.[8]

Several groups have now published algorithms for the calculation of first derivatives of spin-restricted SCF wave-functions of doublets, triplets, open-shell singlets, quartets, and multiconfigurational (MCSCF) wave-functions.[9] The time taken in calculating the derivatives is between one and two times that for a SCF calculation, but the use of these in efficient optimization algorithms such as that of Davidon–Fletcher–Powell[6] is expected to be more efficient and reliable than the commonly used method of cyclic optimization of individual geometrical variables.

Electron Correlation.—Handy *et al.*[10] have suggested that whilst the 1970's will be viewed as the decade in which variational SCF wave-functions of up to almost Hartree–Fock quality came to be applied to a broad range of chemical problems, the 80's will see the routine application of methods which go beyond the Hartree–Fock approximation. The major problem is the calculation of the electron-correlation energy of a system, defined as the difference between the energy calculated at the Hartree–Fock level and that obtained from the exact solution of the non-relativistic Schroedinger equation. An approach has been outlined for a 'standard' direct configuration interaction (CI) procedure which includes all single and double excitations and which is expected to recover about 95% of the correlation energy for a small molecule such as water, and rather less for larger molecules. The price to be paid, however, is considerably increased computing time, which may show a dependence on the basis set of between n^5 and n^6 (where n is the size of the basis set), compared with the usual n^4 dependence of single-configuration *ab initio* calculations and the n^2 dependence of the NDO semi-empirical methods! Procedures have recently been developed, based on the graphical unitary group approach to CI,[11] which can reduce the computing time two- or three-fold. Papers from a recent symposium concerned with this problem have appeared.[12]

Improvements in the use of pseudo-core or effective core potentials for elements of high atomic weight have been reported.[13] Morokuma has described his energy, charge, and spin decomposition analysis,[14] which enables the evaluation of charge and spin redistributions at the transition state and along the reaction pathway. The importance of charge-transfer interactions in chemical reactions was emphasized.

Lipscomb *et al.* have given a detailed comparison of the PRDDO method with the STO-3G, CNDO/2, INDO, and several other procedures.[15] PRDDO was about 6 times slower than CNDO/2, and eliminated 90% of the error of the latter as compared with reference SCF calculations. Relative energies were only moderately

[8] (*a*) D. Poppinger, *Chem. Phys. Letters*, 1975, **34,** 332; (*b*) H. B. Schlegel, S. Wolfe, and F. Bernardi, *J. Chem. Phys.*, 1975, **63**, 3632; (*c*) M. Dupuis and H. F. King, *ibid.*, 1978, **68**, 3998; (*d*) A. Komornicki and R. L. Jaffe, *ibid.* 1979, **71**, 2150.

[9] (*a*) J. D. Goddard, N. C. Handy, and H. F. Schaefer, *J. Chem. Phys.*, 1979, **71**, 1525; (*b*) S. Kato and K. Morokuma, *Chem. Phys. Letters*, 1979, **65**, 19.

[10] N. C. Handy, J. D. Goddard, and H. F. Schaefer, *J. Chem. Phys.*, 1979, **71**, 426.

[11] (*a*) P. E. M. Siegbahn, *J. Chem. Phys.*, 1979, **70**, 5391; (*b*) B. R. Brooks and H. F. Schaefer, *ibid.*, 1979, **70**, 5092.

[12] *Internat. J. Quantum Chem.*, 1978, Vol. 14.

[13] P. A. Christiansen, Y. S. Lee, and K. S. Pitzer, *J. Chem. Phys.*, 1979, **71**, 4445.

[14] S. Nagase and K. Morokuma, *J. Amer. Chem. Soc.*, 1978, **100**, 1666.

[15] T. A. Halgren, D. A. Kleier, J. H. Hall, L. D. Brown, and W. N. Lipscomb, *J. Amer. Chem. Soc.*, 1978, **100**, 6595.

well predicted by PRDDO. Dewar and Ford report that the MNDO method,[6] whilst only slightly slower than the CNDO/2 method, is about three times more accurate than PRDDO or STO-3G for the calculation of relative energies and has about the same accuracy for the calculation of ionization potentials and dipole moments. Schweig *et al.*[16] have developed the LNDO/S method. It is parameterized on the electronic transition energies (ETEs) and vertical ionization potentials (VIPs) of a number of small molecules, and not surprisingly it gives very good estimates of these two quantities for a large number of hydrocarbons, appearing to be superior to the CNDO/S method. The HAM/3 method has given rise[17] to several polemics concerning the rigour of its theoretical foundations. It provides a fast and apparently reliable method of predicting ionization energies.

The MINDO/3 method has now been tested with some success for a range of properties, although a number of defects have been reported.[18]

LCAO–MO theory continues to be combined with force-field calculations. An interesting combination of extended Hückel theory with an empirical force field is a useful adjunct to existing molecular-mechanics methods, and represents an improved method for the study of arenes.[19]

3 Electronic Structure and Geometries of Molecules

Although the success of SCF–MO methods in reproducing molecular geometries has been amply demonstrated, there remain a few problem areas. Pople and co-workers[20] have investigated the effect of electron correlation at the *ab initio* level. Using the MP2/6-31G* method, they found that the only major improvement was in bonds to electronegative elements such as F and O.

A number of groups have analysed substituent effects in organic molecules. Wiberg[21] has shown that substituent effects in CH_3—X can be separated into σ- and π-components in a manner analogous to those in aromatic compounds, and calculations have been carried out which show that strain energies in small ring systems are decreased by σ-donating and π-withdrawing or π-donating substituents (*e.g.* Li or O).[22]

The role of *d*-orbitals in species such as H_3NO or H_3PO has been re-investigated.[23] The results of calculations using a basis set of double-zeta quality and with the inclusion of electron correlation suggest that these molecules are best considered as bipolar species, and that there is no justification in describing the *d*-functions as valence orbitals, although they are necessary if molecular energies are to be predicted accurately.

[16] G. Lauer, K. W. Schulte, and A. Schweig, *J. Amer. Chem. Soc.*, 1978, **100**, 4925.
[17] L. Asbrink, C. Fridh, E. Lindholm, and S. de Bruijn, *Chem. Phys. Letters*, 1979, **66**, 43.
[18] (*a*) T. J. Zielinski, D. L. Brean, and R. Rein, *J. Amer. Chem. Soc.*, 1978, **100**, 6266; G. Klopman, P. Andreozzi, A. J. Hopfinger, O. Kikuchi, and M. J. S. Dewar, *ibid.*, 1978, **100**, 6267; (*b*) S. P. McManus and M. R. Smith, *Tetrahedron Letters*, 1978, 1897; (*c*) S. Bantle and R. Ahlrichs, *Chem. Phys. Letters*, 1978, **53**, 148; (*d*) G. Frenking, H. Goetz, and F. Marschner, *J. Amer. Chem. Soc.*, 1978, **100**, 5295.
[19] D. A. Dougherty and K. Mislow, *J. Amer. Chem. Soc.*, 1979, **101**, 1401.
[20] D. J. DeFrees, B. A. Levi, S. K. Pollack, W. J. Hehre, J. S. Binkley, and J. A. Pople, *J. Amer. Chem. Soc.*, 1979, **101**, 4085.
[21] K. B. Wiberg, *J. Amer. Chem. Soc.*, 1979, **101**, 2204.
[22] J. D. Dill, A. Greenberg, and J. F. Liebman, *J. Amer. Chem. Soc.*, 1979, **101**, 6814.
[23] H. Wallmeier and W. Kutzelnigg, *J. Amer. Chem. Soc.*, 1979, **101**, 2804.

Neutral Species.—The replacement of H by Li or of C by Si has been rich grazing ground for theoretical studies. A 'state of the art' calculation on dilithiomethane re-affirms it to have nearly degenerate singlet and triplet ground states, in both tetrahedral and planar configurations.[24] Schleyer's group continue to predict unusual structures for lithiated species, *e.g.* (1)—(3).[25] They used the popular method of carrying out geometry optimizations with a STO-3G basis set and evaluating relative energies with a 4-31G basis set (STO-3G/4-31G). They found

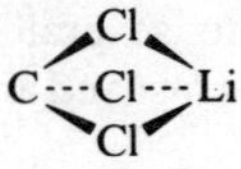
(1)

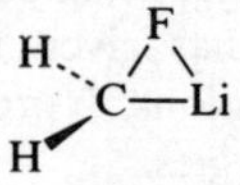
(2)

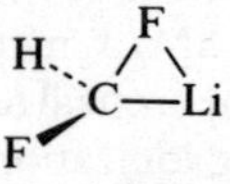
(3)

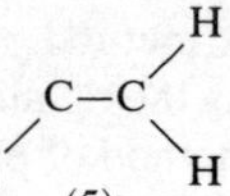
(5)

that (3) is best considered as a carbene complex with LiF, the lone pair on the carbene overlapping with an empty orbital on the metal and back donation occurring from the fluorine into the unoccupied carbene *p*-orbital. There is some support for the structure of (3) from n.m.r. studies.[26] Dilithioethylene (4) has been studied, using a double-zeta basis set plus CI,[27] the conclusion being that the twisted triplet is the most stable state. The planar triplet is 5 kJ mol^{-1} and the planar singlet is 123 kJ mol^{-1} higher in energy. Calculations at the STO-3G/4-31G level have also been reported for a number of lithio derivatives of allene, propyne, and cyclopropene.[28] In most cases the most stable isomer has a non-classical structure, involving Li bridging.

The reaction between acetylene and an aluminium atom is predicted to be exothermic by 84 kJ mol^{-1}.[29] These studies, at the *ab initio* level, using basis sets of double-zeta quality, suggest that the π-complex is not bonding and that instead the most stable isomer involves a hydrogen shift to form (5). Interest in the ability of silicon to replace carbon in organic molecules is unabated. The analogue of ethylene, Si_2H_4, was calculated,[30] using a STO 4-31G basis, to favour the silylene structure (6) rather than the double-bonded form (7). A similar conclusion was reached by Schaefer and co-workers[31] concerning $CSiH_4$. Using a very large orbital basis set, and what they describe as almost a complete CI, they found that (8) and (9) were essentially degenerate in energy.

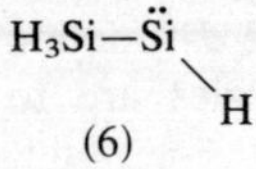
(6)

$H_2Si{=}SiH_2$
(7)

$H_2Si{=}CH_2$
(8)

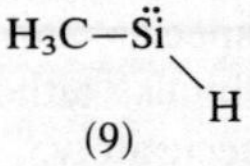
(9)

(10)

Ab initio calculations[32] suggest that the silylene (10) has a singlet ground state and is a relatively stable species, unlike its carbon analogue, which has a triplet ground

[24] W. D. Laidig and H. F. Schaefer, *J. Amer. Chem. Soc.*, 1978, **100**, 5922.
[25] T. Clark and P. von R. Schleyer, (*a*) *J.C.S. Chem. Comm.*, 1979, 883; (*b*) *Tetrahedron Letters*, 1979, 4963; (*c*) *J. Amer. Chem. Soc.*, 1979, **101**, 7747.
[26] D. Seebach, H. Siegel, K. Mullen, and K. Hiltbrunner, *Angew. Chem. Internat. Edn.*, 1979, **18**, 784.
[27] W. D. Laidig and H. F. Schaefer, *J. Amer. Chem. Soc.*, 1979, **101**, 7184.
[28] E. D. Jemmis, J. Chandrasekhar, and P. von R. Schleyer, *J. Amer. Chem. Soc.*, 1979, **101**, 2848.
[29] M. Trenary, M. E. Casida, B. R. Brooks, and H. F. Schaefer, *J. Amer. Chem. Soc.*, 1979, **101**, 1638.
[30] L. C. Snyder and Z. R. Wasserman, *J. Amer. Chem. Soc.*, 1979, **101**, 5222.
[31] H. F. Schaefer, *Accounts Chem. Res.*, 1979, **12**, 288.
[32] J. C. Barthelat, G. Trinquier, and G. Bertrand, *J. Amer. Chem. Soc.*, 1979, **101**, 3785.

state. Another molecule of great interest, silabenzene (11), has been calculated by Schlegel *et al.*[33] to have as much as $\frac{2}{3}$ of the resonance energy of benzene itself. Whether this will be reflected in its properties remains to be seen. Further papers have appeared[34] on cyclobutadiene (12). The vibrational frequencies of the rectangular ground-state singlet have been calculated, using an *ab initio* method,[34a] and were found to be in good agreement with those reported for matrix-trapped (12). Lipkowitz[34b] has studied the possibility that complexes of (12) with CO, HCN, or benzene might be formed. CNDO/2 predicts the formation of all three complexes to be considerably exothermic, whereas STO-3G predicts that none of the complexes is stable. In a similar study using MNDO,[34c] the complex with CO was found to form exothermically, but on relaxing the C_{4v} symmetry restraint the system was found to collapse to bicyclopentenone. Schweig and Thiel report that the optimized geometry of tetra-t-butyltetrahedrane, using MNDO, has only T symmetry, with the implication that the molecule is chiral. However, a very low barrier to racemization was predicted.[35]

Higher annulenes have been studied by the semi-empirical MINDO/3 method,[36] which was found to over-estimate the amount of bond alternation. Possible reasons for this were discussed, including the inadequate treatment of electron correlation in the method.

A remarkable series of anti van't Hoff compounds have been postulated, on the basis of STO 4-31G calculations.[37] By substituting two geminal hydrogens on a carbon atom with the cyclic —BH—BH— unit, Schleyer and his co-workers have suggested that compounds (13)—(15) exist preferentially in planar, perpendicular, and planar forms respectively. Isomers with conventional van't Hoff stereochemistry were calculated to be as much as 84 kJ mol^{-1} higher in energy! Although little optimism was expressed for the synthesis of (13)—(15), a more hopeful candidate (16) was suggested, which might be expected to display similar properties.

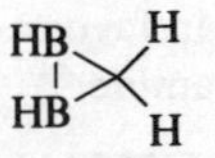
(13)

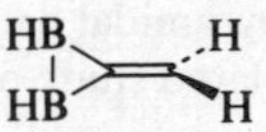
(14)

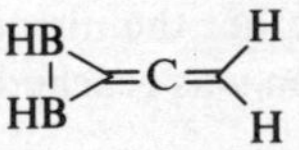
(15)

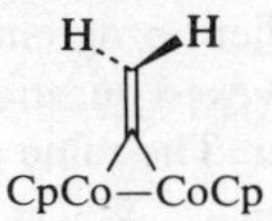
(16)

The benzynes continue to attract attention.[38] The vibrational spectrum of *o*-benzyne has been calculated, using MNDO,[38a] and *ab initio* GVB calculations suggest that the *meta*- and *para*-isomers have singlet ground states and exist as biradicals.[38b] Although the calculations suggested that bicyclic isomers are higher in energy, Washburn has apparently characterized bicyclo[3.1.0]hexatriene (*m*-benzyne).[38c]

33 H. B. Schlegel, B. Coleman, and M. Jones, *J. Amer. Chem. Soc.*, 1978, **100**, 6499.

34 (*a*) L. J. Schaad, B. A. Hess, and C. S. Ewig, *J. Amer. Chem. Soc.*, 1979, **101**, 2281, (*b*) K. B. Lipkowitz, *ibid.*, 1978, **100**, 7535; (*c*) A. Schweig and W. Thiel, *Tetrahedron Letters*, 1978, 1841.

35 A. Schweig and W. Thiel, *J. Amer. Chem. Soc.*, 1979, **101**, 4742.

36 (*a*) H. Vogler, *J. Mol. Struct.*, 1979, **51**, 289; (*b*) H. Baumann, *J. Amer. Chem. Soc.*, 1978, **100**, 7196.

37 K. Krogh-Jespersen, D. Cremer, D. Poppinger, J. A. Pople, P. von R. Schleyer, and J. Chandrasekhar, *J. Amer. Chem. Soc.*, 1979, **101**, 4843.

38 (*a*) M. J. S. Dewar, G. P. Ford, and H. S. Rzepa, *J. Mol. Struct.*, 1979, **51**, 275; (*b*) J. O. Noell and M. D. Newton, *J. Amer. Chem. Soc.*, 1979, **101**, 51; (*c*) W. N. Washburn, R. Zahler, and I. Chen, *ibid.*, 1978, **100**, 5863.

The structures of crucial intermediates involved in ozonolysis have been investigated. The 'Criegee intermediate' (17) is calculated to have a high barrier to *syn*/*anti* isomerization (134 kJ mol^{-1}), with the *syn* form favoured by 12—17 kJ mol^{-1}. This is consistent with many experimental observations. Two isomeric species (18) and (19) have been calculated to be considerably lower in energy (by 139 and 113 kJ mol^{-1} respectively), and the possible involvement of such species in ozonolysis cannot be excluded.[39] A closely related study by Harding and Goddard[40] came to very similar conclusions.

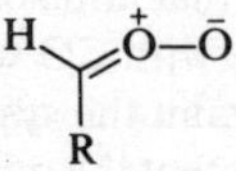
(17)

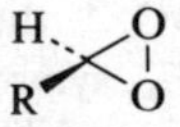
(18)

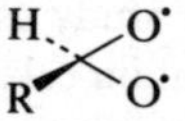
(19)

Cremer[41] has published an important series of papers on the conformations of primary (20) and final ozonides (21). *Ab initio* calculations find the rings to be puckered in each case. The oxygen envelope (E) is the most stable for the primary ozonide, and an explanation is offered for the contrasting stereochemical behaviour of *cis*- and *trans*-alkenes upon ozonolysis.

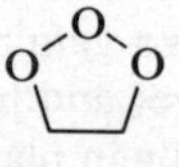
(20)

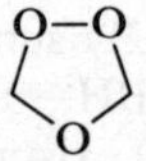
(21)

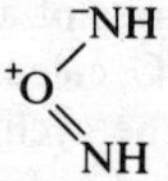
(22)

The novel dipolar species (22) was predicted by STO 4-31G calculations to be a good candidate for detection, particularly if the hydrogens are replaced by electron-withdrawing substituents.[42] A study of several amides, using the TEXAS *ab initio* gradient program with complete geometry optimization,[43] appears to have settled the vexed question of whether the nitrogen is pyramidal or planar in favour of the latter. The same conclusion was reached in a related study of formamide.[44]

Conformations.—Pople and his co-workers have applied their GAUSSIAN 70 *ab initio* program to the study of conformational analysis, with some success. One unexpected result was their prediction that certain cyclic alkenes such as (23) are puckered and not planar.[45] Conformational effects in pyranosides were modelled with dimethoxymethane. Although complete geometry optimization was not possible at the time of the study, the 4-31G method correctly predicted[46] that the molecule has the preferred conformation (24). This is in excellent agreement with the anomeric effects observed in pyranosides. This study has subsequently been

39 D. Cremer, *J. Amer. Chem. Soc.*, 1979, **101**, 7199; K. Karlstöm, S. Engström, and B. Jönsson, *Chem. Phys. Letters*, 1979, **67**, 343.
40 L. B. Harding and W. A. Goddard, *J. Amer. Chem. Soc.*, 1978, **100**, 7180.
41 D. Cremer, *J. Chem. Phys.*, 1979, **70**, 1898, 1911, 1928.
42 R. Cimiraglia, M. Persico, and J. Tomasi, *Theor. Chim. Acta*, 1978, **49**, 13.
43 G. Fogarasi, P. Pulay, F. Török, and J. E. Boggs, *J. Mol. Struct.*, 1979, **57**, 259.
44 N. R. Carlsen, L. Radom, N. V. Riggs, and W. R. Rodwell, *J. Amer. Chem. Soc.*, 1979, **101**, 2234.
45 H.-U. Wagner, G. Szeimies, J. Chandrasekhar, P. von R. Schleyer, J. A. Pople, and J. S. Binkley, *J. Amer. Chem. Soc.*, 1978, **100**, 1210.
46 G. A. Jeffrey, J. A. Pople, J. S. Binkley, and S. Vishveshwara, *J. Amer. Chem. Soc.*, 1978, **100**, 373.

extended to methoxymethyl fluoride and chloride, as models for the study of glycosyl halides.[47] Fluoride was found to have a greater anomeric effect than chloride, with OH having an intermediate value.

(23)

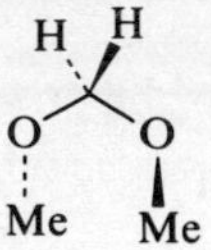
(24)

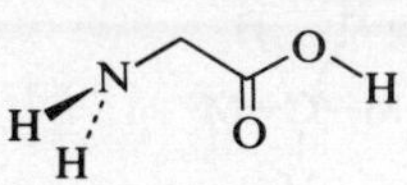
(25)

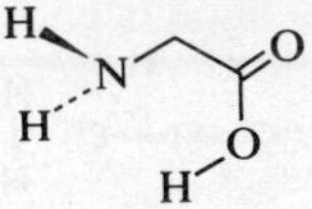
(26)

An apparent disagreement between the predicted (STO 4-31G) conformation of glycine and microwave results appears to have been resolved.[48] Complete geometry optimization at the 4-21G level, using the TEXAS program, agrees with earlier calculations in predicting (25) to be more stable than (26) by 9 kJ mol^{-1}, although the microwave results had suggested that the predominant conformer was (26). However, since only (26) is predicted to have a large dipole moment [6.54 D *vs.* 1. 1 D for (25)], it seems probable that only microwave transitions from (26) were detected experimentally. Pople and co-workers[49] report 4-31G calculations which suggest that the zwitterionic form of glycine is a shallow minimum in the potential-energy surface, a barrier of only 2 kJ mol^{-1} separating it from (26), which is 121 kJ mol^{-1} more stable. On the basis of these results, a hydration energy of at least 209 kJ mol^{-1} has been proposed.

Other conformational studies have been carried out on vinyl thioethers,[50a] esters,[50b] peroxyformic acid,[50c] n-butane,[50d] substituted thiophens,[50e] and biacetyl.[50f]

Charged Species.—Schleyer *et al.*[51] have studied ions of the type $CH_3M_2^+$ (M = Li, BeH, Na, or MgH), which are interesting as models for electrophilic substitution reactions. Whilst the most stable structure for CH_5^+ has C_s symmetry (27), those when M is Li or BeH have D_{3h} symmetry (28), suggesting a fine balance between retention and inversion of configuration in S_E2 reactions.

The combination of using MINDO/3 for rapid optimization of the geometry and an *ab initio* 4-31G basis for energy calculations has been successful in a study of numerous carbo-cations.[52] Corrections for zero-point energies were made using thermodynamic quantities evaluated from MINDO/3 calculations. Hehre *et al.*,[53] on the basis of 4-31G/6-31G* calculations, have proposed that two forms of the cyclobutylcarbinyl cation may be stable; *i.e.* one form, with C_s symmetry, (29) and an alternative isomer which is structurally very similar but does not have the plane of symmetry, and which is twisted to some extent. Since this latter species is only

[47] G. A. Jeffrey and J. H. Yates, *J. Amer. Chem. Soc.*, 1979, **101** 820.
[48] H. L. Sellers and L. Schäfer, *J. Amer. Chem. Soc.*, 1978, **100**, 7728.
[49] Y.-C. Tse, M. D. Newton, S. Vishveshwara, and J. A. Pople, *J. Amer. Chem. Soc.*, 1978, **100**, 4329.
[50] (*a*) J. Kao, *J. Amer. Chem. Soc.*, 1978, **100**, 4685; (*b*) J. R. Larson, N. D. Epiotis, and F. Bernardi, *ibid.*, 1978, **100**, 5713; (*c*) R. D. Bach, C. L. Willis, and T. J. Lang, *Tetrahedron*, 1979, **35**, 1239; (*d*) M. R. Peterson and I. G. Csizmadia, *J. Amer. Chem. Soc.*, 1978, **100**, 6911; (*e*) J. Kao and L. Radom, *ibid.*, 1979, **101**, 311; (*f*) J. Tyrell, *ibid.*, 1979, **101**, 3766.
[51] E. D. Jemmis, J. Chandrasekhar and P. von R. Schleyer, *J. Amer. Chem. Soc.*, 1979, **101**, 527.
[52] H. J. Köhler and H. Lischka, *J. Amer. Chem. Soc.*, 1979, **101**, 3479.
[53] B. A. Levi, E. S. Blurock, and W. J. Hehre, *J. Amer. Chem. Soc.*, 1979, **101**, 5537.

2 kJ mol^{-1} higher in energy than (29), it is difficult to be certain that this result is not an artifact of either the method or the geometry optimization. Using the same method, it was found that cyclobutyl cation was not a stable species, collapsing without activation to (29).

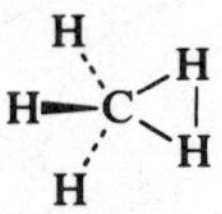
(27)

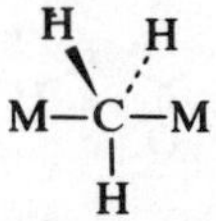
(28)

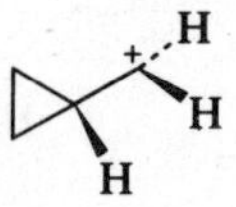
(29)

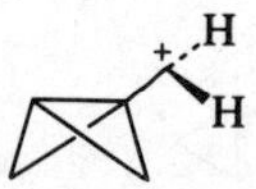
(30)

On the basis of STO-3G calculations, but without geometry optimization, Greenberg[54] has made the fascinating prediction that the central bond in bicyclobutane is comparable to the vinyl group in its ability to stabilize a carbo-cation centre. The rotational barrier about the C—C bond in (30) was calculated to be as high as 133 kJ mol^{-1}. Further studies, using a higher quality basis set and complete geometry optimization, would be desirable. Substitution of a methyl group in the terminal position was calculated (STO-3G) to stabilize an allyl cation by 71 kJ mol^{-1}, whereas substitution on the central carbon atom has a much smaller effect (21 kJ mol^{-1}).[55] The stabilization was found to be related to the rate of solvolysis.

The cyclobutadienyl dication may naively be expected to be aromatic and planar, on the basis of the Hückel $4n+2$ rule. It has been reported that, in fact, a puckered form is 31 kJ mol^{-1} more stable, as a consequence of destabilizing 1,3-interactions and the results of orbital mixing and reorientation.[56]

Chandrasekhar and Schleyer[57] have studied 1,2-shifts in cyclic carbonium ions. Both the MINDO/3 and STO-3G methods predict that the activation energy increases with decreasing ring size [*e.g.* by 64 kJ mol^{-1} (MINDO/3) or 93 kJ mol^{-1} (STO-3G) for cyclopropyl cation relative to ethyl cation]. The effect has been attributed to increasingly poor orbital overlap in the smaller rings. A re-investigation of the protonation of benzene, using the MINDO/3 and STO-3G methods, suggests that previously found differences between the two methods may have been a result of inadequate optimization of the geometry.[58] The structure corresponding to a π-complex was found to represent the transition state for a 1,2 hydrogen shift.

Another interesting prediction[59] is of two-fold aromaticity for (31). It has been suggested that this ion, which is known from mass spectral studies, is especially stable by virtue of (*a*) having six π-electrons and (*b*) having two σ-electrons occupying the symmetric Walsh orbital.

Lischka and Kohler[60] have published a detailed comparison of the results of MINDO/3 and *ab initio* CEPA/PNO calculations for small carbo-cations. Both methods predict non-classical bridging structures to be slightly more stable than the classical forms. MINDO/3, however, was found to predict excessive stability for cyclic halonium ions (32). McLafferty and co-workers have discovered some

54 A. Greenberg, *Tetrahedron Letters*, 1978, 3509.
55 H. Mayer, W. Förner, and P. von R. Schleyer, *J. Amer. Chem Soc.*, 1979, **101**, 6032.
56 K. Krogh-Jespersen, P. von R. Schleyer, J. A. Pople, and D. Cremer, *J. Amer. Chem. Soc.*, 1978, **100**, 4301.
57 J. Chandrasekhar and P. von R. Schleyer, *Tetrahedron Letters*, 1979, 4057.
58 T. Sordo, J. Bertrán, and E. Canadell, *J.C.S. Perkin II*, 1979, 1486.
59 J. Chandrasekhar, E. D. Jemmis, and P. von R. Schleyer, *Tetrahedron Letters*, 1979, 3707.
60 H. Lischka and H. J. Köhler, *J. Amer. Chem. Soc.*, 1978, **100**, 5297.

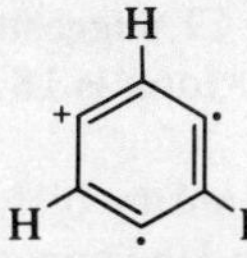
(31)

(32)

$H_3C{-}S^+$
(33)

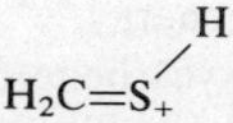
(34)

interesting differences in the behaviour of oxygen and sulphur.[61] Both (33) and (34) were deduced to be stable on the basis of mass spectral studies, and STO 4-31G calculations verified this, predicting (33) to be a triplet. In contrast, the methoxy cation is predicted to be unstable, and does not seem to have been detected experimentally, although there is evidence of a weak complex between H_2 and HCO^+.

Radom and co-workers have calculated that eleven isomers of $C_2H_4O^{+\cdot}$ are stable, of which five as yet unknown species are proposed as reasonable candidates for identification.[62] The most stable isomer corresponded to ionized vinyl alcohol. There have been several investigations of $C_2H_5O^+$ as a model for the electrophilic addition of HO^+ to an alkene. Various methods[63] are in agreement that the bridging ion (35) is more stable than the open form (36) by 92 kJ mol^{-1} (*ab initio*/CEPA), 30 kJ mol^{-1} (STO 4-31G), 21 kJ mol^{-1} (MNDO), or 100 kJ mol^{-1} (MINDO/3). A high barrier has been predicted for the isomerization of (35) to (36) (25 kJ mol^{-1})[63] and an alternative isomer (37) is predicted by all the methods to be considerably more stable.[64]

(35)

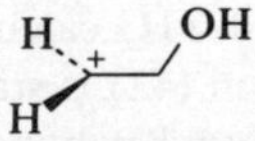
(36)

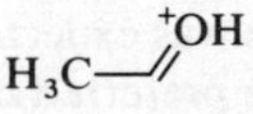
(37)

Carbenes, Triplets, and Excited States.—A review on singlet–triplet separations in carbenes and biradicals has appeared.[65] There is now sufficient confidence in the reliability of the calculated singlet–triplet energy separation in methylene that, when an anomalous experimental result is reported, it is confidently concluded that the experimental interpretation, and not the calculations, must be erroneous![66] The theory has now reached the stage where a small difference in the latest experimental value ($\simeq$33 kJ mol^{-1}) and the latest calculations ($\simeq$42 kJ mol^{-1}) can be attributed to the neglect of relativistic effects in the latter.[66]

The triplet states of a number of carbenes[67a] and nitrenes[67b] have been shown to be stabilized relative to the singlet by σ-donating substituents (*e.g.* Li) and to be destabilized by π-donors (*e.g.* F, OR, NR_2, or CR_2^-). Cyanocarbene has proved rather mysterious. Experimental evidence suggests that it is linear, *i.e.* (38), and calculations at the STO 4-31G/small CI level appeared to support this conclusion.[67c]

61 J. D. Dill and F. W. McLafferty, *J. Amer. Chem. Soc.*, 1979,, **101**, 6526.
62 W. J. Bouma, J. K. MacLeod, and L. Radom, *J. Amer. Chem. Soc.*, 1979, **101**, 5540.
63 A. C. Hopkinson, M. H. Lien, I. G. Csizmadia, and G. Yates, *Theor. Chim. Acta*, 1978, **43**, 97.
64 H. Lischka and H. J. Köhler, *Chem. Phys. Letters*, 1979, **63**, 326.
65 W. T. Borden and E. R. Davidson, *Ann. Rev. Phys. Chem.*, 1979, **30**, 125.
66 D. Feller and E. R. Davidson, *Chem. Phys. Letters*, 1980, **69**, 201.
67 (*a*) N. C. Baird and K. F. Taylor, *J. Amer. Chem. Soc.*, 1978, **100**, 1333; (*b*) J. F. Harrison, R. C. Liedtke, and J. F. Liebman, *ibid.*, 1979, **101**, 7162; (*c*) J. F. Harrison, A. Dendramis, and G. E. Leroi, *ibid.*, 1978, **100**, 4352; (*d*) M. F. Zandler, J. D. Goddard, and H. F. Schaefer, *ibid.*, 1979, **101**, 1072.

A re-investigation, using a more extensive basis set and a larger CI, suggests that the species is bent, with an angle of about 135 °, and that the linear form is 18 kJ mol^{-1} higher in energy.[67d]

Hydroxycarbene (39) has been implicated in the photochemistry of formaldehyde.[31] A significant delay between the decay of the S_1 state and the appearance of CO photoproduct suggested that an intermediate might be involved. Goddard and Schaefer,[68] on the basis of extensive CI calculations, found that (39) was energetically accessible, although there appears to be no low barrier to the conversion of (39) into H_2 and CO. It was concluded[31] that, although (39) may be 'visited' in the course of photodissociation of H_2CO, it does not play an important role.

H—C=C=N

(38)

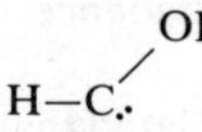

(39)

(40)

Cyclopropenylidene (40) is of interest because the singlet state can be regarded as aromatic in the Hückel $4n+2$ sense. Indeed, CI studies suggest that configurations which violate the $4n+2$ rule contribute relatively less to the wave-function than they do in other carbenes.[69] If the equilibrium geometry is distorted by lengthening the C—C single bond, a singlet–triplet crossing was found to occur, and was felt to be significant in rationalizing the non-stereospecific addition of singlet carbenes of this type to alkenes.[69]

Cyclopropylchlorocarbene is apparently less selective (more reactive) towards alkenes than was expected.[70] STO 4-31G calculations offer an interesting rationalization. The preferred conformation (41) is stabilized by interaction of the vacant carbenic *p*-orbital with the cyclopropyl σ-bonds, but this does not allow sterically favourable overlap with the π-orbitals of an approaching alkene. This problem does not occur with the alternative conformation (42), which, however, is predicted to be 40 kJ mol^{-1} higher in energy, and would be expected to show less selectivity.

(41)

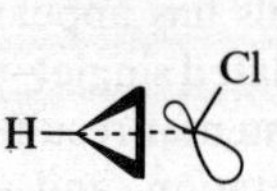

(42)

A 'state of the art' double-zeta basis set/CI calculation on trimethylenemethane (43) suggests that the elusive ($^3A_2'-^1B_1$) energy separation is not less than 59 kJ mol^{-1},[71a] while a much less extensive calculation on Closs's biradical (44) suggests that, in this case, the singlet and triplet energies are almost equal.[71b]

Salem has reviewed the phenomenon of sudden polarization.[72] Calculations using a large basis set and extensive CI suggest that if one of the CH_2 groups in the

68 J. D. Goddard and H. F. Schaefer, *J. Chem. Phys.*, 1979, **71**, 5117.
69 R. Shepard, A. Banerjee, and J. Simons, *J. Amer. Chem. Soc.*, 1979, **101**, 6174.
70 R. A. Moss, M. Vezza, W. Guo, R. C. Munjal, K. N. Houk, and N. G. Rondan, *J. Amer. Chem. Soc.*, 1979, **101**, 5088.
71 (*a*) D. M. Hood, H. F. Schaefer, and R. M. Pitzer, *J. Amer. Chem. Soc.*, 1978, **100**, 8009; (*b*) M. P. Conrad, R. M. Pitzer, and H. F. Schaefer, *ibid.*, 1979, **101**, 2245.
72 L. Salem, *Accounts Chem. Res.*, 1979, **12**, 87.

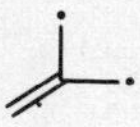

(43)

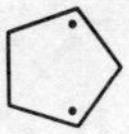

(44)

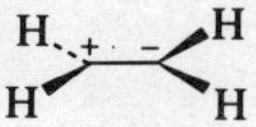

(45)

orthogonal S_1 state of ethylene becomes pyramidal, a large amount of charge separation will occur,[73] *e.g.* as in (45), by as much as 0.5 e for 10 ° of bending. Salem speculates that such an effect could be important in the mechanism of vision (Scheme 1).[72]

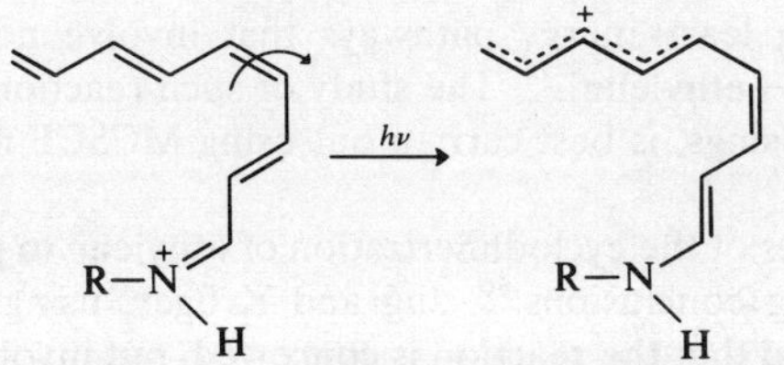

Scheme 1

4 Dynamic Processes and Reaction Hypersurfaces

An instructive and entertaining paper by Müller, dealing with reaction hypersurfaces, is recommended to anyone who has had difficulties in locating a transition state! Several new ways of locating stationary points in a potential surface are reviewed, and illustrated with examples.[74]

The ability to predict accurately the rate constants of organic reactions entirely from first principles has come closer to realization.[75] Morokuma and co-workers[75a] have calculated rate constants and kinetic isotope effects for the reaction of hydrogen radicals with ethylene, using a STO 4-31G basis set, and obtained excellent agreement with experiment. In general, however, such accuracy is not to be expected, although the calculation of kinetic isotope effects *a priori*, using SCF–MO methods, for a variety of reactions has been shown to give reasonably good results.[75b,c]

Pericyclic Reactions.—Radom and co-workers[76] have found that the calculated STO-3G/4-31G barrier to the 'forbidden' suprafacial 1,3-shift in propene is slightly lower than that for the 'allowed' 1,3-antarafacial shift. In the transition state located for the former, the migrating hydrogen atom is essentially directly bonded to the central carbon atom, resulting in a biradical-like structure (46). There remains doubt whether (46) truly represents a transition state for a 1,3-hydrogen shift or whether it is the transition state for propene–cyclopropane isomerization. The same method was found to predict the barrier to the 1,5-shift in pentadiene to be 259 kJ mol^{-1}; much lower than for the 1,3-shift, but still in error by a large amount. An excellent account of theoretical calculations concerned with 1,2-shifts has appeared.[31]

[73] B. R. Brooks and H. F. Schaefer, *J. Amer. Chem. Soc.*, 1979, **101**, 307.
[74] K. Müller, *Angew. Chem. Internat. Edn.*, 1980, **19**, 1.
[75] (*a*) S. Nagase, T. Fueno, and K. Morokuma, *J. Amer. Chem. Soc.*, 1979, **101**, 5849; (*b*) S. B. Brown, M. J. S. Dewar, G. P. Ford, D. J. Nelson, and H. S. Rzepa, *ibid.*, 1978, **100**, 7832; (*c*) D. J. McLennan, *Aust. J. Chem.*, 1979, **32**, 1869.
[76] W. J. Bouma, M. A. Vincent, and L. Radom, *Int. J. Quantum Chem.*, 1978, **14**, 767.

(46)

Reudenberg and co-workers have concluded[77] that the dimerization of singlet ($^1A_1$) methylene to give ethylene in its excited ($^1A_g{}^*$) state occurs with no barrier, despite being formally an orbital-symmetry-forbidden reaction! Calculations show that several other reactions that are symmetry-forbidden occur with no activation energy by adopting least-energy pathways that involve no symmetry (*e.g.* $^1A_1$ CH_2+H_2; $^1A_1$ CH_2+ethylene).[78] The study of such reactions, which may involve avoided orbital crossings, is best carried out using MCSCF methods, or with very large CI.[79]

Theoretical studies of the cyclodimerization of ethylene to give cyclobutane have produced conflicting conclusions.[80] Jug and Krüger, using their semi-empirical SINDO method, find that the reaction is concerted, but involves an unsymmetrical transition state.[80a] According to CI applied *after* geometry optimization, a biradical should be encountered along the reaction path, but this should not affect the concerted nature of the reaction. This does not preclude the possibility that CI applied *during* geometry optimization would lead to the biradical as a definite minimum on the reaction pathway. In contrast, Burke and Leroy[80b] used the STO-3G/CI method and found that the symmetric $[\pi^2{}_s+\pi^2{}_s]$ forbidden reaction has a lower barrier (345 kJ mol^{-1}) than the $[\pi^2{}_s+\pi^2{}_a]$ allowed reaction (447 kJ mol^{-1}), but that both are much higher in energy than the stepwise biradical pathway (177 kJ mol^{-1}). The Diels–Alder reaction between ethylene and butadiene continues to be controversial.[81] Previous STO 4-31G/CI studies with partial geometry optimization suggested a symmetrical addition, although an alternative pathway involving a biradical intermediate was, at most, 17 kJ mol^{-1} higher in energy.[81a] More recently, Jug and Krüger have suggested that the reaction is concerted, but involves an unsymmetrical transition state.[80a] Dewar *et al.*,[81b] using MINDO/3 and applying CI *during* geometry optimization, found it to be stepwise and to involve a biradical intermediate. Kinetic isotope effects, often cited as evidence for a concerted reaction, were shown to be consistent with this mechanism. However, the MINDO/3 method has been criticized[81a] and a more complete calculation, involving complete optimization of the geometry and extensive CI, might help to resolve this problem. Studies using STO 4-31G find that the reduction of ethylene with *cis*-di-imide is concerted, and involves a symmetrical transition state.[82]

[77] L. M. Cheung, K. R. Sundberg, and K. Ruedenberg, *Int. J. Quantum Chem.*, 1979, **16**, 1103.
[78] (*a*) H. Kollmar and V. Staemmler, *Theor. Chim. Acta*, 1979, **51**, 207; (*b*) B. Zurawski and W. Kutzelnigg, *J. Amer. Chem. Soc.*, 1978, **100**, 2654.
[79] S. Yamabe, T. Minato, and Y. Osamura, *J. Amer. Chem. Soc.*, 1979, **101**, 4525.
[80] (*a*) K. Jug and H. W. Krüger, *Theor. Chim. Acta*, 1979, **52**, 19; (*b*) L. A. Burke and G. Leroy, *Bull. Soc. chim. belges.*, 1979, **88**, 379.
[81] (*a*) R. E. Townshend, G. Ramunni, G. Segal, W. J. Hehre, and L. Salem, *J. Amer. Chem. Soc.*, 1976, **98**, 2190; (*b*) M. J. S. Dewar, S. Olivella, and H. S. Rzepa, *ibid.*, 1978, **100**, 5650.
[82] (*a*) D. J. Pasto and D. M. Chipman, *J. Amer. Chem. Soc.*, 1979, **101**, 2290; (*b*) E. Flood and P. N. Skancke, *Chem. Phys. Letters*, 1978, **54**, 53.

Other Reactions.—Nucleophilic addition to a triple bond is known to proceed with *trans* stereochemistry. Two theoretical studies, using the *ab initio* approach, have used the addition of H^- to acetylene as a model for the reaction.[83] The calculated structure of the transition state (47) shows that the nucleophile and the developing negative charge are indeed *trans* with respect to each other. It was also demonstrated that acetylene is less sensitive than ethylene to changes in the angle of approach of the nucleophile. Although the preferred angle is about 127 °, and not 60 ° (as implied by Baldwin's rules), deformation to achieve the latter angle is more facile in the case of acetylene than ethylene. Very similar transition states were reported for nucleophilic addition to nitrile oxides[84a] and for the addition of a hydrogen atom to acetylene.[84b]

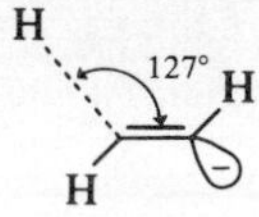

(47)

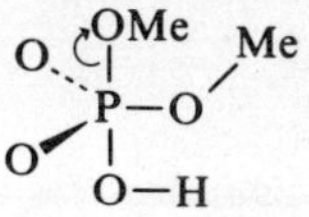

(48)

The reaction between peroxyformic acid and alkenes or imines has been studied, using a STO-4G basis set, with the conlusion that an intermediate adduct may be involved.[85] Since no calculations of the reaction path were carried out, and transition states were not located as saddle points in the potential surface, this conclusion can only be qualitative. The same is true of a study of the photochemical transformation of oxaziridine to formamide,[86] in which the N—O bond is thought to cleave first, followed by a decay to the S_0 ground state and subsequent migration of hydrogen. Bigot *et al.* have come to similar conclusions, on the basis of rather more detailed calculations.[87] A thorough STO-3G study of the HF-catalysed ring-opening of oxiran[88] suggests that initial protonation on oxygen is followed by rearrangement to give the *cis*-fluorohydrin. It was predicted that the gas-phase reaction should occur with no formation of the *trans*-isomer.

Gorenstein and co-workers[89] have studied the hydrolysis of dimethyl phosphate by hydroxide ion. Using a STO-3G basis set, which does not include *d*-functions, they concluded that the energy of the dimethoxyphosphorane transition state (48) was critically dependent upon the orientation of the basal ester bonds, but not of the leaving group.

[83] (*a*) C. E. Dykstra, A. J. Arduengo, and T. Fukunaga, *J. Amer. Chem. Soc.*, 1978, **100**, 6007; (*b*) R. W. Strozier, P. Caramella, and K. N. Houk, *ibid.*, 1979, **101**, 1340.
[84] (*a*) G. Leroy, M. T. Nguyen, M. Sana, K. J. Dignam, and A. F. Hegarty, *J. Amer Chem. Soc.*, 1979, **101**, 1988; (*b*) S. Nagase and C. W. Kern, *ibid.*, 1979, **101**, 2544.
[85] A. Ažman, J. Koller, and B. Plesničar, *J. Amer. Chem. Soc.*, 1979, **101**, 1107.
[86] E. Oliveros, M. Riviere, J. P. Malrieu, and C. Teichteil, *J. Amer. Chem. Soc.*, 1979, **101**, 318.
[87] B. Bigot, D. Roux, A. Sevin, and A. Devaquet, *J. Amer. Chem. Soc.*, 1979, **101**, 2560.
[88] G. Algona, E. Scrocco, and J. Tomasi, *Theor. Chim. Acta*, 1979, **51**, 11.
[89] D. G. Gorenstein, B. A. Luxon, and J. B. Findlay, *J. Amer. Chem. Soc.*, 1979, **101**, 5869.

4 Reaction Mechanisms

Part (i) Pericyclic Reactions

By R. J. BUSHBY

Department of Organic Chemistry, The University, Leeds LS2 9JT

1 Introduction

The pericyclic reactions of singlet oxygen[1] and of the photoexcited states of organic anions[2] have been reviewed.

The effect of substituents on pericyclic reactions has long remained a source of difficulty. This year two interesting approaches to the problem have been described. Carpenter and Wilcox[3] have used a simple HMO model. They have shown that for electrocyclic reactions, 1,5-hydrogen shifts, cyclopropane stereomutations, and reactions involving carbon–carbon bond homolysis, there is a linear relationship (1) between $\Delta\Delta H^{+}$ (the measured effect of the substituent on the enthalpy of activation in kcal mol^{-1}) and $\Delta\Delta E_{\pi}$ (the calculated difference between the effect of the substituent on the HMO π-energy of the substrate and that of a suitable model for the transition state, both in units of β. Möbius cyclobutadiene is taken as a model for the transition state in the conrotatory opening of cyclobutene).

$$\Delta\Delta H^{+} = 19.31\Delta\Delta E_{\pi} \qquad (1)$$

In most cases, deviations from linearity are small. For example the effects of phenyl substituents on the enthalpy of activation for opening the cyclobutenes (2), (3), and (4) [relative to the parent system (1)] are calculated from equation (1) to be −7.0, −12.0, and +7.0 kcal mol^{-1} respectively whereas the measured values are −6.3, −11.2, and +6.4 kcal mol^{-1}. The treatment is limited to hydrocarbon systems and breaks down for reactions, such as 3,3-sigmatropic rearrangements, where the substituents themselves drastically alter the position and nature of the transition state. This variable nature of the transition state in Cope/Claisen rearrangements is

(1)

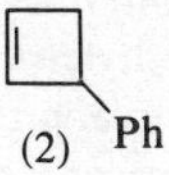

(2)

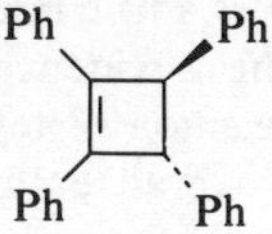

(3)

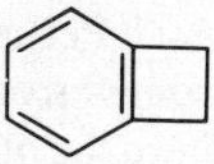

(4)

[1] A. A. Frimer, *Chem. Rev.*, 1979, **79**, 359.
[2] M. A. Fox, *Chem. Rev.*, 1979, **79**, 253.
[3] C. F. Wilcox and B. K. Carpenter, *J. Amer. Chem. Soc.*, 1979, **101**, 3897.

deliberately built into the model developed by Gajewski.[4] He makes use of the simplest equation for an energy surface with a saddle point; namely equation (2), in which, for 3,3-sigmatropic shifts, x and y refer to the fraction of 1,6-bond formation and 3,4-bond breaking, and a, b, c, and d are constants.

$$\Delta G = \mathrm{a}x + \mathrm{b}y + \mathrm{c}xy + \mathrm{d} \tag{2}$$

Application of suitable boundary conditions and solution for the position of the saddle point gives rise to equation (3), in which $\Delta G^{\ddagger}$ is the desired free energy of activation, $\Delta G^{\ddagger}$(BB) and $\Delta G^{\ddagger}$(BM) are the calculated free energies for forming the (appropriately substituted) bond-broken and bond-made species (5) and (6), ΔG_r is the free energy of reaction, and p is an empirically derived parameter which is found to be constant for a given series of reactions.

$$\Delta G^{\ddagger} = \frac{\Delta G^{\ddagger}(\mathrm{BB}) * \Delta G^{\ddagger}(\mathrm{BM}) * p}{\Delta G^{\ddagger}(\mathrm{BB}) + \Delta G^{\ddagger}(\mathrm{BM}) - (\Delta G_r/p)} \tag{3}$$

For a value of $p = 1.5$, this equation leads to predicted free energies of activation for the 3,3-sigmatropic rearrangements of the dienes (7), (8), and (9) of 41.2, 36.3, and 30.1 kcal mol^{-1} respectively, whereas the experimental values are 41, 35.5, and 31 kcal mol^{-1}. A similar treatment was applied successfully to the Diels–Alder reaction of cyano-substituted dienophiles.

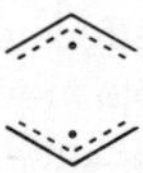
(5)

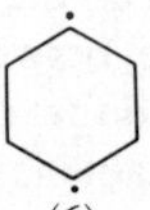
(6)

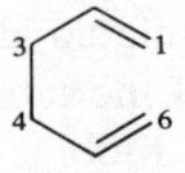

(7)

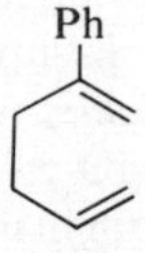

(8)

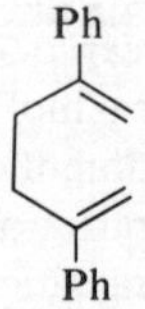

(9)

2 Cycloadditions and Cycloreversions

Charge-transfer complexes are frequently formed between compounds which undergo cycloaddition reactions, but it is difficult to decide whether these are intermediates [equation (4)] or just a dead end [equation (5)].

$$\text{Reactants} \rightleftarrows \text{Charge-transfer complex} \rightarrow \text{Cyclo-adducts} \tag{4}$$

$$\text{Charge-transfer complex} \rightleftarrows \text{Reactants} \rightarrow \text{Cyclo-adducts} \tag{5}$$

Hence one study of the addition of tetracyanoethylene to vinyl ethers suggests that the charge-transfer complex *is* a true intermediate[5] and another, very similar, study that it *is not.*[6] Perhaps the best evidence for the intermediacy of a charge-transfer complex comes from the reaction of tetracyanoethylene with 9,10-dimethylanthracene, where the observed 'overall' enthalpy of activation is negative.[7] This can only be reconciled with equation (4), and a more recent study of the reaction of tetracyanoethylene with a range of anthracene derivatives has shown a correlation

[4] J. J. Gajewski, *J. Amer. Chem. Soc.*, 1979, **101**, 4393.
[5] M. Sasaki, H. Tsuzuki, and M. Okamoto, *J. Org. Chem.*, 1979, **44**, 652.
[6] J. von Jouanne, H. Kelm, and R. Huisgen, *J. Amer. Chem. Soc.*, 1979, **101**, 151.
[7] V. D. Kiselev and J. G. Miller, *J. Amer. Chem. Soc.*, 1975, **97**, 4036.

between the 'overall' free energy of activation and that for formation of a charge-transfer complex. This has also been interpreted as evidence that the charge-transfer complex is a true intermediate.[8]

The formation of zwitterionic intermediates in certain cycloaddition reactions has continued to attract attention, and Moore[9] has discussed the use of furanone azides for the independent generation of such intermediates. For example, the furanone (10) can be used to generate the zwitterion (11), which is thought to be an intermediate in the addition of cyanochloroketen to benzaldehyde.[10] Evidence has also been provided for a zwitterionic intermediate in the addition of tetra-cyanoethylene to thioenol ethers,[11] and the zwitterionic intermediate (12), formed by the addition of bromoketen diethyl acetal to an electron-poor olefin, is trapped internally to give (after work-up) the cyclopropane (13).[12]

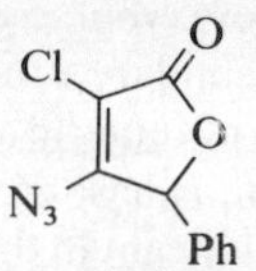

(10)

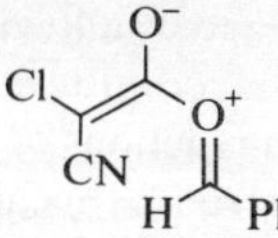

(11)

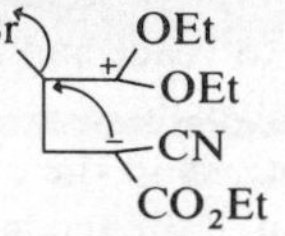

(12)

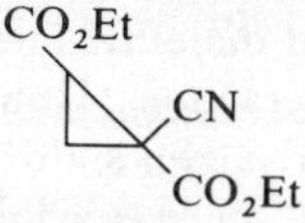

(13)

Several investigations have been reported of the 2+2, Lewis-acid-catalysed addition of propiolate esters to alkenes.[13] Whilst the mechanism of this reaction is still uncertain, the observed stereospecificity, as well as the parallels with keten reactivity, suggest a concerted $\pi 2s + \pi 2a$ process. [Perhaps the keten analogy is best seen if the intermediate complex is written as in formula (14)]. However, the reaction with norbornene gives both normal (15) and rearranged (16) products, showing that a zwitterionic intermediate can be formed when the carbonium ion is sufficiently stable. Some 2+4 cycloaddition reactions are also catalysed by Lewis acids; when the chiral menthol derivative (17) is employed, quite high asymmetric induction can be obtained.[14]

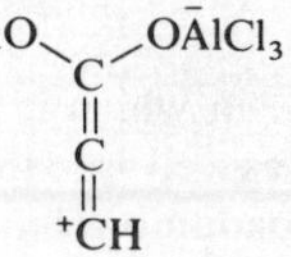

(14)

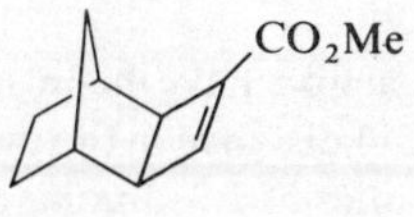

(15)

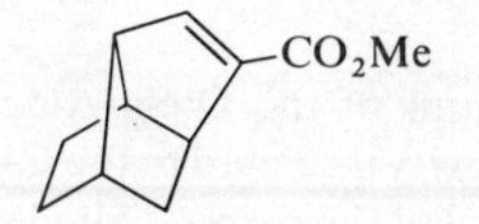

(16)

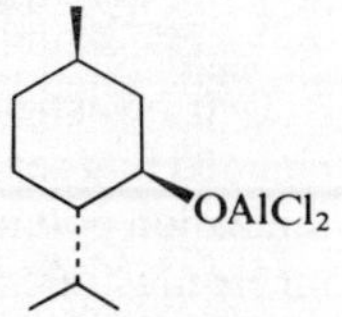

(17)

[8] M. Lofti and R. M. G. Roberts, *Tetrahedron*, 1979, **35**, 2131, 2137.
[9] H. W. Moore, *Accounts Chem. Res.*, 1979, **12**, 125.
[10] H. W. Moore, F. Mercer, D. Kunert, and P. Albaugh, *J. Amer. Chem. Soc.*, 1979, **101**, 5435; but compare O. Krabbenhoft, *J. Org. Chem.*, 1978, **43**, 1305, and E. Schaumann and J. Ehlers, *Chem. Ber.*, 1979, **112**, 1000.
[11] R. Huisgen and H. Graf, *J. Org. Chem.*, 1979, **44**, 2594, 2595.
[12] H. K. Hall, A. B. Padias, A. Deutschmann, and I. J. Westerman, *J. Org. Chem.*, 1979, **44**, 2038.
[13] B. B. Snider and D. M. Roush, *J. Amer. Chem. Soc.*, 1979, **101**, 1906; B. B. Snider, D. J. Rodini, R. S. E. Conn, and S. Sealfon, *ibid.*, p. 5283; R. D. Clark and K. G. Untch, *J. Org. Chem.*, 1979, **44**, 248; H. Fienemann and H. M. R. Hoffmann, *ibid*, p. 2802.
[14] S. Hashimoto, N. Komeshina, and K. Koga, *J.C.S. Chem. Comm.*, 1979, 437.

The energetics for the gas-phase retro-Diels–Alder reactions of 3,6-dihydro-2*H*-pyrans seem to exclude the possibility that biradicals are intermediates.[15] For example, the observed energy of activation for the parent compound (18) is 49.7 kcal mol^{-1}, whereas formation of biradical (19) would require *ca.* 66.0 and of the biradical (20) *ca.* 69.3 kcal mol^{-1}.

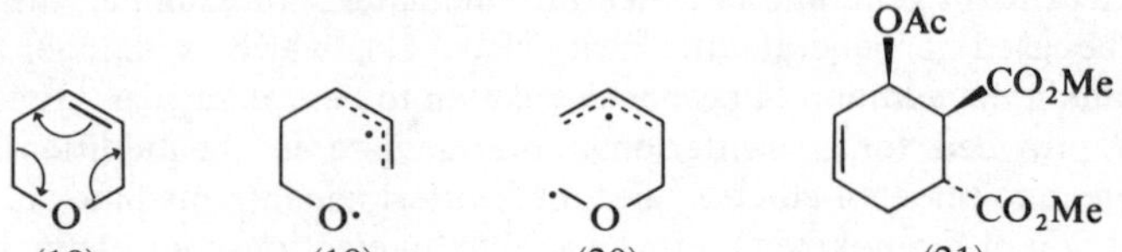

(18) (19) (20) (21)

The stereochemistry of the product (21) obtained by allowing 1-acetoxybutadiene and dimethyl fumarate to react had been explained in terms of secondary orbital interactions. However, investigation of addition reactions of a range of fumaric acid derivatives shows exceptions to the expected stereochemical pattern. It has, therefore, been concluded that, if the secondary orbital effects are ever significant in these systems, then they are small and easily over-ridden by other factors.[16] On the other hand, the preference of N=N dienophiles for the *syn* face of propellane (22) [leading to adducts such as (23)] and of C=C dienophiles for the *anti* face [leading to adducts such as (24)] has been plausibly explained in terms of a secondary orbital interaction between the antisymmetrical combination of the lone-pair orbitals of the N=N and the π^*-orbitals of C=O, which attracts this dienophile to the *syn* side.[17]

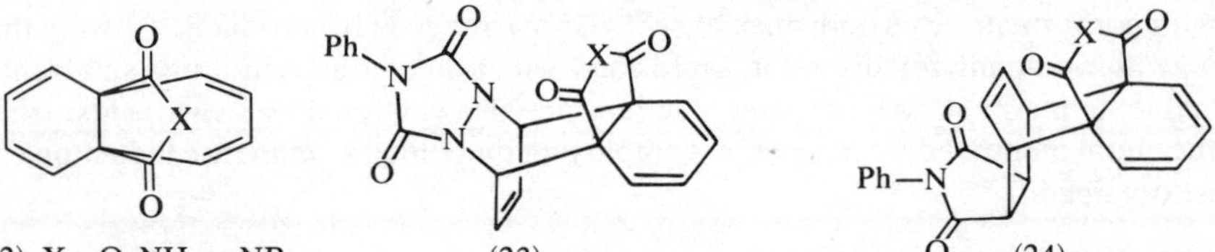

(22) X = O, NH, or NR (23) (24)

Two investigations of the Diels–Alder reaction have been made in which it is claimed that reactivity is related to the C-1–C-4 distance in the diene as well as to the ionization potential of the diene.[18] Similar studies of 1,3-dipolar cycloadditions have nicely illustrated how the electron-rich/-poor nature of the 1,3-dipole influences the relationship between ionization potential and reactivity of the alkene. Whereas diazomethane [19a] is of 'type 1' in Sustmann's classification (reaction being controlled by the HOMO of the 1,3-dipole, and the rate *increasing* roughly as the ionization potential of the alkene *increases*) in diazoacetic ester (25b) 'type 2' behaviour is

[15] H. M. Frey, R. Pottinger, H. A. J. Carless, and D. J. Lingley, *J.C.S. Perkin II*, 1979, 1460; H. M. Frey and S. P. Lodge, *ibid.*, p. 1463.
[16] W. B. T. Cruse, I. Fleming, P. T. Gallagher, and O. Kennard, *J. Chem. Res.* (*S*), 1979, 372.
[17] M. Kaftory, M. Peled, and D. Ginsburg, *Helv. Chim. Acta*, 1979, **62**, 1326.
[18] H.-D. Scharf, H. Plum, J. Fleischhauer, and W. Schleker, *Chem. Ber.*, 1979, **112**, 862; R. Sustmann, M. Böhm, and J. Sauer, *ibid.*, p. 883.
[19] (*a*) J. Geittner, R. Huisgen, and R. Sustmann, *Tetrahedron Letters*, 1977, 881; (*b*) W. Bihlmaier, R. Huisgen, H.-U. Reissig, and S. Voss, *ibid.*, 1979, 2621.

observed [interactions with the HOMO and the LUMO of the 1,3-dipole are both important, leading to a U-shaped plot of log (rate) *vs.* (ionization potential of the alkene)] and the introduction of further electron-withdrawing groups, as in dimethyl diazomalonate (25c) or methyl diazo(phenylsulphonyl)acetate (25d), leads to behaviour intermediate between 'type 2' and 'type 3'[19b] (Sustmann 'type 3' behaviour involves domination by the LUMO of the 1,3-dipole and a rate of reaction which *increases* as the ionization potential of the alkene *decreases*). A similar difference is seen between benzene nitrile oxide (26a), which shows 'type 2' behaviour, and the more electron-deficient benzenesulphonyl nitrile oxide (26b), which shows behaviour intermediate between 'type 2' and 'type 3'.[20]

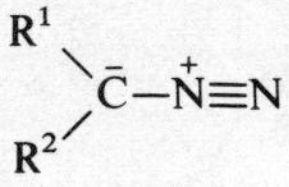

(25) a; $R^1 = R^2 = H$
b; $R^1 = H, R^2 = CO_2Me$
c; $R^1 = R^2 = CO_2Me$
d; $R^1 = CO_2Me, R^2 = SO_2Ph$

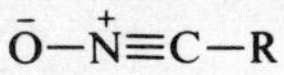

(26) a; $R = Ph$
b; $R = SO_2Ph$

Retro-1,3-dipolar additions[21] and the cycloaddition reactions of triazines[22] have been reviewed. Katritzky and co-workers have continued their extended studies of aromatic ring 1,3-dipoles. 1-Substituted-3-oxido-pyridiniums (27) give adducts (28) with alkenes, and it has been shown that reactivity and selectivity can be rationalized in terms of FMO theory.[23]

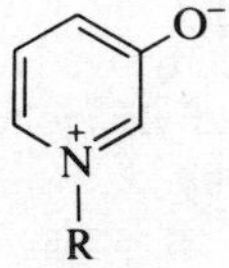

(27)

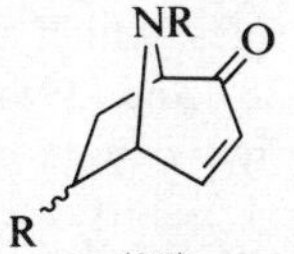

(28)

(29)

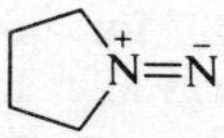

(30)

Ab initio MO calculations (STO-3G and 4-31G) on the thermal elimination of nitrogen from Δ^1-pyrazolines suggest a non-synchronous breaking of the two carbon–nitrogen bonds and (probably) a nitrogen-containing biradical intermediate.[24] The pyrolysis of tetrahydropyridazines (29) and *N*-pyrrolidine nitrenes (30), as well as the thermal rearrangement of cyclobutanes, has generally been interpreted in terms of a common 1,4-biradical, although there is also evidence for a direct retro-(2+2+2) fragmentation of the diazines.[25]

[20] P. A. Wade and H. R. Hinney, *Tetrahedron Letters*, 1979, 139.
[21] G. Bianchi, C. De Micheli, and R. Gandolfi, *Angew. Chem. Internat. Edn.*, 1979, **18**, 721.
[22] J. Bourgois, M. Bourgois, and F. Texier, *Bull. Soc. chim. France*, 1978, **II**, 485.
[23] A. R. Katritzky, N. Dennis, M. Chaillet, C. Larrieu, and M. El Mouhtadi, *J.C.S. Perkin I*, 1979, 408, and related papers.
[24] P. C. Hiberty and Y. Jean, *J. Amer. Chem. Soc.*, 1979, **101**, 2538.
[25] P. B. Dervan, T. Uyehara, and D. S. Santilli, *J. Amer. Chem. Soc.*, 1979, **101**, 2069; P. B. Dervan and T. Uyehara, *ibid.*, p. 2076; P. B. Dervan and D. S. Santilli, *ibid.*, p. 3663.

3 Sigmatropic Rearrangements

The stereochemistry of 2,3-sigmatropic rearrangements has continued to attract attention[26,27] and has been reviewed,[28] as has anchimeric assistance by silicon and its relevance to the mechanism of dyotropic rearrangements.[29]

The phototranspositions of cyano-thiophens [for example the conversion of compound (31) into its isomer (32)] have been shown to involve 5-thiabicyclo-[2.1.0]pent-2-ene intermediates [(33) and (34) in the example given] which

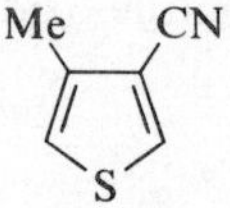

(31)

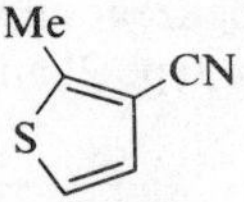

(32)

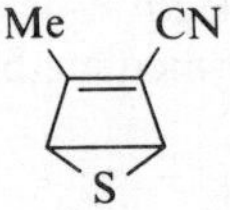

(33)

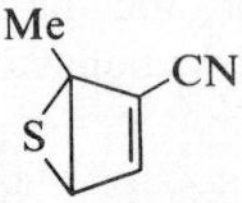

(34)

interconvert *via* a sulphur 'walk'. In some cases it has been shown that these intermediates can be trapped as their 2+4 furan adducts or, as in the case of compound (33), isolated from the reaction mixture so that the sulphur 'walk' can be studied in isolation.[30] An interesting, related carbon 'walk' [bicyclopentene (35) to its isomer (36)] has also been reported.[31] This thermal 1,3-shift involves inversion at the migrating centre, as expected for an orbital-symmetry-controlled process. However, the 1,3-shift in compound (35), the related 1,5-shifts in compounds (37) and (38) [32] and the 1,7-shift in compound (39)[33] all apparently involve inversion at the migrating centre, whether this is 'allowed' or not. The simplest view of these results is that the stereochemistries are determined by 'least motions' factors and that the distinction between 'allowed' and 'disallowed' processes is found not in the stereochemistries but in the activation energies, which are 21.7 37.1, *ca.* 35, and 28.8 kcal mol^{-1} for compounds (35), (37), (38), and (39) respectively. However, by using a specifically deuteriated 3,7-dimethyl-7-methoxymethylcycloheptatriene, Baldwin claims to have shown that the 1,5-shift occurs by a two-step, (C-7 epimerization)+(1,5-shift with retention), 'allowed' route.[32b] It is also found that

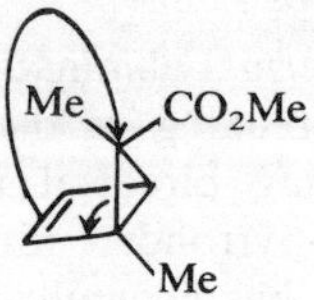

(35)

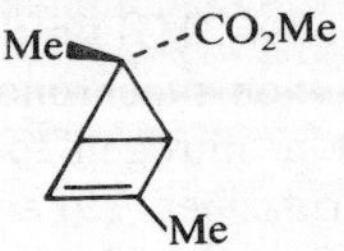

(36)

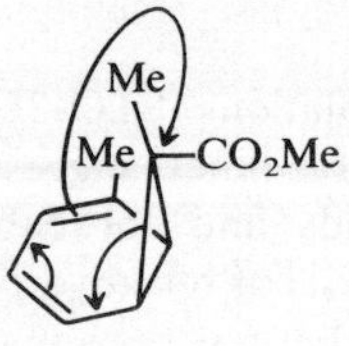

(37)

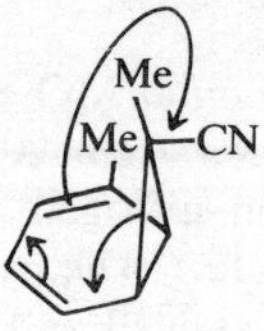

(38)

[26] E. Vedejs, M. J. Arnost, and J. P. Hagen, *J. Org. Chem.*, 1979, **44**, 3230; V. Ceré, C. Paolucci, S. Pollicino, E. Sandri, and A. Fava, *ibid.*, p. 4128.
[27] E. Vedejs, M. J. Arco, D. W. Powell, J. M. Renga, and S. P. Singer, *J. Org. Chem.*, 1978, **43**, 4831; V. Ceré, C. Paolucci, S. Pollicino, E. Sandri, and A. Favo, *ibid.*, p. 4826.
[28] R. W. Hoffmann, *Angew. Chem. Internat. Edn.*, 1979, **18**, 563.
[29] M. T. Reetz, *Angew. Chem. Internat. Edn.*, 1979, **18**, 173.
[30] J. A. Barltrop, A. C. Day, and E. Irving, *J.C.S. Chem. Comm.*, 1979, 881, 966.
[31] F.-G. Klärner and F. Adamsky, *Angew. Chem. Internat. Edn.*, 1979, **18**, 674.
[32] (*a*) F.-G.Klärner, S. Yaslak, and M. Wette, *Chem. Ber.*, 1979, **112**, 1168; (*b*) J. E. Baldwin and B. M. Broline, *J. Amer. Chem. Soc.*, 1978, **100**, 4599.

the photochemical rearrangements of compounds (37) and (38), like their thermal analogues, involve overall inversion of the migrating centre.[32,34] Here epimerization at C-7 certainly competes with the 1,5-shift, and the results have been interpreted in terms of a triplet biradical intermediate (40) which can either close or in which rotation about the C-6–C-7 bond occurs.

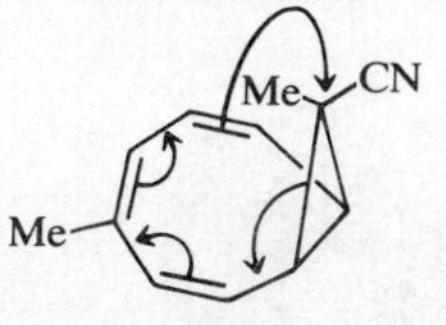

(39)

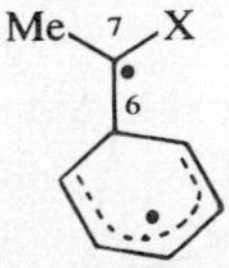

(40) X = CO_2Me or CN

Migratory aptitude in 1,5-sigmatropic shifts seems to be determined by a number of factors.[35–38] Dolbier and his co-workers have shown that the isoindenes (41) [normally encountered as short-lived intermediates in the rearrangement of indenes[36]] can be generated by eliminating N_2O from the corresponding bicyclo-azoxy-system (42).[35,39] They have studied the relative migratory aptitudes of the

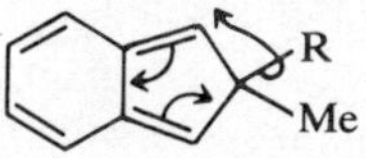

(41)

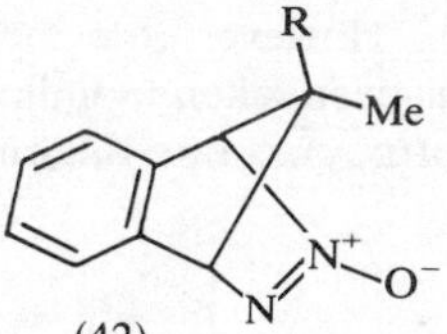

(42)

substituents R in these compounds, seeking for evidence of a radical-pair mechanism. Whilst, on quantitative grounds, they demonstrated that the extent of radical development cannot be very great, they did find a correlation between migratory aptitude and the stability of the radical (R·). In other systems, secondary orbital interactions seem to be important; for example, in the migration of carbonyl groups. Whereas 1,5-shifts of alkyl groups in cyclopentadienes frequently require temperatures above 300 °C, the rate of rearrangement of the formyl-substituted compound (43) is rapid ($k = 90\ s^{-1}$) at room temperature.[37] Here the transition state can be stabilized by an interaction between ψ_3 of the pentadienyl system and π^* of the formyl group (44). A complementary situation is seen in the rapid 1,5-shift of the carbanionic centre in compound (45), where a similar interaction may be written between ψ_4 of the pentadienyl system and a filled p-orbital of the carbanion (46).[38]

[33] F.-G. Klärner and M. Wette, *Chem. Ber.*, 1978, **111**, 282.
[34] F.-G. Klärner and S. Yaslak, *Chem. Ber.*, 1979, **112**, 2286.
[35] W. R. Dolbier, K. E. Anapolle, L. McCullagh, K. Matsui, J. M. Riemann, and D. Rolison, *J. Org. Chem.*, 1979, **44**, 2845.
[36] D. J. Field and D. W. Jones, *J.C.S. Perkin I*, 1979, 1273.
[37] R. J. Bushby and D. W. Jones, *J.C.S. Chem. Comm.*, 1979, 688.
[38] G. Boche and D. Martens, *Chem. Ber.*, 1979, **112**, 175.
[39] W. R. Dolbier, K. Matsui, H. J. Dewey, D. V. Horák, and J. Michl, *J. Amer. Chem. Soc.*, 1979, **101**, 2136.

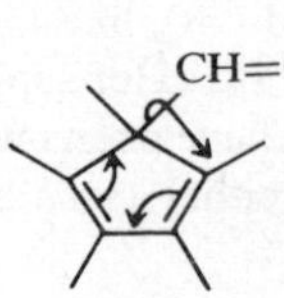

(43)

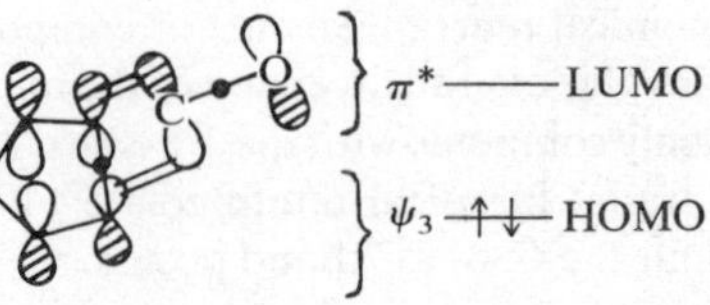

(44)

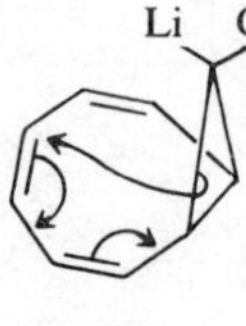

(45)

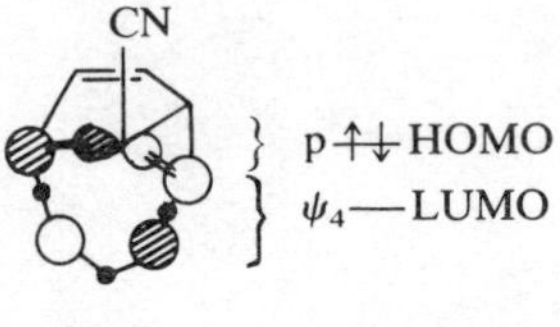

(46)

There have been several attempts to demonstrate carbon analogues (47) of the Claisen rearrangement.[40–43] Such reactions can be facilitated by incorporating a cyclopropane ring into the system, as in compound (48).[41] The general difficulty of these reactions and the need to add a base catalyst had been attributed to a mechanism in which the rate-determining step is prototropic rearrangement of the intermediate (49). However, the lack of scrambling of the isotopic label in compound (50) has been taken to indicate that the first step (3,3-shift), and not the second step (prototropy), is rate-determining.[43]

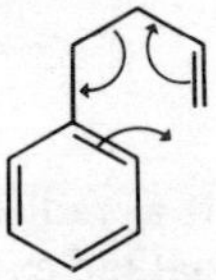

(47)

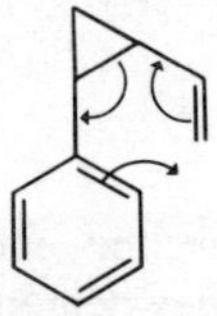

(48)

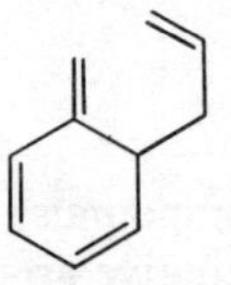

(49)

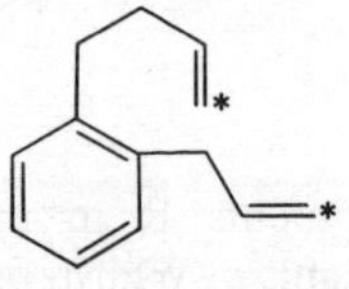

(50)

Evidence for a 1,4-diyl (52) in the Cope rearrangement of the en-yn-ene (51) has been provided by isolating products from the 1,2-silicon shift shown in (52).[44] However, secondary isotope effects for aromatic[45] and aliphatic[46] Claisen rearrangements do not support the notion of a 1,4-diyl in these reactions. Rather, they suggest a non-synchronous concerted process in which bond breaking is in advance of bond making. As expected, they indicate that the transition state comes early in the more exothermic aliphatic rearrangement, but later in the aromatic case.

[40] G. Maas and M. Regitz, *Angew. Chem. Internat. Edn.*, 1977, **16**, 711.
[41] E. N. Marvell and C. Lin, *J. Amer. Chem. Soc.*, 1978, **100**, 877; E. N. Marvell and S. W. Almond, *Tetrahedron Letters*, 1979, 2777, 2779.
[42] J. B. Lambert, D. M. Fabricius, and J. A. Hoard, *J. Org. Chem.*, 1979, **44**, 1480.
[43] J. B. Lambert, D. M. Fabricius, and J. J. Napoli, *J. Amer. Chem. Soc.*, 1979, **101**, 1793.
[44] G. C. Johnson, J. J. Stofko, T. P. Lockhart, D. W. Brown, and R. G. Bergman, *J. Org. Chem.*, 1979, **44**, 4215.
[45] K. D. McMichael and G. L. Korver, *J. Amer. Chem. Soc.*, 1979, **101**, 2746.
[46] J. J. Gajewski and N. D. Conrad, *J. Amer. Chem. Soc.*, 1979, **101**, 2747, 6693.

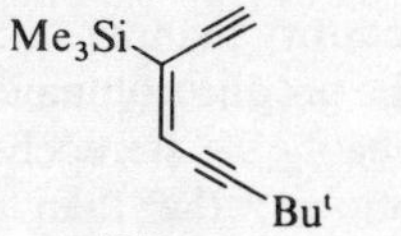

(51)

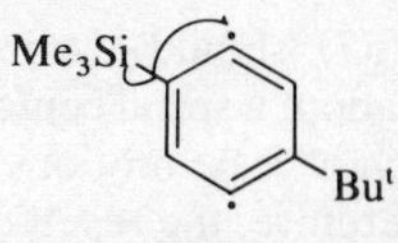

(52)

Stereochemical and regioselective aspects of the photochemical di-π-methane rearrangement have continued to excite interest,[47] and the interconversion of compounds (53) and (54) has provided the first example of a *thermal* reaction of this type.[48]

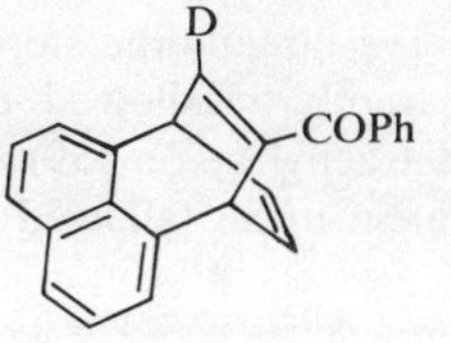

(53)

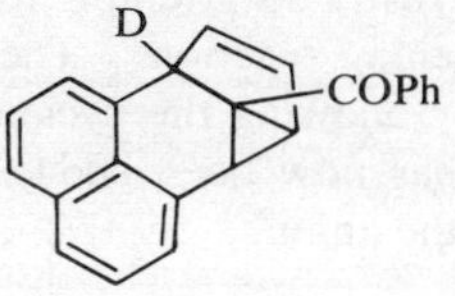

(54)

4 Electrocyclic Reactions

The electrocyclic reactions of 1,5-dipoles have been reviewed[49] and those of radical cations discussed in some detail.[50]

The 16π-electron vinylogous heptafulvene (55)[51] cyclizes thermally in a conrotatory manner to give compound (56). Whilst this is the stereochemistry expected for an 'orbital-symmetry-controlled' process, the related 14π-electron and 12π-electron compounds (57)[52] and (58)[53] also cyclize in a conrotatory sense, despite the fact

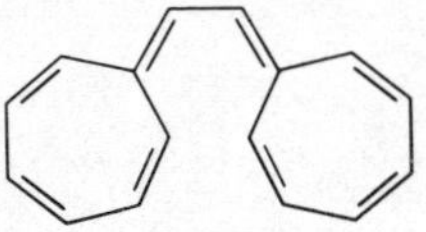

(55)

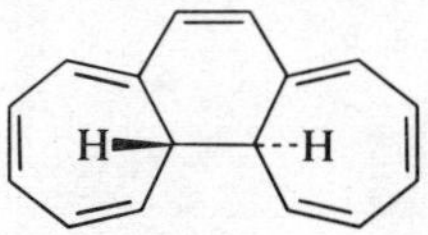

(56)

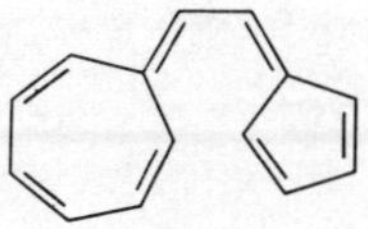

(57)

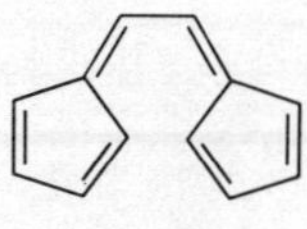

(58)

[47] H. E. Zimmerman and D. R. Diehl, *J. Amer. Chem. Soc.*, 1979, **101**, 1841; H. E. Zimmerman, T. P. Gannett, and G. E. Keck, *J. Org. Chem.*, 1979, **44**, 1982; R. L. Coffin, W. W. Cox, R. G. Carlson, and R. S. Givens, *J. Amer. Chem. Soc.*, 1979, **101**, 3261; L. A. Paquette, A. Y. Ku, C. Santiago, M. D. Rozeboom, and K. N. Houk, *ibid.*, p. 5972; A. Y. Ku, L. A. Paquette, M. D. Rozeboom, and K. N. Houk, *ibid.*, p. 5981; A. Padwa and T. Brookhart, *J. Org. Chem.*, 1979, **44**, 4021.

[48] M. Demuth, U. Burger, H. W. Mueller, and K. Schaffner, *J. Amer. Chem. Soc.*, 1979, **101**, 6763; M. Demuth, C. O. Bender, S. E. Braslavsky, H. Görner, U. Burger, W. Amrein, and K. Schaffner, *Helv. Chim. Acta*, 1979, **62**, 847.

[49] E. C. Taylor and I. J. Turchi, *Chem. Rev.*, 1979, **79**, 181.

[50] E. Haselbach, T. Bally, and Z. Lanyiova, *Helv. Chim. Acta*, 1979, **62**, 577; E. Haselbach, T. Bally, Z. Lanyiova, and P. Baeryschi, *ibid.*, p. 583.

[51] H. Bingmann, L. Knothe, D. Hunkler, and H. Prinzbach, *Tetrahedron Letters*, 1979, 4053.

[52] H. Prinzbach, H. Babsch, and D. Hunkler, *Tetrahedron Letters*, 1978, 649.

[53] H. Prinzbach, H. Sauter, and B. Gallenkamp, *Chem. Ber.*, 1977, **110**, 1382.

that compound (57) 'should' cyclize in a disrotatory manner. Since all three substrates probably adopt a spiral conformation, the products obtained in each case are those expected on a 'least motions' basis. Probably the stereochemistries of these reactions are, therefore, the result of 'least motions' rather than 'orbital symmetry' control. There may, however, be some evidence for a weak alternation between what is 'allowed' and what 'disallowed' in the Arrhenius activation energies, which are 20.0, 24.4, and 22.1 kcal mol^{-1} for compounds (55), (57), and (58) respectively.

It has been claimed that the cyclopropyl anion opens photochemically in a disrotatory manner.[54] However, the tendency of the products to undergo (photochemical) *cis–trans* isomerization renders the result open to some doubt. Similar problems have frustrated attempts to observe directly the stereochemistry of the thermal ring-opening reaction.[55] The apparently 'disallowed' cyclization product (60) obtained by allowing the cyclononatetraenyl anion (59) to react with electrophiles, RX, has now been shown to arise in an 'allowed' manner from the *trans*-isomer of the anion.[56]

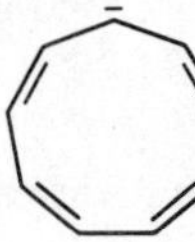

(59)

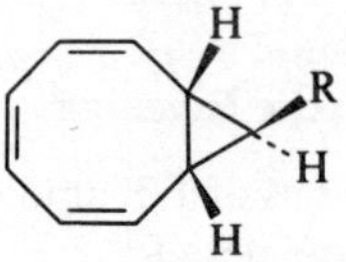

(60)

[54] M. A. Fox, *J. Amer. Chem. Soc.*, 1979, **101**, 4008.
[55] G. Boche, K. Buckl, D. Martens, D. R. Schneider, and H.-U. Wagner, *Chem. Ber.*, 1979, **112**, 2961.
[56] G. Boche, M. Bernheim, D. Lawaldt, and R. Ruisinger, *Tetrahedron Letters*, 1979, 4285.

4 Reaction Mechanisms

Part (ii) Polar Reactions

By H. R. HUDSON

Department of Chemistry, The Polytechnic of North London, Holloway Road, London N7 8DB

1 Introduction

The mechanisms of polar organic reactions are inevitably influenced by the medium in which they occur. Considerable effort is therefore being directed to the investigation of intrinsic properties of solvent-free reactions by the study of ion–molecule reactions in the gas phase.[1] At the same time, attempts are being made to rationalize the polar effects of solvents on rates, equilibria, heats and free energies of solution or transfer, and spectroscopic properties, through linear solvation-energy relationships.[2] Earlier work in this field has been reviewed.[3] The effects of changes of medium on rate processes and structures of transition states have also been examined from the viewpoint of thermodynamic transfer functions,[4] and the effects of micellar catalysts on the rates and mechanisms of a variety of reaction types have been studied.[5] The following report deals with some novel aspects of the year's work on the basic types of polar organic reaction and reaction intermediate. A fuller account is published elsewhere.[6]

2 Solvolytic Reactions

Solvolysis of Saturated Alkyl Substrates.—A novel approach to the analysis of solvolytic reaction mechanisms has been based on a multi-parameter optimization procedure.[7] The method was used to interpret solvolytic data for the substitution and elimination reactions of cyclopentyl *p*-bromobenzenesulphonate (1) and its α-^{2}H,

1 'Gas Phase Ion Chemistry,' ed. M. T. Bowers, Academic Press, New York, 1979.

2 M. J. Kamlet and R. W. Taft, *J.C.S. Perkin II*, 1979, 337; M. J. Kamlet, M. E. Jones, R. W. Taft, and J.-L. Abboud, *ibid.*, p. 342; M. J. Kamlet and R. W. Taft, *ibid.*, p. 349; R. W. Taft and M. J. Kamlet, *ibid.*, p. 1723; M. J. Kamlet, A. Solomonovici, and R. W. Taft, *J. Amer. Chem. Soc.*, 1979, **101**, 3734; M. J. Kamlet, T. N. Hall, J. Boykin, and R. W. Taft, *J. Org. Chem.*, 1979, **44**, 2599.

3 C. Reichardt, *Angew. Chem. Internat. Edn.*, 1979, **18**, 98.

4 E. Buncel and H. Wilson, *Accounts Chem. Res.*, 1979, **12**, 42.

5 D. C. Bredt, N. Utrapiromsuk, and S. S. Jaglan, *J. Org. Chem.*, 1979, **44**, 136; G. B. van de Langkruis and J. B. F. N. Engberts, *ibid.*, p. 141; A. Pillersdorf and J. Katzhendler, *ibid.*, p. 549; W. Tagaki, D. Fukushima, T. Eiki, and Y. Yano, *ibid.*, p. 555; A. Pillersdorf and J. Katzhendler, *ibid.*, p. 934; A. Finiels and P. Geneste, *ibid.*, p. 2036; M. Croce and Y. Okamoto, *ibid.*, p. 2100; Y. Yano, Y. Yoshida, A. Kurashima, Y. Tamura, and W. Tagaki, *J.C.S. Perkin II*, 1979, 1128; T. Kunitake, Y. Okahata, S. Tanamachi, and R. Ando, *Bull. Chem. Soc. Japan*, 1979, **52**, 1967; H. Fujii, T. Kawai, and H. Nishikawa, *ibid.*, p. 1978; T. Kunitake and T. Sakamoto, *ibid.*, p. 2624.

6 'Organic Reaction Mechanisms, 1979', ed. A. C. Knipe and W. E. Watts, Wiley, London, 1980.

7 V. J. Shiner, jnr., D. A. Nollen, and K. Humski, *J. Org. Chem.*, 1979, **44**, 2108.

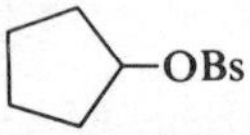

(1)

cis-β-^{2}H, and β-$^2H_4$ analogues in eight different solvent systems consisting of ethanol–water, trifluoroethanol(TFE)–water, and hexafluoroisopropyl alcohol(HFIP)–water. Assuming the possibility of product formation from both contact and solvent-separated ion-pairs, the total number of mechanistic parameters required to define the system is 78. This number was reduced to 46 by assuming that isotope effects are independent of the solvent used and are identical for certain steps of the reaction, and that *anti-/syn*-elimination ratios are identical under specified conditions. To fit the reaction parameters to 95 experimental observations, the simplex method of optimization was used (see, *e.g.*, S. N. Deeming and S. L. Morgan, *Analyt. Chem.*, 1973, **45**, 278A; 1974, **46**, 1170), this being, apparently, the first application of the method to the correlation of rates, product distributions, and isotope effects. The outcome of the analysis was that significant fractions of ion-pair return were shown to occur in most solvents, nucleophilic solvation was not apparently involved, and the secondary β-^{2}H effects on elimination of a proton from the tight ion-pair suggested a twisted-envelope transition-state structure for this process. The interpretation differs from that of Bentley and Schleyer, who consider borderline solvolyses of secondary sulphonates to occur *via* the rate-determining formation of nucleophilically solvated ion-pairs for which return is kinetically insignificant [*Ann. Reports* (*B*), 1976, **73**, 55; 1974, **71**, 116].

The importance of nucleophilic solvation for simple s-alkyl cations has been stressed in a comparative study of heats of ionization (ΔH_i) for alkyl chlorides in SbF_5 and their corresponding free energies of limiting solvolysis in ethanol.[8] The results confirm the view that rates of limiting solvolysis are a good measure of stabilities of carbonium ions. For simple secondary alkyl chlorides, the solvolytic data correlated well with ΔH_i after a correction for nucleophilic solvation had been made. In the case of secondary or tertiary 2-norbornyl substrates (2), a good correlation was obtained for the *exo*-isomer, indicating that the solvolytic transition state is closely related to the resulting carbo-cation. The result is in keeping with σ-delocalization in this case, and it is interesting to note that a similarly good correlation was not found for the *endo*-isomer. Other evidence in favour of a steric interpretation of high *exo-/endo*-rate ratios in 2-norbornyl solvolyses has been provided by the relative rates for some 2-substituted *exo-* or *endo*-5,6-trimethylene-2-norbornyl *p*-nitrobenzenesulphonates (3), in which the 5,6-trimethylene bridge provides steric interference to σ-interaction in the *exo*-isomer.[9] *exo-/endo*-Rate ratios of 118 (2-phenyl), 286 [2-(5′-coumaranyl)], and 420 (2-methyl) were nevertheless observed.

Evidence has been presented for unexpected examples of nucleophilic solvent assistance in the solvolysis of t-butyl halides. The conclusions were based on a comparison of reaction rates with those of adamantyl substrates in aqueous ethanol and HFIP, and indicated a mechanistic change for the t-butyl halides as the medium

[8] E. M. Arnett, C. Petro, and P. von R. Schleyer, *J. Amer. Chem. Soc.*, 1979, **101**, 522.
[9] H. C. Brown, C. G. Rao, and D. L. Vander Jagt, *J. Amer. Chem. Soc.*, 1979, **101** 1780.

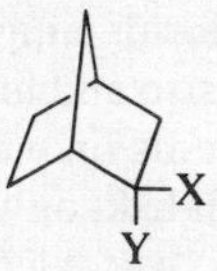

(2) X = Cl, Y = H or substituent
Y = Cl, X = H or substituent

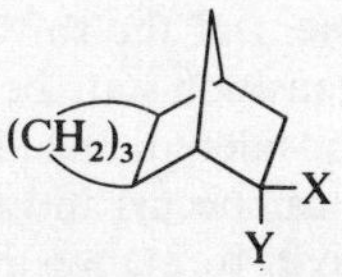

(3) X = OBs, Y = substituent
Y = OBs, X = substituent

became more nucleophilic.[10] It was also shown that trifluoroacetic acid (TFA) is more nucleophilic than HFIP towards simple s-alkyl tosylates, although trifluoroacetolyses of the latter are normally regarded as good models for S_N1 (limiting) reactions.

A new scale of solvent nucleophilicities, based on the two-term Grunwald–Winstein equation, has been set up by the use of first-order rate coefficients obtained for the solvolysis of triethyloxonium hexafluorophosphate in various organic or aqueous–organic solvent mixtures.[11] By the use of a positively charged substrate, contributions from the solvent ionizing power (Y) are reduced to less than 10% of those for an initially neutral substrate, and can be estimated from data available for solvolysis of t-butyldimethylsulphonium ion. The new 'N_{KL}' values [based on equation (1)] were applied to previously studied solvolyses of alkyl tosylates and

$$N_{KL} = \log(k/k_0)_{Et_3O^+} - 0.55\,Y \qquad (1)$$

alkyl chlorides, and were shown to be applicable to both RX- and RX^+-type substrates. The reactivity selectivity principle (r.s.p.) has continued to attract attention in the study of solvolytic reactions, although there are circumstances under which it does not hold. Some possible reasons for its failure have been discussed in terms of More O'Ferrall plots [*cf. J. Chem. Soc.* (*B*), 1970, 274], which illustrate the possibility of movement of the transition state over a potential surface, without significant change in the nucleophile–carbon distance.[12] The lack of a relationship between reactivity and selectivity in the reactions of octyl, 1-methylheptyl, and benzyl derivatives with *m*-chloroaniline and ethanol has led to the conclusion that selectivity values derived from two unrelated nucleophiles (*i.e.* those not having a common nucleophilic atom) cannot be utilized as a measure of structure of the transition state.[13] Selectivity was also found to be dependent on the nature of the substrate and the solvent. The results were rationalized by perturbation molecular orbital theory, and it was suggested that the differential HOMO–LUMO gap governs selectivity. The range of applicability of the r.s.p. to S_N2 reactions in which a series of related nucleophiles and related leaving groups is involved is still not clear. In the solvolytic reactions of 1-adamantyl bromide, picrate, 2,4-dinitrophenolate, and tosylate in aqueous ethanol, the variation of selectivity with substrate and with solvent composition was attributed to the differing abilities of the various ion-pair intermediates to select a molecule of water or of ethanol in the formation of a solvent-separated ion-pair.[14]

[10] T. W. Bentley, C. T. Bowen, W. Parker, and C. I. F. Watt, *J. Amer. Chem. Soc.*, 1979, **101**, 2486.
[11] D. N. Kevill, and G. M. L. Lin, *J. Amer. Chem. Soc.*, 1979, **101**, 3916.
[12] J. M. Harris, S. G. Shafer, J. R. Moffatt, and A. R. Becker, *J. Amer. Chem. Soc.*, 1979, **101**, 3295.
[13] Y. Karton and A. Pross, *J.C.S. Perkin II*, 1979, 857.
[14] P. R. Luton and M. C. Whiting, *J.C.S. Perkin II*, 1979, 646.

A new treatment of the solvolysis of t-butyl chloride employs $Me_4N^+ Cl^-$ as a model for the transition state because of its similar size and high polarity.[15] Twelve solvents were considered, including hydroxylic solvents (in which both substitution and elimination can occur) and aprotic media (which lead only to elimination). The free energy of cavity formation in each solvent, $|\bar{G}_{cav}|_s$, was calculated for the reactant R, the activated complex AC, and the model M; by taking methanol as the reference solvent, relative cavity terms, $|\bar{G}_{cav}|_s^{s_1}$, were obtained. These were combined with free energies of transfer from gas phase to solvent, $|\bar{G}_t|_s^g$, to give the relative solvent–solute interaction terms, $|G_{int}|_s^{s_1}$. The linear correlation obtained between $|G_{int}|_s^{s_1}$ terms for AC and those for M indicates that the charge separation in the transition state remains approximately constant for all the solvents studied, whilst the slope (0.73) confirms the highly polar nature of the activated complex.

Solvolysis of Vinyl Substrates.—The first comparison of titrimetric (k_t) and polarimetric (k_α) rate constants for the solvolysis of a vinyl substrate has been made for 9-(α-bromoanisylidene)-10-hydroxy-9,10-dihydroanthracene (4) in 2,6-lutidine-buffered TFE at 49.5 °C.[16] The extent of rate depression by a common ion and the identity of k_t° and k_α values showed that a free vinyl cation (5) was involved. Product formation from ion-pairs, and ion-pair return with racemization, in TFE were excluded. Some caution must, however, be exercised in the application of the k_α/k_t probe to systems such as this in which alternative rate-determining ionization steps are available. Although C—Br heterolysis was rate-determining in good ionizing and relatively non-acidic solvents (TFE or 80% ethanol), it was found that the initial site of reaction was solvent-dependent.[17] In acetic acid, for example, the initial cleavage was of the bond between C-10 and oxygen, and alternative reaction pathways were introduced (Scheme 1). In either case, the carbo-cation that is formed is non-chiral, and optical activity is lost.

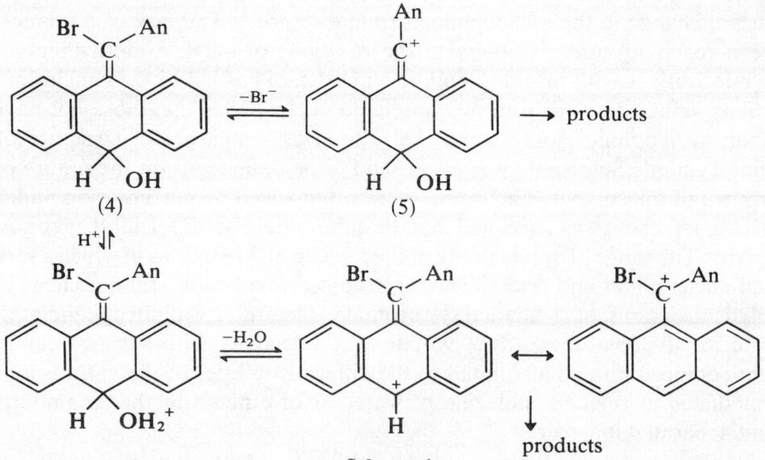

Scheme 1

[15] J. J. M. Ramos, R. Reisse, and M. H. Abraham, *Canad. J. Chem.*, 1979, **57**, 500.
[16] Z. Rappoport and J. Greenblatt, *J. Amer. Chem. Soc.*, 1979, **101**, 1343.
[17] Z. Rappoport and J. Greenblatt, *J. Amer. Chem. Soc.*, 1979, **101**, 3967.

The solvolysis of (bromocyclopropylmethylene)cyclopropane (6) in aqueous ethanol of TFE has been shown to proceed *via* a vinyl cation (7), in which stabilization by both the cyclopropyl and the cyclopropylidene rings is possible.[18] The effects of added salts (KBr, $LiClO_4$, or LiCl), together with product analyses, indicated a reaction scheme which involves rearrangement to the 2-cyclopropylcyclobutenyl (8) and 4-cyclopropylhomopropargyl (9) cations, and in which

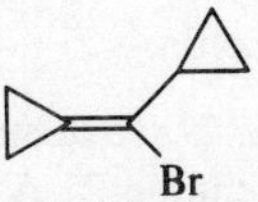

(6)

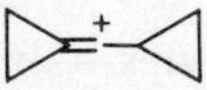
(7)

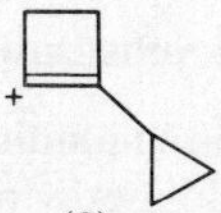
(8)

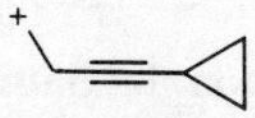
(9)

intimate ion-pairs [(7), (8), and (9)], solvent-separated ion-pairs [(7) and (8)], and dissociated ions [(7) and (8)] may occur. In aqueous ethanol, Grunwald–Winstein *m* values of 0.92 or 1.00 (the highest so far obtained for a vinyl derivative) showed the solvolysis to be almost a pure k_c process. Rearrangement among cyclobutenyl, cyclopropylidene, and homopropargyl cations has also been reported in the solvolyses of a series of 2- or 3-substituted cyclobutenyl nonaflates (10—12; R = Me or aryl).[19] The rates of solvolysis of cyclic vinyl nonaflates have previously been shown to increase with ring size in the range C_5—C_9, a reflection of the preferred linear geometry of the vinyl cation. Much higher reactivity for the four-membered cyclobutenyl derivatives is therefore attributed to the formation of stabilized non-classical intermediates, (13)—(15) respectively.

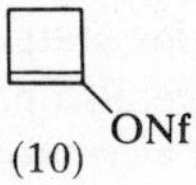

(10)

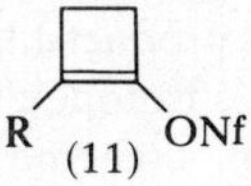

(11)

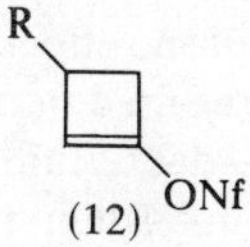

(12)

(13)

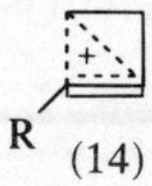

(14)

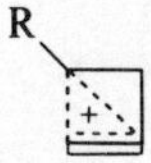

(15)

An unusual example of complete rearrangement with substitution in the reaction of a substituted 1-cyclohexenyl substrate has been provided by the solvolysis of the steroidal derivative 5α-cholest-1-en-1-yl triflate (16).[20] The reaction in a sodium-acetate-buffered mixture of acetone and water yielded substitution products (17)—(20), presumably *via* an allyl cation intermediate (21), which was formed in a concerted ionization and migration process.

[18] R. Kopp and M. Hanack, *Chem. Ber.*, 1979, **112**, 2453.
[19] M. Hanack, E. J. Carnahan, A. Krowczynski, W. Schoberth, L. R. Subramanian, and K. Subramanian, *J. Amer. Chem. Soc.*, 1979, **101**, 100.
[20] G. Ortar and E. Morera, *Tetrahedron Letters*, 1979, 4881.

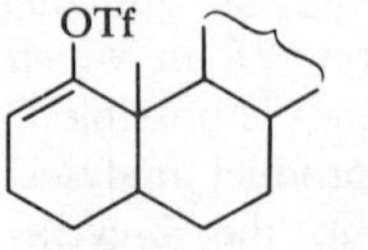

(16)

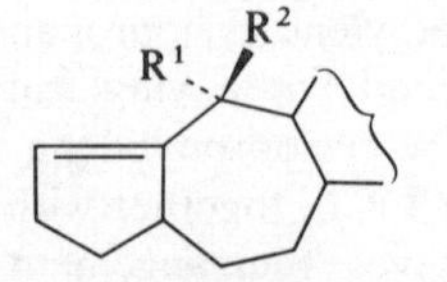

(17) $R^1 = OH, R^2 = Me$
(18) $R^1 = Me, R^2 = OH$

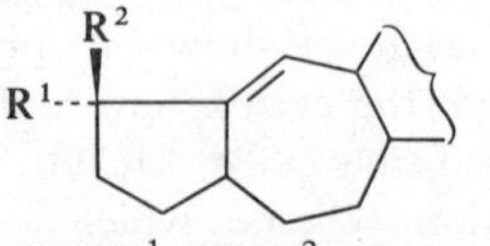

(19) $R^1 = H, R^2 = OH$
(20) $R^1 = OH, R^2 = H$

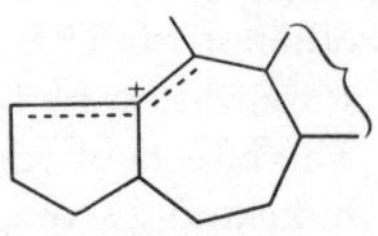

(21)

3 Nucleophilic Substitution in Non-solvolytic Reactions

Substitution in Saturated Aliphatic Substrates.—In the S_N2 reaction of thiophenoxide ion with benzyldimethylphenylammonium ion (Scheme 2; $Ar^1 = Ar^2 = Ar^3 = Ph$) in DMF at 0 °C, an unusually large secondary α-deuterium kinetic isotope effect was observed (1.179 ± 0.007, *i.e.* 1.086 ± 0.003 per α-D).[21] This is significantly higher than the normal values for S_N2 reactions ($k_H/k_D = 1.082$, or 1.04 per α-D), and indicates that maximum values are more dependent on the bulk of the leaving group than was previously supposed. Such values should therefore be established for each leaving group before being used as a criterion of mechanism. In an analogous

$$Ar^1S^- + Ar^2CH_2\overset{+}{N}Me_2Ar^3 \rightarrow \left[Ar^1S\cdots\underset{Ar^2}{\overset{H\quad H}{C}}\cdots NMe_2Ar^3 \right]^{\ddagger} \rightarrow Ar^1SCH_2Ar^2 + Me_2NAr^3$$

Scheme 2

study (Scheme 2; Ar^1=*p*-$MeOC_6H_4$, *p*-MeC_6H_4, Ph, or *p*-ClC_6H_4; Ar^2=Ph; Ar^3= *p*-$MeOC_6H_4$, Ph, or *p*-ClC_6H_4), heavy atom (N) kinetic isotope effects, secondary α-deuterium isotope effects, and Hammett ρ values showed that a change to a better leaving group resulted in a more product-like transition state.[22]

Secondary α-deuterium kinetic isotope effects have also been used to study the symmetrical Finkelstein reaction (Scheme 3), but were complicated in dipolar

$$XC_6H_4YCZ_2Cl + Et_4NCl^* \rightleftharpoons XC_6H_4YCZ_2Cl^* + Et_4NCl$$

X = H, *p*-Cl, *p*-Me, *p*-MeO, *p*-NO_2, *m*-Cl, or *m*-Me

Y = O, S, CH$\overset{t}{=}$CH, none (*i.e.* benzyl), or CH_2

Z = H or D

Scheme 3

aprotic solvents by the formation of a substrate–nucleophile complex.[23] Specific association was demonstrated (Scheme 4) by the observation of a low-field shift of absorption in the ^{1}H n.m.r. spectrum, and it was seen to be strongest in the case of the most electronegative anion, *i.e.* chloride.[24] The effect was observed in acetonitrile, DMSO, and DMF but was retarded by the addition of methanol because of competitive hydrogen-bonding. Complex formation was thought to involve

[21] K. C. Westaway and S. F. Ali, *Canad. J. Chem.*, 1979, **57**, 1089.
[22] K. C. Westaway and S. F. Ali, *Canad. J. Chem.*, 1979, **57**, 1354.
[23] J. Hayami, N. Hihara, N. Tanaka, and A. Kaji, *Bull. Chem. Soc. Japan*, 1979, **52**, 831.
[24] J. Hayami, T. Koyanagi, N. Hihara, and A. Kaji, *Bull. Chem. Soc. Japan*, 1978, **51**, 891, and cited papers.

$$ArXCH_2Cl + Y^- \rightleftharpoons ArXCH_2ClY^-$$

Ar = Ph or substituted Ph
X = none, S, SO, SO_2, or CO
Y = Cl, Br, or I (as R_4NY; R = Et or Bu)

Scheme 4

hydrogen-bonding of the nucleophile to the 'acidic' α-methylene protons, thus locating it in the close neighbourhood of the rear side of the C—Cl bond, in a site of potential S_N2 attack. The overall observed secondary α-deuterium isotope effects were therefore dissected into equilibrium isotope effects of bimolecular association (K_H/K_D) and kinetic isotope effects for unimolecular scrambling of the complex (k_1^H/k_1^D), such that $k_2^H/k_2^D = (k_1^H/k_1^D) \times (K_H/K_D)$.[23] For activated systems, the analysis leads to a fairly large and uniform kinetic isotope effect (1.127—1.147), approaching that for limiting solvolysis, whereas the value for the 2-arylethyl chlorides is nearer to unity. The investigation was also extended to substrates of extremely low reactivity, *viz. p*-nitroarylsulphonyl-chloromethanes, *p*-nitroaryl-sulphinyl-chloromethanes, and 2-chloro-1-ethanones.[25] For the halogeno-ketones the association mechanism leads to an α-deuterium effect of *ca.* 1, suggesting the intermediacy of a very tight ion-pair. The positive entropies of activation for these highly deactivated substrates do not accord with steric hindrance to reaction, but the overall interpretation is still not clear. The association–scramble mechanism is a possibility that requires further investigation.

The intriguing possibility of an S_N2 reaction with retention of configuration has been theoretically predicted for substitution in small ring compounds (P. D. Gillespie, and I. Ugi, *Angew. Chem.*, 1971, **83**, 493; W. D. Stohrer and K. R. Schneider, *Chem. Ber.*, 1976, **109**, 285). Unfortunately, an attempt to demonstrate such a process by diazotization of *syn*- or *anti*-11-aminotricyclo[4.4.1.$0^{1,6}$]undec-3-ene (22) in the presence of sodium bromide failed.[26] As reported for the decomposition of *exo*-7-norcaranediazonium ions [*Ann. Reports* (*B*), 1978, **75**, 57], retentive substitution results when disrotatory cyclopropyl–allyl rearrangement is initiated, but it cannot be completed, for reasons of ring strain. A partly opened cyclopropyl cation is therefore most probably involved, as described previously.

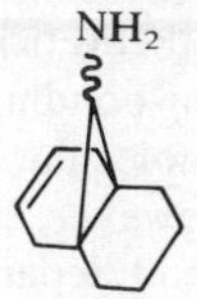

(22)

Preponderant retention of configuration has long been known to occur in the formation of alkyl chloride by the thermal decomposition of an optically active s-alkyl chloroformate in the liquid phase. A mechanism which involves the formation and collapse of an ion-pair intermediate is most likely. Extensive alkyl rearrangement in the case of primary alkyl chloroformates shows that these

[25] J. Hayami, T. Koyanagi, and A. Kaji, *Bull. Chem. Soc. Japan*, 1979, **52**, 1441.
[26] W. Kirmse and T. Engbert, *Angew. Chem. Internat. Edn.*, 1979, **18**, 228.

decompositions also have considerable carbonium-ion character.[27] Products are obtained resulting from competitive 1,2-hydride, 1,2-alkyl, and 1,3-hydride shifts, the latter being thought to proceed *via* protonated cyclopropane intermediates.

An example has been reported of what is claimed to be a directly observable model of an S_N2 transition state (23).[28] The central carbon atom in the structure shown is

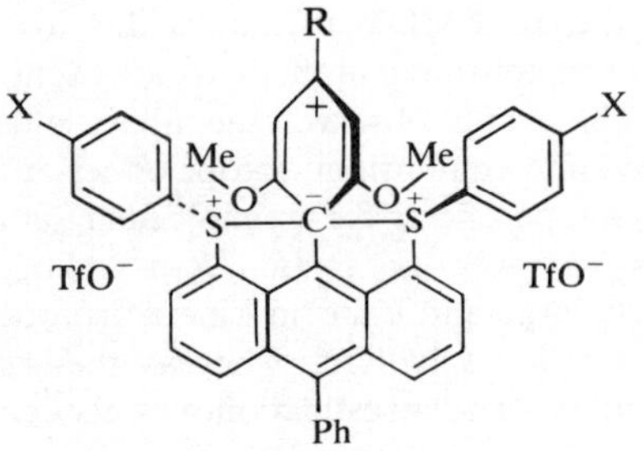

(23) X = Me or F
R = H or Bu^t

associated with five ligands, which contribute ten electrons to the bonding, unlike pentaco-ordinate 'onium species (*e.g.* CH_5^+), in which only eight electrons are involved. The species was prepared in liquid SO_2 or SO_2–SO_2ClF at −80 °C, and was distinguished from other non-symmetrical possibilities, in which one of the C—S bonds is absent, by ^1H and ^{19}F n.m.r. spectroscopy.

Allylic Substitution.—Although the precise timing of bond-making and bond-breaking in the S_N2' process has been the subject of controversy, the stereochemical outcome is open to direct investigation. Both *syn*- and *anti*-attack have been reported previously, according to conditions. It has now been shown[29] that the attack of diethylamine on (−)-(*R*)- and (+)-(*S*)-3-chloro-(*Z*)-[1-^{2}H]but-1-ene proceeds with >95% *syn*-stereospecificity (Scheme 5). This is a relatively simple allylic system, which has no built-in conformational preference for *syn*-attack, and yet the degree of stereospecificity is much higher than that reported previously for the amminolysis of (*R*)- or (*S*)-[1-^{2}H]alk-1-en-3-yl 2,6-dichlorobenzoates [*Ann. Reports* (*B*), 1978, **75**, 51]. Most theoretical analyses predict a preference for *syn*-attack, and the role of hydrogen-bonding between the attacking and leaving groups is still not clear. An intramolecular displacement with concerted allylic rearrangement by carbanion has, however, been shown to proceed with an *anti*-relationship between the attacking and departing groups (Scheme 6),[30] this paralleling a previous result with sulphide as the internal nucleophile. It is likely that the nature of the counter-ion, the leaving group, and the medium are all important factors in determining the course of the reaction.

The reactions of α-, β-, and γ-methylallyl chlorides with silver nitrate in acetonitrile have been shown to exhibit kinetics which are first-order in substrate and of

27 H. R. Hudson, A. J. Koplick, and D. J. Poulton, *J.C.S. Perkin II*, 1979, 57.
28 T. R. Forbus, jnr., and J. C. Martin, *J. Amer. Chem. Soc.*, 1979, **101**, 5057.
29 R. M. Magid and O. S. Fruchey, *J. Amer. Chem. Soc.*, 1979, **101**, 2107.
30 G. Stork and A. R. Schoofs, *J. Amer. Chem. Soc.*, 1979, **101**, 5081.

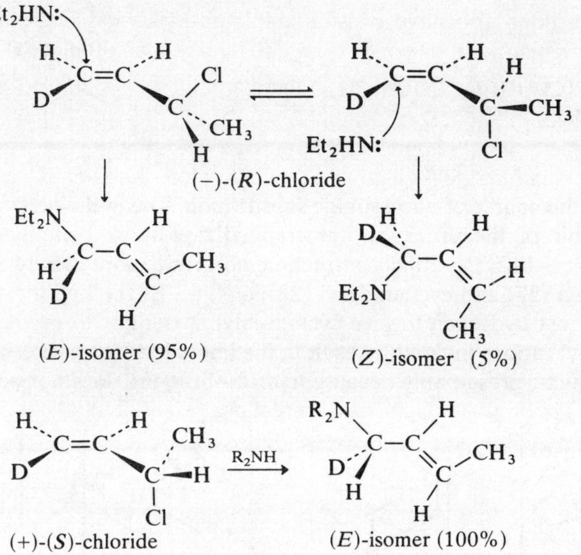

Scheme 5

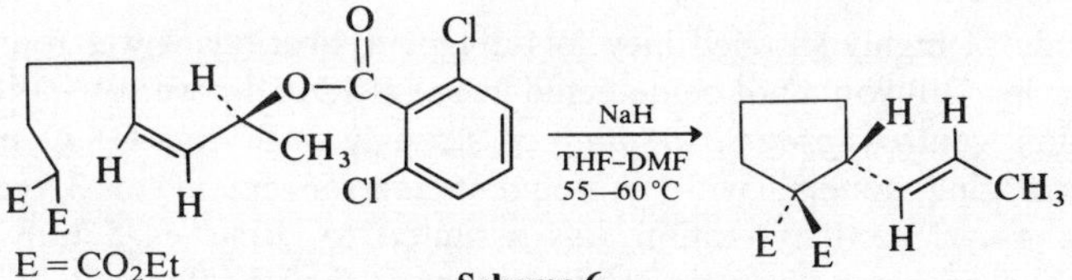

Scheme 6

order 1.5 in silver nitrate.[31] The results indicate that either nucleophilic assistance by nitrate ion is involved in a rate-determining attack on a pre-formed ion-pair intermediate, or that synchronous S_N2' attack is more important in the presence of silver ion than was previously believed for similar anionic substitutions in its absence.

4 Carbo-cations

Detailed information on the structures and properties of carbo-cations continues to be obtained from studies in super-acid media and in the gas phase. Also reported this year are the first high-resolution ^{13}C n.m.r. spectra of carbo-cations in the solid state, determined for diethoxycarbonium hexachloroantimonate (24) and heptamethylbenzenonium tetrachloroaluminate (25) by 'magic-angle' spinning.[32]

Preparation and Properties in Super-acid Media.—The ^{1}H and ^{13}C n.m.r. spectra of secondary cycloalkyl cations derived from 1-chloro-cycloalkanes (C_8—C_{11}) in SbF_5/SO_2ClF–SO_2F_2 at *ca.* −140 °C reveal hydrido-bridged structures (26)—(29).[33] These correspond to those cycloalkyl systems in which transannular hydride shifts

[31] D. N. Kevill and C. R. Degenhardt, *J. Amer. Chem. Soc.*, 1979, **101**, 1465.
[32] J. R. Lyerla, C. S. Yannoni, D. Bruck, and C. A. Fyfe, *J. Amer. Chem. Soc.*, 1979, **101**, 4770.
[33] R. P. Kirchen and T. S. Sorensen, *J. Amer. Chem. Soc.*, 1979, **101**, 3240, and cited references.

$[CH_3CH_2O\overset{+}{C}HOCH_2CH_3]\ SbCl_6^-$
(24)

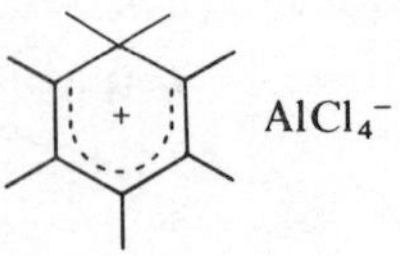

(25)

occur in the course of nucleophilic substitution. The cyclo-octyl cation (26) was the least stable of the series, and rearranged rapidly to 1-methylcycloheptyl ($t_{\frac{1}{2}}$ *ca.* 17 min at −142 °C). Ring contraction occurred more slowly in the case of the cyclononyl (27) and cycloundecyl (28) ions, whilst the hydrido-bridged cyclodecyl ion (29) lost hydrogen to give cyclodecalyl. Attempts to prepare cycloheptyl and cyclohexyl cations invariably result in the immediate formation of the ring-contracted products, presumably because hydrido-bridging does not occur in these cases.

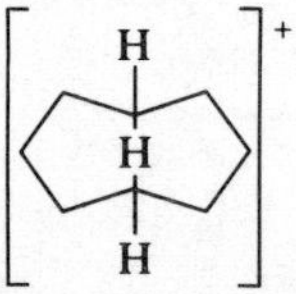
(26)

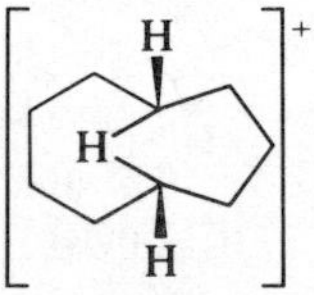
(27)

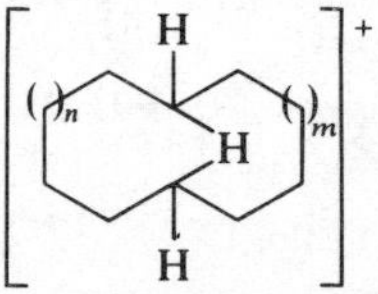
(28)

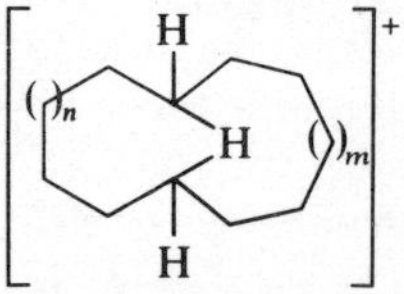
(29)

In a study of highly strained bicyclo[1.1.1]pentyl cations it was found that the parent species (30) could not be detected in SbF_5/SO_2ClF, even at −140 °C.[34] The carbo-cation centre appears to interact strongly with the C—C σ-bonds of the bicyclic ring system, with consequent rearrangement to 3-cyclopentenyl (31). The 4-cyclopentenyl cation was assumed to occur as a transient intermediate. The presence of a phenyl substituent at position 2, *i.e.* as in (32), had a stabilizing effect on the carbo-cation, which could then be observed at −140 °C, although rearrangement to the allylic structure (33) was quantitative at −30 °C.

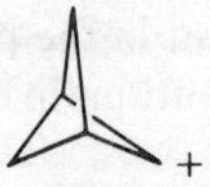
(30)

(31)

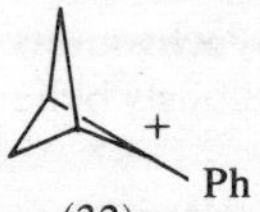
(32)

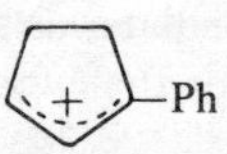
(33)

The first examples of the previously unknown chloromethylhalonium ions have been reported.[35] Bis(chloromethyl)chloronium ion (34) was obtained by the interaction of dichloromethane with an excess of SbF_5 in SO_2ClF at −130 °C, and it was found to be a powerful chloromethylating agent which reacted with sulphur dioxide, monohalogeno- and dihalogeno-alkanes, and ethers as shown (Scheme 7).

Gas-phase Studies.—Heats of formation (ΔH_f) for simple alkyl cations (Me^+, Et^+, $i\text{-}Pr^+$, and $t\text{-}Bu^+$) have been determined *via* photoelectron spectroscopy of the corresponding radicals.[36] The value for t-butyl (162.9 ± 1.2 kcal mol^{-1}) agrees well with a recent figure derived by photoionization mass spectrometry of isobutane and

[34] G. A. Olah, G. K. Surya Prakash, and G. Liang, *J. Amer. Chem. Soc.*, 1979, **101**, 3932.
[35] G. A. Olah and M. R. Bruce, *J. Amer. Chem. Soc*, 1979, **101**, 4765.
[36] F. A. Houle and J. L. Beauchamp, *J. Amer. Chem. Soc.*, 1979, **101**, 4067.

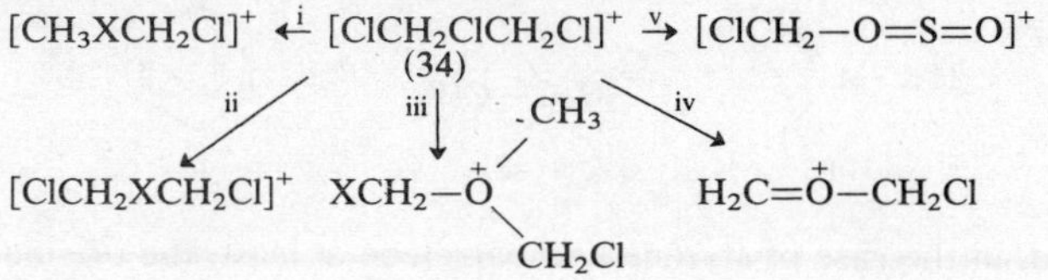

Reagents: i, CH_3X; ii, $ClCH_2X$; iii, XCH_2OCH_3; iv, $ClCH_2OCH_2Cl$; v, SO_2

Scheme 7

the t-butyl halides (162.1 ± 8 kcal mol^{-1}),[37] and indicates that a currently accepted value is 6—7 kcal mol^{-1} too high (*cf.* F. P. Lossing and G. P. Semeluk, *Canad. J. Chem.*, 1970, **48**, 955). A new value for ΔH_f for (2-norbornyl$^+$) (182 ± 2 kcal mol^{-1}), determined from a study of proton-transfer equilibria, agrees well with previous data and shows this cation to be more stable than other cyclic secondary cations by *ca.* 6 kcal mol^{-1}.[38] The 2-adamantyl cation (39) has been detected as a stable species in the gas phase for the first time.[39] Dissociative ionization of the corresponding bromides gave the 1-adamantyl (40) and 2-adamantyl (39) ions, which did not rearrange, and which were distinguished by collision activation spectroscopy. Bromide-transfer reactions have confirmed the order of stabilities indicated for ions (35)—(40) and have shown that (37) and (40) differ in stability by no more than 4 kcal mol^{-1}.[40] The 1-adamantyl cation is therefore to be regarded as a normal tertiary cation, but 2-adamantyl appears to be more stable than might be expected for a classical secondary cation.

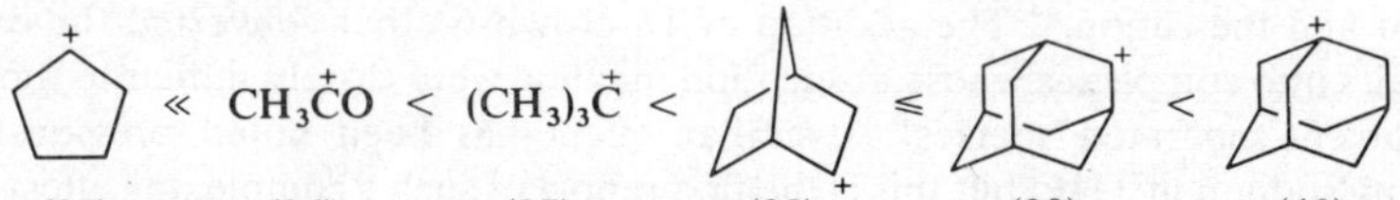

(35) (36) (37) (38) (39) (40)

Photoionization mass spectrometry has shown that two non-interconverting halonium ions, (41) and (42), can exist in the gase phase, these being produced by photoionization of the corresponding 1,1-dihalogeno- and 1,2-dihalogeno-alkanes respectively.[41] Heats of formation were calculated for each species, and it was shown that the cyclic isomer (42) is more stable by 1.4 kcal mol^{-1} if X is Br but less stable by 5.6 kcal mol^{-1} if X is Cl. The relative stabilities of other 'onium ions (X = OH, SH, NH_2, or PH_2) were found by the use of ion cyclotron resonance spectroscopy, which enabled the preferred direction of exchange [equation (2)] to be determined. Other

$$C_2H_4X^+ + HY \rightleftharpoons C_2H_4Y^+ + HX \qquad (2)$$

gas-phase studies have included a photoionization investigation of the structures of the $C_8H_8^+$ cations obtained from the isomeric compounds styrene, cyclo-octatetraene, and barrelene (43).[42] Rearrangement to a common structure was not

[37] R. G. McLoughlin and J. C. Traeger, *J. Amer. Chem. Soc.*, 1979, **101**, 5791.
[38] P. P. S. Saluja and P. Kebarle, *J. Amer. Chem. Soc.*, 1979, **101**, 1084.
[39] C. Wesdemiotis, M. Schilling, and H. Schwarz, *Angew. Chem. Internat. Edn.*, 1979, **18**, 950.
[40] R. Houriet and H. Schwarz, *Angew. Chem. Internat. Edn.*, 1979, **18**, 951.
[41] D. W. Berman, V. Anicich, and J. L. Beauchamp, *J. Amer. Chem. Soc.*, 1979, **101**, 1239.
[42] R. C. Dunbar, M. S. Kim, and G. A. Olah, *J. Amer. Chem. Soc.*, 1979, **101**, 1368.

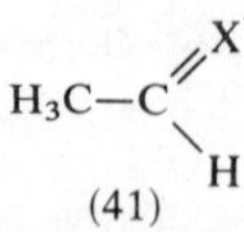

(41)

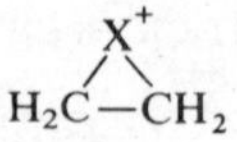

(42)

(43)

observed, and, in the case of styrene and barrelene at least, the original structure appeared to be retained. Protonation, however, caused relatively rapid rearrangement to the most stable gas-phase form of $C_8H_9^+$, which is thought to be the styryl cation.

The first clear evidence for the existence of cyclopropylcarbinyl and cyclobutyl cations in the gase phase has been provided by allowing multi-tritiated cyclobutane to undergo β-decay [equation (3)] in the presence of water or ammonia in the dilute

$$\text{c-}C_4H_{8-n}T_n \xrightarrow{\beta\text{-decay}} \text{c-}C_4H_{7-n}T_n + {}^3He + \beta^- \quad (3)$$

gaseous state.[43] In addition to products derived from skeletal rearrangement, cyclobutanol (or cyclobutylamine) and cyclopropylcarbinol (or cyclopropylamine) were formed, presumably by interaction of the nucleophilic water or ammonia with the corresponding cyclic cations.

5 Carbanions

Spectrophotometric studies of (triphenylmethyl)lithium and (triphenylmethyl)potassium in dimethoxyethane (DME), THF, and diethyl ether have shown that contact ion-pairs or solvent-separated ion-pairs may exist, according to the solvent and the cation.[44] The addition of 18-crown-6 ether converted the contact ion-pairs into complexes whose absorption maxima were closely similar to those for the solvent-separated species. A similar effect has been noted previously for fluorenylsodium in THF, but this is the first report of such a complexing effect when the counter-ion is lithium. Ion-pairing has been shown to increase the stability of the fluoradenide ion (FD^-) (44) with respect to its electron-transfer equilibrium with nitrobenzene.[45] The ion was generated in the presence of methoxide ion in methanol, which can stabilize both MeO^- and $PhNO_2^{\cdot-}$ by hydrogen-bonding. Although hydrogen-bonding between methanol and the FD^- ion might also occur, this is a highly delocalized carbanion, and it is energetically favourable to replace such solvation by ion-pairing, provided that a large cation is available. Ion-pairs were shown to be formed with cations such as R_4N^+ or Cs^+, and with smaller cations in the presence of crown ethers.

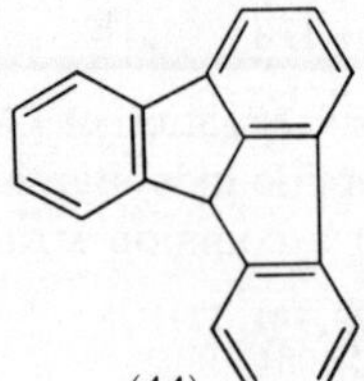

(44)

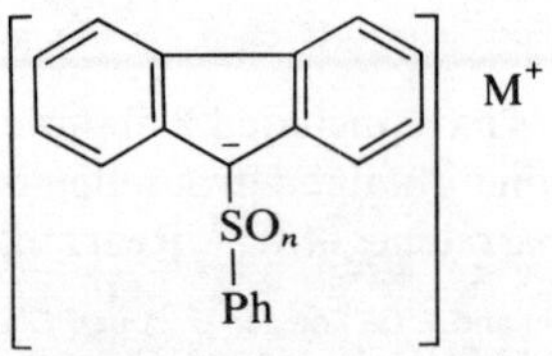

(45) $n = 0, 1,$ or 2

[43] F. Cacace and M. Speranza, *J. Amer. Chem. Soc.*, 1979, **101**, 1587.
[44] E. Buncel and B. Menon, *J. Org. Chem.*, 1979, **44**, 317.
[45] R. D. Guthrie and N. S. Cho, *J. Amer. Chem. Soc.*, 1979, **101**, 4698.

To obtain information on the role of neighbouring sulphur as a stabilizing group for carbanions, the oxidation number of sulphur was varied in a series of alkali-metal fluorenyl carbanions (45) that were substituted with phenyl-sulphur groups in the 9-position.[46] At +20 to −78 °C the electronic absorption spectrum of the sulphenyl carbanion (45; $n = 0$) corresponded to the solvent-separated species $PhSFl^-\|M^+$ or to the free anion, and a stabilization due to the phenylsulphenyl substituent of *ca.* 5.5 kcal mol^{-1} was calculated. The sulphinyl carbanion (45; $n = 1$) existed as contact ion-pairs even in MTHF at 77 K, and its greater stability was explained on the basis of electrostatic interaction between the sulphinyl oxygen and the counter-cation. A completely different type of absorption spectrum in the case of the sulphonyl derivative (45; $n = 2$) was attributed to intramolecular charge transfer, which places the negative charge on the sulphonyl oxygen atoms in the excited state. Proton n.m.r. spectra, oxidation potentials, and electron-transfer reactions for the three types of carbanion were also reported, and it was concluded that the stabilizing influence of the phenyl-sulphur groups can be correlated with their inductive effects. The role of p_π–d_π interaction was not clear. Selenium-stabilized carbanions have also been reported as intermediates in the reactions of enolates with 3-methylallyl bromide.[47]

New information on rotational barriers for allyl carbanions in solution has been obtained from n.m.r. exchange rates for the terminal protons of allyl–alkali-metal compounds in THF.[48] The results provide the first reliable evidence on the simple allyl anion, and give a new lower limit for $\Delta G^{\ddagger}$ of 18 kcal mol^{-1}, determined for allylcaesium. Quantitative data were also given for the rotational barriers in alkyl-substituted allyl carbanions.

Gas-phase Studies.—A new method for the generation of carbanions in the gas phase involves the reaction of the corresponding trimethylsilyl derivative with fluoride ion (Scheme 8).[49] This has the advantage over methods based on proton

$$F^- + CH_3C{\equiv}CSiMe_3 \rightarrow CH_3C{\equiv}C^- + FSiMe_3$$
$$F^- + HC{\equiv}CCH_2SiMe_3 \rightarrow HC{\equiv}CCH_2^- + FSiMe_3$$

Scheme 8

abstraction that the desired anion is produced unambiguously. The possibility of base-catalysed rearrangement of the anion is also obviated. Reaction with molecular oxygen in either its triplet or singlet state appears to be a potentially valuable method for the gas-phase degradation of organic anions and for the determination of their structures.[50] Specific cleavage is induced, reflecting the site of negative charge; *e.g.*, the 2,4-hexadienyl anion reacted with triplet oxygen at the non-terminal negative sites to give enolate anions (Scheme 9).

$$CH_3{-}\overset{\delta-}{C}H{\cdots}CH{\cdots}\overset{\delta-}{C}H{\cdots}CH{\cdots}\overset{\delta-}{C}H_2 + {}^3O_2 \rightarrow \begin{cases} CH_3CH{=}CHCHO + {}^-OCH{=}CH_2 \\ CH_3CH{=}CH{-}O^- + OCHCH{=}CH_2 \\ CH_3CHO + {}^-OCH{=}CHCH{=}CH_2 \end{cases}$$

Scheme 9

[46] S. Nishi and M. Matsuda, *J. Amer. Chem. Soc.*, 1979, **101**, 4632.
[47] H. J. Reich and M. L. Cohen, *J. Amer. Chem. Soc.*, 1979, **101**, 1307.
[48] T. B. Thompson and W. T. Ford, *J. Amer. Chem. Soc.*, 1979, **101**, 5459.
[49] C. H. Depuy, V. M. Bierbaum, L. A. Flippin, J. J. Grabowski, G. K. King, and R. J. Schmitt, *J. Amer. Chem. Soc.*, 1979, **101**, 6443.
[50] R. J. Schmitt, V. M. Bierbaum, and C. H. Depuy, *J. Amer. Chem. Soc.*, 1979, **101**, 6443.

6 β-Elimination

The structures of *E*2 transition states have been discussed in the context of elimination reactions of substituted benzyldimethylcarbinyl chlorides in the presence of methanol, methoxide, and mercaptide ions.[51,52] Hammett ρ values of -0.9 for methanolysis and of $+1.2$ for the formation of $ArCH{=}CMe_2$ from the methoxide-induced eliminations of the 3,5-dichloro- and *p*-nitro-derivatives[51] can be rationalized on the basis of a transition state which may vary between the paenecarbenium and paenecarbanion extremes [(46)—(48)]. A similar variation in

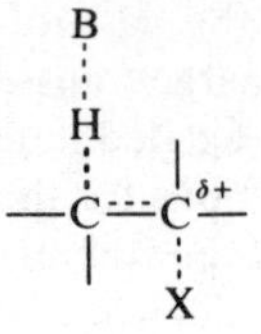
(46)

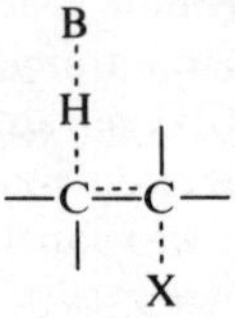
(47)

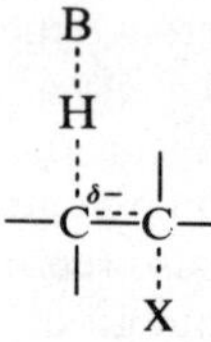
(48)

transition-state structure is consistent with the U-shaped Hammett plot which was obtained for mercaptide-ion-induced eliminations, and in this case a competing $(E2)_{ip}$ mechanism might also be involved.[52] The *E*2C/*E*2H hypothesis [*cf. Ann. Reports* (*B*), 1977, **74**, 82] was discussed in some detail, particularly with respect to the relative rates of eliminations induced by EtS^- and MeO^-, but was considered to provide a less satisfactory explanation of the experimental results.

The influence of attacking base on positional orientation has been studied for various s-alkyl substrates [equation (4)] in their reactions with oxyanions, nitrogen bases, or carbanions in DMSO. Previous investigations had shown that the relative proportions of terminal and internal alkenes obtained from 2-iodobutane (49; X = I, R = Me) were controlled by base strength rather than size in the case of attack by oxyanions, unless extremely bulky species such as 2,6-di-t-butylphenoxide were used.[53] In general, a stronger base yields more of the alk-1-ene, and a similar variation with base strength has been observed for amide anions derived from substituted anilines, amides, and sulphonamides. The effects of methide ions were more dependent on base type, and ion-pairing was seen to be an important factor in eliminations promoted by dimsyl anion. The level of complexity of the base at which its steric influence becomes significant was found to depend on the size of the β-substituent, R, but it was independent of the leaving group, X.[54] For $R{=}Pr^i$, steric effects became apparent with t-butoxide. Anomalous results in the formation of 4,4-dimethylpent-2-ene from (49; $R = Bu^t$, X = OTs) were interpreted in terms of a *syn*-elimination which is forced on the system by the unfavourable steric interactions which would occur between the t-butyl and tosylate moieties in the transition state for *anti*-elimination.

$$\underset{(49)}{RCH_2CHXCH_3} \xrightarrow{\text{base}} RCH{=}CHCH_3 + RCH_2CH{=}CH_2 \qquad (4)$$

[51] J. F. Bunnett and S. Sridharan, *J. Org. Chem.*, 1979, **44**, 1458.
[52] J. F. Bunnett, S. Sridharan, and W. P. Cavin, *J. Org. Chem.*, 1979, **44**, 1463.
[53] R. A. Bartsch, D. K. Roberts, and B. R. Cho, *J. Org. Chem.*, 1979, **44**, 4105.
[54] R. A. Bartsch, R. A. Read, D. T. Larsen, D. K. Roberts, K. J. Scott, and B. R. Cho, *J. Amer. Chem. Soc.*, 1979, **101**, 1176.

'Complex bases' such as $NaNH_2$–Bu^tONa are known to favour *syn*-elimination from *trans*-1,2-dibromo-alkanes (C_5—C_7) (P. Caubere, *Accounts Chem. Res.*, 1974, **7**, 301). Investigations with compounds containing two different halogens have now shown that the normal order of leaving-group ability for dehydrohalogenations ($I > Br > Cl \gg F$) is reversed in reactions of this type; *e.g.*, *trans*-1-bromo-2-fluoro- and *trans*-1-chloro-2-fluoro-cyclohexane each gave 85% yields of 1-bromo- or 1-chloro-cyclohexene, with none of the corresponding fluoro-alkene.[55] The result was attributed to the dominant role played by metal–halogen interaction in the *syn*-transition state (50; B = base, M = metal ion) as opposed to the importance of carbon–halogen bond strength in conventional β-eliminations. Comparisons of rates of *syn*- and *anti*-elimination have been made for the sterically constrained *trans*-2,3-dihalogeno- (51) and *cis*-2,3-dihalogeno-2,3-dihydrobenzofurans (52) respectively;

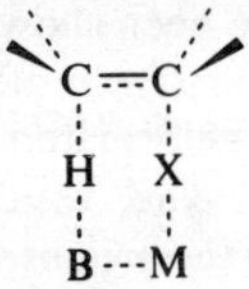
(50)

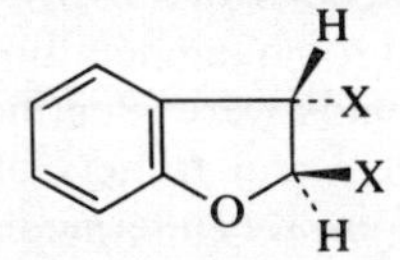
(51)

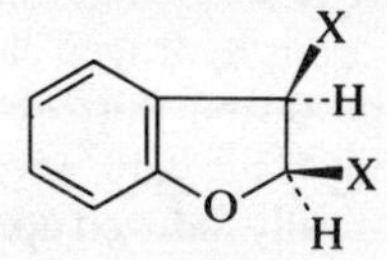
(52)

the *anti*-elimination was favoured by a factor of *ca.* 33 000 for sodium ethoxide and of *ca.* 10 000 for potassium t-butoxide.[56] Together with the kinetic effects of a 3-deuterio- or a 5-chloro-substituent in the *trans*-2,3-dibromo-compound, the results suggested that the transition state for *syn*-elimination is affected significantly by association of the base but not by the basicity of the medium. The *syn*/*anti* dichotomy has also been investigated for the alkoxide-catalysed eliminations of two homologous series of positionally isomeric alkyl-trimethylammonium salts $RCH_2CHXC_5H_{11}$ and $RCHXCH_2C_5H_{11}$ (R = H, Me, Et, Pr^n, Pr^i, or Bu^t; $X = \overset{+}{N}Me_3$ Cl^-).[57] Overall rates were dissected into their *syn*- and *anti*-components, and it was concluded that the steric factors are more important in controlling the stereochemical (*syn*- or *anti*-) outcome, whilst polar effects are overriding in determining the orientation of products (Hofmann *vs.* Saytzeff).

7 Electrophilic Addition

A claim that the kinetic effects of methyl substitution in styrene on its rate of bromination more closely resemble those of sulphenylation than of hydration [*cf. Ann. Reports* (*B*), 1978, **75**, 64], and that a bridged bromonium ion is therefore involved in the rate-determining step,[58] has been challenged on the grounds that the overall rate constants in protic solvents include terms for both free bromine and tribromide ions [equation (5)].[59] The use of corrected values for k_{Br_2}, together with

$$k_{exp}(1 + K[Br^-]) = k_{Br_2} + Kk_{Br_3^-}[Br^-] \qquad (5)$$

[55] J. G. Lee and R. A. Bartsch, *J. Amer. Chem. Soc.*, 1979, **101**, 228.
[56] E. Baciocchi and G. V. Sebastiani, *J. Org. Chem.*, 1979, **44**, 28.
[57] J. Zavada and M. Pankova, *Coll. Czech. Chem. Comm.*, 1979, **44**, 1273.
[58] G. H. Schmid, *J. Org. Chem.*, 1978, **43**, 777.
[59] M. F. Ruasse, J. E. Dubois, and A. Argile, *J. Org. Chem.*, 1979, **44**, 1173.

an enlarged body of data, led to the conclusion that α- and β-methyl substituent effects are both large for additions to $RCH{=}CH_2$ where R is an alkyl group, but that the α-effect is small and the β-effect is large where R is phenyl, as in hydration. An open carbenium-ion-like transition state is therefore indicated for the bromination of styrene, as reported previously. The basis of comparison with arenesulphenylation has also been criticized on the grounds that log *k versus* log *k* correlations are poor, and that steric interactions in the rate-determining steps of the two processes are so different that a linear free-energy relationship over a wide range of substituents cannot be expected.[60]

Relative rates of reaction ($k_{C\equiv C}/k_{C=C}$) for pairs of phenyl-substituted alkynes and alkenes with diphenylmethyl or 1-phenylethyl cations are mainly in the range 0.2—3.5, and are consistent with a transition state having an open cation structure in each case.[61] Steric factors appear to dominate the effect of structure on reactivity.

What are described as 'bizarre reactions' of but-2-yne have been shown to occur with anhydrous hydrogen chloride at room temperature.[62] Apart from the expected addition products, thirteen compounds were identified as chlorinated and non-chlorinated cyclic and acyclic dimers and trimers of but-2-yne, resulting from electrophilically induced cycloadditions. Key intermediates in the proposed reaction scheme were formed as shown in Scheme 10. Cyclodimerization products having 1,3-dialkyl-1,3-dibromocyclobutane structures were likewise obtained by the interaction of anhydrous hydrogen bromide with alkynes.[63]

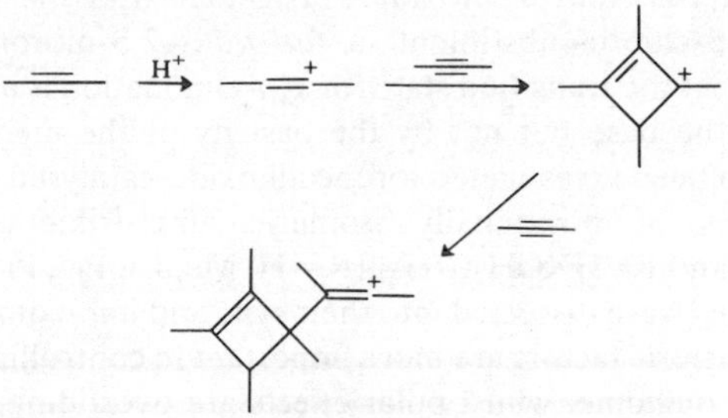

Scheme 10

8 Nucleophilic Addition to Alkenes

Suitably activated carbon–carbon double bonds may undergo nucleophilic addition by a two-step process, the first step of which consists in the rate-determining formation of an association complex. The rate of attack is frequently too fast for conventional kinetic studies, but temperature-jump and stopped-flow techniques have now been successfully applied to the reactions of substituted benzylidene-Meldrum's acids (53) and substituted benzylidene-*NN'*-dimethylbarbituric acids (54) with amines in acetonitrile or chloroform.[64] The zwitterionic structure assigned to the association complex (55) is in agreement with its lower stability in chloroform

[60] M. F. Ruasse, A. Argile, E. Bienvenue-Goëtz, and J. E. Dubois, *J. Org. Chem.*, 1979, **44**, 2758.
[61] F. Marcuzzi, G. Melloni, and G. Modena, *J. Org. Chem.*, 1979, **44**, 3022.
[62] H. Schneider and K. Griesbaum, *J. Org. Chem.*, 1979, **44**, 3316.
[63] K. Griesbaum, W. Seiter, H. Schneider, M. El Abed, and Z. Rehman, *Annalen*, 1979, 1137.
[64] B. Schreiber, H. Martinek, P. Wolschann, and P. Schuster, *J. Amer. Chem. Soc.*, 1979, **101**, 4708.

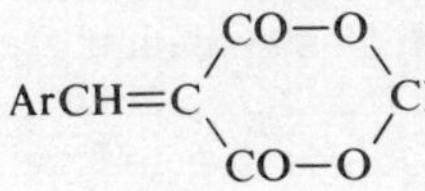
(53)

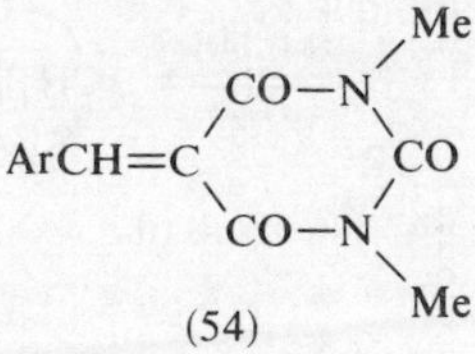
(54)

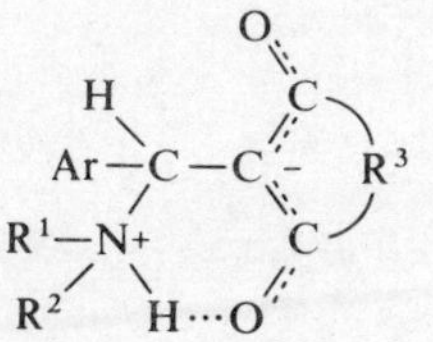
(55) $R^3 = OCMe_2O$ or N(Me)CON(Me)

and with the fact that tertiary amines, which provide no hydrogen atom for intramolecular hydrogen-bonding in the complex, associate more weakly than primary or secondary amines. Electron-withdrawing substituents in the aromatic ring afford stabilization.

Some unusual examples of nucleophilic addition are provided by the base-catalysed reactions of certain tricylic derivatives (56), which yield ethers (57) by intramolecular attack of the carbon–carbon double bond by alkoxide ion derived from the neighbouring hydroxyl group.[65] In those cases in which $R^2 + R^3 = O$ (*i.e.* a carbonyl group is present), the reaction was thought to proceed by virtue of the stabilizing effect of an intermediate homo-enolate (58), but steric compression appears to be the sole driving force for reaction when R^2 and R^3 are H and/or MeO.

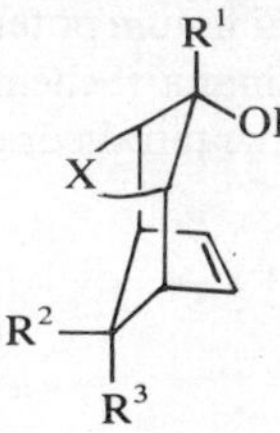
(56)

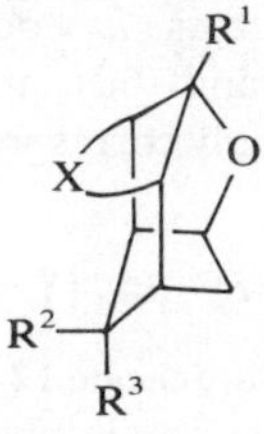
(57)

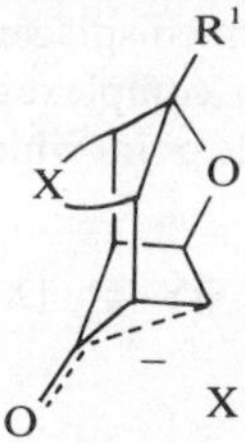
(58)

X = CH=CH, $(CH_2)_2$, or $(CH_2)_3$

9 Nucleophilic Reactions of Carbonyl Compounds

A new correlation for the reactivities of carbonyl compounds with nucleophiles (BH_4^-, CN^-, SO_3^{2-}, NH_2OH, and RS^-) has been derived for twenty-six ketones [equation (6); k is the rate constant for a given ketone; k_0 is the rate constant for cyclohexanone as standard].[66] The slope of the line is approximately unity, and the relationship can therefore be simplified to $\log (k/k_0) = B$.

$$\log k = A \log k_0 + B \qquad (6)$$

The intramolecular transesterifications of some ω-hydroxyalkyl thionobenzoates and thionoacetates have been shown to give rise to a novel class of thermally stable anionic tetrahedral intermediate (Scheme 11).[67] The compounds were obtained as stable solids which were insoluble in aprotic solvents such as acetonitrile or dichloromethane but which reacted with water to yield the starting materials.

[65] R. A. Pfund and C. Ganter, *Helv. Chim. Acta*, 1979, **62**, 228.
[66] A. Finiels and P. Geneste, *J. Org. Chem.*, 1979, **44**, 1577.
[67] F. Khouri and M. K. Kaloustian, *J. Amer. Chem. Soc.*, 1979, **101**, 2249.

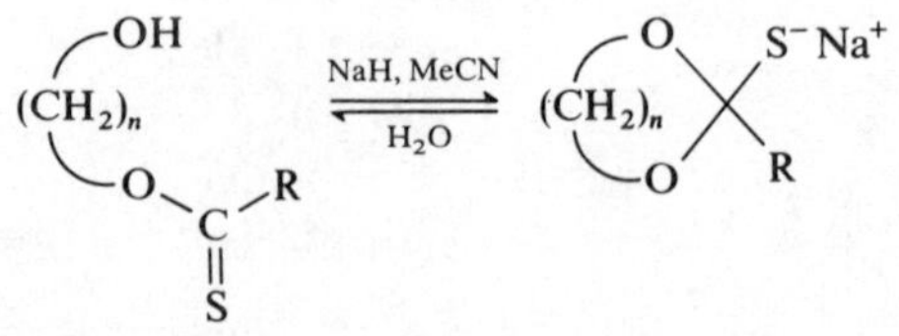

R = Ph or Me, $n = 2$ or 3

Scheme 11

Gas-phase Reactions.—Pulsed ion cyclotron resonance spectroscopy (ICR) has been used to investigate the behaviour of acyl compounds towards nucleophiles in the gas phase.

$$RCOY + X^- \rightarrow RCOX + Y^- \qquad (7)$$

Acyl halides undergo displacement reactions [equations (7)], although a number of competitive processes may also occur, *viz.* proton transfer, base-induced complexation, and base-induced elimination.[68] Nucleophilicity of the attacking group decreased in the order $MeO^- \geqslant F^- > CN^- > SH^- > Cl^-$, which is the same as for gas-phase basicities and is essentially the same as is found in protic solvents. This contrasts with the behaviour of nucleophiles in displacements at sp^3 carbon, for which the order of nucleophilicity is very susceptible to solvation effects. Reaction rates suggested that displacement occurs *via* a double-minimum potential well, with loose association complexes, (59) and (60), at the minima (Scheme 12) and an intervening saddle point which possibly corresponds to the tetrahedral intermediate structure.

$$X^- + RCOY \rightleftharpoons [X \cdots RCOY]^- \rightleftharpoons [Y \cdots RCOX]^- \rightleftharpoons Y^- + RCOX$$

Scheme 12

The relatively unstable character of tetrahedral intermediates in the absence of solvation has been made clear by the gas-phase interactions of phenyl acetate with a range of nucleophiles (OH^-, MeO^-, CN^-, SH^-, MeS^-, and PhO^-).[69] In spite of the reluctance of the phenyl ring to undergo S_N2 attack, this mode of reaction [equation (8)] occurred exclusively, in preference to the displacement of phenoxide by the $B_{AC}2$ mechanism [equation (9)]. However, reaction by the $B_{AC}2$ route occurred

$$X^- + PhOCOCH_3 \rightarrow XPh + {}^-OCOCH_3 \qquad (8)$$

$$X^- + PhOCOCH_3 \rightarrow XCOCH_3 + PhO^- \qquad (9)$$

when partially solvated nucleophiles (obtained, *e.g.*, from the reaction $X^- + HCO_2Me \rightarrow X^- \cdots\cdots HOMe + CO$) were used. It is clear that charge delocalization by even a single solvent molecule is extremely important for the stabilization of transition states for processes proceeding *via* tetrahedral intermediates.

[68] O. I. Asubiojo and J. I. Brauman, *J. Amer. Chem. Soc.*, 1979, **101**, 3715.
[69] E. K. Fukuda and R. T. McIver, jnr., *J. Amer. Chem. Soc.*, 1979, **101**, 2498.

4 Reaction Mechanisms

Part (iii) Free-Radical Reactions

By A. G. DAVIES

Department of Chemistry, University College London, 20 Gordon Street, London WC1H 0AJ

1 General

A useful introductory text on organic free radicals has been published.[1] Other general reviews include articles on the mechanisms of free-radical reactions,[2] anchimerically assisted bond homolysis,[3] vibrational spectroscopy of free radicals,[4] fluorine-containing free radicals,[5] directive effects in gas-phase radical addition reactions,[6] homolytic aromatic *ipso*-substitution reactions,[7] and the capto-dative stabilization of radicals carrying both an electron-acceptor and an electron-donor group.[8]

An interesting development in technique is the application of e.s.r. spectroscopy to the direct observation of the radicals that are formed during the pyrolysis of hydrocarbons.[9] The compounds, in benzene solution, are passed slowly through the cavity at temperatures up to 566 °C and pressures up to 140 kg cm^{-2}, usually well above their critical temperatures, but the high density of the fluid quenches the angular momentum of molecular rotation, and sharp, well-resolved spectra are observed. As yet, the technique has been applied mainly to radicals of the benzyl type.

2 Carbon-centred Radicals

Electron spin resonance spectroscopy continues to be the technique that is used most widely to investigate the structures of free radicals.

In an sp^2-hybridized alkyl radical (1) the α-carbon atom is in the nodal plane of the *p*-orbital containing the unpaired electron, and hence the hyperfine coupling to this carbon is low (about 40 G). On the other hand, an sp^3-hybridized radical (2) has a finite *s*-electron density at the carbon nucleus, and the hyperfine coupling is large. The temperature dependence of $a(^{13}C)$ can further be used to determine the out-of-plane bending frequency of the radical.

[1] D. C. Nonhebel, J. M. Tedder, and J. C. Walton, 'Radicals', Cambridge University Press, 1979.
[2] C. Rüchardt, *Uspekhi Khim.*, 1978, **47**, 2014.
[3] M. T. Reetz, *Angew. Chem. Internat. Edn.*, 1978, **17**, 594.
[4] R. E. Hester, *Adv. Infrared Raman Spectrosc.*, 1978, **4**, 1.
[5] 'Fluorine-containing Free Radicals', ed. J. W. Root, A.C.S. Symposium Series, 1978, Vol. 66.
[6] J. M. Tedder and J. C. Walton, *Adv. Phys. Org. Chem.*, 1978, **16**, 51.
[7] J. C. Traynham, *Chem. Rev.*, 1979, **79**, 323.
[8] H. G. Viehe, R. Merényi, L. Stella, and Z. Janousek, *Angew. Chem. Internat. Edn.*, 1979, **18**, 917.
[9] R. Livingston, H. Zeldes, and M. S. Conradi, *J. Amer. Chem. Soc.*, 1979, **101**, 4312.

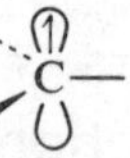
(1)

(2)

In the ethyl[10] and isopropyl radicals,[11] the value of $a(^{13}C_\alpha)$ increases monotonically with temperature. These radicals have a planar structure at the minimum of the potential-energy curve, and the out-of-plane bending frequencies which are derived (541 and 380 cm^{-1}, respectively) are in good agreement with the values obtained from i.r. spectroscopy.[12]

In contrast, the t-butyl radical [$a(^{13}C_\alpha)$ *ca.* 45 G] appears to be most stable when it is bent by 11.5° from planarity. The trineopentylmethyl radical[13] [$a(^{13}C_\alpha)$ *ca.* 40.5 G] shows the lowest value of $a(^{13}C_\alpha)$ of any t-alkyl radical, and is presumably forced nearer to planarity by steric repulsion between the neopentyl groups. Below −85 °C, the spectrum shows that the hydrogen atoms on the β-carbon atoms become non-equivalent as the neopentyl groups are frozen into specific conformations. Steric repulsion between pairs of these radicals also confers on them a high persistence in solution: in benzene, at 20 °C, their half-life is 11 minutes.

Steric effects are also the basis of the persistence of $(RS)_3C^\bullet$ radicals.[14] Hexakis(trifluoromethylthiyl)ethane dissociates at room temperature.[15] The C—C bond length is 1.7 nm (compared with 1.54 nm in ethane) and the bond dissociation energy is only 13.7 kJ mol^{-1}. The value of $a(^{13}C_\alpha)$ in the radical (3) is 40.09 G, confirming that the radical is approximately planar [*cf.* (1)]; in contrast, the radical $(MeO)_3C^\bullet$ has $a(^{13}C) = 154$ G, showing that it is pyramidal [*cf.* (2)].

$$(CF_3S)_3C{-}C(SCF_3)_3 \rightleftharpoons 2(CF_3S)_3C\cdot$$

(3)

The value of $a(H_\beta)$ in an alkyl group can similarly give evidence regarding the conformation about the C_α—C_β bond. Hyperconjugative coupling will be largest when the β-C—H bond eclipses the semi-occupied *p*-orbital, *i.e.* (4), and least when it is in the nodal plane, as shown in (5).

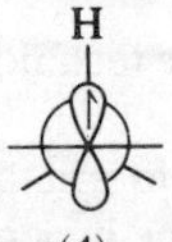

(4)

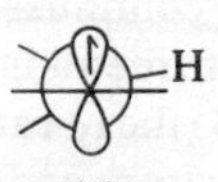

(5)

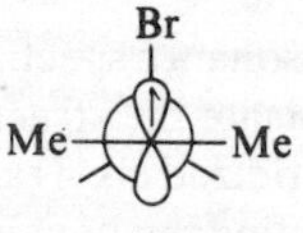

(6)

By this type of argument, the β-bromoalkyl radical $Me_2\dot{C}CH_2Br$, about which there has been some controversy, has been shown to have the preferred conformation (6).[16] Other spectra that can be observed when isobutyl bromide in an adamantane matrix is subjected to *X*-radiolysis apparently result from the interaction of the radicals $Me_2CHCH_2^\bullet$ and $Me_3C^\bullet$ with Br^-.[16,17]

[10] D. Griller, P. R. Marriott, and K. F. Preston, *J. Chem. Phys.*, 1979, **71**, 3703.
[11] D. Griller and K. F. Preston, *J. Amer. Chem. Soc.*, 1979, **101**, 1875.
[12] J. Pacansky, D. E. Home, G. P. Gardini, and J. Bargon, *J. Phys. Chem.*, 1977, **81**, 2149.
[13] K. Schluetter and A. Berndt, *Tetrahedron Letters*, 1979, 929.
[14] R. Schecker, U. Henkel, and D. Seeback, *Chem. Ber.*, 1977, **110**, 2880.
[15] A. Haas, K. Schlosser, and S. Steenken, *J. Amer. Chem. Soc.*, 1979, **101**, 6282.
[16] M. C. R. Symons and I. G. Smith, *J.C.S. Perkin II*, 1979, 1362.
[17] R. V. Lloyd, D. E. Wood, and M. T. Rogers, *J. Amer. Chem. Soc.*, 1974, **96**, 7130.

The first well-defined e.s.r. spectrum of an alkane radical cation (7) has been reported from the γ-irradiation of 2,2,3,3,-tetramethylbutane in the presence of a carbon tetrahalide as an electron scavenger, at 77 K.[18] It is suggested that the two Me_3C groups, which are held together by a one-electron bond, are approximately planar (7); six C—H bonds lie parallel with the singly occupied σ-orbital and, by hyperconjugation, show a large hyperfine coupling constant of 32 G, and the other twelve hydrogen atoms show $a = 4.5$ G. In the absence of the carbon tetrahalide, electron return gives the alkane in an excited state, and fragmentation occurs.

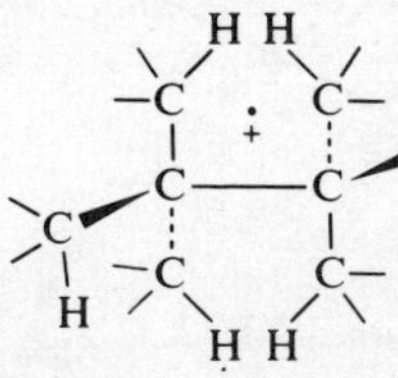

(7)

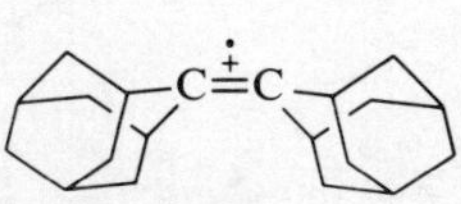

(8)

The first long-lived mono-olefin radical cation (8) has also been observed,[19] from the electrolytic oxidation of adamantylideneadamantane in an e.s.r. cell. It is thought to owe its persistence to the fact that the β-carbon atoms are all located at bridgeheads, from which loss of a proton is forbidden by Bredt's rule.

E.s.r. spectroscopy has also been used to show the existence of two isomeric forms of the pentadienyl radical that are not interconvertible on the time-scale of the technique. Photolysis of di-t-butyl peroxide in the presence of penta-1,4-diene above −60 °C shows the presence of only one species, which has been identified as the (*EE*)-isomer (9), but at −120 °C the predominant radical that is formed is the (*EZ*)-isomer (10).[20,21]

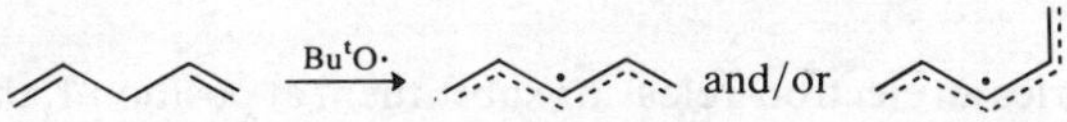

(9) (10)

One of the more satisfying aspects of Hückel MO theory is its success in interpreting quantitatively the e.s.r. spectra that are observed when a substituent interacts with the π-electron system of an annulene radical. It is inevitable that most studies should have been carried out on the benzene radical anions ($C_6H_5R^{\cdot-}$), but the cyclopentadienyl radicals ($C_5H_4R^{\cdot}$) would have certain advantages because the π-electron system is concentrated over fewer nuclei (so that the observed effects are often larger) and because the radicals are uncharged, and the unpaired electron resides in a bonding rather than an antibonding orbital.

There has previously been no satisfactory route to the cyclopentadienyl radicals, because the substituted cyclopentadienes C_5H_5R react with alkoxyl radicals by addition rather than hydrogen abstraction (except when R is a trialkylsilyl group).[22]

18 M. C. R. Symons and I. G. Smith, *J. Chem. Res.* (*S*), 1979, 382; J. T. Wang and F. Williams, *Chem. Phys. Letters*, 1979, **67**, 202; M. C. R. Symons, *ibid.*, 1980, **69**, 198.
19 S. F. Nelson and C. R. Kessel, *J. Amer. Chem. Soc.*, 1979, **101**, 2503.
20 D. Griller, K. U. Ingold, and J. C. Walton, *J. Amer. Chem. Soc.*, 1979, **101**, 758.
21 R. Sustman and H. Schmidt, *Chem. Ber.*, 1979, **112**, 1440.
22 M. Kira, M. Watanabe, and H. Sakurai, *J. Amer. Chem. Soc.*, 1977, **99**, 7780.

The general routes to the radicals $C_5H_4R^{\cdot}$ shown in Scheme 1 have now been developed, and the e.s.r. spectra can be analysed to identify the substituent effects of the groups R.[23,24]

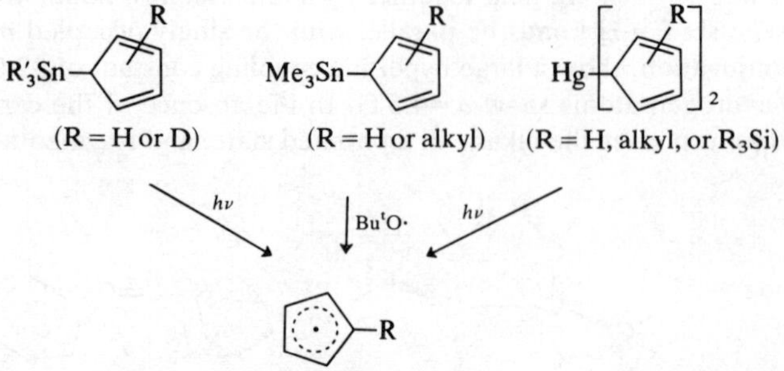

Scheme 1

The forms of the degenerate symmetrical (ψ_S) and antisymmetrical (ψ_A) wave-functions for the cyclopentadienyl radical are shown in (11) and (12). These are equally populated by the unpaired electron in the $C_5H_5\cdot$ radical, resulting in an equal spin density at each carbon centre and an e.s.r. spectrum which consists of a binomial sextet; $a(5H) = 5.98$ G.

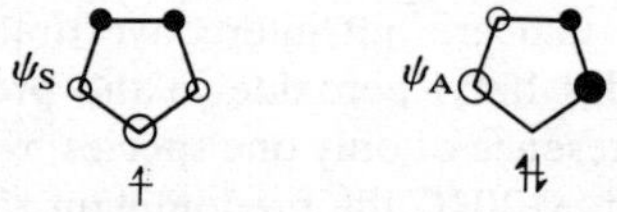

(11) (12)

If the ring carries an electron-releasing substituent at position 1, the ψ_S MO will be destabilized because of the high electron density at C-1, but ψ_A will be unaffected because it has a node passing through this position. Two electrons will therefore enter the ψ_A orbital, and the unpaired electron will occupy the ψ_S orbital. The value of $a(H_j)$ will be given by $c_j^2(5 \times 5.98)$ G, where c_j is the Hückel coefficient at position j; this predicts that $a(H2, 5) = 1.1$ G and $a(H3, 4) = 7.8$ G.

Conversely, an electron-attracting substituent should stabilize ψ_S, which will be doubly occupied, leaving the unpaired electron in ψ_A, so that $a(H2, 5) = 10.8$ G and $a(H3, 4) = 4.1$ G.

When R is Me, Et, Pr^i, or Bu^t, the e.s.r. spectrum that is observed shows that the unpaired electron is wholly in the ψ_S orbital; for example, for $C_5H_4Me\cdot$ radical, $a(H2, 5) = 1.10$ G and $a(H3, 4) = 7.73$ G at 24 °C. This is compatible with the accepted behaviour of alkyl groups acting as electron-releasing substituents. On the other hand, the spectrum of $C_5H_4SiMe_3\cdot$ shows that the Me_3Si group weakly attracts electrons, leaving the unpaired electron principally (but not wholly) in the ψ_A orbital.

[23] A. G. Davies and M.-W. Tse, *J.C.S. Chem. Comm.*, 1978, 353; P. J. Barker, A. G. Davies, and J. D. Fisher, *ibid.*, 1979, 587.

[24] M. Kira, M. Watanabe, and H. Sakurai, *Chem. Letters*, 1979, 973.

Penta-alkyldisilyl substituents R_3SiR_2Si, however, act as weakly electron-releasing groups in the cyclopentadienyl radical, whereas in the benzene radical anion they act as electron-attracting substituents.

The spectrum of the $C_5H_4D\cdot$ radical shows that the ψ_A MO is higher in energy than ψ_S. This is ascribed not to a Coulombic effect but to the difference in the vibrational frequencies of the C—H and C—D bonds acting upon the resonance integral of the ψ_S MO,[24,25] or upon the various Jahn–Teller-distorted isomers of the radical.[26]

The above correlations between structure, stability, and e.s.r. spectra all relate to carbon-centred π-radicals (1). Relatively little is known about such relationships for σ-radicals, partly because the e.s.r. spectra of two of the most important classes of these radicals, namely the vinyl and aryl radicals, are difficult to observe in fluid solution, and the spectra of most of the small number of acyl σ-radicals (13) that have been observed in the past have shown broad lines, with no resolvable structure.[27]

The intramolecular interactions, and the conformations which they impose, are similar in the acyl radicals and their parent aldehydes. Further, the n.m.r. couplings, $^3J(CH\text{–}CHO)$, in the aldehydes (14) and the e.s.r. couplings, $a(H_\beta)$, in the acyl radicals (15) are similar functions of the dihedral angle θ, being largest when $\theta = 0°$ [β-C—H bond *trans* to the σ-C(O)—H bond or singly occupied σ-orbital] and least when θ is 90—270°.

When R is a small alkyl group (C_1—C_5), the most stable conformation of the aldehydes RCH_2CHO and of the acyl radicals $RCH_2\dot{C}O$ is that in which the R—C bond eclipses the carbonyl group [θ in (14) and (15) is 120 or 240°]. The n.m.r. coupling constant $^3J(CH\text{–}CHO)$ is small (<1.8 Hz) and the e.s.r. coupling is unresolvable, accounting for the broad signals that have been observed previously. If R is larger than C_5, θ is reduced, probably because of steric hindrance between R and the carbonyl group; the n.m.r. coupling increases (to *ca.* 2 Hz) and the e.s.r. coupling becomes resolvable (2–3 G). In the extreme, di-t-butylacetaldehyde, which is most stable in the conformation (16) in which the β-C—H bond is *trans* to the aldehydic C—H bond, shows a large n.m.r. coupling, $^3J(CH\text{–}CHO) = 6$ Hz, and the corresponding acyl radical, with a similar conformation ($\theta = 0$), shows $a(H_\beta) = ca.$ 11 G.[28]

In 2,3-unsaturated aldehydes, interaction between the π-system of the carbonyl group and that of the double bond stabilizes the molecule in the s-*trans* conformation, and $^3J(CH\text{–}CHO) = 6$ Hz. The corresponding unsaturated acyl radicals adopt the same conformation (17), in which $\theta = 60°$ and $a(H_\beta) = 19$ G.

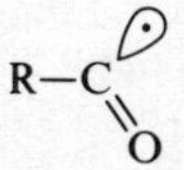

(13)

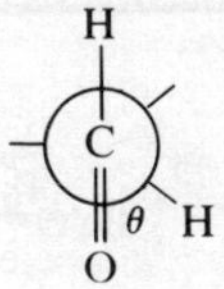

(14)

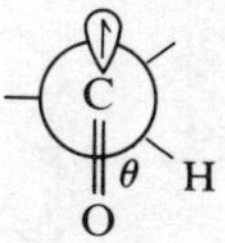

(15)

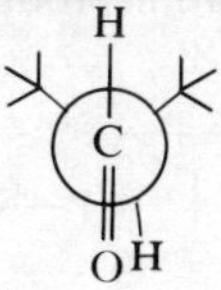

(16)

[25] P. J. Barker and A. G. Davies, *J.C.S. Chem. Comm.*, 1979, 815.
[26] T. Clark, J. Chandrasekhar, and P. von R. Schleyer, *J.C.S. Chem. Comm.*, 1980, 266.
[27] H. Paul and H. Fischer, *Helv. Chim. Acta*, 1973, **56**, 1575.
[28] A. G. Davies and R. Sutcliffe, *J.C.S. Chem. Comm.*, 1979, 473; *J.C.S. Perkin II*, 1980, 819.

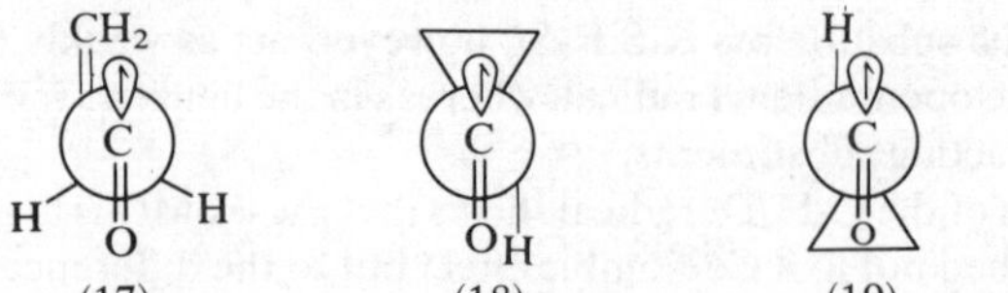

(17) (18) (19)

Similarly, the interaction of the π-system of the carbonyl group with the Walsh π-orbitals of the cyclopropyl ring restricts rotation about the C_α—C_β bond in cyclopropylcarbaldehyde, but now the two conformations have approximately equal stability. Below −90 °C, the aldehyde reacts with t-butoxyl radicals to show the spectra of the two radicals (18), in which $\theta = 0°$ and $a(H_\beta) = 18.2$ G, and (19), where $\theta = 180°$, and $a(H_\beta) = 0.5$ G. Above about −90 °C, rotation about the C_α—C_β bond is rapid on the e.s.r. time-scale, and a single time-averaged spectrum is observed; $a(H_\beta) = ca.$ 9 G. Simulation of the spectra over the region of intermediate exchange rates gives a barrier to rotation of 17.5 kJ mol^{-1}, close to the value of 18.4 ± 1.7 kJ mol^{-1} that has been obtained by microwave spectroscopy for the parent aldehyde.[29]

Similar considerations may be expected to apply to the σ-imidoyl radicals (21). Previously, their spectra had been observed only from the radicals obtained from the corresponding imines (20; R^2=alkyl).[30] A wider variety of imidoyl radicals (21a)

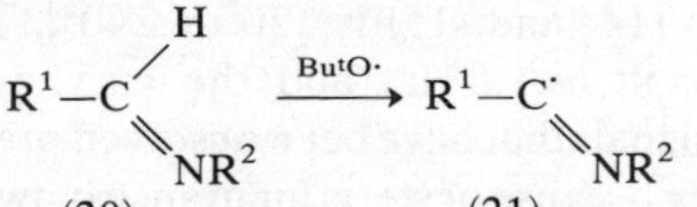

(20) (21)

$$X^\cdot + :C{=}NR^2 \rightarrow X\dot{C}{=}NR^2$$

(21a)

have now been prepared by the addition of radicals X· [*e.g.* Bu^{t}O·, Me_3SiO·, $(EtO)_2\dot{P}O$, MeS·, and Me_3Si·] to alkyl isocyanides (R^2 = Me, Bu^n, Bu^t, Me_3Si, *etc.*).[31]

When X is Bu^{t}O or Bu^{t}S, the imidoyl radicals undergo cleavage within X to give the radicals Bu^t· and OCNR^2 or SCNR^2, as appropriate, but when X is MeS, Et_3Si, or Ph, the N—R^2 bond breaks to give $\dot{R}^2$ and XC≡N.

An unusual case of a radical existing in two distinct electronic configurations has been detected in the reaction of t-butoxyl radicals with the phosphorus(III) isocyanate (22)[32] (the proposed π and σ-N configurations of the succinimidyl radical are a second example[33]). Two e.s.r. spectra are observed, one being ascribed to the conventional trigonal-bipyramidal phosphoranyl radical (23) and the other to its isomerization product (24), where the electron occupies a σ-orbital on the isocyanate

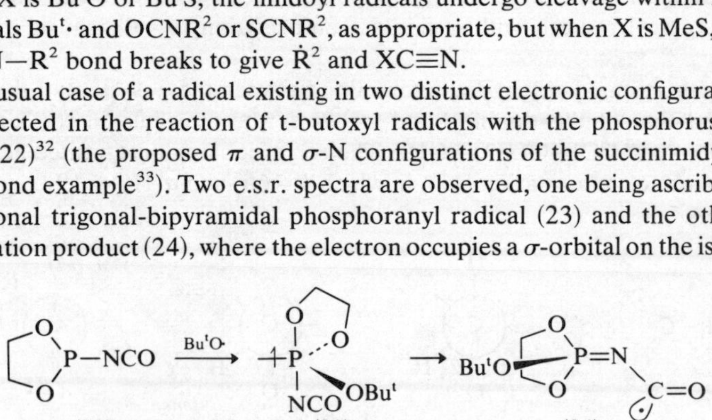

(22) (23) (24)

[29] P. M. Blum, A. G. Davies, and R. Sutcliffe, *J.C.S. Chem. Comm.*, 1979, 217.
[30] W. C. Danen and C. T. West, *J. Amer. Chem. Soc.*, 1973, **95**, 6872.
[31] P. M. Blum and B. P. Roberts, *J.C.S. Perkin II*, 1978, 1313.
[32] J. A. Baban and B. P. Roberts, *J.C.S. Chem. Comm.*, 1979, 537.
[33] See *Ann. Reports (B)*, 1978, **75**, 74.

ligand. The large value of the $^{13}C_{\alpha}$ hyperfine coupling (136.1 G) and the low g value (2.0009) are characteristic of acyl radicals.

3 Sulphur-centred Radicals

The e.s.r. spectra of the RS· radicals and the RO· radicals (which were discussed in last year's Report[34]) cannot be detected unless a specific electronic interaction breaks the degeneracy of the doubly occupied and singly occupied orbitals on the heteroatoms. In the ROO· radicals, this is accomplished by interaction with an electron pair on the β-oxygen atom. A similar interaction occurs in the RSS· radicals, and the spectrum of Bu^{t}SS· ($g = 2.025$, $\Delta H_{pp} = 4$ G) has now been detected from the photolysis of t-butylthiosulphenyl chloride, Bu^{t}SSCl. At −86 °C, the rate constant for its second-order decay (2×10^8 l mol^{-1} s^{-1}) is about 10^8 times greater than that of the Bu^{t}OO· radical.[35]

The degeneracy can also be broken by nitrogen, and a variety of dialkylaminothiyl radicals, for example Et_2NS· [a(N) = 10.7 G, a(4H) = 6.1 G, $g = 2.0156$], have recently been identified.[36] On the other hand, the persistent radicals that are observed when tetrasulphur tetranitride is photolysed in the presence of alkenes are not the sulphur-centred radicals (25), as was originally believed,[37] but are instead nitrogen-centred radicals [*e.g.* (26); a(N) = 13.11 G, a(2H) = 3.49 G, $a(2\,^{33}S)$ = 2.89 G, $g = 2.0064$].[38]

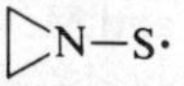

(25)

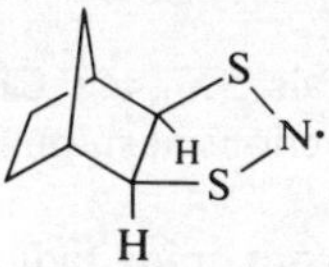

(26)

Again, the degeneracy can be broken by interaction with the π-electron system of an aromatic ring, and the arylthiyl radicals ArS· have been detected when the products of the photolysis of diaryl disulphides were condensed on a cold finger at 77 K.[39]

Alkylthiyl radicals can react with alkenes either by addition to the double bond or by abstraction of allylic hydrogen. In the reaction between methylthiyl radicals and cyclopentene, the balance between these two processes is strongly temperature-dependent. At −100 °C, only the spectrum of the adduct (27) is observed; at −60 °C, only the cyclopentenyl radical (28) is detected.[40]

Sulphides also fairly readily lose an electron to reagents such as HO·,[41] NO^+,[42] Bu$^t\overset{+}{O}$H,[43] or $NH_3^{+\cdot}$,[44] or electrochemically,[43] to give the radical cations $R_2S^{+\cdot}$.[45] These

[34] See *Ann. Reports* (*B*), 1978, **75**, 76.
[35] J. E. Bennett and G. Brunton, *J.C.S. Chem. Comm.*, 1979, 62.
[36] W. C. Danen and D. D. Newkirk, *J. Amer. Chem. Soc.*, 1976, **98**, 516; B. Maillard and K. U. Ingold, *ibid.*, p. 520; J. A. Baban and B. P. Roberts, *J.C.S. Perkin II*, 1978, 678.
[37] S. A. Fairhurst, W. R. McIlwaine, and L. H. Sutcliffe, *J. Magn. Resonance*, 1979, **35**, 121.
[38] S. Rolfe, D. Griller, K. U. Ingold, and L. H. Sutcliffe, *J. Org. Chem.*, 1979, **44**, 3515.
[39] W. Moerke, A. Jezierski, and H. Singer, *Z. Chem.*, 1979, **19**, 147.
[40] L. Lunazzi, G. Placucci, and L. Grossi, *J.C.S. Chem. Comm.*, 1979, 533.
[41] K.-D. Asmus, D. Bannemann, C.-H. Fischer, and D. Weltwisch, *J. Amer. Chem. Soc.*, 1979, **101**, 5322.
[42] T. L. Wolford and P. B. Rousch, *J. Amer. Chem. Soc.*, 1978, **100**, 6416.
[43] W. B. Gara, J. R. M. Giles, and B. P. Roberts, *J.C.S. Perkin II*, 1979, 1444.
[44] B. C. Gilbert and P. R. Marriott, *J.C.S. Perkin II*, 1979, 1425.
[45] Reviewed by K.-D. Asmus, *Accounts Chem. Res.*, 1979, **12**, 436.

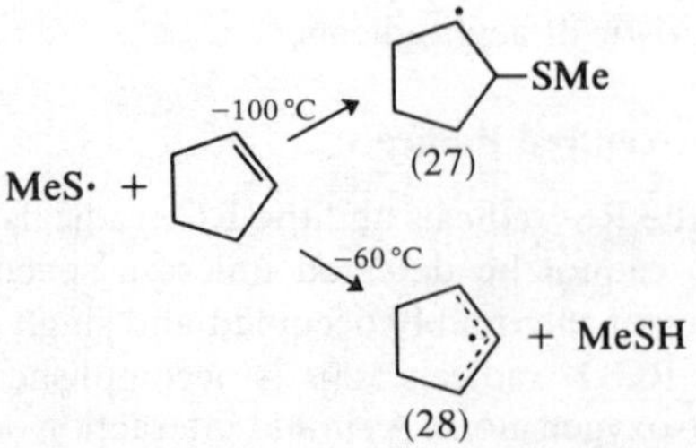

may be stabilized sterically or electronically as the monomers; *e.g.*, $Bu^t_2S^{+\cdot}$, $(Me_2N)_2S^{+\cdot}$, and $Ph\overset{+\cdot}{S}Me$. Otherwise they react (inter- or intra-molecularly) with a second sulphur centre, giving the dimer cation radicals $(R_2SSR_2)^{+\cdot}$, with an equilibrium constant of 10^3—10^4 l mol^{-1} [*e.g.* $Me_2SSMe_2^{+\cdot}$; $a(12H) = 6.3$ G]. The unpaired electron is probably contained in a σ^*-orbital, with the optical absorption due to a $\sigma \rightarrow \sigma^*$ transition (29).

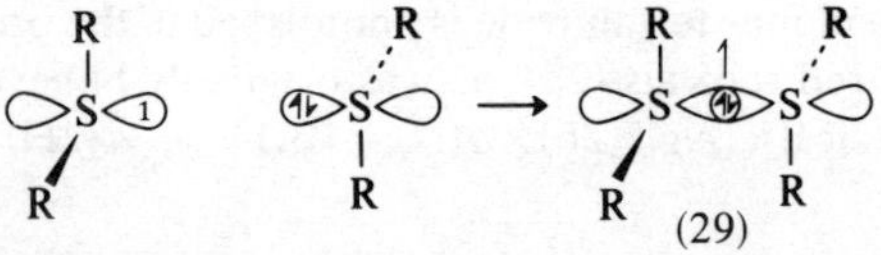

(29)

The sulphuranyl radicals $^{\cdot}SX_3$, *e.g.* $CF_3S\dot{S}Me_2$ and $Me_3SiO\dot{S}Me_2$, which occur as intermediates in S_H2 reactions at sulphur, probably also have the unpaired electron confined in a σ^*-orbital.[43,46]

A variety of sulphur-containing radicals are known in which the sulphur carries oxygen ligands, but the unpaired electron appears to be centred mainly on sulphur (*e.g.* $R_2N\dot{S}O$, $R\dot{S}O_2$, and $R\dot{S}O$).[47,48] The structures of the arenesulphonyl radicals $Ar\dot{S}O_2$ have been re-investigated. These are σ-radicals, with the unpaired electron in the plane of the benzene ring, and (like the benzoyl radical) the hyperfine coupling constants are in the sequence a(H-m) > a(H-o), a(H-p), and not $o > p > m$, as had previously been supposed.[48]

4 Nitrogen-centred Radicals

The detection of $R_2N^{\cdot}$ radicals by e.s.r. spectroscopy is not beset with the problems associated with orbital degeneracy that are encountered with $RO^{\cdot}$ and $RS^{\cdot}$ radicals, and the variety of nitrogen-centred radicals which have been studied is increasing rapidly. The reactions often involve addition of a radical to a multiple bond to nitrogen. Thus aminyl radicals, *e.g.* (30), have been prepared by the addition of alkyl radicals to imines[49] and hydrazyl radicals, *e.g.* (31), by addition to azodicarboxylates.[50] Azo-compounds which carry an electronegative substituent X in the α-position similarly undergo addition by trialkylmetallic radicals $(R_3M^{\cdot})$, but this is

[46] J. R. M. Giles and B. P. Roberts, *J.C.S. Chem. Comm.*, 1978, 623.
[47] C. Chatgilialoglu, B. C. Gilbert, C. M. Kirk, and R. O. C. Norman, *J.C.S. Perkin II*, 1979, 1084; B. C. Gilbert and P. R. Marriott, *ibid.*, p. 1425.
[48] C. Chatgilialoglu, B. C. Gilbert, and R. O. C. Norman, *J.C.S. Perkin II*, 1979, 770.
[49] B. P. Roberts and J. N. Winter, *J.C.S. Chem. Comm.*, 1978, 960.
[50] B. P. Roberts and J. N. Winter, *Tetrahedron Letters*, 1979, 3575.

$$Me_3CN{=}CH_2 + R^{\bullet} \rightarrow Me_3C\dot{N}CH_2R \quad (30) \qquad EtO_2CN{=}NCO_2Et + Bu^{t}\cdot \rightarrow EtO_2C\underset{}{N}(Bu^{t}){-}\dot{N}CO_2Et \quad (31)$$

followed by elimination of R_3MX, then β-scission of the resulting nitrogen-centred radical to give an azine ($M = R_3Si$ or R_3Sn; X = Cl, PhS, or PhO), as shown in Scheme 2.[51]

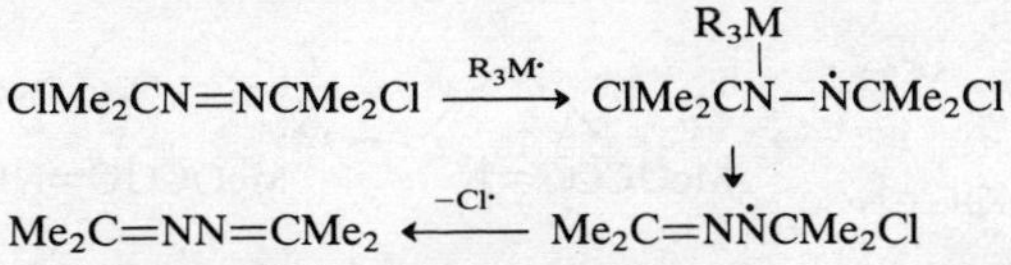

Scheme 2

Photolysis of the azines gives the iminyl radicals (32), which can also be formed by the photolysis of di-t-butyl peroxide in the presence of primary or secondary alkyl azides. The iminyl radicals recombine to form the azines at diffusion-controlled rates.[52]

$$Bu^{t}O^{\bullet} + H{-}CR_2{-}NN_2 \rightarrow Bu^{t}OH + N_2 + \underset{(32)}{R_2C{=}N^{\bullet}} \underset{h\nu}{\rightleftharpoons} R_2C{=}NN{=}CR_2$$

Iminyl radicals can also be prepared by addition to a nitrile. The 4-cyanobutyl radical (33) undergoes irreversible ring closure with a rate constant of $4.5 \times 10^2\ s^{-1}$ at 259 K,[52] but photolysis of cyclobutanone azine (34) shows only the spectrum of the

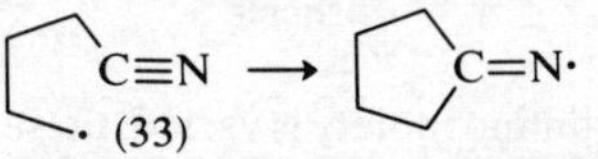

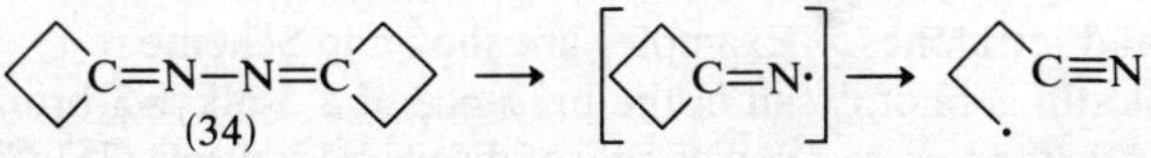

ring-opened radical. A reversible ring-closure of this type has been proposed to account for the 1,4-homolytic transfer of a cyanide group,[53] as shown in Scheme 3, and an intermolecular addition provides the basis for the peroxide-initiated cyanation of hydrocarbons by methyl cyanoformate (Scheme 4).[54]

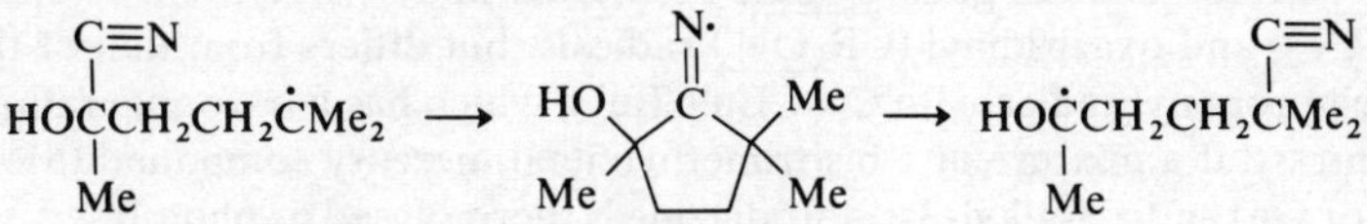

Scheme 3

C. Grugel and W. P. Neumann, *Annalen*, 1979, 1675.
B. P. Roberts and J. N. Winter, *J.C.S. Perkin II*, 1979, 1353.
D. W. Watt, *J. Amer. Chem. Soc.*, 1976, **98**, 271.
D. D. Tanner and P. M. Rahimi, *J. Org. Chem.*, 1979, **44**, 1674.

$$\mathrm{Bu^tO^\bullet + RH \rightarrow Bu^tOH + R^\bullet}$$

$$\mathrm{R^\bullet + MeOCOC{\equiv}N \rightarrow MeOCO\underset{R}{C}{-}N^\bullet \rightarrow MeO\dot{C}O + RC{\equiv}N \quad (>70\%)}$$

Scheme 4

Under the same conditions, 2,4-dimethylpentane gave an azacyclopentene (35) by a sequence of reactions that includes the cycloaddition of an alkyl radical to the nitrogen of an imine.

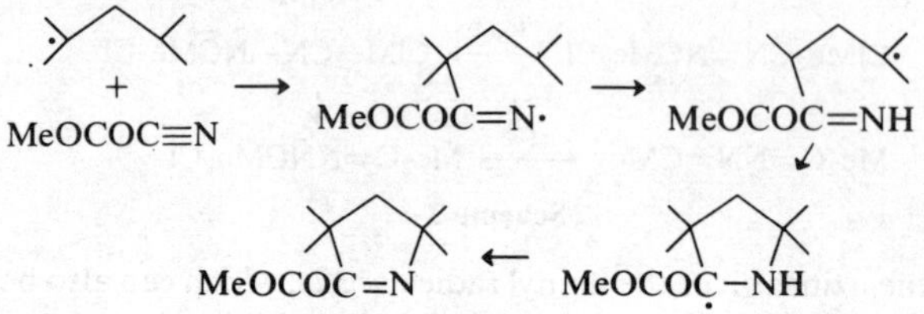

(35)

For use in chemical synthesis, iminyl radicals can conveniently be generated by the oxidation of imino-oxyacetic acids with persulphate, or thermolysis of the corresponding t-butyl peroxy-esters, *e.g.* as shown in Scheme 5.[55]

$$\mathrm{Ph_2C{=}NOCH_2CO_2H \xrightarrow{K_2S_2O_8} Ph_2C{=}NO\dot{C}H_2 \xrightarrow{-CH_2O} Ph_2C{=}N^\bullet \rightarrow (Ph_2C{=}N)_2}$$

$$\mathrm{Ph_2C{=}NOCH_2CO_3Bu^t \xrightarrow{\Delta} Ph_2C{=}NO\dot{C}H_2}$$

Scheme 5

If the structure of the arylimino moiety is varied, these reactions can provide useful routes to a variety of compounds such as phenanthrolines, quinolines, pyridines, pyrimidines, and tetralones.[55] Examples are shown in Scheme 6.

If hexabutylditin is photolysed in the presence of a 3-alkyl-3-bromodiazine, the bromine is abstracted to give a new family of diazirinyl radicals (36).[56] Correlation of the results from ^{13}C labelling of the ring atom with INDO calculations shows that these are π-radicals in which $a(^{13}\mathrm{C})$ has a negative sign and arises principally by spin polarization from nitrogen through the C—N bonds. The radical is therefore best represented by the valence-bond structures (37a) and (37b), with (38) making no significant contribution.

This structure is analogous to that of the aziridinyl ($\mathrm{\overline{CR_2CR_2N^\bullet}}$), diaziridinyl ($\mathrm{\overline{CR_2NRN^\bullet}}$), and oxaziridinyl ($\mathrm{\overline{CR_2ON^\bullet}}$) radicals, but differs from that of the tri-t-butylcyclopropenyl radical ($\mathrm{\overline{Bu^tC{=}CBu^tCBu^{t\bullet}}}$), which has a σ-configuration.

In contrast, if a mixture of a bis(trimethylmetal)mercury compound $\mathrm{(Me_3M)_2Hg}$ (M = Si or Ge) and a 3-alkyl-3-chlorodiazine is thermolysed or photolysed, the $\mathrm{R_3M}$

[55] A. R. Forrester, M. Gill, D. J. Meyer, J. S. Sadd, and R. H. Thompson, *J.C.S. Perkin I*, 1979, 606; A. R. Forrester, M. Gill, J. S. Sadd, and R. H. Thompson, *ibid.*, p. 612; A. R. Forrester, M. Gill, and R. H. Thompson, *ibid.*, pp. 616, 621; A. R. Forrester, M. Gill, R. J. Napier, and R. H. Thompson, *ibid.*, p. 632; A. R. Forrester, M. Gill, C. J. Meyer, and R. H. Thompson, *ibid.*, p. 637.

[56] Y. Maeda and K. U. Ingold, *J. Amer. Chem. Soc.*, 1979, **101**, 837.

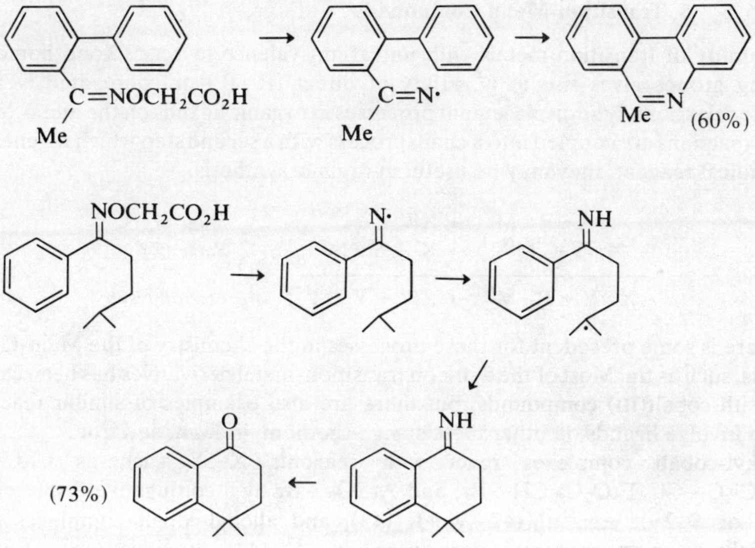

Scheme 6

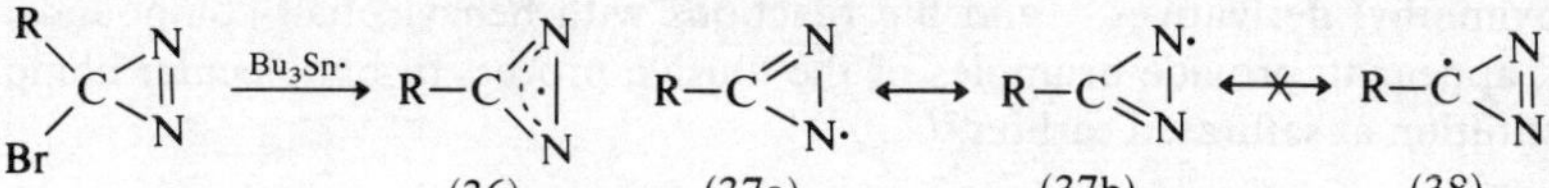

(36) (37a) (37b) (38)

radical is added at nitrogen to give a 2-metallated diaziridinyl radical (39).[57] When M = Si, the two nitrogen atoms are non-equivalent, but when M = Ge they are rendered equivalent, presumably by a rapid 1,2-migration of Me_3Ge. These diaziridinyl radicals may then react further to give *N*-metallated 1,2,3,5-tetrazinyl radicals (40), which are in equilibrium with their corresponding dimers.

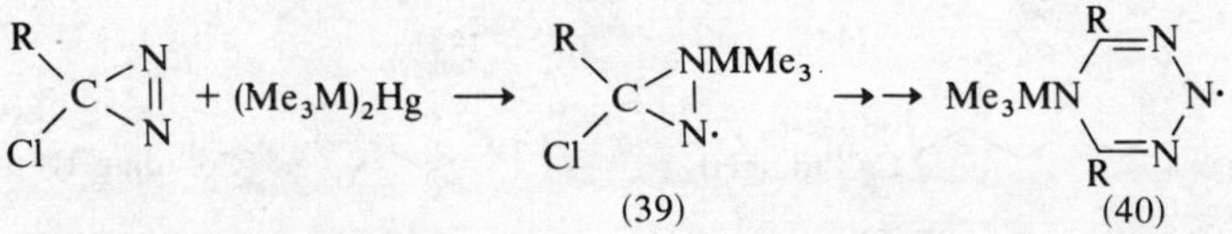

(39) (40)

Trialkylsilyl radicals similarly add to the nitrogen multiple-bond system of a variety of organic azides to give either 1,3- or 3,3-triazenyl radicals.[52] The available evidence suggests that these are the 3,3-triazenyls (41), in which the unpaired electron is mainly associated with the central nitrogen, in a σ-orbital that is in the N—N—N plane.

$$R^1_3Si^{\bullet} + R^2N_3 \rightarrow R^1_3Si(R^2)N{-}N{=}N^{\bullet} \text{ or } R^2N{=}N{-}\dot{N}SiR^1_3$$

(41)

[57] C. Grugel and W. P. Neumann, *Annalen*, 1979, 870.

5 Transition-Metal Compounds

The ability of transition metals with univariant valence to act as good homolytic leaving groups gives rise to a variety of direct (R=R′) or conjugated (R≠R′) bimolecular homolytic displacement processes at organic ligands on the metal. When these reactions are coupled into a chain process with a second step which regenerates the radical reagent, they may be useful in organic synthesis.

$$X^{\bullet} + R{-}M^{N} \rightarrow XR' + {}^{\bullet}M^{N-1} \qquad \textit{Stage (1)}$$
$$X{-}Y + {}^{\bullet}M^{N-1} \rightarrow X^{\bullet} + Y{-}M^{N} \qquad \textit{Stage (2)}$$
$$X{-}Y + R{-}M^{N} \rightarrow XR' + Y{-}M^{N} \qquad \textit{Overall result}$$

There is some precedent for these processes in the chemistry of the Main-Group metals, such as tin. Most of the work on transition-metal derivatives has been carried out with cobalt(III) compounds, but there are also examples of similar reactions which involve ligands on other metals; *e.g.*, rhodium, iridium, and iron.

Alkyl-cobalt complexes react with reagents X—Y such as $Cl_3C{-}Cl$, $(NC)Cl_2C{-}Cl$, $(EtO_2C)_2CH{-}Br$, and $ArSO_2{-}Br$ by a conjugative displacement (S_H2' or $S_H2\gamma$); *e.g.*,[58,59] (42) yields (43), and allenyl-cobalt complexes react similarly to give propargyl derivatives; *e.g.*,[57] (44) provides (45). Butenyl compounds, *e.g.* (46), undergo ring-closure, giving a useful route to cyclopropylmethyl derivatives,[60] and the reactions with benzylcobalt compounds, *e.g.* (47), appear to provide examples of the elusive process of bimolecular homolytic substitution at saturated carbon.[61]

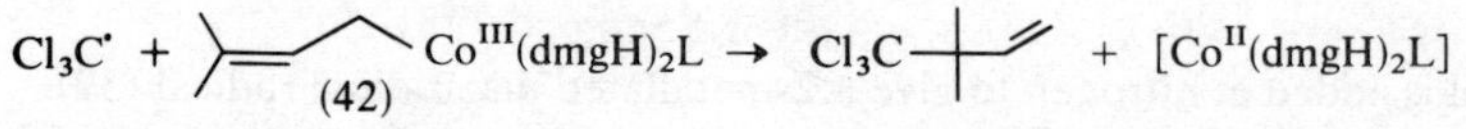

(43) (dmgH = dimethylglyoximato)

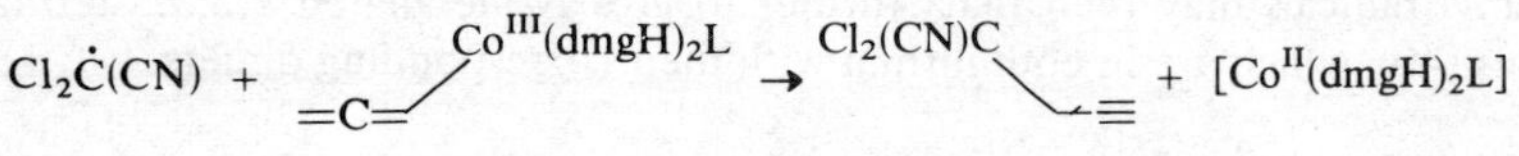

(44) (45)

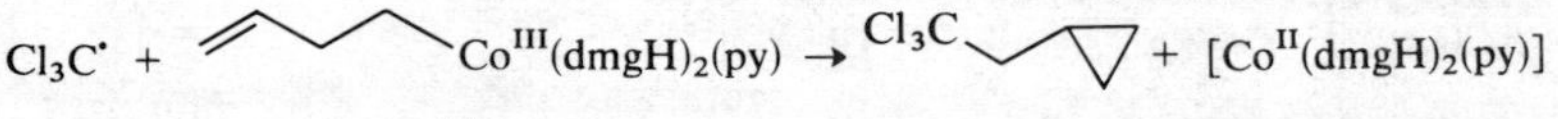

(46)

$$Cl_3C^{\bullet} + [ArCH_2Co^{III}(dmgH)(imid)] \rightarrow ArCH_2CCl_3 + [Co^{II}(dmgH)_2(imid)]$$

(47)

(imid = imidazole)

[58] A. Bury, C. J. Cooksey, T. Funabiki, B. D. Gupta, and M. D. Johnson, *J.C.S. Perkin II*, 1979, 1050.
[59] A. E. Crease, B. D. Gupta, M. D. Johnson, A. Bialkowska, K. N. V. Duong, and A. Gaudemar, *J.C.S. Perkin I*, 1979, 2611.
[60] A. Bury, M. R. Ashcroft, and M. D. Johnson, *J. Amer. Chem. Soc.*, 1978, **100**, 3217.
[61] B. D. Gupta, T. Funabiki, and M. D. Johnson, *J.C.S. Chem. Comm.*, 1977, 653.

5 Arynes, Carbenes, Nitrenes, and Related Species

By D. W. KNIGHT

Department of Chemistry, The University, Nottingham, NG7 2RD

1 Arynes

Ab initio calculations have established that *o*-, *m*-, and *p*-benzynes have singlet ground states, although some doubts still exist in the case of *p*-benzyne.[1] The total energies of the isomers increase in the order $o < m < p$, with the equilibrium structures of the latter two isomers showing considerable diradical character, in contrast to the aryne nature of *o*-benzyne. Although the bicyclic structure of *p*-benzyne may represent an energy minimum, this is far above the monocyclic structure. An analysis of the products arising from the thermolysis of hexa-1,5-diyn-3-enes, *e.g.* (1), suggests that the reaction proceeds *via* 1,4-dehydrobenzyne diradicals, which rearrange to 1,3-dehydrobenzenes by [1,2] shifts of the trimethylsilyl groups, presumably reflecting the greater stability of *m*-benzynes.[2] Treatment of the indene derivative (2) with a strong base at 0 °C gives a large number of products, whose formation can be rationalized by assuming the intermediacy of 2-chloro-1,3-dehydronaphthalene in both the bicyclic (diradical) and tricyclic forms.[3]

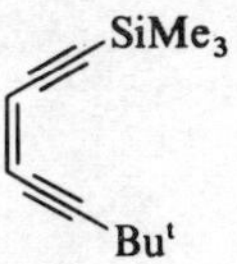

(1)

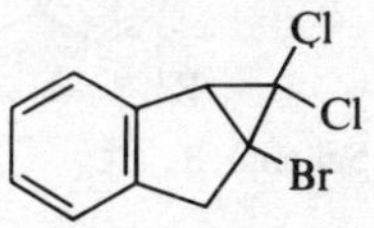

(2)

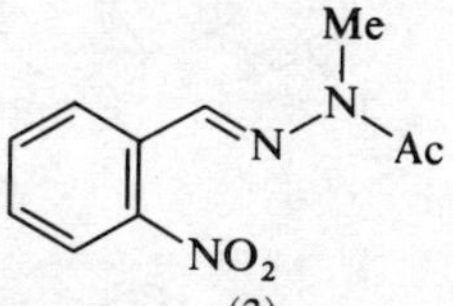

(3)

o-Benzyne can be generated photochemically from the *o*-nitrobenzaldehyde derivative (3),[4] and a full report has appeared on the generation of polymer-linked benzyne.[5]

Ab initio calculations suggest that the high electrophilicity of benzyne is due to a significant lowering of its LUMO energy level relative to the LUMO of linear but-2-yne, while the HOMO energies of the two molecules are approximately the same.[6] The synthetic utility of this electrophilicity continues to be exploited. Thus, a

[1] J. O. Noell and M. D. Newton, *J. Amer. Chem. Soc.*, 1979, **101**, 51.
[2] G. C. Johnson, J. J. Stofko, Jr., T. P. Lockhart, D. W. Brown, and R. G. Bergman, *J. Org. Chem.*, 1979, **44**, 4215.
[3] W. E. Billups, J. D. Buynak, and D. Butler, *J. Org. Chem.*, 1979, **44**, 4218.
[4] Y. Maki, T. Furuta, and M. Suzuki, *J.C.S. Perkin I*, 1979, 553.
[5] S. Mazur and P. Jayalekshmy, *J. Amer. Chem. Soc.*, 1979, **101**, 677; *cf.* J. Rebek, Jr., and J. E. Trend, *ibid.*, p. 737.
[6] N. G. Rondan, L. N. Domelsmith, K. N. Houk, A. T. Brown, and R. H. Levin, *Tetrahedron Letters*, 1979, 3237.

Japanese group have reported a potentially useful route to polycyclic alkaloids (Scheme 1)[7] while a new route to pyrrolo[3,4-*c*]pyridines probably proceeds *via* a pyridyne intermediate (Scheme 2).[8]

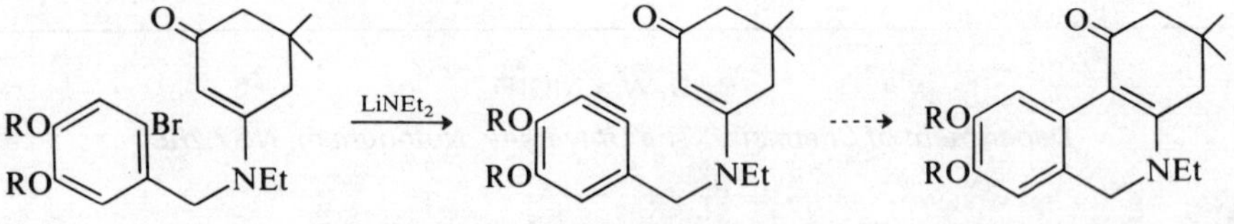

Scheme 1

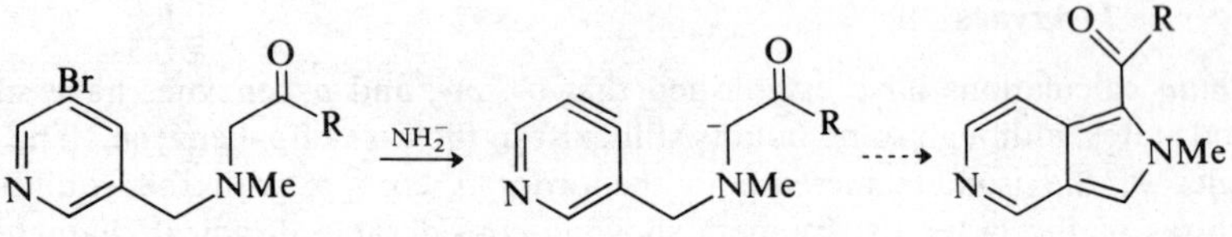

Scheme 2

Phthalide enolates add to benzyne to provide a simple if not very efficient route to anthraquinones (Scheme 3)[9] and, in a similar type of reaction, enolates of cyclic $\alpha\beta$-unsaturated ketones condense with benzyne (Scheme 4), affording ultimately the tricyclic alcohols, *e.g.* (4), which are useful as precursors of benzocyclo-octenones.[10]

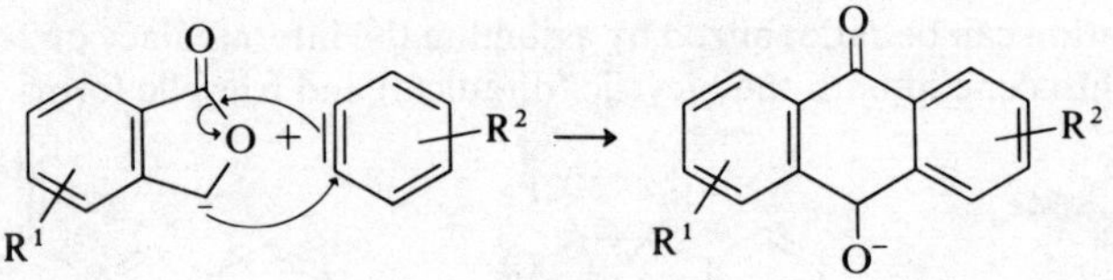

Scheme 3

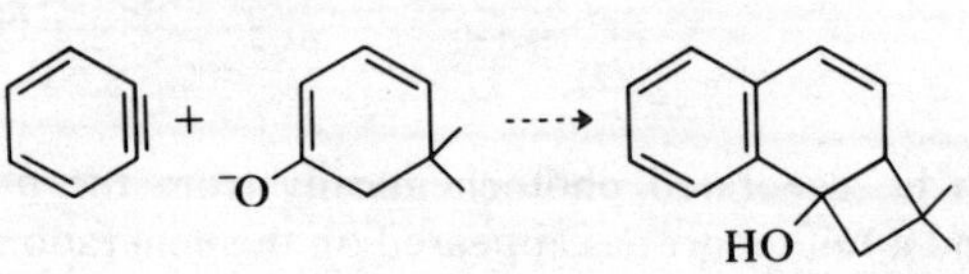

(4)

Scheme 4

Some evidence of the existence of the heteroaryne species (5) is the isolation of (6), its (formal) Diels–Alder adduct with anthracene.[11]

[7] H. Iida, Y. Yuasa, and C. Kibayashi, *J. Org. Chem.*, 1979, **44**, 1074; *cf. ibid.*, p. 1236.
[8] I. Ahmed, G. W. H. Cheeseman, and B. Jaques, *Tetrahedron*, 1979, **35**, 1145; see also M. Mallet and G. Quéguiner, *ibid.*, p. 1625.
[9] P. G. Sammes and D. J. Dodsworth, *J.C.S. Chem. Comm.*, 1979, 33.
[10] M. Essiz, G. Guillaumet, J.-J. Brunet, and P. Caubère, *J.C.S. Chem. Comm.*, 1979, 276; see also M. L. Viriot, *J. Chem. Res.*, (*S*) 1979, 324; H. Iida, Y. Yuasa, and C. Kibayashi, *J. Org. Chem.*, 1979, **44**, 3985.
[11] M. G. Reinecke and H. H. Ballard, *Tetrahedron Letters*, 1979, 4981.

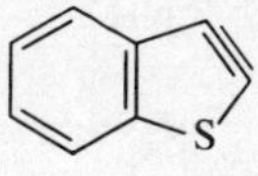

(5)

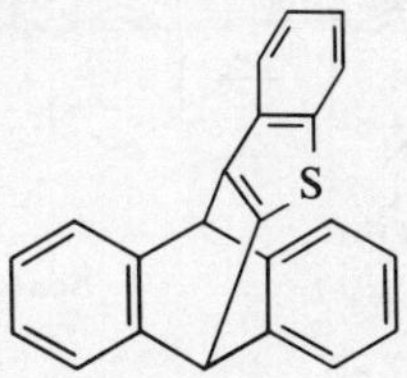

(6)

2 Nitrenes

A useful review has appeared concerning advances in the synthetic applications of aryl and heteroaryl nitrenes.[12] Chapman *et al.* have summarized recent studies on the low-temperature photochemistry of C_6H_5N species;[13] it now seems clear that, under these conditions, triplet phenylnitrene and triplet 2-pyridylmethylene interconvert *via* 1-aza-1,2,4,6-cycloheptatetraene.

Ab initio calculations predict that an energetically favourable pathway in the reduction of ethylene by *trans*-di-imide proceeds by way of ethylaminonitrene.[14]

The reaction between cyanamide and iodobenzene diacetate produces cyanonitrene, which can be trapped by a variety of reagents.[15]

The nitrene (7), obtained by oxidation of the corresponding arylsulphenamide, adds to electron-rich olefins to give aziridines in 38—64% yield.[16] The stereoselectivities of the additions imply that the singlet and triplet states of (7) are in equilibrium and that the addition of triplet (7) to olefins is reversible. A milder method for the generation of phthalimidonitrene is by thermolysis, in refluxing benzene, of the sulphimide (8).[17] The nitrene so produced is a relatively stable, singlet species, which adds to olefins stereoselectively to give aziridines.

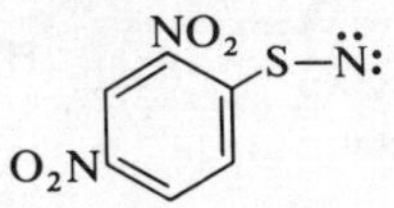

(7)

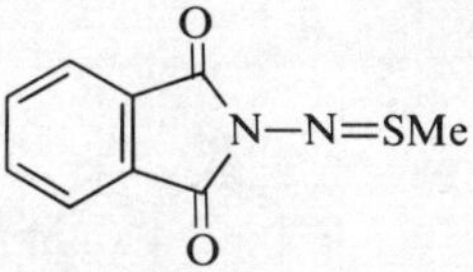

(8)

A study of the reactions between alcohols and ethoxycarbonylnitrene, generated thermally or photolytically from ethyl azidoformate, indicates that only the singlet species is capable of inserting into the O—H bonds of alcohols.[18]

The photolysis of pyridine *N*-oxides, leading to formyl-pyrroles (Scheme 5), gives considerably improved yields when performed in the presence of copper(II) salts.[19]

[12] B. Iddon, O. Meth-Cohn, E. F. V. Scriven, H. Suschitzky, and P. T. Gallagher, *Angew. Chem. Internat. Edn.*, 1979, **18**, 900.

[13] O. L. Chapman, R. S. Sheridan, and J.-P. LeRoux, *Rec. Trav. chim.*, 1979, **98**, 334; O. L. Chapman and R. S. Sheridan, *J. Amer. Chem. Soc.*, 1979, **101**, 3690.

[14] D. J. Pasto and D. M. Chipman, *J. Amer. Chem. Soc.*, 1979, **101**, 2290.

[15] J. E. G. Kemp, D. Ellis, and M. D. Closier, *Tetrahedron Letters*, 1979, 3781.

[16] R. S. Atkinson and B. D. Judkins, *J.C.S. Chem. Comm.*, 1979, 832, 833.

[17] M. Edwards, T. L. Gilchrist, C. J. Harris, and C. W. Rees, *J. Chem. Res.* (*S*), 1979, 114.

[18] N. Torimoto, T. Shingaki, and T. Nagai, *J. Org. Chem.*, 1979, **44**, 2585; see also L. D'Epifanio, L. Pellacani, and P. A. Tardella, *ibid.*, p. 3605.

[19] F. Bellamy and J. Streith, *J. Chem. Res.* (*S*), 1979, 18.

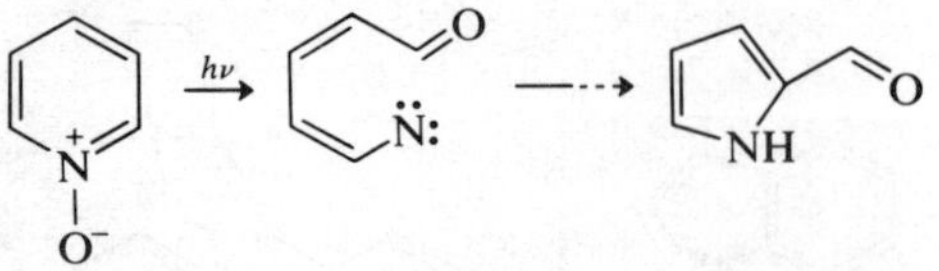

Scheme 5

A highly practical route to pentaco-ordinate phosphoranes consists of treatment of a suitable azide with a phosphine[20] (Scheme 6).

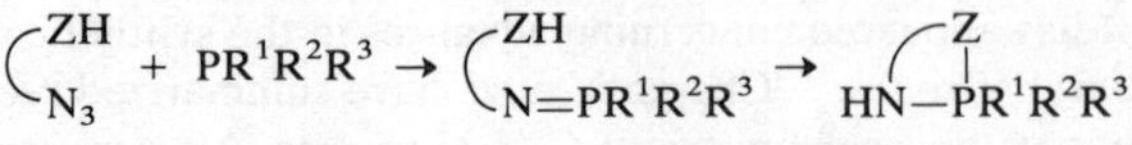

Z = O or NH

Scheme 6

Previous studies with imidoylnitrenes (9; R = H) revealed that a major reaction pathway was attack by the nitrene at the *ortho*-position of the aromatic ring followed by re-aromatization. It has now been found that, when the *ortho*-positions are blocked, *e.g.* (9; R = Me), a rather different set of rearrangements occurs *via* 3a*H*-benzimidazoles (10), leading to cyclopenta[*d*]pyrimidines by sequential [1,5] vinyl-, imidoyl-, and hydride-shifts (Scheme 7).[21] Also isolated from thermally generated (9; R = Me) were low yields of benzimidazoles, *e.g.* (11), which could have arisen from (10) by extremely rare but thermally allowed [1, 9] alkyl shifts.

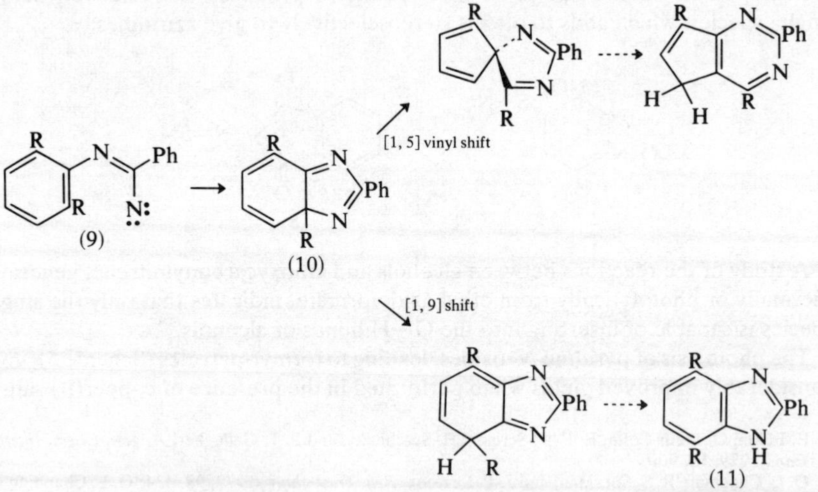

Scheme 7

[20] J. I. G. Cadogan, I. Gosney, E. Henry, T. Naisby, B. Nay, N. J. Stewart, and N. J. Tweddle, *J.C.S. Chem. Comm.*, 1979, 189; J. I. G. Cadogan, N. J. Stewart, and N. J. Tweddle, *ibid.*, p. 191.

[21] T. L. Gilchrist, C. J. Moody, and C. W. Rees, *J.C.S. Perkin I*, 1979, 1871; T. L. Gilchrist, P. F. Gordon, D. F. Pipe, and C. W. Rees, *ibid.*, p. 2303.

Some evidence that intramolecular insertion of an arylnitrene into thiophenes proceeds *via* a ring-opened form is the isolation of one such compound (12) (see Scheme 8);[22] however, it is still uncertain that (12) is a true intermediate, and not a by-product.

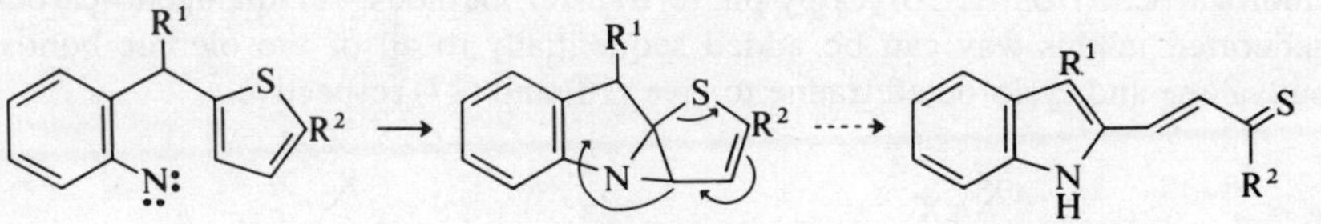

(12)

Scheme 8

A second example of a *bis*-nitrene metal complex, *cis*-$[Mo(NPh)_2(S_2CNEt_2)_2]$, has been reported. *X*-Ray analysis at −130 °C reveals that the two phenylnitrene ligands are not identical, one being nearly linear while the second is bent at nitrogen (∠MoNC = 139.4 °), with a longer Mo—N bond than the former.[23]

Phenylphosphinidine (PhP̈:), thermally generated from $(PhP)_5$, inserts into Ge–X bonds (X = H, C, N, O, or P) to provide a route to germylphosphines.[24]

3 Carbenes

The stable, isolable salt (13) serves as a useful source of methylene for the stereospecific conversion of olefins into cyclopropanes in good yields.[25]

The formation of cyclopropanes from carbenes and olefins, when thought of in Frontier Orbital terms, is generally carried out with an electrophilic carbene (*e.g.*, $:CCl_2$), where the dominant orbital interaction is between the LUMO of the carbene and the HOMO of the olefin, and less commonly using a nucleophilic carbene, *e.g.* $:C(OMe)_2$. This spectrum of reactivity has been summarized,[26] and it has also been demonstrated that methoxychlorocarbene (14) is an ambiphilic species in that, although its additions to electron-rich olefins may be characterized as electrophilic, (14) is a nucleophilic species in the reactions with electron-poor olefins.[27] The preparation of (14) by thermolysis of diazirine (15), both in solution and in the gas phase, involves a direct two-bond cleavage, to the exclusion of the alternative pathway *via* one-bond rupture to give a diazo intermediate; a process which operates when various other substituents (*e.g.*Ph or alkyl) are present.[28]

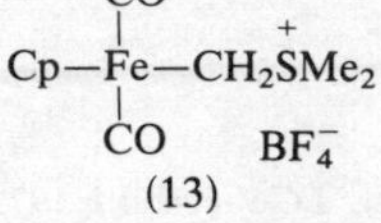

(13)

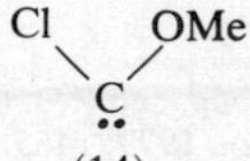

(14)

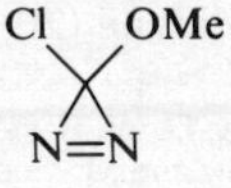

(15)

[22] G. Jones, C. Keates, I. Kladko, and P. Radley, *Tetrahedron Letters*, 1979, 1445; see also, G. Jones and W. H. McKinley, *J.C.S. Perkin I*, 1979, 599.

[23] B. L. Haymore, E. A. Maatta, and R. A. D. Wentworth, *J. Amer. Chem. Soc.*, 1979, **101**, 2063.

[24] J. Escudié, C. Couret, and J. Satgé, *Rec. Trav. chim.*, 1979, **98**, 461.

[25] S. Brandt and P. Helquist, *J. Amer. Chem. Soc.*, 1979, **101**, 6473.

[26] R. A. Moss and R. C. Munjal, *Tetrahedron Letters*, 1979, 4721.

[27] R. A. Moss, M. Fedorynski, and W.-C. Shieh, *J. Amer. Chem. Soc.*, 1979, **101**, 4736; N. P. Smith and I. D. R. Stevens, *J.C.S. Perkin II*, 1979, 1298.

[28] N. P. Smith and I. D. R. Stevens, *J.C.S. Perkin II*, 1979, 213.

The generation of dibromocarbene from bromoform under phase-transfer conditions is successful when cetyltrimethylammonium bromide is used as catalyst, whereas benzyltriethylammonium chloride favours the formation of $^{-}CBr_3$ only.[29] A careful choice of reagents is also necessary for the generation of bromochlorocarbene from $HCBr_2Cl$ by phase-transfer methods.[30] Dihalogeno-carbenes generated in this way can be added sequentially to all of the olefinic bonds in bullvalene and cyclo-octatetraene to give (16) and (17) respectively.[31]

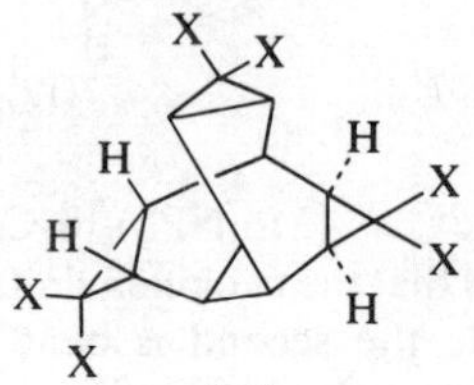

(16) X = Cl or Br

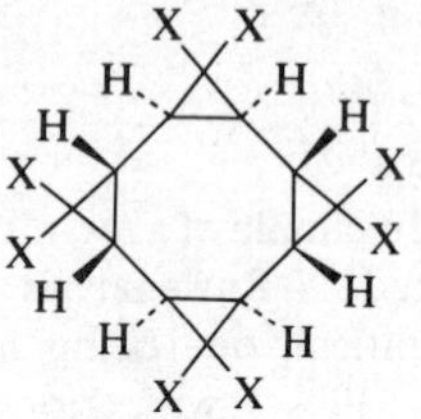

(17)

The preparation and chemistry of 1,2,2-trifluoroethylidene has been investigated; reactions with alkanes lead mainly to C—H insertion products, with the usual order of reactivity ($3^y > 2^y > 1^y$) being observed.[32] Halogenocarbenes, thermally generated from chloroform, add to pyrrole and some methyl-pyrroles to give surprisingly high yields of chloro-pyridines.[33]

Interest in Simmons–Smith-type reactions continues. Conventional conditions using *gem*-dihalides and zinc or copper-based catalysts are usually only successful when applied to electron-rich olefins. It has now been found that, by employing a nickel catalyst, *i.e.* $[Ni(PPh_3)_4]$, in the presence of sodium iodide and zinc, electron-poor olefins (*e.g.* vinyl ketones or acrylates) can be efficiently converted into cyclopropanes.[34] When copper powder is used in these reactions, the addition is highly stereoselective, indicating the intermediacy of organo-coppers rather than free carbenes.[35] Indeed, *ab initio* calculations[36] support this view, and suggest that the true intermediates in the Simmons–Smith reaction may have considerable metallocarbenium character. Such species have been directly observed, using ^{13}C n.m.r.[37] Furthermore, in the reaction between dibromocyclopropane (18) and methyl-lithium, products are found which attest to the intermediacy of (19) as well as of cyclopropylidene (20).[38]

29 M. S. Baird, A. G. W. Baxter, B. R. J. Devlin, and R. J. G. Searle, *J.C.S. Chem. Comm.*, 1979, 210.
30 E. V. Dehmlow and M. Slopianka, *Annalen*, 1979, 1465.
31 E. V. Dehmlow and M. Lissel, *Annalen*, 1979, 181.
32 R. N. Haszeldine, R. Rowland, J. G. Speight, and A. E. Tipping, *J.C.S. Perkin I*, 1979, 1943; R. N. Haszeldine, C. Parkinson, P. J. Robinson, and W. J. Williams, *J.C.S. Perkin II*, 1979, 954.
33 R. E. Busby, M. Iqbal, M. A. Khan, J. Parrick, and C. J. G. Shaw, *J.C.S. Perkin I*, 1979, 1578.
34 H. Kanai and N. Hiraki, *Chem. Letters*, 1979, 761.
35 N. Kawabata, I. Kamemura, and M. Naka, *J. Amer. Chem. Soc.*, 1979, **101**, 2139; N. Kawabata, M. Tanimoto, and S. Fujiwara, *Tetrahedron*, 1979, **35**, 1919.
36 T. Clark and P. von R. Schleyer, *J.C.S. Chem. Comm.*, 1979, 883; *Tetrahedron Letters*, 1979, 4963.
37 D. Seebach, H. Siegel, K. Müllen, and K. Hiltbrunner, *Angew. Chem. Internat. Edn.*, 1979, **18**, 784; H. Siegel, K. Hiltbrunner, and D. Seebach, *ibid.*, p. 785.
38 M. S. Baird and A. G. W. Baxter, *J.C.S. Perkin I*, 1979, 2317.

(18) (19) (20)

Cyclopropylidenes usually undergo ring-opening reactions to give allenes unless these would be highly strained or unless insertion pathways are readily available. An example of the latter phenomenon is the formation of (22) and (23) from the *anti*-3-vinylbicyclo[4.1.0] heptane (21),[39] by insertion into the C-3—H and C-5—H bonds respectively. Similar treatment of the *syn* isomer of (21) at low temperatures results in addition to the vinyl group. The dual pathway in the reactions of cyclopropylidene (24) to give allenes (25) and insertion products (26) has been investigated. A Hammett plot gives $\rho = -0.5$, indicating little build-up of charge at the benzylic carbon during insertion, assuming that changes in the *para*-substituent have little effect on the rate of formation of allene.[40]

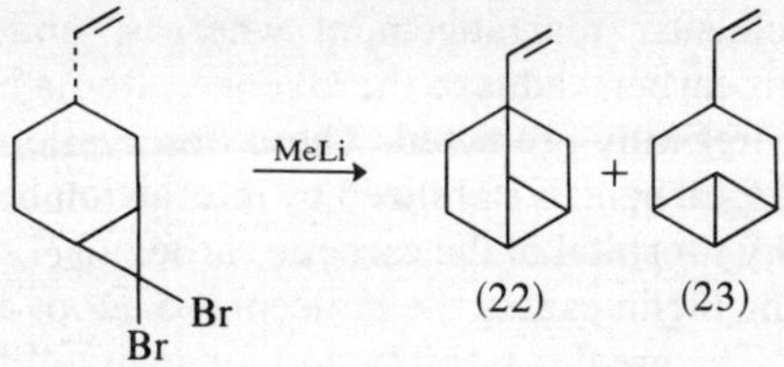

(21) (22) (23)

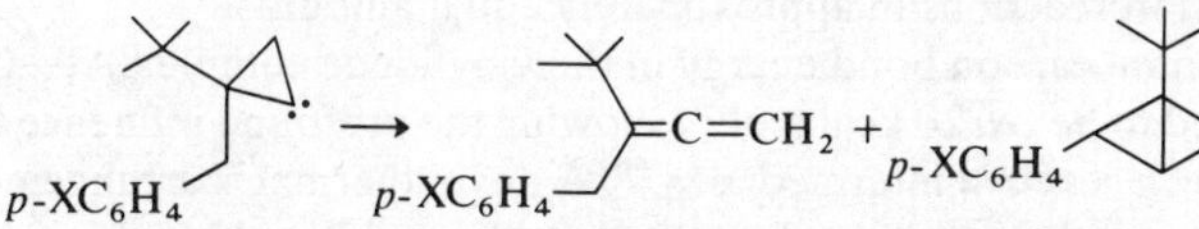

(24) (25) (26)

Vinylcyclopropylidenes usually undergo [1,3] shifts to give 3-cyclopentenylidenes; it has now been suggested that the *cis*-dienylcyclopropylidene (27) reacts by a novel [1,5] shift to give (28), ultimately isolated as cycloheptatriene (29).[41] Alternative mechanistic rationales (*e.g.* insertion into vinylic C—H or into the terminal double bond) seem to be energetically unfavourable.

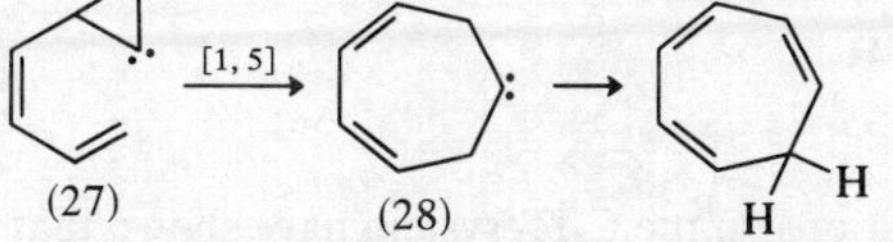

(27) (28) (29)

[39] M. S. Baird, *J.C.S. Perkin I*, 1979, 1020; see also R. M. Cory, L. P. J. Burton, and R. G. Pecherle, *Synth. Comm.*, 1979, **9**, 735; M. S. Baird, P. Sadler, J. Hatem, J.-P. Zahra, and B. Waegell, *J.C.S. Chem. Comm.*, 1979, 452.

[40] M.S. Baird, *J.C.S. Chem. Comm.*, 1979, 776.

[41] U. H. Brinker and I. Fleischhauer, *Angew. Chem. Internat. Edn.*, 1979, **18**, 396.

MINDO/3 calculations indicate that the initial step in the transformation of cyclobutylidene into methylenecyclopropane involves electrophilic attack by the empty *p*-orbital of the carbene carbon onto the methylene group at C-3 to produce a bicyclobutane-type intermediate, a formulation which represents a 'non-classical' carbene.[42]

2-Vinylcyclobutylidenes (30), in contrast to vinylcyclopropylidenes (*cf.* ref. 41), do not undergo carbene–carbene rearrangements but instead give mainly allylidenecyclopropanes (31) by C-2—C-3 bond cleavage.[43]

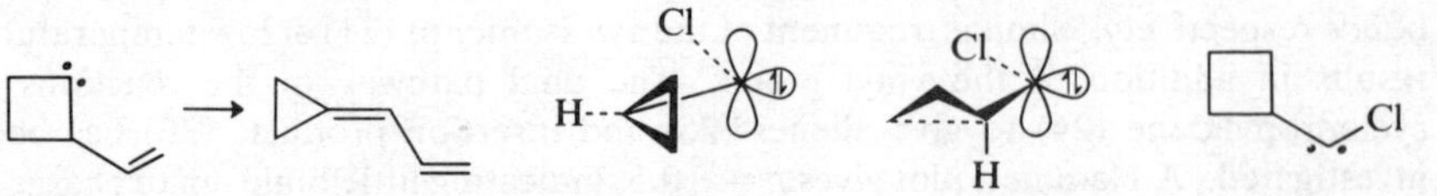

(30) (31) (32) (33) (34)

Isopropylidenechlorocarbene, when generated in the presence of olefins, undergoes mainly intramolecular rearrangement whereas, under similar conditions, cyclopropylidenechlorocarbene adds to the alkenes, although lower selectivities are found than those theoretically predicted. These observations can be explained by assuming that the latter carbene is stabilized by interaction between the cyclopropyl σ-bonds and the empty *p*-orbital of the carbene carbon (32), and that this stability is lost when the incoming olefin causes the molecule to adopt a less crowded, twisted conformation (33).[44] The greater stability and longevity of (32) contrasts with the cyclobutyl analogue (34), which undergoes intramolecular rearrangement and cyclopropanation reactions in approximately equal amounts.[45]

The aluminium–carbon bond energy in the vinylidene complex Al—C=CH_2 has been calculated to be *ca.* 20 kcal mol^{-1}, showing the profound influence that a metal can have on such reactive intermediates.[46] A novel thermal rearrangement, termed the 'α-alkynone cyclization', involves the formation of 2-cyclopentenones (37) from α-alkynones (35), probably *via* alkylidenecarbenes (36).[47] A general route to alkylidenes is by a facile loss of nitrogen from diazoalkenes, these being prepared by Wadsworth–Emmons reactions between dimethyl diazomethylphosphonate and aldehydes or alkyl aryl ketones.[48]

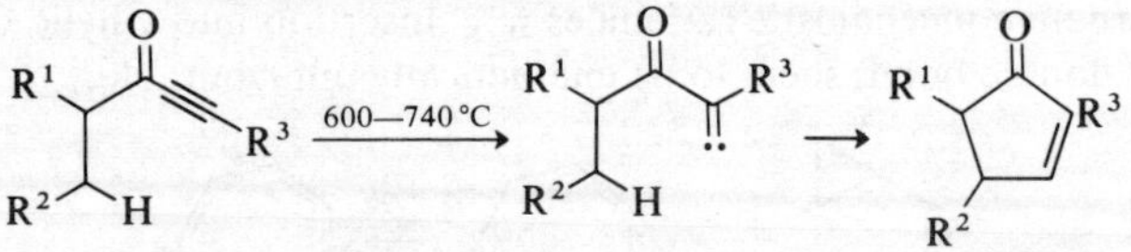

(35) (36) (37)

Ab initio calculations on the C_4H_2 system have shown that the two arrangements butatrienylidene (38) and ethynylvinylidene (39) are of almost the same energy, and

[42] W. W. Schoeller, *J. Amer. Chem. Soc.*, 1979, **101**, 4811.
[43] U. H. Brinker and L. König, *J. Amer. Chem. Soc.*, 1979, **101**, 4738.
[44] R. A. Moss, M. Vezza, W. Guo, R. C. Munjal, K. N. Houk, and N. G. Rondan, *J. Amer. Chem. Soc.*, 1979, **101**, 5088.
[45] R. A. Moss, M. E. Fantina, and R. C. Munjal, *Tetrahedron Letters*, 1979, 1277.
[46] M. Trenary, M. E. Casida, B. R. Brooks, and H. F. Schaefer, III, *J. Amer. Chem. Soc.*, 1979, **101**, 1638.
[47] M. Karpf and A. S. Dreiding, *Helv. Chim. Acta.*, 1979, **62**, 852.
[48] J. C. Gilbert, U. Weerasooriya, and D. Giamalva, *Tetrahedron Letters*, 1979, 4619.

that the energy difference between them and the stable structure buta-1,3-diene is similar to the difference between vinylidene and acetylene.[49] A brief report has appeared on the generation of the 4,4-dimethyl derivative of (38) by base-induced γ-elimination of triflate from (40).[50]

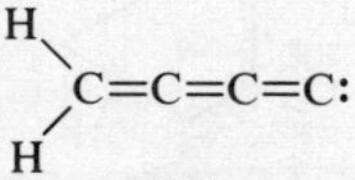
(38)

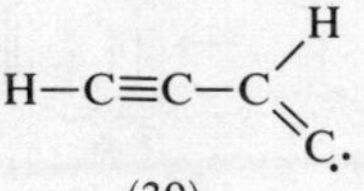
(39)

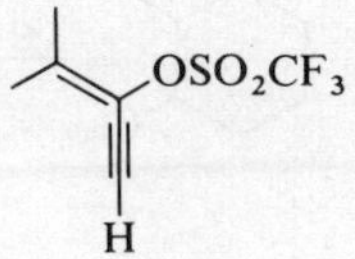
(40)

An authoritative review has appeared on [1,2]-hydrogen shifts involving carbenes and nitrenes.[51] A detailed study of such [1,2] rearrangements in non-rigid cyclohexylidenes, *e.g.* (41), amplifies previous work in leading to the conclusion that, for such conformations, an axial hydrogen is more likely to migrate, as its bonding orbital is nearly parallel with the empty *p*-orbital of the carbene and perpendicular to the filled non-bonding orbital.[52]

In contrast to earlier findings with methylene, the energy gap between the singlet and triplet ground states in simple alkylcarbenes (42) appears[53] to be considerably less than 20 kcal mol^{-1}.

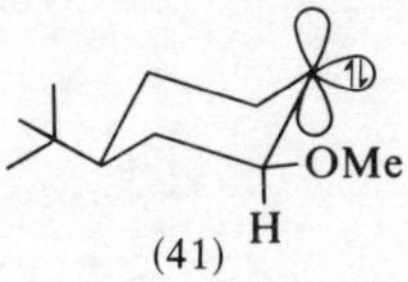
(41)

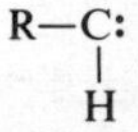
(42) R = Bu^t or Pr^i

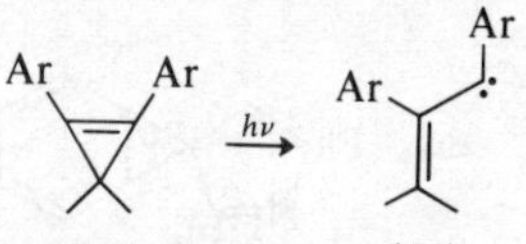
(43) (44)

Padwa[54] has discussed the probable intermediacy and subsequent reactions of vinylcarbenes (44) in photochemical reactions of cyclopropenes (43). Photochemically generated vinylsulphinylcarbene (45) undergoes a novel rearrangement to give vinyl-sulphines (46), which can be trapped or isolated as the corresponding $\alpha\beta$-unsaturated ketone.[55] 1,1-Dichloroallyl-lithium serves as a useful source of chlorovinylcarbene (47), which reacts with simple olefins to give 1-chloro-1-vinylcyclopropanes in 18—42% yield.[56]

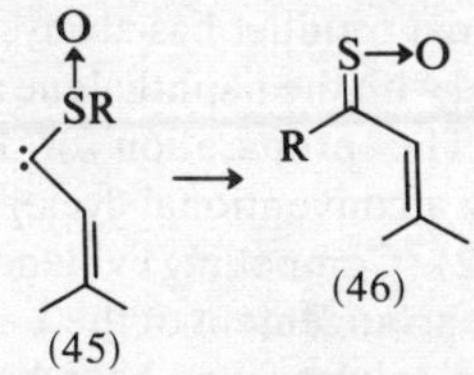
(45) (46)

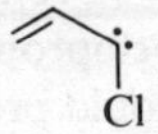
(47)

[49] C. E. Dykstra, C. A. Parsons, and C. L. Oates, *J. Amer. Chem. Soc.*, 1979, **101**, 1962.
[50] P. J. Stang and T. E. Fisk, *J. Amer. Chem. Soc.*, 1979, **101**, 4772; *Synthesis*, 1979, 438.
[51] H. F. Schaefer, III, *Accounts Chem. Res.*, 1979, **12**, 288.
[52] L. S. Press and H. Shechter, *J. Amer. Chem. Soc.*, 1979, **101**, 509.
[53] K.-T. Chang and H. Shechter, *J. Amer. Chem. Soc.*, 1979, **101**, 5082.
[54] A. Padwa, *Accounts Chem. Res.*, 1979, **12**, 310; A. Padwa, T. J. Blacklock, and R. Loza, *Tetrahedron Letters*, 1979, 219, 1617.
[55] M. Franck-Neumann and J. L. Lohmann, *Tetrahedron Letters*, 1979, 2397.
[56] R. A. Moss and R. C. Munjal, *Synthesis*, 1979, 425.

In a rather odd reaction, 1-chloro-2-alkyl-cyclohexenes react with organolithiums to give bicyclo[4.1.0]heptanes (Scheme 9), probably *via* the carbene (48) and the cyclopropene (49).[57]

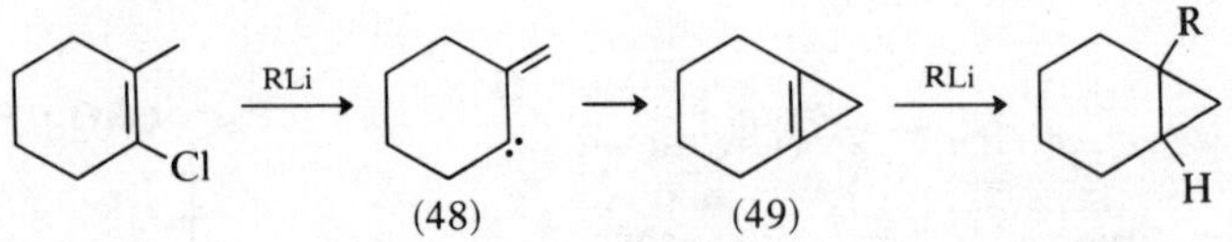

Scheme 9

Reactions of (presumably) singlet pyrrolylidene (50) with substituted benzenes proceed *via* two pathways, depending on the electronic properties of the substituent (Scheme 10);[58] not surprisingly, triplet (50) gives only the substitution products (52) without the participation of spironorcaradiene (51).

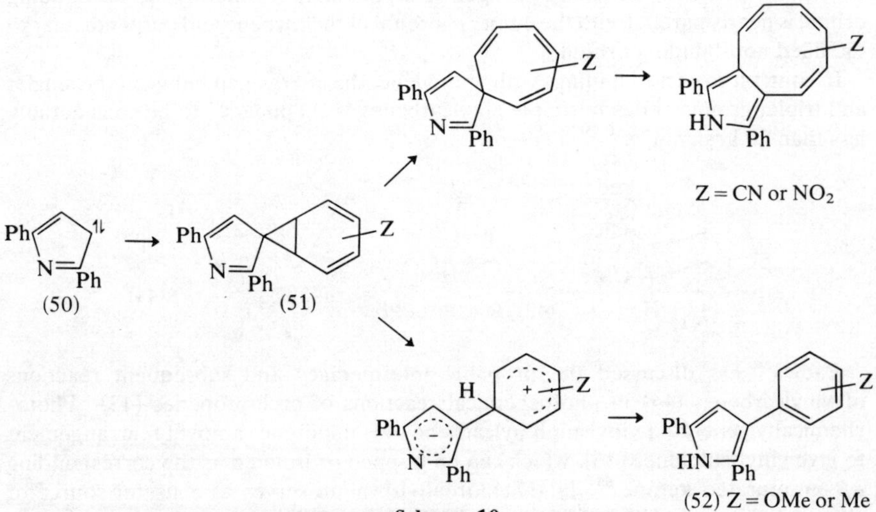

Scheme 10

The thermal rearrangement of benzocyclopropane to ethynylcyclopentadiene (Scheme 11) is well known, but the final product has always eluded proper characterization. This has now been achieved with the naphthalene analogue (53), obtained from naphtho[2,3]cyclopropane.[59] The preparation of benzvalene *via* cyclopentadienylcarbene could proceed by a conventional cyclopropanation reaction or by an overall [1,4] addition (Scheme 12). Compelling evidence for the latter pathway is the isolation of only (55) from the rearrangement of the 1-methyl analogue (54).[60]

Full details for the preparation of *p*-tolylcarbene have been published, together with some useful notes on the generation of carbenes.[61] Arylcarbenes, photo-

[57] P. G. Gassman, J. J. Valcho, and G. S. Proehl, *J. Amer. Chem. Soc.*, 1979, **101**, 231.
[58] M. Nagarajan and H. Shechter, *J. Amer. Chem. Soc.*, 1979, **101**, 2198.
[59] C. Wentrup, E. Wentrup-Byrne, P. Müller, and J. Becker, *Tetrahedron Letters*, 1979, 4249.
[60] U. Burger and G. Gandillon, *Tetrahedron Letters*, 1979, 4281.
[61] C. M. Dougherty and R. A. Olofson, *Org. Synth.*, 1978, **58**, 37.

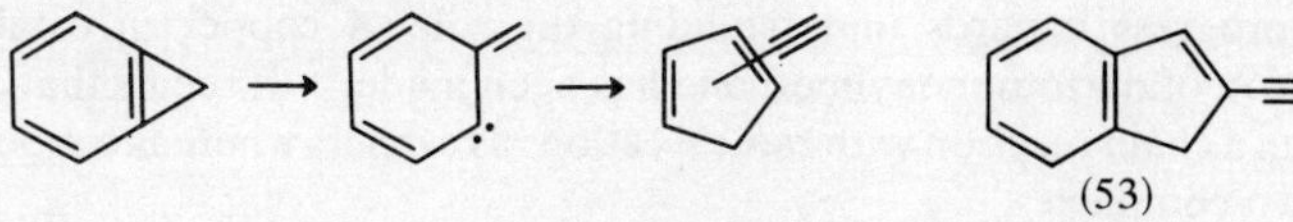

(53)

Scheme 11

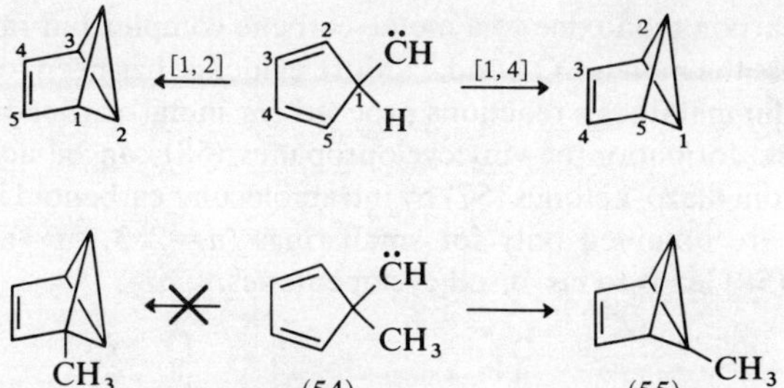

(54) (55)

Scheme 12

chemically generated in rigid matrices of alkanes, preferentially insert into primary C—H bonds, in contrast to the propensity for reaction with secondary and tertiary C—H bonds in liquid alkanes at room temperature. This effect is probably due to the steric constraints imposed by the rigid matrix.[62] These constraints can also cause the reactivity of arylcarbenes to be dependent on the precursor. Thus, determinations of singlet/triplet ratios by an analysis of the products arising from carbenes generated in matrices at low temperatures (*e.g.*, −196 °C) can be unreliable.[63] Thermally generated σ-substituted arylcarbenes such as (56) mainly undergo intramolecular insertion reactions to give five- and six-membered rings (Scheme 13);[64] σ-alkoxy-arylcarbenes also react by a formal double proton abstraction to give the corresponding vinyl ethers,[65] although labelling studies reveal that this reaction does not occur *via* a geminal abstraction followed by a [1,2] hydride shift.

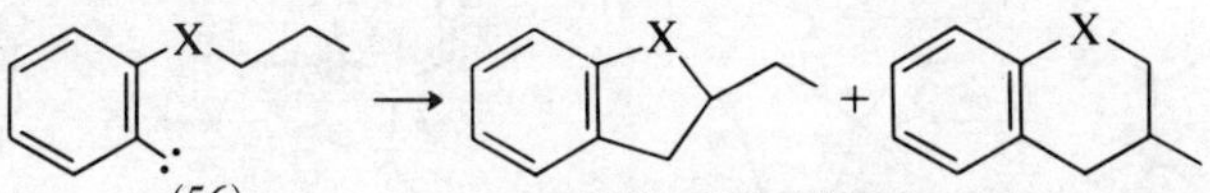

(56) X = CH_2, NR, O, or S

Scheme 13

It is known that the bivalent carbon in *p*-tolylcarbene is capable of migration to give styrene and benzocyclobutene. Good evidence has been found for the occurrence of a similar process in icosahedral carbaboranes;[66] it is interesting to note that other analogies between benzene and carbaboranes have been made in the past.

[62] H. Tomioka, *J. Amer. Chem. Soc.*, 1979, **101**, 256.
[63] H. Tomioka, G. W. Griffin, and K. Nishiyama, *J. Amer. Chem. Soc.*, 1979, **101**, 6009.
[64] W. D. Crow and H. McNab, *Austral. J. Chem.*, 1979, **32**, 89, 99, 123.
[65] W. D. Crow and H. McNab, *Austral. J. Chem.*, 1979, **32**, 111.
[66] S. Chari, G. K. Agopian, and M. Jones, Jr., *J. Amer. Chem. Soc.*, 1979, **101**, 6125.

Further progress towards understanding the role of copper in catalysing the decomposition of diazodiphenylmethane has been made.[67] It seems that copper(II) is involved in a chain reaction with radical cations as carriers whereas copper(I) reacts *via* unknown complexes.

The mechanism of the reactions between carbenes and metal carbonyls to give ketens has remained obscure for many years. It now seems that this occurs not by addition of carbon monoxide to a metal–carbene complex but rather by addition of carbene across the metal–CO bond.[68] More evidence has been found to support the idea that olefin metathesis reactions proceed *via* metal–carbenoid complexes.[69]

The specific formation of vinylcyclopropanes (58) can be achieved under mild conditions from diazo-ketones (57) by intramolecular carbenoid insertion reactions; good yields are obtained only for small rings ($n = 2, 3,$ or 4).[70] Thermolysis of compounds (58) leads to *cis*-fused cyclopentenes.

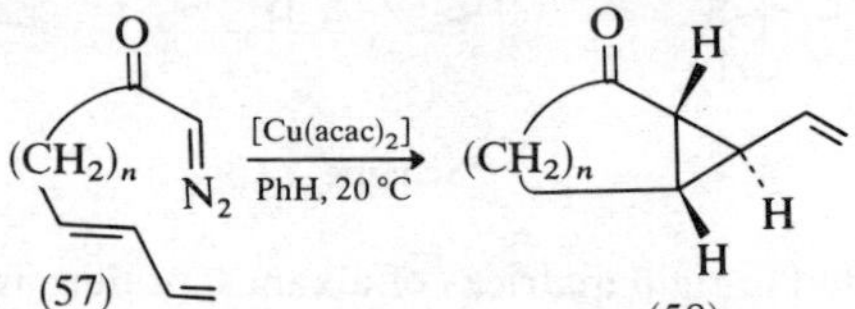

(57) (58)

Surprisingly, the effects of conformation on the Wolff rearrangement of α-diazo-ketones to ketens have not until recently been investigated. It has now been shown[71] that under aprotic, thermal, or photochemical conditions, an *S-Z* conformation with the migrating group *trans* to the nitrogen is necessary for keten formation, suggesting that the Wolff rearrangement is a concerted reaction rather than one involving ketocarbenes (Scheme 14). Set against this conclusion are some further studies[72] on the photo-Wolff reaction of α-diazo-ketones, using ^{13}C labelling, which indicate the occurrence of a carbene–carbene rearrangement *via* an oxeten intermediate, in line with previous work.

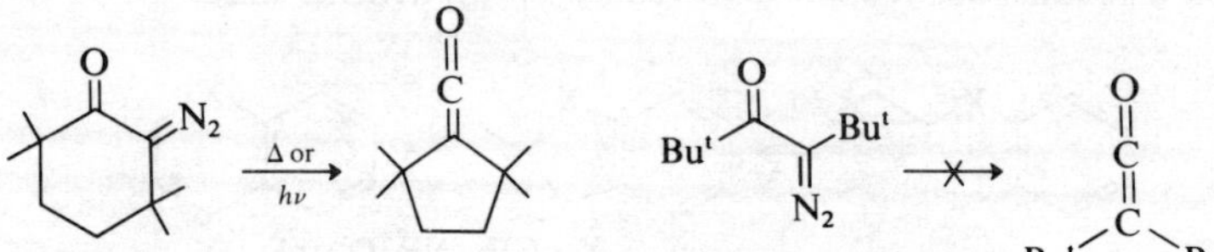

Scheme 14

It has long been known that ketocarbenes (59; R = H) react by insertion into the *endo*-C—H bond at C-5. When this position is blocked by an acetate group, *i.e.* (59;

[67] D. Bethell, K. L. Handoo, S. A. Fairhurst, and L. H. Sutcliffe, *J.C.S. Perkin II*, 1979, 707, D. Bethell, M. F. Eeles, and K. L. Handoo, *ibid.*, p. 714.
[68] W. A. Herrmann, J. Plank, M. L. Ziegler, and K. Weidenhammer, *J. Amer. Chem. Soc.*, 1979, **101**, 3133.
[69] R. H. Grubbs and C. R. Hoppin, *J. Amer. Chem. Soc.*, 1979, **101**, 1499; F. N. Tebbe, G. W. Parshall, and D. W. Ovenall, *ibid.*, p. 5074; see also C. O'Donohoe, J. K. A. Clarke, and J. J. Rooney, *J.C.S. Chem. Comm.*, 1979, 648.
[70] T. Hudlicky, J. P. Sheth, V. Gee, and D. Barnvos, *Tetrahedron Letters*, 1979, 4889.
[71] F. Kaplan and M. L. Mitchell, *Tetrahedron Letters*, 1979, 759.
[72] K.-P. Zeller, *Chem. Ber.*, 1979, **112**, 678; *Annalen*, 1979, 2036; *Monatsh.*, 1979, **110**, 393.

R = OAc), insertion into the C—R bond is replaced by reaction with the carbonyl group of the acetate followed by rearrangement to (60).[73] When R is OMe, the carbene inserts into a C—H bond of the methoxy-group to give (61).

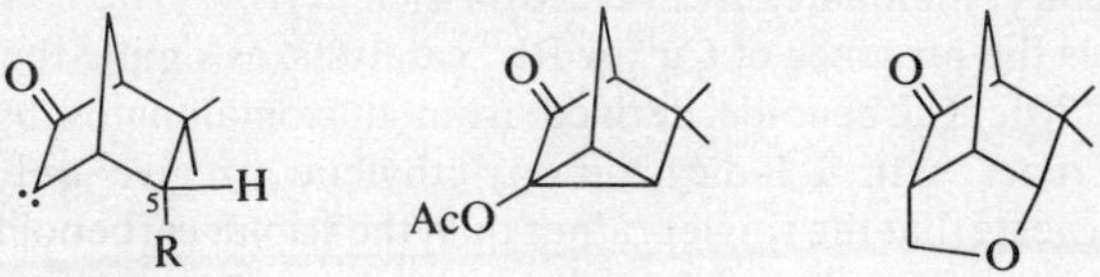

(59) (60) (61)

α-Iminocarbene (62), when photochemically generated in methanol, largely undergoes rearrangement to the ketenimine (63) before being trapped by the solvent (Scheme 15).[74] The intermediate (63) can be trapped by other reagents; for example, with benzylideneanilide, when the β-lactam imide (64) can be isolated.

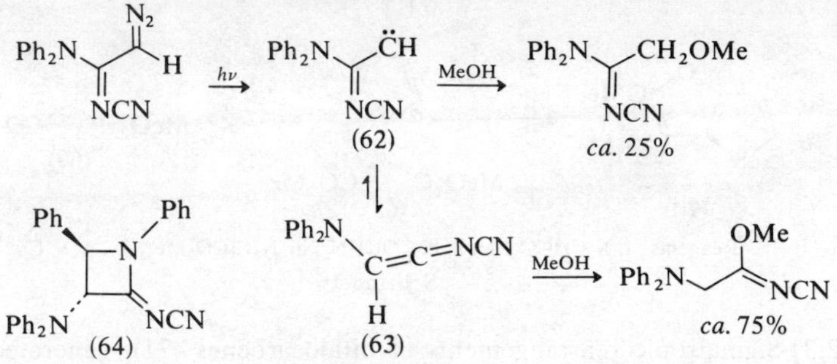

Scheme 15

The reaction of carbene (65) with alcohols leads to a number of products, the most notable of which is the methoxysilane (66); its formation may involve a silene intermediate.[75]

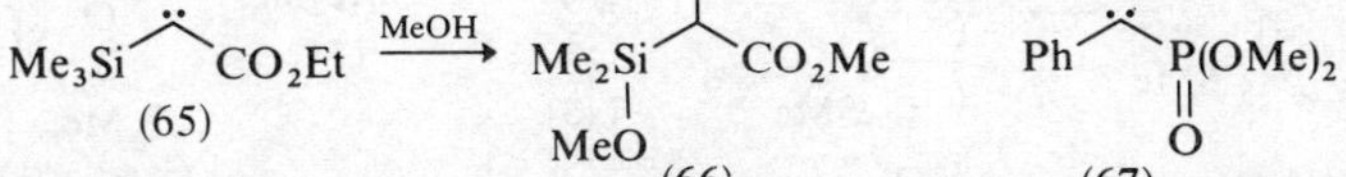

(65) (66) (67)

Photolytically generated phosphorylcarbenes (67) insert into the hydroxy-group of simple alcohols at 27 °C but into the C—H bonds at −196 °C As the sensitized photolytic generation of (67) at 27 °C leads to an increase in products of C—H insertion, these results may be rationalized by assuming that O—H insertion products arise *via* the singlet carbene (*cf.* ref. 18) whereas C—H insertion products involve the intermediacy of the triplet carbene. The triplet carbene probably reacts by proton abstraction followed by combination of the resulting radicals.[76]

73 P. Yates and G. F. Hambly, *Canad. J. Chem.*, 1979, **57**, 1668.
74 B. Arnold and M. Regitz, *Angew. Chem. Internat. Edn.*, 1979, **18**, 320.
75 W. Ando, A. Sekiguchi, T. Hagiwara, T. Migita, V. Chowdhry, F. H. Westheimer, S. L. Kammula, M. Green, and M. Jones, Jr., *J. Amer. Chem. Soc.*, 1979, **101**, 6393.
76 H. Tomioka, T. Inagaki, S. Nakamura, and Y. Izawa, *J.C.S. Perkin I*, 1979, 130.

Further reports have appeared on the reactions of thiophen with various diazo-compounds in the presence of rhodium(II) (Scheme 16).[77] The remarkably stable ylide (68) serves as a good source of bis-methoxycarbonyl-carbene (69), which can be used to prepare aryl malonates from electron-rich aryls.[78] The conversion of (68) into (69) requires the presence of Cu^{II} or Rh^{II} catalysts, as simple thermolysis of the ylide leads to (70). Carbenoids derived from diazomalonates by Cu^{II}-induced decomposition react with 1,1-dicyclopropylethylene to give only the addition product; this suggests that the singlet rather than the triplet carbenoid is involved, as the latter would be expected to give rise to rearrangement products.[79]

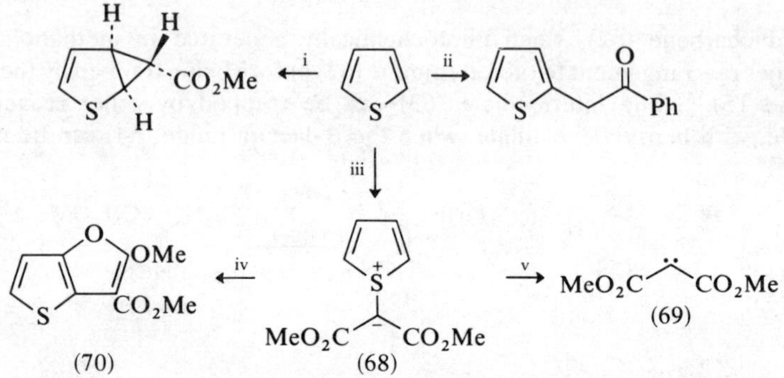

Reagents: i, N_2CHCO_2Me; ii, $PhCOCHN_2$; iii, $N_2C(CO_2Me)_2$; iv, Δ; v, Cu^{II}

Scheme 16

[2,3]-Sigmatropic rearrangements of dithiocarbenes (71), generated from the corresponding tosylhydrazone by base-induced elimination, give only the *trans* product (72), probably due to the intermediate adopting a 'folded envelope' conformation in which the substituent R occupies a pseudo-equatorial position.[80]

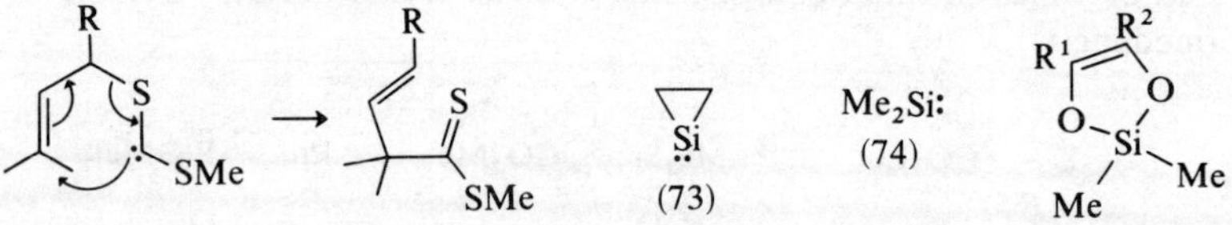

(71) (72) (75)

Theoretical calculations indicate that, in contrast to cyclopropylidene, which is considerably *less* stable than its neutral, acyclic isomers (*i.e.* allene, propyne), silacyclopropylidene (73) is much *more* stable than 2-sila-allene and 2-silapropyne.[81] Also, a singlet ground state is predicted for (73), again in contrast to cyclopropylidene, which has a triplet ground state. The elusive dimethylsilylene (74) has been detected by u.v. and i.r. spectroscopy in hydrocarbon glasses at −196 °C; the

77 R. J. Gillespie and A. E. A. Porter, *J.C.S. Perkin I*, 1979, 2624; R. J. Gillespie, J. Murray-Rust, P. Murray-Rust, and A. E. A. Porter, *J.C.S. Chem. Comm.*, 1979, 366.
78 R. J. Gillespie and A. E. A. Porter, *J.C.S. Chem. Comm.*, 1979, 50.
79 M. E. Alonso and M. Gómez. *Tetrahedron Letters*, 1979, 2763.
80 T. Nakai and K. Mikami, *Chem. Letters*, 1979, 1081.
81 J.-C. Barthelat, G. Trinquier, and G. Bertrand, *J. Amer. Chem. Soc.*, 1979, **101**, 3785.

species appears to be stable under these conditions, and can be trapped by various reagents to give *ca.* 60% yields of the expected products.[82] Dimethylsilylene (74) can also be generated by photolysis of $(Me_2Si)_6$, and it adds to α-diketones to give silacyclopent-4-enes such as (75).[83] Further support for the existence of silylenes is the isolation of (77) from the co-thermolysis of (76) with anthracene; presumably the silylenes dimerize to disilenes prior to reaction with the anthracene. If the latter is replaced by 2,3-dimethylbuta-1,3-diene, then the 1-silacyclopentene (79) is isolated, the silylene probably being trapped as the silacyclopropane (78) before it can dimerize.[84]

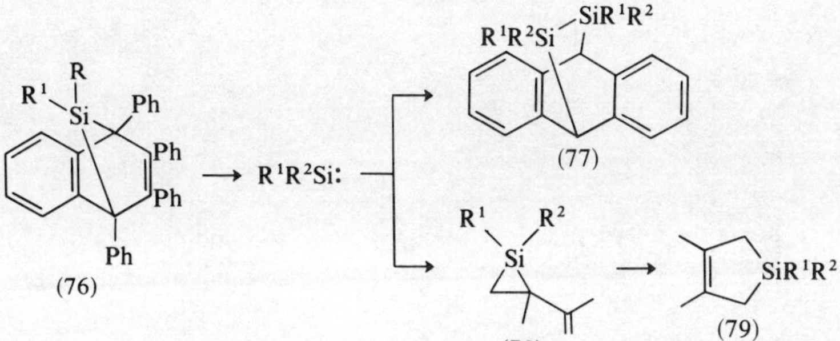

(76) (77) (78) (79)

82 T. J. Drahnak, J. Michl, and R. West, *J. Amer. Chem. Soc.*, 1979, **101**, 5427.
83 W. Ando and M. Ikeno, *J.C.S. Chem. Comm.*, 1979, 655.
84 Y. Nakadaira, T. Kobayashi, T. Otsuka, and H. Sakurai, *J. Amer. Chem. Soc.*, 1979, **101**, 486; H. Sakurai, Y. Nakadaira, and T. Kobayashi, *ibid.*, p. 487.

6 Electro-organic Chemistry

By R. LINES
Raychem Limited, Faraday Road, Dorcan, Swindon SN3 5HH

1 Introduction

This review is organized, as in previous years, according to the nature of the species which is believed to participate in the initial electrode reaction. Of general interest is a report by a Swedish group[1] on the advantages and use of a capillary-gap cell for large laboratory-scale organic electrosyntheses.

2 Anodic Processes

The Anodic Oxidation of Carboxylates.—A procedure for the conversion of α-(phenylthio)carboxylic acids into the corresponding aldehydes or acetals (Scheme 1) under mild conditions has been described by a Japanese group.[2]

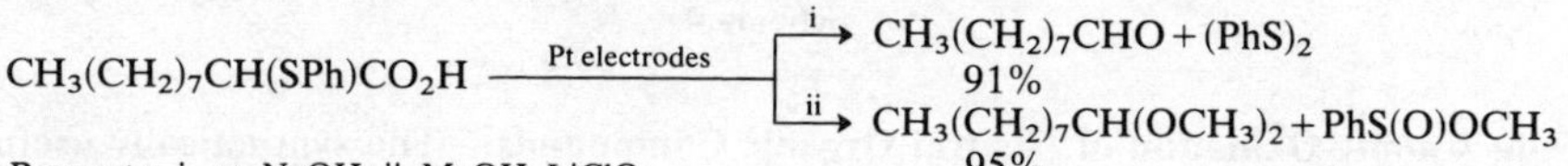

Reagents: i, aq. NaOH; ii, MeOH, $LiClO_4$

Scheme 1

The Kolbe reaction has also been used to advantage for the preparation of cyclic and semi-cyclic *N*-acyl *NO*-acetals.[3] The definition of semi-cyclic and cyclic *N*-acyl *NO*-acetals has been given by Nyberg, quoted in ref. 3. Such compounds are precursors of *N*-acyl-immonium ions, which in turn are key intermediates in the synthesis of some amino-acids and alkaloids. The starting materials are *N*-acyl-prolines or *N*-acyl-pipecolic acids, and these are smoothly converted into the acetals in high yield by anodic oxidation in alcoholic solution (Scheme 2).

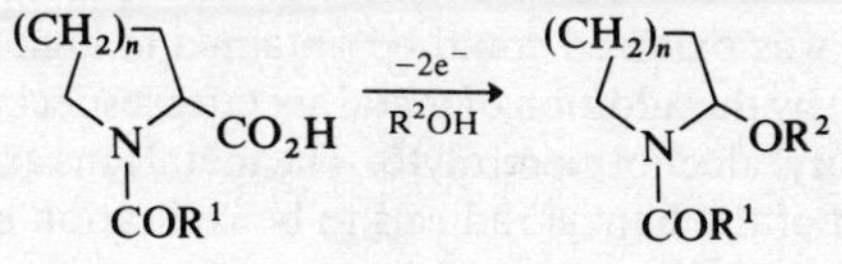

Scheme 2

1 L. Eberson, J. Hlavaty, L. Jönsson, K. Nyberg, R. Servin, H. Sternrup, and L.-G. Wistrand, *Acta Chem. Scand. (B)*, 1979, **33**, 133.
2 J. Nokami, M. Kawada, and R. Okawara, *Tetrahedron Letters*, 1979, 1045.
3 T. Iwasaki, H. Horikawa, K. Matsumoto, and M. Miyoshi, *J. Org. Chem.*, 1979, **44**, 1552.

The finding that malonic acid derivatives can function as ketone synthons has been applied by a Japanese group[4] to the synthesis of jasmones in 43% overall yield, starting from diethyl malonate (Scheme 3).

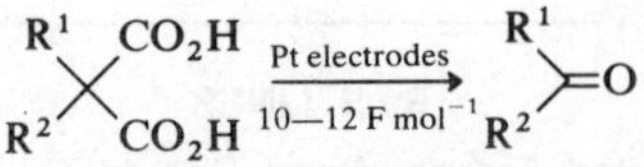

Scheme 3

The synthetic utility following the electrolytic decarboxylation of 1-alkyl-2-cycloalkene-1-carboxylic acids[5] is well demonstrated by the preparation of *dl*-muscone (1; $R^1 = Me$, $n = 12$), outlined in Scheme 4. The reaction involves a regiospecific acetoxylation at the α-position of the unsaturated acids, and this is used to advantage in the synthesis of *dl*-muscone by serving as a means of introducing a methyl group at the β-position of the cyclopentadecanone precursor.

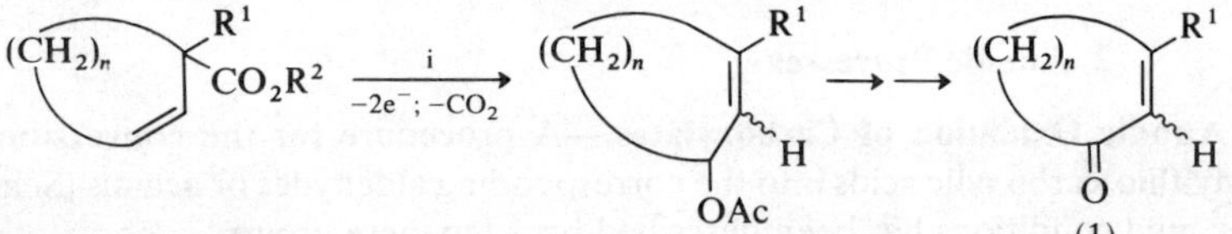

(1)

Reagents: i, HOAc, Bu^tOH

Scheme 4

The Anodic Oxidation of Neutral Organic Compounds.—The synthetically useful organic intermediates benzoquinone bis-acetals have been found to be readily prepared by the anodic oxidation of 2-substituted 1,4-dimethoxybenzenes in methanolic KOH solution.[6] Many of these compounds were also found to undergo selective monohydrolysis at the acetal remote from the 2-substituent to give a good yield of the equally useful mono-acetals. The anodic oxidation of methoxy-naphthalenes has also been studied in methanolic KOH solution[7] and found to be a convenient route to methoxylated naphthalenes and 1,2-naphthoquinone mono-acetals.

The electro-acetoxylation of 4-methylphenyl acetate to 4-acetoxybenzyl acetate has formed the basis of a synthesis of vanillin described by Torii and co-workers.[8] A detailed study of the electro-acetoxylation conditions revealed that the products in which the side-chain was oxidized could be obtained in high yield (90%) and with high selectivity (88%) by the addition of metal acetates, especially copper(II) acetate, to the acetic acid–t-butyl alcohol electrolyte. The metal ions are thought[8] to promote the smooth oxidation of the benzyl radicals to benzyl cation intermediates.

[4] J. Nokami, T. Yamamoto, M. Kawada, M. Izumi, N. Ochi, and R. Okawara, *Tetrahedron Letters*, 1979, 1047.
[5] S. Torii, T. Inokuchi, K. Mizuguchi, and M. Yamazaki, *J. Org. Chem.*, 1979, **44**, 2303.
[6] D. R. Henton, B. L. Chenard, and J. S. Swenton, *J.C.S. Chem. Comm.*, 1979, 326.
[7] M. G. Dolson, D. K. Jackson, and J. S. Swenton, *J.C.S. Chem. Comm.*, 1979, 327.
[8] S. Torii, H. Tanaka, T. Siroi, and M. Akada, *J. Org. Chem.*, 1979, **44**, 3305.

Methanolic NaCN appears to be an excellent medium for the anodic cyanation of pyrroles[9] and methyl-naphthalenes.[10] For a series of 1-alkyl-pyrroles,[9] substitution was found to take place preferentially at a vacant 2- or 5-position, while the 1,2,5-trialkyl-derivatives underwent cyanation in the side-chain to form dialkyl-pyrrole-2-acetonitriles. An interesting feature of this reaction is that, in contrast to the anodic cyanation of cyclohexene in the same reaction medium [see *Ann. Reports* (*B*), 1977, **74**, 155], no isocyanides or methoxylated products were observed.[9] Oxidation of some methyl-naphthalenes in methanolic NaCN resulted in a variety of naphthonitriles.[10] The orientation of attack was found to be directed by the substituent effect of the methyl groups rather than by any steric factors. In many cases the yields of products from the cyanation of pyrroles and methyl-naphthalenes are sufficient to make the methods preparatively attractive; *e.g.*, Scheme 5.

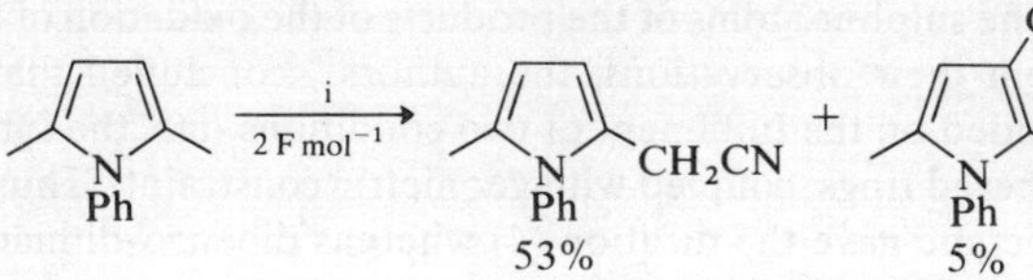

Reagents: i, MeOH, NaCN

Scheme 5

A novel anodic dimerization has been discovered[11] in which the enamine (2) undergoes coupling between the β-carbon atom and the 4-position of the phenyl ring (see Scheme 6). Until now, all of the reported oxidative dimerizations of enamines have occurred predominantly at the ethylenic carbon β to the nitrogen atom.

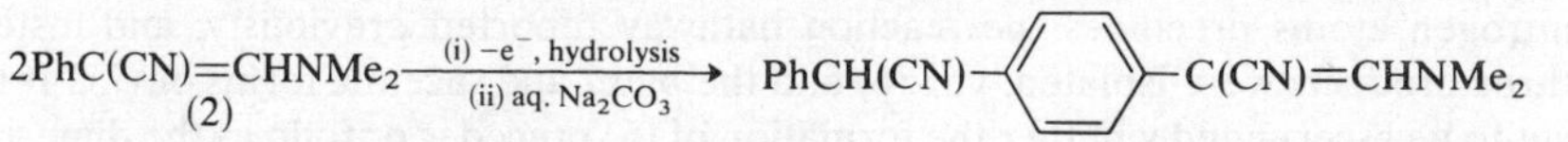

Scheme 6

Another novel observation, this time concerning a cleavage reaction, relates to the stereoselective oxidation of (*dl*)-4,4′-dihydroxyhydrobenzoin to 4-hydroxy-benzaldehyde at a mild steel anode.[12] Under the electrolysis conditions (aqueous NaOH, pH 13.5, 2—3 V *vs.* SCE), the *meso*-form was found to be completely electro-inactive. An explanation advanced by the author[12] is that a complex between the oxide coating on the steel anode and the hydroxy-groups of the glycol is a necessary prerequisite for charge transfer. For the *meso*-compound the phenyl rings prevent a favourable alignment of the hydroxy-groups, thus inhibiting oxidation.

Some preliminary findings have been published on the oxidation of a series of strained polycyclic hydrocarbons.[13] An excellent correlation was found between the $E_{\frac{1}{2}}$ potentials and the HOMO's of the compounds studied. For these compounds the HOMO's are associated with a carbon–carbon rather than a carbon–hydrogen bond

[9] K. Yoshida, *J. Amer. Chem. Soc.*, 1979, **101**, 2116.
[10] K. Yoshida and S. Nagase, *J. Amer. Chem. Soc.*, 1979, **101**, 4268.
[11] M. Masui, T. Michida, C. Ueda, and H. Ohmori, *Chem. and Ind.*, 1978, 922.
[12] J. H. Wagenknecht, *J. Electrochem. Soc.*, 1979, **126**, 983.
[13] P. G. Gassman and R. Yamaguchi, *J. Amer. Chem. Soc.*, 1979, **101**, 1308.

and indeed, on oxidation in pyridine–methanol, tricyclo[4.1.1.0.$0^{2,7}$]heptane, for example, gave predominantly the acetal (3).

A study has been made of the oxidation of a series of mono- and di-thioethers and of a number of mesocyclic dithioethers.[14] The latter compounds underwent facile,

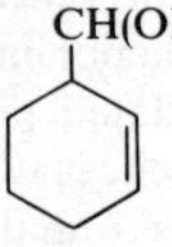

(3)

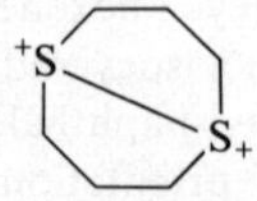

(4)

reversible oxidation, in contrast to the irreversible oxidation of the non-cyclic derivatives. The difference in behaviour was attributed[14] to transannular interactions between the sulphur atoms of the products of the oxidation of the mesocyclic compounds. From their observations, the authors[14] concluded that transannular interaction depended on the fulfilment of two conditions, *viz.* the formation of two five- or six-membered rings, coupled with geometric constraint. Thus, for example, 1,5-dithiacyclo-octane gave the dication (4) whereas dibenzo-dithiacyclododecane underwent irreversible oxidation.

3 Cathodic Processes

The Cathodic Reduction of Organic Cations.—Following last years report of the formation of a pyrrolodiazepine from the electrolysis of a dihydrodiazepine salt [see *Ann. Reports* (*B*), 1978, **75**, 139], the reduction of the *NN′*-disubstituted dihydrodiazepine salt (5) has been studied.[15] The presence of the substituents on the nitrogen atoms precludes the reaction pathway reported previously, and instead three products were isolated, viz. (6) and the *meso* and racemic forms of (7). It has yet to be ascertained whether the formation of (6) precedes or follows the dimerization of (5) to (7).

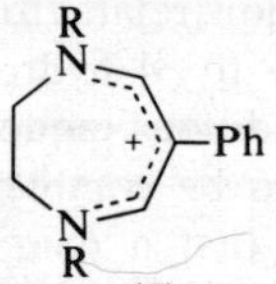

(5)

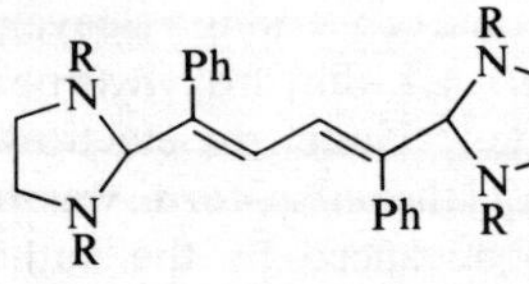

(6)

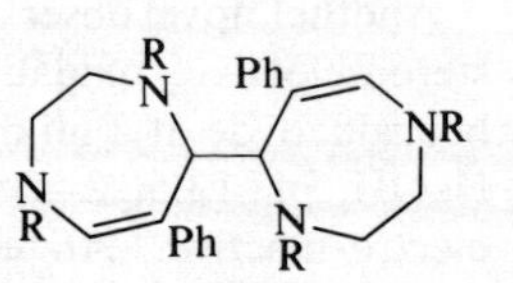

(7)

The Cathodic Reduction of Neutral Organic Compounds.—The cleavage of diaryl sulphones in aprotic solvents to the corresponding aryl sulphinate ion and aromatic hydrocarbon has, until now, been thought to proceed entirely *via* charged intermediates, *i.e.* anion radicals and dianions. Evidence has now been presented to suggest that radical intermediates are involved.[16] Thus, reduction of (8) in DMF gave (9) by cleavage of a carbon–sulphur bond in the anion radical of (8), followed by an intramolecular radical substitution on the adjacent phenyl ring.

[14] G. S. Wilson, D. D. Swanson, J. T. Klug, R. S. Glass, M. D. Ryan, and W. K. Musker, *J. Amer. Chem. Soc.*, 1979, **101**, 1040.
[15] D. Lloyd, C. A. Vincent, and D. J. Walton, *Bull. Soc. chim. belges*, 1979, **88**, 113.
[16] J. Grimshaw and J. Trocha-Grimshaw, *J.C.S. Perkin I*, 1979, 799.

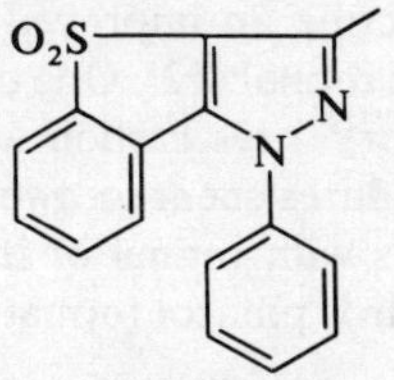

(8)

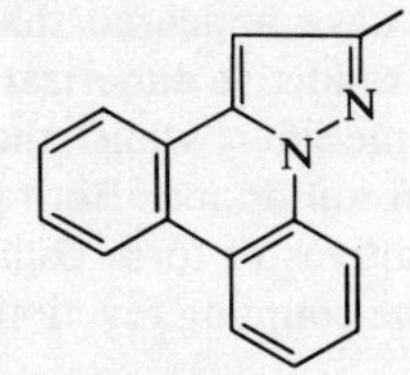

(9)

The electroreduction of a number of $\alpha\omega$-9,9′-dianthryl-ethanes $[\pi(CH_2)_n\pi]$ has been studied,[17] with a view to determining how the interactions between the aromatic centres affected the reversible electrode potentials. The results showed that strong interactions operated for the compounds with $n = 0$ and $n = 2$, but decreased as the length of the methylene chain increased. At $n = 6$ the two successive charge transfers (anion radical to dianion) merged into a single wave, with a peak width close to that of a reversible one-electron transfer, confirming the independence of the electro-active centres. In a separate study of bianthryl (*i.e.* $n = 0$),[18] fragmentation into anthracene and 9,10-dihydroanthracene was observed upon electrolysis at the potential of the second wave and above. This represents the first example of the direct electrochemical cleavage of a biaryl into the corresponding monomer.

A very useful example of the selective protection of the primary or secondary hydroxy-group in a diol has been presented by van der Stouwe and Schafer.[19] The authors found that tritylone ethers can be cathodically cleaved in high yield and under mild conditions to the corresponding alcohols. By taking advantage of the different reduction potentials of the tritylone and *p*-cyanobenzyl ether protecting groups, the selective protection of the hydroxy-groups of pentane-1,4-diol was demonstrated.

The electroreduction of some cyclic ethers, namely epoxides of structure (10), has also shown interesting behaviour.[20] The direct reduction of the epoxides in DMF at mercury afforded products (alchohols and hydrocarbons) by a two-electron pathway.

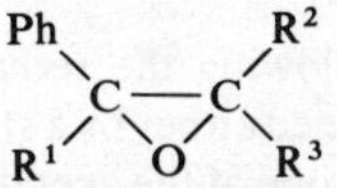

(10)

The same products were also obtained by indirect reduction using electrogenerated reducing agents, and autocatalysis was observed in some cases. Thus, for (10; $R^2 = Ph$), ring opening followed by elimination of OH^- generated *trans*-stilbene, which was then able to form a stable anion radical and thereby, once a suitable concentration of it had been attained, undergo homogeneous electron exchange with the substrate.

[17] K. Itaya and A. J. Bard, *Z. Phys. Chem.* (*Frankfurt*), 1978, **112**, 1.
[18] O. Hammerich and J.-M. Saveant, *J.C.S. Chem. Comm.*, 1979, 938.
[19] C. van der Stouwe and H. J. Schafer, *Tetrahedron Letters*, 1979, 2643.
[20] K. Boujlel and J. Simonet, *Electrochim. Acta*, 1979, **24**, 481.

Two papers[21,22] have appeared that describe an improved synthesis of retinal pinacol (11) by the reductive dimerization of retinal (12). One of the methods[22] uses $CrCl_3 \cdot 6H_2O$ as a modifier, whilst the other[21] uses carbon acids such as diethyl malonate as the proton donor. Both procedures seem to owe their success to the abilities of the additives to form complexes with retinal or its anion radical, and thereby to direct the coupling reaction towards pinacol formation.

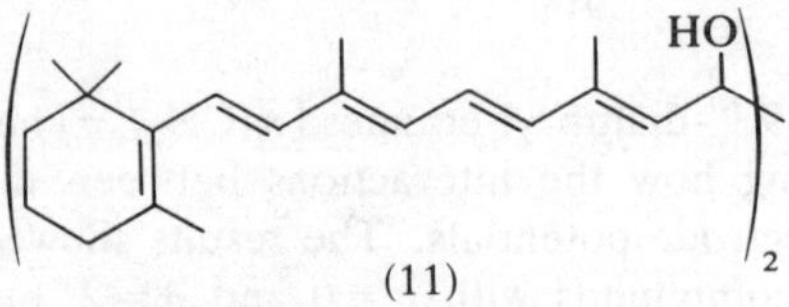

(11)

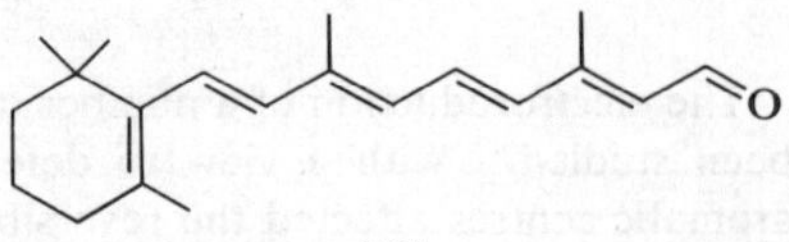

(12)

Retinal pinacol (11) provides an entry into the C_{40} carotenoids and, as such, an efficient method of electrochemically cleaving allylic hydroxy-groups would be of great synthetic value. The leaving ability of such a group is, however, rather low. It seems appropriate, then, that the methyl ester of abscisic acid (13) has been chosen as a model compound for the study of the conditions required for hydroxyl cleavage.[23] The hydroxy-group attached to C-6 is activated both by the enone and by the unsaturated side-chain ester. Contrary to expectations, it was found that, upon reduction, the bicyclic ester (14) was obtained in good yield. It appears that the hydroxy-group, by preventing conjugation throughout the molecule, allows selective transfer of electrons to the side-chain. The radical anion that is so formed isomerizes and undergoes rapid cyclization.

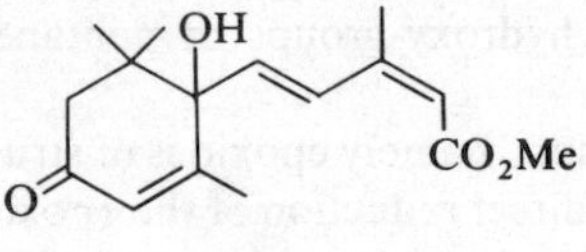

(13)

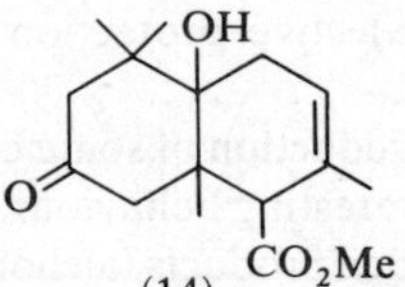

(14)

The formation of alkenes following the reductive elimination of acetate from vicinal diacetates can be rationalized either by a stepwise or a concerted process. The observation of hydrogenolysis of one of the acetate groups of diacetoxystilbene upon reduction in aqueous DMF[24] has provided good evidence for the stepwise mechanism. In contrast with dry DMF, where diphenylacetylene is the major product, the aqueous medium is sufficiently acidic that the intermediate carbanion is protonated. Reduction of 9,10-diacyloxy-phenanthrenes led to an unexpected acyl–oxygen fission.[24] Phenanthraquinone and the appropriate carboxylic acid were isolated.

Following a previous report of the formation of *NO*-dialkyl-hydroxylamines from the electrochemical reduction of nitro-compounds in the presence of alkyl halides

[21] L. A. Powell and R. M. Wightman, *J. Amer. Chem. Soc.*, 1979, **101**, 4412.
[22] D. W. Sopher and J. H. P. Utley, *J.C.S. Chem. Comm.*, 1979, 1087.
[23] B. Terem and J. H. P. Utley, *Electrochim. Acta*, 1979, **24**, 1081.
[24] C. Adams, N. M. Kamkar, and J. H. P. Utley, *J.C.S. Perkin II*, 1979, 1767.

[see *Ann. Reports* (*B*), 1977, **74**, 158], a description of the efficient synthesis of *NO*-diacetyl-hydroxylamines has been published.[25] The method consists of directly reducing aliphatic or aromatic nitro-(and nitroso-)compounds in the presence of acetic anhydride. Isolated yields range from 40 to 80%.

Another efficient transformation concerns the selective reduction of aliphatic alcohols to alkanes *via* the methanesulphonate esters.[26] The method appears effective even when the alcohol contains other functional groups, such as ester, alkene, nitrile, or epoxide.

A variety of 1,1-dichloro-alkenes have been synthesized,[27] in good yields, by the reduction of diethyl trichloromethylphosphonate and the appropriate carbonyl compounds (Scheme 7). The method has the advantage of avoiding the strongly basic and expensive alkyl-lithium reagents.

$$Cl_3CP(O)(OEt)_2 \xrightarrow[-1.2\ V\ vs.\ SCE]{Hg\ cathode} Cl_2\bar{C}P(O)(OEt)_2 \xrightarrow{R^1R^2CO} R^1R^2C{=}CCl_2$$

Scheme 7

The electroreduction of some *N*-ethyl-maleimides has led to the discovery of a hydrodimerization reaction in aqueous solution.[28] Normally, aprotic conditions are required for the formation of hydro-dimers from diactivated alkenes, where protonation of the intermediate anion radicals competes with their dimerization. In the present example, a radical–radical coupling mechanism is thought to operate.

Two papers have appeared on the synthesis of substituted barbaralanes.[29,30] The first paper[29] reports on the reduction of bridged 1,5-benzodiazepines (15) in acetonitrile to give substituted barbaralanes (16) in good yield. In the second paper,[30] substituted barbaralanes, *e.g.* (17), were obtained by the reduction of bicyclo[3.3.1]nonadienediones (18), in the presence of acetic anhydride, at the potential of the second wave. Reduction of the substrates at the first wave gave dimeric products only.

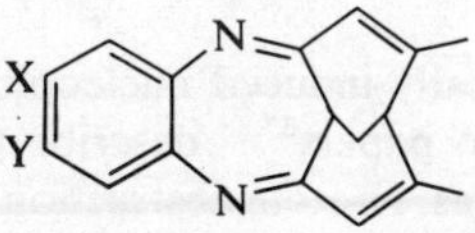

(15)

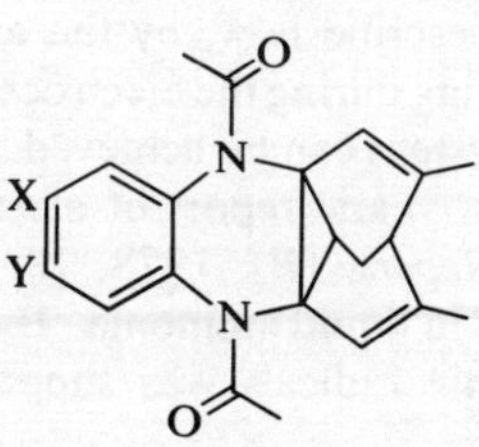

(16)

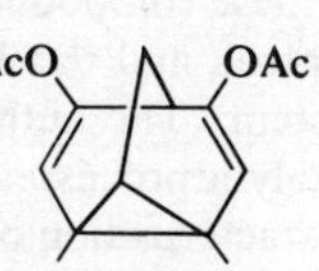

(17)

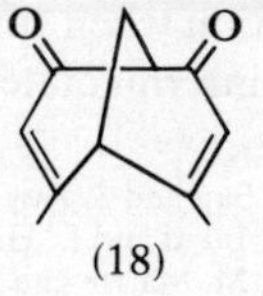

(18)

X, Y = H, Cl, or Me

[25] L. Christensen and P. E. Iversen, *Acta Chem. Scand.* (*B*), 1979, **33**, 352.
[26] T. Shono, Y. Matsumura, K. Tsubata, and Y. Sugihara, *Tetrahedron Letters*, 1979, 2157.
[27] F. Karrenbrock, H. J. Schafer, and I. Langer, *Tetrahedron Letters*, 1979, 2915.
[28] H. Zoutendam and P. T. Kissinger, *J. Org. Chem.*, 1979, **44**, 758.
[29] J. M. Mellor, B. S. Pons, and J. H. A. Stibbard, *J.C.S. Chem. Comm.*, 1979, 761.
[30] J. M. Mellor, B. S. Pons, and J. H. A. Stibbard, *J.C.S. Chem. Comm.*, 1979, 759.

Evidence has been presented for the reversible formation of the tetranegative ion of the aromatic hydrocarbon decacyclene.[31] Cyclic voltammetry of the hydrocarbon in dimethoxyethane at 0 °C gave four reversible one-electron waves, the fourth being located at −3.35 V (*v.s.* Ag|Ag$^+$). This ion represents the most reduced form of an aromatic hydrocarbon yet reported.

Both electrogenerated anions[32] and radical dianions[33] have been cited as electron-transfer reagents in solution. 1-Ethyl-4-methoxycarbonylpyridinium iodide[32] furnished the anion upon reduction, whilst the radical dianion was generated[33] from the conjugated phenol (19) in the presence of a strong base. In each case, a solution electron transfer was demonstrated by cyclic voltammetry for the reduction of a benzylic chloride.

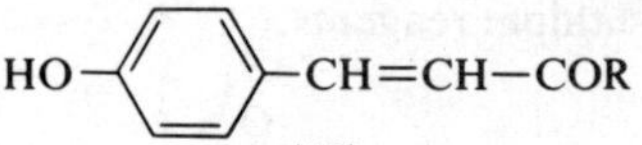

(19)

Fritz and Kornrumpf,[34] by the electrolysis of CF_2Br_2 at platinum electrodes in dichloromethane containing Bu_4NBr, are the first investigators to obtain the carbene CF_2 electrochemically. Current efficiencies of up to 60% were recorded for the trapping of the carbene by various alkenes.

Contrary to some statements that the reduction of allyl bromides and iodides proceeds in two distinct one-electron steps, compelling evidence has now been presented[35] to show that in fact the reduction follows a single two-electron pathway. Detailed electro-analytical studies of various allyl halides in acetonitrile revealed that peaks that had previously been attributed to the separate reduction of the allyl radical to the anion were the result of specific surface interactions of the allyl halides with the electrode materials. The reductions can thus be formalized as an e_1ce_2 process, where $E^\circ_2 < E^\circ_1$.

The reduction of organic halides, this time of benzene and pyridine, is also the subject of a comprehensive paper by French researchers.[36] Initial electron transfer to these compounds is slow, thus placing the electrochemical reaction under kinetic control, and this limits the amount of information that can be obtained from the system. The authors[36] describe how, by the analysis of the homogeneous redox catalytic processes occurring during the electrochemical reductions, a more complete characterization of the system can be achieved.

Following on from last years' report of electrochemically induced nucleophilic substitution [see *Ann. Reports* (*B*), 1978, **75**, 141], two papers[37,38] describe the reaction when conducted in liquid ammonia. By this means, H-atom abstraction by the intermediate aromatic radicals was suppressed, so that attention could be

[31] T. Saji and S. Aoyagui, *J. Electroanalyt. Chem. Interfacial Electrochem.*, 1979, **102**, 139.
[32] H. Lund and L. H. Kristensen, *Acta Chem. Scand.* (*B*), 1979, **33**, 495.
[33] A. M. Martre and J. Simonet, *J. Electroanalyt. Chem. Interfacial Electrochem.*, 1979, **97**, 287.
[34] H. P. Fritz and W. Kornrumpf, *J. Electroanalyt. Chem. Interfacial Electrochem.*, 1979, **100**, 217.
[35] A. J. Bard and A. Merz, *J. Amer. Chem. Soc.*, 1979, **101**, 2959.
[36] C. P. Andrieux, C. Blooman, J. M. Dumas-Bouchiat, and J.-M. Saveant, *J. Amer. Chem. Soc.*, 1979, **101**, 3431.
[37] C. Amatore, J. Chaussard, J. Pinson, J.-M. Saveant, and A. Thiebault, *J. Amer. Chem. Soc.*, 1979, **101**, 6012.
[38] C. Amatore and J.-M. Saveant, *J. Electroanalyt. Chem. Interfacial Electrochem.*, 1979, **103**, 303.

focussed on the other major competing process to substitution, *viz.* electron transfer to the radicals. A detailed analysis of the kinetics of the competitive electron-transfer process, using cyclic voltammetry, gave good evidence for the correctness of the $S_{RN}1$ mechanism (Scheme 8).

$$\begin{array}{rcll} ArX + e^- & \rightleftharpoons & ArX^{\cdot-} & (E_1^\circ = \text{electrolysis potential}) \\ ArX^{\cdot-} & \rightarrow & Ar^{\cdot} + X^- & \\ Ar^{\cdot} + Y^- & \rightarrow & ArY^{\cdot-} & \\ ArY^{\cdot-} - e^- & \rightleftharpoons & ArY & (E_2^\circ, \text{ if } E_1^\circ > E_2^\circ) \end{array}$$

Scheme 8

4 Miscellaneous

Interest continues to be shown in modifying electrode surfaces by applying polymeric films. Two communications have appeared on the electrochemical polymerization of pyrroles.[39,40] Oxidation of pyrrole[39] in acetonitrile containing Et_4NBF_4 gave a tough film, of conductivity $100\ \Omega^{-1}\ cm^{-1}$. Elemental analysis suggested a structure containing polymerized pyrrole units and BF_4^- ions in the ratio 4:1. The polymer could be modified further by, *e.g.*, nitration. Polymer films of *N*-methylpyrrole[40] exhibited much lower conductivities ($10^{-3}\ \Omega^{-1}\ cm^{-1}$) and thus, by co-electrolysis of solutions of pyrrole and *N*-methylpyrrole, films of intermediate conductivities could be obtained. The potentialities of using the charge-transfer complex TTF/TCNQ (tetrathiafulvalene/tetracyanoquinodimethane) as an electrode material in aqueous solution have been explored by Jaeger and Bard.[41] The complex was found to be stable over a potential span of 0.9 V, and the oxidation and reduction of redox couples, *e.g.* Cu^{II}/Cu^{I}, could be carried out. Beyond the potential limits, insoluble compounds of TTF or TCNQ with the ions of the supporting electrolyte were formed.

A further application of the poly(4-nitrostyrene)-coated electrode referred to in last years' report [see *Ann. Reports* (*B*), 1978, **75**, 145] has been found for the reduction of 1,2-dibromo-1,2-diphenylethane to *trans*-stilbene.[42] Like the reduction of oxygen on this electrode, electron transfer to the dibromide *via* the charged polymer was found to take place 0.4 V less cathodically than at a clean platinum surface. The electrode was also found to be sufficiently stable for preparative-scale electrolyses. Catalyst turnover was of the order of 10^4.

Chemically modified electrodes, as well as being able to catalyse charge transfers, can also alter isomer ratios, a case in point being the selective chlorination of anisole at a cyclodextrin-modified electrode.[43] At a graphite anode the chlorination of anisole in aqueous NaCl gave a *para* : *ortho* ratio of about 5:1. Chemically bound α-cyclodextrin on the graphite changed the ratio to 18:1, while adsorbed cyclodextrin on the electrode surface increased the ratio to >25:1. In these cases it is thought[43] that the *ortho*-positions of anisole are blocked by the cyclodextrin structure, leading to preferential *para*-chlorination.

39 A. F. Diaz and K. K. Kanazawa, *J.C.S. Chem. Comm.*, 1979, 635.
40 K. K. Kanazawa, A. F. Diaz, R. H. Geiss, W. D. Gill, J. F. Kwak, J. A. Logan, J. F. Rabolt, and G. B. Street, *J.C.S. Chem. Comm.*, 1979, 854.
41 C. D. Jaeger and A. J. Bard, *J. Amer. Chem. Soc.*, 1979, **101**, 1690.
42 J. B. Kerr and L. K. Miller, *J. Electroanalyt. Chem. Interfacial Electrochem.*, 1979, **101**, 263.
43 T. Matsue, M. Fujihira, and T. Osa, *J. Electrochem. Soc.*, 1979, **126**, 500.

Turning to more conventional electrolytic procedures, the almost quantitative conversion of 3-substituted sydnones into their 4-chloro- or 4-bromo-derivatives has been discovered.[44] The success of the method appears to lie with the mild conditions of the electrolysis (MeOH, NaBr or Me_4NCl), which minimizes any tendency for ring opening. Attempts at fluorination and iodination of 3-substituted sydnones by this method were unsuccessful.

A preparatively significant synthesis of esters and ketones from primary and secondary alcohols has been accomplished by the use of iodide ions as catalytic current carriers.[45] Typically, electrolysis of an aqueous solution of the alcohol and KI (4:1 ratio) yields the ketone or ester in up to 84% yield. Although the authors[45] suggest the iodonium ion as the key intermediate (Scheme 9), it seems unlikely that this species would be generated in an aqueous environment and at the low anode potential (0.6—0.8 V *vs.* SCE).

$$2RCH_2OH \xrightarrow[I^-]{0.8\text{ V } vs.\text{ SCE}} RC(=O)CH_2R$$

Scheme 9

Liquid SO_2, as an electrochemical solvent, has been re-examined by two groups.[46,47] Liquid SO_2 has many advantages from an electrochemical viewpoint, *e.g.* high dielectric constant (*ca.* 20), low nucleophilicity, and high solvating ability for covalent compounds. Contamination by water can be a problem, however, and the solubility of most inorganic salts in the solvent is low. Anhydrous liquid SO_2 can be obtained by vacuum-line techniques[46] or by using $AlCl_3$ as a supporting electrolyte.[47] In the latter case the solutions are oxidizing, and the cation radicals of a number of aromatic hydrocarbons can be obtained directly. Further electrochemical oxidation to the dications is also possible. In contrast, anhydrous liquid SO_2 containing $Bu_4N^+ClO_4^-$ was found to have a very wide anodic range[46] (*ca.* 3.4 V *vs.* SCE), and the dications of a variety of aromatic compounds could easily be generated.

Halide ions, particularly bromide (as sodium bromide), have been found to be excellent promoters of the electrolytic cross-coupling of disulphides with imides[48] and with dialkyl (or diaryl) phosphites[49] (Scheme 10). The mechanism is thought to involve electrogenerated bromine, which reacts with either the imide or the phosphite to give an intermediate (*N*-bromo-imide or phosphorobromide) that is capable of reacting with the disulphides and regenerating bromide ions.

$$2R^1(CO)_2NH \xrightarrow[2\text{ F mol}^{-1}]{i} 2R^1(CO)_2NSR + H_2$$

$$2(R^1O)_2P(O)H \xrightarrow[2\text{ F mol}^{-1}]{i} 2(R^1O)_2P(O)SR + H_2$$

Reagents: i, RSSR, NaBr

Scheme 10

[44] H.-J. Tien, T. Nonaka, and T. Sekine, *Chem. Letters*, 1979, 283.
[45] T. Shono, Y. Matsumura, J. Hayashi, and M. Mizoguchi, *Tetrahedron Letters*, 1979, 165.
[46] L. A. Tinker and A. J. Bard, *J. Amer. Chem. Soc.*, 1979, **101**, 2316.
[47] P.-C. Lacaze, J.-E. Dubois, and M. Delamar, *J. Electroanalyt. Chem. Interfacial Electrochem.*, 1979, **102**, 135.
[48] S. Torii, H. Tanaka, and M. Ukida, *J. Org. Chem.*, 1979, **44**, 1554.
[49] S. Torii, H. Tanaka, and N. Sayo, *J. Org. Chem.*, 1979, **44**, 2938

Finally, an unusual mixed-valence ferrocene compound (20) has been synthesized by the reaction of ferrocenyl-lithium with boron trifluoride.[50] Structure determinations revealed that the complex contains one sterically and one electronically distinct ferrocene group; these are not coincident. The electronic absorption spectrum showed bands characteristic of inter-valence charge-transfer transitions, and these were attributed[50] to a through-space rather than to a through-bond mechanism. Cyclic voltammetry of the complex in dichloromethane gave one reversible reduction wave and three reversible one-electron oxidation waves (Scheme 11).

$$[(Fc)B\{(Fc)_3^-\}] \xrightleftharpoons{+e^-} [(Fc^+)B\{(Fc)_3^-\}] \xrightleftharpoons{-e^-} [(Fc^+)B(Fc)_3] \xrightleftharpoons{-e^-} [(20)]^{2+} \xrightleftharpoons{-e^-} [(20)]^{3+}$$

(20)

$$Fc = (\pi\text{-}C_{10}H_9)_2Fe$$

Scheme 11

[50] D. O. Cowan, P. Shu, F. L. Hedberg, M. Rossi, and T. J. Kistenmacher, *J. Amer. Chem. Soc.*, 1979, **101**, 1304.

7 Photochemistry

By A. COX

Department of Chemistry and Molecular Sciences, University of Warwick, Coventry CV4 7AL

A review has appeared[1] surveying some of the more complex adiabatic photoreactions of organic molecules such as fragmentations, electrocyclic rearrangements, and geometrical isomerizations.

The effect of pulsed infrared radiation on *cis*-3,4-dichlorocyclobutene has been examined.[2] Despite the fact that the experimental conditions used were probably sufficient to raise the substrate to vibrational levels considerably in excess of the normal lowest reaction channel, *cis,trans*-1,4-dichlorobuta-1,3-diene was the only product detected, clearly showing an adherence to the Woodward–Hoffmann Rules for this thermally-allowed transformation. Further investigations into the field of cyclopropene photochemistry have been discussed,[3] with particular reference to 1-methyl-2,3-diphenyl-3-isobutenylcyclopropene and 1,2-diphenyl-3-methyl-3-isobutenylcyclopropene. The results are interpreted in terms of the participation of a biradical rather than a carbene pathway in both the direct and triplet reactions. Interconversions of different vinylcyclopropenes are seen as an incipient di-π-methane rearrangement involving a biradical mechanism. It has been shown[4] that, in contrast to its thermal processes, the photoreactions of bicyclo[4.2.0]octa-2,4,7-trienes are substituent-dependent. Thus, irradiation of the 2,5-dimethyl-3,4-diphenyl-substituted compound leads to a mixture of aromatic and acetylenic products whose formation can be rationalized in terms of thermal equilibria and further photoprocesses. The corresponding tetraphenyl-substituted triene, however, yields only the 1,2,3,4-tetraphenylcyclo-octatetraene.

A good deal of interest is still being shown in the di-π-methane rearrangement. An investigation of some 5-methylenenorborn-2-enes has revealed[5] that, in common with many other rigid 1,4-diene systems, di-π-methane rearrangements can occur from both singlet and triplet excited states. The efficiency of the triplet reaction is highly substituent-dependent and it is demonstrated that the structural constraints of the product control the regiospecificity. Direct irradiation of (−)-3-(*p*-methoxyphenyl)-3-methyl-1,1-diphenylpent-1-ene has been found[6] to lead to a mixture of (−)-*cis*- and (+)-*trans*-2-ethyl-2-methyl-3,3-diphenyl-1-(*p*-methoxyphenyl)cyclopropane, in which the methane carbon has undergone an inversion

[1] N. J. Turro, J. McVey, V. Ramamurthy, and P. Lechtken, *Angew. Chem. Internat. Edn.*, 1979, **18**, 572.
[2] W. C. Danen, D. F. Koster, and R. N. Zitter, *J. Amer. Chem. Soc.*, 1979, **101**, 4281.
[3] H. E. Zimmerman and M. C. Hovey, *J. Org. Chem.*, 1979, **44**, 2331.
[4] R. N. Warrener, I. W. McCay, R. Y. S. Tan, and R. A. Russell, *Tetrahedron Letters*, 1979, 3183.
[5] Z. Goldschmidt and M. Shefi, *J. Org. Chem.*, 1979, **44**, 1604.
[6] H. E. Zimmerman, T. P. Gannett, and G. E. Keck, *J. Org. Chem.*, 1979, **44**, 1982.

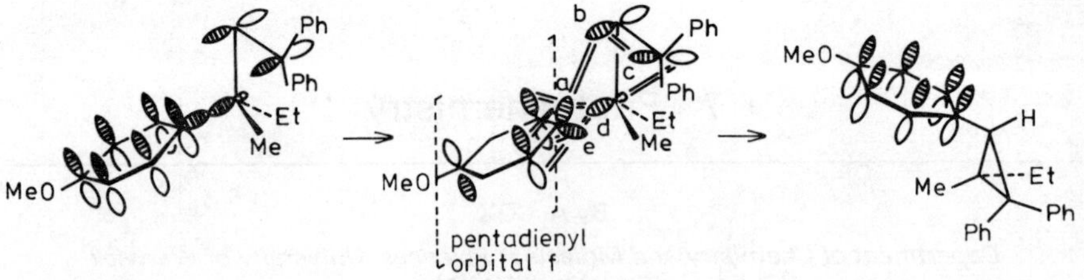

Scheme 1

(Scheme 1). This stereochemistry is in accordance with the predicted allowed stereochemistry arising from a Möbius transition state.

The photoreactions of a series of 5,6-benzo-2-azabicyclo[2.2.2]octa-7,8-dien-3-ones have been used[7] as a probe to study structural and electronic substituent effects on the di-π-methane rearrangement. A phenyl substituent is found to reverse the previously observed regioselectivity and a solvent effect is also apparent. The synthesis of a cyclopropane sesquiterpenoid, taylorione, has been reported,[8] using the di-π-methane rearrangement. This represents only the second example of the use of this reaction in natural product synthesis.

An investigation of the photochemistry of cross-conjugated trienes has shown[9] that whereas irradiation of the dicyano-compound (1) leads to a 'toluene' type of product, the corresponding tetraphenyl compound (2) reacts by a vinyl–vinyl bridging process (Scheme 2). Using SCF–CI calculations along various reaction co-ordinates, a theoretical analysis of the system has been made. This suggests that, in the excited state, phenyl-substitution on the *exo*-methylene group alters vinyl–vinyl bridging from a forbidden to an allowed process.

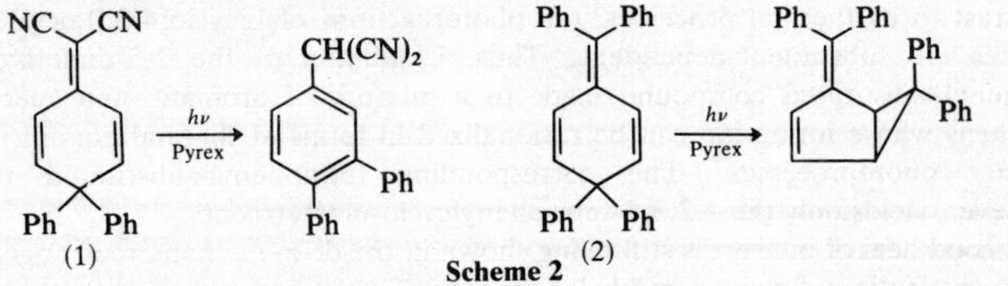

Scheme 2

The photochemistry of 2-halogeno-1,1-diphenylethylenes has been examined[10] and the product distribution found to depend on both the halogen and the solvent. Cleavage of the C–X bond may occur, to give a radical pair, and may be followed by electron transfer, to give an ion pair. This leads to competing ionic and radical behaviour. Products derived from a stabilized triplet state are also possible, and the proportions of the three pathways are dependent on the halogen. A report has appeared[11] of the first example of an olefin analogue of the Type II photoelimination

[7] M. Kuzuya, M. Ishikawa, T. Okuda, and H. Hart, *Tetrahedron Letters*, 1979, 523.
[8] G. Pattenden and D. Whybrow, *Tetrahedron Letters*, 1979, 1885.
[9] H. E. Zimmerman and D. R. Diehl, *J. Amer. Chem. Soc.*, 1979, **101**, 1841.
[10] B. Sket and M. Zupan, *J.C.S. Perkin I*, 1979, 752.
[11] H. Aoyama, T. Hasegawa, M. Okazaki, and Y. Omote, *J.C.S. Perkin I*, 1979, 263.

of ketones. Using a low-pressure mercury arc, photocyclization of *NN*-dibenzylmethacrylamide (3a) to the β-lactam (4) and photodealkylation of *NN*-di-isopropylmethacrylamide (3b) to *N*-isopropylisobutyramide both proceed *via* dimethylketen and the corresponding amine. The first step in both of these transformations is intramolecular hydrogen abstraction by the β-carbon (Scheme 3).

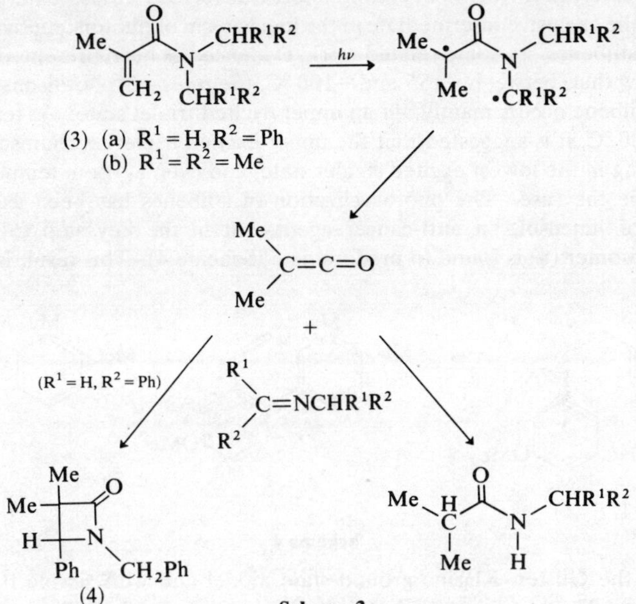

Scheme 3

Chemical evidence has now been presented[12] which adds further weight to an earlier report[13] that direct irradiation of *cis*-1-phenylcyclohexene affords the corresponding *trans*-isomer. At −75 °C and in methanol, irradiation of *cis*-1-phenylcyclohexene gives rise to a dimer (5) thought to arise by a Woodward–Hoffmann-allowed concerted $[\pi^4{}_s + \pi^2{}_s]$ addition of *cis*-1-phenylcyclohexene to the endocyclic *trans* double-bond of *trans*-1-phenylcyclohexene.

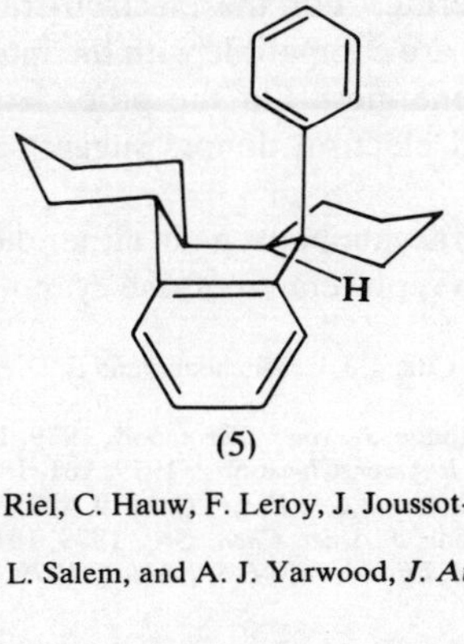

(5)

[12] W. G. Dauben, H. C. H. A. van Riel, C. Hauw, F. Leroy, J. Joussot-Dubien, and R. Bonneau, *J. Amer. Chem. Soc.*, 1979, **101**, 1901.
[13] R. Bonneau, J. Joussot-Dubien, L. Salem, and A. J. Yarwood, *J. Amer. Chem. Soc.*, 1976, **98**, 4329.

It has been observed[14] that the quantum yields of the direct *trans–cis* photoisomerization of some *p*-, *m*-, and *m,m'*-bromine-substituted stilbenes show larger quenching effects by azulene than do the *trans/cis*-photostationary compositions. Together with measurements of the temperature-dependence of the quantum yield of fluorescence of each *trans*-isomer, it is concluded that the positional dependence previously inferred for the heavy-atom effect is incorrect. It also emerges that $^1p^*$ cannot be the exclusive intermediate in the mechanism of photoisomerization of the *cis*-bromostilbenes. In some related work, evidence has been presented[15] supporting the view that, between −155 and −100 °C, the *trans–cis* photoisomerization of 4-bromostilbene occurs mainly *via* an upper excited triplet state. At temperatures above −100 °C, it is suggested that the upper excited triplet mechanism competes with twisting in the lowest excited singlet state, and even at room temperature this may still be the case. The photocyclization of stilbenes has been used[16] in the synthesis of juncusol, an anti-cancer agent, but in the key step (6) → (7), the unwanted isomer (8) is found to predominate (Scheme 4). This result is consistent

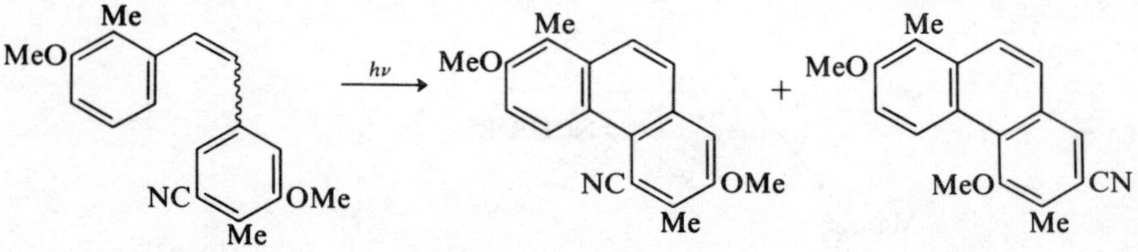

(6) (6:1, *E*:*Z*) (7) (8)

Scheme 4

with both the Güsten–Klasinc ground-state model and with simple frontier MO calculations. The *trans* ⇄ *cis* photointerconversion of 1,2-di-(2'-naphthyl)ethylenes and the photocyclization of the *cis*-isomer to 4a,4b-dihydrophenanthrene derivatives has been investigated[17] over the temperature range +25 to −180 °C. For the *cis*-isomers, the presence of conformer equilibria has been established spectroscopically, and the results also indicate the presence of potential barriers on the pathway between the excited-state starting materials and the products. CIDNP effects have been used[18] to examine the photosensitized photoisomerization of electron-donor and electron-acceptor styrenes, and two different mechanisms of isomerization seem to operate. For the electron-transfer-induced isomerization of donor olefins, the effects are compatible with the interconversion of two different radical ions, of different energies. On the other hand, the effects for acceptor styrenes with photoexcited electron donors suggest a single intermediate in the isomerization.

Irradiation of 1,2,4,5-tetracyanobenzene in either diethyl ether, tetrahydrofuran, or tetrahydropyran leads to replacement of one cyano-group by an ether residue.[19]

[14] J. Saltiel, A. Marinari, D. W.-L. Chang, J. C. Mitchener, and E. D. Megarity, *J. Amer. Chem. Soc.*, 1979, **101**, 2982.
[15] H. Görner and D. Schulte-Frohlinde, *J. Amer. Chem. Soc.*, 1979, **101**, 4388.
[16] A. S. Kende and D. P. Curran, *J. Amer. Chem. Soc.*, 1979, **101**, 1857.
[17] T. Wismonski-Knittel and E. Fischer, *J.C.S. Perkin II*, 1979, 449.
[18] H. D. Roth and M. L. M. Schilling, *J. Amer. Chem. Soc.*, 1979, **101**, 1898.
[19] M. Ohashi, K. Tsujimoto, and Y. Furukawa, *J.C.S. Perkin I*, 1979, 1147.

Three possible mechanisms are considered. One of these in which a tetrahydrofuran radical cation reacts with a neighbouring tetrahydrofuran molecule was definitely eliminated on kinetic grounds. However, no decision was possible between a path involving charge-transfer exciplexes and a zwitterionic mechanism. A novel photochemical aromatic amination has been reported[20] by irradiation of a solution of 9,10-dicyanoanthracene and an amine in acetonitrile. The initial step appears to be transfer of an electron from amine to anthracene *via* an encounter complex or exciplex. Water is also apparently necessary to hydrolyse the intermediate imine. Evidence has been presented[21] on the mechanism of the diene-induced photodechlorination of chloro-aromatics. The reaction involves both exciplex and triplex formation and also a protonation step, and is shown to proceed *via* a long-lived ion-radical intermediate. A study has been made[22] of the orbital topology in a photochemical carbanionic ring-opening, using the anion generated from *cis,trans*-2,3-diphenylcyclopropanecarbonitrile by treatment with lithium di-isopropylamide. The results show that the process is stereochemically non-random but fail to distinguish between disrotatory photochemical opening of an anion and thermal opening of a radical. Irradiation of naphthalene in trifluoroacetic acid has been observed[23] to lead to bimolecular hydrogen exchange *via* a singlet-state process. Consistent with the observed small kinetic isotope effect, the incoming hydrogen is found to be lost about one hundred times more readily than one originally present, and it is concluded that a π-complex is formed. The three-membered ether chain CH_2–O–CH_2 has been used[24] to hold two anthracene chromophores in close proximity while the photobehaviour of the entire system is studied at low temperature. The spectroscopic properties of a series of derivatives of bis-9-anthrylmethyl ethers and their photocyclodimerization reactions have been studied in this way.

A study has been made[25] of the photochemistry of a series of bicyclic $\beta\gamma$-cyclopropyl ketones. The behaviour of these ketones suggests a stepwise sequence of events involving opening of the three-membered ring by α-cleavage, followed by rearrangement of the initial cyclopropylcarbinyl radical under stereoelectronic control to a homoallylic radical. The photochemistry of $\beta\gamma$-cyclopropyl ketones derived from bicyclo[4.2.1]nona-2,4,7-trien-9-one and possessing structural features which, under irradiation, tend to promote decarbonylation has been examined.[26] Relative rates of decarbonylation, 1,3-acyl shifts, and α-cleavage are established; for the compounds chosen, this sequence is found to be in agreement with qualitative information in other systems.

Photoisomerization about a carbon–oxygen partial double bond has been observed[27] (Scheme 5). Irradiation of (9) in FSO_3H at $-70\,°C$, promotes partial isomerization to give a photostationary state comprising (10) (73%) and (11) (27%). It has been suggested that this same photoinduced stereomutation might occur with protonated carbonyl compounds, and that this could be a way for protonated

[20] M. Ohashi, H. Kudo, and S. Yamada, *J. Amer. Chem. Soc.*, 1979, **101**, 2201.
[21] W. K. Smothers, K. S. Schanze, and J. Saltiel, *J. Amer. Chem. Soc.*, 1979, **101**, 1895.
[22] M. A. Fox, *J. Amer. Chem. Soc.*, 1979, **101**, 4008.
[23] N. J. Bunce, Y. Kumer, and L. Ravanal, *J. Org. Chem.*, 1979, **44**, 2612.
[24] A. Castellan, J.-M. Lacoste, and H. Bouas-Laurent, *J.C.S. Perkin II*, 1979, 411.
[25] I. M. Takakis and W. C. Agosta, *J. Amer. Chem. Soc.*, 1979, **101**, 2383.
[26] I. M. Takakis and W. C. Agosta, *J. Org. Chem.*, 1979, **44**, 1294.
[27] R. F. Childs and M. E. Hagar, *J. Amer. Chem. Soc.*, 1979, **101**, 1052.

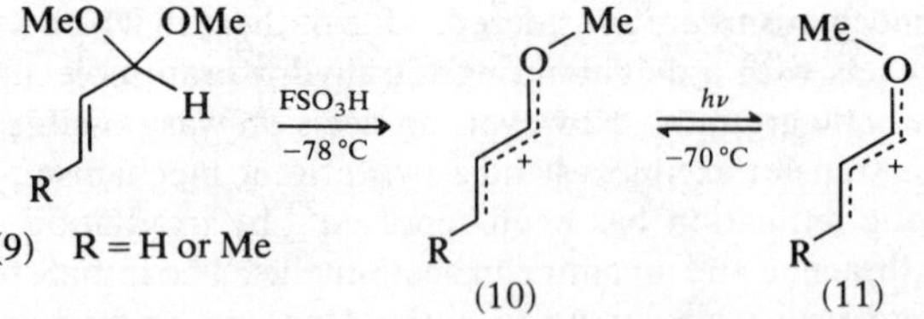

(9) R = H or Me (10) (11)

Scheme 5

carbonyls to relax to their ground state. In a related paper[28] the photoisomerization of protonated 4,4-dimethylcyclohex-2-enone in FSO_3H at −85 °C is reported. Incorporation of the chromophore into a ring system renders isomerization about the double bond more difficult, and protonated 6,6-dimethylbicyclo[3.1.0] hexan-2-one is produced. The effect of methyl substituents on the stereochemistry of the photoinduced addition of methanol to seven- and eight-membered cycloalk-2-enones has been studied.[29] It is concluded that the photoinduced addition of the nucleophile to *cis*-cyclohex-2-enones involves photoisomerization to the *trans*-isomer followed by *syn*-addition of the nucleophile. Proton transfer is probably the rate-determining step. *trans*-Fused $[2s+2a]$ addition products are formed[30] in the photocycloaddition of 17β-hydroxyandrosta-4,6-dien-3-one acetate to the electron-deficient olefin methyl acetate. This breakdown in stereospecificity of cycloaddition confirms a theoretical prediction[31] based on a calculation of the potential energy surface for $^3(\pi,\pi^*)$ cycloadditions which indicates that, if the enone and olefin are a good donor–acceptor pair, then non-Woodward–Hoffmann addition should occur. The regioselectivity of enone photodimerization appears to be profoundly influenced by irradiating in potassium dodecanoate micelles.[32] Thus 3-alkylpentenones are observed to give ratios of head-to-head to head-to-tail dimers of up to 98 : 2. The efficiency of the dimerization is also greatly increased.

Photolysis of 2-arenesulphonyloxy-cyclohex-2-enones leads[33] to 2-hydroxy-3-phenylcyclohex-2-enones in about 40% yield. The reaction, which probably proceeds *via* a triplet state, is not of the photo-Fries type, but may involve S–O cleavage as the primary photochemical step. However, a mechanism involving cyclization to a biradical is also a possibility. A new method for the preparation of bicyclo-[3.2.0]heptan-2-ones has appeared[34] and is especially interesting as, on thermal fragmentation, these compounds lead cleanly to cyclopentenones (Scheme 6). The

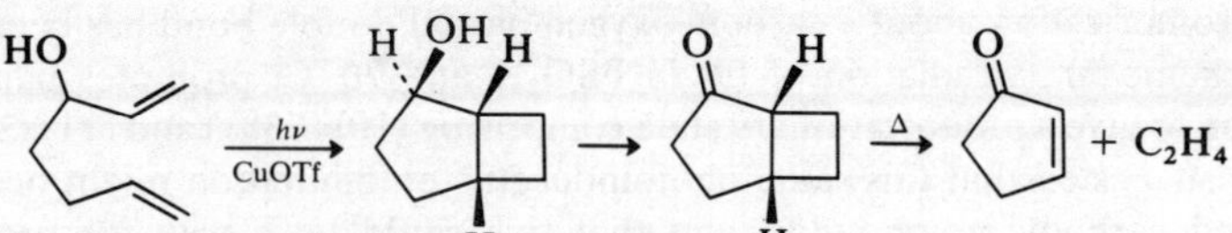

Scheme 6

[28] R. F. Childs, K. E. Hine, and F. A. Hung, *Canad. J. Chem.*, 1979, **57**, 1442.
[29] H. Hart, B. Chen, and M. Jeffares, *J. Org. Chem.*, 1979, **44**, 2722.
[30] G. R. Lenz and L. Swenton, *J.C.S. Chem. Comm.*, 1979, 444.
[31] S. Shaik and N. D. Epiotis, *J. Amer. Chem. Soc.*, 1978, **100**, 18.
[32] K.-H. Lee and P. de Mayo, *J.C.S. Chem. Comm.*, 1979, 493.
[33] A. Feigenbaum, J.-P. Pete, and D. Scholler, *Tetrahedron Letters*, 1979, 537.
[34] R. G. Salomon, D. J. Coughlin, and E. M. Easler, *J. Amer. Chem. Soc.*, 1979, **101**, 3961.

key reaction is the photobicyclization of hepta-1,6-dien-3-ols, using copper(I) trifluoromethanesulphonate as the catalyst. In the absence of copper(I), this reaction is suppressed.

Hitherto it has generally been accepted that whereas electroisomeric linearly conjugated cyclohexa-2,4-dienones of predominantly n,π^* character afford isomeric dienylketens, those of predominantly π,π^* character lead to isomeric bicyclo[3.1.0]hex-3-en-2-ones. However, a recent paper[35] now suggests that bicyclic ketones may also be the products of n,π^* states, and a study of a family of methylated *o*-quinol acetates indicates that the reaction channel depends on such factors as reaction medium and substitution pattern. A paper describing the photoinduced cycloaddition of sterically hindered *p*-quinones to diphenylketen has appeared.[36] This reaction, the first example of its type involving diphenylketen, affords a mixture of a spiro-oxetanone and a cyclobutanone. Both products are considered to arise from the $^3(\pi,\pi^*)$ state of the quinone, and the predominance of the oxetanone product is attributed to the degree of dipole–dipole interaction between the excited state of the quinone carbonyl and the ground state of the diphenylketen. In a related paper,[37] the photoreaction of *p*-benzoquinone with unsymmetrical keten dimers is discussed. A temperature-dependent photo-CIDNP study has revealed[38] that, on irradiation, certain $\beta\gamma$-unsaturated ketones yield radical pairs (Scheme 7) from either the S_1 or T_2 state, depending upon the reaction temperature. Before collapsing to products, one reaction channel that is open to the radicals is a 1,3-acyl shift.

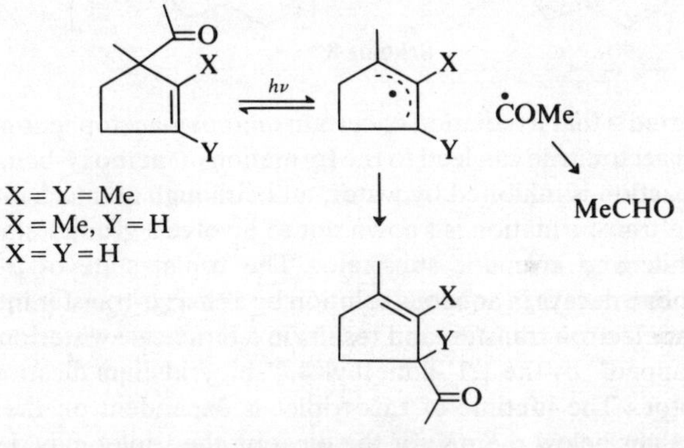

Scheme 7

As part of a series of studies in the field of $\beta\gamma$-unsaturated ketones, the stereochemistry of the oxa-di-π-methane rearrangement of 2-substituted-2-(cyclopent-1-enyl)cyclopentanones has been investigated.[39] The main conclusions drawn are that the triplet-sensitized oxa-di-π-methane rearrangement involves [1,2]-suprafacial carbonyl migration and that this proceeds with inversion of configuration at the

[35] G. Quinkert, F. Cech, E. Kleiner, and D. Rehm, *Angew. Chem. Internat. Edn.*, 1979, **18**, 557.
[36] K. Ogino, T. Matsumoto, and S. Kozuka, *J.C.S. Chem. Comm.*, 1979, 643.
[37] K. Ogino, T. Matsumoto, T. Kawai, and S. Kozuka, *J.C.S. Chem. Comm.*, 1979, 644.
[38] A. Henne, N. P. Y. Siew, and K. Schaffner, *J. Amer. Chem. Soc.*, 1979, **101**, 3671.
[39] R. L. Coffin, W. W. Čox, R. G. Carlson, and R. S. Givens, *J. Amer. Chem. Soc.*, 1979, **101**, 3261.

methane carbon. [1,3]-Rearrangement of the singlet state seems to occur suprafacially, to give a substituted bicyclo[5.3.0]dec-1-en-6-one.

Flash photolysis of various methylated indan-2-ones indicates[40] the presence of a transient, identified as an *o*-xylylene, and which in the case of 1,1,3,3-tetramethylindan-2-one has been shown to be 7,7,8,8-tetramethyl-*o*-xylylene. Thermal decay of these transient non-planar xylylenes may proceed *via* the hitherto unknown antarafacial [1,5]-shift to give styrenes. An examination of the photochemistry of the 2-spirocyclopropylindan-1-one (12) in a rigid matrix at 77 K affords[41] (*E*)-2-ethylideneindan-1-one, formed exclusively from the $^3(n,\pi^*)$ state in a monophotonic process; no reaction, however, occurs in fluid solution. It is suggested that, since the rate of ring-closure in fluid solution is rapid, under these conditions the biradical simply returns to the starting material (Scheme 8).

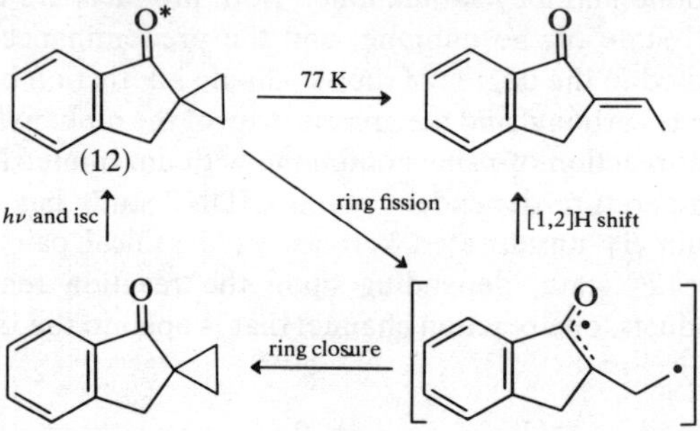

Scheme 8

It has been reported[42] that irradiation of certain methoxy-acetophenones together with cyanide ion in acetonitrile can lead to the formation of methoxy-benzonitriles in high yield. The reaction is inhibited by water, and although no mechanism has yet been suggested the transformation is known not to involve a ground-state complex between nucleophile and aromatic substrate. The triplet state of β-(dimethylamino)propiophenone decays in aqueous solution by a charge-transfer interaction to give intramolecular electron transfer, and results in a biradical zwitterionic product, which has been trapped[43] by the 1,1′-dimethyl-4,4′-bipyridylium dication, acting as an electron acceptor. The lifetime of this triplet is dependent on the pH of the medium, and at values below the pK_a of the dication the triplet may decay before acid–base equilibrium has been established. CIDNP studies of the photoreduction of 4-substituted acetophenones by *NN*-dimethylanilines suggest the presence of two pathways for formation of radical ion-pairs.[44] Triplet ion-pairs are formed by excitation of the ketone followed by electron abstraction from the ground-state amine, and electron-transfer from the singlet excited amine to the ground-state ketone gives a singlet ion-pair. The influence of mercaptans on the photoreduction of various aromatic ketones by alcohols has been studied.[45] Retardation of the

40 K. K. De Fonseka, J. J. McCullough, and A. J. Yarwood, *J. Amer. Chem. Soc.*, 1979, **101**, 3277.
41 R. O. Loutfy and P. Yates, *J. Amer. Chem. Soc.*, 1979, **101**, 4694.
42 A. L. Colb, *J. Amer. Chem. Soc.*, 1979, **101**, 3416.
43 M. V. Encinas and J. C. Scaiano, *J. Amer. Chem. Soc.*, 1979, **101**, 2146.
44 B. M. P. Hendriks, R. I. Walter, and H. Fischer, *J. Amer. Chem. Soc.*, 1979, **101**, 2378.
45 S. G. Cohen, A. W. Rose, P. G. Stone, and A. Ehret, *J. Amer. Chem. Soc.*, 1979, **101**, 1827.

process occurring between triplet benzophenone and propan-2-ol is largely due to hydrogen-transfer reactions from mercaptan to the 2-hydroxy-2-propyl radical and from the benzophenone ketyl radical to the thiyl radical. For 2-mercaptomesitylene the decrease in rate using deuteriated propan-2-ol is greater than in the protio system.

A study of the photochemistry of α-benzoyl-ω-azido-ketones of the type $PhCO(CH_2)_nN_3$ has revealed[46] the existence of a competition between the carbonyl and azide (nitrene) photoprocesses, namely γ-hydrogen abstraction to give Type II products and energy transfer from the triplet ketone to the azide to form nitrene products. The positions of the two chromophores on the molecular skeleton have been varied systematically, so that the rate constants for interaction between the two groups could be compared with their distances apart. The rate variations for $n = 3$ to 5 have been interpreted in terms of the strain present in medium-sized rings.

A hydrophilic polymer-immobilized photosensitizer that is useful for photo-oxygenations in aqueous solution has been described.[47] It is prepared by copolymerization of chloromethylstyrene and the monomethacrylate ester of ethylene glycol as cross-linking agent. This is then heated with Rose Bengal in dry *NN*-dimethylformamide, giving the hydrophilic sensitizer (HP)–Rose Bengal. Singlet oxygenation of cyclo-octa-1,3,5-triene leads[48] to (2+4) adducts (Scheme 9). This

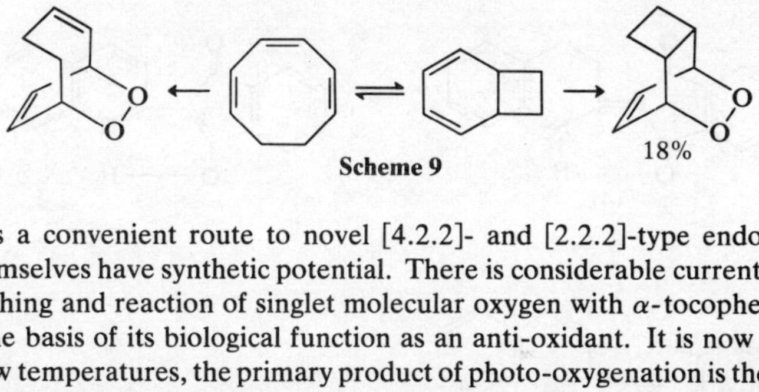

Scheme 9

represents a convenient route to novel [4.2.2]- and [2.2.2]-type endoperoxides, which themselves have synthetic potential. There is considerable current interest in the quenching and reaction of singlet molecular oxygen with α-tocopherol, as this may be the basis of its biological function as an anti-oxidant. It is now reported[49] that, at low temperatures, the primary product of photo-oxygenation is the relatively stable hydroperoxy-dienone (13). Although simple acetylenes do not react with singlet oxygen that has been generated by using Methylene Blue or Rose Bengal in methanol, photosensitized oxygenation of aromatic acetylenes has been shown[50] to yield benzils (Scheme 10). An electron-transfer mechanism similar to that suggested for the photosensitized oxygenation of alkenes has been advanced, and the reactivity

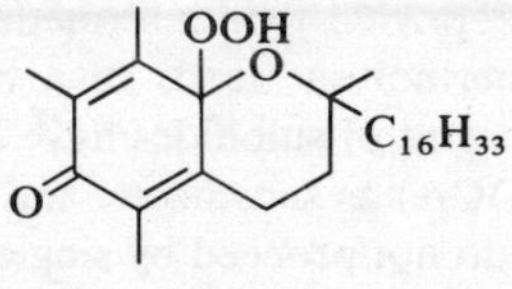

(13)

[46] P. J. Wagner and B. J. Scheve, *J. Amer. Chem. Soc.*, 1979, **101**, 378.
[47] A. P. Schaap, A. L. Thayer, K. A. Zaklika, and P. C. Valenti, *J. Amer. Chem. Soc.*, 1979, **101**, 4016.
[48] W. Adam and I. Erden, *Tetrahedron Letters*, 1979, 2781.
[49] R. L. Cough, B. G. Yee, and C. S. Foote, *J. Amer. Chem. Soc.*, 1979, **101**, 683.
[50] N. Berenjian, P. de Mayo, F. H. Phoenix, and A. C. Weedon, *Tetrahedron Letters*, 1979, 4179.

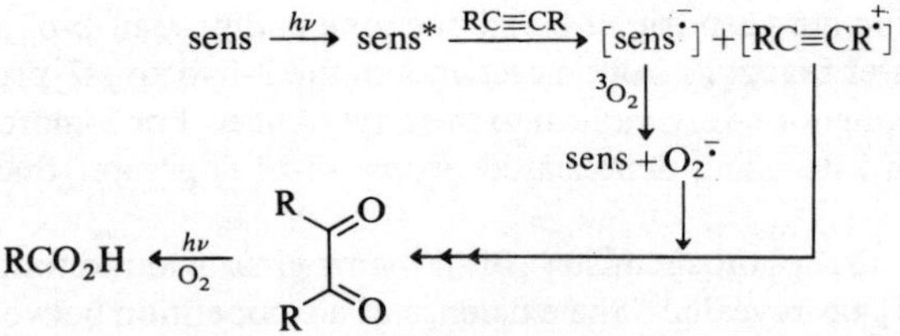

Scheme 10

of the acetylenes is in accordance with the predictions of the Weller equation. Results have now been published[51] which cast doubt on earlier suggestions concerning the dye-sensitized photo-oxygenation of α-oxo-carboxylic acids, in which decarboxylation occurs. Measurements of the rates at which α-oxo-acids react with singlet oxygen in competition with 1,3-diphenylisobenzofuran indicate that reaction with the triplet dye is more efficient than with singlet oxygen. Consequently, singlet oxygen appears not to play a significant role. The Rose-Bengal- or Methylene-Blue-sensitized photo-oxidation of 2-(2-quinolyl)indane-1,3-dione in solution has been shown[52] to lead to a mixture of products including phthalic acid, quinoline-2-carbaldehyde, and quinoline-2-carboxylic acid (Scheme 11). Of the two forms of the

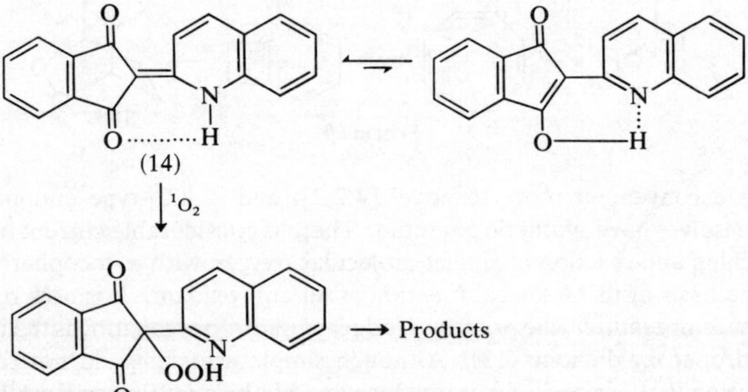

Scheme 11

quinophthalone derivative which exist in equilibrium in solution, (14) is the more susceptible to photo-oxidation. On oxidation, an unstable peroxide is produced by an ene reaction, but a perepoxide-type intermediate is also a possibility. Further photolysis, or possibly thermolysis, leads to a breakdown to products. Photo-oxygenations of different types of sulphides have been carried out, using both dye and dicyanoanthracene (DCA) as sensitizers, and it appears[53] that although those reactions involving DCA do not proceed by singlet oxygen, both involve the same intermediate. A superoxide may participate. Photosensitized oxygenations of various diazo-compounds in the presence of oxygen-atom acceptors have shown[54]

[51] R. S. Davidson, D. Goodwin, and G. Smith, *J.C.S. Chem. Comm.* 1979, 463.
[52] N. Kuramoto and T. Kitao, *J.C.S. Chem. Comm.*, 1979, 379.
[53] W. Ando, T. Nagashima, K. Saito, and S. Kohmoto, *J.C.S. Chem. Comm.*, 1979, 154.
[54] W. Ando, H. Miyazaki, and S. Kohmoto *Tetrahedron Letters*, 1979, 1317.

that the products depend on the electrophilicity of the oxenoid intermediate. Thus the yields of diphenyl sulphoxide formed in the photosensitized oxygenation of benzoylphenyldiazomethane are very different from those obtained from diphenyldiazomethane and diazofluorenone. A simple device for moving the position of a keto-group with concomitant introduction of a double bond has been announced.[55] Thus dye-sensitized oxygenation of vinylsilanes leads to a hydroperoxide, which can be transformed into an allylic alcohol (Scheme 12). This transformation depends on

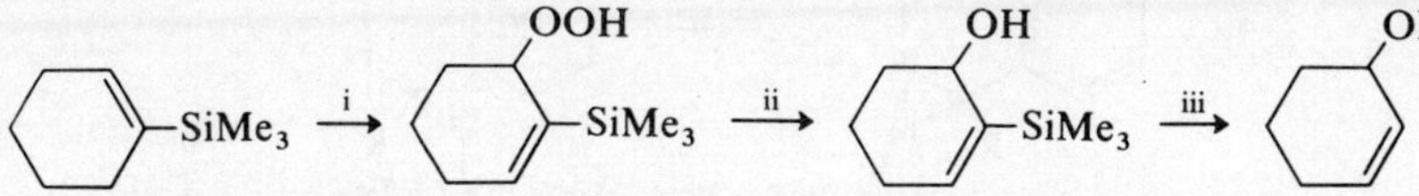

Reagents: i, $^{1}O_2$; ii, $NaBH_4$; iii, $Bu^n{}_4N^+F^-$, dry MeCN

Scheme 12

the fact that $\alpha\beta$-epoxysilanes ring-open with a regioselectivity opposite to that followed by epoxides lacking carbon–metal bonds. A new reaction has been reported[56] in which carbonyl oxides from diazo-compounds oxidize silyl ketones to esters. Thus irradiation of a solution of diphenyldiazomethane, phenyltrimethylsilyl ketone, and *meso*-tetraphenylporphin under oxygen leads to a high yield of trimethylsilyl benzoate. A plausible mechanism is shown in Scheme 13.

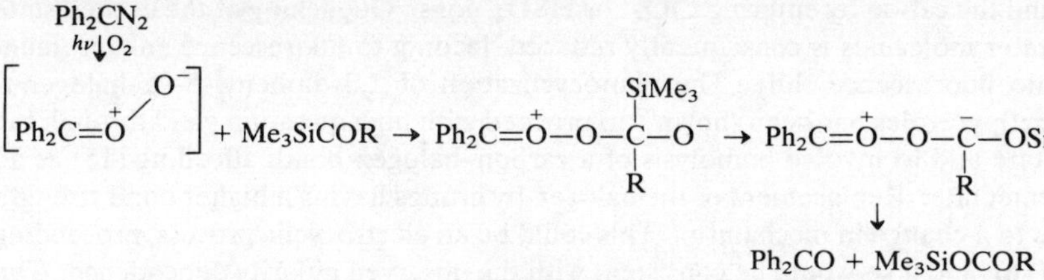

Scheme 13

Irradiation of 4-cyanopyridine in 2,3-dimethylbut-2-ene affords[57] a mixture of 4-(1′,1′,2′-trimethylprop-2′-enyl)pyridine and 4-(2′,3′-dimethylbut-2′-enyl)pyridine; 2-cyanoquinoline undergoes an analogous reaction. It is suggested that the addition proceeds by electron transfer from the π-system of the olefin to the π-system of the aromatic base. However, since no products of photocyclization are apparent, monophotonic hydrogen abstraction from the olefin by the n,π^* state of the base followed by cross-dimerization of the radicals is also a possibility. Formation of 1*H*-1,3-benzodiazepines in the photolysis of isoquinoline *N*-imides has been announced,[58] and probably occurs *via* ring-expansion of an aziridine intermediate (Scheme 14). Although this process is known for aromatic amine *N*-oxides, it is the first example reported involving aromatic amine *N*-imides. A new high-yield synthesis of quinolines has been published.[59] This transformation involves an

[55] W. E. Fristad, T. R. Bailey, L. A. Paquette, R. Gleiter, and M. C. Böhm, *J. Amer. Chem. Soc.*, 1979, **101**, 4420.
[56] A. Sekiguchi, Y. Kabe, and W. Ando, *J.C.S. Chem. Comm.*, 1979, 233.
[57] T. Caronna, S. Morrocchi, P. Traldi, B. M. Vittimberga, *J.C.S. Chem. Comm.*, 1979, 64.
[58] T. Tsuchiya, M. Enkaku, J. Kurita, and H. Sawanishi, *J.C.S. Chem. Comm.*, 1979, 534.
[59] P. de Mayo, L. K. Sydnes, and G. Wenska, *J.C.S. Chem. Comm.*, 1979, 499.

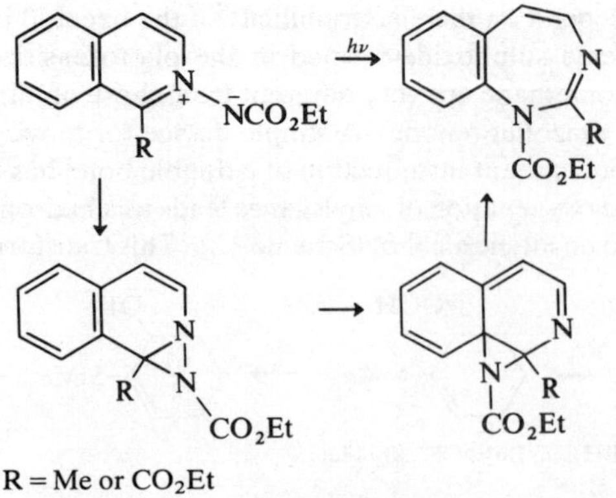

Scheme 14

intramolecular cycloaddition of substituted thioanilides and is thought to proceed *via* a $^1(\pi,\pi^*)$ state. High concentrations of $HClO_4$ or H_2SO_4 increase the fluorescence of a variety of quinolinium ions in the Hammett acidity region, but without any shift in the fluorescence or absorption wavelength.[60] These observations have been interpreted in terms of the formation of an adduct on substitution of the solvation sphere around the cation by entering ClO_4^- or HSO_4^- ions. Quenching of the excited state by water molecules is consequently reduced, leading to fluorescence enhancement but no fluorescence shift. The photocyclization of 1,3-diphenyl-5-(2'-halogenophenyl)pyrazoles has been shown[61] to proceed with high quantum yield through the S_0^v state and to involve homolysis of a carbon–halogen bond, affording (15) as an intermediate. Replacement of the halogen by groups having a higher bond strength leads to a change in mechanism. This could be an electrocyclic process, proceeding *via* (16), and if so would be consistent with the observed polarity dependence. The

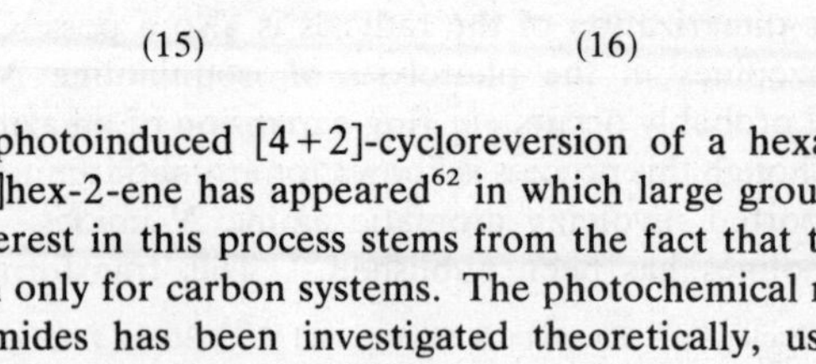

(15) (16)

first report of a photoinduced [4+2]-cycloreversion of a hexasubstituted 2,3-diazabicyclo[3.1.0]hex-2-ene has appeared[62] in which large groups are present at C-4 and C-6. Interest in this process stems from the fact that the $[\pi_s^4 + \pi_a^4]$ has hitherto been seen only for carbon systems. The photochemical rearrangement of oxaziridines to amides has been investigated theoretically, using an *ab initio*

[60] S. C. Chao, J. Tretzel, and F. W. Schneider, *J. Amer. Chem. Soc.*, 1979, **101**, 134.
[61] J. Grimshaw and A. P. de Silva, *J.C.S. Chem. Comm.*, 1979, 193.
[62] G. Ege, K. Gilbert, and B. Hahn, *Tetrahedron Letters*, 1979, 1571.

method.[63] The lowest energy transition results from an $n_{\cdot\pi^{\cdot}} \rightarrow \sigma^*_{NO}$ excitation, promoting rupture of the N–O bond. Migration of hydrogen occurs in a separate step, and probably does so on the ground-state energy surface. Irradiation of 6-methylflavone in aqueous methanol containing sodium sulphite affords[64] two dimers. The mechanism of this process is unclear, but it is suggested that, following excitation of the flavone, an electron is transferred from SO_3^{2-} to give a radical anion; after abstraction of a proton from the solvent, this dimerizes. The photoaddition reactions of the electron-deficient alkene thiochromone 1,1-dioxide with benzene and some simple derivatives have been investigated[65] and an attempt has been made to establish a link between the ionization potential of the aromatic and the course of the phototransformation. It has been shown that the presence of electron-donating groups suppresses the formation of 2:1 adducts, the degree of charge transfer involved possibly promoting dissociation to the ground state. If the ionization potential of the aromatic lies closer to that of thiochromone 1,1-dioxide, the extent of formation of 2:1 adduct correlates well with the difference between the ionization potentials of the arene and the dioxide. Irradiation of 2-methylthianaphthalene 1,1-dioxide affords[66] *anti*-head-to-head (HH) and *anti*-head-to-tail (HT) dimers. It has been suggested that the (HH) dimer results from a monomeric excited triplet and that an excimer is the precursor of the (HT) dimer. An external heavy-atom effect was detected, and this has been interpreted in terms of a spin–orbital perturbation on the intersystem crossing in the 1,4-biradical.

[63] E. Oliveros, M. Riviere, J. P. Malrieu, and C. Teichteil, *J. Amer. Chem. Soc.*, 1979, **101**, 318.
[64] I. Yokoe, M. Taguchi, Y. Shirataki, and M. Komatsu, *J.C.S. Chem. Comm.*, 1979, 333.
[65] I. W. J. Still and T. S. Leong, *Tetrahedron Letters*, 1979, 1097.
[66] M. J. Hopkinson, W. W. Schloman, B. F. Plummer, E. Wenkert, and M. Raju, *J. Amer. Chem. Soc.*, 1979, **101**, 2157.

8 Aliphatic Compounds
Part (i) Hydrocarbons

By K. J. TOYNE
Department of Chemistry, The University of Hull, Hull HU6 7RX

1 Alkanes

Alkynes and some alkenes are reduced to alkanes in high yields by a complex species prepared from sodium borohydride and a cobalt(II) salt (*e.g.* $CoCl_2 \cdot 6H_2O$, $NaBH_4$, and EtOH).[1] Monosubstituted terminal alkenes are most easily reduced, and, although disubstituted alkenes react more slowly, they are also reduced efficiently; more highly substituted alkenes are virtually inert. This high steric selectivity permits the less substituted double bond in dienes to be reduced.

Tri-n-butyltin hydride has been used for many years as an excellent reagent for the hydrodehalogenation of alkyl halides, but a major disadvantage in its use is the difficulty in separating the products from the tin residues. Two simple procedures have been reported which overcome this problem. The first method[2] is based on converting the soluble organotin halide R_3SnX (X = Cl, Br, or I) that is produced in the reaction into the insoluble organotin fluoride R_3SnF, by washing the reaction mixture (in a non-polar solvent) with aqueous potassium fluoride. The organotin fluoride is easily removed by filtration, and the organic layer contains the reduction product. The second method[3] advocates a two-phase solvent system of hexane–acetonitrile in the work-up procedure. The tri-n-butyltin species are soluble in hexane, and the required product partitions preferentially into acetonitrile.

A variation on a standard route to alkanes by decarboxylation has been described (Scheme 1);[4] yields of 50–70% are obtained for primary or secondary alkyl groups

$$\mathrm{RCOCl} \xrightarrow{\mathrm{i}} \mathrm{RH}$$

$$\underset{(1)}{\mathrm{Pr_3Si\cdot}} + \mathrm{RCOCl} \rightarrow \mathrm{Pr_3SiCl} + \mathrm{RCO\cdot}$$

$$\mathrm{RCO\cdot} \rightarrow \mathrm{R\cdot} + \mathrm{CO}$$

$$\mathrm{R\cdot} + \mathrm{Pr_3SiH} \rightarrow \mathrm{RH} + \mathrm{Pr_3Si\cdot}$$

Reagents: i, Pr_3SiH, $(Bu^tO)_2$, 140—170 °C, in ampoules

Scheme 1

[1] S.-K. Chung, *J. Org. Chem.*, 1979, **44**, 1014.
[2] J. E. Leibner and J. Jacobus, *J. Org. Chem.*, 1979, **44**, 449.
[3] J. M. Berge and S. M. Roberts, *Synthesis*, 1979, 471.
[4] N. C. Billingham, R. A. Jackson, and F. Malek, *J.C.S. Perkin I*, 1979, 1137.

but the method is unsatisfactory for tertiary, benzylic, or phenyl groups. A radical chain reaction has been proposed in which the tripropylsilyl radical (1) (generated *via* the t-butoxyl radical) initiates the formation of the alkyl radical (Scheme 1), which then abstracts hydrogen, giving the alkane and regenerating (1).

Electrochemical reduction of methanesulphonates of alcohols (ROH) to alkanes (RH) can be achieved even when functional groups such as ester, nitrile, epoxide, olefin, or aromatic nucleus are present.[5] A divided cell is used, with a lead cathode and a platinum anode, and isolated yields of 70% are reported.

2 Alkenes

Synthesis.—Allylic acetates and allylic phenyl ethers are converted into alk-1-enes in high yields by using a palladium catalyst and ammonium formate, as shown in Scheme 2 for geranyl acetate (2) or linalyl acetate (3).[6]

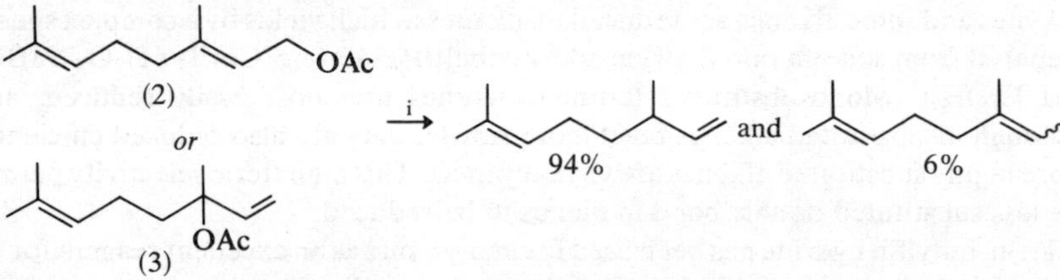

Reagents: i, HCO_2NH_4, $[PdCl_2(PPh_3)_2]$, dioxan, reflux

Scheme 2

Terminal alkenes can be prepared in high yield from alkyl halides that contain one carbon atom less *via* the formation of alkyl selenides (Scheme 3).[7] Methylation of the selenium atom generates a good leaving group, which is eliminated on treatment with base.

$$R^1CH_2Br \xrightarrow{i} R^1CH_2CH_2SeR^2 \xrightarrow{ii,iii} R^1CH{=}CH_2$$

Reagents: i, R^2SeCH_2Li, THF, HMPT ($R^2 = $ Me or Ph); ii, $MeSO_3F$ or MeI, $AgBF_4$; iii, Bu^tOK, DMSO

Scheme 3

A versatile synthesis of terminal alkenes makes use of the fluoride-ion-induced elimination of β-trimethylsilylsulphones in refluxing THF. An attractive feature of this method is that the alkylated β-silylsulphones can be prepared from (4) or from (5) (Scheme 4).[8] A second alkylation of (6) and subsequent elimination gives a 1,1-disubstituted ethene; depending upon the nature of R^1 and R^2, a variety of dienes can be prepared. 1,1-Dialkyl-ethenes can also be prepared directly from terminal alkenes (Scheme 4, reaction iv),[9] and although the reaction uses organo-

[5] T. Shono, Y. Matsumura, K. Tsubata, and Y. Sugihara, *Tetrahedron Letters*, 1979, 2157.
[6] J. Tsuji and T. Yamakawa, *Tetrahedron Letters*, 1979, 613.
[7] S. Halazy and A. Krief, *Tetrahedron Letters*, 1979, 4233.
[8] P. J. Kocienski, *Tetrahedron Letters*, 1979, 2649.
[9] J. J. Barber, C. Willis, and G. M. Whitesides, *J. Org. Chem.*, 1979, **44**, 3603.

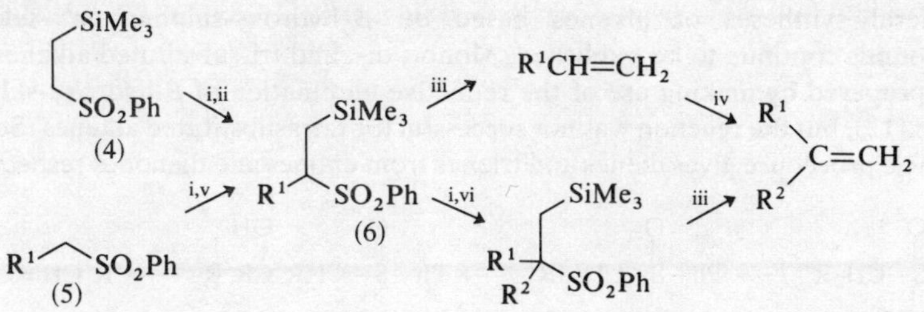

Reagents: i, BuLi, THF; ii, R^1Br; iii, $Bu_4NF{\cdot}3H_2O$, THF, heat; iv, $[Cp_2TiCl_2]$, AlR^2_3, CH_2Cl_2; v, Me_3SiCH_2I; vi, R^2Br

Scheme 4

aluminium and titanium reagents inefficiently, it tolerates the presence of certain functional groups and achieves a transformation in one step which is difficult to accomplish in other ways.

Several Wittig-like reactions involving the conversion of a carbonyl function into an alkene have been reported. The lithium derivative of aldehyde tosylhydrazones (7) and the α-lithio species (8) are generated and a condensation–fragmentation reaction gives the alkene (Scheme 5).[10] In many cases this procedure is more

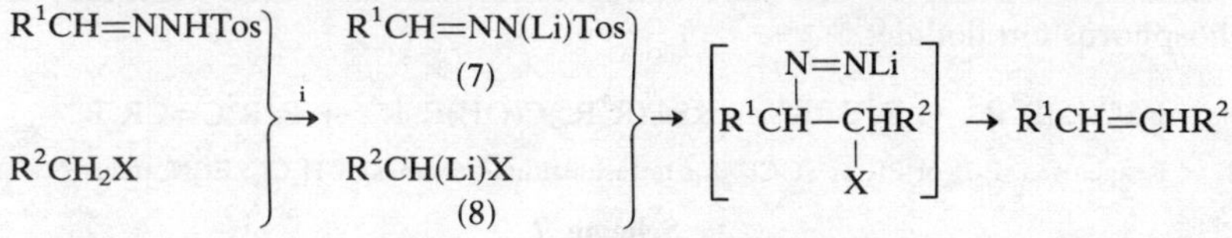

X = *e.g.*, SO_2R^3, SPh, or CN

Reagents: i, Pr^i_2NLi, THF, at −20 °C

Scheme 5

convenient than the Wittig reaction but it offers less stereochemical control; using α-branched nitriles, trisubstituted alkenes can be prepared. Terminal alkenes can also be prepared from ketones, using (chloromethyl)trimethylsilane and triphenylphosphine, which produce methylenetriphenylphosphorane, $Ph_3P{=}CH_2$ (9), as the reactive intermediate.[11] The reaction is carried out at 150 °C in sealed tubes and, although it is not suitable for ketones that have α-hydrogen atoms, because of competing formation of silyl enol ether, it is a simple and clean procedure. A silylated derivative of (9), prepared from methyltriphenylphosphonium bromide and (iodomethyl)trimethylsilane, leads to an allylsilane, $R^1R^2C{=}CHCH_2SiMe_3$ (10), after reaction with a carbonyl compound. Compound (10) reacts with a variety of alkylating agents in the presence of titanium(IV) chloride to give $R^1R^2CZCH{=}CH_2$ (11; Z = *e.g.*, CH_2CH_2OH, R^3, COMe, CH_2CH_2COMe) or, on protodesilylation, (10) gives (11; Z = H).[12]

[10] E. Vedejs, J. M. Dolphin, and W. T. Stolle, *J. Amer. Chem. Soc.*, 1979, **101**, 249.
[11] A. Sekiguchi and W. Ando, *J. Org. Chem.*, 1979, **44**, 413.
[12] I. Fleming and I. Paterson, *Synthesis*, 1979, 446.

Several syntheses of alkenes based on β-hydroxy-sulphur or -selenium compounds continue to be published. Mono-, di-, and tri-substituted alkenes have been prepared by making use of the reductive elimination of β-hydroxy-sulphoximines (12), but the reaction was not successful for tetrasubstituted alkenes (Scheme 6).[13] The procedure gives dienes and trienes from enones and dienones respectively,

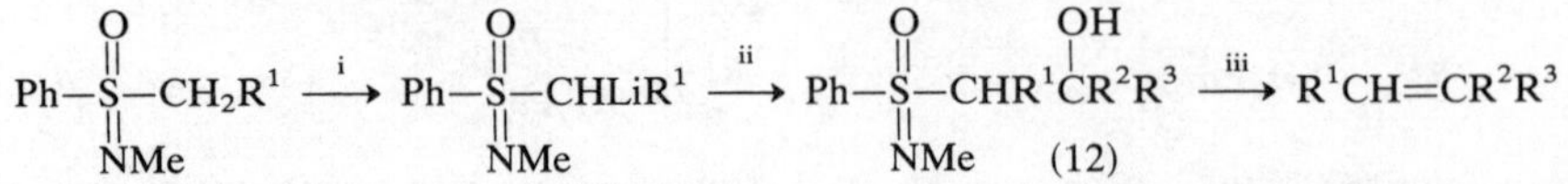

Reagents: i, BuLi, THF; ii, R^2COR^3; iii, Al(Hg), THF, H_2O, HOAc

Scheme 6

and in some cases gives better yields than the Wittig reaction, although the latter reaction is more generally useful. The preparation of di-, tri-, and tetra-substituted alkenes has also been achieved from compounds similar to (12) by using a mild elimination of β-hydroxy-sulphides, which are readily prepared from α-thioalkyllithium and carbonyl compounds (Scheme 7).[14] The regiospecific reaction involves *anti*-elimination of the hydroxy- and sulphenyl groups, probably *via* a thiiranium ion as an intermediate, which is then attacked by iodide ion. The similar deoxygenation of oxirans to alkenes of the same configuration has also been achieved, using diphosphorus tetraiodide.[15]

$$RSC(Li)R^1R^2 + R^3COR^4 \rightarrow RSC(R^1R^2)C(OH)R^3R^4 \xrightarrow{i} R^1R^2C{=}CR^3R^4$$

Reagents: i, P_2I_4 or PI_3 or $SOCl_2$ (for tetrasubstituted alkenes), CH_2Cl_2, Et_3N, reflux.

Scheme 7

α-Lithio-selenides prepared from alkyl aryl selenides react with aldehydes and ketones to give β-hydroxy-selenides, which are readily converted into alkenes by an *anti*-reductive elimination similar to those described above. The mild conditions used for the elimination are compatible with nearly all functional groups except the hydroxy-group of an alcohol (Scheme 8).[16] The α-lithio-selenides can also be alkylated and then converted into selenoxides, which decompose to alkenes.

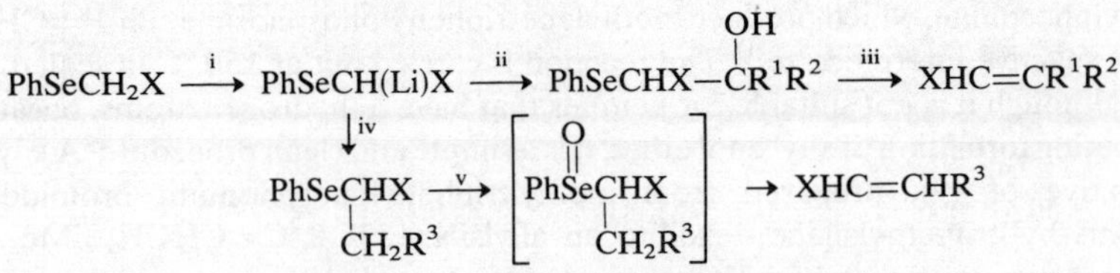

X = *e.g.*, Ar, vinyl, ethynyl, or OMe

Reagents: i, Pr^i_2NLi or lithium 2,2,6,6-tetramethylpiperidide, THF; ii, R^1COR^2; iii, $MeSO_2Cl$, Et_3N, CH_2Cl_2; iv, R^3CH_2Y (Y = Br or I); v, $NaIO_4$, or H_2O_2, or O_3

Scheme 8

[13] C. R. Johnson and R. A. Kirchhoff, *J. Amer. Chem. Soc.*, 1979, **101**, 3602.
[14] J. N. Denis, W. Dumont, and A. Krief, *Tetrahedron Letters*, 1979, 4111.
[15] H. Suzuki, T. Fuchita, A. Iwasa, and T. Mishina, *Synthesis*, 1978, 905.
[16] H. J. Reich, F. Chow, and S. K. Shah, *J. Amer. Chem. Soc.*, 1979, **101**, 6638.

Techniques for the preparation of α-lithio-selenoxides have been developed, and these reagents react with most aldehydes and ketones to give β-hydroxy-selenoxides, which can be reduced to β-hydroxy-selenides (and subsequently converted into alkenes; see Scheme 8) or thermally decomposed to allyl alcohols (Scheme 9).[17] Alkylation of (13) and thermal decomposition of the product also leads to alkenes.

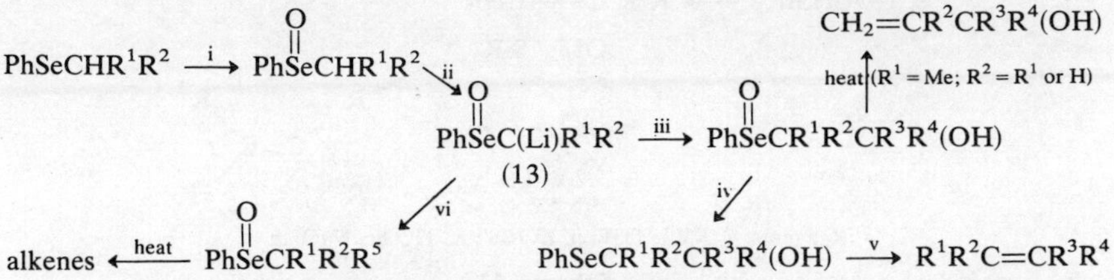

Reagents: i, O_3 or *m*-chloroperbenzoic acid; ii, Pr^i_2NLi, THF; iii, R^3COR^4; iv, HOAc, NaI, and $NaHSO_3$, *or* $(MeO)_3P$; v, $MeSO_2Cl$, Et_3N, CH_2Cl_2 (see Scheme 8); vi, R^5X

Scheme 9

A more detailed account has appeared of a procedure reported in 1975 for converting aromatic and aliphatic esters into isopropenyl compounds by reaction with methylenetriphenylphosphorane or its derivatives (Scheme 10).[18] An excess of n-alkylidenetriphenylphosphorane converts the ester into a branched alkene such that the oxygen atoms in the original ester are replaced by the phosphorane moiety and the double bond remains at the position of the original carbonyl group.

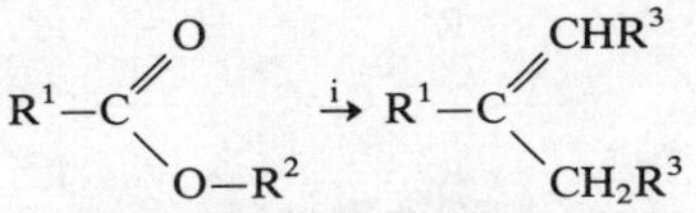

Reagents: i, NaH, DMSO, $Ph_3\overset{+}{P}CH_2R^3\ I^-$

Scheme 10

A further account of a convenient method reported last year[19a] uses a terminal alkyne ($R^1C{\equiv}CH$) and a Grignard reagent (R^2MgBr) to prepare 2,2-disubstituted alkenyl-copper intermediates (14), which are then allowed to react with electrophiles

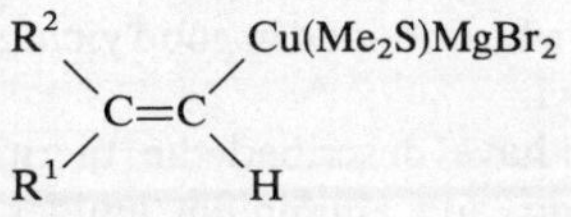

(14)

to give trisubstituted alkenes stereoselectively.[20] The sequence can be adapted to produce compounds with repetitive trisubstituted alkene units, as seen in many naturally occurring compounds such as terpenoids.[21]

Trisubstituted alkenes have also been prepared stereoselectively from α-trialkylsilyl-substituted ketones and alkyl-lithium reagents (Scheme 11).[22] The

[17] H. J. Reich, S. K. Shah, and F. Chow, *J. Amer. Chem. Soc.*, 1979, **101**, 6648.
[18] A. P. Uijttewaal, F. L. Jonkers, and A. van der Gen, *J. Org. Chem.*, 1979, **44**, 3157.
[19] K. J. Toyne, *Ann. Reports (B)*, 1978, **75**, (*a*) p. 161, (*b*) p. 163, (*c*) p. 166.
[20] A. Marfat, P. R. McGuirk, and P. Helquist, *J. Org. Chem.*, 1979, **44**, 3888.
[21] A. Marfat, P. R. McGuirk, and P. Helquist, *J. Org. Chem.*, 1979, **44**, 1345.
[22] M. Obayashi, K. Utimoto, and H. Nozaki, *Bull. Chem. Soc. Japan*, 1979, **52**, 1760.

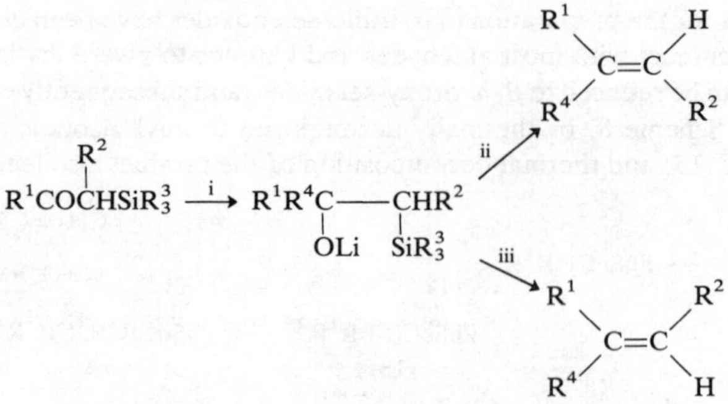

Reagents: i, R^4Li, THF; ii, $KOBu^t$; iii, HOAc, NaOAc

Scheme 11

elimination of the silyl and oxy-functions occurs mainly by a *syn*-process under basic conditions or mainly by an *anti*-mechanism on treatment with acid.

Alkenyl halides undergo stereoselective cross-coupling with organolithium compounds to give alkenes and with thiolate anions to give alkenyl sulphides (Scheme 12).[23] Although stoicheiometric and catalytic processes already exist for the preparation of alkenes from alkenyl halides, the new methods, which have been used with alkyl, aryl, and heterocyclic lithium compounds in both stoicheiometric and catalytic conditions, offer advantages in certain cases.

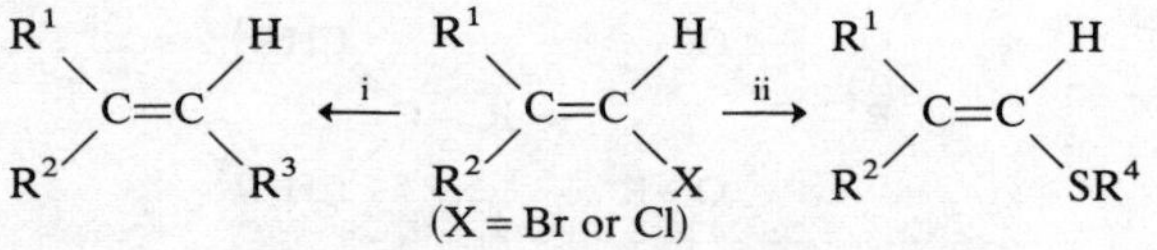

Reagents: i, R^3Li, $[Pd(PPh_3)_4]$; ii, R^4SLi, $[Pd(PPh_3)_4]$

Scheme 12

threo-3-Hydroxycarboxylic acids* (15), obtained from metallated carboxylic acids and aldehydes, can be converted into (*Z*)- or (*E*)-alkenes as shown in Scheme 13.[24] The ratio of (*E*)-/(*Z*)-isomers from each procedure is >99/1 and <3/97, so that practically pure material can be obtained in good yield. In the examples reported, R^1 is Ph or OPh and R^2 is alkyl.

Two related papers[25,26] have described the thermal decomposition, at 150—250 °C, of magnesium, zinc, and aluminium amides $R^1MNR^2{}_2$ (M = metal) into hydrocarbon (R^1H), alkene (from R^2), and a residue which, on hydrolysis, yields primary amine (R^2NH_2)[25] and, secondly, the thermal decomposition, at 195—340 °C, of magnesium, zinc, and aluminium alkoxides R^1MOR^2 (M = metal) into hydrocarbon (R^1H), alkene (from R^2), and metal oxide (MO).[26] The former reaction

* The use of '*threo*' by the authors of ref. 24 is not consistent with the normal IUPAC convention, but has been retained. *Senior Reporter.*

23 S. Murahashi, M. Yamamura, K. Yanagisawa, N. Mita, and K. Kondo, *J. Org. Chem.*, 1979, **44**, 2408.
24 J. Mulzer, A. Pointner, A. Chucholowski, and G. Brüntrup, *J.C.S. Chem. Comm.*, 1979, 52.
25 E. C. Ashby and G. F. Willard, *J. Org. Chem.*, 1978, **43**, 4750.
26 E. C. Ashby, G. F. Willard, and A. B. Goel, *J. Org. Chem.*, 1979, **44**, 1221.

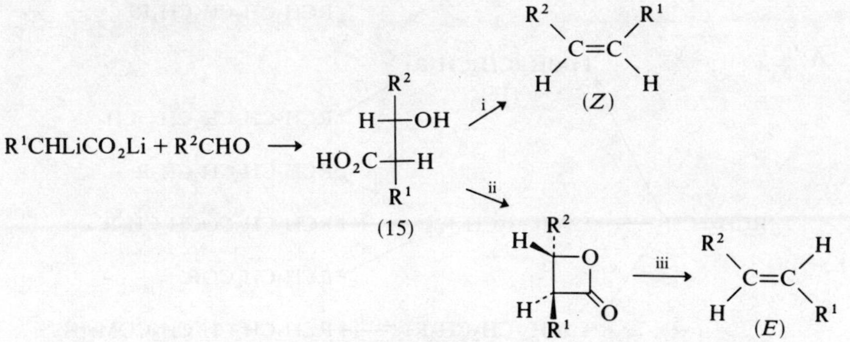

Reagents: i, PPh_3, $EtO_2CN{=}NCO_2Et$; ii, $PhSO_2Cl$, pyridine; iii, heat, to eliminate CO_2

Scheme 13

allows the conversion of a secondary amine into a primary amine and an alkene, and it is an alternative to the Hoffman and the Cope eliminations; the latter reaction is an alternative to Chugaev and acetate pyrolysis procedures for the dehydration of alcohols.

Reactions.—The hydroalumination of alkenes continues to provide the basis for several important synthetic procedures, and the difficulty in removing the catalyst from the organoaluminium product when $LiAlH_4$–$TiCl_4$ is used for hydroalumination can now be avoided. Heterogeneous immobilized titanium catalysts such as SiO_2-supported titanium (IV) chloride and polystyrene-supported titanocene dichlorides have been shown to be excellent catalysts, which are easily recovered from the reagents by filtration and can be re-used many times.[27]

Lithium alkyl-trihydroaluminates (16)[28] and lithium tetra-alkyl-aluminates (17)[29] undergo alkyl–alkyl coupling upon treatment with copper(II) acetate, and an alkyl–allyl cross-coupling is also possible[28] (Scheme 14, reagents ii and iii); a similar reaction, catalysed by copper(I) chloride, using allyl halide has been reported previously.[19b] The alkyl–alkyl coupling reaction therefore provides a new preparative route to alkanes by the dimerization of alkenes, and also enables a variety of dienes to be prepared by coupling the mono-hydroalumination product of a diene (*e.g.* hexa-1,4-diene to dodeca-2,10-diene). When the reaction of organoaluminates with copper(II) acetate is carried out under a carbon monoxide atmosphere, ketones are formed in about 20—30% yield (reagent v),[29] but better routes to aldehydes and ketones from organoaluminates have been reported.[30,31] The reaction of (18) and acrolein or methyl vinyl ketone in the presence of stoicheiometric amounts of copper(II) acetate gives aldehydes or methyl ketones in excellent yield[30] (Scheme 14; reagent vii), and the reaction of (17) with acid chlorides or anhydrides in the presence of catalytic amounts of copper(I) chloride gives ketones (reagent vi).[31] Trialkylboranes have been used previously in similar reactions, but only one of the three alkyl groups on boron is used. The hydroalumination methods appear to give better

[27] F. Sato, H. Ishikawa, Y. Takahashi, M. Miura, and M. Sato, *Tetrahedron Letters*, 1979, 3745.
[28] K. Isagawa, M. Ohige, K. Tatsumi, and Y. Otsuji, *Chem. Letters*., 1978, 1155.
[29] F. Sato, Y. Mori, and M. Sato. *Chem. Letters*, 1978, 1337.
[30] F. Sato, T. Oikawa, and M. Sato, *Chem. Letters*, 1979, 167.
[31] F. Sato, H. Kodama, Y. Tomuro, and M. Sato, *Chem. Letters*, 1979, 623.

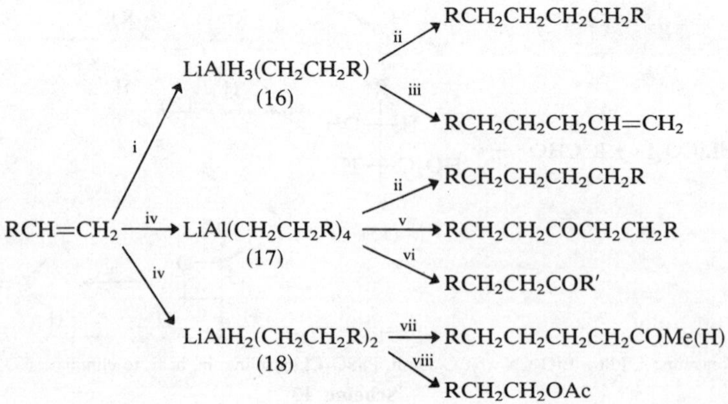

Reagents: i, $LiAlH_4$, $[Cp_2TiCl_2]$; ii, $[Cu(OAc)_2]$, THF; iii, $CH_2{=}CHCH_2Hal$, $[Cu(OAc)_2]$; iv, $LiAlH_4$–$TiCl_4$; v, CO, $[Cu(OAc)_2]$; vi, R′COCl or $(R'CO)_2O$, CuCl; vii, $CH_2{=}CHCOMe(H)$, $[Cu(OAc)_2]$; viii, $[Pb(OAc)_4]$

Scheme 14

yields and to permit the selective introduction of the carbonyl group to one double bond of a non-conjugated diene. Another hydroalumination sequence, which is an attractive alternative to hydroboronation methods, uses the reaction of (18) with lead(IV) acetate to give primary alkyl acetates (Scheme 14; reagent viii);[32] once again, the preparation of an unsaturated acetate or an $\alpha\omega$-diacetate is possible.

Organoaluminium compounds can also be obtained from complexes produced by hydrozirconation of alkenes and alkynes (Scheme 15).[33] Alkyl, alkenyl, or acyl groups in (19) are transferred rapidly and stereospecifically to aluminium chloride to give organoaluminium dichlorides, which can be acylated to yield saturated and $\alpha\beta$-unsaturated ketones; in a similar way, alkenyl-alanes can be prepared from (19). This work illustrates how transmetallation of a readily prepared transition-metal complex can be exploited to produce a more reactive organometallic species.

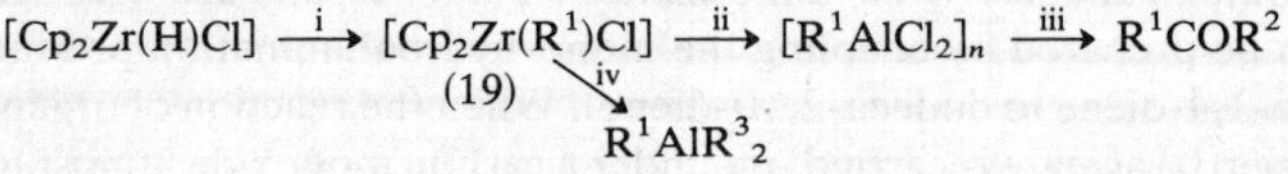

Reagents: i, alkene or alkyne; ii, $AlCl_3$, CH_2Cl_2; iii, R^2COCl, CH_2Cl_2; iv, $R^3{}_2AlCl$, CH_2Cl_2 (for R^1 = alkenyl)

Scheme 15

A standard route to cyclopropanes is by the Simmons–Smith cycloaddition to alkenes. A new reaction which is superficially similar to this procedure but which is more versatile uses the reaction of alkenes with organic *gem*-dihalides and copper powder in an aromatic solvent to give cyclopropane derivatives in good yield.[34] The reaction is electrophilic, apparently proceeding *via* an organocopper intermediate, and the cyclopropane products retain the stereochemical configuration of the alkene

[32] F. Sato, Y. Mori, and M. Sato, *Tetrahedron Letters*, 1979, 1405.
[33] D. B. Carr and J. Schwartz, *J. Amer. Chem Soc.*, 1979, **101**, 3521.
[34] N. Kawabata, I. Kamemura, and M. Naka, *J. Amer. Chem. Soc.*, 1979, **101**, 2139.

from which they are made. A variety of organic *gem*-dihalides can be used, including trihalogenomethanes and dibromoacetic esters, which give monohalogenocyclopropanes and alkoxycarbonyl-cyclopropanes respectively. Another procedure for the preparation of monohalogeno-cyclopropanes from olefins has also been described.[35]

Not only may alkenes be used for preparing three-membered rings, but their importance in the synthesis of four- and five-membered ring systems is illustrated by two interesting papers.[36,37] Cycloaddition of dichloroketen to alkenes (for reaction with alkynes, see Scheme 31) gives cyclobutane derivatives, which may then be ring-expanded to cyclopentanes. Dichloroketen reacts with mono-, di-, tri-, and tetra-substituted alkenes to give 2,2-dichlorocyclobutanones, and the chlorine atoms in the product can be reductively removed (Scheme 16).[36] The cycloaddition of chloroketen has been accomplished in a similar manner.[36] The 2,2-dichlorocyclobutanones undergo a highly regioselective ring-expansion reaction with diazomethane to give 2,2-dichlorocyclopentanones, which, in their turn, can either be completely hydrodechlorinated to cyclopentanones or be partially hydrodechlorinated to monochlorocyclopentanones, and so converted into cyclopentenones or into cyclopentenes.[37] With other diazoalkanes (*e.g.* $MeCHN_2$), α-alkyl-substituted cyclopentanones can be prepared.

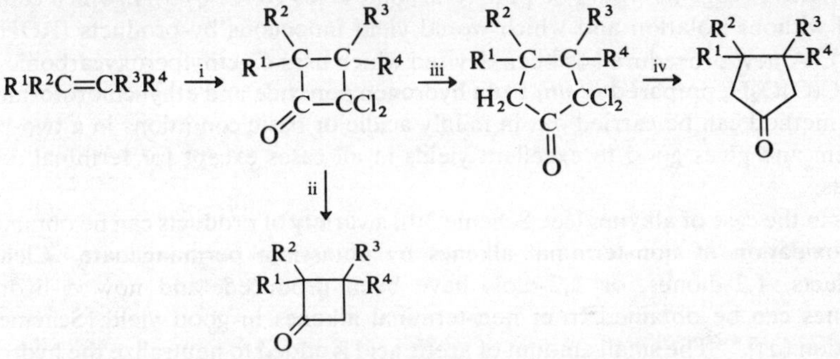

Reagents: i, CCl_3COCl, Zn, Et_2O; ii, $Bu^n{}_3SnH$ or Zn, HOAc; iii, CH_2N_2; iv, Zn, HOAc

Scheme 16

Another route to cyclobutane derivatives uses the reaction of methyl propiolate with mono- and 1,2-di-substituted ethenes in the presence of aluminium trichloride or ethylaluminium dichloride, which appears to act both as a Lewis acid and a proton scavenger [Scheme 17; reaction (*a*)].[38] However, with monosubstituted ethenes, both possible cyclobutenes are formed, as well as some of the ene adduct. The reaction with 1,1-disubstituted, trisubstituted, and tetrasubstituted alkenes followed a different course, and only the ene adducts were obtained [Scheme 17; reaction (*b*)].

The preparation of reactive oxirans by direct epoxidation has been achieved, using *m*-chloroperbenzoic acid in a two-phase system, and the method has been successful

[35] N. Kawabata, M. Tanimoto, and S. Fujiwara, *Tetrahedron*, 1979, **35**, 1919.
[36] D. A. Bak and W. T. Brady, *J. Org. Chem.*, 1979, **44**, 107.
[37] A. E. Greene and J.-P. Deprés, *J. Amer. Chem. Soc.*, 1979, **101**, 4003.
[38] B. B. Snider, D. J. Rodini, R. S. E. Conn, and S. Sealfon, *J. Amer. Chem. Soc.*, 1979, **101**, 5283.

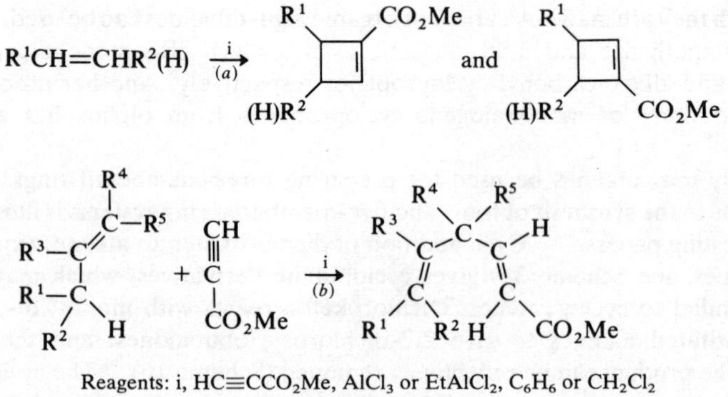

Reagents: i, $HC{\equiv}CCO_2Me$, $AlCl_3$ or $EtAlCl_2$, C_6H_6 or CH_2Cl_2

Scheme 17

in the synthesis of *cis*-, *trans*-, and aryl-oxirans.[39] An aqueous phosphate buffer is used to prevent the reaction of the oxirans with acid. Hydrogen peroxide alone is not able to epoxidize a non-conjugated double bond, but when the OOH group is linked to a multiple bond, as in percarboxylic acids, the reactivity of the reagent is increased. Alternative reagents would be peroxycarbonic acids $ROC(O)O_2H$, which could be used without isolation and which would yield innocuous by-products (ROH and CO_2). A new procedure has been devised which uses *O*-ethylperoxycarbonic acid, $EtOC(O)O_2H$, prepared *in situ* from hydrogen peroxide and ethyl chloroformate.[40] The method can be carried out in mildly acidic or basic conditions in a two-phase system and gives good to excellent yields in all cases except for terminal double bonds.

As in the case of alkynes (see Scheme 30), a variety of products can be obtained in the oxidation of non-terminal alkenes by potassium permanganate. Cleavage products, 1,2-diones, or 1,2-diols have been produced, and now α-hydroxy-ketones can be obtained from non-terminal alkenes in good yield [Scheme 18; reaction (*a*)].[41] The small amount of acetic acid is added to neutralize the hydroxide ions that are formed during the reaction. Different products can also be obtained if the homogeneous reaction mixture from the oxidation of alkenes with potassium

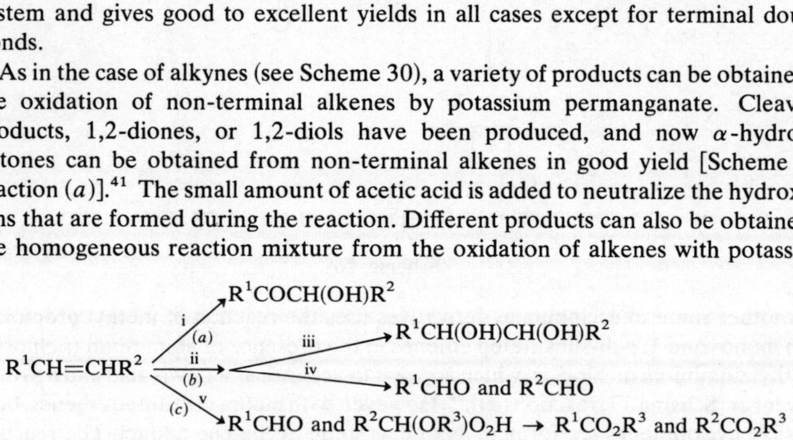

Reagents: i, $KMnO_4$, Me_2CO, H_2O, 2—5% HOAc; ii, $KMnO_4$, $(PhCH_2)Et_3NCl$, CH_2Cl_2; iii, 3% NaOH; iv, aq. AcOH–AcONa; v, O_3, R^3OH, HCl, at −50 to −70 °C and then −10 to +20 °C

Scheme 18

39 M. Imuta and H. Ziffer, *J. Org. Chem.*, 1979, **44**, 1351.
40 R. D. Bach, M. W. Klein, R. A. Ryntz, and J. W. Holubka, *J. Org. Chem.*, 1979, **44**, 2569.
41 N. S. Srinivasan and D. G. Lee, *Synthesis*, 1979, 520.

permanganate solubilized in CH_2Cl_2 is decomposed by aqueous solutions of different pH [Scheme 18; reaction (*b*)].[42] The reaction yields mainly 1,2-diol when quenched with sodium hydroxide solution whereas aqueous sodium acetate–acetic acid can give the aldehyde exclusively.

1,2-Disubstituted ethenes can be ozonolysed directly to esters in good yield if the reaction is carried out in the presence of anhydrous hydrochloric acid in an alcohol [Scheme 18; reaction (*c*)].[43] The Criegee fragments [*i.e.* (20) and (21), or with R^1 and R^2 interchanged] are formed initially, and (21) is dehydrated to an ester and (20) is converted into the acetal, which on further treatment with ozone gives an ester.

Further methods for preparing α-phenylseleno-ketones from alkenes are available (Scheme 19),[44] and subsequent alkylation of (22) at C-1, followed by reductive or oxidative removal of the phenylseleno-group, allows a variety of ketones and $\alpha\beta$-unsaturated ketones, respectively, to be prepared (see also ref. 19*c*). The hydroxyselenation of alkenes has been achieved, in excellent yield, with two new, readily available reagents, *N*-phenylselenophthalimide (N-PSP) and *N*-phenylselenosuccinimide (N-PSS) (Scheme 19).[45] By using this procedure, or an alternative method,[46] the hydroxyselenation of certain dienes leads to oxygen heterocycles such as tetrahydrofuran and tetrahydropyran derivatives, the ring size of the product being determined by the nature of the diene and by the relative thermodynamic stability of the ring.

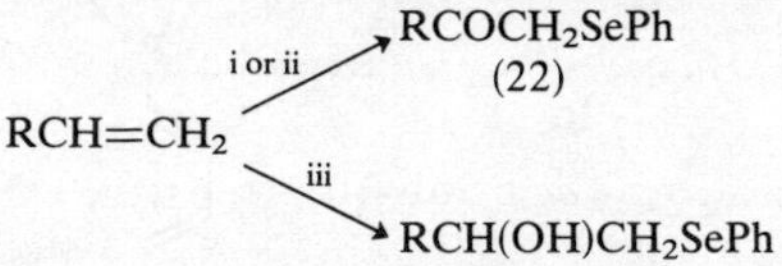

Reagents: i, Bu^tOOH, $(PhSe)_2$, CCl_4, heat to give (PhSeOSePh); ii, $2(PhSe)_2$, $(PhSeO)_2O$ to give (PhSeOSePh); iii, N-PSP or N-PSS, toluene-*p*-sulphonic acid, CH_2Cl_2, H_2O

Scheme 19

Several terminal alkenes have been converted into methyl secondary and tertiary alkyl-lithium reagents in two stages (Scheme 20).[47] Last year, the same authors reported a radical-catalysed addition of thiophenol to terminal alkenes and they have now described an acid-catalysed addition which proceeds with the opposite orientation. In each case, the alkyl phenyl sulphides could be cleaved with lithium in THF.

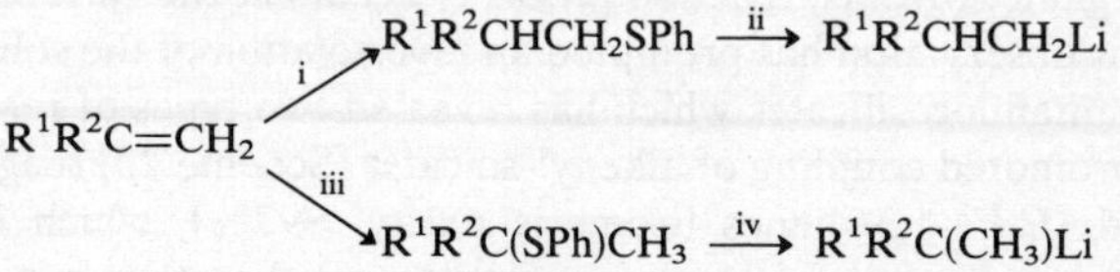

Reagents: i, PhSH, $h\nu$ or $\alpha\alpha$-azoisobutyronitrile; ii, lithium-naphthalene, THF; iii, PhSH, 70% $HClO_4$; iv, Li, THF

Scheme 20

[42] T. Ogino and K. Mochizuki, *Chem. Letters*, 1979, 443.
[43] J. Neumeister, H. Keul, M. P. Saxena, and K. Griesbaum, *Angew. Chem. Internat. Edn.*, 1978, **17**, 939.
[44] M. Shimizu, R. Takeda, and I. Kuwajima, *Tetrahedron Letters*, 1979, 419.
[45] K. C. Nicolaou, D. A. Claremon, W. E. Barnette, and S. P. Seitz, *J. Amer. Chem. Soc.*, 1979, **101**, 3704.
[46] R. M. Scarborough, A. B. Smith, W. E. Barnette, and K. C. Nicolaou, *J. Org. Chem.*, 1979, **44**, 1742.
[47] C. G. Screttas and M. Micha-Screttas, *J. Org. Chem.*, 1979, **44**, 713.

3 Dienes

4-Hydroxytricyclo[4.2.1.$0^{2,5}$]non-7-en-3-one (23), which is readily obtainable from cyclopentadiene and maleic anhydride, has been used as the basis for the preparation of stereochemically defined 1,4-disubstituted 1,3-dienes (Schemes 21).[48] For example, alkylation of (24) and reduction of the carbonyl group followed by esterification gives a sample of compound (25) of known stereochemistry. Mild flash vacuum pyrolysis of (25) generates cyclopentadiene and a 3,4-disubstituted cyclobutene, which undergoes spontaneous conrotatory ring-opening to give diene (26), of predictable stereochemistry, in almost quantitative yield. Although the full sequence is rather long, its value has been demonstrated by the synthesis of dienes with oxygen, sulphur, and alkyl groups and with a chiral directing substituent.

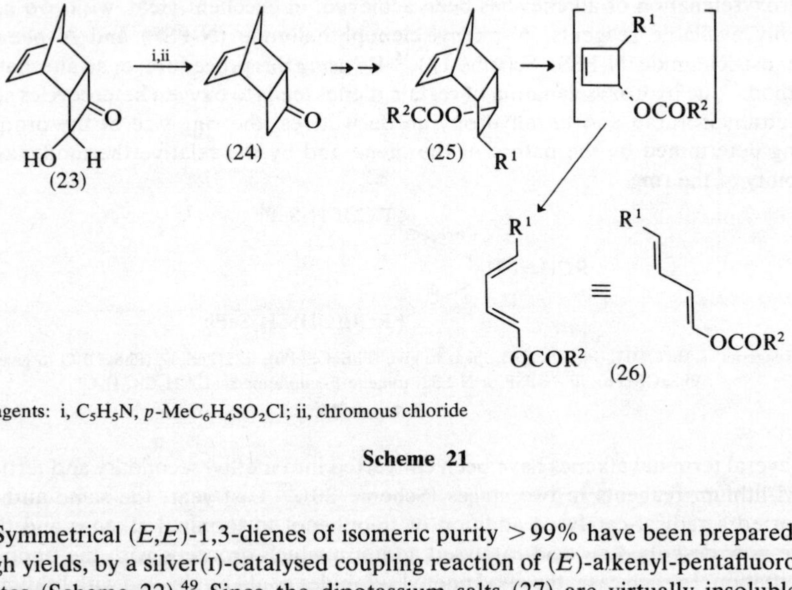

Reagents: i, C_5H_5N, p-$MeC_6H_4SO_2Cl$; ii, chromous chloride

Scheme 21

Symmetrical (E,E)-1,3-dienes of isomeric purity >99% have been prepared, in high yields, by a silver(I)-catalysed coupling reaction of (E)-alkenyl-pentafluorosilicates (Scheme 22).[49] Since the dipotassium salts (27) are virtually insoluble in common organic solvents, reaction probably occurs at the surface of the insoluble silicate. This observation has prompted an investigation of the solid-state reactions of organopentafluorosilicates which has revealed two types of reaction.[50] One is a copper(I)-promoted coupling of alkenyl-silicates [Scheme 22; reagent iii(c)] to give symmetrical (E,E)-1,3-dienes (isomeric purity >92%), which is similar to the silver(I)-catalysed reaction mentioned above, and the other is a 'protonolysis' of alkenyl-silicates (Schemes 22; reagent iv). The 'protonolysis' product is also obtained as the major product from alkyl-pentafluorosilicates.

The preparation of allylic alcohols from oxirans has been reported previously; an extension of this reaction is to start with an allyl alcohol and so produce an enediol

48 B. M. Trost, S. A. Godleski, and J. Ippen, *J. Org. Chem.*, 1978, **43**, 4559.
49 K. Tamao, H. Matsumoto, T. Kakui, and M. Kumada, *Tetrahedron Letters*, 1979, 1137.
50 J. Yoshida, K. Tamao, T. Kakui, and M. Kumada, *Tetrahedron Letters*, 1979, 1141.

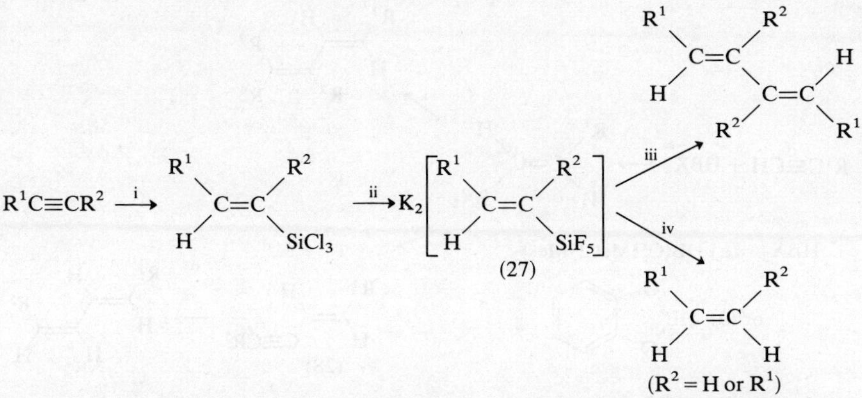

Reagents: i, $HSiCl_3$, $[H_2PtCl_6]$; ii, KF, H_2O or EtOH; iii, (*a*) AgF, MeCN, *or* (*b*) $AgNO_3$, H_2O, ether, *or* (*c*) CuCl (solid state), 200—300 °C, 3—20 mmHg; iv, $CuF_2{\cdot}2H_2O$, heat (solid state)

Scheme 22

which can be converted into a 1,3-diene (Scheme 23).[51] Epoxidation of the allylic alcohol and silylation of the alcohol is followed by ring-opening of the oxiran and desilylation, and finally both hydroxy-groups are removed *via* the dibromide.

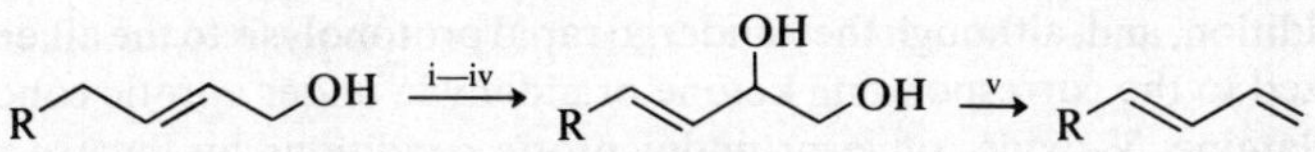

Reagents: i, $[VO(acac)_2]$, Bu^tOOH; ii, Me_3SiCl, $(Me_3Si)_2NH$, C_5H_5N; iii, diethylaluminium 2,2,6,6-tetramethylpiperidide; iv, KF; v, PBr_3, CuBr, Zn

Scheme 23

1-Alkenyl-boranes prepared from terminal alkynes react with (*E*)-1-alkenyl halides or 1-alkynyl halides to give conjugated (*E,E*)-dienes or (*E*)-enynes, respectively, by regio- and stereo-specific reactions; these occur in good yield in the presence of base and tetrakis(triphenylphosphine)palladium as catalyst (Scheme 24, reagents i and ii).[52] Attempts to prepare (*E,Z*)-dienes directly by using a (*Z*)-1-alkenyl halide were unsatisfactory, but the (*E*)-enynes (28) can be readily converted into the corresponding (*E,Z*)-dienes *via* hydroboronation–protonolysis. 1-Alkenyl-boranes from terminal and internal alkynes can similarly be coupled with aryl halides to provide a new stereoselective synthesis of arylated alkenes (29) in high yield.[53]

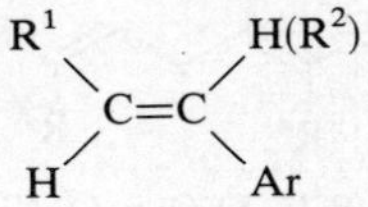

(29)

[51] A. Yasuda, S. Tanaka, H. Yamamoto, and H. Nozaki, *Bull. Chem. Soc. Japan*, 1979, **52**, 1752.
[52] N. Miyaura, K. Yamada, and A. Suzuki, *Tetrahedron Letters*, 1979, 3437.
[53] N. Miyaura and A. Suzuki, *J.C.S. Chem Comm.*, 1979, 866.

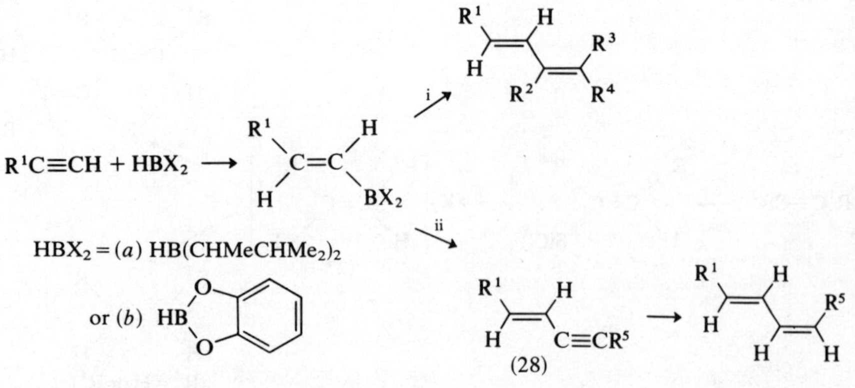

Reagents: i, 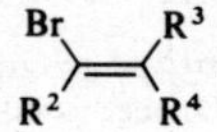 ($R^3 = H$ or R^4), $[Pd(PPh_3)_4]$, aq. NaOH or NaOEt, EtOH; ii, $R^5C{\equiv}CBr$, $[Pd(PPh_3)_4]$, aq. NaOH or NaOEt, EtOH [for reagent (*a*)]

Scheme 24

The hydroboronation of internal or terminal alkynes by 9-borabicyclo[3.3.1]nonane can be readily controlled to give either vinyl-bora- or *gem*-dibora-intermediates.[54] The vinyl-bora-intermediates are formed exclusively by *cis*-addition, and, although they undergo rapid protonolysis to the alkene, they can be oxidized to the corresponding ketone or aldehyde under aprotic conditions with trimethylamine *N*-oxide, or even under protic conditions by inverse addition to buffered hydrogen peroxide.

1,4-Dienes can be prepared in excellent yield by a rapid, regiospecific, general method which starts with an $\alpha\beta\gamma\delta$-unsaturated ketone (Scheme 25).[55] Addition of an alkyl-lithium to the carbonyl group gives an unsaturated alkoxide (30), which is reduced in the same reaction vessel to the corresponding 1,4-diene. The position of

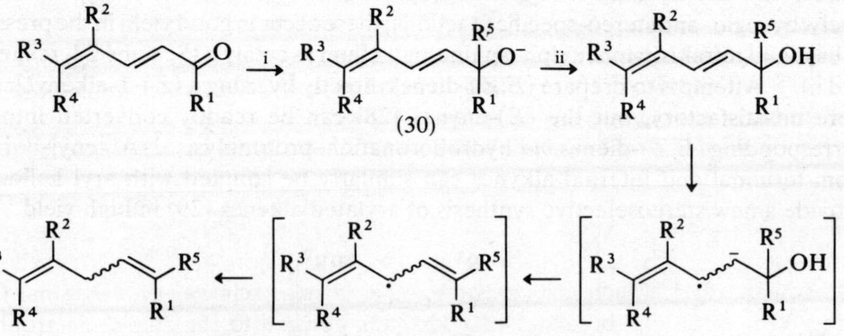

Reagents: i, R^5Li, Et_2O; ii, Li, NH_3, EtOH or Bu^tOH

Scheme 25

[54] H. C. Brown, C. G. Scouten, and R. Liotta, *J. Amer. Chem. Soc.*, 1979, **101**, 96.
[55] J. S. R. Zilenovski and S. S. Hall, *J. Org. Chem.*, 1979, **44**, 1159.

the double bond is predictable, although the product is a mixture of (*E*)- and (*Z*)-isomers at the newly developed double bond.

Another procedure also gives 1,4-dienes without isomerization to 1,3-dienes.[56] The sequence starts with a terminal or internal alkyne, although an unsymmetrical alkyne will give a mixture of products (Scheme 26). Initially the alkyne inserts into a Pd–halogen bond of the palladium complex catalyst and then an allyl halide inserts into the Pd–vinyl bond so formed; subsequent eliminaton of a palladium–halogen species completes the catalytic cycle. The overall reaction represents the addition of an allyl halide to a triple bond, and provides a convenient synthesis of substituted vinyl halides as well as of 1,4-dienes.

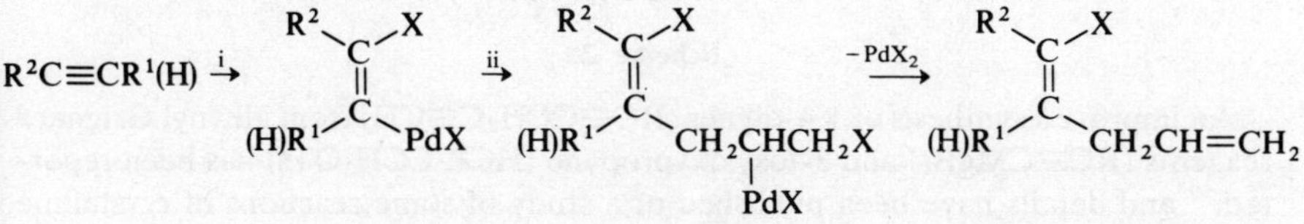

X = Cl or Br

Reagents: i, $[PdX_2(PhCN)_2]$; ii, $CH_2{=}CHCH_2X$

Scheme 26

A new procedure, which is complementary to that described above for the preparation of (*E*)-enynes (28), allows the preparation of conjugated (*Z*)-enynes in good yield, and so provides a route to (*Z,Z*)-dienes (Scheme 27).[57]

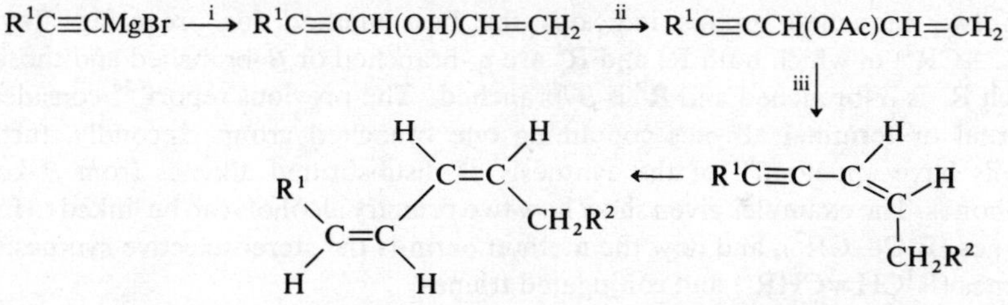

Reagents: i, $CH_2{=}CHCHO$; ii, Ac_2O, pyridine; iii, R^2MgX, CuCl, THF, H^+

Scheme 27

An attractive route to terminal enynes and to 1,3-dienes which introduces the four-carbon-atom conjugated unit in one stage is the ring-opening of pyridazine 1-oxide by Grignard reagents (Scheme 28).[58] Alkyl Grignard reagents give 1,3-dienes as a mixture of geometrical isomers by substitution whereas aryl or alkynyl Grignard reagents, which are less nucleophilic, lead exclusively to terminal (*E*)-enynes. The latter can be metallated and then allowed to react with electrophilic reagents to give internal enynes, and these can then be reduced to conjugated dienes of predictable geometry.

[56] K. Kaneda, T. Uchiyama, Y. Fujiwara, T. Imanaka, and S. Teranishi, *J. Org. Chem.*, 1979, **44**, 55.
[57] G. Cassani, P. Massardo, and P. Piccardi, *Tetrahedron Letters*, 1979, 633.
[58] L. Crombie, N. A. Kerton, and G. Pattenden, *J.C.S. Perkin I*, 1979, 2136.

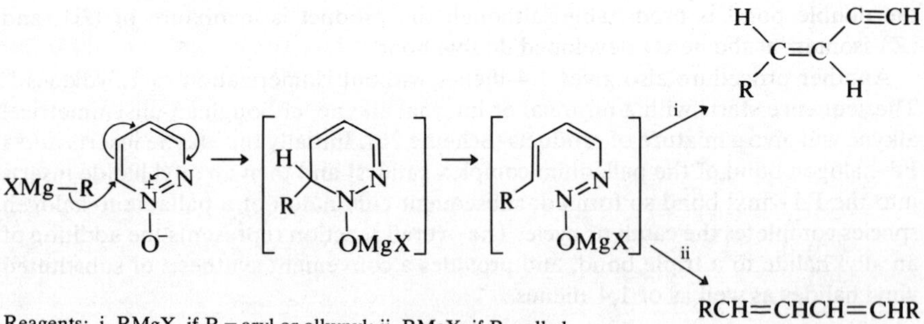

Reagents: i, RMgX, if R = aryl or alkynyl; ii, RMgX, if R = alkyl

Scheme 28

An improved synthesis of 1,4-diynes (RC≡CCH$_2$C≡CH) from alkynyl Grignard reagents (RC≡CMgBr) and 3-tosyloxypropyne (HC≡CCH$_2$OTs) has been reported,[59] and details have been published of a study of some reactions of crystalline dienyl anions prepared from conjugated and non-conjugated dienes.[60]

4 Alkynes

Synthesis.—A volume in the series on 'The Chemistry of Functional Groups' describes the chemistry of the carbon–carbon triple bond[61] and will take its place with other volumes in the series as an important reference work.

Two papers this year are linked with publications referred to in the previous *Annual Report.* First, a survey[62a] of the applicability of several syntheses of alkynes with two branched groups indicates the best method for preparing alkynes ($R^1C{\equiv}CR^2$) in which both R^1 and R^2 are α-branched or β-branched and those in which R^1 is α-branched and R^2 is β-branched. The previous report[62b] considered internal or terminal alkynes containing one branched group. Secondly, further details have appeared[63] of the synthesis of disubstituted alkynes from β-keto-sulphones. The examples given show how two primary alcohols can be linked to form alkynes ($R^1C{\equiv}CR^2$), and now the method permits the stereoselective synthesis of alkenes ($R^1CH{=}CHR^2$) and conjugated trienes.

A new route to terminal alkynes containing one carbon atom more than the primary alkyl halide that is the starting material gives excellent yields (Scheme 29).[64] The alkyl halide is alkylated by dichloromethyl-lithium in hexamethylphosphoric triamide (HMPT), and dehydrochlorination of (31) is achieved by the slow addition of n-butyl-lithium, to produce the lithium acetylenide, which can be protonated or modified further, as shown in Scheme 29. The method can be used for the synthesis of alkynes with a branched R group, but in these cases the primary alkyl iodide must be used.

[59] H. D. Verkruijsse and M. Hasselaar, *Synthesis*, 1979, 292.
[60] H. Yasuda, Y. Ohnuma, M. Yamauchi, H. Tani, and A. Nakamura, *Bull. Chem. Soc. Japan*, 1979, **52**, 2036.
[61] 'The Chemistry of the Carbon–Carbon Triple Bond', ed. S. Patai, Wiley/Interscience, New York, 1978.
[62] F. Bernadou, D. Mesnard, and L. Miginiac, (*a*) *J. Chem. Res.* 1979, (*S*) 190, (*M*) 2201; (*b*) *ibid.*, 1978, (*S*) 106, (*M*) 1501.
[63] B. Lythgoe and I. Waterhouse, *J.C.S. Perkin I*, 1979, 2429.
[64] J. Villieras, P. Perriot, and J. F. Normant, *Synthesis*, 1979, 502.

$$RCH_2X \xrightarrow{i} \underset{(31)}{RCH_2CHCl_2} \xrightarrow{ii} RC{\equiv}CH \text{ or } RC{\equiv}CCH_2OEt$$

$X = Br$ or I

Reagents: i, $LiCHCl_2$, HMPT, H_3O^+; ii, Bu^nLi, THF, H_3O^+ *or* $ClCH_2OEt$

Scheme 29

Reactions.—Many papers have discussed methods to obtain various products by the oxidation of alkynes. Four reports are mentioned here, of which two deal with oxidation of the triple bond and two with oxidation at the α-carbon atoms.

Oxidation of internal alkynes by potassium permanganate can occur in three different ways; (i) to give α-diketones, (ii) by cleavage of the initially formed α-diketone to give carboxylic acids, or (iii) by cleavage of the enol of the α-diketone to give carboxylic acids, with the loss of one carbon atom. Conditions suitable for each of these reactions are shown in Scheme 30.[65] Both of the oxidative cleavages require the presence of aqueous potassium permanganate, and it has been suggested that the symmetrical cleavage occurs by a more polar transition state, which is stabilized by solvation in aqueous solvents, whereas the oxidation of enols does not involve the formation of highly charged intermediates.

$$RCH_2C{\equiv}CCH_2R \rightarrow \underset{(32)}{RCH_2COCOCH_2R} \rightarrow \underset{(33)}{RCH_2CO_2H}$$

$$\downarrow$$

$$RCH_2COC(OH){=}CHR \rightarrow RCH_2COCO_2H + RCHO$$

$$\downarrow$$

$$\underset{(34)}{RCH_2CO_2H + RCO_2H + CO_2}$$

Reagents for preparing (32): $KMnO_4$, anhyd. CH_2Cl_2, phase-transfer assisted[65] *or* $KMnO_4$, acetone, H_2O, $NaHCO_3$, $MgSO_4$;[66] for (33): aq. $KMnO_4$;[65] for (34): aq. $KMnO_4$, CH_2Cl_2, phase-transfer assisted[65]

Scheme 30

Alternative conditions for the preparation of α-diketones are potassium permanganate in aqueous acetone (Scheme 30).[66] In the simple, general method sodium bicarbonate and magnesium sulphate are used as a buffer (initially pH 7.0—7.5), and so they neutralize the hydroxide ions produced during the reaction.

Several chromium(VI) reagents have been compared with regard to their ability to effect α-oxidation of alkynes, and the chromium trioxide–pyridine complex or pyridinium chlorochromate were found to be the best reagents for preparing conjugated ynones.[67] Oxidation at the α-carbon has also been achieved by using selenium dioxide and t-butyl hydroperoxide;[68] in contrast to the behaviour of alkenes, internal alkynes give (35). A comparison of the oxidation of several alkynes shows that an α-methylene or an α-methine atom are almost identical in their

65 D. G. Lee and V. S. Chang, *J. Org. Chem.*, 1979, **44**, 2726.
66 N. S. Srinivasan and D. G. Lee, *J. Org. Chem.*, 1979, **44**, 1574.
67 W. B. Sheats, L. K. Olli, R. Stout, J. T. Lundeen, R. Justus, and W. G. Nigh, *J. Org. Chem.*, 1979, **44**, 4075.
68 B. Chabaud and K. B. Sharpless, *J. Org. Chem.*, 1979, **44**, 4202.

reactivity, and both are much more reactive than methyl. When the alkyne has α-methylene and α-methine atoms, *e.g.* $CH_3CH_2C{\equiv}CCH(CH_3)_2$, the enynone $CH_3COC{\equiv}CC(CH_3){=}CH_2$ is the major product, which is presumably formed by dehydration at the diol or ketol stage.

$$R^1R^2C(OH)C{\equiv}CC(OH)R^3R^4$$

(35)

(36) (37)

The reaction of anhydrous hydrogen bromide with alkynes in the liquid phase has been shown to produce a variety of products in addition to those expected to arise by the normal ionic or radical addition;[69] some terminal alkynes ($RCH_2C{\equiv}CH$) also give cyclodimerization products, such as (36) and (37), and cyclobutene products were detected in the reaction of internal alkynes.

Cyclobutene derivatives have also been made by the cycloaddition of dichloroketen to alkynes (Scheme 31;[70] see Scheme 16 for a similar reaction with alkenes). This general and useful method for the synthesis of cyclobutenones uses dichloroketen (generated *in situ* from trichloroacetyl chloride) in the presence of the alkyne. The reaction is successful with terminal and internal alkynes, and enynes react by cycloaddition to the triple bond.

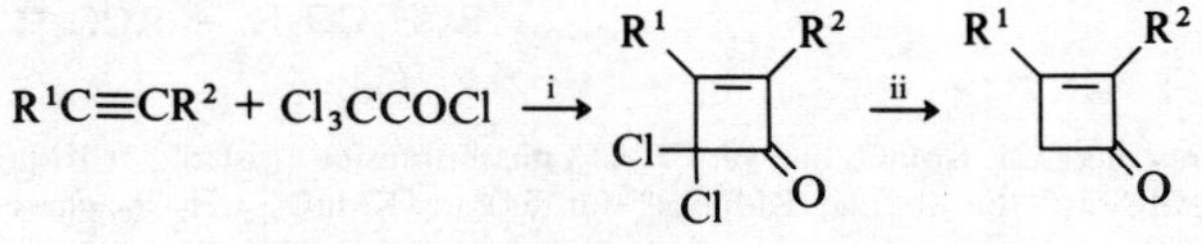

Reagents: i, Zn(Cu), $POCl_3$; ii, Zn, HOAc

Scheme 31

Two papers report syntheses of 1,3-dialkyl-butenynes by 'head-to-tail' dimerization of terminal alkynes (Scheme 32).[71,72] One procedure[71] uses a rhodium(I) catalyst at room temperature to give (38) as the main product, along with smaller amounts of (39). The second procedure[72] uses di-isobutylzinc and the catalyst

$$RC{\equiv}CH \xrightarrow{\text{i or ii}} RC({=}CH_2)C{\equiv}CR \quad (+\ RCH{=}CHC{\equiv}CR)$$

(38) (39)

Reagents: i, $[RhCl(PPh_3)_3]$, room temperature (ref. 71); ii, (*a*) $ZnBu^i_2$, $[Ni(mesal)_2]$, PPh_3, heptane, reflux, (*b*) dilute H_2SO_4 (ref. 72)

Scheme 32

69 K. Griesbaum, W. Seiter, H. Schneider, M. El Abed, and Z. Rehman, *Annalen*, 1979, 1137.
70 A. Hassner and J. Dillon, *Synthesis*, 1979, 689.
71 L. Carlton and G. Read, *J.C.S. Perkin I*, 1978, 1631.
72 G. Giacomelli, F. Marcacci, A. M. Caporusso, and L. Lardicci, *Tetrahedron Letters*, 1979, 3217.

bis-[(*N*-methylsalicylaldimine)]nickel [Ni(mesal)$_2$]. The yields decrease with increased size of the alkyl group and with α- or β-branching, but this can be overcome by adding triphenylphosphine.

Finally, some 1,3-disubstitution reactions of propyne are worthy of mention because they permit such a large variety of alkyne derivatives to be prepared. Dilithiopropyne can be prepared from propyne and then used in reactions with various electrophiles (Scheme 33).[73] Regiospecific reaction occurs first at the propargylic carbon (allowing the formation of alk-1-ynes; reagents ii and iii) and, when alkylation at C-3 is followed by addition of various electrophiles, chain extension at both C-1 and C-3 of propyne is possible by a 'one-flask' procedure (reagents iv—vii). Terminal alkynes, terminal dialkynes, internal alkynes, and various functionalized internal alkynes can all be prepared as shown in Scheme 33, and $\alpha\omega$-dilithiodialkynes (the intermediates with reagent iii) will also react with electrophiles and permit further extension of the chain. Clearly, 1,3-dilithiopropyne is a most useful synthon in alkyne chemistry.

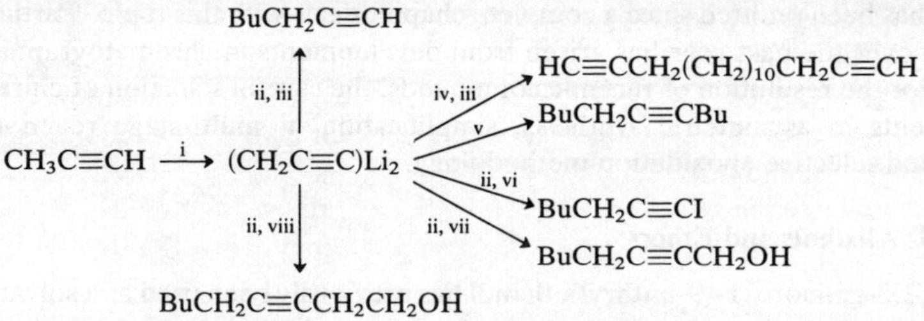

Reagents: i, BuLi, hexane, *NNNN*-tetramethylethylenediamine, at −60 °C; ii, BuBr; iii, H$^+$; iv, Br(CH$_2$)$_{10}$Br; v, BuBr, ether and then BuBr, HMPT; vi, I$_2$; vii, (CH$_2$O)$_n$; viii, CH$_2$CH$_2$O (epoxide)

Scheme 33

5 Allenes

A new general route has been presented for the synthesis of allenes from terminal alkynes (Scheme 34),[74] and the procedure has been used on acetylenic alcohols, ethers, and acetates as well as on unsubstituted terminal alkynes. The reaction probably involves the Mannich base as an intermediate, but the mechanism of the reaction has not been established.

$$RC\equiv CH \xrightarrow{i} RCH{=}C{=}CH_2$$

Reagents: i, CH$_2$O, HNPr$^i_2$, CuBr, dioxan, reflux

Scheme 34

[73] S. Bhanu and F. Scheinmann, *J.C.S. Perkin I*, 1979, 1218.
[74] P. Crabbé, H. Fillion, D. André, and J.-L. Luche, *J.C.S. Chem. Comm.*, 1979, 859.

8 Aliphatic Compounds
Part (ii) Other Aliphatic Compounds

By A. R. TATCHELL

School of Chemistry, Thames Polytechnic, Wellington Street, London SE18 6PF

An attempt has been made in this Report to provide some continuity with previous years, though with a slightly different order of sections; a section on phosphorus compounds has been omitted since a complete chapter deals with this topic. Particular interest over the past year has arisen from developments in chromatographic procedures for the resolution of racemic compounds, the careful selection of chiral inducing agents in asymmetric synthesis, simplification of multi-stage reaction sequences, and selective epoxidation methodology.

1 Alcohols and Ethers

The chiral 2,2,2-trifluoro-1-(9-anthryl)ethanol has previously been used as a solvating agent for the determination of enantiomeric purity and absolute configurations by n.m.r. spectroscopy of the derived diastereoisomeric solvates. Consideration of the stability differences of these solvates has led to the design of a chiral stationary phase that has enabled enantiomers of a wide range of functionality to be separated by h.p.l.c.[1] The stationary phase (1) was prepared from Porasil by treatment first with (3-mercaptopropyl)trimethoxysilane and then with (−)-(*R*)-2,2,2-trifluoro-1-[9-(10-α-bromomethyl)anthryl]ethanol. More than fifty resolutions have been effected, and it has been possible to develop a simple concept of the interactions between the solute enantiomers and the stationary phase, for assessment of the probability of a successful resolution and a prediction of elution order.

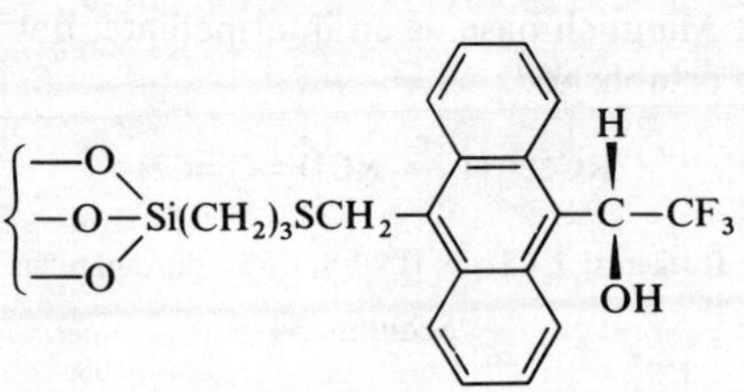

(1)

[1] W. H. Pirkle and D. W. House, *J. Org. Chem.*, 1979, **44**, 1957.

Full details have been reported of a general method that has been developed for the preparation of 1-deuterio primary alcohols, of high enantiomeric purity, by using chiral trialkylboranes as reducing agents.[2] The most effective reagent was *B*-(3-pinanyl)-9-borabicyclo[3.3.1]nonane, derived from (+)-α-pinene and 9-BBN, which then rapidly reduced [1-$^2H_1$]benzaldehyde to (+)-(*S*)-[1-$^2H_1$]benzyl alcohol; using optically pure (+)-α-pinene, none of the (*R*)-enantiomer could be detected by n.m.r. if one used a chiral shift reagent. With deuteriated β-([2-^2H]-3-pinanyl)-9-BBN a range of saturated, unsaturated, and aromatic aldehydes were converted into the corresponding (*R*)-1-deuterio primary alcohols, frequently in enantiomeric yields of greater than 90%. A critical study of the synthesis, configurations, and optical rotations of enantiomerically pure (+)-(*R*)-[2-$^2H_1$]-2-phenylethanol, (+)-(*R*)-[1,1,2-$^2H_3$]-2-phenylethanol, and variously deuteriated phenylethanes has been reported.[3]

Arsonated polystyrene resins have been employed as catalysts for the epoxidation of olefins with aqueous hydrogen peroxide;[4] a useful expoxidizing reagent, *O*-ethylperoxycarbonic acid, formed *in situ* from ethyl chloroformate and hydrogen peroxide under buffered conditions, has also been described.[5] Insertion of a methylene group into the carbonyl group of aldehydes and ketones to form oxirans, using lithiated derivatives of *N*-(toluene-*p*-sulphonyl)sulphimides has much to commend it.[6] Scheme 1 illustrates a specific example.

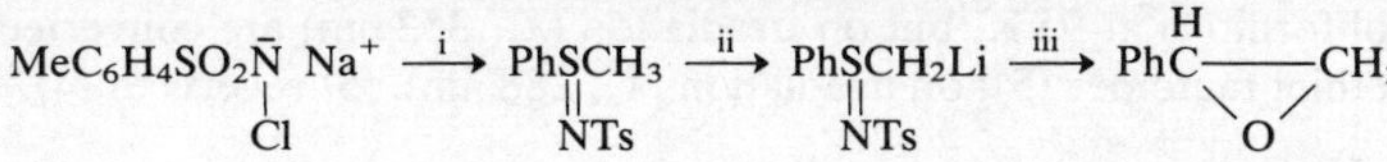

Reagents: i, $PhSCH_3$, CH_2Cl_2, phase-transfer catalyst; ii, BuLi, DMSO; iii, PhCHO

Scheme 1

Epoxidation of polyunsaturated substrates (*e.g.* arachidonic acid, eicosa-*cis*-5,8,11,14-tetraenoic acid) with peroxy-acid reagents is largely non-selective; thus a methodology leading to specific epoxidation of the 5,6- and of the 14,15-double bonds has been developed.[7] The 5,6-epoxide was obtained by treatment of the acid (as its potassium salt) with potassium tri-iodide, isolation of an unstable iodo-δ-lactone, followed by nucleophilic replacement of iodide with hydroxyl; the hydroxy δ-lactone then collapses to give the 5,6-epoxide. The 14,15-epoxide was obtained by conversion of the carboxyl group into the corresponding peroxy-acid, which on standing slowly underwent exclusive intramolecular oxygen transfer *via* a favourable fifteen-membered cyclic intermediate.

The ring-opening of epoxides with the anion from trimethylsilylacetonitrile affords a route to γ-trimethylsiloxy-nitriles, which may then be readily converted into γ-lactones.[8] Reductive ring-opening of epoxides with $Li(OBu^t)_3AlH$ in THF is usually very slow, but may be remarkably facilitated by the addition of molar quantities of triethylborane.[9]

[2] M. M. Midland, S. Greer, A. Tramontano, and S. A. Zderic, *J. Amer. Chem. Soc.*, 1979, **101**, 2352.
[3] R. L. Elsenbaumer and H. S. Mosher, *J. Org. Chem.*, 1979, **44**, 600.
[4] S. E. Jacobson, F. Mares, and P. M. Zambri, *J. Amer. Chem. Soc.*, 1979, **101**, 6946.
[5] R. D. Bach, M. W. Klein, R. A. Ryntz, and J. W. Holubka, *J. Org. Chem.*, 1979, **44**, 2569.
[6] C. R. Johnson, K. Mori, and A. Nakanishi, *J. Org. Chem.*, 1979, **44**, 2065.
[7] E. J. Corey, H. Niwa, and J. R. Falck, *J. Amer. Chem. Soc.*, 1979, **101**, 1586.
[8] I. Matsuda, S. Murata, and Y. Ishii, *J.C.S. Perkin I*, 1979, 26.
[9] H. C. Brown and S. Krishnamurthy, *J. Org. Chem.*, 1979, **44**, 3678.

2 Aldehydes and Ketones

X-Ray photoelectron spectroscopy has been used to measure the O_{1s} bonding energies of enolizable and non-enolizable dicarbonyl compounds.[10] Essentially, the aim was to determine whether the enol form exists in two asymmetric C_s forms, rapidly interconverting *via* the symmetric C_{2v}, or whether it exists entirely in the latter state. Since the method enables determinations to be made on a time scale of 10^{-16} s, some clear evidence on molecular symmetry was anticipated.

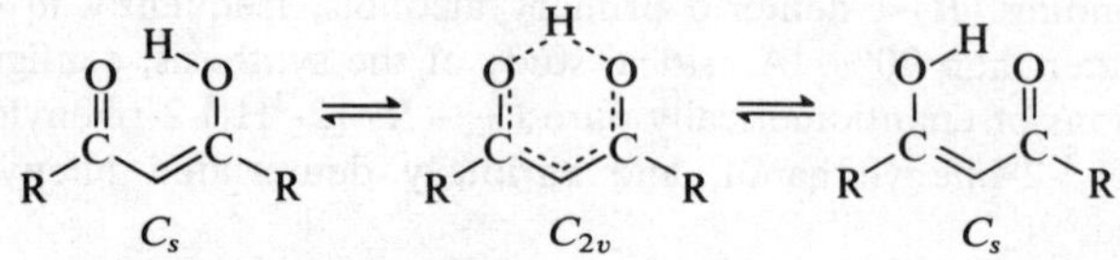

Two dominant ionizations arising from the two non-equivalent oxygens of the C_s forms were found for malonaldehyde, hexafluoroacetylacetone, tropolone, 9-hydroxyphenalenone (2), and 6-hydroxy-2-formylfulvene (3). The last example is of particular interest since the oxygen–hydrogen–oxygen exchange process would be expected to be extremely rapid, owing to the close proximity of the atoms concerned and their nearly linear relationship. β-Thioxo-ketones exist exclusively in the (*Z*)-enol form (4) at 95 K, but on irradiation (λ_{ex} 353 nm) are converted into the (*Z*)-enethiol tautomer (5); on irradiation (λ_{ex} 288 nm), (5) reverts to (4).[11]

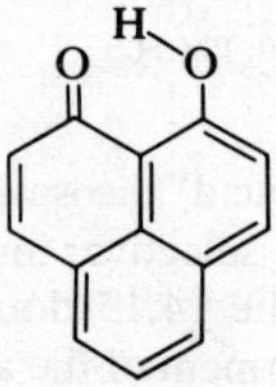
(2)

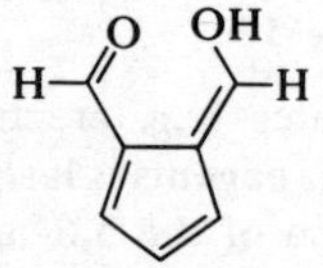
(3)

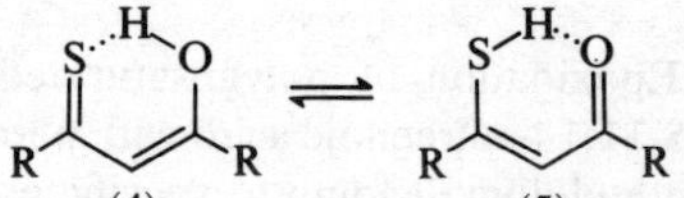
(4) (5)

The formation of (*S*)-2-methoxy-2-phenylpropionaldehyde (7) (see Scheme 2) in 97 ± 2% enantiomeric excess, as judged after conversion into the corresponding methyl ester and subsequent n.m.r. spectroscopy with added chiral reagents,[12] and of (*R*)-2-methoxy-2-phenylacetaldehyde (10) (see Scheme 3) in greater than 70% enantiomeric excess[13] illustrates the value of careful selection of the chiral inducing substrate [(6) or (8)] and the necessity of considering the probable transition state [*e.g.* (9)] of the reaction. It might be expected that both procedures will be exploited to prepare (in very high optical purity) a range of chiral α-hydroxy-aldehydes and their derived acids and alcohols.

The asymmetric reduction of unsymmetrical ketones to chiral secondary alcohols has been an area of activity for over a decade. Chiral ligands that have been popular favourites for complexing with metal hydrides are derived from amines or alcohols,

[10] R. S. Brown, A. Tse, T. Nakashima, and R. C. Haddon, *J. Amer. Chem. Soc.*, 1979, **101**, 3157.
[11] L. Carlsen and F. Duus, *J.C.S. Perkin II*, 1979, 1532.
[12] E. L. Eliel and W. J. Frazee, *J. Org. Chem.*, 1979, **44**, 3598.
[13] L. Colombo, C. Gennari, C. Scolastico, G. Guanti, and E. Narisano, *J.C.S. Chem. Comm.*, 1979, 591.

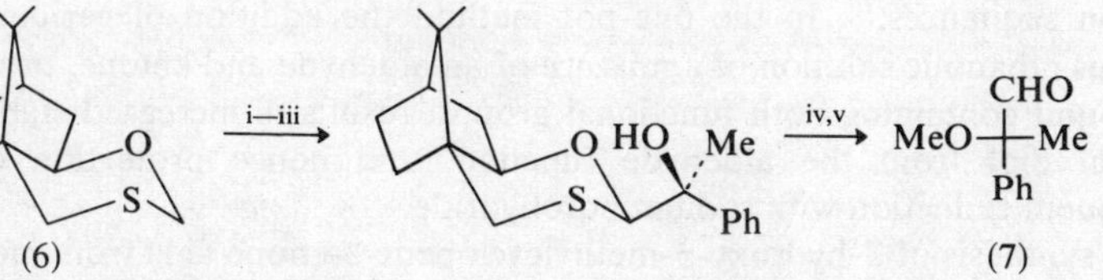

Reagents: i, Bu^sLi, PhCHO; ii, $(COCl)_2$, DMSO, Et_3N; iii, MeMgI; iv, NaH, MeI; v, hydrolysis

Scheme 2

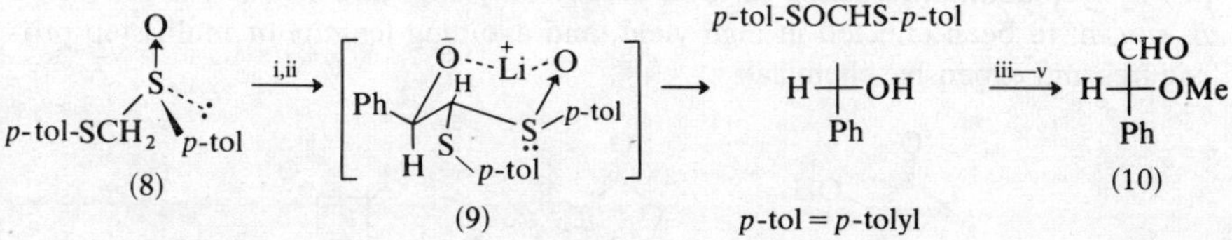

Reagents: i, BuLi; ii, PhCHO, at −78 °C; iii, Me_2SO_4, H_2O, CH_2Cl_2, Bu^nOH; iv, NaI, I_2, Ph_3P; v, I_2, $NaHCO_3$, H_2O, dioxan

Scheme 3

or from carbohydrates and the alkaloids. Currently, the enantiomeric excesses in such reductions need to be high if they are to merit attention. For example, the asymmetric reduction of a range of acetylenic ketones, utilizing the complex $[LiAlH(OR^*)(OAr)_2]$ which arises from $LiAlH_4$ and *N*-methylephedrine (R^*OH), modified by the further addition of 3,5-dimethylphenol (ArOH), gives enantiomeric excesses in the range 75—90%.[14] Reduction of ketones with optically pure β-hydroxysulphoximine–borane complexes has also been reported.[15] Attention has now been turned to improving the enantiomeric excess arising from the addition of an alkyl group to the carbonyl group of aldehydes and ketones for the formation of chiral secondary and tertiary alcohols. In one method[16] the lithium (or sodium) tetra-alkylaluminate was treated with either (−)-*N*-methylephedrine, (−)-quinine, or (+)-cinchonine (R^*OH) to give the chiral reagent $[LiAl(R)_3(OR^*)]$, which then reacted with carbonyl compounds to give chiral secondary and tertiary alcohols of useful optical purity. By employing chiral ligands derived from (*S*)-proline,[17] reaction with alkyl-lithium (and also dialkylmagnesium) gave chiral reagents, which then effected asymmetric addition to the carbonyl group of aldehydes to yield secondary alcohols in high optical yields (up to 95%).

Although several methods for the selective reduction of aldehydes in the presence of ketones are available, and indeed a further variant using lithium borohydride adsorbed on molecular-sieve zeolites has been shown to be effective,[18] there are no efficient and simple methods for a selective reduction in the opposite sense. An attractive solution to the problem has utilized the lanthanoids as catalysts in the

[14] J.-P. Vigneron and V. Bloy, *Tetrahedron Letters*, 1979, 2683.
[15] C. R. Johnson and C. J. Stark, Jr., *Tetrahedron Letters*, 1979, 4713.
[16] G. Boireau, D. Abenhaim, and E. Henry-Basch, *Tetrahedron*, 1979, **35**, 1457.
[17] T. Mukaiyama, K. Soai, T. Sato, H. Shimizu, and K. Suzuki, *J. Amer. Chem. Soc.*, 1979, **101**, 1455.
[18] P. A. Risbood and D. M. Ruthven, *J. Org. Chem.*, 1979, **44**, 3969.

reaction sequences.[19] In the one-pot method the addition of cerium(III) to an aqueous ethanolic solution of a mixture of an aldehyde and ketone, or indeed to a compound containing both functional groups, results in increased stability of the geminal diol from the aldehyde function, and hence protection during the subsequent reduction with sodium borohydride.

The synthesis of 2-hydroxy-3-methylcyclopent-2-enone (11) from methyl acrylate,[20] of dihydrojasmone (12; $R=C_5H_{11}$) and *cis*-jasmone (12; $R=CH_2CH\overset{z}{=}CHEt$) from undecane-2,5-dione and (*Z*)-undec-8-ene-2,5-dione respectively,[21] and of squaric acid (13) and some new derivatives,[22] utilizing a [2+2]-cycloaddition reaction of tetra-alkoxy-ethylenes and oxy-ketens generated *in situ*, have been effected in high yield, and avoiding lengthy or multi-step procedures and expensive chemicals.

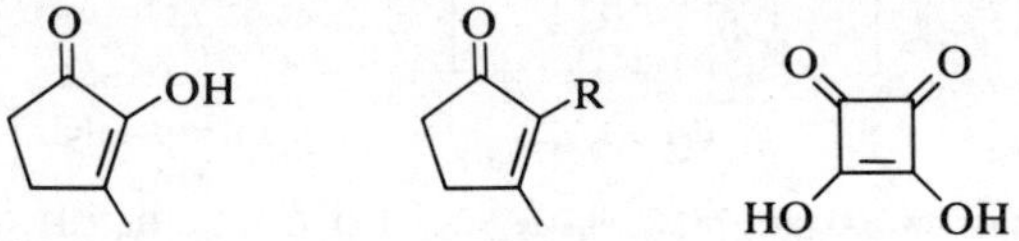

(11) (12) (13)

3 Carboxylic Acids

A flexible method which (in principle) allows for the preparation of all the chiral methyl chiral lactic acids, having all the possible permutations of the isotopes of hydrogen, has been developed.[23] The strategy for one such isomer is shown in Scheme 4. The conversion of (14) into (15) proceeds *via* a σ-vinyl complex; the process by which this is formed is completely stereospecific, as is the subsequent cleavage of the metal–vinyl bond.

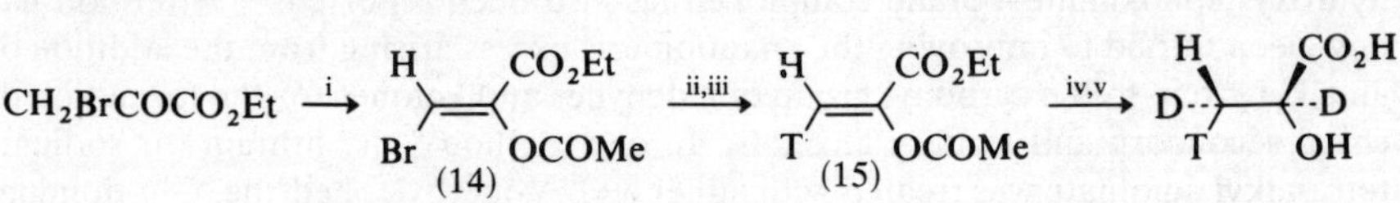

Reagents: i, $(MeCO)_2O$; ii, $[Pd(PPh_3)_4]$; iii, CF_3CO_2T, $(CF_3CO)_2O$; iv, D_2, $[Rh\{(R)\text{-prophos}\}nbd]^+$ ClO_4^-; v, ^-OH, HCl, crystallize from Et_2O–Pr^i_2O

Scheme 4

The stereospecific chain extension of an aldehyde to give a 5-substituted-2-methyl-(2*Z*,4*E*)-pentadienoic acid rather than the more stable (2*E*,4*E*)-isomer has been devised[24a] because of its relevance to the construction of the ansa bridge in the rifamycins,[24b] although the method clearly has more general applications (Scheme 5).

[19] J. L. Luche and A. L. Gemal, *J. Amer. Chem. Soc.*, 1979, **101**, 5848; *J. Org. Chem.*, 1979, **44**, 4187.
[20] R. C. Cookson, and S. A. Smith, *J.C.S. Perkin I*, 1979, 2447.
[21] C. S. Subramaniam, P. J. Thomas, V. R. Mamdapur, and M. S. Chadha, *J.C.S. Perkin I*, 1979, 2346.
[22] D. Belluš, *J. Org. Chem.*, 1979, **44**, 1208.
[23] M. D. Fryzuk and B. Bosnich, *J. Amer. Chem. Soc.*, 1979, **101**, 3043.
[24] (*a*) E. J. Corey and G. Schmidt, *Tetrahedron Letters*, 1979, 2317; (*b*) W. Oppolzer and V. Prelog, *Helv. Chim. Acta*, 1973, **56**, 2287.

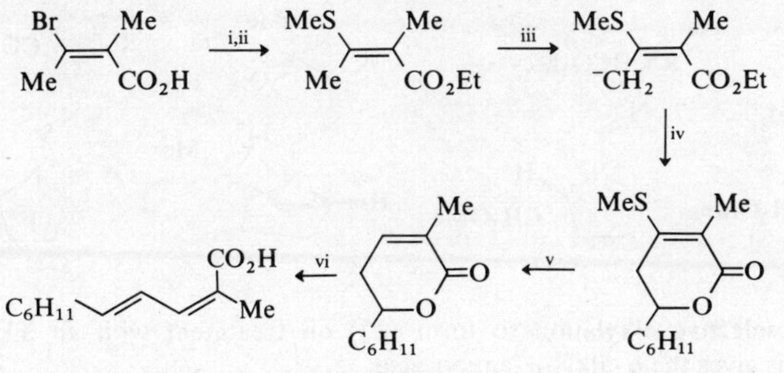

Reagents: i, EtI, K_2CO_3, DMF, ii, NaSMe, iii, $LiNPr^i_2$; iv, $C_6H_{11}CHO$; v, Raney nickel; vi, $KOBu^t$

Scheme 5

Full details are now available[25] of the formation of chiral α-hydroxy-acids (89—98% optical purity) by stereoselective bromolactonization of, for example, (−)-(*S*)-*N*-tigloylproline, followed by debromination and hydrolysis.

An efficient one-pot procedure for the exclusive formation of either the *erythro*- or of the *threo*-2-amino-3-hydroxy-acids is an example of the application of the stereoselective reaction between lithiated derivatives and carbonyl compounds.[26] For the (±)-*erythro*-isomer (17), *NN*-bis(trimethylsilyl)glycine trimethylsilyl ester (16), as its anion, was treated appropriately with the carbonyl compound, whereas for the *threo*-isomer (±)-threonine (18) the anion was derived from *N*-carbobenzoxyglycine ethyl ester (19); in this latter case the initial product was an oxazolidone derivative, which was then hydrolysed under acidic conditions. Scheme 6 illustrates both reactions, using acetaldehyde.

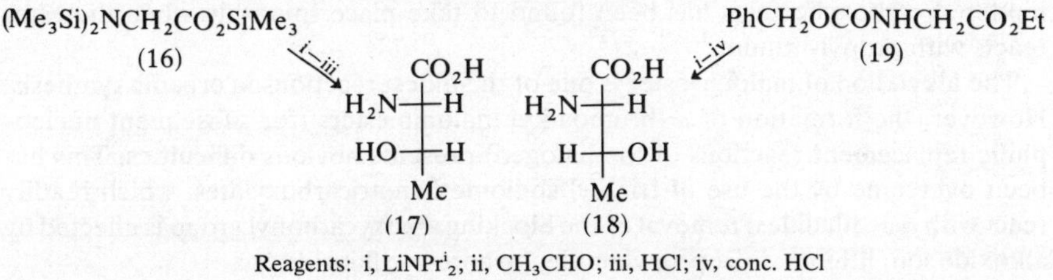

Reagents: i, $LiNPr^i_2$; ii, CH_3CHO; iii, HCl; iv, conc. HCl

Scheme 6

An asymmetric synthesis of α-alkyl-α-amino-acids, using (−)-(*S*)-1-dimethoxymethyl-2-methoxymethyl-pyrrolidine (20), which is readily prepared from (*S*)-proline, as the chiral auxiliary reagent, has been achieved with moderate enantiomeric excess.[27] The racemic acid reacts with the reagent to form a mixture of the diastereoisomeric amidine esters (21). Deprotonation with lithium di-isopropylamide gives a derivative (22), which then undergoes regiospecific and

[25] S.-S. Jew, S. Terashima, and K. Koga, *Tetrahedron*, 1979, **35**, 2337, 2345.
[26] A. Shanzer, L. Somekh, and D. Butina, *J. Org. Chem.*, 1979, **44**, 3967.
[27] M. Kolb and J. Barth, *Tetrahedron Letters*, 1979, 2999.

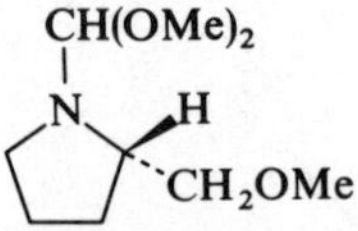

(20)

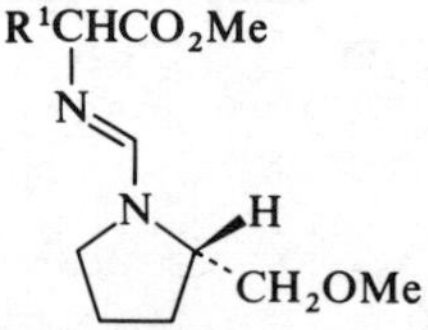

(21)

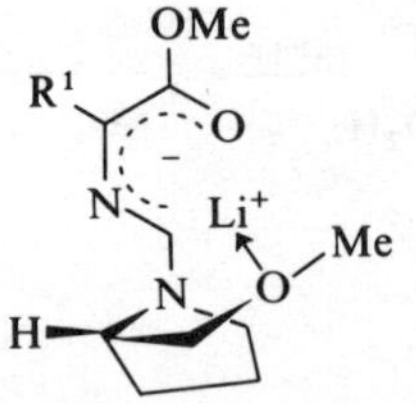

(22)

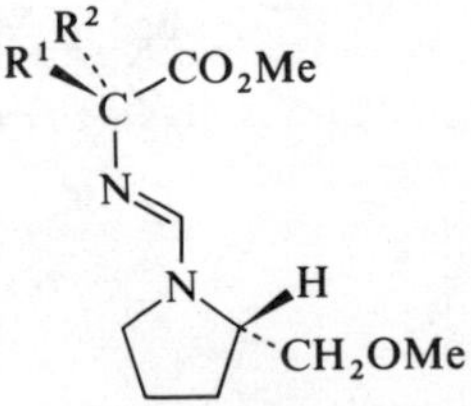

(23)

diastereoselective alkylation to form (23) on treatment with an alkyl halide; hydrolysis gives the α-alkyl-α-amino-acid.

Novel approaches to the synthesis[28] of (*R*)-6-methyltryptophan (a potential sweetening agent) and of ($\pm$)-gabaculine[29] (5-amino-cyclohexa-1,3-dienoic acid) have been devised to enable these compounds to be prepared in more useful amounts.

The resolution, and order of elution, of the enantiomers of racemic α-, β-, and γ-amino-acids, as their corresponding *N*-trifluoroacetyl isopropyl esters, by g.l.c., using *N*-dodecanoyl-L-valine-t-butylamide as the stationary phase, has provided a basis for a provisional hypothesis of selective intermolecular solute–solvent interactions. Coupled with a related study on the *N*-trifluoroacetyl-*O*-acyl derivatives of 2-amino-alkan-1-ols, a better appreciation of the influence of the function of the size of the *O*-acyl substituent on retention values has been demonstrated.[30]

4 Esters and Lactones

Simple procedures whereby sterically hindered acids may be esterified in high yields with unactivated alkyl halides, which involve the use of an anion-exchange resin in either a bi- or a tri-phase system, have been described.[31] Re-conversion into the hindered carboxylic acids has been found to take place smoothly when the ester reacts with propyl-lithium.[32]

The alkylation of malonic ester is one of the oldest reactions in organic synthesis. However, the formation of ω-bromoalkyl malonic esters (for subsequent nucleophilic replacement reactions of the halogen) presents obvious difficulties. This has been overcome by the use of trialkyl sodiomethanetricarboxylates, which readily react with $\alpha\omega$-dihalides; removal of the blocking alkoxycarbonyl group is effected by alkoxide ion, lithium di-isopropylamide, or boron trifluoride.[33]

β-Keto-esters of the type $R^1COCH_2CO_2R^2$ also occupy a position of central importance in the history of organic synthesis. New and convenient procedures for their preparation, in high yield, by the acylation of malonic acid derivatives,[34] or

[28] U. Hengartner, D. Valentine, K. J. Johnson, M. E. Larscheid, F. Piggott, F. Scheidl, J. W. Scott, R. C. Sun, J. M. Townsend, and T. H. Williams, *J. Org. Chem.*, 1979, **44**, 3741.
[29] B. M. Trost and E. Keinan, *J. Org. Chem.*, 1979, **44**, 3451; J.-P. Francois and M. W. Gittos, *Synth. Comm.*, 1979, **9**, 741.
[30] B. Feibush, A. Balan, B. Altman, and E. Gil-Av, *J.C.S. Perkin II*, 1979, 1230.
[31] G. M. Moore, T. A. Foglia, and T. J. McGahan, *J. Org. Chem.*, 1979, **44**, 2425.
[32] C. Lion, J.-E. Dubois, J. A. MacPhee, and Y. Bonzougou, *Tetrahedron*, 1979, **35**, 2077.
[33] H. C. Padgett, I. G. Csendes, and H. Rapoport, *J. Org. Chem.*, 1979, **44**, 3492, 4173.
[34] D. W. Brooks, L. D.-L. Lu, and S. Masamune, *Angew. Chem. Internat. Edn.*, 1979, **18**, 72; J. W. F. K. Barnick, J. L. van der Baan, and F. Bickelhaupt, *Synthesis*, 1979, 787; W. Wierenga and H. I. Skulnick, *J. Org. Chem.*, 1979, **44**, 310.

from the conversion of α-diazo-β-hydroxy-esters, catalysed by rhodium(II) acetate,[35] offer promise to supplement the methods already available.

The absolute configurations of glycidic esters are required both for a study of the stereochemical consequences of their rearrangement with Lewis acids to the corresponding formyl esters[36] and for a study of the stereoselectivity of epoxidation of an $\alpha\beta$-unsaturated ester having a chiral *O*-alkyl group.[37] The specific example that has currently been selected for both these aspects is ethyl (+)-(2*S*,3*R*)-3-methyl-3-phenylglycidate (24); the configuration at C-3 has been unambiguously determined by the sequence of conversions shown in Scheme 7.

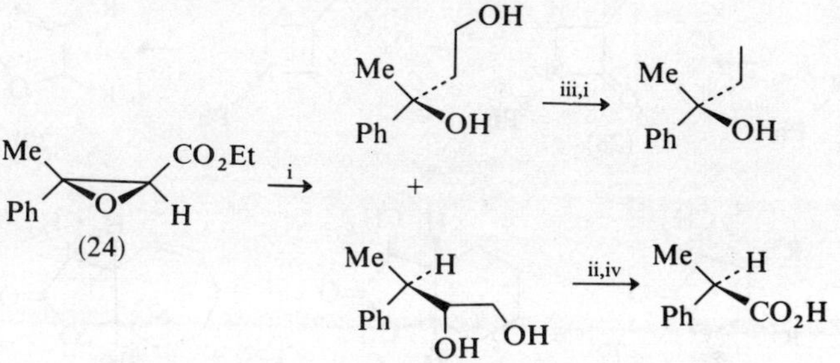

Reagents: i, $LiAlH_4$; ii, HIO_4, $NaHSO_3$; iii, TsCl; iv, $KMnO_4$

Scheme 7

The Diels–Alder reaction of dimethylfulvene with methyl acrylate provides an adduct (25) which, when treated with lithium di-isopropylamide and an alkyl halide, gives the product (26), in which the alkyl group occupies the *exo*-position. Pyrolysis effects the retro-reaction to yield methyl 2-alkyl-prop-2-enoate.[38] A synthetic route to 2-alkyl-but-2-eno-lactones (27) has been developed from this procedure and its utility demonstrated by the synthesis of some relevant naturally occurring lactones.[39]

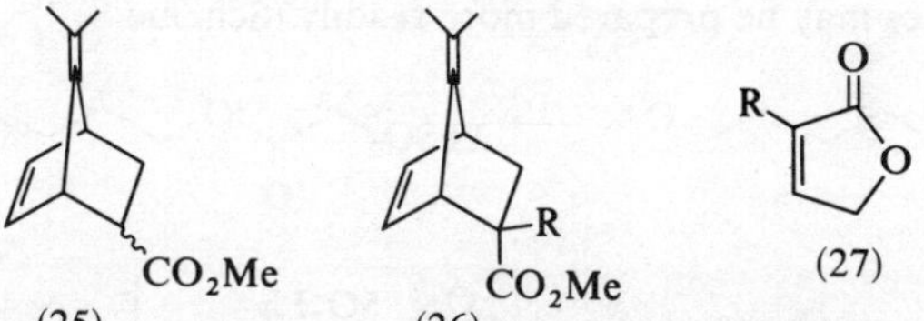

(25) (26) (27)

The formation of α-methylene-γ-butyrolactones from 1-phenyl-β-propane lactam offers a synthetic route that should be of wide applicability.[40a] In essence, the sequence involves the formation of 3-alkylideneazetidin-2-one (28), sequential isomerization and epoxidation of the double bond, and finally an acid-catalysed

[35] R. Pellicciari, R. Fringuelli, P. Ceccherelli, and E. Sisani, *J.C.S. Chem. Comm.*, 1979, 959.
[36] J. M. Domagala and R. D. Bach, *J. Org. Chem.*, 1979, **44**, 2429, 3168.
[37] S. L. Abidi and J. L. Wolfhagen, *J. Org. Chem.*, 1979, **44**, 433.
[38] R. Kimara, A. Ichihara, and S. Sakamura, *Synthesis*, 1979, 516.
[39] A. Ichihara, N. Nio, Y. Terayama, R. Kimura, and S. Sakamura, *Tetrahedron Letters*, 1979, 3731.
[40] (*a*) S. Kano, T. Ebata, K. Funaki, and S. Shibuya, *J. Org. Chem.*, 1979, **44**, 3946, (*b*) L. G. Mueller and R. G. Lawton, *ibid.*, p. 4741.

rearrangement to the anilinomethyl-butenolide (29). Catalytic hydrogenation over Raney nickel followed by a Hofmann degradation of the derived quaternary methiodide gives the α-methylene-γ-butyrolactone (30). When cyclic ketones are employed in the formation of the alkylidene derivative, the ultimate product is the *cis*-fused ring system, *e.g.* (31). The *trans*-α-methylene-γ-lactone system(32) has been obtained by an acid-catalysed rearrangement of t-butyl 2-*exo*-(bicyclo-[4.1.0]hept-2-en-7-yl)propenoate.[40b]

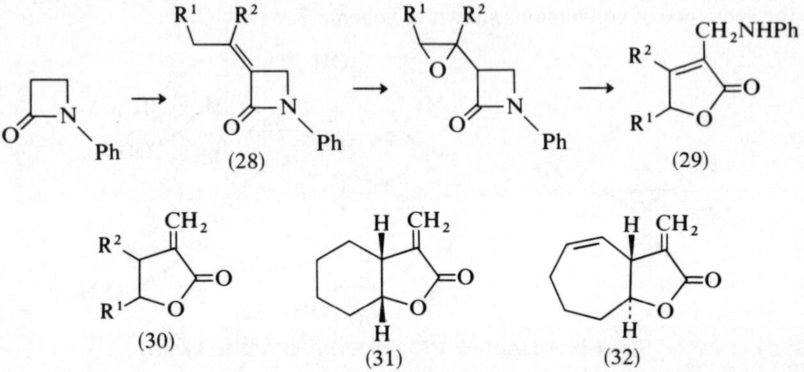

The formation of large ring lactones, by the tetrakis(triphenylphosphine)palladium-catalysed cyclization of allylic acetates having a sulphone function at a more remote position in the carbon chain, which was noted in earlier Reports, has now been applied to the formation of medium-sized ring lactones. The cyclization process allows for the alternative formation of products of two ring-sizes; however, a preference for the larger ring in the 8/6, 9/7, and 10/8 possibilities is observed. This finding, which is contrary to expectations on the basis that the formation of the smaller ring is kinetically more favoured, means that these less accessible lactones may be prepared more readily (Scheme 8).[41]

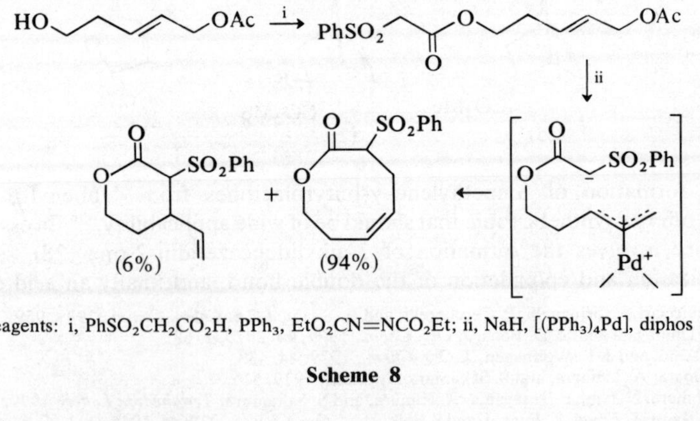

Reagents: i, $PhSO_2CH_2CO_2H$, PPh_3, $EtO_2CN{=}NCO_2Et$; ii, NaH, $[(PPh_3)_4Pd]$, diphos

Scheme 8

[41] B. M. Trost and T. R. Verhoven, *J. Amer. Chem. Soc.*, 1979, **101**, 1595.

5 Amides and Lactams

N-Alkyl- and *N*-aryl-amides may be prepared in a one-pot reaction directly from aliphatic and aromatic acids; the species (33) has been proposed as the intermediate (Scheme 9). The reaction, which proceeds to give almost quantitative yields of product, has been adapted for the synthesis of five-, six-, and seven-membered ring lactams from ω-amino-carboxylic acids.[42]

$$o\text{-NCSC}_6\text{H}_4\text{NO}_2 \xrightarrow{\text{i}} \text{Bu}_3\overset{+}{\text{P}}\text{SC}_6\text{H}_4\text{NO}_2 \xrightarrow{\text{ii}} \underset{(33)}{[\text{R}^1\text{CO}_2\overset{+}{\text{P}}\text{Bu}_3]} \xrightarrow{\text{iii}} \text{R}^1\text{CONHR}^2$$

Reagents: i, Bu_3P; ii, R^1CO_2H; iii, R^2NH_2

Scheme 9

Liquid chromatographic separation of diastereoisomeric amides derived from either racemic carboxylic acids or lactones and optically pure amines, followed by hydrolysis, provides an efficient preparative method for optically pure acids and lactones.[43]

Two β-lactam syntheses have been reported. The first involves cyclization of the 1,3-dianion from an α-phenyl-thioacetamide derivative with di-iodomethane (Scheme 10); subsequently, oxidation at sulphur gave separable diastereoisomeric sulphoxides, or alkylation at C-3 was effected after formation of an anion by reaction with butyl-lithium.[44] In the second synthesis the C-4—N bond was formed under milder conditions than those hitherto reported for the cyclization of β-halogeno-amides (34; X = Cl), and hence provided a reaction sequence that was compatible with the presence of substituents on C-3 which could result in facile racemization; the

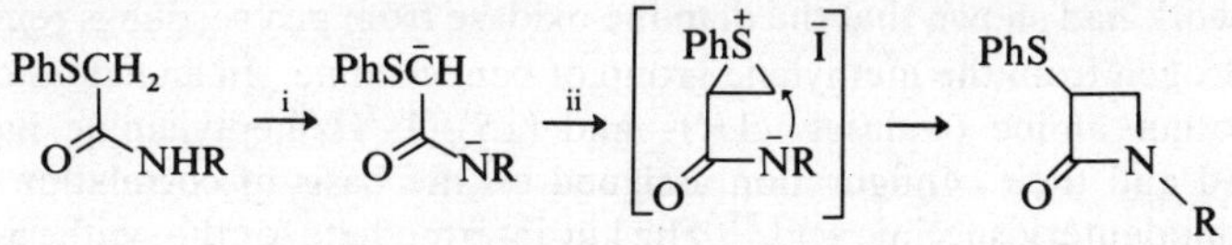

Reagents: i, NaH; ii, CH_2I_2

Scheme 10

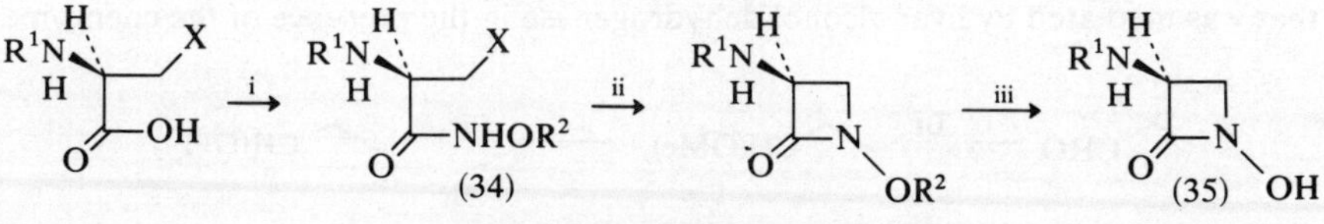

$R^1 = CO_2Bu^t$, $R^2 = CH_2Ph$

Reagents: (X = Cl) i, 1-ethyl-3-(3-dimethylaminopropyl)carbodiimide (WSC), DMF, NH_2OCH_2Ph; ii, NaH, DMF; iii, H_2, Pd/C
(X = OH) i, WSC, DMF, NH_2OCH_2Ph; ii, Ph_3P, $EtO_2CN{=}NCO_2Et$; iii, H_2, Pd/C

Scheme 11

[42] P. A. Grieco, D. S. Clark, and G. P. Withers, *J. Org. Chem.*, 1979, **44**, 2945.
[43] H. Helmchen, G. Nill, D. Flockerzi, W. Schuhle, and M. S. K. Yousset, *Angew. Chem. Internat. Edn.*, 1979, **18**, 62, 63, 65.
[44] K. Hirai and Y. Iwano, *Tetrahedron Letters*, 1979, 2031.

sequence provided a route to substituted *N*-hydroxy-2-azetidinones (35).[45] The success of the method depends on the selective ionization of the N—H bond in the hydroxamic acid derivative (34). Indeed, the reaction was developed for direct use with the β-hydroxy-acid derivative (34; X = OH), as a convenient alternative to the β-halogeno-amide route (Scheme 11).

Three novel fused β-lactam compounds (36) have been isolated from media after culture of *Streptomyces clavuligerus*, and they have been shown to be structurally related to the previously reported clavulanic acid (37).[46]

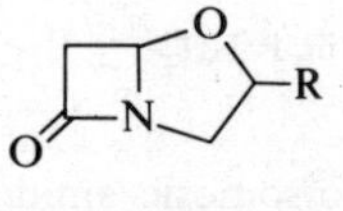

(36) R = CH_2OH, CH_2OCHO, or CO_2H

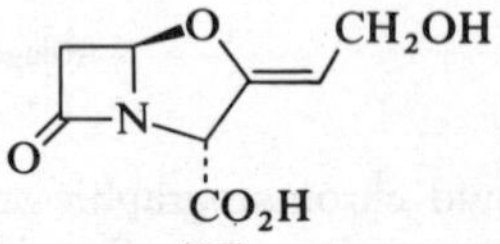

(37)

6 Amines and Imines

The conversion of a primary amine into an aldehyde by amine oxidases may proceed *via* the removal of either the *pro-R*-hydrogen or the *pro-S*-hydrogen (Scheme 12).

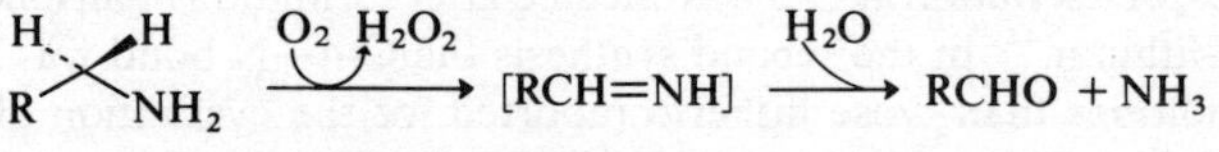

Scheme 12

Previous work had shown that the diamine oxidase from pea seedlings removes the *pro-S*-hydrogen from the methylene group of benzylamine. In an extension of this work to other amine oxidases, (1*R*)- and (1*S*)-[1-$^3H_1$]heptylamine have been synthesized and their configuration assigned on the basis of correlation with the known monodeuteriosuccinic acid.[47] The key intermediate for the synthesis of these labelled amines was (*Z*)-hept-3-enal (38), prepared from acrolein (Scheme 13). Deuterium and tritium labelling was introduced to give the corresponding alcohols (39) (see Scheme 14) by stereospecific transfer from labelled ethanol in a reaction that was mediated by liver alcohol dehydrogenase in the presence of the coenzyme

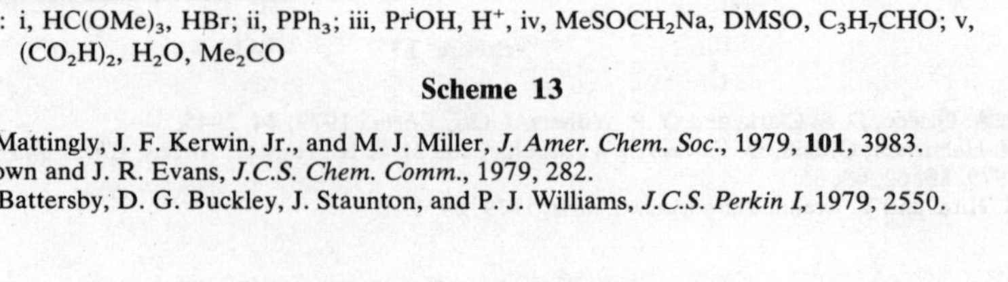

Reagents: i, $HC(OMe)_3$, HBr; ii, PPh_3; iii, Pr^iOH, H^+, iv, $MeSOCH_2Na$, DMSO, C_3H_7CHO; v, $(CO_2H)_2$, H_2O, Me_2CO

Scheme 13

[45] P. G. Mattingly, J. F. Kerwin, Jr., and M. J. Miller, *J. Amer. Chem. Soc.*, 1979, **101**, 3983.
[46] D. Brown and J. R. Evans, *J.C.S. Chem. Comm.*, 1979, 282.
[47] A. R. Battersby, D. G. Buckley, J. Staunton, and P. J. Williams, *J.C.S. Perkin I*, 1979, 2550.

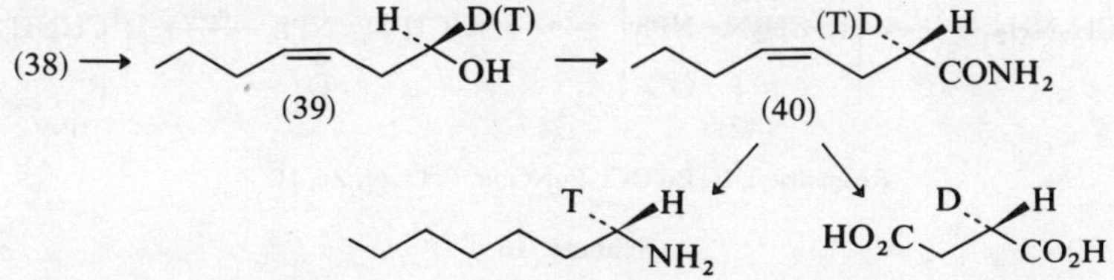

Scheme 14

NAD$^+$. Standard conversions (toluene-*p*-sulphonate into nitrile; then into amide) gave the labelled amides (40), of inverted configuration. The deuteriated acid, on ozonolysis, gave 94 ± 5% optically pure (+)-(2*S*)-[2-$^2H_1$]succinic acid; the tritiated amide, by way of hydrogenation, hydrolysis, and Schmidt degradation, gave (1*S*)-[1-$^3H_1$]heptylamine. The enantiomeric amine was formed by a similar route, but using the tritiated analogue of (38) {(*Z*)-[1-$^3H_1$]hept-3-enal}, formed by the reaction of (38) with sodium borotriti-ide followed by oxidation with chromium trioxide in pyridine, wherein the favourable ^{3}H isotope effect was clearly beneficial. The configurationally authenticated amines, when studied with the monoamine oxidase from rat liver mitochondria, unequivocally demonstrated that the *pro-R*-hydrogen was stereospecifically removed during oxidative deamination.

(*R*)-α-Phenylalkyl-amines have been formed in high optical purity (76—96%) by the stereoselective alkylation (Grignard reagent) of the hydrazone (41) derived from an aryl-aldehyde and (−)-*N*-aminoephedrine (Scheme 15).[48] The stereoselectivity was considered to be due to the (*E*)-stereochemistry of the hydrazone and complexation of the hydrazone with the magnesium of the Grignard reagent before attack by the alkyl group.

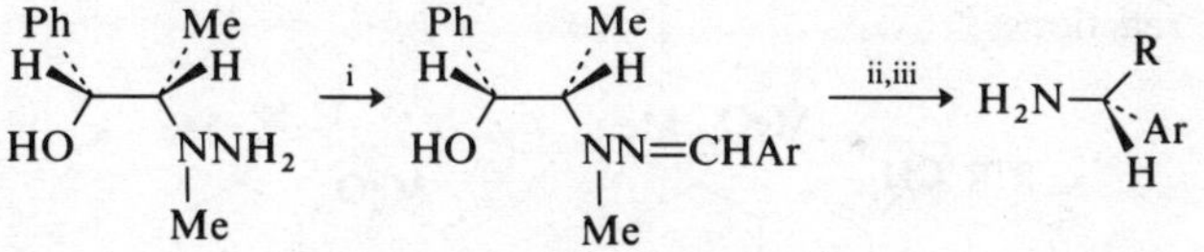

(41)

Reagents: i, ArCHO; ii, RMgX; iii, Pd/C

Scheme 15

The reductive alkylation of azoxy-compounds affords a route whereby α-alkylation of a primary amine may be efficiently carried out (Scheme 16).[49] The intermediate (42) was seen to arise from the homolysis of the di-iodo-amine (formed from iodination under oxidizing conditions) followed by trapping with nitrosobenzene. The same intermediate could equally well arise from homolysis of the monoiodo-amine, trapping, and further iodination. Characteristically, the examples selected for illustrating the value of the method were drawn from simple aliphatic amines, steroidal amines, and amino-glycosides.

[48] H. Takahashi, K. Tomita, and H. Otomasu, *J.C.S. Chem. Comm.*, 1979, 668.
[49] D. H. R. Barton, G. Lamotte, W. B. Motherwell, and S. C. Narang, *J.C.S. Perkin I*, 1979, 2030.

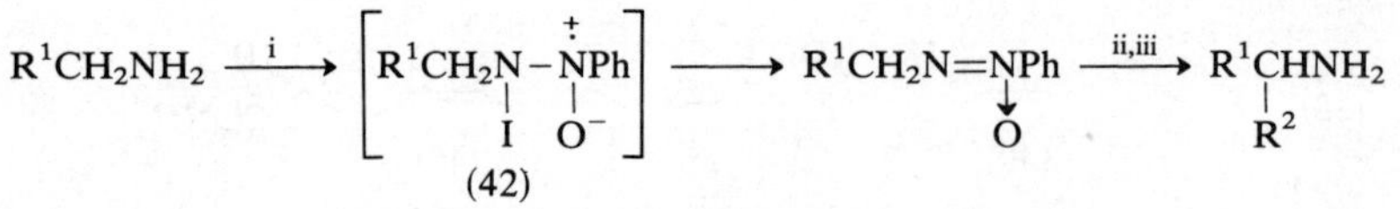

Reagents: i, I_2, Bu^tOCl, PhNO; ii, R^2Li; iii, Zn, H^+

Scheme 16

Full details are now available for the conversion of primary amines into iodides,[50a] bromides and chlorides,[50b] fluorides,[50c] esters,[50d] aldehydes,[50e] and hydrocarbons.[50f] With each conversion, a range of alkyl-, arylalkyl, aryl- and heteroaryl-amine systems were studied, and the limitations of the method with respect to each of these systems noted.

(+)-2-Amino-1-methoxymenth-8-ene reacts with (±)-aldehydes that are chiral at the α-position to give diastereoisomeric aldimines (43). The ratio of diastereoisomers, and hence the optical purity of the aldehyde, may be assessed with good accuracy from integration of the signals corresponding to the pair of doublets from the proton attached to C-1 of the aldimine.[51] This methoxy-amine may also be used as a chiral inducing agent since the aldimine (43; $R^1 = H$, $R^2 =$ Hex or Me), after conversion into the lithiated derivative, readily undergoes a stereoselective reaction at C-2 with a range of electrophiles; subsequent hydrolysis yields the chiral α-alkyl-aldehyde.[52] Similar alkylations of the aldimine derived from the tertiary amine (44) show somewhat lower stereoselectivity. In the case of aldimines derived from (−)-(*S*)-phenylethylamine, where the absence of a chelatable group might be expected to reduce the stereoselectivity of an alkylation reaction, the yields and optical purities of the chiral aldehydes finally isolated are most encouraging.[53] In both these alkylation reactions, interesting concepts of the transition states and of the conformation of the intermediates have been suggested, to explain the stereoselectivity of the reactions.

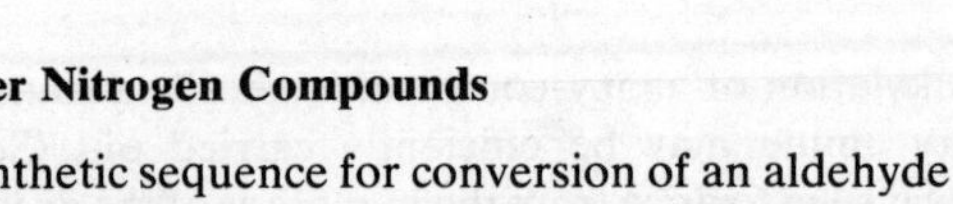

(43) (44)

7 Other Nitrogen Compounds

A conventional synthetic sequence for conversion of an aldehyde (R^1CHO) into the nitrile (R^1R^2CHCN) may proceed through as many as eight stages, in overall yields

50 (*a*) A. R. Katritzky, N. F. Eweiss, and P.-L. Nie, *J.C.S. Perkin I*, 1979, 433; (*b*) A. R. Katritzky, U. Gruntz, A. A. Ikizler, D. H. Kenny, and B. P. Leddy, *ibid.*, p. 436; (*c*) A. R. Katritzky, A. Chermprapai, and R. C. Patel, *J.C.S. Chem. Comm.*, 1979, 238; (*d*) A. R. Katritzky, U. Gruntz, D. H. Kenny, M. C. Rezende, and H. Shiekh, *J.C.S. Perkin I*, 1979, 430; (*e*) A. R. Katritzky, M. J. Cook, A. Ikizler, and G. H. Millet, *ibid.*, p. 2500; (*f*) A. R. Katritzky, J. Lewis, and P.-L. Nie, *ibid.*, p. 442.

51 A. I. Meyers and Z. Brich, *J.C.S. Chem. Comm.*, 1979, 566.

52 A. I. Meyers, Z. Brich, G. W. Erickson, and S. G. Traynor, *J.C.S. Chem. Comm.*, 1979, 567.

53 R. R. Fraser, F. Akiyama, and J. Banville, *Tetrahedron Letters*, 1979, 3929; R. R. Fraser and J. Banville, *J.C.S. Chem. Comm.*, 1979, 47.

which rarely exceed 12%. An ingenious continuous process has been devised which allows for the quenching of the intermediate tertiary halide, by allowing the reaction mixture to flow into a phase-transfer catalytic system, to give finally the nitrile in >90% yield.[54]

Alkyl and aryl isocyanates are formed, in excellent yield, by heating dry 2,4,6-triphenylpyridine *N*-acyl-imines (45) under reduced pressure. The acyl-imines are prepared either from 1-amino-2,4,6-triphenylpyridinium perchlorate (46) or from 2,4,6-triphenylpyrylium perchlorate, as shown in Scheme 17.[55] Alkyl isocyanates (and also substituted ureas and urethanes) may also be synthesized, in wide variety, from the reagent (48),[56] which, incidentally, has also been used for the preparation of *N*-t-butoxycarbonyl-amino-acid esters.[57]

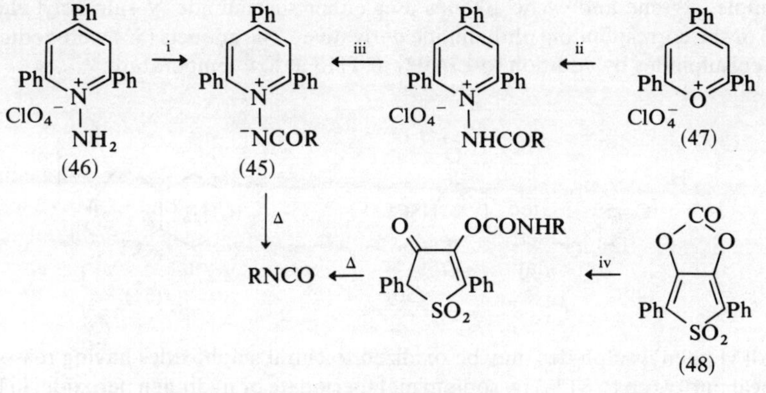

Reagents: i, $RCOCl$, K_2CO_3; ii, $RCONHNH_2$; iii, KOH; iv, RNH_2

Scheme 17

The coupling reaction between lithium acetylides and 2-chloro-2-nitropropane gives high yields of nitro-acetylenes, which may then be elaborated to $\alpha\beta$-unsaturated ketones (Scheme 18).[58]

$$RC{\equiv}CLi \xrightarrow{i} RC{\equiv}CCMe_2NO_2 \xrightarrow{ii} [RCOCH_2CMe_2NO_2] \longrightarrow RCOCH{=}CMe_2$$

Reagents: i, $ClCMe_2NO_2$; ii, Hg^{2+}, H_2SO_4

Scheme 18

A nitro-group sited in an aliphatic or alicyclic system may be replaced with hydrogen by reaction with the sodium salt of methanethiol in HMPT or DMSO by a sequence which proceeds *via* a radical mechanism. With β-aryl-nitro-alkanes, where the aryl residue is unsubstituted, replacement with hydrogen occurs similarly in both HMPT and DMSO; when the aryl residue has electron-withdrawing

[54] J. A. Foulkes and J. Hutton, *Synth. Comm.*, 1979, **9**, 625.
[55] A. R. Katritzky, L. Lewis, and P-L. Nie, *J.C.S. Perkin I*, 1979, 446.
[56] H. Schmidt, O. Hollitzer, A. Seewald, and W. Steglich, *Chem. Ber.*, 1979, **112**, 727.
[57] G. Schnorrenberg and W. Steglich, *Angew. Chem. Internat. Edn.*, 1979, **18**, 307.
[58] M. Jawdosiuk, M. Makasza, B. Mudryk, and G. A. Russell, *J.C.S. Chem. Comm.*, 1979, 488.

substituents, the reaction becomes solvent-dependent, and replacement of the nitro-group with the thiomethyl group predominates in DMSO. With branched-chain nuclear-substituted β-aryl-nitro-alkanes, rearrangment of the carbon skeleton side-chain during reaction provides a rationale of the mechanism of this interesting reaction.[59]

8 Sulphur Compounds

Application of flash vacuum pyrolysis and microwave spectroscopic techniques has enabled the structure of the lachrymatory factor of the onion to be finally identified as (*Z*)-propanethial *S*-oxide (49).[60]

A new method for the formation of a range of episulphides, in high yield, from, for example, styrene and cyclic alkenes uses either succinimide *N*-sulphenyl chloride (50) or the corresponding phthalimide derivative. The adducts (51) were reduced to the episulphides by addition to $LiAlH_4$ in THF at low temperature.[61]

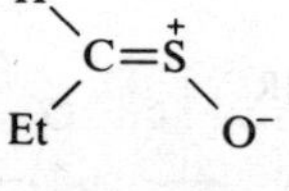

(49)

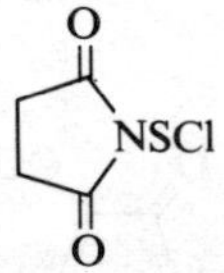

(50)

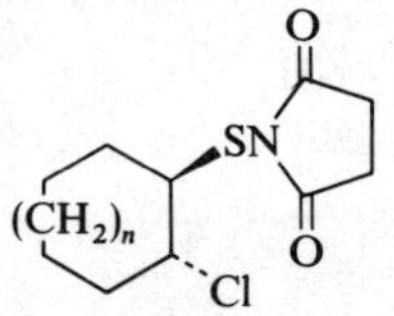

(51)

Alkyl phenyl sulphides may be oxidized to chiral sulphoxides having reasonable optical purity (up to 81%) by sodium metaperiodate or hydrogen peroxide, in buffer solution, with bovine serum albumin. Prolonged oxidation gave lower chemical yields of sulphoxide, but usually of greater optical purity. This was seen to arise from a two-stage oxidation process, the first stage being an oxidation to the chiral sulphoxide having excess of one enantiomer and the second a preferential oxidation of the less abundant enantiomer to the sulphone.[62] Prediction of the configuration of the sulphoxides from a consideration of the nature of the substituents is at present premature.

Racemic 2-alkyl-tetrahydrothiopyran-4-ones (±)-(53), readily formed from methyl acrylate (52) as shown in Scheme 19, give the optically pure *cis*- and *trans*-2-alkyl-tetrahydrothiopyran-4-ols (54) and (55), each having (*S*) configuration at C-4, when submitted to horse liver alcohol dehydrogenase (HLADH)-mediated reduction. These products, which may be readily separated by chromatography, were converted into the optically pure (3*S*)-alcohols (56) by desulphurization. The latter are not obtained in an optically pure state by the enzymic reduction of the corresponding acyclic ketones.[63]

[59] N. Kornblum, S. C. Carlson, and R. C. Smith, *J. Amer. Chem. Soc.*, 1979, **101**, 647; N. Kornblum, J. Widmer, and S. C. Carlson, *ibid.*, p. 658.

[60] E. Block, R. E. Penn, and L. K. Revelle, *J. Amer. Chem. Soc.*, 1979, **101**, 2200.

[61] M. V. Bombala and S. V. Ley, *J.C.S. Perkin I*, 1979, 3013.

[62] T. Sugimoto, T. Kokubo, J. Miyazaki, S. Tanimoto, and M. Okano, *J.C.S. Chem. Comm.*, 1979, 402, 1052.

[63] J. Davies and J. B. Jones, *J. Amer. Chem. Soc.*, 1979, **101**, 5405.

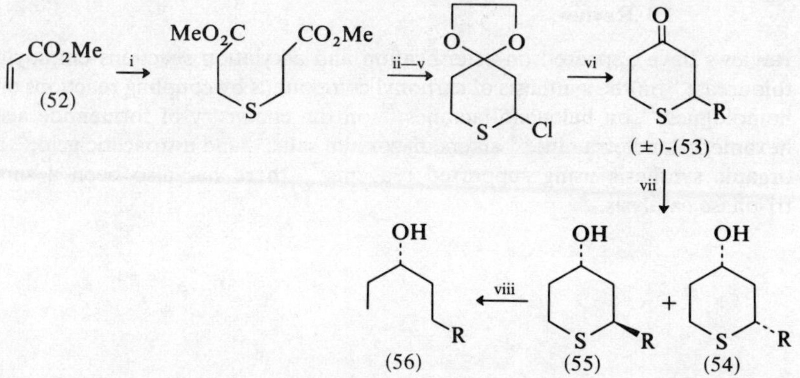

Reagents: i, H_2S; ii, NaH; iii, H^+, Δ; iv, $(HOCH_2)_2$, H^+; v, *N*-chlorosuccinimide; vi, RMgX, H^+; vii, HLADH, NAD^+; viii, Ni

Scheme 19

9 Alkyl Halides

The nucleophilic substitution reactions of alkyl halides are the most widely used and the most effective monitor of the efficiency of bi- and tri-phase catalytic systems. Catalysts currently being further studied for value in this type of reaction are principally, though not exclusively, quaternary ammonium and phosphonium salts immobilized on a polystyrene matrix,[64] or on silica gel.[65] Polymer-bound acyclic poly(oxyethylene) monomethyl ethers have been used for generation of dichlorocarbene and for Michael addition and Wittig reactions.[66] Apart from surveying the preparative value of these systems, attention is being directed to correlating such factors as, for example, the distance between the anionic and cationic centres, the length of the carbon chain linking the ionic site to the polymer matrix (which influences the degree of protrusion of the site into the reaction medium), and the porosity of the resin. Alumina has also been shown to catalyse nucleophilic substitutions and the oxidation of alcohols by permanganate.[67]

Polymer-immobilized quaternary ammonium fluorides have been effectively used for the introduction of fluorine (by replacement of sulphonyloxy-groups) into the steroid nucleus and into monosaccharide residues.[68] The synthetic value of lithium organocuprates is likely to be enhanced by recent studies on polymer-bound iodo(triarylphosphine)copper(I) in contact with a solution of the alkyl- or aryl-lithium.[69]

[64] S. L. Regen, J. C. K. Heh, and J. McLick, *J. Org. Chem.*, 1979, **44**, 1961; S. L. Regen and J. J. Besse, *J. Amer. Chem. Soc.*, 1979, **101**, 4059; H. Molinari, F. Montanari, S. Quici, and P. Tundo, *ibid.*, p. 3920; M. S. Chiles and P. C. Reeves, *Tetrahedron Letters*, 1979, 3367.
[65] P. Tundo and P. Venturello, *J. Amer. Chem Soc.*, 1979, **101**, 6606.
[66] S. Yanagida, K. Takahashi, and M. Okahara, *J. Org. Chem.*, 1979, **44**, 1099.
[67] S.-J. Liaw, S. Quici, and S. L. Regen, *J. Org. Chem.*, 1979, **44**, 2029; S. Quici and S. L. Regen, *ibid.*, p. 3436; S. L. Regen, S. Quici, and M. D. Ryan, *J. Amer. Chem. Soc.*, 1979, **101**, 7629.
[68] S. Colonna, A. Re, G. Gelbard, and E. Cesarotti, *J.C.S. Perkin I*, 1979, 2248.
[69] R. H. Schwartz and J. S. Filippo, Jr., *J. Org. Chem.*, 1979, **44**, 2705.

10 Reviews

Reviews have appeared on esterification and alkylation reactions employing isothioureas;[70] on the synthesis of carbonyl compounds by coupling reactions of lower homologues;[71] on halogeno-lactones;[72] on the chemistry of formamide acetals,[73] hexamethylenetetramine,[74] alkenediazonium salts,[75] and nitroacetic acid;[76] and on organic synthesis using supported reagents;[77] there has also been a survey of tri-phase catalysis.[78]

[70] L. J. Mathias, *Synthesis*, 1979, 561.
[71] S. F. Martin, *Synthesis*, 1979, 633.
[72] M. D. Dowle and D. I. Davies, *Chem. Soc. Rev.*, 1979, **8**, 171.
[73] R. F. Abdulla and R. S. Brinkmeyer, *Tetrahedron*, 1979, **35**, 1675.
[74] N. Blažević, D. Kolbah, B. Belin, V. Šunjić, and F. Kadfež, *Synthesis*, 1979, 161.
[75] K. Bott, *Angew. Chem. Internat. Edn.*, 1979, **18**, 259.
[76] M. T. Shipchandler, *Synthesis*, 1979, 666.
[77] A. McKillop and D. W. Young, *Synthesis*, 1979, 422, 481.
[78] S. L. Regen, *Angew. Chem. Internat. Edn.*, 1979, **18**, 421.

9 Alicyclic Chemistry

By A. COX
Department of Chemistry and Molecular Sciences, University of Warwick Coventry CV4 7AL

1 Introduction

Reviews have been published on the potential of electrophilic cyclopropanes in synthesis,[1] on the preparation of squaric acid and derivatives,[2] and on methods available for the synthesis of bicyclo[3.3.1]nonanes.[3]

2 Synthesis

Monocycles.—A new facile synthesis of cyclopropyl-alkadienes has appeared.[4] The reaction of diazo-compounds with (*E,E*)-octa-1,3,5-triene and with (*Z*)- and (*E*)-hexa-1,3,5-trienes gives 1-pyrazolines, resulting from regiospecific 1,3-dipolar addition of the diazo-compound to the triene at the terminal double bond. Thermal or photochemical decomposition of the pyrazoline leads exclusively to the desired product, in high yield. This method provides access to a large number of different cyclopropyl-alkadienes. At moderate temperatures, the reaction of olefins with *gem*-dihalides in an aromatic solvent and in the presence of copper powder leads to the formation of cyclopropanes.[5] Similarly, trihalogeno-methanes afford mono-halogeno-cyclopropanes, and dibromoacetic esters give alkoxycarbonyl-cyclopropane derivatives. This reaction displays high stereospecificity in that *cis*- and *trans*-olefins usually lead to cyclopropanes having *cis*- and *trans*-substituents. The aromatic hydrocarbon solvent appears to play an important role in the formation of the organo-copper intermediate, with which it probably forms a complex.

The *cis-/trans-* ratio of the dicyanocyclopropanes produced in a catalytic two-phase system, using a tetra-alkylammonium salt derived from 2-chloro-alkane-nitriles with 1-cyano-alkenes, has been shown[6] to depend on whether the catalyst is present or not. This difference is believed to originate from the fact that, in the absence of catalyst, both intermediate carbanions are present in the interfacial region whereas in the catalytic reaction they are in the organic phase. This causes two different transition-state interactions.

A new ring-enlargement route to 2-alkynyl-cyclobutanones has appeared.[7] The key step involves treatment of a crude mixture of tosylates of 1-hydroxy-

[1] S. Danishefsky, *Accounts Chem. Res.*, 1979, **12**, 66.
[2] A. H. Schmidt and W. Reid, *Synthesis*, 1978, 869.
[3] J. A. Peters, *Synthesis*, 1979, 321.
[4] M. Schneider and A. Rau, *Angew. Chem. Internat. Edn.*, 1979, **18**, 231.
[5] N. Kawabata, I. Kamemura, and M. Naka, *J. Amer. Chem. Soc.*, 1979, **101**, 2139.
[6] A. Jonczyk and A. Kwast, *Tetrahedron Letters*, 1979, 541.
[7] J. Salaün, A. Fadel, and J. M. Conia, *Tetrahedron Letters*, 1979, 1429.

cyclopropylcarbinol that have the tertiary hydroxy-group protected by a β-methoxyethoxymethyl group with LiCl in THF. A mixture of the 2-alkynyl-cyclobutanone and an enol ether is formed. Treatment of this with anhydrous zinc chloride in methylene dichloride causes the enol ether to rearrange to the desired cyclobutanone. Cycloaddition of dichloroketen to acetylenes has been reported[8] to be a general synthesis of cyclobutenones. Both mono- and di-substituted acetylenes participate successfully, and enynes react preferentially at the triple bond. The reaction of bis-dithioacetals with methyl-lithium in THF containing tetramethylethylenediamine (TMEDA) at 0 °C under argon causes a base-induced ring-closure to occur.[9] This constitutes a new route to functionalized cyclobutanes and cyclopentanes, and may prove to be especially useful in view of the known synthetic versatility of sulphur in organic chemistry (see Scheme 1).

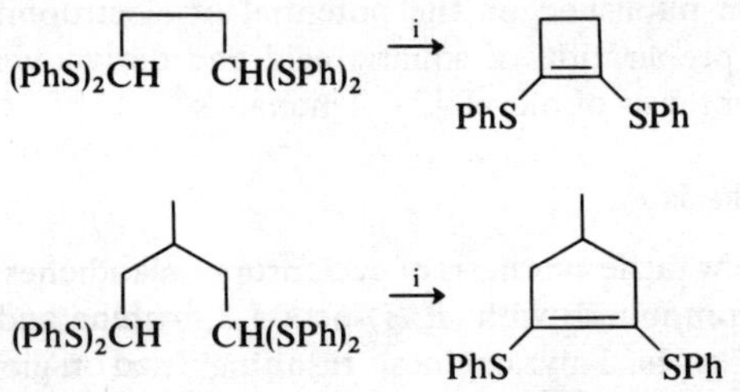

Reagents: i, MeLi, THF, TMEDA, at 0 °C

Scheme 1

A new synthesis of cyclopentanones involves cycloaddition of dichloroketen to olefins.[10] Following ring-expansion of the product $\alpha\alpha$-dichlorocyclobutanones, using diazomethane, 2,2-dichlorocyclopentanones are produced in good yields. The reaction conditions are mild, and, most significantly, the intermediates offer the possibility of other synthetic transformations, such as simple dechlorination using zinc, or some more elaborate reaction. Palladium acetate has been observed[11] to catalyse the cyclization of 4,4-disubstituted 1,6-dienes to give good yields of substituted cyclopentenes. Varying the substituents shows that chelate action by these groups is unimportant.

Thermolysis of pent-1-yn-3-ones in the gas phase has been observed to lead to substituted cyclopent-2-enones,[12] and provides a new, simple tool for the preparation of monocyclic, bicyclic, and spiro compounds containing this structural moiety. The reaction involves formation of a new carbon–carbon bond between an acetylenic and a non-activated carbon atom followed by [1,2]migration of one of the substituents at the triple bond. This may be explained by the formation of an alkylidenecarbene intermediate, which inserts into a C—H bond that is five carbon atoms away, at the non-acetylenic part of the ketone. Another paper[13] reports the

[8] A. Hassner and J. Dillon, *Synthesis*, 1979, 689.
[9] T. Cohen, D. Ouellette, and W. M. Daniewski, *Tetrahedron Letters*, 1979, 5063.
[10] A. E. Greene and J.-P. Deprés, *J. Amer. Chem. Soc.*, 1979, **101**, 4003.
[11] R. Grigg, T. R. B. Mitchell, and A. Ramasubbu, *J. C. S. Chem. Comm.*, 1979, 669.
[12] M. Karpf and A. S. Dreiding, *Helv. Chim. Acta*, 1979, **62**, 852.
[13] H. Sakurai, A. Shirahata, and A. Hosomi, *Angew. Chem. Internat. Edn.*, 1979, **18**, 163.

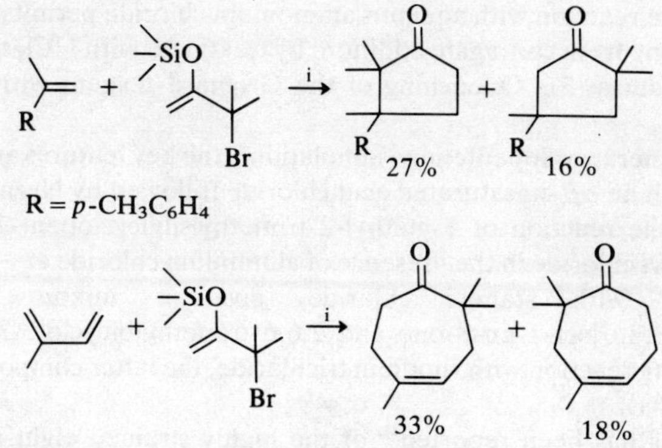

Reagents: i, $ZnCl_2$, CH_2Cl_2

Scheme 2

use of 3-bromo-3-methyl-2-trimethylsiloxybut-1-ene in the synthesis of five- and seven-membered cyclic ketones (Scheme 2). Yields varying between 10 and 97% are reported for a range of different alkenes and alkadienes.

A new method for annulation to form cyclopentene which promises to have many applications in synthesis has been announced.[14] It is based on the conjugate addition of a synthon for the β-oxo-carbanion (1) to produce 1,6-dicarbonyl compounds; the chosen synthon is the acetal-containing Grignard reagent (2). As shown in Scheme 3,

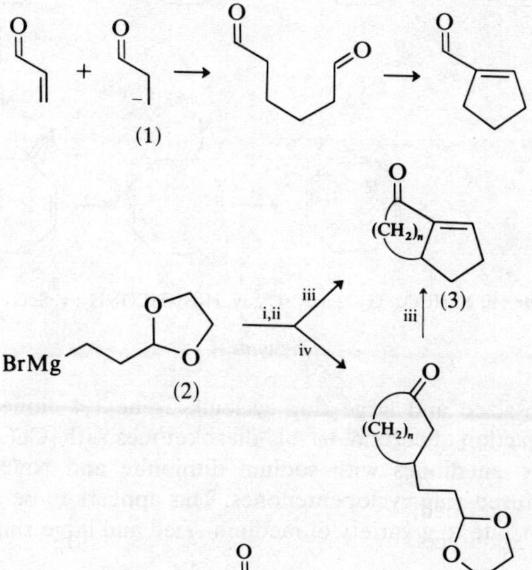

Reagents: i, CuBr, Me_2S, THF, at −78 °C; ii, $(CH_2)_n$ 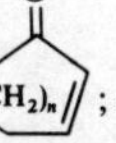 ; iii, HCl, H_2O; iv, NH_4Cl, H_2O

Scheme 3

[14] A. Marfat and P. Helquist, *Tetrahedron Letters*, 1979, 4217.

quenching of the reaction with aqueous ammonium chloride permits isolation of the products resulting from conjugate addition; by treatment with HCl, these lead to the annulation products (3). Quenching of the Grignard reagent with HCl gives (3) directly.

In another general cyclopentenone annulation[15] the key features are acylation of a vinyl-silane with an $\alpha\beta$-unsaturated acid chloride followed by Nazarov cyclization. For example, the reaction of 3-methyl-2-trimethylsilylcyclopent-1-ene with $\beta\beta$-dimethylacryloyl chloride in the presence of aluminium chloride at −78 °C, followed by treatment with stannic chloride, gives a mixture of 2,6,6-trimethylbicyclo[3.3.0]oct-1-en-8-one and 2,6,6-trimethylbicyclo[3.3.0]oct-1(5)-en-8-one. Following reaction with rhodium trichloride, the latter compound is obtained exclusively (55%).

The synthesis has been reported[16] of the highly strained eight-membered-ring compound cyclo-octa-1,5-dien-3-yne (Scheme 4). The compound begins to decompose at about −15 °C, and at 27 °C it has a half-life of about 2 hours. This short lifetime is consequent upon facile di- and oligo-merization and also upon a low thermal barrier for transformation into cyclo-octatetraene. The molecule exists in a C_2 conformation, and is thus a relatively rare case of a chiral hydrocarbon without an asymmetric carbon atom.

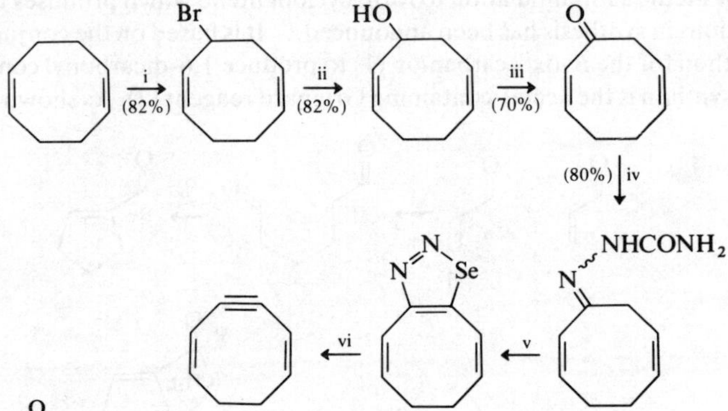

Reagents: i,  NBr ; ii, $NaHCO_3$, H_2O; iii, CrO_3; iv, $H_2NNHCONH_2$; v, SeO_2; vi, 170 °C

Scheme 4

Some medium-sized and large ring cycloalk-2-ene-1,4-diones have been prepared[17] by the reaction of certain $\alpha\omega$-bis-diazoketones with [Cu(acac)$_2$]. Successive treatment of the enediones with sodium dithionite and NaOH leads to their conversion into fused-ring cyclopentenones. This appears to be a general process, offering a direct route to a variety of medium-sized and large rings.

15 W. E. Fristad, D. S. Dime, T. R. Bailey, and L. A. Paquette, *Tetrahedron Letters*, 1979, 1999.
16 H. Meier, T. Echter, and H. Petersen, *Angew. Chem. Internat. Edn.*, 1978, **12**, 942.
17 S. Kulkowit and M. A. McKervey, *J. C. S. Chem. Comm.*, 1978, 1069.

Polycycles.—A new synthesis has been described[18] of variously substituted spiro[2.4]heptadienes from fulvenes, using *S*- and *N*-ylides. Irradiation of the isopropylidene diazomalonate (4) leads to the corresponding spiro-activated cyclopropanes in high yield, and this forms the key step in a new synthesis of these compounds.[19] The cyclopropanation of *cis*- and *trans*-but-2-ene is stereospecific, and high stereoselectivity is reported with such alkenes as 3- and 4-t-butyldimethylsiloxycyclopent-1-ene and with 1,3,3-trimethylcyclohex-1-ene. Insertion into C—H bonds was rarely observed, and the simplicity and scope of the method suggest that it will be a powerful tool in synthesis.

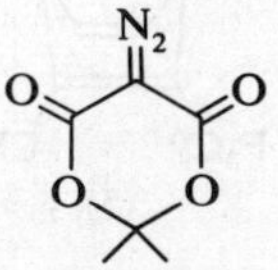

(4)

With a view to devising a route to bicyclo[3.1.0]hexatriene, the elimination reactions of 4,6-dibromobicyclo[3.1.0]hex-2-ene have been studied.[20] Depending upon the base used, either bromobenzene or 6-substituted fulvenes are formed. However, labelling experiments show that bicyclo[3.1.0]hexatriene is an intermediate on the reaction pathway to the fulvene. The chemistry of the triene is as anticipated for an azulenoid hydrocarbon, but, so far, trapping experiments using diphenylisobenzofuran have failed, and only nucleophilic addition products have been found.

1,6-Dimethylbicyclo[4.4.1]undeca-2,4,7,9-tetraene, a vinylogue of norbornadiene, has been synthesized.[21] The u.v. spectrum shows a bathochromic shift of 17 nm compared with cyclohepta-1,3-diene, and this can be attributed either to planar fixation of the diene units or to some homoconjugation. The compound is also thermally very labile, and at room temperature it is converted into 5,7-dimethyl-*cis*-bicyclo[5.4.0]undeca-2,4,8,10-tetraene in 100% yield in a process which, although formally a [1,5]-sigmatropic shift of a butadiene bridge, probably proceeds as in Scheme 5. In a study of the intramolecular Diels–Alder reaction, the equilibrium constant for the transformation of 3-methyleneocta-1,7-diene into bicyclo-[3.3.1]non-1-ene has been determined[22] as $K(427\,°\text{C}) = 0.28$. However, an

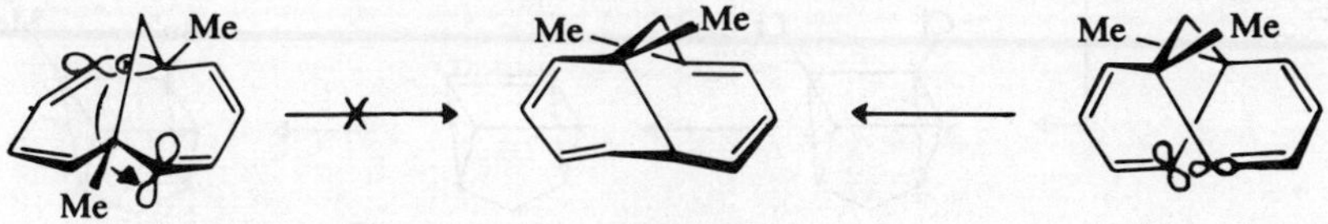

Scheme 5

[18] W.-D. Schröber and W. Friedrichsen, *Annalen*, 1978, 1648.
[19] T. Livinghouse and R. V. Stevens, *J. Amer. Chem. Soc.*, 1978, **100**, 6479.
[20] W. N. Washburn, R. Zahler, and I. Chen, *J. Amer. Chem. Soc.*, 1978, **100**, 5863.
[21] J. Frank, W. Grimme, and J. Lex, *Angew. Chem. Internat. Edn.*, 1978, **17**, 943.
[22] K. J. Shea and S. Wise, *Tetrahedron Letters*, 1979, 1011.

unfavourably high entropy of reaction precludes a high yield of product. In an effort to enhance the synthetic potential, electron-withdrawing groups have been placed on the dienophile; in all cases examined, activated triene esters and ketones undergo cycloaddition under milder conditions and in higher yield than unactivated hydrocarbons. Another inside–outside bicyclic molecule, 10,11-bistrifluoromethyl-*i,o*-bicyclo[7.2.2]trideca-10,12-diene (5), has been reported.[23] It is prepared by the

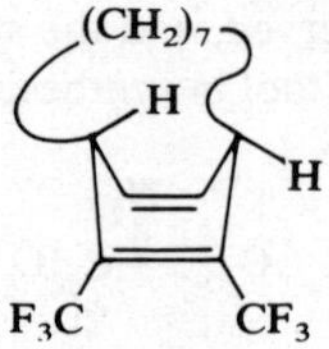

(5)

thermal reaction between *cis,trans*-cycloundeca-1,3-diene and hexafluorobut-2-yne, and its structure has been proved by *X*-ray analysis of the complex with iron tricarbonyl. It has been suggested[24] that the highly strained bridgehead olefin tetracyclo[3.2.0.0^{2,7}.0^{4,6}]hept-1(7)-ene (6) is formed as an intermediate when 1-chloroquadricyclane is treated with a base such as n-butyl-lithium. Addition of methyl bromide to the reaction gives n-butylmethylquadricyclane, probably according to Scheme 6. Support comes from the observation that, if 1-chloroquadricyclane in THF is allowed to react with a mixture of anthracene and lithium 2,2,6,6-tetramethylpiperidide, the Diels–Alder adduct of (6) and anthracene can be isolated in 40% yield. The corresponding adduct from (7) was not, however, detected.

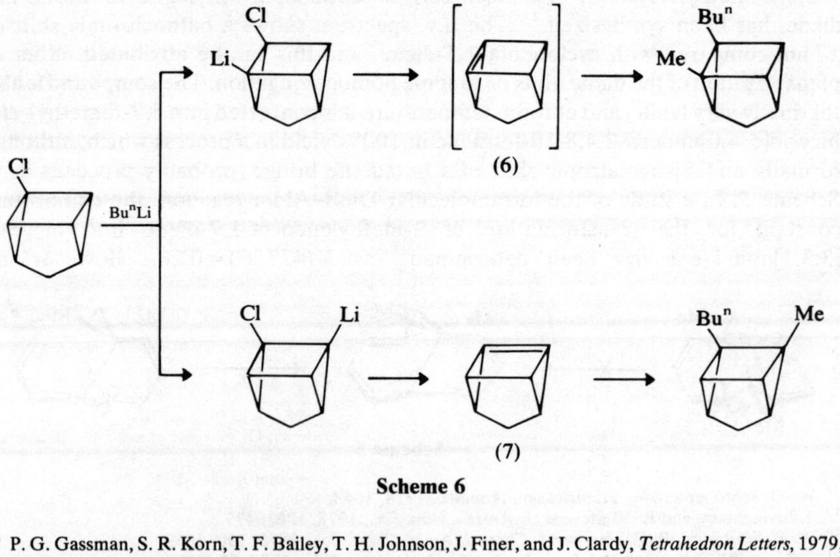

Scheme 6

[23] P. G. Gassman, S. R. Korn, T. F. Bailey, T. H. Johnson, J. Finer, and J. Clardy, *Tetrahedron Letters*, 1979, 3401.

[24] J. Harnisch, O. Baumgärtel, G. Szeimies, M. van Meerssche, G. Germain, and J.-P. Declercq, *J. Amer. Chem. Soc.*, 1979, **101**, 3370.

Experimental evidence has been described[25] attesting to the existence of tricyclo[3.1.0.0^{2,6}]hex-1(6)-ene (8), a homologue of tricyclo[4.1.0.0^{2,7}]hept-1(7)-ene, as a reactive species. 1-Chlorotricyclo[3.1.0.0^{2,6}]hexane is observed to react with n-butyl-lithium to give 1-n-butyltricyclo[3.1.0.0^{2,6}]hexane. This and related observations have been interpreted in terms of lithiation of the chlorocarbon to give (9). Following elimination of LiCl, (8) is formed, which in turn adds the lithium base to its highly strained double bond.

(8)

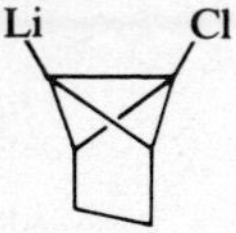

(9)

An improved synthesis of 4e-chloroadamantylideneadamantane* has been reported.[26] This involves treatment of adamantylideneadamantane with phenylsulphenyl chloride to afford the product directly, probably *via* the intermediacy of a thiiranium salt. A paper has also appeared[27] that describes the systematic classification and nomenclature of the subset of diamond hydrocarbons known as 'polymantanes', using a graph-theoretical approach.

As part of an investigation of structural aspects of planoid deformation of the tetraco-ordinate carbon atom, 13-oxa-14-oxopentacyclo[5.5.2.1.0^{4,15}.0^{10,15}]-pentadecane, a bridged tetraquinacane, has been synthesized.[28] Since this compound has specific functionality at two bridgeheads, it is well suited for the preparation of new bridgehead olefins and of annulenes with a central carbon atom. Deslongchamps' full paper has now appeared,[29] giving experimental details for a synthesis of triquinacene and of 2,3-dihydrotriquinacen-2-one, starting from Thiele's acid. Hitherto, the synthesis of molecules constructed of repeating alicyclic units has largely been restricted to those homologous series such as the adamantanes and related families. This has been largely because thermodynamic factors have made them readily available by Lewis-acid-catalysed cationic rearrangements. A new approach, based on a two-fold cation–olefin cyclization, has now been announced,[30] and the synthesis of *syn*-[3.2.1]^{2}geminane (octahydro-3*H*,8*H*-3,5a:8,10a-dimethanoheptalene) (10) and [2.2.2]^{2}geminane (octahydro-2,4a:6,8a-diethanonaphthalene) (11) described (see Scheme 7).

Some evidence has been advanced[31] for the existence of naphtharadialene, 1,2,3,4,5,6,7,8-octakis(methylene)-Δ^9-octalin, as a reactive intermediate. Pyrolysis of either 1,4,5,8-tetrakis(chloromethyl)-2,3,6,7-tetramethylnaphthalene or the

* Chlorine is equatorial in the cyclohexylidene ring. *Senior Reporter.*

25 U. Szeimies-Seebach, J. Harnisch, G. Szeimies, M. Van Meerssche, G. Germain, and J.-P. Declercq, *Angew. Chem. Internat. Edn.*, 1978, **17**, 848.
26 J. Bolster, R. M. Kellogg, E. W. Meijer, and H. Wynberg, *Tetrahedron Letters*, 1979, 285.
27 A. T. Balaban and P. von R. Schleyer, *Tetrahedron*, 1979, **34**, 3599.
28 R. Keese, A. Pfenninger, and A. Roesle, *Helv. Chim. Acta*, 1979, **62**, 326.
29 P. Deslongchamps, U. O. Cheriyan, Y. Lambert, J.-C. Mercier, L. Ruest, R. Russo, and P. Soucy, *Canad. J. Chem.*, 1978, **56**, 1687.
30 H. Park, P. F. King, and L. A. Paquette, *J. Amer. Chem. Soc.*, 1979, **101**, 4773.
31 H. Hart, M. Jeffares, A. Teuerstein, and D. L. Ward, *J. Amer. Chem. Soc.*, 1978, **100**, 8012.

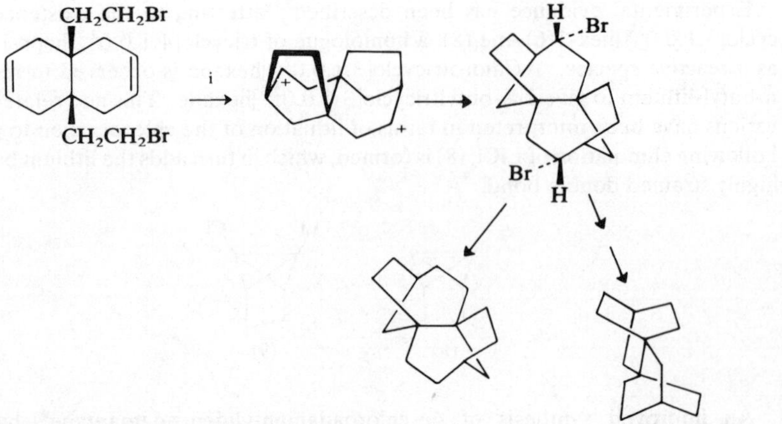

Reagents: i, silica gel, stir at room temperature for 48 h (10) (11)

Scheme 7

corresponding 2,3,6,7-tetrakis(chloromethyl)-1,4,5,8-tetramethylnaphthalene both lead to (12) *via*, it is suggested, naphtharadialene. Support for this postulate comes from the observation that pyrolysis of 2,3-bis(chloromethyl)-1,4,5,6,7,8-hexamethylnaphthalene gives a good yield of the relatively stable (13).

The synthesis of the bridgehead alkene bicyclo[4.2.2]deca-1,5-diene has been reported.[32] Its Raman spectrum shows a band at 1620 cm^{-1}, and at 25 °C its half-life for conversion into 2,5-dimethylenebicyclo[4.2.0]octane is reported to be 314 min.

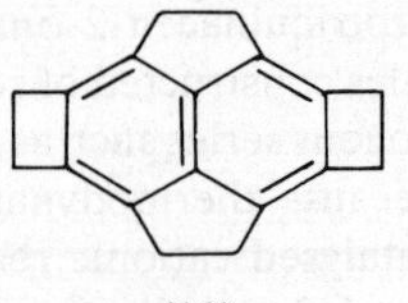

(12)

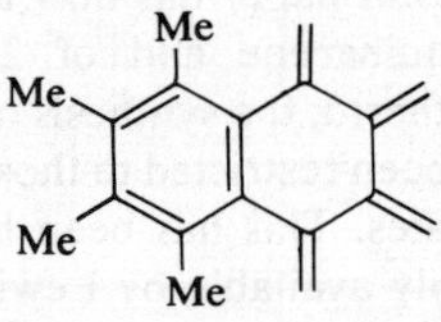

(13)

3 Stereochemistry

Using a variety of force fields, calculations have been performed[33] on tetra-t-butyltetrahedrane; the results show the molecule to possess T symmetry and to be chiral. The four t-butyl groups are twisted by about 14°, and all twelve methyl groups by about 2–6° in the same direction from a staggered T_d conformation. Enantiomerization is, however, expected to be facile, since the structure having T_d symmetry is a possible transition state and lies 8–21 kJ mol^{-1} above the ground state. A study of the conformations of cyclohexanecarboxaldehyde in the gas phase, using microwave spectroscopy, has revealed[34] the presence of two stable conformers. In both of these the ring assumes a chair conformation, but different torsional angles are

32 J. R. Wiseman and J. J. Vanderbilt, *J. Amer. Chem. Soc.*, 1978, **100**, 7730.
33 W. D. Hounshell and K. Mislow, *Tetrahedron Letters*, 1979, 1205.
34 P. N. Kao and P. H. Turner, *J. Amer. Chem. Soc.*, 1979, **101**, 4497.

displayed about the C-1—CHO bond. The *gauche* conformer, in which the carbonyl group eclipses the C-1—C-2 ring bond, has been shown to be 3.0 ± 0.8 kJ mol^{-1} more stable than the *cis* conformer, in which the carbonyl group eclipses the axial C-1—H bond. These results are in good agreement with those obtained theoretically and also by n.m.r. spectroscopy.

The first example has appeared[35] of an electrochemical reduction showing a separate cyclic voltammetric peak for each of two interconverting conformers. At room temperature, solutions of *trans*-1,2-dibromocyclohexane in DMF and containing tetraethylammonium bromide show a single reduction peak, due to the totally irreversible, two-electron reduction of the rapidly interconverting diaxial and diequatorial conformers. Lowering the temperature causes a second peak to appear, attributable to reduction of the longer-lived diequatorial conformer. At −90 °C the relative heights of the two peaks due to the diaxial and diequatorial conformers are independent of the scan rate, so that peak heights reflect the equilibrium concentrations of the two conformers and *K* is directly derivable.

Low-temperature ^{13}C n.m.r. spectroscopy has revealed[36] that, in those conformations of 2,2-dimethyl- and 2,2,6,6-tetramethyl-cyclohexanes having the 1-substituent in an axial position, an additional destabilization exists as compared with monosubstituted cyclohexanes. The effect is consequent upon the equatorial methyl group being pushed towards the axial substituent by the reflex effect of the axial methyl group. This view is supported by considerations involving a force-field calculation which correctly predicts the experimentally determined conformational equilibria. In order to assess the effect of bulky substituents on the conformation of cyclohexa-1,4-diene, the crystal and molecular structures of *trans*-1,4-dihydro-4-tritylbiphenyl have been determined.[37] These studies agree with previously published n.m.r. data that the cyclohexadiene ring is relatively flat ($\alpha = 171.8°$),* with the trityl group located pseudo-axially, and they stand in contrast to suggestions obtained from space-filling models that there is a pseudo-equatorial environment. Investigation of *cis*-1,4-dihydro-4-tritylbiphenyl shows[38] that the ring is only slightly puckered ($\alpha_{mean} \approx 175°$) but somewhat distorted ($\alpha_{C\text{-}1\text{—}Ph} \approx 177°$, whereas $\alpha_{C\text{-}4\text{—}CPh_3} \approx 173°$), and that both substituents are pseudo-equatorial. This suggests that a trityl group does not lock a cyclohexadiene ring into a boat geometry. A planar conformation of the ring appears to be favoured in which the controlling factor is packing of the substituents about the central trityl carbon atom. The first absolute configurational assignments of two chiral [8]annulenes, namely 1,2,3-trimethyl- and 1,2,3,4-tetramethyl-cyclo-octatetraenes, have been reported;[39] they are (14) and (15) respectively. The chiroptical properties of the molecules have been determined, and these show that there are four essentially unconjugated ethylene chromophores, coupled principally by electrostatic potential between the $\pi \rightarrow \pi^*$ transition dipoles.

* α is the angle between the planes of the two double bonds.

35 A. J. Klein and D. H. Evans, *J. Amer. Chem. Soc.*, 1979, **101**, 757.
36 H.-J. Schneider and W. Freitag, *Chem. Ber.*, 1979, **112**, 16.
37 M. C. Grossel, A. K. Cheetham, and J. M. Newsam, *Tetrahedron Letters*, 1978, 5229.
38 M. C. Grossel, A. K. Cheetham, D. A. O. Hope, K. P. Lam, and M. J. Perkins, *Tetrahedron Letters*, 1979, 1351.
39 J. M. Gardlik, L. K. Johnson, L. A. Paquette, B. A. Solheim, J. P. Springer, and J. Clardy, *J. Amer. Chem. Soc.*, 1979, **101**, 1615.

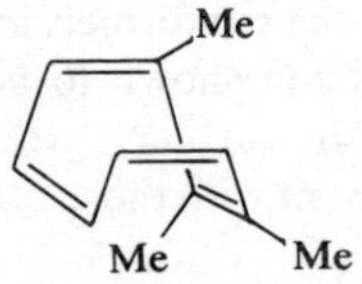

(14)

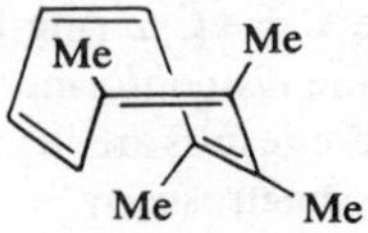

(15)

A new way of resolving chiral cyclo-octatetraenes has appeared,[40] using a Diels–Alder reaction of chiral triazolinediones. Enantiomerically pure (−)-*endo*-bornyl-1,2,4-triazolinedione is prepared and allowed to react with racemic 1,2,3-trimethylcyclo-octatetraene, giving a mixture of two stereoisomeric urazoles. Repeated crystallization separates the two diastereomers, from which the parent polyolefin can be obtained in a high state of enantiomeric purity by a hydrolysis–oxidation procedure. The importance of this procedure stems from the significance of the cyclo-octatetraene nucleus as the annulene ring system that is most nearly ideally suited for probing the asymmetry of cyclic polyolefins.

Iterative force-field calculations have been used[41] to investigate the conformations of cycloundecane, cyclotridecane, and cyclopentadecane. At −150 °C, cycloundecane is found to be a mixture of the [12323]* and [335] conformations, with the latter predominating. Cyclotridecane is predicted to exist mainly in the [13333] conformation and cyclopentadecane, at low temperatures, probably prefers the highly symmetrical [33333] conformation. Using iterative force-field calculations, the strain-energy barrier for pseudorotation of the [3333] conformation of lowest energy of cyclododecane has been found[42] to be 33.1 kJ mol^{-1}. This result is in agreement with the free-energy barrier obtained from n.m.r. data. Evidence is now available,[43] from dynamic n.m.r. spectroscopy and from iterative force-field calculations, on the conformations of *cis*-cyclododecene. Low-temperature ^{13}C n.m.r. suggests the presence of two conformations, in the ratio 2 : 1, with $\Delta G^{\ddagger}(-120\,°C) \approx$ 29 kJ mol^{-1}. The calculations indicate that the conformation of lowest energy is [1_{ene}2333] and the next higher one is [1_{ene}2342]. If it is assumed that the only mechanism of exchange is interconversion of these two conformers and their mirror images, then the barrier observed spectroscopically can be ascribed to the separation of these two conformations. A similar method has been used[44] to determine the conformations of cyclododecyne. The results show the major conformation to be symmetrical, and to have a [3_{yne}333] structure, which may be regarded as being derived from the conformation of lowest energy of cyclododecane, *i.e.* the [3333] conformation; the minor conformation could not be identified. Assuming that the only exchange mechanism is interconversion of [3_{yne}333] and [4_{yne}332], the conformational barrier at −95 °C has been assessed at -33 ± 1.2 kJ mol^{-1}.

* The description of these conformers uses the notation of Dale (J. Dale, *Acta Chem. Scand.*, 1973, **27**, 1115.)

[40] J. M. Gardlik and L. A. Paquette, *Tetrahedron Letters*, 1979, 3597.
[41] F. A. L. Anet and T. N. Rawdah, *J. Amer. Chem. Soc.*, 1978, **100**, 7810.
[42] F. A. L. Anet and T. N. Rawdah, *J. Amer. Chem. Soc.*, 1978, **100**, 7166.
[43] F. A. L. Anet and T. N. Rawdah, *Tetrahedron Letters*, 1979, 1943.
[44] F. A. L. Anet and T. N. Rawdah, *J. Amer. Chem. Soc.*, 1979, **101**, 1887.

Low-resolution microwave spectroscopy has been used[45] to obtain the conformations of some bicyclic monoterpenes based on the bicyclo[3.1.0]hexane skeleton. For thujone and some related compounds, a boat-like conformation has been established for the bicyclic system, and the puckering angle has also been derived. The synthesis has been reported[46] of the first optically active anti-Bredt-rule compound that has known absolute configuration. The compound, (−)-(5*S*)-bicyclo[3.3.1]non-1-ene (16), $[\alpha]_D = -720°$ (by calculation), can be regarded as a methylene-bridge derivative of (−)-(*R*)-*trans*-cyclo-octene.

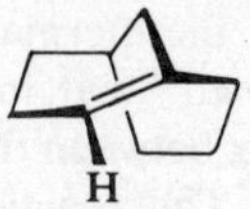

(16)

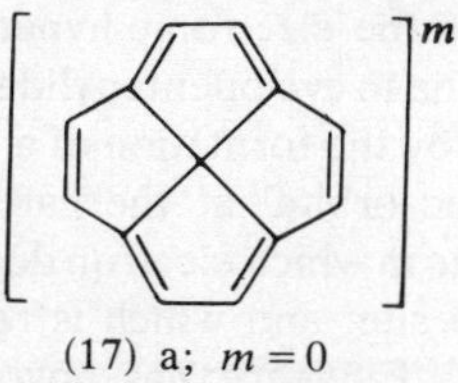

(17) a; $m = 0$
b; $m = 2-$
c; $m = 2+$

Pure enantiomers of tricyclo[6.4.0.0^{4,9}]dodecane have been prepared and found to have easily measurable optical rotations ($[\alpha]_D = -30.0°$), despite the fact that, according to Brewster's rules, they ought to be almost optically inactive.[47] It has been suggested that the reason for the high optical rotation is that all three cyclohexane rings exist in slightly twisted chair conformations. In order to throw light on whether the central tetraco-ordinate carbon atom that is enclosed by a [12]annulene may be planar, MINDO/3 calculations have been carried out[48] on (17a—c). The results show that all structures attempt to avoid a planar conformation and that those having a planar central carbon atom are of much higher energy than those that do not. The strain is distributed over all carbon centres, and this leads to highly directed π-orbitals at C-1, C-4, C-7, and C-10. These observations have some important implications for synthetic endeavours directed at molecules of this type.

4 Structural Properties and Orbital Interactions

Using *ab initio* SCF calculations, some theoretical studies have been made[49] on the singlet and triplet cyclopropylidene–allene system. These have shown that triplet cyclopropylidene (18) is lower in energy than singlet cyclopropylidene (19) by 35.1 kJ mol^{-1} and that triplet allene is a bent planar molecule that is 33.4 kJ mol^{-1} lower in energy than the corresponding structure with a linear geometry. The barrier height for disrotatory opening of (19) is 75.3 kJ mol^{-1} and for (18) is 79.5 kJ mol^{-1}. The structure of $C_4H_7^+$ has been the subject of further investigations,[50] and the existence of a second minimum-energy form (20) has been revealed. This should be regarded as being made up equally from the classical forms (21) and (22).

[45] Z. Kisiel and A. C. Legon, *J. Amer. Chem. Soc.*, 1978, **100**, 8166.
[46] M. Nakazaki, K. Naemura, and S. Nakahara, *J. C. S. Chem. Comm.*, 1979, 82.
[47] H. Buding and H. Musso, *Angew. Chem. Internat. Edn.*, 1978, **17**, 851.
[48] M. C. Böhm, R. Gleiter, and P. Schang, *Tetrahedron Letters*, 1979, 2575.
[49] D. J. Pasto, M. Haley, and D. M. Chipman, *J. Amer. Chem. Soc.*, 1978, **100**, 5272.
[50] B. A. Levi, E. S. Blurock, and W. J. Hehre, *J. Amer. Chem. Soc.*, 1979, **101**, 5537.

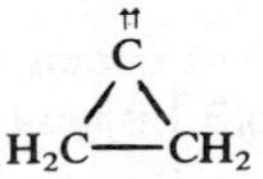

(18)

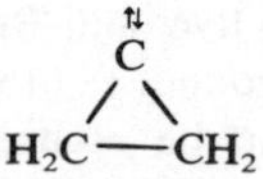

(19)

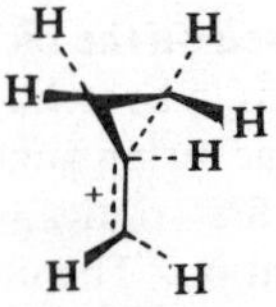

(20)

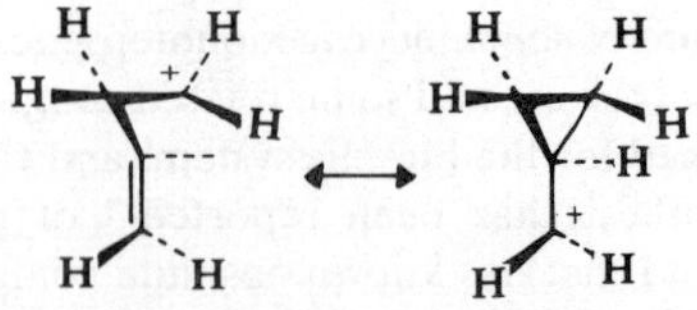

(21) (22)

An exploration of the electronic hypersurface of the thermal transformation of vinylcyclopropylidene to cyclopentenylidene has shown[51] that, in its singlet state, the reaction is initiated by the formation of a π-complex between the double bond and the empty *p* atomic orbital at the carbene site. The reaction proceeds *via* a non-classical carbene in which electron density is shifted from the initial double bond toward the carbene site, and which is relatively unstable toward deformation to cyclopentenylidene. Evidence has now been presented[52] arguing strongly that cyclobutadiene is definitely not square (D_{4h}) but is probably rectangular. This conclusion is based on the appearance of four bands below 1700 cm^{-1} in the i.r. spectrum of cyclobutadiene produced from several precursors, and is consistent with a rectangular geometry for both cyclobutadiene and tetradeuteriocyclobutadiene, of point group D_{2h}. In a related paper[53] supporting these conclusions, an *ab initio* SCF method has been used to calculate the vibrational frequencies of cyclobutadiene, assuming a rectangular geometry for the ground state. However, the authors do point out that the electronic energy obtained by this method does exceed a better, published value.

As part of a study of $p_\sigma p_\sigma$ bonding, the π-face bonding of cyclobutadiene and CO, HCN, and benzene has been studied.[54] Total energies obtained by the CNDO/2 method have been found to be unreliable for several possible intermolecular geometries when compared with minimal *ab initio* results, which all show repulsive interactions except at long range, for CO. The results suggest that benzene, CO, and HCN should not be considered as ligands of cyclobutadiene.

(2,2′-Bipyridyl)(tetramethylcyclobutadiene)nickel(O), a member of a previously unknown class of complexes, has been prepared.[55] These complexes are of interest[56] in relation to the nickel-catalysed synthesis of cyclo-octatetraene. A quantum-mechanical study has been undertaken,[57] using the MINDO/3 technique, to determine the electronic hypersurface for the conversion of cyclobutylidene into methylenecyclopropane. The results show that the facile reaction is initiated by formation of the non-classical carbene (23), which has a bicyclobutane-like structure and in which electron density is shifted towards the carbene site. Ring opening

[51] W. W. Schoeller and U. H. Brinker, *J. Amer. Chem. Soc.*, 1978, **100**, 6012.
[52] S. Masamune, F. A. Souto-Bachiller, T. Machiguchi, and J. E. Bertie, *J. Amer. Chem. Soc.*, 1978, **100**, 4889.
[53] L. J. Schaad, B. A. Hess, and C. S. Ewig, *J. Amer. Chem. Soc.*, 1979, **101**, 2281.
[54] K. B. Lipkowitz, *J. Amer. Chem. Soc.*, 1978, **100**, 7535.
[55] U. Griebsch and H. Hoberg, *Angew. Chem. Internat. Edn.*, 1978, **17**, 950.
[56] P. W. Jolly and G. Wilke, 'The Organic Chemistry of Nickel', Academic Press, New York, 1975, Ch. 2, p. 94.
[57] W. W. Schoeller, *J. Amer. Chem. Soc.*, 1979, **101**, 4811.

[(23) → (24)] is controlled by orbital-symmetry considerations and proceeds in a conrotatory fashion relative to the residue. Since there is a strong separation of charge in the transition state, electron-donating substituents at C-4 and/or C-2 should enhance the rate.

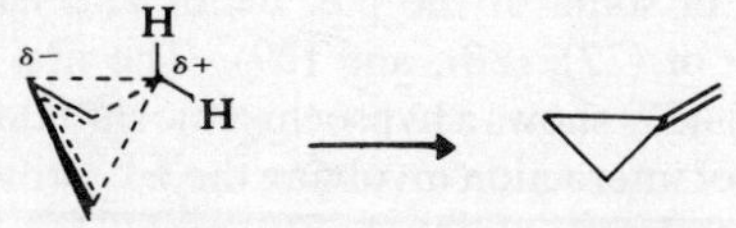

(23) (24)

An investigation of the electronic structure of 1,2- and 1,3-dimethylenecyclobutane has been carried out,[58] using photoelectron spectroscopy in conjunction with *ab initio* MO calculations. In the 1,3-derivative, the through-bond interaction dominates the through-space interaction of the two localized π-orbitals, leading to an inverted sequence of the orbitals. The splitting of the π MO's in the 1,2-derivative, on the other hand, is explained by the fact that the π–π interaction typical of conjugated dienes is effectively reduced by a through-space interaction across the ring between the π- and pseudo-π-orbitals of CH_2. Calculations have also been made[59] of the potential surface of D_{2h} 1,3-dimethylenecyclobutadiene. These show that the lowest singlet is ^1A_g but that the ground state is $^3B_{2u}$.

The 1,6-dimethylcyclodecyl cation has been generated from the corresponding alcohol, using FSO_3H–SbF_5 in SO_2ClF, at −120 °C. A study of this species, using ^1H and ^{13}C n.m.r., has shown[60] that it exists as a symmetrical μ-hydrido-bridged structure. It has been suggested that the dominant factor in promoting such a 'tertiary' ion is the release of steric strain in achieving relatively strainless decalin-type geometry from the strained cyclodecyl geometry. The electronic structure and photoelectron spectrum of cyclododeca-1,5,9-triyne have been determined,[61] and suggest that the molecule adopts a conformation that is a quasi-chair, having D_3 symmetry. Substantial interaction appears to exist between the non-conjugated acetylenic π-orbitals.

The difference between $^1J_{C\text{-}1\text{-}C\text{-}3}$ in (25) and $^1J_{C\text{-}1\text{-}C\text{-}3}$ in (26) is very large (−5.4 Hz), and it has been concluded[62] that substituents and/or steric influences exert a powerful effect on the value of $^1J_{C\text{-}1\text{-}C\text{-}3}$ in bicyclobutanes. Using the Muller–Pritchard relationship between $^1J_{CH}$ and *s* character gives 36.9% *s* character in C-1–H of (25) and 41.0% in bicyclobutane itself. $^1J_{C\text{-}1\text{-}C\text{-}3}$ is −17.49 Hz, and any of

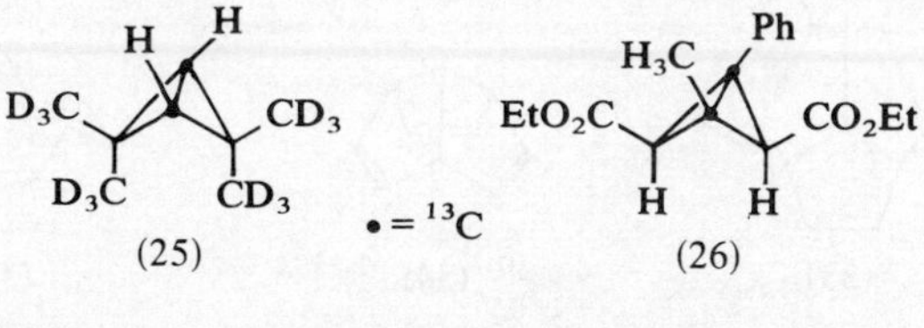

(25) (26)

[58] P. Hemmersbach, M. Klessinger, and P. Bruckmann, *J. Amer. Chem. Soc.*, 1979, **100**, 6344.
[59] E. R. Davidson, W. T. Borden, and J. Smith, *J. Amer. Chem. Soc.*, 1978, **100**, 3299.
[60] R. P. Kirchen and T. S. Sorensen, *J. C. S. Chem. Comm.*, 1978, 769.
[61] K. N. Houk, R. W. Strozier, C. Santiago, R. W. Gandour, and K. P. C. Vollhardt, *J. Amer. Chem. Soc.*, 1979, **101**, 5183.
[62] H. Finkelmeier and W. Lüttke, *J. Amer. Chem. Soc.*, 1978, **100**, 6261.

the available coupling constant–hybridization relationships give a negative value for s^2. Since this leads to an imaginary *s* character for its orbitals, it follows that none of the available equations can be applied to the central bond.

A study of the conformationally rigid α-diketones (30), (31), and (32) has permitted[63] assignment of some of the p.e. bands and has also revealed a close similarity to the spectra of (27), (28), and (29). The first band ($\pi_+^* \leftarrow n_+$) in the electronic spectra of (30)–(32) shows a hypsochromic shift through the series, and has been explained in terms of interaction involving the π_+^*-orbital of the C_2O_2 fragment and the π-orbitals of the olefinic moiety in (30) and (31).

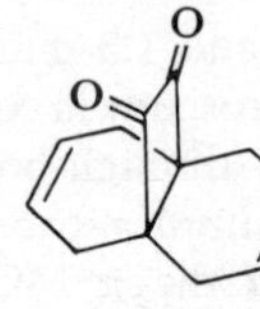

(27)

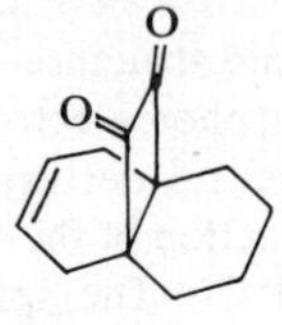

(28)

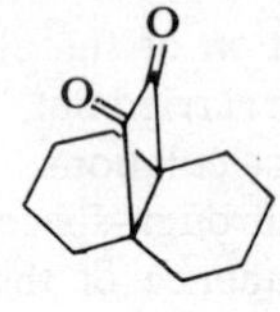

(29)

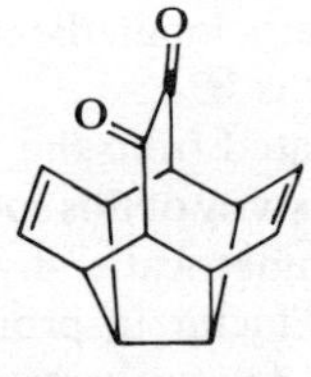

(30)

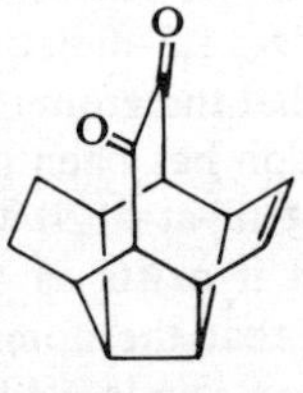

(31)

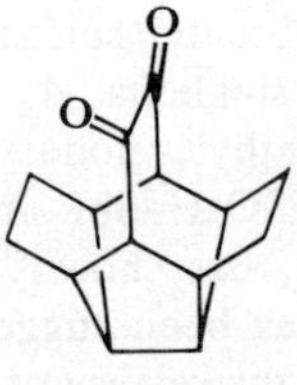

(32)

A quantitative assessment of *pp*–σ overlap has appeared[64] for C_{16}-hexaquinacene (33). Relative to other related molecules, *e.g.* (34) and (35), models suggest that (33) has an enhanced sphericality, resulting in an increased degree of in-plane alignment of the $p\pi$-orbital triad. However, the study shows that, although the geometry is favourable, the longer inter-atomic distances have a counterbalancing effect on the magnitude of the *p*–*p* overlap integral.

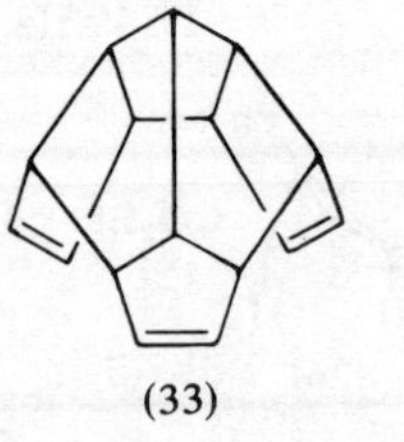

(33)

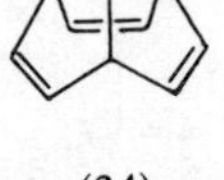

(34)

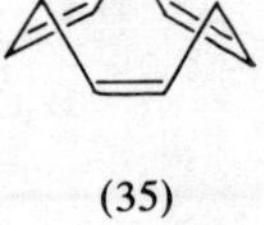

(35)

[63] R. Bartetzko, R. Gleiter, J. L. Muthard, and L. A. Paquette, *J. Amer. Chem. Soc.*, 1978, **100**, 5589.
[64] G. G. Christoph, J. L. Muthard, L. A. Paquette, M. C. Böhm, and R. Gleiter, *J. Amer. Chem. Soc.*, 1978, **100**, 7782.

5 Reactions

Metal-promoted Reactions.—The reaction of 2,2,3,3-tetradeuteriomethylenecyclopropane (36) with dimethyl fumarate has been carried out in the presence of bis(acrylonitrile)nickel(O) in benzene solution (Scheme 8). Spectroscopic analysis of the products shows[65] that the deuterium atoms are completely statistically scrambled over three carbons in cyclo-adduct (37). It was concluded that a trimethylenemethane-nickel(O) complex of type (38), or the related rapidly equilibrating σ-complex, is responsible for the formation of such adducts.

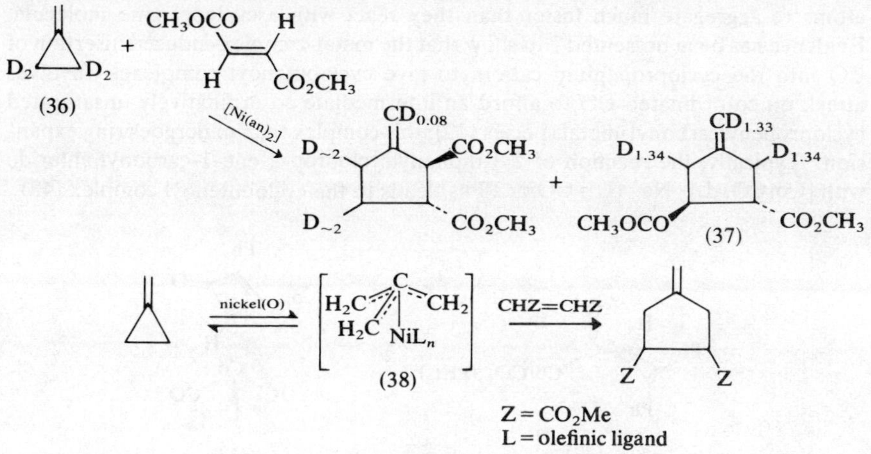

Scheme 8

Co-deposition of 6,6-dimethylfulvene and nickel vapour at −196 °C has been shown[66] crystallographically to lead to a mixture of 4,4,8,8-tetramethyl-1,4,5,8-tetrahydro-*s*-indacene and its 1,4,7,8-tetrahydro-isomer, both of which are formally derivable by a thermally forbidden [6+6] cycloaddition followed by [1,5]-hydrogen shifts. It has been suggested that the mechanism involves a stepwise dimerization on a nickel atom template. The eighteen-electron tetrahedral complex (39) appears to be the initial product, which then gives (40); finally, coupling of the two π-allyl

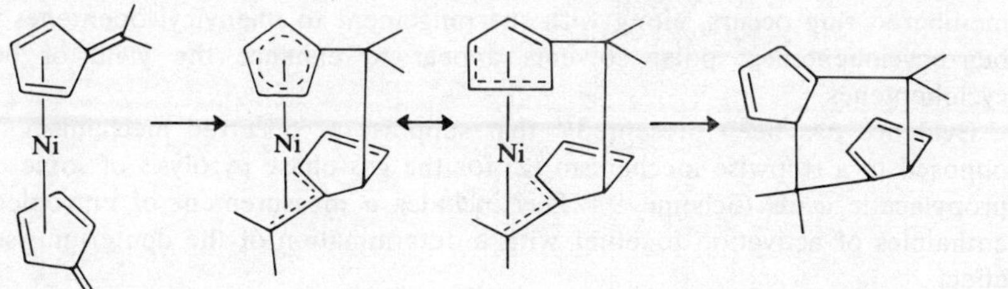

(39) (40) (41)

[65] R. Noyori, M. Yamakawa, and M. Takaya, *Tetrahedron Letters*, 1978, 4823.
[66] N. Hao, J. F. Sawyer, B. G. Sayer, and M. J. McGlinchey, *J. Amer. Chem. Soc.*, 1979, **101**, 2203.

moieties leads to (41), which rearranges to the product. That the reaction takes this route rather than leading to the twenty-electron nickelocene system is presumably a consequence of thermodynamic factors.

An investigation of reactions between metal atoms and organocyclopropanes has revealed[67] that co-condensation of chromium and quadricyclane followed by a 1 hour warm-up of the matrix promotes an efficient isomerization to norbornadiene. Together with other observations on related systems, the intervention of any carbon–carbon bond-insertion reaction is precluded, and the observed isomerization is thought to be heterogeneously catalysed. It has been concluded that the metal atoms re-aggregate much faster than they react with a cyclopropane molecule. Evidence has been presented[68] to show that the metal-complex-induced insertion of CO into the cyclopropenium cation, to give cyclobutenoyl complexes, involves attack on co-ordinated CO to afford an intermediate co-ordinatively unsaturated cyclopropenylcarbonyl-metal species (42); this complex then undergoes ring expansion. Typically, the reaction of 2,3-diphenylcycloprop-2-ene-1-carbonyl chloride with $[Co(CO)_3L]^- Na^+$ (L = CO or PPh_3) leads to the cyclobutenoyl complex (43).

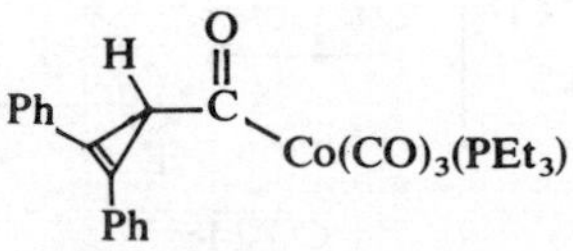

(42)

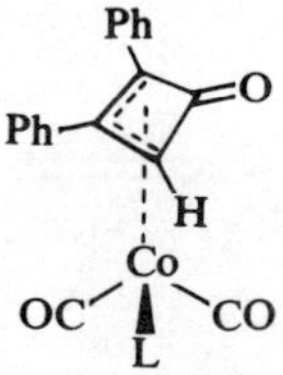

(43)

Thermally Promoted Reactions.—Two distinct secondary deuterium isotope effects have been measured[69] in the stereomutations of *cis*- and *trans*-1-cyano-2-phenylcyclopropane. Replacement of hydrogen by deuterium at C-1 gives a normal secondary isotope effect of about 1.07 ± 0.02. However, a value of 1.13 ± 0.02 is obtained if the hydrogen at C-3 is replaced. Such a large effect suggests a high degree of interaction between C-1 and C-2 and with the H—C-3 bond in the transition state. A study has also been made[70] of the thermal reactions of some 1-aryl-2-vinylcyclopropanes. Between 100 and 200 °C, *cis–trans* isomerization of the three-membered ring occurs, along with rearrangement to phenylcyclopentenes or to benzocycloheptenes; polar solvents appear to enhance the yield of benzocycloheptenes.

Evidence has been presented[71] that supports a concerted mechanism (1) as opposed to a stepwise mechanism (2) for the gas-phase pyrolysis of some cyclopropylacetic acids (Scheme 9). This includes a measurement of entropies and enthalpies of activation together with a determination of the deuterium isotope effect.

[67] J. A. Gladysz, J. G. Fulcher, R. C. Ugolick, A. T. L. Hanlan, and A. B. Bocarsly, *J. Amer. Chem. Soc.*, 1979, **101**, 3388.
[68] C. E. Chidsey, W. A. Donaldson, R. P. Hughes, and P. F. Sherwin, *J. Amer. Chem. Soc.*, 1979, **101**, 233.
[69] J. E. Baldwin and C. G. Carter, *J. Amer. Chem. Soc.*, 1979, **101**, 1325.
[70] G. Maas, *Chem. Ber.*, 1979, **112**, 3241.
[71] D. B. Bigley and C. L. Fetter, *J. C. S. Perkin II*, 1979, 122.

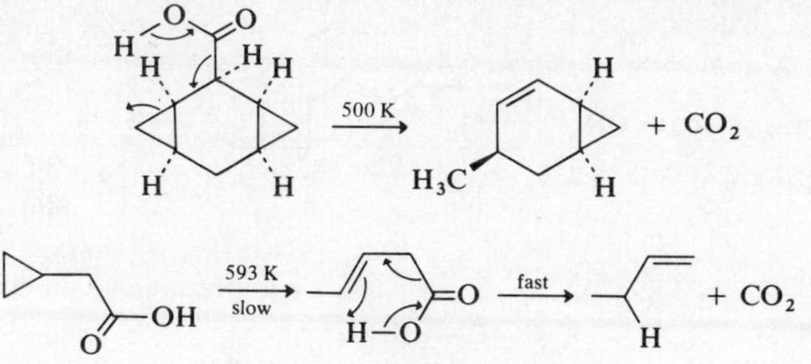

(1)

(2)

Scheme 9

In a study of the thermal processes of 2-vinylcyclobutylidene (44), no cyclohex-3-enylidene [analogous to the cyclopent-3-enylidene produced from 2-vinylcyclopropylidene (45)] was detected.[72] Instead, 2-vinylcyclobutylidene that has been generated by flash pyrolysis of the sodium salt of 2-vinylcyclobutanone tosylhydrazone gives a mixture of vinylmethylenecyclopropane and allylidenecyclopropane, from cleavage of the C-3—C-4 and C-2—C-3 bonds, respectively. Although conformational effects and different methods of generation may in part be responsible for the difference in behaviour between (44) and (45), thermochemical studies show (45) to be less strained than (44) by 42—54 kJ mol^{-1}.

(44) (45)

The vinylcyclobutane–cyclohexene rearrangement is known to occur at elevated temperatures but examples that occur at room temperature are rare. However, it has been reported[73] that the reaction of 1,1-dicyclopropylbuta-1,3-diene with TCNE gives 1-(2′,2′-dicyclopropylvinyl)-2,2,3,3-tetracyanocyclobutane, and if allowed to react for longer periods, 3,3-dicyclopropyl-4,4,5,5-tetracyanocyclohex-1-ene (46). Although (46) is formally the product of a (4+2) cycloaddition, these observations can also be interpreted in terms of a vinylcyclobutane–cyclohexene rearrangement. Following solvent and substituent studies, the rearrangement was concluded to be a heterolytic process involving a zwitterionic intermediate (47). In a study[74] of the thermolysis of t-butyl cubaneperoxycarboxylate, measurements of the rate and activation parameters show that the cubyl radical is formed 220-fold less rapidly than the t-butyl radical under the same conditions. It has been suggested that this is a consequence of the hybridization of the exocyclic bonding orbital of the bridgehead carbon atom and the polar character of the endocyclic carbon–carbon bonding orbitals.

On pyrolysis, 1,3-dimethylenecyclopentane (48) has been shown[75] to behave similarly to 1,2-dimethylenecyclobutane. Thus, at 370.5 °C, (48) undergoes random

[72] U. H. Brinker and L. König, *J. Amer. Chem. Soc.*, 1979, **101**, 4738.
[73] N. Shimizu and S. Nishida, *J. C. S. Chem. Comm.*, 1978, 931.
[74] T.-Y. Luh and L. M. Stock, *J. Org. Chem.*, 1978, **43**, 3271.
[75] J. J. Gajewski and J. Salazar, *J. Amer. Chem. Soc.*, 1979, **101**, 2739.

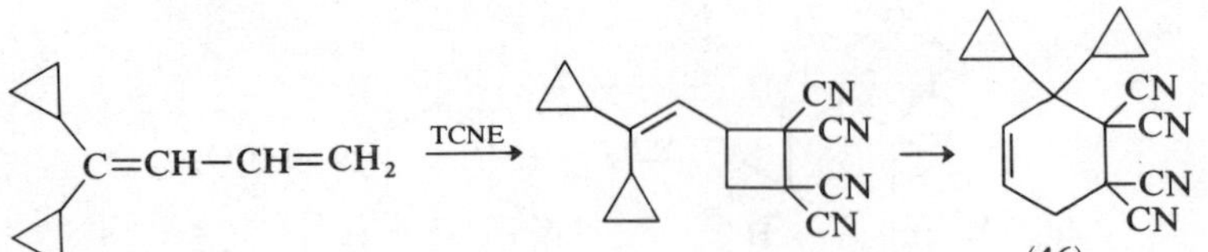

(46)

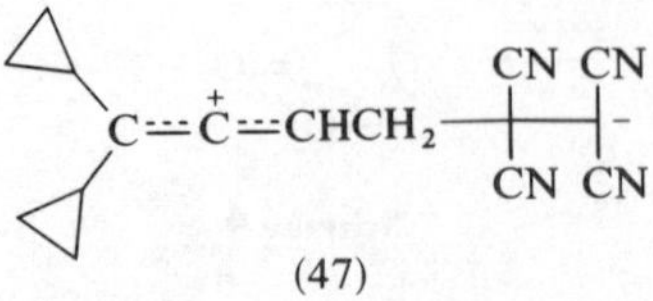

(47)

scrambling of the methylene groups with an activation energy of approximately 222 kJ mol^{-1}, implying that there is cleavage to 2,2′-bisallylmethane followed by random closure (Scheme 10). The ring-opening was shown to be almost stereospecific, and this has been attributed to steric effects.

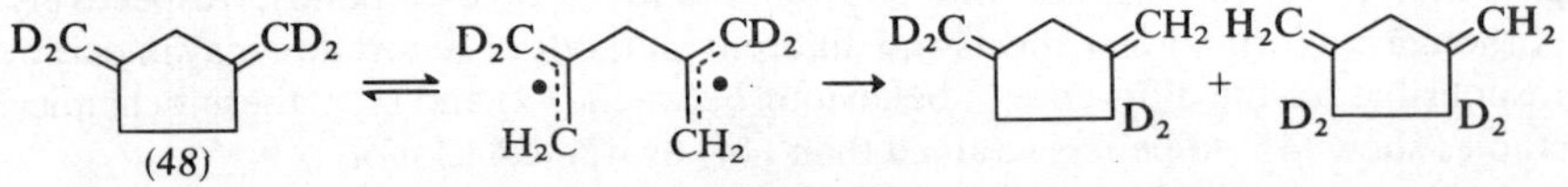

(48)

Scheme 10

A companion paper reports[76] that pyrolysis of (−)-(4*R*,5*R*)-*trans*-4,5-dimethyl-1,3-dimethylenecyclopentane gives (+)-(4*S*)-4-methyl-(*E*)-1-ethylidene-3-methylenecyclopentane, with 11.9% preservation of optical purity. Along with other results, this implies that the 2,2′-bisallylmethane biradical undergoes closure, with stereospecificity, to the [1,3]-shift product, with partial inversion of configuration of the migrating carbon. This stereospecificity has been suggested to arise as a consequence of stereospecific generation of a biradical followed by ring-closure (*via* a least-motion path) which is faster than bond rotation.

In a study that was mounted to investigate the processes involved in the thermal decomposition of cyclohexene, 3,3,6,6-tetradeuteriocyclohexene has been pyrolysed.[77] Four primary unimolecular processes, *i.e.* retro-Diels–Alder, elimination of D_2, elimination of HD, and elimination of H_2, appear to be involved, in the ratio 20:1:0.1:0.02. If concerted, the last two processes are symmetry-forbidden, and have an activation energy ~25 kJ mol^{-1} larger than the symmetry-allowed elimination of D_2.

The first example of a cyclic conjugated ene-allene, *cis*-cyclodeca-1,2,4-triene, has been isolated; on thermolysis at 100 °C this gives *trans*-bicyclo[4.4.0]deca-2,4-diene quantitatively.[78] This reaction appears to proceed *via* a stereospecific [1,5]-hydrogen migration to give *trans,cis,cis*-cyclodeca-1,3,5-triene as an intermediate, and as such it stands in contrast to thermal reactions of the acyclic ene-allenes, whose degree of stereoselectivity is much smaller.

[76] J. J. Gajewski and J. Salazar, *J. Amer. Chem. Soc.*, 1979, **101**, 2740.
[77] D. C. Tardy, R. Ireton, and A. S. Gordon, *J. Amer. Chem. Soc.*, 1979, **101**, 1508.
[78] D. E. Minter, G. J. Fonken, and F. T. Cook, *Tetrahedron Letters*, 1979, 711.

The optically active systems 2,7-dimethyl[norcaradiene⇌cycloheptatriene]-7-carbonitrile and the corresponding carboxylate have been used[79] to investigate the stereochemistry of the thermal norcaradiene–norcaradiene rearrangement. Contrary to the predictions of the Woodward–Hoffmann rules, the rearrangement proceeds highly stereoselectively and with inversion at C-7. The results have been interpreted in terms of biradical transition states.

Kinetic data have been published[80] on the Cope rearrangement of some *cis*-1,2-dialkenyl-cyclopropanes. The steric effects of substituents clearly support the view that the Cope reaction is a concerted $[\sigma^2_s + \pi^2_s + \pi^2_s]$ process, involving a boat-like conformation in the transition state. In order to probe further the influence of molecular conformation on the direction of the Cope rearrangement, the thermal behaviour of (49), a compound that is free to adopt one of two possible conformations, has been studied.[81] The results indicate that (49) cleanly rearranges in hot benzene *via* transposition of the weaker cyclopropane cross-link to give (50), which, under the conditions employed, rearranges further.

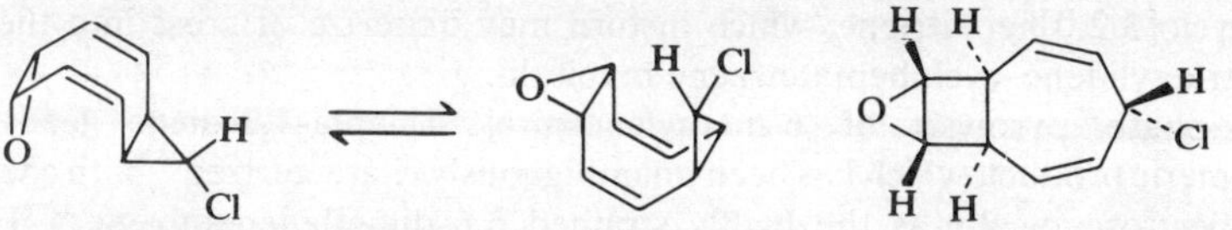

(49a) (49b) (50)

The thermal isomerization of 6-difluoromethylenebicyclo[3.2.0]hept-2-ene (51) occurs[82] exclusively through two competing 1,3-sigmatropic processes, (53) formally arising by shift of C-7 from C-1 to C-3 and (54) by shift of C-1 from C-7 to C-8. Absence of the Cope-derived product (55) would appear to be due to conformational processes in the initially formed radical (52) in that there is a significantly greater barrier for passage of the CF_2 end of the allyl radical in (52) past the CH_2 group at C-4 than for the comparable rotation of the analogous protio-radical.

The thermal reaction of bicyclo[5.1.0]octa-2,5-diene (56) with SO_2 affords[83] 7-thiabicyclo[4.2.1]nona-2,4-diene 7,7-dioxide (Scheme 11). An additional mechanism, involving prior isomerization of a double bond in (56), is also possible since a facile reaction occurs between 1,2-homotropylidene and dry SO_2 in [$^2H_8$]toluene, giving an almost quantitative yield of the same sulphone.[84]

A new member of the C_7H_6 energy-surface, *i.e.* 7-norbornadienylidene,* has been generated by pyrolysis of the corresponding *N*-nitrosourea.[85] It is a compound of special interest, as it is the premier example of the foiled carbenes. Depending on the temperature, 7-norbornadienylidene may lose carbon to give benzene or rearrange

* Bicyclo[2.2.1]hepta-2,5-dien-7-ylidene. *Senior Reporter.*

79 F.-G. Klärner, S. Yaslak, and M. Wette, *Chem. Ber.*, 1979, **112**, 1168.
80 M. P. Schneider and A. Rau, *J. Amer. Chem. Soc.*, 1979, **101**, 4426.
81 A. G. Anastassiou and R. L. Mahaffey, *Tetrahedron Letters*, 1979, 3349.
82 W. R. Dolbier, C. A. Piedrahita, and B. H. Al-Sader, *Tetrahedron Letters*, 1979, 2957.
83 J. Dalling, J. H. Gall, and D. D. MacNicol, *Tetrahedron Letters*, 1979, 4789.
84 D. D. MacNicol, personal communication.
85 W. T. Brown and W. M. Jones, *J. Org. Chem.*, 1979, **44**, 3090.

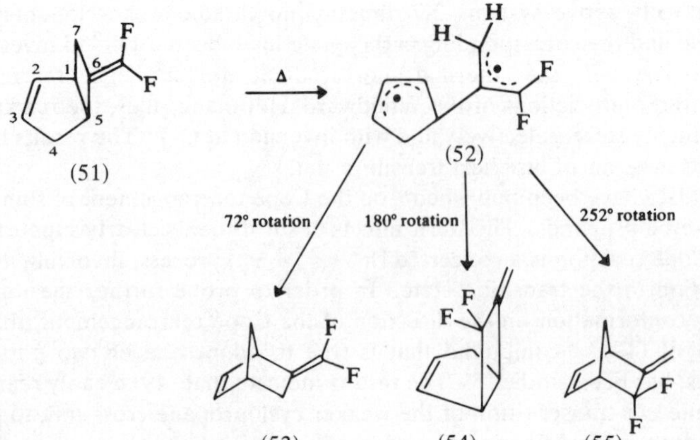

(51) (52) (53) (54) (55)

to bicyclo[3.2.0]heptatriene, which in turn may dimerize or cross into the cyclo-heptatrienylidene–cycloheptatetraene manifold.

Gas-phase pyrolysis of 6-methylenespiro[2.4]hepta-1,3-diene leads to a monomeric product, which has been unambiguously characterized[86] both chemically and spectroscopically as the highly strained 6,6-dimethylenefulvene.* The u.v. spectrum is interesting in that, despite strain caused by the cyclopropane ring, it is very similar to that of dialkyl-fulvenes. These results cast further doubt on an earlier claim[87] to the synthesis of this compound.

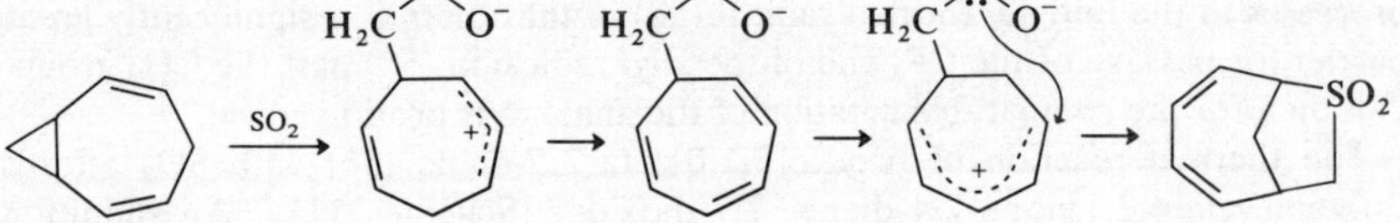

Scheme 11

A report has appeared[88] on the possibility of preferential double inversion *via* a π-spiropentane in the geometrical isomerization of 1,2,4-trimethylspiropentane. The results show that the *trans*-isomers undergo conrotatory double inversion and that the *cis*-isomers undergo disrotatory double inversion. Double inversion appears to be favoured by a factor of three over single inversion, and the mutual direction of rotation is dependent on steric and dynamic effects rather than electronic ones.

* 5-Cyclopropylidenecyclopenta-1,3-diene. *Senior Reporter.*

[86] R. D. Miller and D. Kaufmann, *J. C. S. Chem. Comm.*, 1978, 496.
[87] H. Schaltegger and M. Neuenschwander, *Chimia* (*Switz.*), 1962, **16**, 231.
[88] J. J. Gajewski, R. J. Weber, and M. J. Chang, *J. Amer. Chem. Soc.*, 1979, **101**, 2100.

In an investigation of the reactivity of cyclopentadiene and di-*exo*-methylene compounds in Diels–Alder reactions, the frontier-orbital method has been used[89] to show that the distance between C-1 and C-4 of the diene is significant for the reaction. Reactivity has been shown to be inversely dependent on this distance, and a linear dependence between this distance and log k_2 has been demonstrated experimentally.

As part of a study of the effect of pressure on the [2+2→4] cycloaddition of tetracyanoethylene to enol ethers, the general volume profile has been measured for a number of different enol ethers. The results obtained[90] accord with a mechanism involving a zwitterionic intermediate and a 'dead-end' charge-transfer complex.

The timing of bond changes on the way from reactants to products has been investigated[91] for the allyl-cation route to seven-membered rings. Using the sodium-iodide-induced debromination of $\alpha\alpha'$-dibromo-ketones, in the presence of copper and a conjugated diene, it has been found that cycloaddition of the W conformation of 1,3-dimethyl-2-oxyallyl cations to electron-rich cyclic conjugated dienes is accompanied by configurational loss. It was concluded that a push–pull mechanism is involved in which the first σ-bond is formed in an electrophilic step and the second σ-bond in a nucleophilic step. Electronic as well as steric factors may play a part.

A novel type of thermal pericyclic reaction has been announced.[92] The reaction of hex-1-yne and n-hexane (1 : 10) in a high-pressure/high-temperature flow reactor leads to alkenes as major products; these are thought to have risen according to Scheme 12. This new reaction may have importance for vinylation and for the fragmentation of alkanes to alkenes.

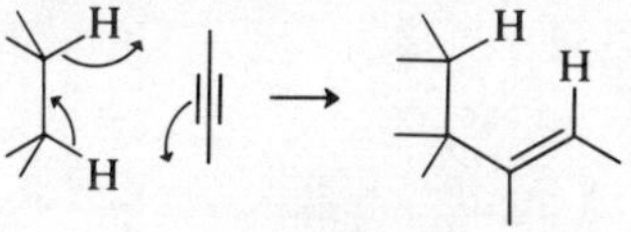

Scheme 12

By using a labelled *ortho*-allyl substituent, it has been possible to examine the first step of the carbon analogue of the Claisen rearrangement quite separately from the second step.[93] The results suggest that the first step is rate-determining. Above 350 °C, a high-yield ene cyclization to form nine-membered rings occurs, and it can be attributed to a steric acceleration arising from loss of degrees of freedom on placement of the ene components in *ortho*-related side-chains on a benzene ring.

A theoretical study has been made[94] of the thermal interconversion of tetra-t-butyltetrahedrane and tetra-t-butylcyclobutadiene. The transition state is found to be early on the reaction co-ordinate, and the activation energy has been estimated to be 84 kJ mol^{-1}.

[89] R. Sustmann, M. Böhm, and J. Sauer, *Chem. Ber.*, 1979, **112**, 883.
[90] J. von Jouanne, H. Kelm, and R. Huisgen, *J. Amer. Chem. Soc.*, 1979, **101**, 151.
[91] D. I. Rawson, B. K. Carpenter, and H. M. R. Hoffmann, *J. Amer. Chem. Soc.*, 1979, **101**, 1786.
[92] J. Metzger and P. Köll, *Angew. Chem. Internat. Edn.*, 1979, **18**, 71.
[93] J. B. Lambert, D. M. Fabricius, and J. J. Napoli, *J. Amer. Chem. Soc.*, 1979, **101**, 1793.
[94] A. Schweig and W. Thiel, *J. Amer. Chem. Soc.*, 1979, **101**, 4742.

[illegible]

[illegible] addition and subsequent fragmentation [illegible] a chain mechanism [illegible] for pericyclic reactions [illegible]

Another type of [illegible] has been established. [illegible]

[illegible]

Scheme 17

[illegible]

[illegible]

10 Aromatic Compounds

By R. BOLTON

Department of Chemistry, Bedford College (University of London), London NW1 4NS

1 General and Theoretical Considerations

The criteria of homoaromatic character have been calculated using *ab initio* STO-2G methods, giving results similar to those obtained using MINDO/3 and frontier orbital methods.[1] A MINDO/3 study of nitrobenzene reproduced the geometry of the nitro-group adequately, but failed to reflect the structure of the aryl system so effectively.[2] Other calculations, using both CNDO/2 and MINDO/3 methods, have been applied to the ^{1}H n.m.r. shifts associated with annulenes. Good agreement with experimental values is found only if [14]annulene is assumed to be a buckled structure, and if [18]annulene is nearly planar.[3]

An interesting example of a 10π-aromatic system is 1,4-dihydro-1,4-diazocine (1). The parent molecule, together with some of the simpler derivatives, shows both chemical behaviour and spectroscopic properties consistent with a delocalized aromatic structure; however, when electron-withdrawing substituents ($-CO_2Me$, $-COPh$, $-SO_2Me$) are attached to the nitrogen atoms, the resulting ten-membered ring does not show aromatic properties.[4]

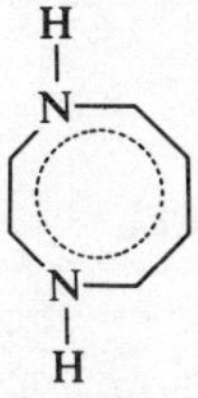

(1)

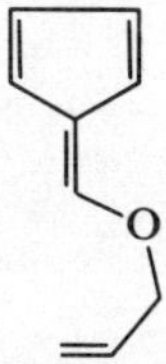

(2)

An analogue to the Claisen rearrangement has been suggested[5] in the reaction of (trifluoromethyl)cyclopentadiene with potassium hydroxide and allyl alcohol, when the observed product, allyl 1-allylcyclopenta-2,4-dienoate, was suggested to result *via* the fulvene (2). Other similar processes, having their analogy in conventional

[1] R. C. Haddon, *J. Org. Chem.*, 1979, **44**, 3608.
[2] L. P. Davis and R. M. Guidry, *Austral. J. Chem.*, 1979, **32**, 1369.
[3] H. Vogler, *J. Mol. Struct.*, 1979, **51**, 289.
[4] (*a*) H.-J. Altenbach, H. Stegelmeier, M. Wilhelm, B. Voss, J. Lex, and E. Vogel, *Angew. Chem., Internat. Edn.*, 1979, **18**, 962; (*b*) M. Breuninger, B. Gallenkamp, K.-H. Müller, H. Fritz, H. Prinzbach, J. J. Daly, and P. Schönholzer, *ibid.*, p. 964.
[5] T. Olsson and O. Wennerstroem, *Acta Chem. Scand.* 1979, **B33**, 256.

benzene chemistry, were also reported. Both ^{1}H and ^{13}C n.m.r. spectroscopy studies have been applied to some 6-(*para*-substituted-phenyl)-fulvenes.[6]

The problems associated with the transmission of electronic effects across aryl systems linked in various ways have received study by semi-empirical HMO and LCAO MO SCF methods, and the results were compared with experimental measurements of the properties of amines and phenols corresponding to the structure $YC_6H_4MC_6H_4X$, in which M is the linkage through which the electronic perturbations are transmitted from the substituent X to the amino- or hydroxyl group ($Y = NH_2$ or OH).[7] Both calculation methods were held to be effective; not surprisingly, the degree and character of the various conjugative interactions were found important.

2 Benzene Derivatives

Electrophilic Substitution.—A mechanistic problem, inherent in each mode of substitution and satisfactorily solved in none of them, is the reality of the π-complex as a discrete intermediate in reaction. The relative roles of both σ- and π-complexes in the structure $C_6H_7^+$, an intermediate in proton-exchange reactions, has been examined by MINDO/3 and *ab initio* calculation methods. Two possible routes appear to be available, so that the π-complex is found not to be a necessary intermediate in the formation of the σ-complex.[8] Calculation therefore offers only equivocal support to theories suggesting the essential intermediacy of π-complexes in aromatic substitution.

The first synthesis of a phenyl carbo-cation has been reported. The formation of this ion from the products of solvolysis of (2a) was demonstrated by the isolation of between 1 and 10% of phenol and phenolic ethers derived from 2,3,6-trimethylphenol, and showing therefore the intermediacy of the 2,3,6-trimethylphenyl carbo-cation.[8a]

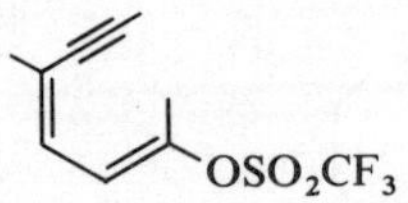

(2a)

Nitration. Two types of anomalous product may result from *ipso*-attack, and throughout the year a number of interesting advances have been made in studies of the chemistry of such products. Suzuki and his colleagues have continued their investigations of the side-chain attack which characterizes attempted nitration processes in fully substituted benzene systems. Thus, the reaction of hexamethylbenzene with nitric acid in acetic acid at low temperatures provides four derivatives of pentamethylbenzyl alcohol (Scheme 1). The same four derivatives are also formed from (3), a possible intermediate; but, although they are formed at the same

[6] A. Otter, H. Muehle, M. Neuenschwander, and H. P. Kellerhals, *Helv. Chim. Acta*, 1979, **62**, 1626.
[7] V. A. Dadali, T. M. Prokep'eva, and Yu.B.Vysotskii, *Zhur. Org. Khim.*, 1979, **15**, 1129.
[8] T. Sordo, J. Bertrán, and E. Canadell, *J.C.S. Perkin II*, 1979, 1486.
[8a] M. Hanack and U. Michel, *Angew. Chem., Internat. Edn.*, 1979, **18**, 870.

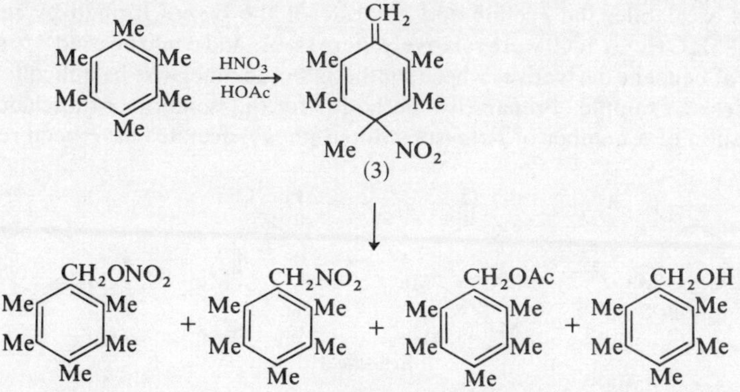

Scheme 1

rate in each total reaction, the product distribution differs in the two processes.[9] The corresponding reaction of derivatives of pentamethylacetophenone and of 4,6-diacetyl-1,2,3,5-tetramethylbenzene showed an interesting trend towards attack at the most crowded methyl groups in the system. This negative steric effect might arise from a contribution by the adjacent carbonyl system in structures such as (4) in a sequence such as Scheme 2 demonstrates, and occurs even when trihalogenomethyl ketones [*e.g.* $Cl_3CC(O)C_6Me_5$] undergo nitration.[10] The problems associated with these reactions have been discussed.[11]

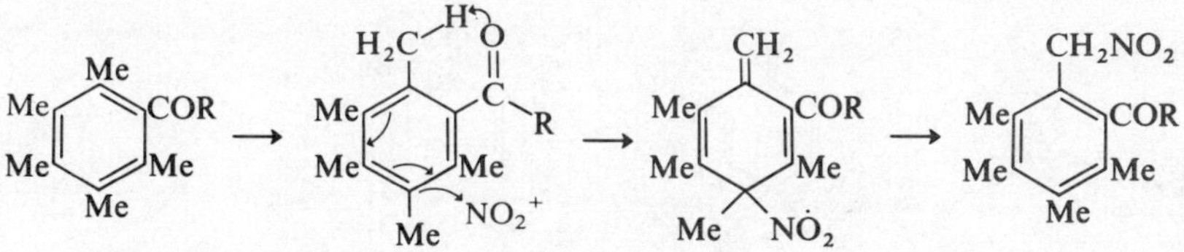

(4) **Scheme 2**

Another facet of *ipso*-substitution is shown by the isolation of adducts such as 4-methyl-4-nitrocyclohexa-2,5-dienone (5) from the nitration of phenols. Myhre and his colleagues have made a study of adducts such as (5) and, in particular, their synthetic uses and their rearrangement under acidic conditions to products involving formal nitrodeprotonation.[12] Although both the dienones and their reduction

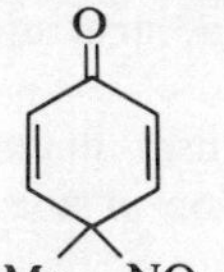

(5)

[9] H. Suzuki, T. Mishina, and T. Hanafusa, *Bull. Chem. Soc. Japan.*, 1979, **52**, 191 (*Chem. Abs.*, 1979, **90**, 203 107w).
[10] H. Suzuki, M. Hashihama, T. Mishina, and T. Hanafusa, *J.C.S. Chem. Comm.*, 1979, 69.
[11] H. Suzuki, *Yuki Gosei Kagaku Kyokaishi*, 1979, **37**, 290 (*Chem. Abs.*, 1979, **91**, 55 649c).
[12] K. S. Feldman, A. McDermott, and P. C. Myhre, *J. Amer. Chem. Soc.*, 1979, **101**, 505.

products are labile, the acetate and chloride of the dienol formed by reduction ($NaBH_4$–MeOH, −15 °C) were relatively more stable and could be used[13] to supply a number of benzene derivatives whose synthesis would otherwise be difficult. Scheme 3 provides an example. Preparative methods[14] for the isolation of these adducts by the nitration of a number of *para*-substituted phenyl acetates have been reported.

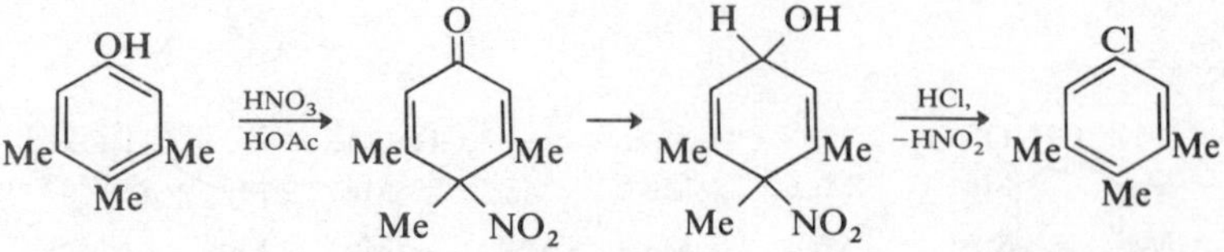

Scheme 3

The nitration of phenol and cresols in sulphuric acid (58—80%) is a quantitative process which involves attack by the nitronium ion at near the encounter-controlled rate. The *ipso*-attack which had earlier been identified[15] in the nitration of *p*-cresol also accounts for the formation of quantities of 2-methyl-6-nitrophenol in the nitration of *o*-methylanisole.[16] While there has been much evidence to suggest that such cyclohexadiene derivatives are formed in heterolytic nitration reactions, polar mechanisms are not essential. 2,6-Di-t-butyl-4-methylphenol gives 2,6-di-t-butyl-4-methyl-4-nitrocyclohexa-2,4-dienone (6) and the further product of addition (7) on treatment with dinitrogen tetroxide at room temperature for 2 h; the conditions suggest a radical mechanism.[17]

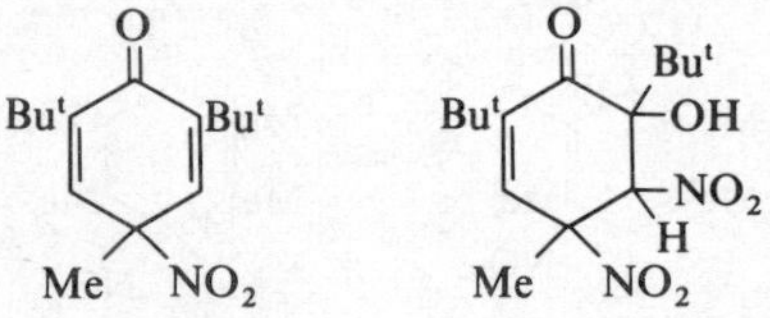

(6) (7)

An original nitration mechanism, observed under conditions in which a classical nitration process might confidently have been expected, was found by Giffney and Ridd[18] in the reaction of nitric acid with *NN*-dimethylaniline in 85% sulphuric acid. The process was catalysed by nitrous acid, when *NNN'N'*-tetramethylbenzidine and its nitration products were identified along with the expected 3- and 4-nitro-*NN*-dimethylaniline. A radical nitration sequence (Scheme 4) was suggested; the isolation of the tetramethylbenzidine structures gave strong circumstantial evidence for the intermediate $PhNMe_2^{+\cdot}NO\cdot$.

The classical nitration mechanism underwent more detailed analysis when Schofield's school used the nitration of anisole and of toluene under encounter-

[13] K. S. Feldman and P. C. Myhre, *J. Amer. Chem. Soc.*, 1979, **101**, 4768.
[14] C. E. Barnes, K. S. Feldman, M. W. Johnson, H. W. H. Lee, and P. C. Myhre, *J. Org. Chem.*, 1979, **44**, 3925.
[15] R. G. Coombes and J. G. Golding, *Tetrahedron Letters*, 1978, 3583.
[16] R. G. Coombes, J. G. Golding, and P. Hadjigeorgiu, *J.C.S. Perkin II*, 1979, 1451.
[17] G. Brunton, H. W. Cruse, K. M. Riches, and A. Whittle, *Tetrahedron Letters*, 1979, 1093.
[18] J. C. Giffney and J. H. Ridd, *J.C.S. Perkin II*, 1979, 618.

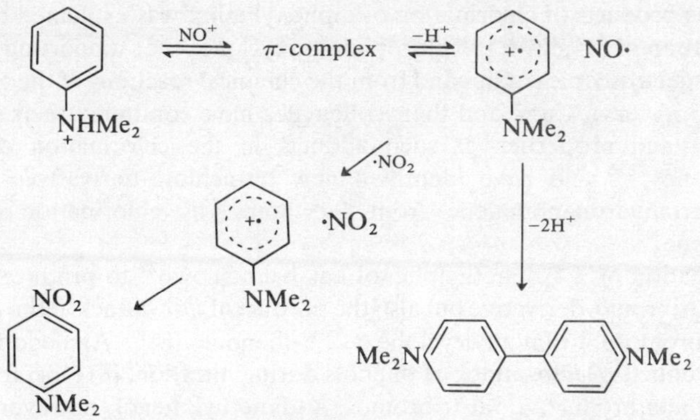

Scheme 4

controlled rates to measure the rate of formation of nitronium ions in aqueous acid media.[19] Other studies of more conventional nitration reactions include the conclusion that the formation of picramide from 2,4-dinitroaniline involved the intermediacy of the *N*-nitro-derivative, since the rate of formation of picramide from *N*-nitro-2,4-dinitroaniline under acid conditions was similar to that from nitric acid and 2,4-dinitroaniline under the same conditions.[20]

The mercury(II)-catalysed nitration of arenes[21] was shown to involve mercuration as its essential, rate-determining step; the orientation of nitration paralleled that of mercuration and differed from that observed in the absence of mercury salt.

The nitration of phenanthrene in acetic anhydride was reported[22] to give substantial quantities of 10-acetoxy-10′-nitro-9,9′-biphenanthryl. The identification of this product, together with the apparently ubiquitous observation of *ipso*-attack in such reaction media, may make it desirable to re-examine the earlier work on the electrophilic substitution of polyannular systems, which took nitration in acetic anhydride or acetic acid as the archetypal process.

Sulphonation. The details of the sulphonation of 2-methylnaphthalene and of all ten dimethyl-naphthalenes have been carefully studied.[23] Steric factors appear to be unusually important, and appreciable amounts of disulphonation occurred in the same ring. Directing effects appear to be unusually weak in these systems compared with those in the benzene system, and suggest that it is an over-simplification to regard the naphthalene system as just an extended benzene system. A number of substituted biphenyls have also been studied.[24]

Halogenation. Since the early work of de la Mare's group, the concept of addition occurring in parallel with heterolytic substitution by chlorine[25] in aromatic systems has been well documented. The identification of much 2,4-dichlorodiphenyl ether

19 R. B. Moodie, K. Schofield, and P. G. Taylor, *J.C.S. Perkin II*, 1979, 133.
20 G. F. Burya and L. R. Andreeva, *Izvest. Akad. Nauk SSSR, Ser. Khim.*, 1979, 1464 (*Chem. Abs.*, 1979, **91**, 156 941x).
21 L. M. Stock and T. L. Wright, *J. Org. Chem.*, 1979, **44**, 3467.
22 D. M. D. Lane, K. E. Richards, G. J. Wright, and A. Fischer, *Austral. J. Chem.*, 1978, **31**, 2737.
23 K. Lammertsma and H. Cerfontain, *J.C.S. Perkin II*, 1979, 673.
24 T. A. Kortekaas and H. Cerfontain, *J.C.S. Perkin II*, 1979, 224.
25 P. B. D. de la Mare, 'Electrophilic Halogenation', Cambridge University Press, 1976.

among the products of chlorination of diphenyl ether was explained by postulating the formation of 3,4,5,6-tetrachlorophenoxycyclohexene; supporting evidence was found in spectroscopic studies and from the chemical reactions of the crude reaction product.[26] Aversa, Cum, and their colleagues have continued their studies of the formation and properties of such adducts in the chlorination of substituted naphthalenes[27,28] and have identified new tetrachloro-derivatives of 1-methyl-1,2,3,4-tetrahydronaphthalene from the homolytic chlorination of 1-methylnaphthalene.[28]

Bromination of 3,4-dimethylphenol has been shown[29] to produce not only the expected tribromo-derivative but also the product of *ipso*-attack at the *para*-position [2,4,6-tribromo-3,4-dimethylcyclohexa-2,5-dienone (8)]. Analogously with the similar products of *ipso*-attack of phenols during nitration, (8) rearranges in acid to give the main product, 2,5,6-tribromo-3,4-dimethylphenol; however, in light, the re-aromatization product is 2,6-dibromo-4-bromomethyl-3-methylphenol (9).

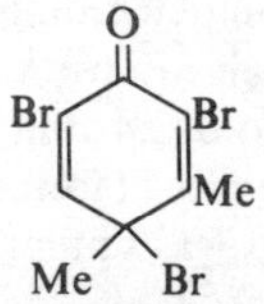

(8)

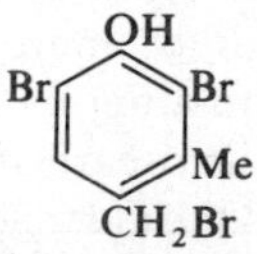

(9)

Side-chain bromination by *N*-bromosuccinimide in tetrachloromethane to give the mono-bromination product does not show a clear preference towards σ or σ^+ in reflecting the variation of yield with the nature of the substituent in fifteen derivatives of toluene; however, the corresponding formation of benzal bromides from the ring-substituted benzyl bromides showed a better agreement with the Brown–Okamoto parameters (ρ^+, -0.88).[30]

The preparative chlorination of polyfluoroaromatic amines by t-butyl hypochlorite proceeds by two distinct stages. The initially formed *N*-chloroaniline derivative may be rearranged to the stable ring-substituted derivative after treatment with catalytic quantities of molecular iodine.[31]

Other Electrophilic Processes. Friedel–Crafts acylation of some polybenzenoid systems has shown interesting results. The electrophilic attack of substituted fluorenes may show orientation effects in which the directing effect of the substituent within one benzene ring acts in opposition to that commonly seen through the stabilizing effect of the second aromatic system upon attack at the 2-position of the fluorene moiety. Earlier reports of such substitution reactions have not been systematic (*e.g.* attack upon 1,4-difluorofluorene[32]) or have not allowed such a competition of effect (*e.g.* 1-methylfluorene),[33] but a beginning has been made[34] of

26 W. D. Watson and H. E. Henries, *J. Org. Chem.*, 1979, **44**, 1155.
27 M. C. Aversa, G. Cum, P. Giannetto, and G. Romeo, *J. Chem. Res.*, 1978, (*S*) 292; (*M*) 3629.
28 M. C. Aversa, G. Cum, P. Giannetto, and G. Romeo, *J. Chem. Res.* 1979, (*S*) 338; (*M*) 3979.
29 J. M. Brittain, P. B. D. de la Mare, N. S. Isaacs, and P. D. McIntyre, *J.C.S. Perkin II*, 1979, 933.
30 W. Offermann and F. Voegtle, *J. Org. Chem.*, 1979, **44**, 710.
31 R. E. Banks, M. G. Barlow, J. C. Hornby, and T. J. Noakes, *J. Fluorine Chem.*, 1979, **13**, 179.
32 M. J. Namkung and T. L. Fletcher, *Canad. J. Chem.*, 1967, **45**, 2569.
33 E. O. Arene and D. A. H. Taylor, *J. Chem. Soc.* (*C*), 1966, 481.

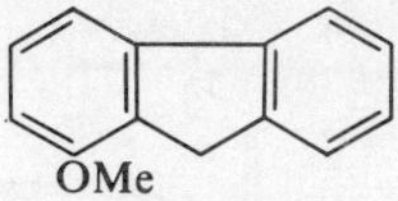

(10)

a systematic study of such behaviour. For example, 1-methoxyfluorene (10) provides a complex mixture of partially demethylated products, of which 45% is 4,7-diacetyl-1-hydroxyfluorene or its methylated or acetylated derivatives; a small amount of 2,4-diacetyl-1-hydroxyfluorene was also detected. Apparently the strong *para*-directing properties of the methoxy-substituent override the tendency of the fluorene system towards 2-substitution, although at least part of this behaviour may result from steric factors.[33,34] However, the selection of substituted fluorenes was not exhaustive, and there is still room for a more systematic study of substituent effects in this system.

Gore and his co-workers have recently turned their attention to the acylation of naphthalene derivatives, with some unexpected results. The acetylation of acenaphthene provides products of both 3- and 5-attack, the relative amounts of which may be altered by changing the effective bulk of the attacking agent either by increasing the size of the acyl fragment or by changing to a more complexing solvent: nitrobenzene assists attack at the less hindered position, as it does in the acylation of naphthalene itself.[35] In contrast, 3-, 4-, and 5-acyl derivatives are identified in the corresponding reaction of 1,8-dimethylnaphthalene and, in addition, some evidence was found of rearrangement to derivatives of 1,7-dimethylnaphthalene.[36] The Perrier method, in which the complex between acyl halide and aluminium halide is pre-formed and dissolved in the reaction solvent before addition of the aromatic substrate, was used in a study of the complicated products of acetylation of 1-chloronaphthalene. All seven isomers were detected; the major product (*ca.* 80%) was 4-acetyl-1-chloronaphthalene unless the reaction solvent was nitrobenzene, when approximately equal amounts of the 4-, 6-, and 7-acetyl-1-chloronaphthalenes were obtained. The 1-chloro-substituent was also found to activate attack at the 4-position by a factor of *ca.* 5, a result at variance with expectation both from studies of the Friedel–Crafts acylation of chlorobenzene and from measurements of the attack of chlorobenzene by other electrophiles. Some indications that the yields of product in the acetylation of some naphthalene derivatives depend upon time suggest that the unexpected effect may arise from idiosyncracies of this particular reaction, rather than from a general property of 1-chloronaphthalene.[37] Rearrangements, already noted in the acenaphthene work as well as in the attack of 1,8-dimethylnaphthalene,[36] may be involved.

Thallation [$Tl(OCOCF_3)_3$–CF_3CO_2H, at 25 °C] has been shown to be more selective than mercuration (11) but otherwise appears to involve a conventional electrophilic (ρ, 4) substitution.[38]

[34] D. R. Buckle, N. J. Morgan, and R. G. Alexander, *J.C.S. Perkin I*, 1979, 3004.
[35] *e.g.*, H. Luther and G. Wächter, *Chem. Ber.*, 1949, **82**, 161 and refs. therein.
[36] P. H. Gore and M. Jehangir, *J.C.S. Perkin I*, 1979, 3007.
[37] P. H. Gore and I. M. Khan, *J.C.S. Perkin I*, 1979, 2779.
[38] P. Y. Kwok, L. M. Stock, and T. L. Wright, *J. Org. Chem.*, 1979, **44**, 2309.

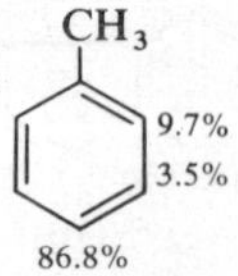

$k_{PhMe}/k_{PhH} = 43.5 \pm 2.0$

(11)

Nucleophilic Substitution.—*The* S_NAr *Mechanism.* The classic mechanism of aromatic nucleophilic substitution continues to receive much attention. Tiecco and his co-workers have reported the displacement of halogen from otherwise unactivated aromatic systems by thiolate nucleophiles in hexamethylphosphoramide at 80 °C; the order of ease of displacement was reported to be I > F > Br > Cl and this, together with the lack of influence of azobenzene (expected to inhibit the $S_{RN}1$ mechanism), was advocated as evidence for the suggested displacement mechanism. However, the differences of rate between iodobenzene, fluorobenzene, and bromobenzene were only a factor of two in the rate constant; chlorobenzene was only one-eighth as rapidly attacked as bromobenzene, and this contributed to the apparently wide spread of rate constants in the attack of the monohalogenobenzenes.[39] It seems still probable that more than one mechanism might prevail in these reaction conditions, for although confirmation of the order of displacement was found by studying dihalogeno-benzenes of the structure XC_6H_4Y, such studies are complicated by the different effects that the X and the Y substituents must have upon the attack which displaces the other substituent.

A very different situation has been found in studies of the effect of complexation of metal upon the reactivity of aryl halides. Early work[40] demonstrated the large activation which complexation with chromium, as in (12), might bring about. Knipe,

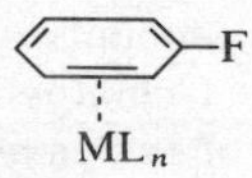

(12) a; $ML_n = Cr(CO)_3$
b; $ML_n = Mo(CO)_3$

McGuinness, and Watts[41] have reported rather more detailed studies of the effect of complexing upon the ease of displacement of fluorine and of chlorine in phenyl halides complexed with various metal ligands. Considerable differences were found by changing the nature of the metal and also the ease of displacement of the halogen atom [F ≫ Cl; k_F/k_{Cl} *ca.* 500 in (12b)] paralleled much more closely that found in 'conventional' processes such as the methoxydehalogenation of pentachloro-halogenobenzenes or of 2,4-dinitro-halogenobenzenes in methanol.[42] Another important demonstration of the effect of complexing by metals was shown in the

[39] P. Cogolli, F. Maiolo, L. Testaferri, M. Tingoli, and M. Tiecco, *J. Org. Chem.*, 1979, **44**, 2642.
[40] B. Nicholls and M. C. Whiting, *J. Chem. Soc.*, 1959, 551; J. F. Bunnett and H. Hermann, *J. Org. Chem.*, 1971, **36**, 4081.
[41] A. C. Knipe, S. J. McGuinness, and W. E. Watts, *J.C.S. Chem. Comm.*, 1979, 842.
[42] *e.g.*, J. Miller, 'Aromatic Nucleophilic Substitution,' Elsevier, Amsterdam, 1968.

methoxydechlorination of 5-chloro-1,10-phenanthroline in 89.8% DMSO–10.2% MeOH mixture. The rate of this process was accelerated[43b] by the presence of a number of transition metals, the effect appearing to be linked with the formal charge associated with the complex. In the preceding paper dealing with the displacement reaction of the uncomplexed chlorophenanthroline, the kinetic behaviour suggested a somewhat more complicated mechanism, which was reported as the coexistence of terms that are first-order and second-order in methoxide ion [equation (1)]:[43a]

$$d[Cl^-]/dt = k[ArCl][OMe^-] + k'[ArCl][OMe^-]^2 \quad (1)$$

Although the point was not given detailed study, it is possible that the second term in the kinetic equation may represent catalysis by sodium ion. Variations of the observed rate constant caused by changes in the concentration of reagent salts have been well known for some time, but detailed measurement of the effect of changes in the gegenion upon rate have only recently been reported.[44] The rates of ethoxy-dehalogenation of iodo- and chloro-2,4-dinitrobenzene in anhydrous ethanol by lithium, sodium, potassium, or caesium ethoxides were measured at 25 °C, and were compared with the results of a similar study with ethyl iodide as substrate. The major salt effects that were found were consistent with two effective reagents; 'free' ethoxide ion and 'paired' ethoxide ion. Conductimetric measurement was used to determine the contribution of each type of reagent in the solutions, and the variations which were observed could be interpreted in terms of the different degrees of association of the different metal ethoxides. Kinetic complications were caused partly because of the difficulty of assessing the contribution by the metal halide that is produced upon the initial pre-equilibrium between ion-pairs and free ethoxide ion, and, at higher concentrations, because of a general divergence which might have reflected the inadequacy of the kinetic form to cope with relatively high concentrations of species in solution.

Another approach to the problem was made by Gold and Toullec,[45] who took the pragmatic approach to define a term J_E [equation (2)] which reflected the ability of solutions of sodium ethoxide in ethanol to donate OEt^- to 2,4-dinitrophenetole

$$J_E = pK + pK_E + \log([\text{Adduct}]/[\text{2,4-Dinitrophenetole}]) \quad (2)$$

dissolved in them. The ion produced, (13), was detected spectroscopically, and the apparent equilibrium constant K_{app} was corrected for changes in ionic strength to give K, which, along with the autoprotolysis constant of ethanol (K_E), provided the

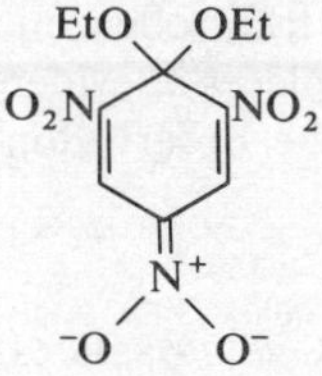

(13)

[43] (*a*) K. Jackson, J. H. Ridd, and M. L. Tobe, *J.C.S. Perkin II*, 1979, 607; (*b*) *ibid.*, p. 611.
[44] P. Jones, R. Harrison, and Lord Wynne-Jones, *J.C.S. Perkin II*, 1979, 1679.
[45] V. Gold and J. Toullec, *J.C.S. Perkin II*, 1979, 596.

parameters necessary for the new function. A similar function, applied to methanol, already exists.[46,47]

The formation of Meisenheimer complexes, of which (13) is an example, remains a source of interest.[48] Gibson and Crampton[49] used spectroscopic, kinetic, and equilibrium measurements to demonstrate that the reaction between hydroxide ion in aqueous media and picryl derivatives (1-X-2,4,6-trinitrobenzenes) is a process which first involves attack at C-3, and that the subsequent rearrangement of hydroxide from the 3-position to the 1-position then provides the conventionally formulated anion. It was felt, although on less rigid evidence, that the corresponding process was less likely in the analogous 2,4-dinitrophenyl systems. Low-temperature flow methods were used in an n.m.r. study of Meisenheimer complexes between aliphatic amines and 2,4,6-trinitroanisole in DMSO–MeOH (1:1).[50] Surprisingly, the products of the two reactions were different, and apparently reflected different stabilities of the two Meisenheimer intermediates. With n-butylamine, the fast formation of (14) was followed by the expected displacement of methoxide ion to give *N*-(n-butyl)picramide. Di-n-butylamine, however, rapidly provided the corresponding complex, which did not then lose OMe^- but instead underwent methanolysis to give 1,1-dimethoxy-2,4,6-trinitrocyclopentadiene anion (15).

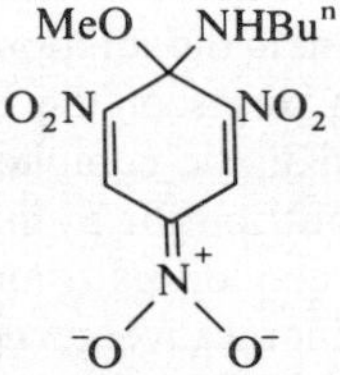
(14)

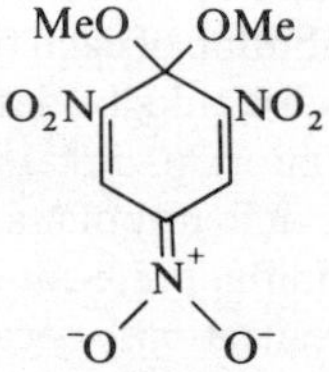
(15)

Stabilities of Meisenheimer complexes have also been measured in the nucleophilic attack of pyridinium ions by nucleophiles[51,52] and in the attack of (poly)nitrocyano-thiophens and -selenophens. In the latter case, addition of methoxide ion occurred more readily with the nitro-group *ortho* and *para* to the site of attack, the order being 2,4-dinitro > 2-nitro-4-cyano > 2-cyano-4-nitro, in describing the ease of attack of carbon bearing hydrogen in either heterocyclic system. The selenophen derivatives, perhaps expectedly, were better able to stabilize such structures than their thiophen analogues.[53] Bernasconi has revised and extended his earlier work on similar complexes between 1,3,5-trinitrobenzene and amines.[54]

One of the earliest compelling pieces of evidence for the two-stage mechanism of nucleophilic displacement was the observation of base catalysis. Hirst and his

46 F. Terrier, *Ann. Chim. (France)*, 1969, **4**, 153.
47 C. H. Rochester, *J. Chem. Soc.*, 1965, 2404.
48 S. Sekiguchi, *Yuki Gosei Kagaku Kyokaishi*, 1978, **36**, 633 (*Chem. Abs.*, 1979, **90**, 71 243a).
49 B. Gibson and M. R. Crampton, *J.C.S. Perkin II*, 1979, 648.
50 C. A. Fyfe, S. W. H. Damji, and A. Koll, *J. Amer. Chem. Soc.*, 1979, **101**, 951.
51 S. W. H. Damji and C. A. Fyfe, *J. Org. Chem.*, 1979, **44**, 1757.
52 S. W. H. Damji, C. A. Fyfe, D. Smith, and F. J. Sharom, *J. Org. Chem.*, 1979, **44**, 1761.
53 F. Terrier. A.-P. Chatrousse, and C. Paulmier, *J. Org. Chem.*, 1979, **44**, 1634.
54 C. F. Bernasconi, M. C. Muller, and P. Schmid, *J. Org. Chem.*, 1979, **44**, 3189.

colleagues[55] have observed that such catalysis apparently occurs in the reaction between aniline and 1-fluoro-2,4-dinitrobenzene in dimethylformamide, in acetonitrile, or in nitromethane, but not in dimethyl sulphoxide (DMSO). 1-Chloro-2,4-dinitrobenzene showed no such effect under any of these conditions. The evidence for catalysis, however, was a kinetic dependence upon the aniline concentration. The order with respect to aniline was greater than unity. The results allow a number of interpretations, one of which might be the instability of the solvents towards aniline. Catalysis by piperidine was shown in the aminodemethoxylation of 2-methoxy-3-nitrothiophen in benzene by piperidine at 10—30 °C. The evidence was again the incursion of a term involving a higher order with respect to the base than that expected in the simple bimolecular process, and the interpretation of the mechanism by which this catalysis occurred draws attention to the Meisenheimer intermediate.[56] A similar observation of catalysis by methoxide ion has already been noted.[56b] In an attempt to quantify the effects of *ortho*-substituents in similar displacement reactions, the rates of aminodebromination of 2-bromo-3,5-dinitrothiophen in methanol by a number of *ortho*-substituted anilines were measured. A multi-parameter equation was used to link the observed rates with steric and electronic properties of the substituents.[57]

Aminodenitration has been reported as a preparative method in the synthesis of acridones from polynitro-benzophenones, the displacement being assisted by the nitro-substituents as well as by the acyl group (Scheme 5).[58] However, the attack by piperidine upon 2,3-dinitrophenol is reported[59] to give 6-piperidino-3-nitrophenol by a sequence (Scheme 6) which involves initially a remote attack *meta* to the group that is ultimately displaced.

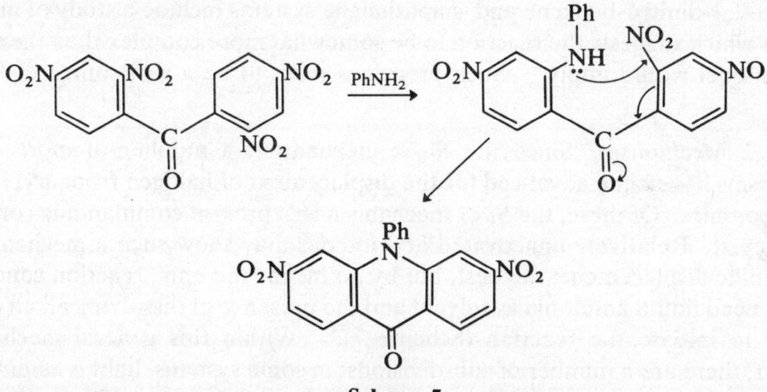

Scheme 5

Among novel systems showing nucleophilic displacement reactions, attack by methoxide ion upon pyrylium ions continues to attract attention. The rates of the rapid reactions, so far as generalizations may be made, are consistent with the substituent effects expected.[60]

[55] T. O. Bamkole, J. Hirst, and I. Onyido, *J.C.S. Perkin II*, 1979, 1317.
[56] G. Consiglio, R. Noto, and D. Spinelli, *J.C.S. Perkin II*, 1979, 222; *J. Org. Chem.*, 1978, **43**, 4038.
[57] G. Consiglio, R. Noto, D. Spinelli, and C. Arnone, *J.C.S. Perkin II*, 1979, 219.
[58] J. H. Gorvin and D. P. Whalley, *J.C.S. Perkin I*, 1979, 1364.
[59] R. E. Maxwell, *J.C.S. Chem. Comm.*, 1979, 428.
[60] G. Doddi, S. Fornarini, G. Illuminati, and F. Stegel, *J. Org. Chem.*, 1979, **44**, 4496.

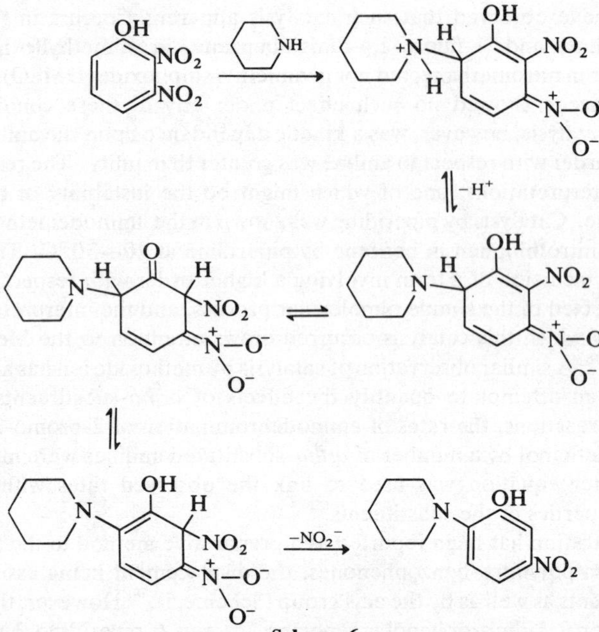

Scheme 6

Other variations in the conditions used for displacing chlorine by hydroxide in the 1-chloro-2,4-dinitro-benzene and -naphthalene systems include a study of micellar catalysis which suggests the reaction to be somewhat more complex than the simple kinetic model would imply.[61] Such processes seem to be a rich source of kinetic study.

The $S_{RN}1$ Mechanism. Since the S_NAr mechanism, a number of more recent mechanisms have been advanced for the displacement of halogen from aryl halides by nucleophiles. Of these, the $S_{RN}1$ mechanism is at present commanding considerable interest. Relatively unactivated aryl iodides may show such a mechanism of nucleophilic displacement; the best, but by no means the only, reaction conditions seem to need liquid ammonia as solvent and the presence of dissolving alkali metals or light to initiate the reaction (Scheme 7).[62] Within this general mechanism, however, there are a number of sub-divisions; in some systems, light is essential for the displacement to occur, whereas in others, light merely speeds the process but is

$$ArX + e^- \rightarrow [ArX]^{\overline{\cdot}}$$
$$[ArX]^{\overline{\cdot}} \rightarrow Ar\cdot + X^-$$
$$Ar\cdot + Y^- \rightarrow [ArY]^{\overline{\cdot}}$$
$$[ArY]^{\overline{\cdot}} + ArX \rightarrow ArY + [ArX]^{\overline{\cdot}}$$

Scheme 7

[61] C. A. Bunton, L. S. Romsted, and G. Savelli, *J. Amer. Chem. Soc.*, 1979, **101**, 1253.
[62] J. F. Bunnett, *Accounts Chem. Res.*, 1978, **11**, 413.

not essential. Thus, Pierini and Rossi[63] found that the displacement of halogen from halogeno-naphthalenes, -phenanthrenes, and -quinolines by $PhSe^-$ and $PhTe^-$ required light catalysis, and would not proceed in the dark, whereas Swartz and Bunnett[64] noticed that the phenylation of phosphanions by such a mechanism is accelerated by light, but that a slower reaction could be discerned in the dark; this presumably reflected the presence of adventitious amounts of radical sources. After the identification of the $S_{RN}1$ mechanism, a systematic study of the requirements and limitations of the mechanism has been begun. Thus, Galli and Bunnett[65] have compared the rates of phenylation of diethylphosphite anion [$(EtO)_2PO^-$] and pinacolone enolate [$Me_3CC(O)CH_2^-$] by six different mono-substituted benzenes PhX, where X = F, Cl, Br, I, SPh, or NMe_3^+, in liquid ammonia at the boiling point. The ratio of relative rates of attack upon the two anionic nucleophiles was 1.36 ± 0.08; this indicated that the important reactant could be Ph· (or a common reagent made from each PhX), but not the anion radical $PhX^{\cdot-}$. This study followed earlier work with Scamehorn[66,67] in which the apparent selectivity between nucleophiles, or the apparent differences in reactivity of iodobenzene and bromobenzene, is not found in competition reactions, and such selectivity would therefore seem to arise from different standing concentrations of radicals under the different reaction conditions.

In addition to these studies of the mechanism of the reaction, a number of reports[68—70] have dealt with the synthetic applications of the reactions. Perhaps one of the most compelling pieces of evidence in favour of the $S_{RN}1$ mechanism in the form suggested by Bunnett is the electrochemical induction of the same processes, in which the electrodes act as sources or sinks of electrons. The application of cyclic voltammetry[71] to such reactions suggests that the loss of halogen from 2-halogeno-quinolines (as the radical anion) increases in the order Cl < Br < I and provides rate constants for these carbon–halogen bond fissions [$10^{-4}k$/s: 1.7 (Cl), 13 (Br), 300 (I) in liquid ammonia at −40 °C] consistent with such processes. Rate constants of the order of 10^8 l mol^{-1} s^{-1} were deduced for the reactions of resulting phenyl radical with various nucleophiles, and values of between 6 and 45×10^6 l mol^{-1} s^{-1} for the exchange reaction which serves to re-initiate the chain process.

Another contribution to the systematic study of these reactions came from Moon and Wolfe,[72] who took the reaction of potassium acetone enolate with 2-chloro-quinoline as the standard reaction and attempted to determine the effects of various solvents upon the course and rate of the reaction. Perhaps predictably, in view of the complex interactions between free-radical and heterolytic processes, the experimentally observed effect was not simply linked to the expected solvent properties derived from studies of wholly heterolytic or wholly homolytic reactions. Thus,

63 A. B. Pierini and R. A. Rossi, *J. Org. Chem.*, 1979, **44**, 4667.
64 J. E. Swartz and J. F. Bunnett, *J. Org. Chem.*, 1979, **44**, 4673.
65 C. Galli and J. F. Bunnett, *J. Amer. Chem. Soc.*, 1979, **101**, 6137.
66 R. G. Scamehorn and J. F. Bunnett, *J. Org. Chem.*, 1977, **42**, 1449, 1457.
67 R. G. Scamehorn and J. F. Bunnett, *J. Org. Chem.*, 1979, **44**, 2604.
68 A. B. Pierini and R. A. Rossi, *J. Organometallic Chem.*, 1979, **168**, 163.
69 R. A. Rossi, R. H. de Rossi, and A. B. Pierini, *J. Org. Chem.*, 1979, **44**, 2662.
70 J. E. Swartz and J. F. Bunnett, *J. Org. Chem.*, 1979, **44**, 340.
71 C. Amatore, J. Chaussard, J. Pinson, J. M. Saveant, and A. Thiebault, *J. Amer. Chem. Soc.*, 1979, **101**, 602.
72 M. P. Moon and J. F. Wolfe, *J. Org. Chem.*, 1979, **44**, 4081.

dimethylformamide or tetrahydrofuran sustained the photochemical process, while 1,2-dimethoxyethane, diethyl ether, benzene, and dimethyl sulphoxide were less satisfactory, although all solvents showed a 'dark' (thermal) reaction concurrent with any light-catalysed process. In tetrahydrofuran, under photochemical conditions, potassioacetone displaced bromine less rapidly than chlorine from the appropriate 2-halogeno-quinoline, and had very little effect upon iodobenzene, whereas lithioacetone or potassioacetophenone had no appreciable effect upon 2-chloroquinoline. Evidently the gegenion is important in these processes, and also there is a subtle interplay between reactivity and the charge distribution in the carbanion.

Diazonium-ion Processes.—Important contributions have been made both to the mechanisms of formation of aryldiazonium ions and to their modes of decomposition. D. L. H. Williams and his co-workers[73] have reported the halide-ion-catalysed reaction between *para*-substituted anilines and nitrous acid in mineral acid media; the reaction seems to involve nitrosyl halides as the true electrophiles, and is also characterized by reversibility of the nitrosation step [equation (3)] as well as by

$$PhNH_2 + NOX \rightarrow Ph\overset{+}{N}H_2NO\ X^- \qquad (3)$$

the incursion of encounter control in the attack of the more activated amines. Challis, following the nitrosation reactions, has observed the reaction of nitrosyl chloride with a number of primary aromatic amines and secondary alkylamines in alkaline (0.1M-NaOH) or substantially neutral (phosphate buffer; pH 6.85) solution. While both primary and secondary amine attack could generally be described by the kinetic form of equation (4):

$$d[ArN_2^+]/dt \text{ (or } d[R_2NNO]/dt) = k[\text{Amine}][\text{NOCl}] \qquad (4)$$

Amines which were more basic than *N*-methyl-4-nitroaniline showed a rate of nitrosation substantially proportionate to the rate of the competing hydrolysis with water and independent of their basicity. This was interpreted as evidence of an encounter-controlled process in which the relative rates of attack upon NOCl by the two possible nucleophiles was determined by concentration considerations alone.[74] Challis and Shuker have also identified an interesting and specific effect by polyhydroxylic species such as glycols which, through the corresponding nitrite monoesters, can bring about nitrosation of amines in alkaline media; this, applied to sugars, may have a significant bearing upon mechanisms of carcinogenesis *in vivo*.[75]

Among studies of the reactions of aryldiazonium cations with nucleophiles, a report of the incorporation of radioactive iodine and of ^{211}At has shown[76] that some considerable selection must occur before the displacement of nitrogen and the formation of the carbon–halogen bond to account for the degree of incorporation of astatine at very low concentrations (*ca.* 10^{-12} mol l^{-1}, 10—35% incorporation). Scheme 8 shows the suggested sequence.

$$ArN_2^+ + X^- \rightleftharpoons [ArN_2^+, X^-] \rightleftharpoons [ArN_2\cdot X\cdot] \rightarrow ArX + N_2$$

Scheme 8

[73] M. R. Crampton, J. T. Thompson, and D. L. H. Williams, *J.C.S. Perkin II*, 1979, 18.
[74] B. C. Challis and D. E. G. Shuker, *J.C.S. Perkin II*, 1979, 1020.
[75] B. C. Challis and D. E. G. Shuker, *J.C.S. Chem. Comm.*, 1979, 315.
[76] G. J. Meyer, K. Roessler, and G. Stoecklin, *J. Amer. Chem. Soc.*, 1979, **101**, 3121.

A considerable contribution appears in a paper by Maurer, Szele, and Zollinger[77a] in which the contributions of kinetic studies to the assessment of various mechanisms are discussed. An unusually clear-sighted analysis of the consistency of kinetic forms with the various mechanistic ideas was applied to the dediazoniation process; the argument deserves perusal even by physical organic chemists who are not directly concerned with such reactions. Scheme 9 shows the process consistent with experimental fact, including the incorporation of nitrogen into the diazonium ion through the reverse reaction ($Ar^+ + N_2 \rightleftharpoons ArN_2^+$).

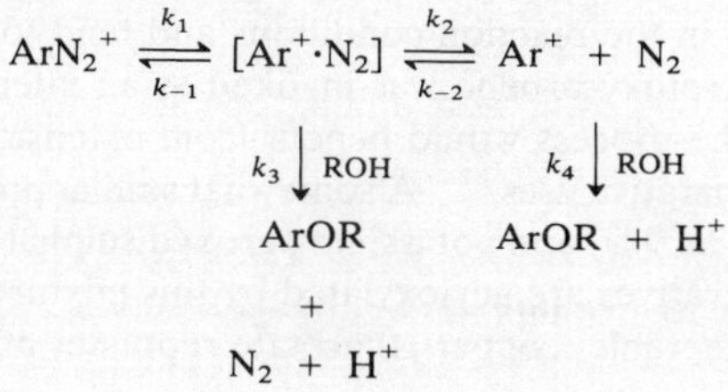

Scheme 9

A second communication[77b] dealt with a kinetic isotope effect which unexpectedly increases with temperature. The diazo-coupling reactions of *NN*-dimethylaniline and *m*-toluidine apparently show this behaviour as a result of two reactions, of the same kinetic order, which both involve molecular complexing between the diazonium ion and the nitrogen atom of the amine; partitioning between the two processes explains this temperature dependence.

3 Preparative Aspects

Benzene Derivatives.—The use of arene complexes in organic synthesis has been reviewed.[78]

Oxidation in various guises has become popular in studies of synthetic methods. The most fundamental use of dehydrogenation reactions is in the formation of aromatic systems from cyclohexyl compounds, and the use of thionyl chloride and zinc chloride as such an aromatization mixture has been reported, although the abstract gives little idea of the generality or limitations of this reagent mixture.[79] Benzyltriethylammonium permanganate has been advocated as an effective oxidizing agent for the conversion of ethers into esters by oxidation at the α-carbon atom.[80] Dibenzyl ether provides benzyl benzoate in 80% yield, and the order of ease of attack of radicals attached to oxygen was shown to fall in the sequence $PhCH_2 >$ Ph-Alkyl $> RCH_2 > RR'CH > Me > Ph$. A radical mechanism would be consistent with these observations.

[77] (*a*) W. Maurer, I. Szele, and H. Zollinger, *Helv. Chim. Acta*, 1979, **62**, 1079; (*b*) J. R. Penton and H. Zollinger, *J.C.S. Chem. Comm.*, 1979, 819.

[78] G. Jaouen, *Org. Chem.* (*New York*), 1978, **33** (*Transition Met. Organomet. Org. Synth., Part 2*) 65 (*Chem. Abs.*, 1979, **91**, 73 717c).

[79] J. Knabe, W. Schmitt, and W. Simon, *Arch. Pharm.* (*Weinheim*), 1979, **312**, 445 (*Chem. Abs.*, 1979, **91**, 107 723u). See also M. Albeck and T. Tamary, *J. Organometallic Chem.*, 1978, **164**, C23 for the corresponding use of tellurium(IV) halides.

[80] H. J. Schmidt and H. J. Schaefer, *Angew. Chem.*, 1979, **91**, 78.

Metal-catalysed processes also were reported; thus, palladium(II) trifluoroacetate oxidizes PhX to a mixture of biaryl and teraryls.[81] Acetoxymethylation has also been brought about by using a mixture of lithium bromide, tellurium(IV) oxide, and acetic acid at temperatures near the boiling point of the mixture. The simpler and less reactive aromatic hydrocarbons, such as benzene, toluene, and *o*- and *p*-xylene, were attacked only at the ring under these conditions. At higher temperatures, however, coupling occurred to give the appropriate diarylmethane ($ArCH_2Ar$), and the more reactive substrates gave more complex products. Mesitylene (1,3,5-trimethylbenzene), for example, gave (16), suggesting that the original products are themselves unstable in the reaction conditions and tend to alkylate the unreacted starting material. Acetoxycarbene was invoked as an intermediate in the original acetoxylation, but the process would benefit from extensions both of mechanistic study and of the preparative uses.[82a] A somewhat similar process involves the use of trifluoroacetic acid (*ca.* 9%) and potassium peroxydisulphate in acetic acid solution. Methylbenzene derivatives are acyloxylated by this mixture, both the ring and the side-chain being vulnerable; copper(II) acetate represses nuclear attack.[82]

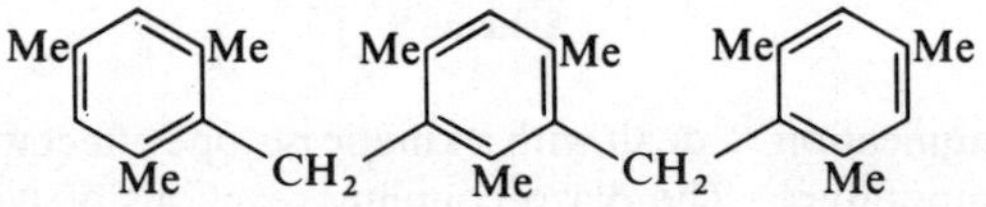

(16)

Oxidative methods are also involved in the coupling of arenes by a mixture of copper(II) chloride and aluminium chloride; in this way, 1-(2′,4′,6′-trimethylphenyl)naphthalene was prepared from mesitylene and naphthalene. The reaction was held to involve radical-cation intermediates.[83] The oxidation of sterically hindered phenols by periodic acid (HIO_4) also was reported,[84] and periodic acid was the oxidizing agent in a selective conversion of unsaturated alcohols into ketones. 2,3-Dichloro-5,6-dicyanobenzoquinone and periodic acid were treated with the allylic alcohol in a mixture of benzene and aqueous hydrochloric acid (0.1M), when a number of compounds of the general formula PhCH=CHCH(R)OH were oxidized to the corresponding ketone in 80—90% yields.[85]

Reduction of carbonyl groups to methylene systems was also achieved under unusually mild conditions for *o*-acyl-phenols, where the ethyl carbonate derivative gave high yields of the corresponding *o*-alkyl-phenol on treatment with sodium borohydride (3 molecular equivalents).[86] Photochemical reduction, apart from the photo-stimulated $S_{RN}1$ processes already reported, has been applied to the trichlorobenzenes.[87] Irradiation of solutions of each trichlorobenzene in methanol or in acetone provides all the dichlorobenzenes and monochlorobenzene, suggesting

[81] M. N. Vargaftik and I. I. Moiseev, *Izvest. Akad. Nauk SSSR, Ser. Khim.*, 1979, 242.
[82] (*a*) J. Bergman and L. Engman, *Tetrahedron Letters*, 1978, 3279; (*b*) L. Jonsson and L. G. Wistrand, *J.C.S. Perkin I*, 1979, 669.
[83] L.-S. Wen and P. Kovacic, *Tetrahedron*, 1978, **34**, 2723.
[84] H. D. Becker and K. Gustafsson, *J. Org. Chem.*, 1979, **44**, 428.
[85] S. Cacchi, F. La Torre, and G. Paolucci, *Synthesis*, 1978, 848.
[86] N. Minami and S. Kijima, *Chem. and Pharm. Bull.* (*Japan*), 1979, **27**, 1490 (*Chem. Abs.*, 1979, **91**, 174 937e).
[87] G. G. Choudry, A. A. M. Roof, and O. Hutzinger, *Tetrahedron Letters*, 1979, 2059.

some scrambling process, and also tetra- and penta-chlorobiphenyls. The mechanism of formation of these coupling products deserves further study of the process whereby two aryl fragments apparently unite rather than abstract hydrogen from the protic solvent.

Light-induced decomposition of the intermediates in the well-known and valuable synthesis of biaryls from the decomposition of aromatic amines by nitrite esters in aromatic solvents[88] is reported to cause a more selective attack of the aryl radical upon the aromatic substrate, without appreciable loss of yield (17—60%).[89]

Demethylation has been brought about most selectively in two special cases. Trimethyliodosilane gives *ca.* 70% yields of the demethylation product in its reaction with (17) or (18);[90] also, 2,3,4,6-tetramethoxybenzaldehyde loses specifically the methyl group attached to oxygen at C-2 on treatment with aluminium chloride in benzene.[91]

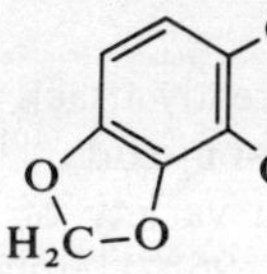

(17)

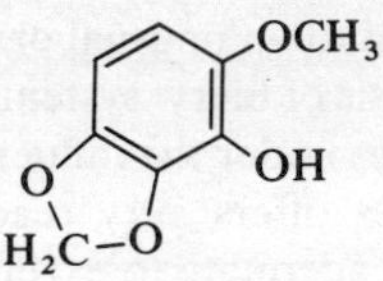

(18)

Amongst other preparative reactions involving diazonium intermediates, the decompositions of triazenes derived from the coupling of a diazotized aniline with piperidine were successfully used to provide aryl fluorides containing ^{18}F. Caesium fluoride was the source of the labelled halogen, and the decomposition was carried out in tetrahydrofuran containing methanesulphonic acid, when up to 50% of the radioisotope was incorporated in the derived aryl fluoride.[92] The synthesis of aryl fluorides by the decomposition of *p*-methyl- or *p*-t-butylbenzenediazonium borofluoride in solvents was studied in some detail. The main products were the expected aryl fluoride and the solvolysis product (*e.g.*, phenol from water); these were obtained in relative amounts which reflected the ratio K_D/α, where K_D was the dissociation constant and α was the degree of dissociation of the ion pairs.[93]

The use of titanium(III)–phenol complexes to promote phenylation by decomposing aryldiazonium ions at the *ortho*-positions has also been advanced.[94]

The synthesis of aryl fluorides by ion-exchange processes, using dipolar aprotic solvents and activated nitro-chloro-benzenes (*e.g.* *o*-chloronitrobenzene), was extended by a number of German patents in which the use of caesium fluoride was claimed to assist the action of potassium fluoride as a source of fluoride ion in these nucleophilic displacements to the extent that *meta*-nitro- or -cyano-chlorobenzenes were sufficiently rapidly attacked to make the process commercially viable. Coupled with this was the observation that 18-crown-6 ether aided the displacement to such

88 Huang Shu, *Acta Chim. Sinica*, 1959, **25**, 171; J. I. G. Cadogan, *J. Chem. Soc.*, 1962, 4257.
89 M. Julliard, C. Siv, G. Vernin, and J. Metzger, *Helv. Chim. Acta*, 1978, **61**, 2941.
90 J. Minamikawa and A. Brossi, *Tetrahedron Letters*, 1978, 3085.
91 E. G. Paul and P. S.-C. Wang, *J. Org. Chem.*, 1979, **44**, 2307.
92 T. J. Tewson and M. J. Welch, *J.C.S. Chem. Comm.*, 1979, 1149.
93 H. G. O. Becker and G. Israel, *J. Prakt. Chem.*, 1979, **321**, 579 (*Chem. Abs.*, 1979, **91**, 192 627h).
94 T. Caronna, F. Ferrario, and S. Servi, *Tetrahedron Letters*, 1979, 657.

an extent that dimethylformamide was a satisfactory solvent, although in the absence of such catalysts the lower boiling point and thermal instability makes this solvent less useful than tetramethylene sulphone.[95—97]

A number of specific processes now follow. The earlier use of the t-butyl group as a blocking group in the synthesis of *ortho*-substituted benzene derivatives (*e.g.* the preparation of 2,2′,6,6′-tetramethylbenzophenone by the de-t-butylation of the corresponding 4,4′-di-t-butyl ketone, itself obtained from 5-t-butyl-1,3-dimethylbenzene)[98] has been used to obtain 1,2- and 1,2,3-polysubstituted benzenes.[99] The corresponding use of derivatives of *p*-t-butyliodobenzene in the Ullmann synthesis of biaryls, followed by the de-t-butylation of the resulting materials, also offers a straightforward route to *ortho*-substituted biphenyls.[100] Doubtless other processes may be employed to obtain biaryls or similar species containing a removable t-butyl group, and this synthetic route has much value where the appropriate starting materials are only available as the result of substitution reactions blocked by a t-butyl group and directed to an unusual orientation.

Another synthesis of biaryl systems involves the ready attack upon *m*-dinitrobenzene by aryl iodides under the influence of copper t-butoxide.[101] This variant of the Ullmann syntheses offers very ready access to a variety of derivatives of 2,6-dinitrobiphenyl in adequate to excellent yield (10—96%) [equation (5)].

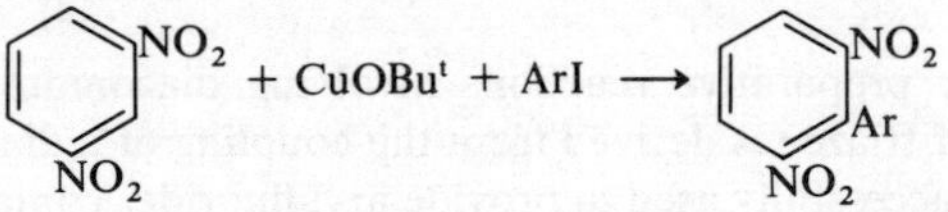

(5)

Polychlorobiphenyls have also been prepared by the reaction of *o*-chloranil and alkynes or benzynes[102] in a method reminiscent of the similar reactions in polyfluoroaromatic chemistry. Stilbenes may also be obtained by a coupling process involving an olefinic system with extended conjugation, either to a carbonyl group or to a benzene ring, an aryl iodide, palladium(II) acetate, and triethylamine. Styrene, acrylic acid, or methyl acrylate each provide the appropriate aryl derivative [equation (6)] very effectively; nitro- and bromo-substituents may be present in the aromatic ring.[103]

$$\mathrm{ArI + CH_2{=}CHX \xrightarrow{Pd(OAc)_2,\ Et_3N} ArCH{=}CHX} \qquad (6)$$

Friedel–Crafts reactions of optically active 3-chlorobutan-1-ol, the corresponding acid, or its esters with aluminium chloride and benzene gave optically specific products which arose from inversion of configuration at the attacking carbon atom, and therefore contributed to a knowledge of the stereochemistry of attack in this

95 H. G. Oeser, K. H. Koenig, and D. Mangold, Ger. Offen. 2 724 366 (*Chem. Abs.*, 1979, **90**, 71 891k).
96 H. G. Oeser, K. H. Koenig, and D. Mangold, Ger. Offen. 2 724 367 (*Chem. Abs.*, 1979, **90**, 103 599g).
97 H. G. Oeser, K. H. Koenig, and D. Mangold, Ger. Offen. 2 803 259 (*Chem. Abs.*, 1979, **91**, 157 418u).
98 K. Maruyama, R. Tanikaga, and R. Goto, *Nippon Kagaku Zasshi*, 1963, **84**, 75 (*Chem. Abs.*, 1964, **60**, 144b).
99 M. Tashiro and T. Yamato, *J.C.S. Perkin I*, 1979, 176.
100 M. Tashiro and T. Yamato, *J. Org. Chem.*, 1979, **44**, 3037.
101 J. Cornforth and T. W. Wallace, *J.C.S. Chem. Comm.*, 1979, 294.
102 J. L. Pyle, R. A. Lunsford, and J. S. Cantrell, *J. Org. Chem.*, 1979, **44**, 2391.
103 J. E. Plevyak, J. E. Dickerson, and R. F. Heck, *J. Org. Chem.*, 1979, **44**, 4078.

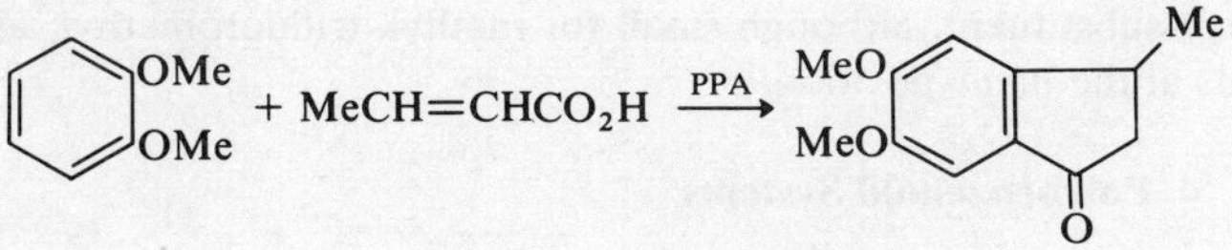

(7)

reaction.[104] Derivatives of 1,2-dialkoxy-benzenes were reported to give 5,6-substituted indanones [equation (7)] upon treatment with polyphosphoric acid (PPA) and crotonic acid, in a reaction with evident parallels to the Friedel–Crafts acylation process,[105] and Suzuki's interest in crowded aromatic ketones led to the synthesis of a number of such species by the further methylation ('pivaloylation') of propionoyl residues already present in the system, using potassium t-butoxide to generate the carbanion, methyl iodide as the alkylating agent, and t-butyl alcohol–benzene mixture as solvent [equation (8)].[106]

$$\mathrm{ArCOCH_2Me} \xrightarrow{\mathrm{KOBu^t,\ MeI}} \mathrm{ArCOCMe_3} \qquad (8)$$

Other reports of the synthesis of such ketones point to a less predictable influence of the order of addition of reagents upon the yields of tetramethyldiacyl-benzenes. Di-acylation can be achieved if any of the tetramethylbenzenes are treated with three molecular proportions of acetyl or propionoyl chloride and six molecular proportions of aluminium chloride — apparently without any of the migration of methyl group that has been noted earlier — but the yield of diacetodurene is 85% if a solution of the hydrocarbon in carbon disulphide is added to the other reagents, but falls to 10% if a mixture of acyl chloride and hydrocarbon in carbon disulphide is added to a suspension of aluminium chloride in the same solvent.[107,*cf.*108]

Polyfluoroaromatic chemistry is exemplified by a study of photosubstitution in which hexafluorobenzene, with benzophenone in methanol, provided 2,3,4,5,6-pentafluorobenzyl alcohol and benzpinacol; pentafluoropyridine reacted analogously to give the 4-substituted product only, suggesting some nucleophilic character to the radical.[109] Nucleophilic reactions of decafluoropyrene were also reported.[110] Methoxydefluorination occurred at the 1-, 3-, 6-, and 8-positions, in agreement with the modified I_π concept,[111] but also with resonance theory and other theoretical calculations. Polychloroaromatic chemistry was advanced by a careful and detailed synthesis of the isomeric monobromo- and -iodo-tetrachloropyridines by unambiguous methods.[112]

A warning on the use of aryl-cadmiums as reagents in the synthesis of aryl ketones was sounded. *meta*-Substituted aryl-cadmiums gave *para*-substituted ketones.[113] The effect was greatest with the methoxy-substituent, but was also evident with the

[104] S. Nakashima, T. Kondo, Y. Nakamoto, and S. Suga, *Bull. Chem. Soc. Japan.*, 1979, **52**, 2377.
[105] H. G. Grant, *Z. Naturforsch.*, 1979, **34b**, 728.
[106] H. Suzuki, Y. Aomori, T. Mishina, and T. Hanafusa, *Bull. Chem. Soc. Japan*, 1978, **51**, 3393.
[107] A. G. Pinkus and N. Kalyanam, *Org. Prep. Proced. Int.*, 1978, **10**, 255 (*Chem. Abs.*, 1979, **90**, 86 938e).
[108] H. Galenkamp and A. C. Faber, *Rec. Trav. chim.*, 1958, **77**, 850.
[109] M. Zupan, B. Sket, and B. Pahor, *Proc. IUPAC Symp. Photochem. 7th.*, 1978, 398 (*Chem. Abs.*, 1979, **90**, 121 098e).
[110] J. Burdon, I. W. Parsons, and H. S. Gill, *J.C.S. Perkin I*, 1979, 1351.
[111] J. Burdon and I. W. Parsons, *J. Amer. Chem. Soc.*, 1977, **99**, 7445.
[112] A. G. Mack, H. Suschitzky, and B. J. Wakefield, *J.C.S. Perkin I*, 1979, 1472.
[113] P. R. Jones and J. G. Shelnut, *J. Org. Chem.*, 1979, **44**, 696.

thiomethoxy-substituent, although small for methyl, trifluoromethyl, and fluorine substituents at the *meta*-position.

4 Polybenzenoid Systems

Fluoranthene chemistry has been reviewed,[114] and a study of selectivity relations in naphthalene, biphenyl, and benzothiophen has contributed considerably to this problem.[115] 9-Phenanthryne has been indicted as the intermediate involved in the reaction of 9-bromophenanthrene with ketone enolate anions in tetrahydrofuran under the influence of sodium amide. The reaction conditions, however, are strongly reminiscent of those in $S_{RN}1$ processes, and such an alternative mechanism deserves consideration.[116]

Newman has reported a new synthesis of fluoro-methylbenz[*a*]anthracenes, the key process of which is the treatment of a dihydro-dihydroxy-derivative with diethylaminosulphur trifluoride (Et_2NSF_3).[117] Correspondingly, *trans*-7,8-dihydroxy-7,8,9,10-tetrahydrobenz[*a*]pyrene can be converted into 7-fluorobenz[*a*]pyrene. Unexpectedly, the product of reaction of the dihydroxytetrahydro-compound (19) with Et_2NSF_3 is a product of ring-contraction, the fluorine atoms occurring in a difluoromethyl substituent outside the ring; aromatization of this product with dichlorodicyanobenzoquinone reverses the ring-contraction to produce the six-membered (aromatic) ring (Scheme 10).[118]

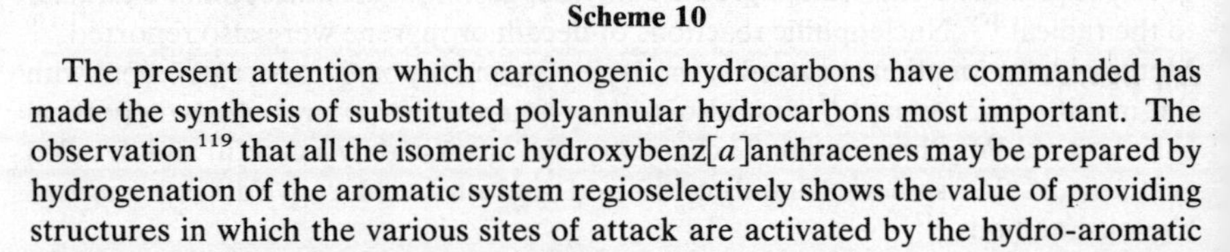

(19)

Scheme 10

The present attention which carcinogenic hydrocarbons have commanded has made the synthesis of substituted polyannular hydrocarbons most important. The observation[119] that all the isomeric hydroxybenz[*a*]anthracenes may be prepared by hydrogenation of the aromatic system regioselectively shows the value of providing structures in which the various sites of attack are activated by the hydro-aromatic fragment of the partially reduced benzanthracene system.

Among work on chemical carcinogens, the present emphasis is upon the bay region of such hydrocarbons. Epoxides such as (20) are held to undergo nucleophilic attack, forming a bond between C-1 and an exocyclic amino-group of a guanine fragment in the DNA structure. Thus, 15,16-dihydro-11-methylcyclopenta-[*a*]phenanthren-17-one (21; R = Me) is carcinogenic whereas the 11-ethyl

[114] N. Campbell and N. H. Wilson, *Commun.-R. Soc. Edinburgh, Phys. Sci.*, 1979, **15**, 193 (*Chem. Abs.*, 1979, **91**, 157 479g).
[115] A. A. Hamffray and R. L. Bruce, *J. Org. Chem.*, 1979, **44**, 724.
[116] M.-L. Viriot, *J. Chem. Res.*, 1979, (*S*) 324; (*M*) 3686.
[117] M. S. Newman and J. M. Khanna, *J. Org. Chem.*, 1979, **44**, 866.
[118] M. S. Newman, V. S. Khanna, and K. Kanakarajan, *J. Amer. Chem. Soc.*, 1979, **101**, 6788.
[119] P. P. Fu, C. Cortez, K. B. Sukumaran, and R. G. Harvey, *J. Org. Chem.*, 1979, **44**, 4265.

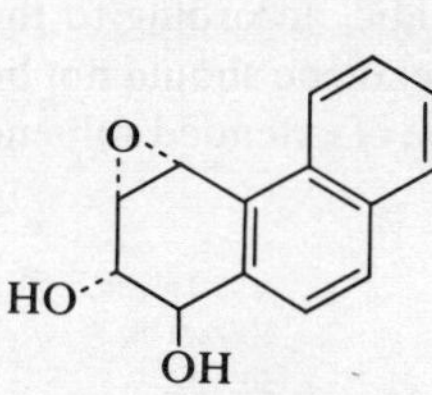

(20)

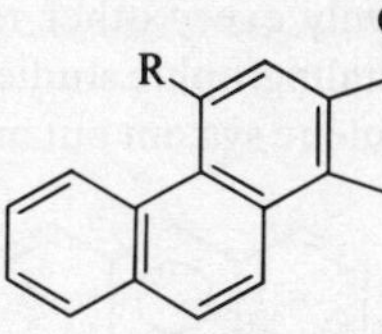

(21)

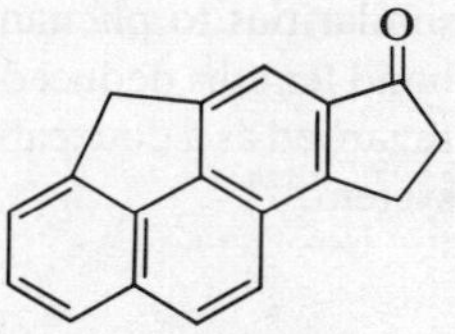

(22)

analogue (21; R = Et) is less effective, and the corresponding butyl derivative is inert. However, the bridged analogue, 15,16-dihydro-1,11-methanocyclopenta[*a*]phenanthren-17-one (22), was prepared by conventional methods from 1,10-methano-1,2,3,4-tetrahydrophenanthren-4-one *via* a Stobbe condensation with diethyl succinate and found to be carcinogenic.[120] While such emphasis on the bay region is justified empirically, and molecular studies of the dihydro-diols and diol epoxides derived from the carcinogenic hydrocarbons show that a covalent bond, if formed between a macromolecule and the appropriate carbon atoms of the polyannular hydrocarbon, must be axial to minimize steric effects,[121] there is no correlation between either S_N1 or S_N2 reactivity of the epoxides of carcinogenic and non-carcinogenic hydrocarbons and their mutagenic properties.[122] Perhaps this problem comes from the real difficulties of providing a nucleophile comparable in both chemical reactivity and steric properties to the macromolecule with which it is compared.

A polyannular system has also been used to demonstrate the mechanism of reaction of trifluoromethyl hypofluorite (CF_3OF). 4-Acetoxypyrene (23) readily gave 4-acetoxy-5-fluoro-4-trifluoromethoxy-4,5-dihydropyrene (24) at −78 °C in dichloromethane; the adduct was unstable at room temperatures, and over a few days it provided 4-acetoxy-5-fluoropyrene. 2-Acetamidopyrene (25) did not, despite the large activating effect of the *ortho*-acetamido-group, undergo an electrophilic attack upon the 1-position. It was therefore deduced that, unlike difluorination, attack by trifluoromethyl hypofluorite is not an electrophilic process.[123]

The synthesis and properties of kekulene (26) were reported in the previous *Annual Report* [Vol. 75, p. 236]; however, studies of the system now suggest

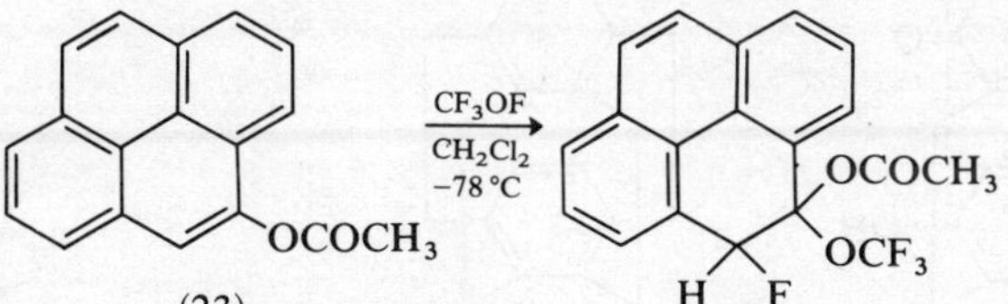

(23) (24)

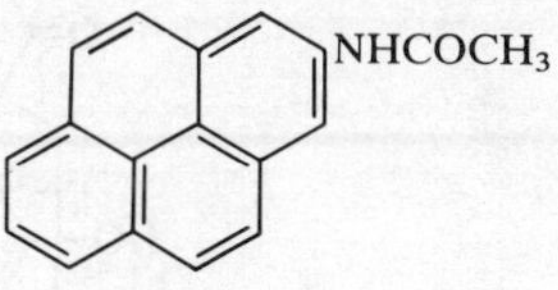

(25)

Scheme 11

[120] T. S. Bhatt, M. C. Coombs, A.-M. Kissonerghis, A. F. D. Clayton, and M. McPartlin, *J.C.S. Chem. Comm.*, 1979, 433.
[121] D. E. Zacharias, J. P. Glusker, P. P. Fu, and R. G. Harvey, *J. Amer. Chem. Soc.*, 1979, **101**, 4043.
[122] T. Okamoto, K. Shudo, N. Miyata, Y. Kitahara, and S. Nagata, *Chem. and Pharm. Bull. (Japan)*, 1978, **26**, 2014.
[123] T. B. Patrick, G. L. Cantrell, and C.-Y. Chang, *J. Amer. Chem. Soc.*, 1979, **101**, 7434.

similarities to phenanthrene. Only every other ring is aromatic, according to the bond lengths deduced from crystallographic studies, so that kekulene should not be regarded as a dodecabenzo-annulene system but more as a type of extended stilbene system.[124]

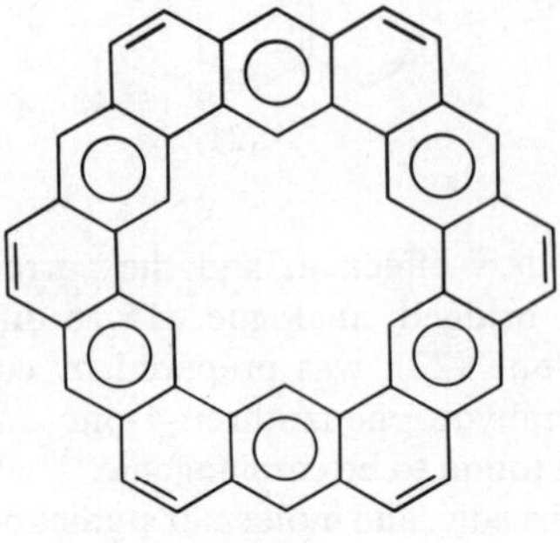

(26)

5 Non-benzenoid Systems and Cyclophanes

Cyclophanes.—A number of new and important cyclophanes have been reported, many of them synthesized by classical routes that usually involve extrusion of groups (photo-extrusion of sulphur dioxide is a frequent method of linking the two aliphatic chains to make the cyclophane structure)[125] or classical methods of ring-closure common to alicyclic chemistry. [2.2](2,7)Naphthalenoparacyclophane has been reported,[126] and the interesting quinhydrone derivative of [2.2.2.2](1,2,4,5)cyclophane (27) is the subject of a second paper.[127] The first preparation of [2.2.2.2]-(1,2,3,4)cyclophane (28) involves building up the third and fourth ring linkages by a Friedel–Crafts reaction of monochlorodimethyl ether ($ClCH_2OMe$) upon a paracyclophane structure, the cyclization finally being achieved through a dibromide structure in the same way that 9,10-dihydrophenanthrene was much earlier obtained from 2,2′-bis(bromomethyl)biphenyl.[128] [5.1]Metacyclophane also has been prepared,[129] and the synthesis of some derivatives of biphenyl linked by a polymethylene chain has produced a new type of strained, non-planar biaryl structure.

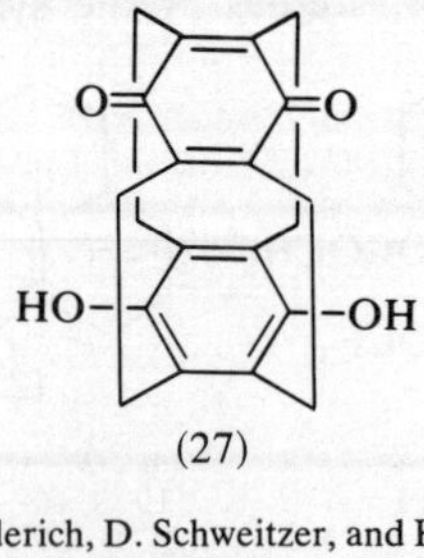

(27)

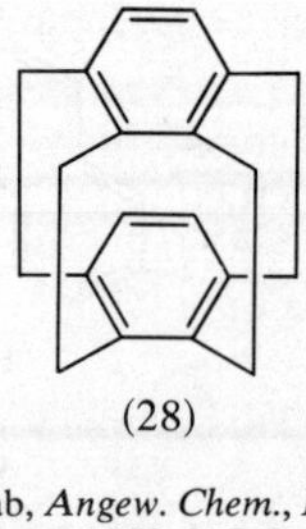

(28)

[124] C. Krieger, F. Diederich, D. Schweitzer, and H. A. Staab, *Angew. Chem., Internat. Edn.*, 1979, **18**, 699.
[125] *e.g.*, G. D. Ewing and V. Boekelheide, *J.C.S. Chem. Comm.*, 1979, 207; R. S. Givens, R. J. Olsen, and P. L. Wylie, *J. Org. Chem.*, 1979, **44**, 1608.
[126] J. R. Davy, M. N. Iskander, and J. A. Reiss, *Tetrahedron Letters*, 1978, 4085.
[127] H. A. Staab and V. M. Schwendemann, *Annalen*, 1979, 1258.
[128] J. Kleinschroth and H. Hopf, *Angew. Chem. Internat. Edn.*, 1979, **18**, 329.
[129] N. Finch and C. W. Gemenden, *J. Org. Chem.*, 1979, **44**, 2804.

The essential step is the ring-closure of derivatives of biphenyl with chains attached to the 2- and 4′-position by the Prelog–Stoll acyloin condensation.[130]

The synthesis of [1.1.1.1]metacyclophane-7,14,21,28-tetraol derivatives (29) properly falls together with the new interest in even larger polyphenol structures.[131] The trivial name 'calixarene' has been advanced[132] for such systems, which are exemplified by (30); this is formally a derivative of [1.1.1.1.1.1.1.1]metacyclophane-7,14,21,28,35,42,49,56-octaol, and it undergoes attack by chloro-2,4-dinitrobenzene, or by camphorsulphonyl chloride, to give products in which up to six of the eight hydroxyl groups have been substituted.

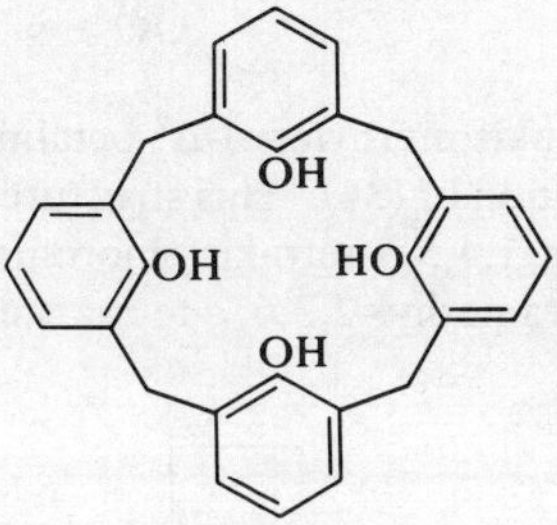

(29)

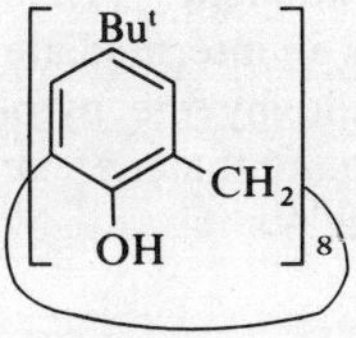

(30)

The photochemistry of cyclophanes also presents some interesting aspects; [6 + 6] reversions have been observed in the photolysis of paracyclophanes and the derived [2.2.2](1,2,4)- and [2.2.2.2](1,2,3,5)-cyclophanes (Scheme 12).[133]

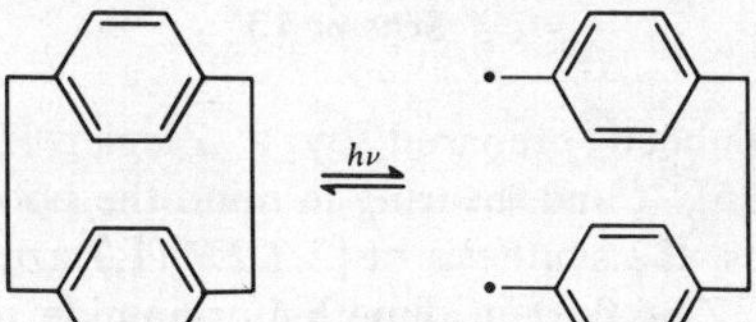

Scheme 12

Annulenes.—Two reviews of annulene chemistry have appeared.[134,135] Interesting annuleno-annulenes have recently been prepared; examples are trisdehydro-[18.10.2][14]annuleno[22]annulene[136] derivatives, which appear to show independent ring currents, and a number of structures described in a paper which reviews the main synthetic methods available to produce these compounds.[137]

Considerable argument has been directed towards the relative stabilities of benzocyclobutene systems and radialene (1,2,3,4,5,6-hexamethylenecyclohexane)

130 M. Nakazaki, K. Yamamoto, S. Isoe, and M. Kobayashi, *J. Org. Chem.*, 1979, **44**, 2160.
131 H. Kaemmerer, G. Happel, V. Boehmer, and D. Rathay, *Monatsh.*, 1978, **109**, 767.
132 R. Muthukrishnan and C. D. Gutsche, *J. Org. Chem.*, 1979, **44**, 3962.
133 G. Kaupp, E. Teufel, and H. Hopf, *Angew. Chem., Internat. Edn.*, 1979, **18**, 215.
134 P. J. Garratt, in 'Comprehensive Organic Chemistry', ed. J. F. Stoddart, Pergamon, Oxford, 1979, vol. 1, p. 361.
135 M. Iyoda, *Kagaku No Ryoiki*, 1978, **32**, 907 (*Chem. Abs.*, 1979, **91**, 56 484g).
136 S. Akiyama and M. Nakagawa, *Tetrahedron Letters*, 1978, 1483.
137 M. Nakagawa, *Angew. Chem., Internat. Edn.*, 1979, **18**, 202.

structures, without a definite result. Thus, maleic anhydride was successfully used to identify the equilibrium between benzocyclobutene (31) and 1,2-bis(methylene)-cyclohexa-3,5-diene (32).[138] Despite the earlier report,[139] tricyclobutabenzene (33) is reported to be stable with respect to decomposition to radialene.[140]

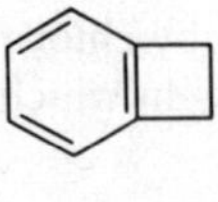

(31)

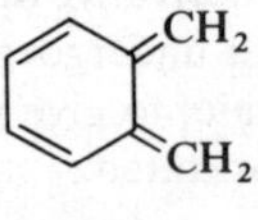

(32)

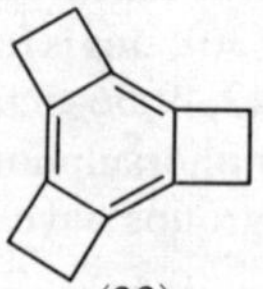

(33)

Naphtharadialene [1,2,3,4,5,6,7,8-octakis(methylene)-Δ^9-octalin] has been described as an intermediate in the production of (34). This structure is obtained in *ca.* 15% yield by the pyrolysis of either 1,4,5,8-tetrakis(chloromethyl)-2,3,6,7-tetramethylnaphthalene or 1,4,5,8-tetramethyl-2,3,6,7-tetrakis(chloromethyl)-naphthalene.[141]

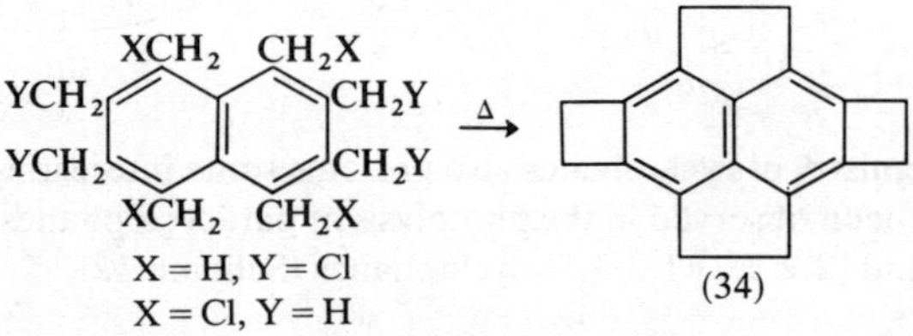

Scheme 13

5,5′-Bisazulenyl has been prepared by a Ziegler–Hafner reaction of the pentamethinium salt (35),[142] and, bearing in mind the poor yields associated with many azulene syntheses, the synthesis of [2.2.2.2](1,3)azulenophane represents a considerable triumph. Zinc dust in dimethylformamide provides 1,2-bis-(1-azulenyl)ethane, which undergoes attack under Mannich reaction conditions at C-3 of the azulene residues to give the precursor to the azulenophane, obtained in 0.1% yield by high-dilution cyclization methods (Scheme 14).[143]

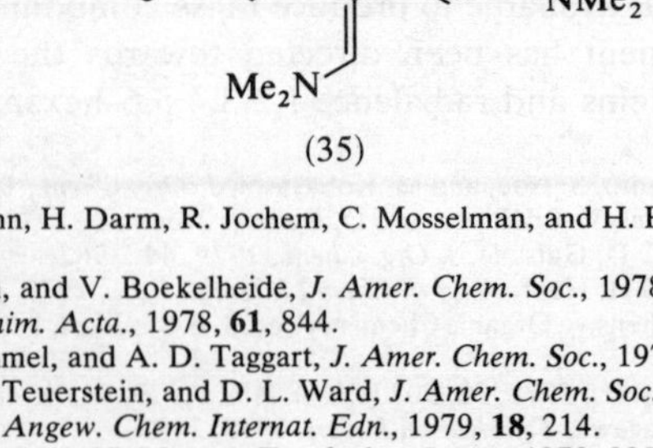

(35)

[138] W. R. Roth, M. Biermann, H. Darm, R. Jochem, C. Mosselman, and H. Hermann, *Chem. Ber.*, 1978, **111**, 3892.
[139] L. G. Harruf, M. Brown, and V. Boekelheide, *J. Amer. Chem. Soc.*, 1978, **100**, 2893; P. Schiess and M. Heitzmann, *Helv. Chim. Acta.*, 1978, **61**, 844.
[140] W. Nutakul, R. P. Thummel, and A. D. Taggart, *J. Amer. Chem. Soc.*, 1979, **101**, 770.
[141] H. Hart, M. Jeffares, A. Teuerstein, and D. L. Ward, *J. Amer. Chem. Soc.*, 1978, **100**, 8012.
[142] M. Hanke and Ch. Jutz, *Angew. Chem. Internat. Edn.*, 1979, **18**, 214.
[143] M. Fujimura, T. Nakazawa, and I. Murata, *Tetrahedron Letters*, 1979, 825.

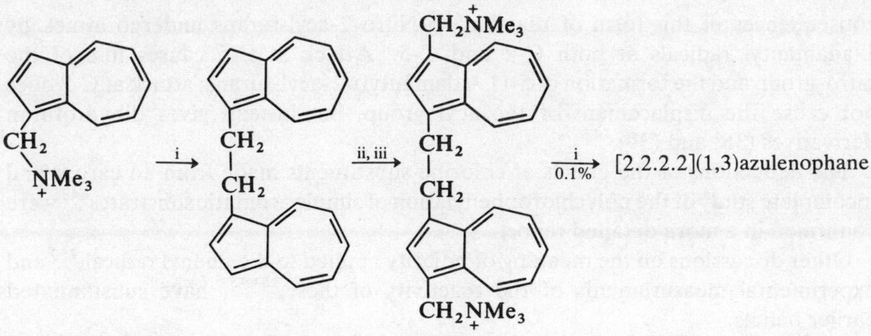

Reagents: i, Zn, DMF; ii, HCHO, $(Me_2N)_2CH_2$; iii, MeI

Scheme 14

Probably the most spectacular of the structures prepared this year is (36), which is evidently the *ne plus ultra* of cyclopropenium ionic structures. The synthetic methods employed[144] are displayed in Scheme 15.

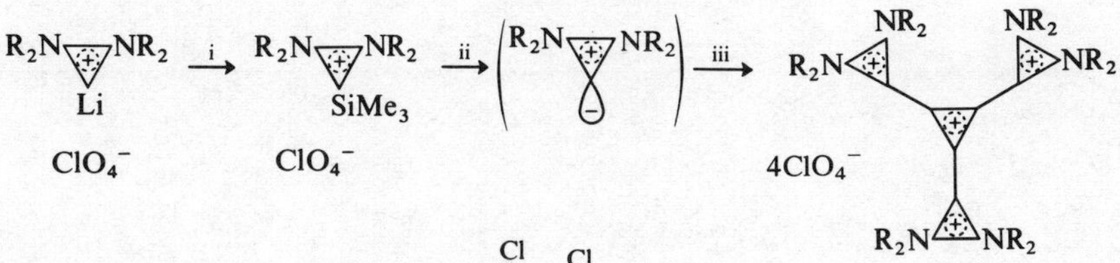

(36)

Reagents: i, $Me_3SiOSO_2CF_3$; ii, KF, THF; iii, 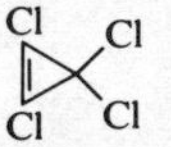 $HClO_4$

Scheme 15

6 Free-radical Aromatic Chemistry

Much of the present interest is directed towards *ipso*-attack. The phenomenon was well known in homolytic chemistry before its observation in electrophilic aromatic substitution, but a revival of interest is demonstrated by the report[145] that the benzyl radical displaces chlorine from *o*-dichlorobenzene and a more detailed investigation of alkyldenitration or alkyldeacetylation caused by the attack upon C-2 by adamantyl radicals of the heterocyclic species (37).[146] Tiecco extended this interest in *ipso*-attack of heterocyclic systems to an elegant demonstration of the various

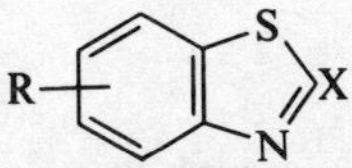

(37) $X = COCH_3$ or NO_2

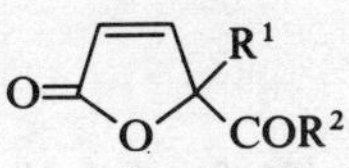

(38)

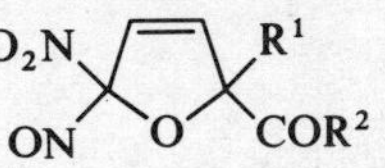

(39)

[144] R. Weiss, M. Hertel, and H. Wolf, *Angew. Chem., Internat. Edn.*, 1979, **18**, 473.

[145] R. Henriquez, W. M. MacKenzie, N. McPhail, and D. C. Nonhebel, *Colloq. Int. C.N.R.S.* 1977 (publ. 1978), No. 278 (*Radicaux Libres Org.*) p. 313 (*Chem. Abs.*, 1979, **91**, 192 439y).

[146] M. Tiecco, in ref. 145, p. 423 (*Chem. Abs.*, 1979, **91**, 192 442u).

consequences of this form of reaction. 5-Nitro-2-acyl-furans undergo attack by 1-adamantyl radicals at both C-2 and C-5. Attack at C-5 causes loss of the nitro-group and the formation of 5-(1′-adamantyl)-2-acyl-furans; attack at C-2 does not cause the displacement of the acyl group, but instead gives dihydrofuran derivatives (38) and (39).[147]

The deductions of the effects of chlorine substituents made from an earlier and incomplete study of the polychlorophenylation of simple aromatic substrates[148] were confirmed in a more detailed report.[149]

Other discussions on the meaning of polarity applied to the phenyl radical,[150] and experimental measurements of the reactivity of these,[151,152] have substantiated earlier beliefs.

[147] P. Cogolli, L. Testaferri, M. Tiecco, and M. Tingoli, *J.C.S. Chem. Comm.*, 1979, 800.
[148] R. Bolton, E. P. Mitchell, and G. H. Williams, *J. Chem. Res.*, 1977, (*S*) 223; (*M*) 2618.
[149] G. Vernin, L. Bouscasse, J. Metzger, and C. Párkányi, *J. Chem. Res.*, 1979, (*S*) 26; (*M*) 0228.
[150] A. A. Zavitsas, G. Hanna, A. Arafat, J. Ogunwole, and L. R. Zavitsas, *Colloq. Int. C.N.R.S.* 1977 (publ. 1978), No. 278 (*Radicaux Libres Org.*) p. 479 (*Chem. Abs.*, 1979, **91**, 174 498n).
[151] J. P. Lorand, R. G. Kryger, and N. R. Herron, in ref. 150, p. 463 (*Chem. Abs.*, 1979, **91**, 174 497m).
[152] T. Migita, K. Takayama, Y. Abe, and M. Kosugi, *J.C.S. Perkin I*, 1979, 1137.

11 Heterocyclic Compounds

By A. J. BOULTON and M. J. COOK
School of Chemical Sciences, University of East Anglia, Norwich NR4 7TJ

1 Heterocycles in Transformations of Functional Groups

1,4-Dihydropyridines have received considerable attention as reducing agents in a number of laboratories. The role of metal ions as catalysts has been discussed,[1] and the specific effect of zinc ions in the reduction of pyridine-2-carbaldehyde by 1-benzyl-1,4-dihydronicotinic acid diethylamide has been described.[2] A chiral 1,4-dihydropyridine that has been worked into a crown ether system (1) delivers hydride in an asymmetric reduction.[3] These reactions are not confined to unsaturated carbon: sulphonium salts are also attacked.[4]

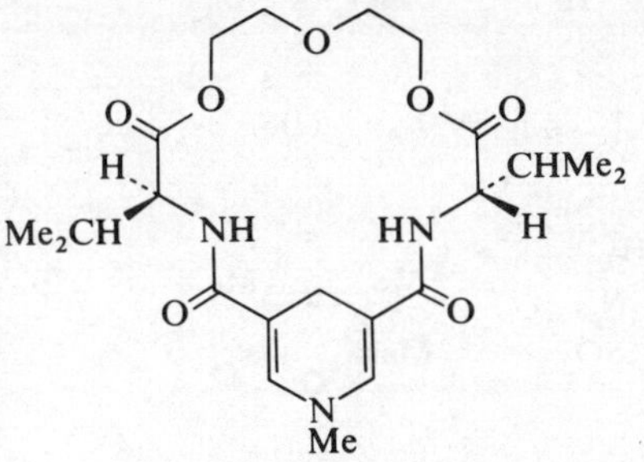

(1)

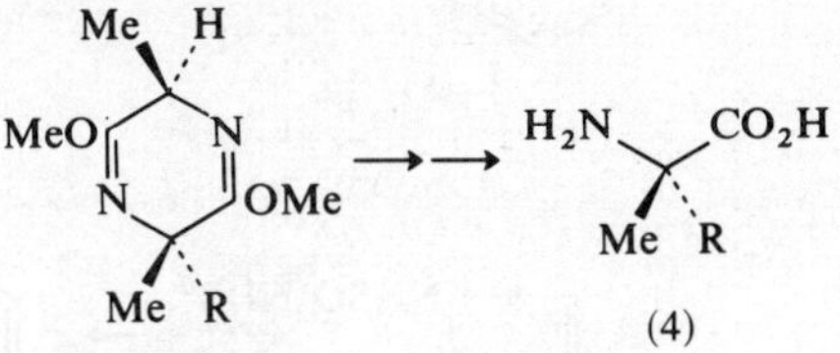

(2) R = H
(3) R = alkyl

(4)

Asymmetric induction, using proline derivatives, has been described by two Japanese groups. In one approach, the basic nitrogen forms the first attachment to the substrate, and the carboxy-group delivers the substituent which forms the new asymmetric centre.[5] In the other, the amino-acid is converted into a range of chiral amine ligands to complex Grignard reagents and organo-lithium compounds during addition to aldehydes.[6] The lactim ether (2) derived from L-alanine can be lithiated, and the *C*-alkylated product (3) then used to form amino-acids (4) in good optical yield.[7] An *N*-(camphorsulphonyl)-oxaziridine has been proposed as a chiral oxidizing agent.[8]

[1] R. A. Gase and U. K. Pandit, *J. Amer. Chem. Soc.*, 1979, **101**, 7059.
[2] R. A. Hood and R. H. Prince, *J.C.S. Chem. Comm.*, 1979, 163.
[3] J. G. de Vries and R. M. Kellogg, *J. Amer. Chem. Soc.*, 1979, **101**, 2759.
[4] T. J. van Bergen, D. M. Hedstrand, W. H. Kruizinger, and R. M. Kellogg, *J. Org. Chem.*, 1979, **44**, 4953.
[5] S. S. Jew, S. Terashima, and K. Koga, *Tetrahedron*, 1979, **35**, 2337, 2345.
[6] T. Mukaiyama, K. Soai, T. Sato, H. Shimizu, and K. Suzuki, *J. Amer. Chem. Soc.*, 1979, **101**, 1455.
[7] U. Schöllkopf, W. Hartwig, and U. Groth, *Angew. Chem. Internat. Edn.*, 1979, **18**, 863.
[8] F. A. Davis, R. Jenkins, S. Q. A. Rizvi, and T. W. Panunto, *J.C.S. Chem. Comm.*, 1979, 600.

The use of pyrylium salts for converting primary amines into other functionalities by nucleophilic displacement of a pyridine ring from the derived pyridinium salts [*cf. Ann. Reports* (*B*), 1977, **74**, 251] has led to a number of further preliminary communications, and some full papers. Particularly noteworthy transformations not hitherto mentioned in these Reports include the preparation of a wide variety of alkyl, aryl, and heteroaryl iodides from the corresponding amines,[9] of aryl[10] and alkyl[11] thiocyanates, and of alkyl fluorides,[12] and the *C*-alkylation of nitronate anions.[13] Steric acceleration to nucleophilic displacement apparently gives the fused pyrylium salts (5) an advantage over the 2,4,6-triphenyl compounds in these amine-substitution reactions.[10] A 'reductive deamination' procedure, which had earlier been applied to benzylic pyridinium salts, by reduction with borohydride to the 1,2-dihydropyridine (6) and thermal decomposition,[14,15] was found to proceed more readily, and with a wider range of substrates (R = alkyl or aryl), when the 1,4-dihydropyridine (7) was heated.[16] The mechanism of the decomposition of 1,2-dihydropyridine (6) was shown by labelling experiments not to involve a retro-ene reaction (*cf.* ref. 14), and probably to be free-radical in nature.[16]

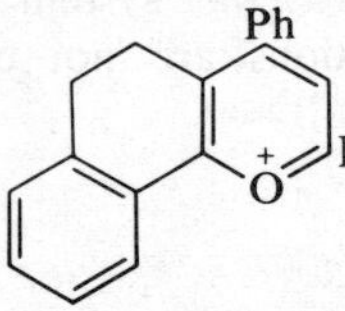

(5)

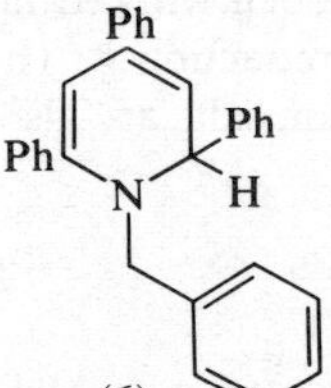

(6)

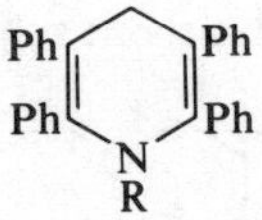

(7)

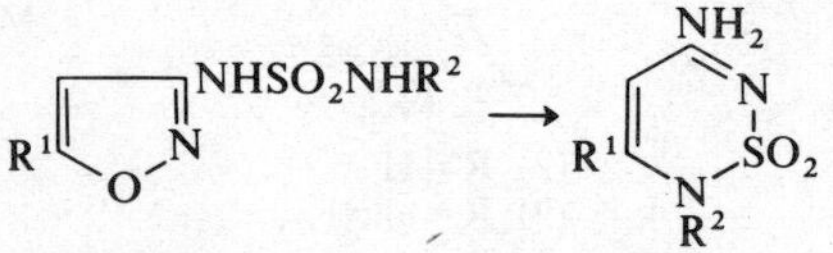

(8) (9)

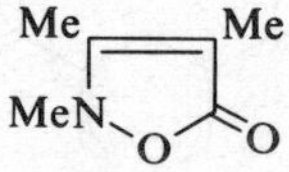

(10)

Isoxazoles have long been employed as protective and directive groups, usually with advantage being taken of the ready reductive cleavage of the N—O bond to generate the reactive enamino-ketone function at an appropriate stage in the sequence. A typical example of this type was reported recently, in the synthesis of the 1,2,6-thiadiazine 1,1-dioxide (9) from (8).[17] Now 2-isoxazolines are found to be perhaps even more versatile, being able to undergo reactions not open to the isoxazoles.[18] Metalation of 2,3,4-trimethylisoxazolin-5-one (10), and its bromination with *N*-bromosuccinimide followed by reaction with triphenylphosphine, both

[9] A. R. Katritzky, N. F. Eweiss, and P.-L. Nie, *J.C.S. Perkin I*, 1979, 433.
[10] A. R. Katritzky and S. S. Thind, *J.C.S. Chem. Comm.*, 1979, 138.
[11] A. R. Katritzky, U. Gruntz, N. Mongelli, and M. C. Rezende, *J.C.S. Perkin I*, 1979, 1953.
[12] A. R. Katritzky, A. Chermprapai, and R. C. Patel, *J.C.S. Chem. Comm.*, 1979, 238.
[13] A. R. Katritzky, G. De Ville, and R. C. Patel, *J.C.S. Chem. Comm.*, 1979, 602.
[14] A. J. Boulton, J. Epsztajn, A. R. Katritzky, and P.-L. Nie, *Tetrahedron Letters*, 1976, 2689.
[15] A. R. Katritzky, J. Lewis, and P.-L. Nie, *J.C.S. Perkin I*, 1979, 442.
[16] A. R. Katritzky, K. Horvath, and B. Plau, *J.C.S. Chem. Comm.*, 1979, 300.
[17] H. A. Albrecht, J. F. Blount, F. M. Konzelmann, and J. T. Plati, *J. Org. Chem.*, 1979, **44**, 4191.
[18] V. Jäger and W. Schwab, *Tetrahedron Letters*, 1978, 3129; V. Jäger, H. Grund, and W. Schwab, *Angew. Chem. Internat. Edn.*, 1979, **18**, 78.

activate the methyl group at the 3-position, and the intermediates show potential for future construction work.[19]

The anion derived from 1,3-dihydrobenzo[*c*]thiophen dioxide has been used by two research groups in the synthesis of polycyclic compounds.[20,21] The principle is that thermal decomposition of the alkenylated product (11) generates an intermediate (12), which can undergo an intramolecular Diels–Alder reaction in the same way as that earlier described [*Ann. Reports* (*B*), 1974, **71**, 345] for similar intermediates from dihydrobenzocyclobutenes.

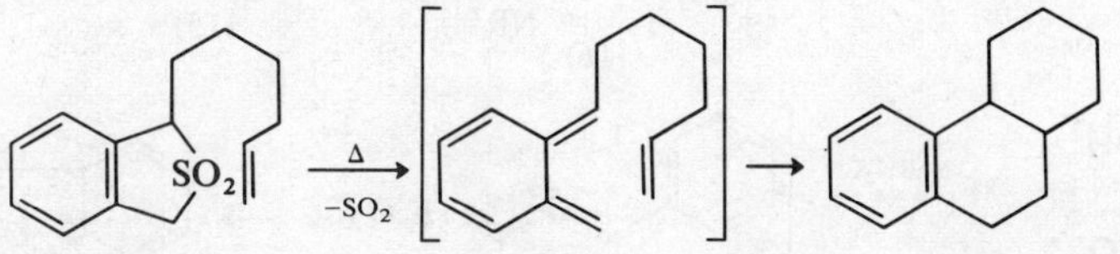

(11) (12)

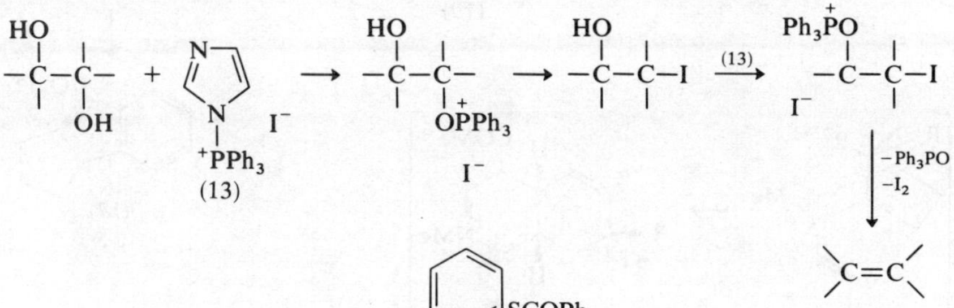

(13)

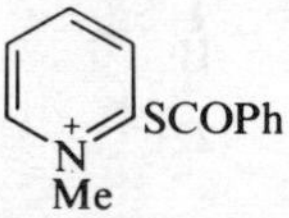

(14)

A mixture of imidazole, triphenylphosphine, and iodine is recommended as a reagent for the reduction of vicinal diols to olefins.[22] It is suggested that the intermediate (13) is formed, which converts a hydroxy-group into an iodo-group, as indicated; the other hydroxy-group is activated, and iodide ion completes the reduction. Another heterocyclic reagent which is conveniently prepared *in situ* is the water-soluble benzoylating agent (14).[23]

2 General Synthetic Methods, Reaction Types, and Physicochemical Studies

In this Section we discuss some aspects of heterocyclic chemistry which are not conveniently categorized by ring size, as in the following Sections.

[19] S. A. Tischler and L. Weiler, *Tetrahedron Letters*, 1979, 4903.
[20] K. C. Nicolaou and W. E. Barnette, *J.C.S. Chem. Comm.*, 1979, 1119.
[21] W. Oppolzer, D. A. Roberts, and T. G. C. Bird, *Helv. Chim. Acta*, 1979, **62**, 2017.
[22] P. J. Garegg and B. Samuelsson, *Synthesis*, 1979, 469.
[23] M. Yamada, Y. Watabe, T. Sakakibara, and R. Sudoh, *J.C.S. Chem. Comm.*, 1979, 179.

Synthesis.—*o*-Formyl-*N*-alkyl-anilines (15), important intermediates for the preparation of many heterocyclic systems, can be prepared by the action of BCl_3 and an isocyanide on the alkyl-aniline; this is an interesting variant on the *o*-acyl-aniline synthesis using nitriles, reported last year [*Ann. Reports* (*B*), 1978, **75**, 241]. The intermediate benzoborazoline derivatives (16) can be isolated.[24]

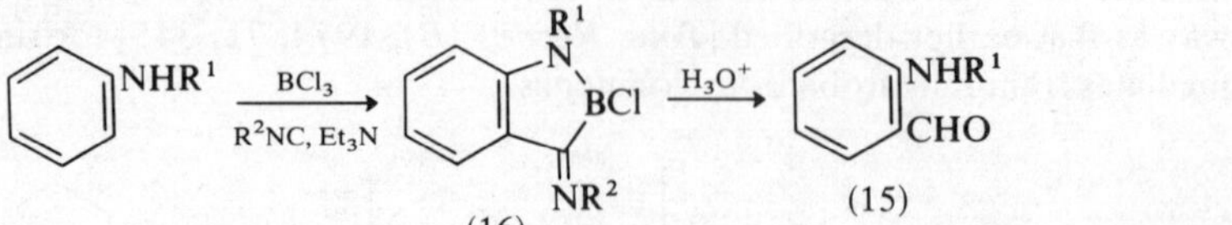

(16) (15)

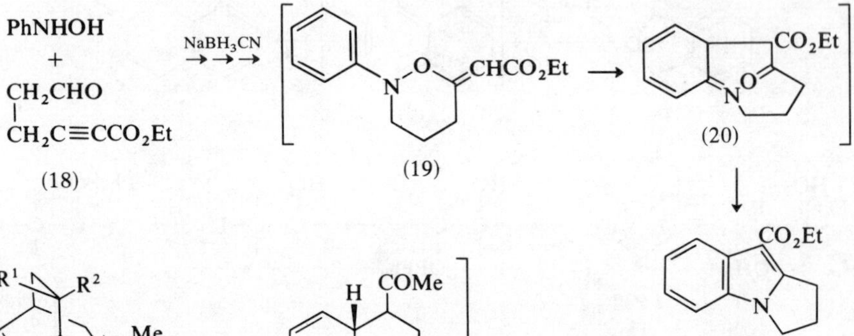

(18) (19) (20) (17)

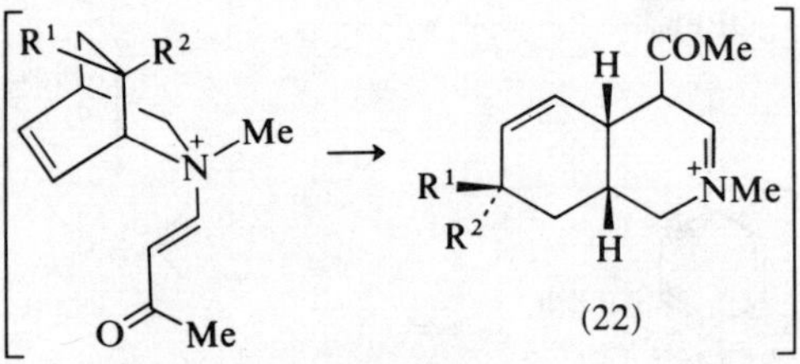

(21) (22)

A number of heterocyclic syntheses were noted which use a hetero-Cope rearrangement at some stage. The important pyrrolo[1,2-*a*]indole system (17) was prepared from phenylhydroxylamine and an acetylenic aldehyde (18) in a multi-step sequence in which the 1,2-oxazine (19) was rearranged to the benzazocine (20). None of the intermediates was isolated.[25] A Diels–Alder addition and Cope rearrangement are concerned in the conversion of a dihydropyridine, through an isoquinuclidine, into a hexahydroisoquinoline, in which the key step is the rearrangement of the (un-isolated) intermediates (21) and (22).[26] Also relevant here are new syntheses of 3-acyl-pyrrolidines,[27] of 3,3′-bi(pyrrolidin-2-one),[28] and of 5,6,7,8-tetrahydroquinoline.[29]

Cycloaddition.—A reaction which was at first incorrectly interpreted[30] has now been straightened out, and some interesting chemistry of azoalkenes (23) has emerged.

24 T. Sugasawa, H. Hamana, T. Toyoda, and M. Adachi, *Synthesis*, 1979, 99.
25 R. M. Coates and C. W. Hutchings, *J. Org. Chem.*, 1979, **44**, 4742.
26 P. S. Mariano, D. Dunaway-Mariano, and P. L. Huesmann, *J. Org. Chem.*, 1979, **44**, 124.
27 L. E. Overman and M. Kakimoto, *J. Amer. Chem. Soc.*, 1979, **101**, 1310.
28 Y. Tamaru, T. Harada, and Z. Yoshida, *J. Amer. Chem. Soc.*, 1978, **100**, 1923.
29 T. Kusumi, K. Yoneda, and H. Kakisawa, *Synthesis*, 1979, 221.
30 S. Sommer, *Chem. Letters*, 1977, 583.

Enamines do not undergo [4+2] but rather [3+2] cycloaddition, giving initial azomethine-imines (24). Proton transfer [to form (25)] and loss of the secondary amine produces *N*-amino-pyrroles (26). One group[31] reports the isolation of the intermediate (24); another[32] that of (25): probably both are correct, as the subtituents chosen by the two sets of workers were quite different. The *N*-amino-pyrroles were found to undergo cycloadditions, to form benzenes, with loss of an amino-nitrene, when acetylenic esters are used, and, reversibly, to form adducts of type (27) with *N*-phenylmaleimide.[33] Cyclo-adducts of (24) were also prepared.[31] On the other hand, dimethyl azodicarboxylate forms the six-membered [4+2] cyclo-adduct (28) with phenylazocyclohexene.[34]

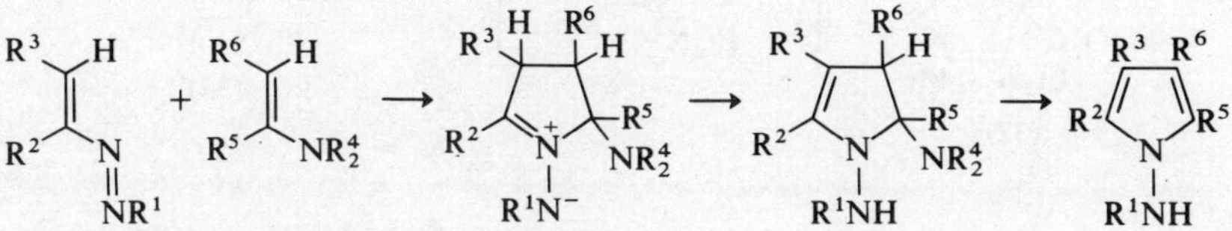

(23) (24) (25) (26)

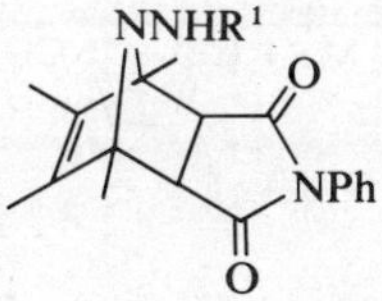

(27)

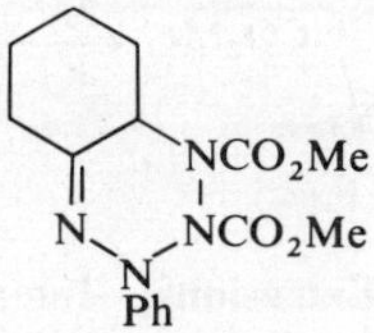

(28)

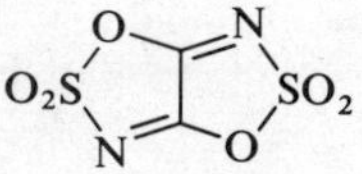

(29)

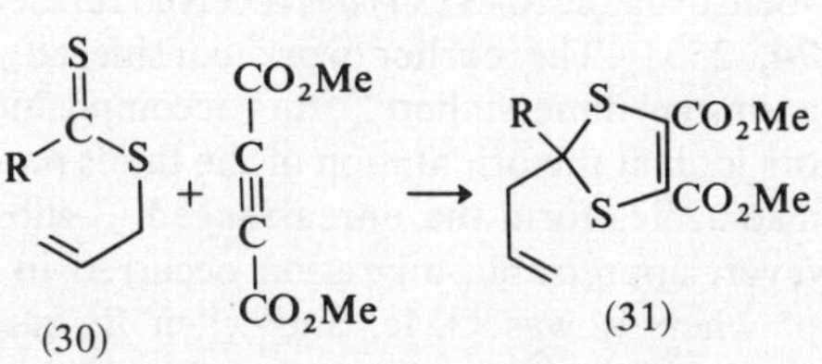

(30) (31)

Sulphur trioxide and cyanogen form the 'criss-cross' cyclo-adduct (29).[35] An interesting cycloaddition mode is shown by allyl and propargyl dithioesters, *e.g.* (30), which with dimethyl acetylenedicarboxylate forms (31).[36]

The reaction between dimethyl diazomalonate and an ynamine is HOMO (ynamine)-/LUMO (dipole)-controlled.[37] The product (32) undergoes an interesting example of a 'conducted tour' migration of one of the methoxycarbonyl

[31] S. Sommer, *Angew. Chem. Internat. Edn.*, 1979, **18**, 695.
[32] A. G. Schultz, W. K. Hagmann, and Ming Shen, *Tetrahedron Letters*, 1979, 2965.
[33] A. G. Schultz and Ming Shen, *Tetrahedron Letters*, 1979, 2969.
[34] S. Sommer and U. Schubert, *Angew. Chem. Internat. Edn.*, 1979, **18**, 696.
[35] H. W. Roesky, N. Amin, G. Remmer, A. Gieren, U. Riemann, and B. Dederer, *Angew. Chem. Internat. Edn.*, 1979, **18**, 223.
[36] V. Drozd and O. A. Popova, *Tetrahedron Letters*, 1979, 4491.
[37] R. Huisgen, H.-U. Reissig, and H. Huber, *J. Amer. Chem. Soc.*, 1979, **101**, 3647.

groups, guided in this instance by diphenylketen. The intermediate (33) can be isolated.[38] The same group have published an interesting short note on the [2+2] and [3+2] cyclo-adducts formed from methyl 3,3-dimethyl-3*H*-pyrazole-5-carboxylate.[39] The reactivity of 3,4-dimethoxyfuran in [4+2] cycloadditions (Diels–Alder reactions) has been reported extensively in two papers from Eugster's group.[40]

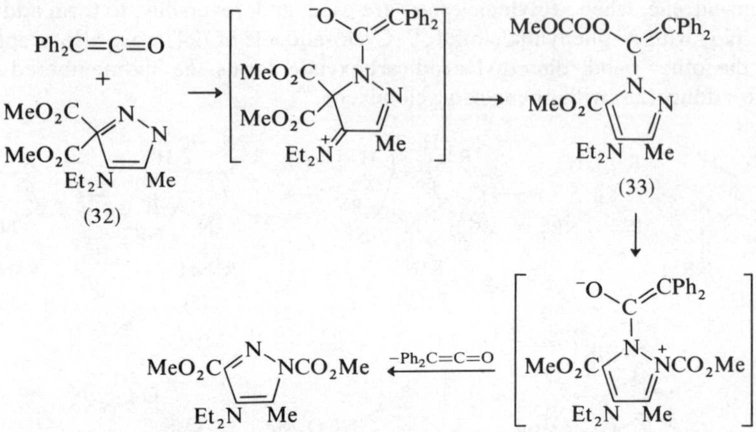

Nitrogen Extrusion and Other Extrusions.—Interest in the elimination of nitrogen from cyclic azo-compounds has been maintained at a high level. The question of the intermediacy of benzothiirens (35) in the thermal and photochemical decompositions of 1,2,3-benzothiadiazoles (34) has received further attention [*cf. Ann. Reports* (*B*), 1977, **74**, 253]. The earlier work considered evidence from the structures of the thianthrene dimers: here[41] the decomposition is conducted in tetralin, and the authors look at the orientation of the thiols produced. In all cases, 6-substituted benzothiadiazoles form the unrearranged (3-substituted) thiols (36) predominantly. However, appreciable migration occurred in the photochemical decomposition, except when R was OMe, and when R was CO_2Me some rearrangement was also found in the thermal reaction. The evidence does not require the benzothiiren to occupy an energy minimum on the reaction profile. The decomposition of benzothiadiazole (34; R = H) has also been studied by variable-temperature photoelectron spectroscopy, which is claimed to be a useful technique

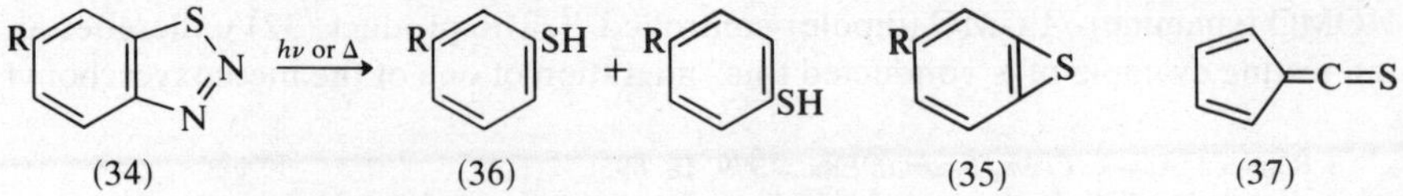

(34) (36) (35) (37)

[38] H.-U. Reissig and R. Huisgen, *J. Amer. Chem. Soc.*, 1979, **101**, 3648.
[39] R. Huisgen and H.-U. Reissig, *J.C.S. Chem. Comm.*, 1979, 568.
[40] P. X. Iten, A. A. Hofmann, and C. H. Eugster, *Helv. Chim. Acta*, 1979, **62**, 2202; A. A. Hofmann, I. Wyrsch-Walraf, P. X. Iten, and C. H. Eugster, *ibid.*, p. 2211.
[41] R. C. White, J. Scoby, and T. D. Roberts, *Tetrahedron Letters*, 1979, 2785.

for the detection of transient species at elevated temperatures. However, here no benzothiiren was observed, only the rearrangement product, *i.e.* the thioketen (37).[42]

Nitrogen is eliminated at high temperatures from the fused cinnoline (38), to give the unstable biphenylene analogue (39).[43] Naphth[1,8-*bc*]azete (40) is formed quantitatively, on irradiation of (41), and oxidation of (41) with *m*-chloroperbenzoic acid gives the sulphoxide of (40) directly.[44] Stereochemical aspects of the thermal[45] and practical aspects of the photochemical[46] decomposition of cyclic azo-compounds have been reported, and *ab initio* calculations on the reaction have been performed.[47]

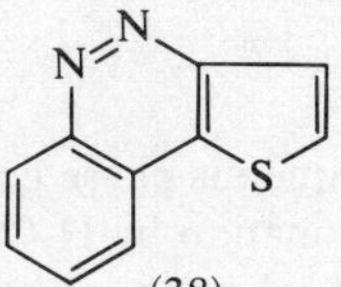

(38)

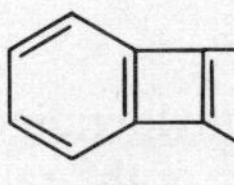

(39)

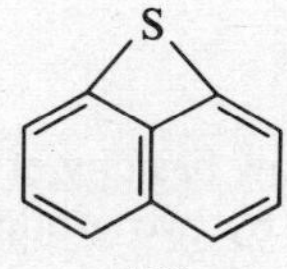

(40)

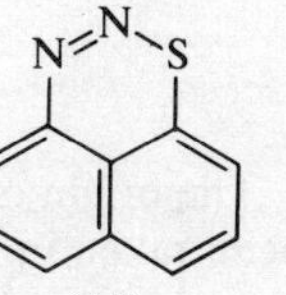

(41)

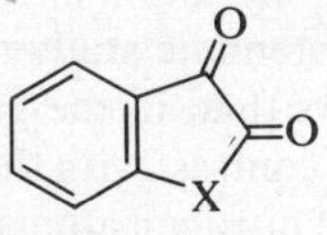

(42) X = O
(43) X = S

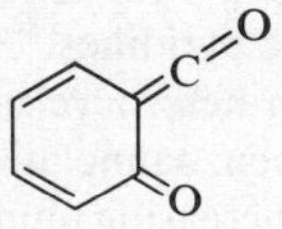

(44)

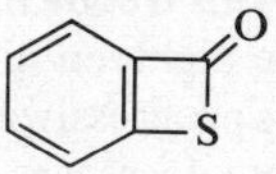

(45)

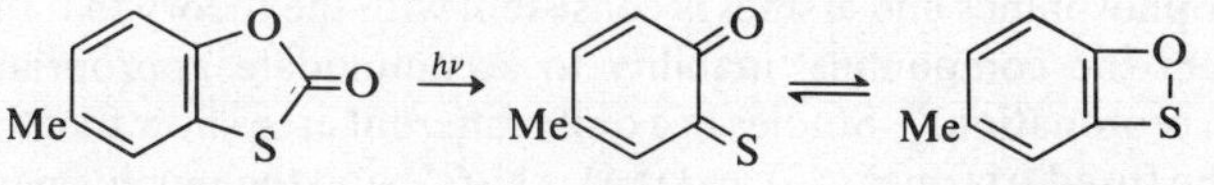

(46) (47) (48)

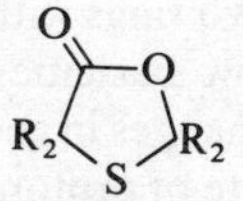

(49)

Elimination of carbon monoxide (at 600 °C) from benzofuran-2,3-dione (42) initially forms the keten (44), but under similar conditions thioisatin (43) gives benzothiet-2-one (45), according to the photoelectron spectra.[42] Irradiation of (46) at 77 K forms the monothio-*o*-quinone (47), which undergoes photo-equilibration with its cyclic isomer (48). Both products proved to be too unstable to be isolated.[48] The primary products of thermal elimination of CO_2 from (49), *i.e.* the thiocarbonyl ylides, cyclized to episulphides before they could be detected.[49]

[42] R. Schulz and A. Schweig, *Tetrahedron Letters*, 1979, 59.
[43] J. W. Barton and D. J. Lapham, *Tetrahedron Letters*, 1979, 3571.
[44] J. Nakayama, T. Fukushima, E. Seki, and M. Hoshimo, *J. Amer. Chem. Soc.*, 1979, **101**, 7684.
[45] P. B. Dervan, T. Uyehara, and D. S. Santilli, *J. Amer. Chem. Soc.*, 1979, **101**, 2069; P. B. Dervan and T. Uyehara, *ibid.*, p. 2076; D. S. Santilli and P. B. Dervan, *ibid.*, p. 3663.
[46] P. S. Engel, C. J. Nalepa, R. A. Leckonby, and W. K. Chae, *J. Amer. Chem. Soc.*, 1979, **101**, 6435.
[47] P. C. Hiberly and Y. Jean, *J. Amer. Chem. Soc.*, 1979, **101**, 2538.
[48] P. de Mayo, A. C. Weedon, and G. S. K. Wong, *J. Org. Chem.*, 1979, **44**, 1977.
[49] T. B. Cameron and H. W. Pinnick, *J. Amer. Chem. Soc.*, 1979, **101**, 4755.

Physicochemical Properties, and Structural Studies.—The basicities of bis-annelated pyridines (50) fall regularly with decreasing ring size.[50] For $n = 4$, the pK_a is 8.09, but the novel highly strained compound with $n = 2$, prepared by two groups this year,[50,51] has $pK_a = 4.40$. From data for pyridine ($pK_a = 5.3$) and compound (51) ($pK_a = 4.85$), it appears that the effects of the annelated four-membered rings are additive, there being a lowering of 0.45 p*K* units per ring. However, the quinoline (52) has a pK_a of 3.99, which is 1.07 units lower than quinoline itself.[52]

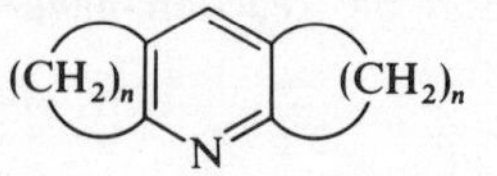

(50)

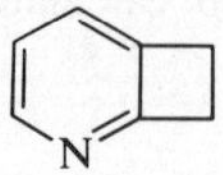

(51)

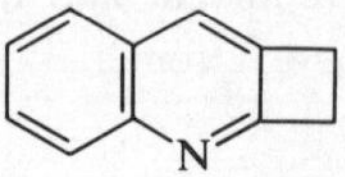

(52)

The origin of the low basicity of 2,6-di-t-butylpyridine in the aqueous phase has been taken up again by two groups. A study of the rate of protonation by H_3O^+ supports the view that the low basicity is due to steric hindrance of solvation (rather than steric compression of the N—H bond of the conjugate acid).[53] This theory is also supported by a complete analysis of the thermodynamics of hydration, which also demonstrates that the hydration properties of both the cation and the free base are at variance with trends in other pyridines.[54] A systematic study of the proton affinities of some eighteen nitrogen heterocycles shows that, in the gas phase, the basicity increases progressively on benzoannelation (in contrast with the trend in the solution phase) but decreases with increasing number of nitrogen atoms in the ring.[55] The basicities of phosphorin and arsenin, both in solution and in the gas phase, are unexpectedly low: a comparison of proton affinities and core ionization data for the two rings with those for phosphines and arsines is consistent with the theory that the low basicities arise from the compounds' inability to accommodate appropriate changes in geometry on protonation.[56] Studies in a quite different area show that the site of protonation of the fused triazines (53) and (54) is N-6,[57] and the involvement of the zwitterionic tautomer (55) in isotopic exchange at C-8 of various purines has been verified.[58]

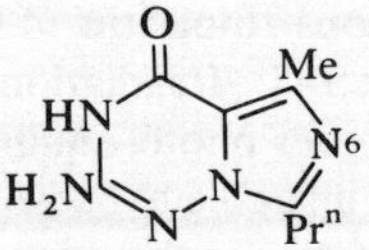

(53)

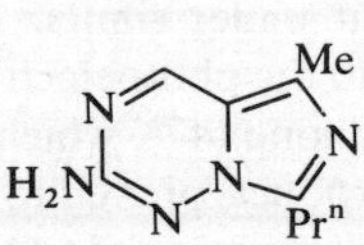

(54)

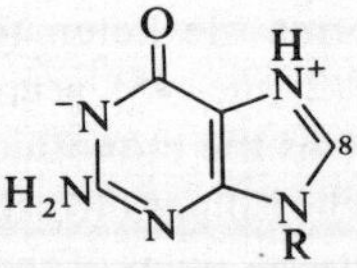

(55)

[50] R. P. Thummel and D. K. Kohli, *Tetrahedron Letters*, 1979, 143.
[51] A. Naimann and K. P. C. Vollhardt, *Angew. Chem. Internat. Edn.*, 1979, **18**, 411.
[52] J. H. Markgraf, J. H. Antin, F. J. Walker, and R. A. Blatchly, *J. Org. Chem.*, 1979, **44**, 3261.
[53] C. F. Bernasconi and D. J. Carré, *J. Amer. Chem. Soc.*, 1979, **101**, 2707.
[54] E. M. Arnett and B. Chawla, *J. Amer. Chem. Soc.*, 1979, **101**, 7141.
[55] M. Meot-Ner, *J. Amer. Chem. Soc.*, 1979, **101**, 2396.
[56] A. J. Ashe, M. K. Bahl, K. D. Bomben, W.-T. Chan, J. K. Gimzewski, P. G. Sitton, and T. D. Thomas, *J. Amer. Chem. Soc.*, 1979, **101**, 1764.
[57] J. P. Riley, F. Heatley, I. H. Hillier, P. Murray-Rust, and J. Murray-Rust, *J.C.S. Perkin II*, 1979, 1327.
[58] J. R. Jones and S. E. Taylor, *J.C.S. Perkin II*, 1979, 1253, 1587.

Three examples of the use of nitrogen n.m.r. illustrate its potential value in heterocyclic chemistry. In the imidazole series, ^{15}N shifts provide information about the active sites in the molecule, demonstrating a degree of charge transfer between the cation and the anion of (56) in $CHCl_3$.[59] Chemical shifts of ^{14}N are superior to ^{13}C shifts for studying tautomeric equilibria of pyridine-2-thione and various 2-amino-pyridines, because of the very large difference in chemical shifts between nitrogen atoms in the alternative environments.[60] Coupling constants $^{15}N-^{1}H$ and $^{15}N-^{13}C$ have been used as a sensitive probe for studying the tautomerism of the uracil anion; the two forms (57) and (58) are very evenly balanced.[61]

A new mass spectrometric technique has been devised for investigating gas-phase tautomerism. It involves comparison of the collision-induced dissociation–mass-analysed ion kinetic energy (CID–MIKE) spectra of cations formed by ethylation (thermodynamically controlled) of the mobile system with those of the protonated forms of *N*- and *O*-ethyl models. The results confirm that 2-hydroxypyridine predominates over 2-pyridone, but that 2-quinolone is favoured over 2-hydr-oxyquinoline, under these conditions.[62]

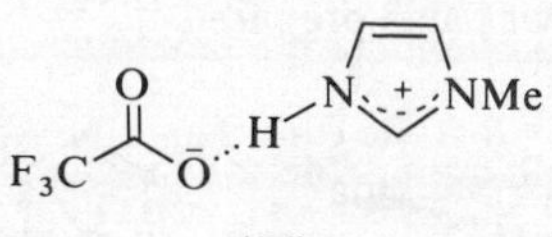

(56)

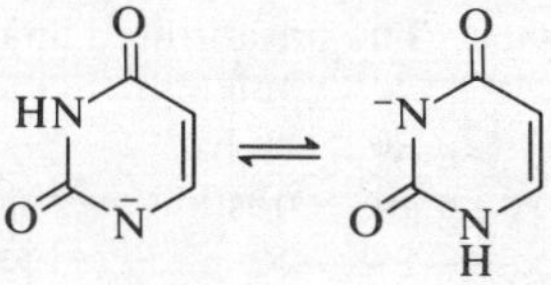

(57) (58)

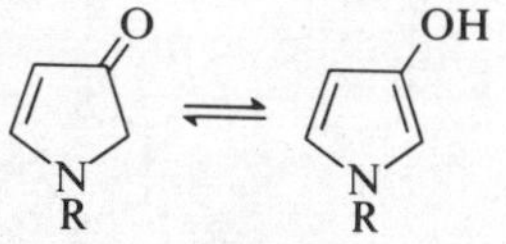

(59)

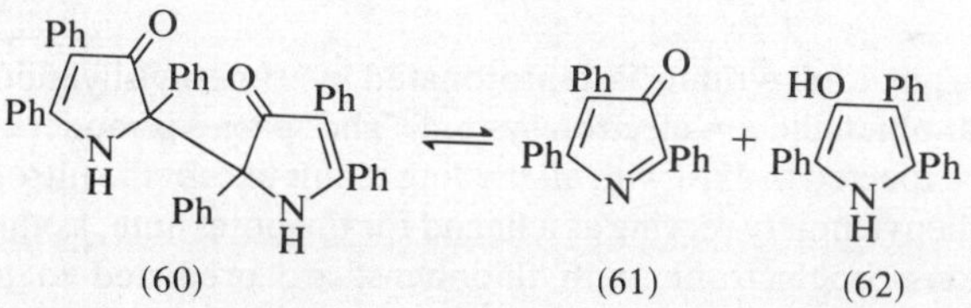

(60) (61) (62)

There has been some activity in the hydroxy-pyrrole (pyrrolinone) area recently. A new route to pyrrolin-2-one has been described, and the 3*H*/5*H* equilibrium investigated: the 5*H*-form predominates.[63] In the 3-series, C=O forms are reported to prevail in some cases, *e.g.* (59; R = Ph or CH_2Ph), but both tautomers (2-*H* and

[59] I. I. Schuster and J. D. Roberts, *J. Org. Chem.*, 1979, **44**, 3864; I. I. Schuster, C. Dyllick-Brenzinger, and J. D. Roberts, *ibid.*, p. 1765.
[60] L. Stefaniak, *Org. Magn. Resonance*, 1979, **12**, 379.
[61] R. L. Lipnick and J. D. Fissekis, *J. Org. Chem.*, 1979, **44**, 1627.
[62] A. Maquestiau, Y. van Haverbeke, R. Flammang, H. Mispreuve, A. R. Katritzky, J. Ellison, J. Frank, and Z. Meszáros, *J.C.S. Chem. Comm.*, 1979, 888.
[63] J. T. Baker and S. Sifniades, *J. Org. Chem.*, 1979, **44**, 2798.

OH) are seen in others, *e.g.* (59; R=Me). Various 4-carboxy-substituted 3-hydroxy-pyrroles favour the OH forms.[64] A compound which had earlier been reported to be 3-hydroxy-2,4,5-triphenylpyrrole has been found to have the dimeric structure (60), and to dissociate reversibly on heating, to give the pyrrolone (61) and the hydroxy-pyrrole (62), or a tautomer thereof.[65] A theoretical study (MINDO/3) of the tautomerism of 3-hydroxypyrrole had very limited success in predicting the experimental results.[66]

In the field of 'hypervalent' sulphur compounds, n.m.r. measurements in the nematic phase add further evidence for the C_{2v} symmetry (or, at least fail to provide evidence for anything different) of compounds of type (63).[67] However, reports of equilibrations involving bond-switching at sulphur have appeared in several journals.[68—70] Perhaps the simplest example to illustrate here is that between the two amines (64) and (65), which is observed directly by n.m.r. When the sulphur is replaced by oxygen the equilibration does not occur.[70] An *X*-ray study of the system (66) appears to indicate that there is little or no bonding between the S and X atoms.[71] Hypervalent sulphur (or hypovalent nitrogen) is evident in the persistent radicals formed by photolysis of S_4N_4–olefin adducts, to which structures (67) have been assigned. The unsaturated analogue (68) was also prepared.[72]

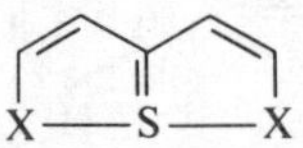
(63)

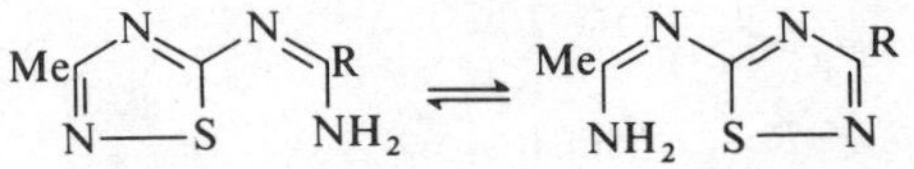
(64) (65)

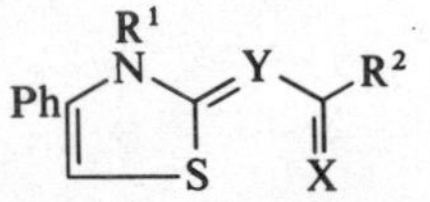
(66) Y=N or CH
X=O or S

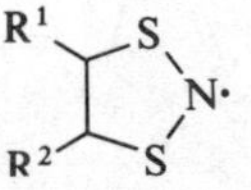
(67)

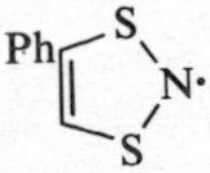
(68)

1,4-Dihydro-1-methylpyridine is deprotonated by trimethylsilylmethylpotassium to yield an organometallic 8π-electron system. The species is moderately stable (at −50 °C), and is expected to exist with all the ring atoms except the nitrogen coplanar, with the pentadienyl moiety serving as a ligand for the potassium, as shown in (69).[73] Phosphole anions, isoelectronic with thiophens, are predicted to possess a de-localized π-electron system, and evidence for considerable π character in the C—P

[64] T. Momose, T. Tanaka, and T. Yokota, *Heterocycles*, 1977, **6**, 1827; T. Momose, T. Tanaka, T. Yokota, N. Nagamoto, and K. Yameda, *Chem. Pharm. Bull.*, 1979, **27**, 1448.
[65] J. P. Freeman and M. J. Haddadin, *Tetrahedron Letters*, 1979, 4813.
[66] A. Karpfen, P. Schuster, and H. Berner, *J. Org. Chem.*, 1979, **44**, 374.
[67] J. P. Jacobsen, J. Hansen, C. T. Pedersen, and T. Pedersen, *J.C.S. Perkin II*, 1979, 1521.
[68] K. Akiba, S. Arai, and F. Iwasaki, *Tetrahedron Letters*, 1978, 4117.
[69] K. Akiba, S. Arai, T. Tsuchiya, Y. Yamamoto, and F. Iwasaki, *Angew. Chem. Internat. Edn.*, 1979, **18**, 166.
[70] K. Akiba, T. Kobayashi, and S. Arai, *J. Amer. Chem. Soc.*, 1979, **101**, 5857.
[71] R. J. S. Beer, D. McMonagle, M. S. S. Siddiqui, A. Hordvik, and K. Jynge, *Tetrahedron*, 1979, **35**, 1199.
[72] S. Rolfe, D. Griller, K. U. Ingold, and L. H. Sutcliffe, *J. Org. Chem.*, 1979, **44**, 3515.
[73] M. Schlosser and P. Schneider, *Angew. Chem. Internat. Edn.*, 1979, **18**, 489.

bonds has been obtained, using ^{31}P n.m.r. The anions, *e.g.* (70), are formed by cleavage of the *P*-phenyl derivatives with potassium, and are sufficiently stable to be unquenched by added ethanol (contrast dialkyl-phosphines, p*K* *ca.* 35).[74] The novel bis-1,2,6-selenadiazine (71), the sulphur analogue of which was reported last year [*Ann. Reports* (*B*), 1978, **75**, 244] shows its proton chemical shift at δ 5.9; *cf.* 4.74 for the thiadiazine. The difference in the magnitudes of the induced paramagnetic ring currents is determined by the HOMO–LUMO energy gap.[75]

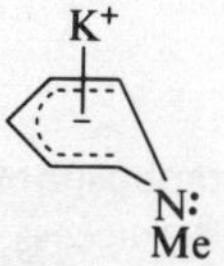

(69)

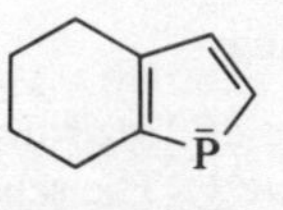

(70)

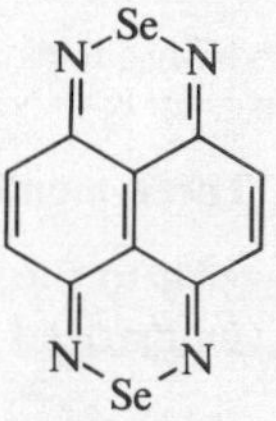

(71)

In the area of conformational analysis, the controversy alluded to last year [*Ann. Reports* (*B*), 1978, **75**, 266], concerning the effect of a second heteroatom on the barrier to inversion of an *N*-methyl group in six-membered rings, is now resolved. Apparently conflicting standpoints have been reconciled, with the statement that a β-heteroatom increases the '*ax–ts* half-barrier', but decreases the '*eq–ts* half-barrier'.[76] Another controversial question has been that of the orientation of the N—H bond in piperidines. The weight of evidence is now heavily in favour of the predominance of the NH-equatorial form, but some loose ends remain. The difference in chemical shifts of geminal protons α to the nitrogen was used by some as a criterion for assigning an NH-axial preference; now, however, the chemical shifts of the α-protons in (72) and (73), drawn here as their predominant conformers, have been found not to be significantly different.[77] An *X*-ray crystal structure confirms the axial t-butyl group on the chair ring in (72), with a small degree of torsion to relieve the strain.[77] A study on bicyclic sulphonium salts has shown that the equilibrium for isomerization of *cis*/*trans* ring fusion in (74) favours the *trans*-fused isomer. However, the ring fusion is predominantly *cis* in the ester (75), and also in

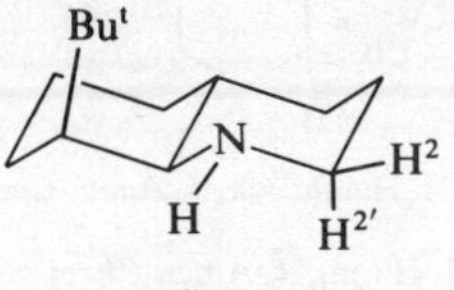

(72)

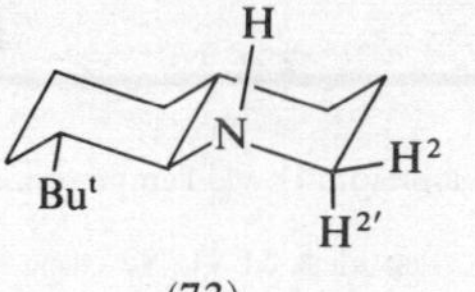

(73)

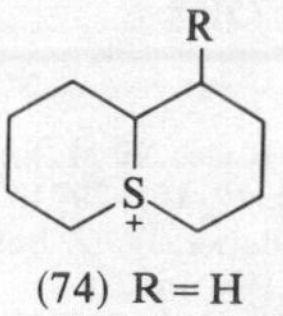

(74) R = H
(75) $R = CO_2Me$

[74] L. D. Quin and W. L. Orton, *J.C.S. Chem. Comm.*, 1979, 401.
[75] M. L. Kaplan, R. C. Haddon, F. C. Schilling, J. H. Marshall, and F. B. Bramwell, *J. Amer. Chem. Soc.*, 1979, **101**, 3306.
[76] A. R. Katritzky, R. C. Patel, and F. G. Riddell, *J.C.S. Chem. Comm.*, 1979, 674.
[77] F. W. Vierhapper and E. L. Eliel, *J. Org. Chem.*, 1979, **44**, 1081.

the ylide derived from the ion (74).[78] Among phosphorus heterocycles, twist conformers are adopted by the following: (76) in both the solid state and solution, (77) in the solid state but not in solution,[79] and (78)[80] and (79),[81] both in solution. In 1-methyl-silacyclohexane the methyl group is mainly axial, according to the ^{1}H n.m.r. spectrum.[82]

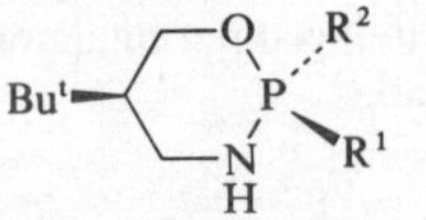

(76) $R^1 = NMe_2$, R^2 is =O
(77) R^1 is =O, $R^2 = NMe_2$

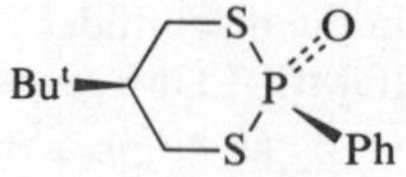

(78)

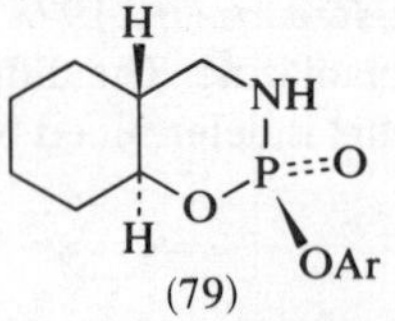

(79)

3 Three-membered Rings

A novel approach to the synthesis of 1-azirines *via* oxazaphospholines has been described, as summarized in Scheme 1. The scheme provides access to azirines not

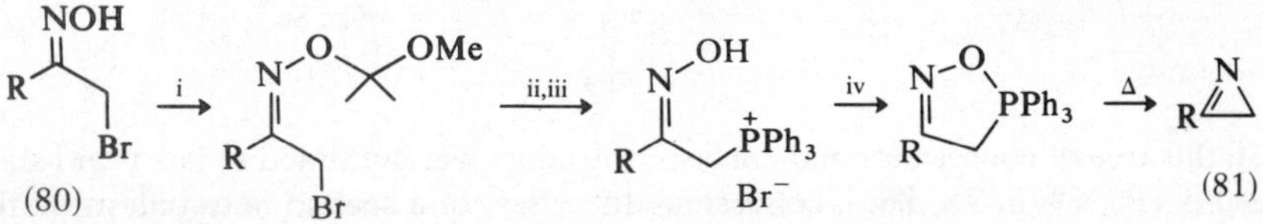

(80) (81)

Reagents: i, $Me_2C(OMe)_2$, H^+; ii, PPh_3; iii, H_3O^+; iv, Et_3N

Scheme 1

readily accessible by the vinyl azide route.[83] *N*-Sulphenyl-aziridines can be prepared by oxidation of sulphenamides in the presence of olefins, and the sulphenyl groups are readily removed by borohydride.[84] The products of two earlier reported ring syntheses have now been shown to be wrongly assigned. Thus, alkyldiazoacetates and *NN'*-di-isopropylcarbodi-imide, in the presence of copper triflate or rhodium(II) acetate, give (82) rather than (83)[85] [*cf. Ann. Reports* (*B*), 1976, **73**, 241], and the problems encountered in separating the supposed phosphiren (84) [*cf. Ann. Reports* (*B*), 1972, **69**, 425] from 1,5-diazabicyclo[4.3.0]non-5-ene hydrobromide have now been explained: the product is in fact (85)![86]

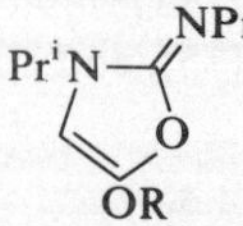

(82)

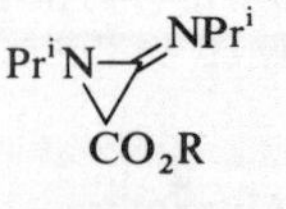

(83)

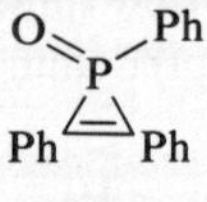

(84)

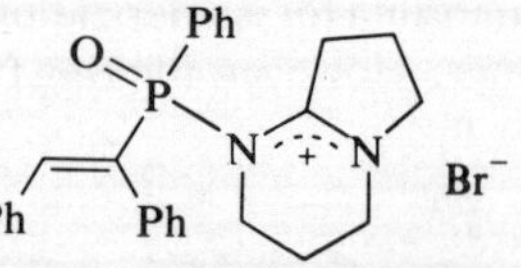

(85)

[78] D. M. Roush, E. M. Price, L. K. Templeton, D. H. Templeton, and C. H. Heathcock, *J. Amer. Chem. Soc.*, 1979, **101**, 2971.
[79] G. S. Bajwa, W. G. Bentrude, N. S. Pantaleo, M. G. Newton, and J. H. Hargis, *J. Amer. Chem. Soc.*, 1979, **101**, 1602.
[80] R. O. Hutchins, B. E. Maryanoff, M. J. Castillo, K. D. Hargrave, and A. T. McPhail, *J. Amer. Chem. Soc.*, 1979, **101**, 1600.
[81] D. G. Gorenstein and R. Rowell, *J. Amer. Chem. Soc.*, 1979, **101**, 4925.
[82] R. Carleer and M. J. O. Anteunis, *Org. Magn. Resonance*, 1979, **12**, 673.
[83] A. Hassner and V. Alexanian, *J. Org. Chem.*, 1979, **44**, 3861.
[84] R. S. Atkinson and B. D. Judkins, *J.C.S. Chem. Comm.*, 1979, 832, 833.
[85] J. Drapier, A. Feron, R. Warin, A. J. Hubert, and P. Teyssié, *Tetrahedron Letters*, 1979, 559.
[86] H. Quast and M. Heuschmann, *J.C.S. Chem. Comm.*, 1979, 390.

A strong theme in the year's organic literature has been the investigation of ring-cleavage reactions, and a number of theoretical treatments have been reported. The Dewar π-complex theory was described in detail nearly 30 years ago, and now a new series of calculations supports a detailed restatement of the relationship between π-complexes and three-membered rings containing O, S, NH, PH, SiH_2, OH^+, SH^+, NH_2^+, *etc.* In particular, the direction of nucleophilic ring-opening of unsymmetrically substituted compounds, which it is argued is a criterion for distinguishing between the π-complex type and a classical ring species, is found to be generally consistent with the MINDO scale of π-complex character. According to the theory, significant π-complex character can be attributed to the protonated ring systems.[87] The photochemical ring-opening of oxiran[88] and of aziridine, diaziridine, and oxaziridine,[89] and the rearrangements of the last system,[90,91] have been the subject of *ab initio* calculations, as also have the oxidation of imines to oxaziridines[92] and the acid-catalysed hydrolysis of benzene oxides.[93] An empirical investigation of the mechanism of hydrolysis of oxaziridine has provided evidence for both *O*- and *N*-protonation of this system.[94]

Some new ring-cleavage reactions have been described, some of possible synthetic use. The ring-opening of aryl-oxirans under certain conditions ($CuSO_4$–C_5H_5N, phosphate buffer at pH 7) gives *cis*-diols. The stereochemistry of the cleavage is reproduced by nucleophiles other than water, which enter at the benzylic position.[95] Ylides are obtained in a novel ring-opening of 2-aryl-3,3-dicyano-oxirans, using pyridine [to give (86)], sulphides, or triphenylphosphine.[96] 1,3-Di-t-butyl-aziridinone, a particularly stable α-lactam, reacts very readily with *m*-chloroperbenzoic acid to give the oxaziridine and CO; *N*-oxidation is postulated to be the first step.[97] The oxaziridine (87), with phenylmagnesium bromide, gives biphenyl, but, with phenyl-lithium, phenol is produced, and also the adduct (88). The analogue (89) follows the latter pathway with both organometallic reagents.[98]

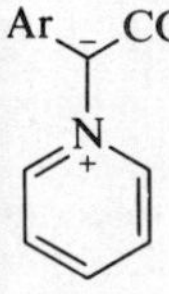

(86)

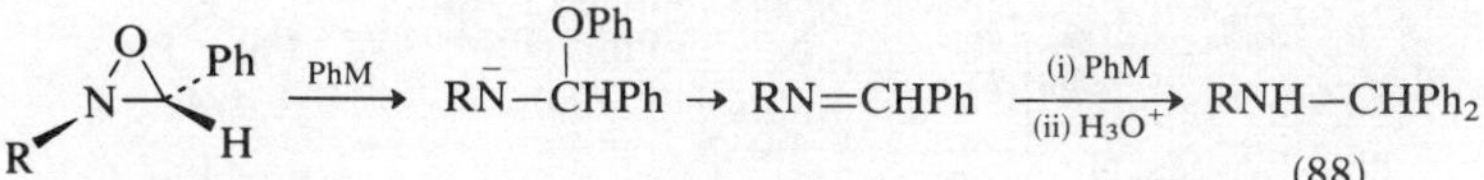

(88)

(87) R = Bu^t
(89) R = $PhSO_2$

(M = Li or MgBr)

[87] M. J. S. Dewar and G. P. Ford, *J. Amer. Chem. Soc.*, 1979, **101**, 783.
[88] B. Bigot, A. Sevin, and A. Devaquet, *J. Amer. Chem. Soc.*, 1979, **101**, 1095, 1101.
[89] J. Sauer, *Tetrahedron*, 1979, **35**, 2109.
[90] B. Bigot, D. Roux, A. Sevin, and A. Devaquet, *J. Amer. Chem. Soc.*, 1979, **101**, 2560.
[91] E. Oliveros, M. Riviere, J. P. Malrieu, and C. Teichteil, *J. Amer. Chem. Soc.*, 1979, **101**, 318.
[92] A. Ažman, J. Koller, and B. Plesničar, *J. Amer. Chem. Soc.*, 1979, **101**, 1107.
[93] J. E. Ferrell and G. H. Loew, *J. Amer. Chem. Soc.*, 1979, **101**, 1385.
[94] A. R. Butler, B. C. Challis, and A. M. Lobo, *J.C.S. Perkin II*, 1979, 1035; A. R. Butler, J. G. White, B. C. Challis, and A. M. Lobo, *ibid.*, p. 1039.
[95] M. Imuta and H. Ziffer, *J. Amer. Chem. Soc.*, 1979, **101**, 3990.
[96] A. Robert, M. T. Thomas, and A. Foucaud, *J.C.S. Chem. Comm.*, 1979, 1048.
[97] T. Hata and M. Watanabe, *J. Amer. Chem. Soc.*, 1979, **101**, 1323.
[98] F. A. Davis, P. A. Mancinelli, K. Balasubramanian, and U. K. Nadir, *J. Amer. Chem. Soc.*, 1979, **101**, 1044.

Another burst of papers from Zürich on the apparently inexhaustible chemistry of 3-dimethylamino-2,2-dimethyl-2*H*-azirine has appeared.[99–102] [see *Ann. Reports* (*B*), 1977, **74**, 255]. A feature of its reaction with heterocumulenes is that alternative pathways are followed, as indicated, depending upon the nucleophilicity of the atoms or groups X and Y in the intermediates (90).[101] However, among thiocyanates the reaction of the phenyl compound is anomalous, giving (91).[99] 3-Phenyl-2*H*-azirines undergo addition of CO with rhodium catalysts to give vinyl isocyanates.[103]

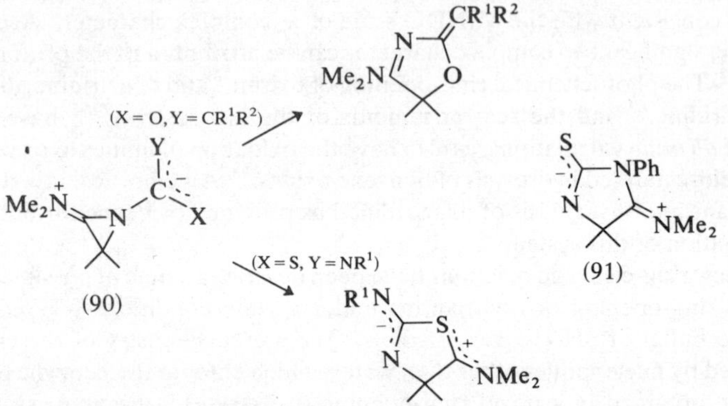

(90) (91)

Thiiran–imines (92) show three different modes of addition to different doubly- or triply-bonded reactants (see Scheme 2). The reactions (*a*) and (*c*) find analogies in the chemistry of diaziridine-imines and *α*-lactams, respectively. The successive

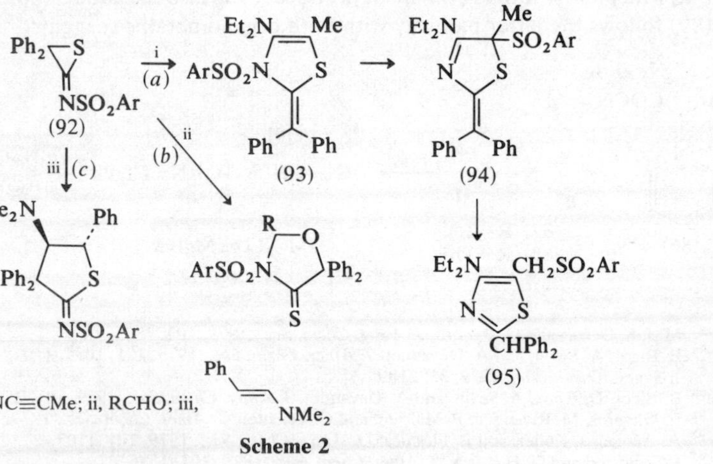

Scheme 2

[99] U. Schmid, H. Heimgartner, and H. Schmid, *Helv. Chim. Acta*, 1979, **62**, 160.

[100] G. Mukherjee-Müller, S. Chaloupka, H. Heimgartner, H. Schmid, H. Link, K. Bernauer, P. Schonholzer, and J. J. Daly, *Helv. Chim. Acta*, 1979, **62**, 768.

[101] J. Lukáč and H. Heimgartner, *Helv. Chim. Acta*, 1979, **62**, 1236.

[102] G. Mukherjee-Müller, H. Heimgartner, and H. Schmid, *Helv. Chim. Acta*, 1979, **62**, 1429.

[103] T. Sakakibara and H. Alper, *J.C.S. Chem. Comm.*, 1979, 458.

rearrangements (93) → (94) → (95) of the ynamine adduct are noteworthy.[104] Di-t-butyldiaziridinone (96) does not form the expected thione with the sulphurizing agent (97); instead, the five-membered heterocycle (98) is produced, which reportedly exists in two stereoisomeric forms.[105]

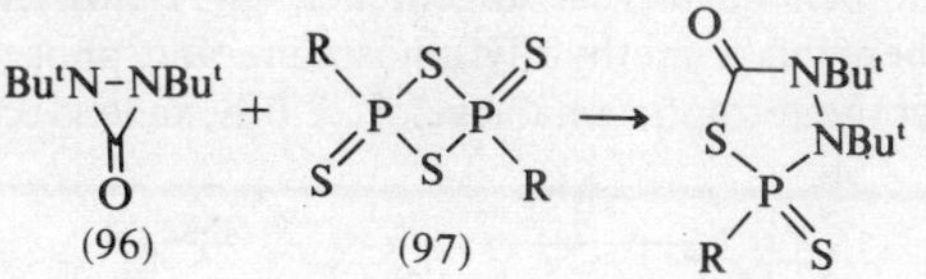

(96) (97) (98) ($R = C_6H_4OMe$-*p*)

4 Four-membered Rings

General.—The sodium salt of dimethyl *N*-(toluene-*p*-sulphonyl)sulphoximine, a nucleophilic methylene-transfer reagent, is known to react with dialkyl ketones to give oxirans. It has now been found that excess of the reagent leads to oxetans, so providing a particularly convenient route to this ring system.[106] Azetidine-2,4-diones (malonimides) result from the photochemical ring-contraction of succinimides, in yields of up to 50%,[107] and stable four-membered-ring nitrones (azetine 1-oxides) (99) have been obtained from the reaction of nitro-alkenes with ynamines.[108] A series of reactions involving isocyanides have been reported, giving imino-azetidines from *N*-phthalimido-aziridines bearing electron-withdrawing groups, in a (3+1) cheletropic reaction with the valence-isomeric azomethine ylides,[109] 2,4-bis(imino)thietans from arylsulphonylimino-thiirans (92),[110] and tris(imino)thietans (100) from arylsulphonyl isothiocyanates.[111] The tris-imine (100) reacts at C-2 with methanol, diethylamine, and ethanethiol, to give open-chain products, but at C-4 with hydrazoic acid, finally forming the tetrazole (101).[112]

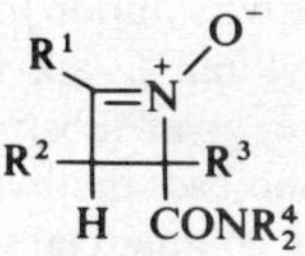

(99)

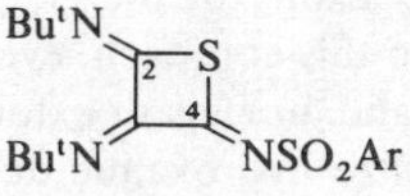

(100)

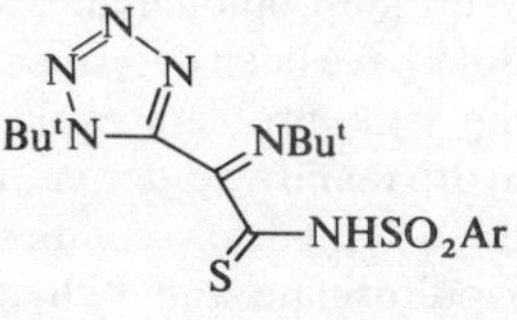

(101)

[104] G. L'abbé, J.-P. Dekerk, S. Toppet, J.-P. Declercq, G. Germain, and M. Van Meerssche, *Tetrahedron Letters*, 1979, 1819.
[105] G. L'abbé, J. Flémal, J.-P. Declercq, G. Germain, and M. Van Meerssche, *Bull. Soc. chim. belges*, 1979, **88**, 737.
[106] S. C. Welch and A. S. C. Prakasa Rao, *J. Amer. Chem. Soc.*, 1979, **101**, 6135.
[107] K. Maruyama, T. Ishitoku, and Y. Kubo, *J. Amer. Chem. Soc.*, 1979, **101**, 3670.
[108] A. D. de Wit, M. L. M. Pennings, W. P. Trompenaars, D. N. Reinhoudt, S. Harkema, and O. Nevestveit, *J.C.S. Chem. Comm.*, 1979, 993.
[109] J. Charrier, H. Person, and A. Foucaud, *Tetrahedron Letters*, 1979, 1381.
[110] G. L'abbé and J.-P. Dekerk, *Tetrahedron Letters*, 1979, 3213.
[111] G. L'abbé, L. Huybrechts, J.-P. Declercq, G. Germain, and M. Van Meerssche, *J.C.S. Chem. Comm.*, 1979, 160.
[112] G. L'abbé, L. Huybrechts, S. Toppet, J.-P. Declercq, G. Germain, and M. Van Meerssche, *Bull. Soc. chim. belges*, 1979, **88**, 297.

The first authentic stable bisdioxetan (102) has been reported. On thermolysis, it forms benzoic anhydride quantitatively.[113] The indole dioxetan (103) has been found to be sufficiently stable to allow its spectroscopic and chemical investigation.[114] A 1,2-disilacyclobutane (104) has been prepared; it undergoes some unusual and interesting thermal and photochemical rearrangements.[115] A further intriguing reaction is that between phenylbis(trimethylsilyl)phosphine and phosgene, which gives as a secondary product the tetraphosphabicyclo[2.2.0]hexane (105).[116]

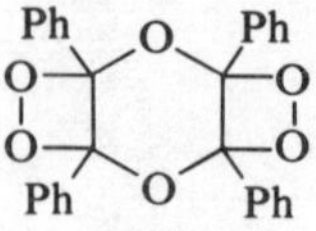

(102)

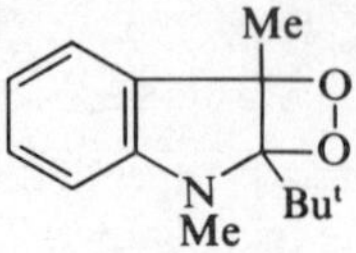

(103)

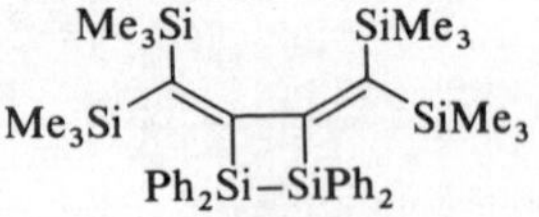

(104)

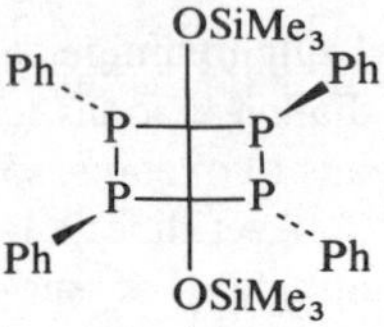

(105)

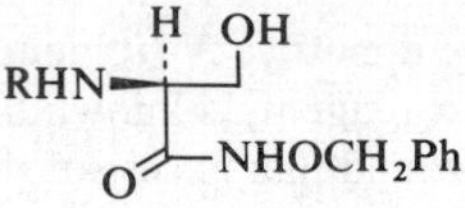

(106)

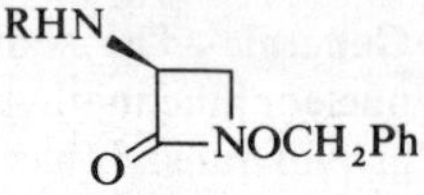

(107)

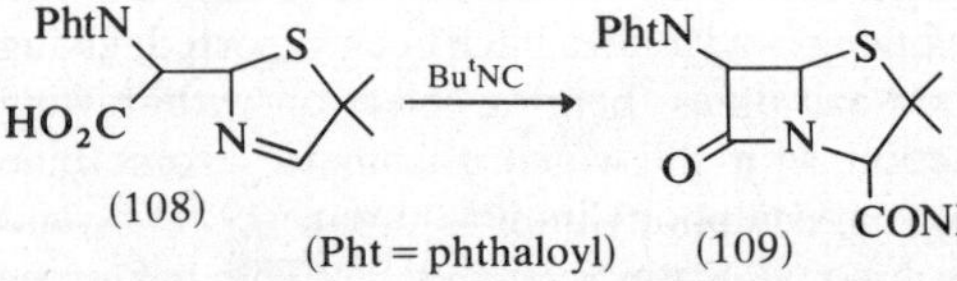

(108) (Pht = phthaloyl) (109)

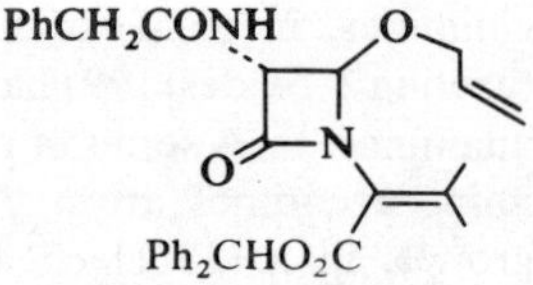

(110)

β-Lactams.—Routes to β-lactams attract attention because of their application to the synthesis of pharmacologically active materials. Base-catalysed cyclization of γ-bromopropionamides suffers from a competing elimination reaction, but conditions have been described which minimize the latter,[117] and an alternative solution to the same problem exploits the acidity of the N—H bond of *O*-acyl and *O*-alkyl hydroxamic acid derivatives. In this approach, cyclization (with base) gives substituted *N*-hydroxy-compounds, and an elegant extension of the method uses diethyl azodiformate and PPh_3 to cyclize hydroxamic derivatives (106) of commercially available *N*-protected L-serine. This affords the cyclic compound (107) with retention of chirality.[118] A further β-lactam-forming reaction converts the thiazoline (108) into the penicillin (109), using t-butyl isocyanide.[119]

The first direct introduction of a substituent at C-6 in cephalosporins and at C-5 in penicillins has been reported. The benzoyloxy-group enters the bridgehead position

[113] W. Adam, C.-C. Cheng, O. Cueto, I. Erden, and K. Zinner, *J. Amer. Chem. Soc.*, 1979, **101**, 4735.
[114] I. Saito, S. Matsugo, and T. Matsuura, *J. Amer. Chem. Soc.*, 1979, **101**, 4757, 7332.
[115] M. Ishikawa, T. Fuchikami, M. Kumada, T. Higuchi, and S. Miyamoto, *J. Amer. Chem. Soc.*, 1979, **101**, 1348.
[116] R. Appel, V. Barth, M. Halstenberg, G. Huttner, and J. von Seyerl, *Angew. Chem. Internat. Edn.*, 1979. **18**, 872.
[117] H. H. Wasserman, D. J. Hlasta, A. W. Tremper, and J. S. Wu, *Tetrahedron Letters*, 1979, 549.
[118] P. G. Mattingly, J. F. Kerwin, and M. J. Miller, *J. Amer. Chem. Soc.*, 1979, **101**, 3983.
[119] A. Schutz and I. Ugi, *J. Chem. Res.*, 1979, (*S*) 157, (*M*) 2064.

in the α-configuration if one uses t-butyl perbenzoate and catalytic cuprous chloride in refluxing benzene.[120] Stereoselective synthesis of 6β-alkyl-penicillanates is achieved by reduction of the known 6α-alkyl-6β-isocyano-derivative with tri-n-butyltin hydride,[121] and the stereocontrolled synthesis of 7α-methoxy-1-oxa-cephems has also been described. In this the methoxy-group is introduced at the 3α-position of the precursor (110), obtained from the appropriate 6-*epi*-penicillin.[122]

There has been a spate of reports on the preparation and properties of further examples in the penem series. The derivatives (111), unsubstituted at the 6-position, have dominated the scene, with syntheses being described that use clavulanic acid[123] and 6-aminopenicillanic acid[124,125] as precursors, or with structures such as (112)[126] and (113)[127] as intermediates. A penem sulphoxide is produced, on *S*-oxidation of a thia-analogue of clavulanic acid.[128] The acids (111) prove more stable than their 6-acylamino-substituted derivatives, and, in marked contrast to penicillanic and cephalosporanic acids (which are also unsubstituted on the β-lactam ring), they show antibiotic activity towards a broad spectrum of bacteria.[126]

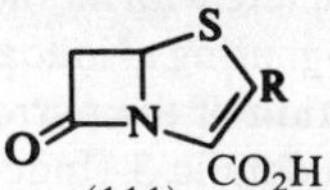

(111)

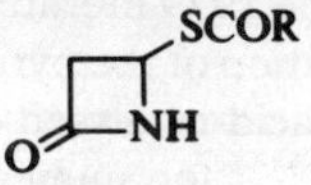

(112)

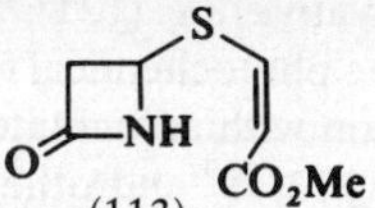

(113)

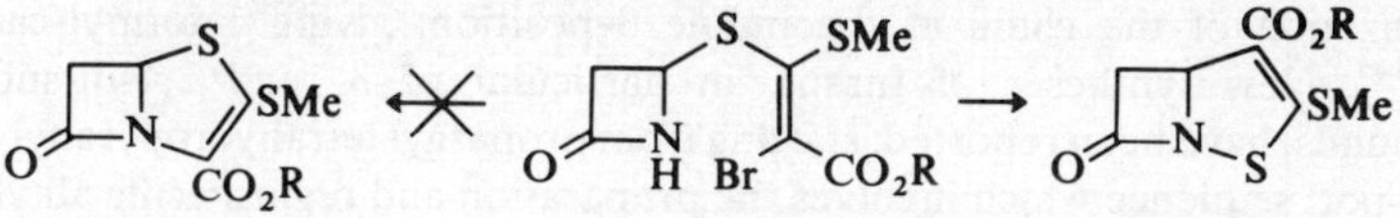

(115) (114) (116)

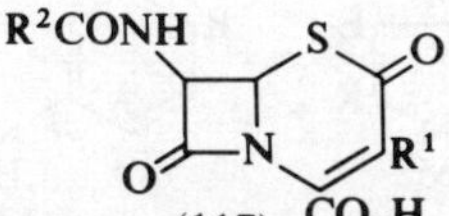

(117)

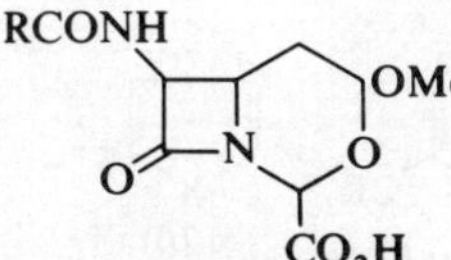

(118)

An unusual Cu^{I}-mediated addition process merits special attention. The reaction of (114) with base in the presence of CuI and PBr_3 was claimed to give the 2-alkylthio-penem (115),[129] but a report in the early 1980 literature has revised this structure in favour of the unexpected isopenem (116). Details of the preparation of authentic (115) are promised.[130] Other 'non-classical' β-lactam structures include

[120] H. Matsumura, T. Yano, M. Ueyama, K. Tori, and W. Nagato, *J.C.S. Chem. Comm.*, 1979, 485.
[121] D. I. John, E. J. Thomas, and N. D. Tyrrell, *J.C.S. Chem. Comm.*, 1979, 345.
[122] S. Uyeo, I. Kikkawa, Y. Hamashima, H. Ona, Y. Nishitani, K. Okada, T. Kubota, K. Ishikura, Y. Ide, K. Nakano, and W. Nagata, *J. Amer. Chem. Soc.*, 1979, **101**, 4403.
[123] P. C. Cherry, C. E. Newall, and N. S. Watson, *J.C.S. Chem. Comm.*, 1979, 663.
[124] C. M. D. Beels, M. S. Abu-Rabie, P. Murray-Rust, and J. Murray-Rust, *J.C.S. Chem. Comm.*, 1979, 665.
[125] I. Ernest, J. Gosteli, and R. B. Woodward, *J. Amer. Chem. Soc.*, 1979, **101**, 6301.
[126] M. Lang, K. Prasad, W. Holick, J. Gosteli, I. Ernest, and R. B. Woodward, *J. Amer. Chem. Soc.*, 1979, **101**, 6296.
[127] H. R. Pfaendler, J. Gosteli, and R. B. Woodward, *J. Amer. Chem. Soc.*, 1979, **101**, 6306.
[128] P. Lombardi, G. Franceschi, and F. Arcamone, *Tetrahedron Letters*, 1979, 3777.
[129] F. DiNinno, E. V. Linek, and B. G. Christensen, *J. Amer. Chem. Soc.*, 1979, **101**, 2210.
[130] S. Oida, A. Yoshida, T. Hayashi, E. Nakayama, S. Sato, and E. Ohki, *Tetrahedron Letters*, 1980, **21**, 619.

the 2-oxocephem system (117). Unfortunately, expectations that increased conjugation to the 2-oxo-group would increase the reactivity of the β-lactam carbonyl, thereby enhancing antibacterial activity, were not fully realized.[131,132] Compounds based on (118) represent the first examples of β-lactams fused to a saturated six-membered ring that have potent antibacterial activity.[133]

5 Five-membered Rings

Synthetic modifications of five-membered-ring heterocycles, using lithiation reactions, have been extended in scope by several groups. Lithium 3-furoate may be lithiated at the 2-position, to give promising intermediates.[134] 2-Furyl-,[135,136] 1-methyl-2-pyrrolyl-,[135] and 1-methyl-2-indolyl-lithium[137] react with a trialkylborane to form the (heterocyclyl)trialkylborate anions (119) (9-R-9-BBN will serve instead of R_3B[135,137]). The ions (119) may be decomposed (by I_2 or *N*-chlorosuccinimide) to give (120), or first be treated with an electrophilic alkylating agent (*e.g.* MeI, or an $\alpha\beta$-unsaturated ketone), then decomposed (by H_2O_2 and NaOH), to give a 2,3-dialkyl derivative, *e.g.* (121).[137] The 1979 literature is replete with indole syntheses, including the photochemical formation of the pyrrole ring, using *o*-iodoaniline in an $S_{RN}1$ reaction with an enolate,[138] acid-catalysed cyclization of the pyrrole (122) to the indole (123),[139] and others[140—144] too numerous to detail. 3-(Indol-2-yl)pyridinium salts, with base, undergo cleavage of the pyridine ring and subsequent re-cyclization of the chain at the indole 3-position, giving 1-formyl-carbazoles (124).[145] New syntheses of furans, in particular of 3- and 2,4-di-substituted compounds, have been reported, starting from propargyl tetrahydropyranyl ether, in a 'one-pot' sequence which involves the preparation and regiospecific alkylation of the carbanion (125).[146] A synthesis of 3-acyl-furans, using sulphur techniques, has been reported.[147]

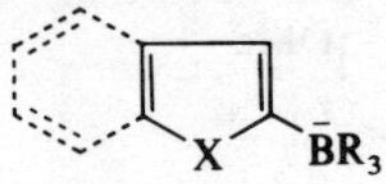

(119) X = O or NMe

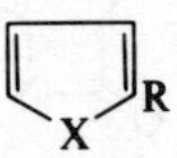

(120)

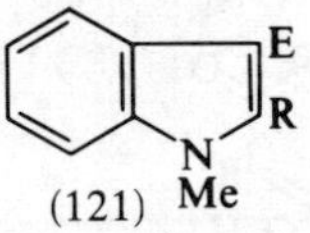

(121)

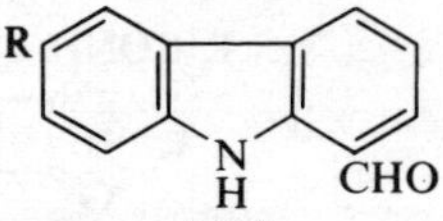

(124)

[131] C. U. Kim, P. F. Misco, and D. N. McGregor, *J. Medicin. Chem.*, 1979, **22**, 743.
[132] I. Ernest, *Helv. Chim. Acta*, 1979, **62**, 2681.
[133] J. G. Gleason, T. F. Buckley, K. G. Holden, D. B. Bryan, and P. Siler, *J.Amer. Chem. Soc.*, 1979, **101**, 4730.
[134] D. W. Knight, *Tetrahedron Letters*, 1979, 469.
[135] E. R. Marinelli and A. B. Levy, *Tetrahedron Letters*, 1979, 2313.
[136] I. Akimoto and A. Suzuki, *Synthesis*, 1979, 146.
[137] A. B. Levy, *Tetrahedron Letters*, 1979, 4021.
[138] R. Beugelmans and G. Roussi, *J.C.S. Chem. Comm.*, 1979, 950.
[139] M. Natsume and H. Muratake, *Tetrahedron Letters*, 1979, 3477.
[140] I. Fleming and M. Woolias, *J.C.S. Perkin I*, 1979, 829.
[141] M. Somei, F. Yamada, and C. Kaneko, *Chem. Letters*, 1979, 943.
[142] G. S. Ponticello and J. J. Baldwin, *J. Org. Chem.*, 1979, **44**, 4003; U. Hengartner, D. Valentine, K. K. Johnson, M. E. Larscheid, F. Pigott, F. Scheidl, J. W. Scott, R. C. Sun, J. M. Townsend, and T. H. Williams, *ibid.*, p. 3741.
[143] Y. Ito, K. Kobayashi, N. Seko, and T. Saegusa, *Chem. Letters*, 1979, 1273.
[144] R. Beugelmans, H. Ginsburg, M. Le Goff, A. Lecas, J. Pusset, and G. Roussi, *Heterocycles*, 1979, **12**, 811.
[145] A. N. Kost, T. V. Stupnikova, R. S. Sagitullin, B. P. Zemskii, and A. K. Sheinkman, *Doklady Akad. Nauk S.S.S.R*, 1979, **244**, 103, [*Proc. Acad. Sci.* (*U.S.S.R.*), 1979, **244**, 14].
[146] M. Stähle and M. Schlosser, *Angew. Chem. Internat. Edn.*, 1979, **18**, 875.
[147] K. Inomata, M. Sumita, and H. Kotake, *Chem. Letters*, 1979, 709.

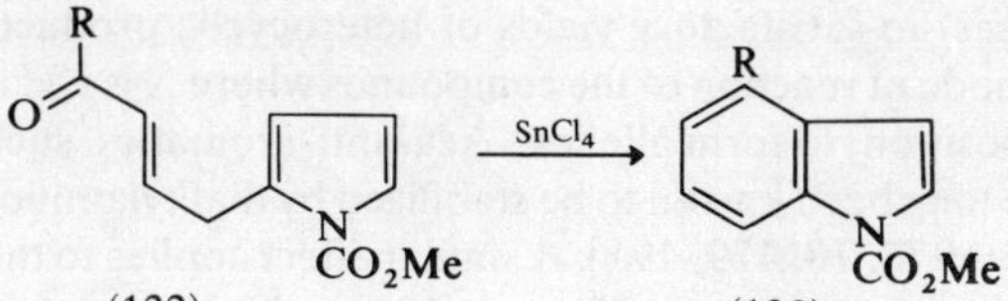

(122) (123)

$LiOCH_2CH{-}\bar{C}H{-}CH{-}OTHP$
Li^+

(125)

Work in several laboratories has been directed towards the synthesis of thiophen-3-acetic and -3-malonic acids, one approach to which involves the thiophenium *S*-ylides [*cf. Ann. Reports* (*B*), 1978, **75**, 254]. The dichloro-compound (126) thermally cyclizes, with loss of HCl, to the thienofuran (127),[148] and with copper catalysis it liberates dicarbomethoxycarbene, which can be trapped.[149] A full paper[150] describes a wide range of these ylides, and also the preparation of the ester (128) and its rearrangement to (129). Another approach to thiophen-3-acetic esters involves cyclization of the dihalogeno-diene (130).[151]

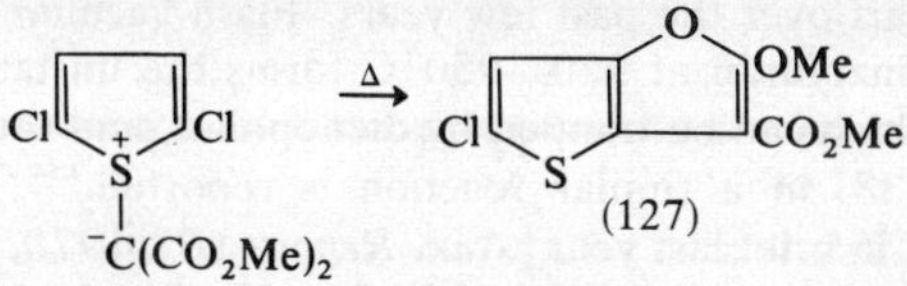

(126) (127)

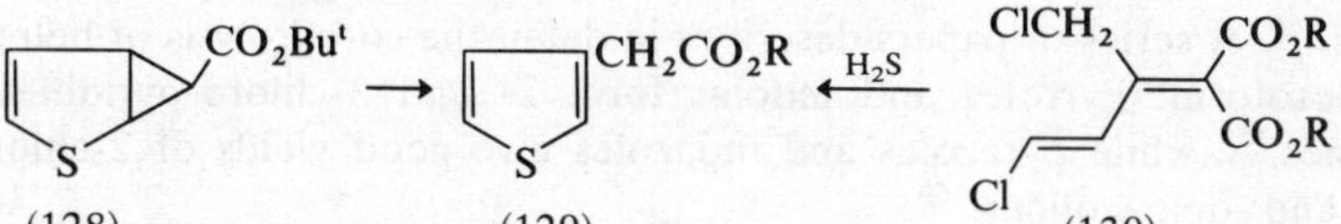

(128) (129) (130)

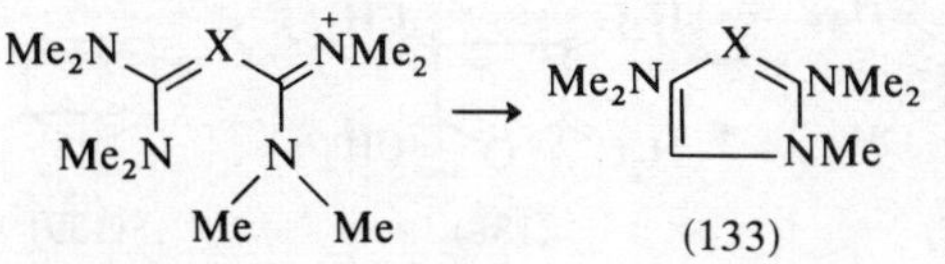

(X = CH or N)

(131) (133)

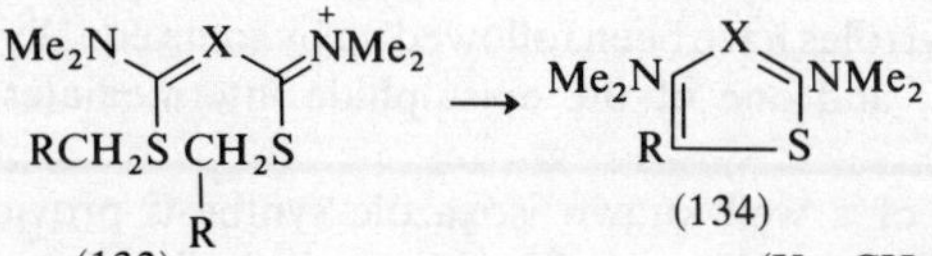

(132) (134)

(X = CH or N; R = H or Me)

In an extremely compressed pair of papers, Gompper and Schneider[152] have described the cyclization of ylides generated by the action of strong base on cations of type (131) and (132). The reaction proceeds with elimination of a molecule of amine

[148] R. J. Gillespie, J. Murray-Rust, P. Murray-Rust, and A. E. A. Porter, *J.C.S. Chem. Comm.*, 1979, 366.
[149] R. J. Gillespie and A. E. A. Porter, *J.C.S. Chem. Comm.*, 1979, 50.
[150] R. J. Gillespie and A. E. A. Porter, *J.C.S. Perkin I*, 1979, 2624.
[151] P. J. Clayton, A. W. Guest, A. W. Taylor, and R. Ramage, *J.C.S. Chem. Comm.*, 1979, 500.
[152] R. Gompper and C. S. Schneider, *Synthesis*, 1979, 213, 215.

or thiol, and results (in some cases) in satisfactory yields of heterocyclic products (133) and (134). An alternative mode of reaction of the compounds where X is CH is deprotonation from this central position, to form allenes. Aza-anti-aromatics, such as azete and aza-pentalenes, have long been known to be stabilized by dialkylamino-substitution [*cf. Ann. Reports* (*B*), 1973, **70**, 479, 491]. A similar effect applies to the highly coloured aza-cyclopentadienium cations (135) and (136), which are made from tetrachloro-2*H*-imidazole and 1,1,3-trichloro-1*H*-isoindole, respectively.[153]

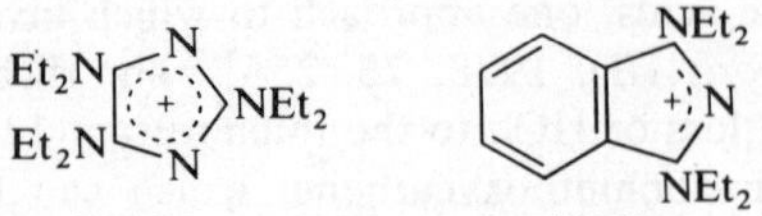

(135) (136)

Techniques for controlling high-temperature reactions, which avoid the conversion of the majority of the organic material into unpromising charcoals, have been developed to a fine art over the past few years. Flash vacuum pyrolysis (FVP) of 4,5,6,7-tetrahydrobenzofuran at 920—950 °C forms the unstable dimethylenedihydrofuran (137), which can be trapped by dienophiles, and the generation of the 'furanoradialene' (138) in a similar reaction is reported.[154] Pyrolysis of isoxazolinones, reported in brief last year [*Ann. Reports* (*B*), 1978, **75**, 258], has been found to produce formonitrile oxide (HCNO) when applied to 4-isonitroso-5-isoxazolinones.[155] Other nitrile oxides are obtained in good yields by FVP of furoxans.[156] A series of papers describes in detail the co-pyrolysis of heterocycles with chloroform. Pyrroles and indoles form 2- and 3-chloro-pyridines[157] and -quinolines,[158] while pyrazoles and indazoles give good yields of 2-chloro-pyrimidines and -quinazolines.[159]

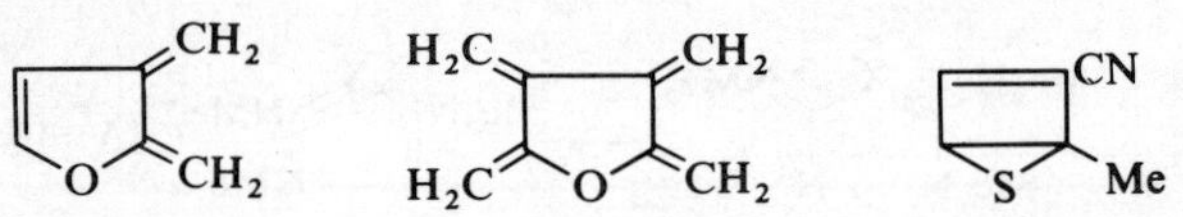

(137) (138) (139)

The reports of recent years on the 'nitrogen-walk' process of photo-rearrangement of cyano-pyrroles have been followed up by some equally elegant studies on the cyano-thiophens, and one of the episulphide intermediates (139) has been isolated.[160]

An extension of a well-known isoxazole synthesis provides a useful route to 1,2,4-oxadiazoles (140) from acyl-amidines, including the interesting and rather

[153] R. Gompper and K. Bichlmayer, *Angew. Chem. Internat. Edn.*, 1979, **18**, 156.
[154] J. Jullien, J. M. Pechine, F. Perez, and J. J. Piade, *Tetrahedron Letters*, (*a*) 1979, 3079; (*b*) 1980, **21**, 611.
[155] C. Wentrup, *Angew. Chem. Internat. Edn.*, 1979, **18**, 467.
[156] W. R. Mitchell and R. M. Paton, *Tetrahedron Letters*, 1979, 2443.
[157] R. E. Busby, M. Iqbal, M. A. Khan, J. Parrick, and C. J. G. Shaw, *J.C.S. Perkin I*, 1979, 1958.
[158] R. E. Busby, S. M. Hussain, M. Iqbal, M. A. Khan, J. Parrick, and C. J. G. Shaw, *J.C.S. Perkin I*, 1979, 2782.
[159] R. E. Busby, J. Parrick, S. M. H. Rizvi, and C. J. G. Shaw, *J.C.S. Perkin I*, 1979, 2786.
[160] J. A. Barltrop, A. C. Day, and E. Irving, *J.C.S. Chem. Comm.*, 1979, 881, 966.

unstable cases where R = H. Replacing the hydroxylamine in Scheme 3 by a hydrazine leads to 1,2,4-triazoles.[161] More details have been published of a novel synthesis of imidazoles from *N*-chloro-amidines, and a further study of the chemical behaviour of the imidazoline intermediates has been made.[162] 4-Methylene-4,5-dihydro-oxazoles (141), intermediates in a synthesis of oxazoles from acetylenic imino-ethers, have been isolated, and can be transformed by electrophilic reagents into oxazoles with a 4-(substituted methyl) group.[163]

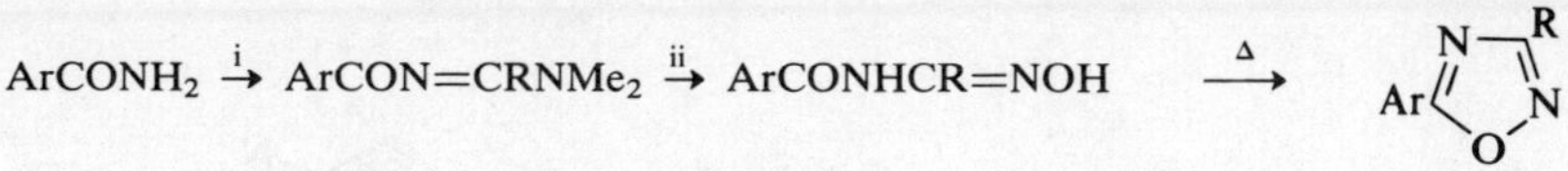

Reagents: i, $(MeO)_2CRNMe_2$; ii, NH_2OH,AcOH (140)

Scheme 3

Although the desire to patent it seems to be slackening, novel aspects of the evergreen Beirut reaction are still uncovered. Unsubstituted quinoxaline di-*N*-oxide is from time to time reported as being produced by the diethylamine-catalysed reaction of benzofuroxan with a variety of unlikely compounds; now the Lebanese groups have shown that this is produced, amongst other things, simply from the amine and the benzofuroxan.[164] With 1-diethylaminobutadiene the enamine (142) is formed.[165]

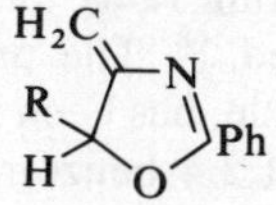

(141)

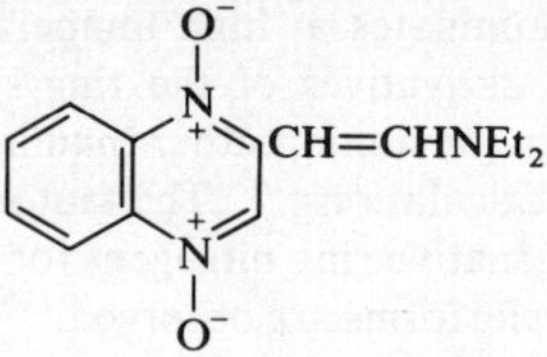

(142)

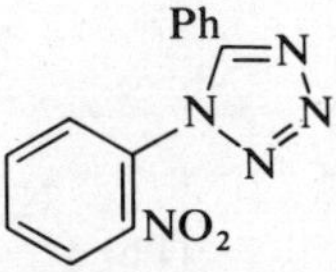

(143)

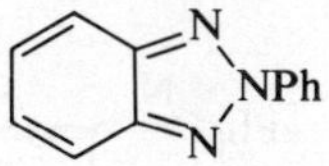

(144)

An interesting synthesis of 2-aryl-benzotriazoles has been described. Starting from precursors such as (143), it is suggested that *o*-nitrophenyl-carbodi-imides are formed. Subsequent rearrangements involving the nitro-group result in the loss of CO_2, with formation of the benzotriazole (144).[166] A nitro-group is again involved, in the base-induced cyclization of *N*-phenacyl-*o*-nitro-*N*-tosylaniline (145). The

161 Y.-I. Lin, S. A. Lang, M. F. Lovell, and N. A. Perkinson, *J. Org. Chem.*, 1979, **44**, 4160.
162 L. Citerio, D. Pocar, R. Stradi, and B. Gioia, *J.C.S. Perkin I*, 1978, 309; L. Citerio, D. Pocar, M. L. Saccarello, and R. Stradi, *Tetrahedron*, 1979, **35**, 2375, 2453.
163 L. E. Overman, S. Tsuboi, and S. Angle, *J. Org. Chem.*, 1979, **44**, 2323.
164 M. Z. Nazer, C. H. Issidorides, and M. J. Haddadin, *Tetrahedron*, 1979, **35**, 681.
165 P. Devi, J. S. Sandhu, and G. Thyagarajan, *J.C.S. Chem. Comm.*, 1979, 710.
166 P. G. Houghton, D. F. Pipe, and C. W. Rees, *J.C.S. Chem. Comm.*, 1979, 771.

benzoyl and tosyl fragments are lost, and 2-ethoxy-1-hydroxybenzimidazole is formed (Scheme 4).[167]

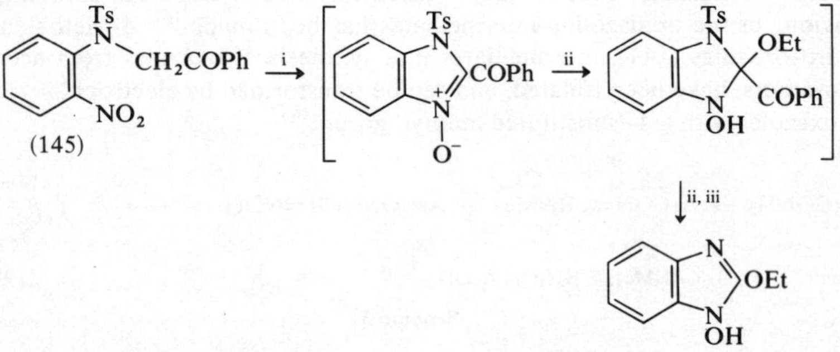

Reagents: i, NaOEt, EtOH; ii, EtO^-; iii, H^+

Scheme 4

A number of complex azide–tetrazole equilibria have been studied. 3-Azido-5-phenyl-*s*-triazole (146) is favoured over the cyclized triazolotetrazole forms (147) and (148), but the (slow) equilibrium of the anion is more finely balanced: in DMSO the cyclized ion (149) is preferred at low temperatures, while the azide [anion of (146)] predominates at high temperatures. An interesting range of isomeric *N*-methylated derivatives of the ring system was prepared.[168] The preference for cyclization in the anion, rather than in the neutral molecule, has been supported by theoretical calculations.[169] The tautomerism of 3-azido-1,2,4-benzotriazine, which has two alternative ring nitrogens for cyclization, has been studied: the open-chain and both cyclic forms are observed.[170] The equilibrium between tetrazole and azide in the system (150) ⇌ (151) shows a linear dependence between the Hammett substituent constant for R, *i.e.* σ, and the equilibrium constant, K (rather than

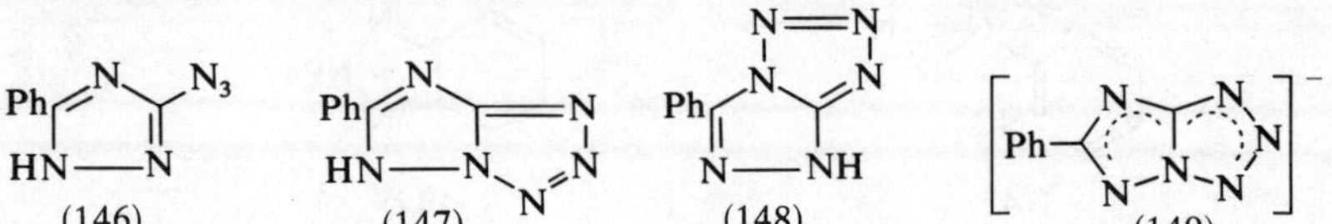

(146) (147) (148) (149)

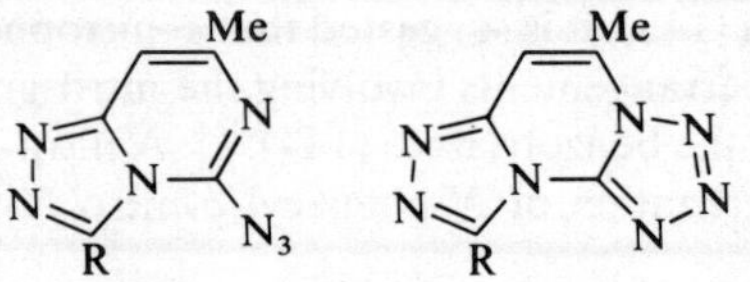

(150) (151)

[167] J. Machin and D. M. Smith, *J.C.S. Perkin I*, 1979, 1371.
[168] R. N. Butler, T. McEvoy, E. Alcalde, R. M. Claramunt, and J. Elguero, *J.C.S. Perkin I*, 1979, 2886.
[169] S. Olivella and J. Vilarrasa, *J. Heterocyclic Chem.*, 1979, **16**, 685.
[170] A. Messmer, G. Hajos, J. Tamas, and A. Neszmelyi, *J. Org. Chem.*, 1979, **44**, 1823.

log *K*).[171] This is a curious result, and it would be interesting to see whether it is found in other systems, or with a wider range of substituents. The structures (152) of the reaction products from *S*-alkylated thioamides (153) and azide ion have been corrected: the acyclic triaza-dienes (154) are formed,[172] as was predicted by Holm.[173]

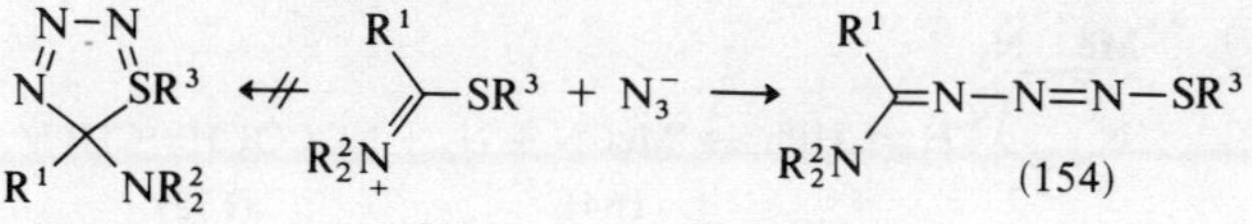

(152) (153) (154)

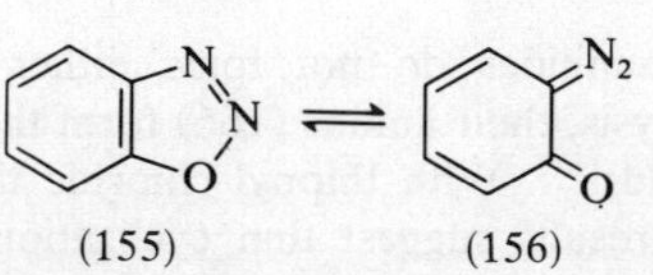

(155) (156)

Variable-temperature photoelectron spectroscopy has been referred to earlier (Section 2). The claim[174] to have observed 1,2,3-benzoxadiazole (155) as the major component of the gas-phase equilibrium with the diazo-phenol (156) by this method is particularly interesting, and confirmation by other physicochemical techniques would be very welcome.

Carbon atoms, formed in the decomposition of diazotetrazole (157), insert into furan with ring-cleavage, to give the acetylenic aldehyde (158). The fate of the carbon was followed by labelling experiments; both C=C addition and C-2—H insertion seem to occur.[175] Ring-cleavage is a common fate of five-membered heterocyclic rings adjacent to which a carbene or nitrene centre is generated, and the literature of 1979 contains a number of examples of this. Thermolysis of the oxazolyldiazomethane (159) gives an acyl-imide (160) [with an obvious structural relationship to (158)], which readily rearranges to the acyl-enamine (161). Under

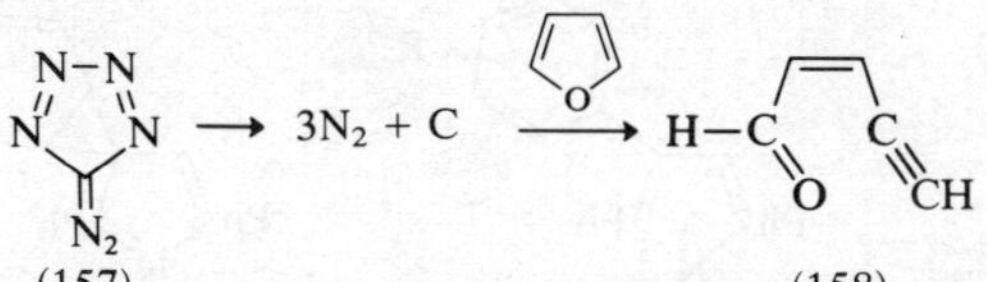

(157) (158)

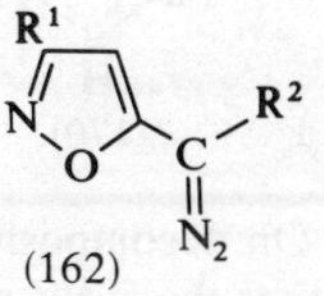

(162)

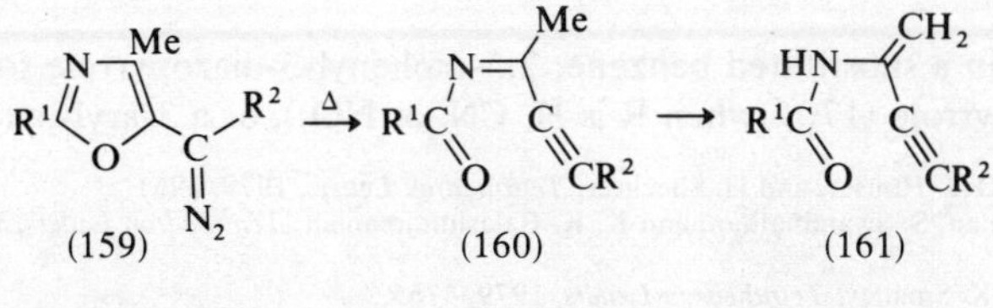

(159) (160) (161)

[171] A. Könnecke, R. Dörre, E. Kleinpeter, and E. Lippmann, *Tetrahedron*, 1979, **35**, 1957.
[172] G. L'abbé, A. Willocx, J.-P. Declercq, G. Germain, and M. Van Meerssche, *Bull. Soc. chim. belges*, 1979, **88**, 107.
[173] A. Holm, *Adv. Heterocyclic Chem.*, 1976, **20**, 173.
[174] R. Schulz and A. Schweig, *Angew. Chem. Internat. Edn.*, 1979, **18**, 692.
[175] S. F. Dyer and P. B. Shevlin, *J. Amer. Chem. Soc.*, 1979, **101**, 1303.

more vigorous conditions fragmentation occurs, to the nitrile R^1CN and an acetylenic ketone, and similar products are formed from the isoxazolyldiazomethane (162). For these reactions, intermediate dehydro-oxazines are suggested.[176] The 4-azido-isoxazole (163), on heating, fragments to give the two cyanides (164) and (165), and nitrogen.[177]

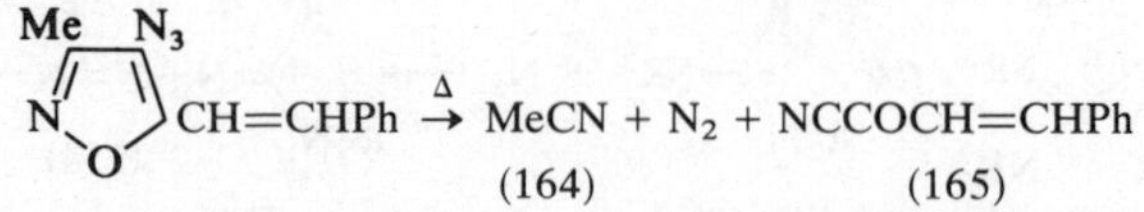

(164) (165)

(163)

Although *o*-azido-benzanilides do not form either arylamino-anthranils or indazolinones on thermolysis, their anions (166) form the 2-aryl-indazolin-3-ones (167) in fair to good yields.[178] With thionyl chloride they give 3-chloro-2-aryl-indazoles.[179] These two results suggest that cyclization requires a lone pair of electrons in the molecular plane, on the nitrogen atom which cyclizes to the azide. However, anions of *o*-azidoacetophenones (168), apparently isoelectronic with the neutral amides, do form the C—N bond, to give the indoxyls (169).[180] Whatever the mechanisms may be, the reactions are useful ones.

Photochemical reactions of anthranils,[181,182] indoxazenes,[182] indazoles,[183] and 2,1-benzisothiazoles[184] have been reported extensively. Ring-cleavage products are generally observed, often with concomitant nucleophilic substitution in the benzene ring.

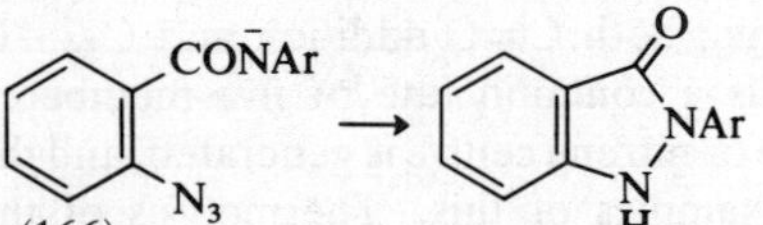

(166) (167)

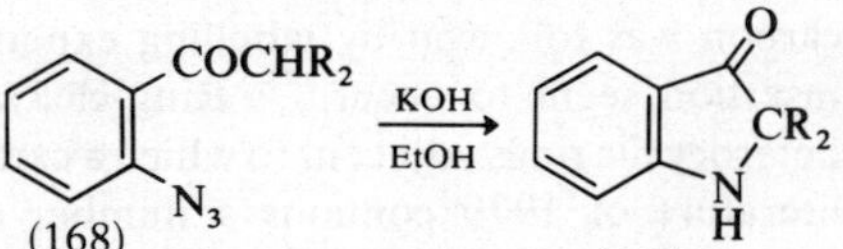

(168) (169)

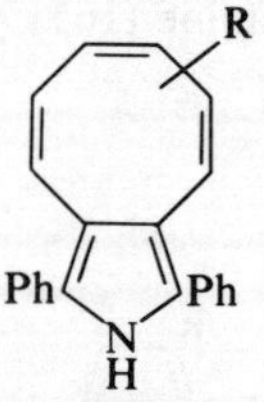

(170)

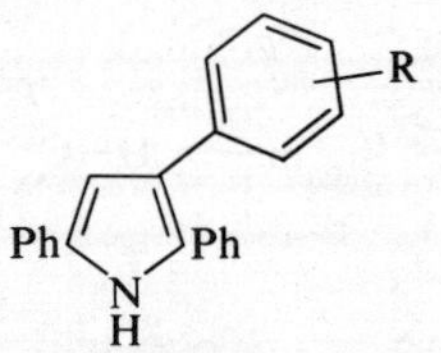

(171)

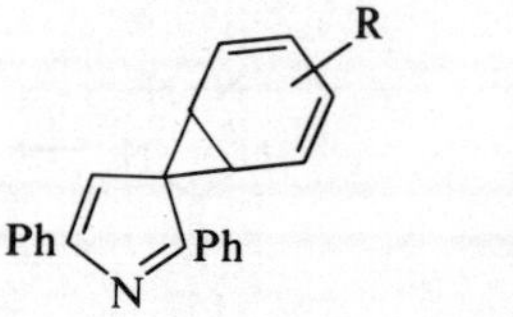

(172)

On decomposition in a substituted benzene, 2,5-diphenyl-3-diazopyrrole forms either the cyclo-octapyrrole (170) (when R is H, CN, or NO_2), or a 3-aryl-pyrrole

[176] S.-I. Hayashi, M. Nair, D. J. Houser, and H. Shechter, *Tetrahedron Letters*, 1979, 2961.
[177] G. Kumar, K. Rajagopalan, S. Swaminathan, and K. K. Balasubramanian, *Tetrahedron Letters*, 1979, 4685.
[178] M. A. Ardakani and R. K. Smalley, *Tetrahedron Letters*, 1979, 4765.
[179] M. A. Ardakani, R. K. Smalley, and R. H. Smith, *Synthesis*, 1979, 308.
[180] M. A. Ardakani and R. K. Smalley, *Tetrahedron Letters*, 1979, 4769.
[181] E. Giovannini and B. F. S. E. de Sousa, *Helv. Chim. Acta*, 1979, **62**, 185, 198.
[182] T. Doppler, H. Schmid, and H.-J. Hansen, *Helv. Chim. Acta*, 1979, **62**, 271, 304, 314.
[183] E. Georgarakis, H. Schmid, and H.-J. Hansen, *Helv. Chim. Acta*, 1979, **62**, 234.
[184] B. Jackson, H. Schmid, and H.-J. Hansen, *Helv. Chim. Acta*, 1979, **62**, 391.

(171) (when R is Me or OMe). An intermediate spiro-norcaradiene (172) is suggested, for the formation of both products.[185]

Cycloaddition reactions of diazo-pyrazoles have been studied widely, and recent work in this area has included the reactions of 3-diazo-pyrazoles and -indazoles and of 4-diazo-1,2,3-triazoles with various ylides[186a] and with isocyanates,[186b] to form a wide variety of condensed heterocyclic systems, usually in good yield.

1-Chlorobenzotriazole has been reported to be a shelf-stable oxidizing agent, but the unfused analogue (173) decomposes very easily. The first step is probably migration of Cl, to form the 4*H*-isomer (174), which can either lose nitrogen to form the chloro-azirene (175), or cleave to benzonitrile and phenylchlorodiazomethane; the observed products are derived from these.[187]

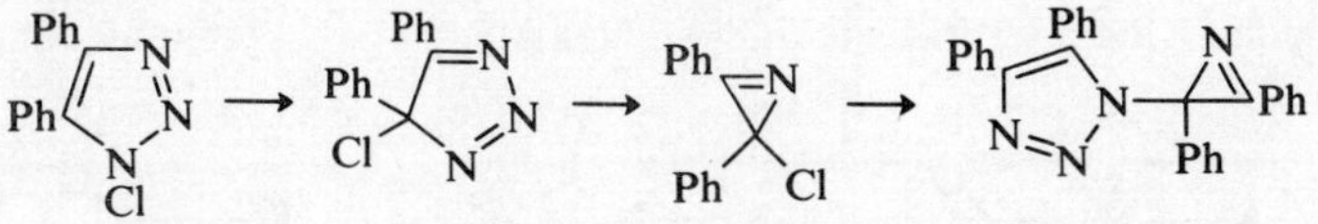

(173) (174) (175)

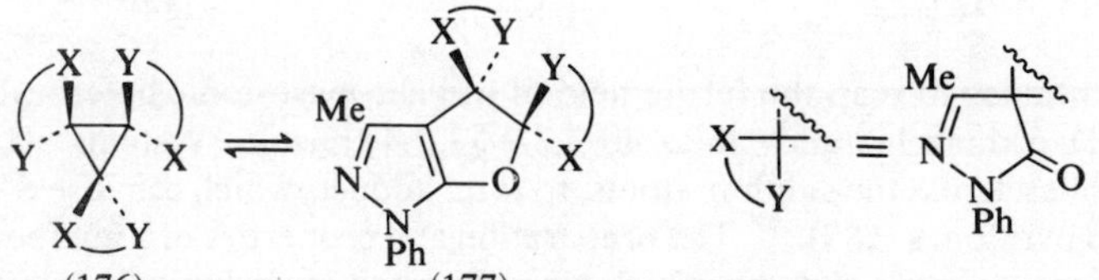

(176) (177)

The spiro-pyrazolinone (176) has been prepared, and in solution it shows an interesting rapid temperature-dependent equilibrium with the isomer (177).[188]

A synthesis of 4-amino-isothiazoles (178) from α-tosyloximino-nitriles has been reported (Scheme 5), and the products were shown to be flexible synthons for a variety of ring systems containing the fused isothiazole nucleus.[189]

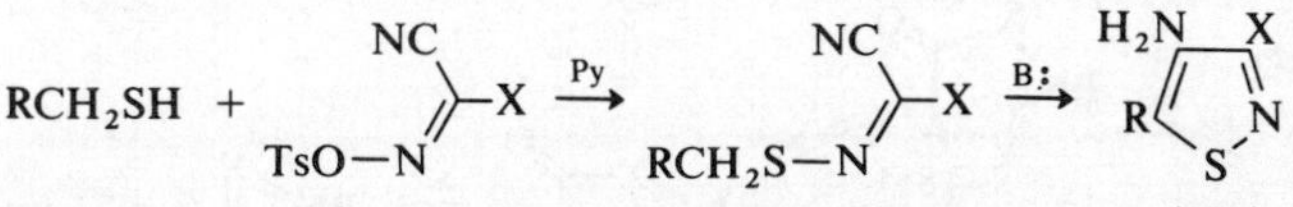

(178)

Scheme 5

Nucleophilic attack at the sulphur atom of heteroaromatic rings is nothing new, but novel aspects continue to be discovered. Isothiazolium and 1,2,5-thiadiazolium

[185] M. Nagarajan and H. Shechter, *J. Amer. Chem. Soc.*, 1979, **101**, 2198.
[186] G. Ege and K. Gilbert, *Tetrahedron Letters*, 1979, (*a*) 1567; (*b*) 4253.
[187] T. C. Gallagher, M. J. Sasse, and R. C. Storr, *J.C.S. Chem. Comm.*, 1979, 419.
[188] G. Mann, L. Hennig, H. Wilde, S. Hauptmann, S. Behrendt, and M. Kretschmer, *Tetrahedron Letters*, 1979, 4645.
[189] K. Gewald and P. Bellmann, *Annalen*, 1979, 1534.

salts (179) incorporate cyanide and methylpropiolate anions with ring expansion: the product (180; X = CH, Z = N) slowly decomposes to the starting materials.[190] The selenophen (181) is ring-cleaved by organolithium reagents.[191]

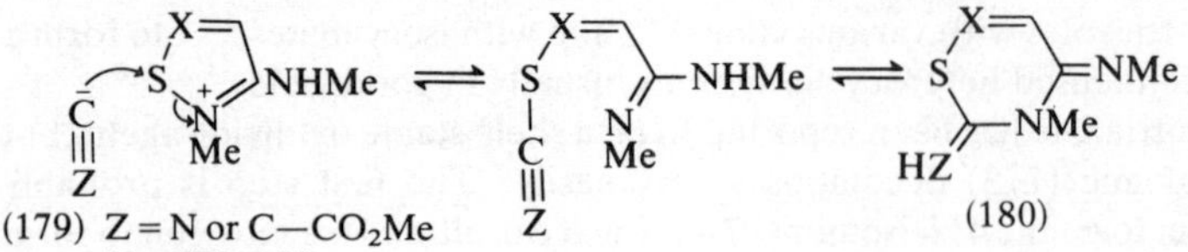

(179) Z = N or C—CO_2Me
X = CH or N

(180)

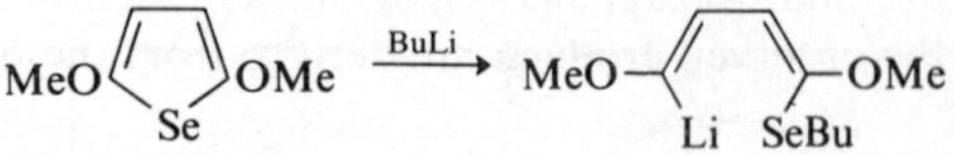

(181)

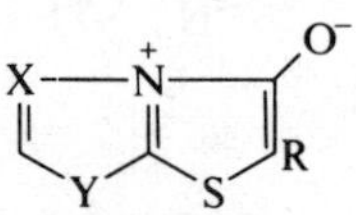

(182)

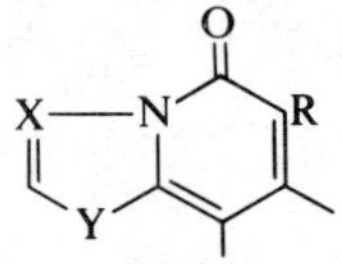

(183)

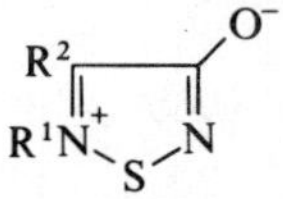

(184)

Potts continues to reap the fertile field of bicyclic meso-ionic heterocycles. The imidazo[2,1-*b*]thiazole and thiazolo[3,2-*b*][1,2,4]triazole systems (182) add dipolarophiles across the sulphur atoms, to form adducts which can lose S or H_2S to form fused pyridones (183).[192] The preparation and properties of a number of novel monocyclic meso-ionic systems, which were reported in preliminary notes in 1976 [*Ann. Reports* (*B*), 1976, **73**, 252], are published in full detail in an extensive series of papers from the Sheffield group.[193] Meso-ionic derivatives (184) of the 1,2,5-thiadiazole system have been prepared.[194]

More work on the photochemistry of meso-ionic monocycles (185) has been reported. Two important types of reaction which have been postulated are the ring-opening [to (186)] and the further cyclization [to (187)], with decomposition *via* an anti-aromatic intermediate (188) (Scheme 6). In the case of the meso-ionic

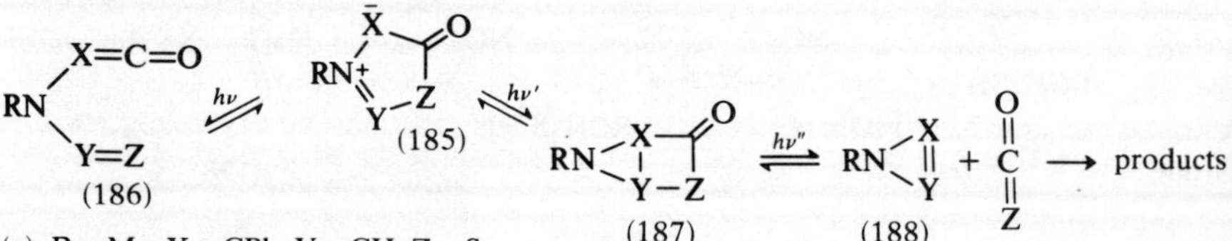

(a) R = Me, X = CPh, Y = CH, Z = S
(b) R = Ph, X = Y = N, Z = O
(c) R = Ar, X = CR′, Y = N, Z = O

Scheme 6

[190] J. Rokach, P. Hamel, Y. Girard, and G. Reader, *Tetrahedron Letters*, 1979, 1281.
[191] S. Gronowitz, A. Hallberg, and T. Frejd, *Tetrahedron*, 1979, **35**, 2607.
[192] K. T. Potts and S. Kanemasa, *J. Org. Chem.*, 1979, **44**, 3803, 3808.
[193] E. Cawkill, R. N. Hanley, G. P. Rowson, I. S. Smith, W. D. Ollis, and C. A. Ramsden, *J.C.S. Perkin I*, 1979, 724, and five succeeding papers.
[194] K. Masuda, J. Adachi, and K. Nomura, *J.C.S. Chem. Comm.*, 1979, 331.

thiazoles (185a), the heterocumulene (186a) can be trapped, but it has not yet been detected spectroscopically.[195] The oxatriazole (185b), on photolysis, forms phenyl azide and also phenyl isocyanate. The azide was at one time thought to arise through the 'cyclic azide' (188b), but Danish authors[196] have shown that such a symmetrical intermediate cannot be on the main pathway to the azide, by irradiating the labelled compound (185b; $Y = {}^{15}N$): the phenyl azide was found to carry the marker practically exclusively on the terminal nitrogen. The formation of the isocyanate also suggests that the phenyl group, with or without its attached nitrogen, moves closer to the ring carbon atom before cleavage.[196] The cyclization–fragmentation (lower route of Scheme 6) is also found in (185a), as shown by the production of carbonyl sulphide: the authors apparently accept the intermediate (188a) here.[195] The thiazole (185a) reacts with singlet oxygen to form (189), presumably *via* the intermediate (190).[195] Sydnones (185c), which hitherto have been reported only to form charge-transfer complexes with tetracyanoethylene, react with it under more forcing conditions, with loss of CO_2, giving the open-chain hydrazones (191).[197]

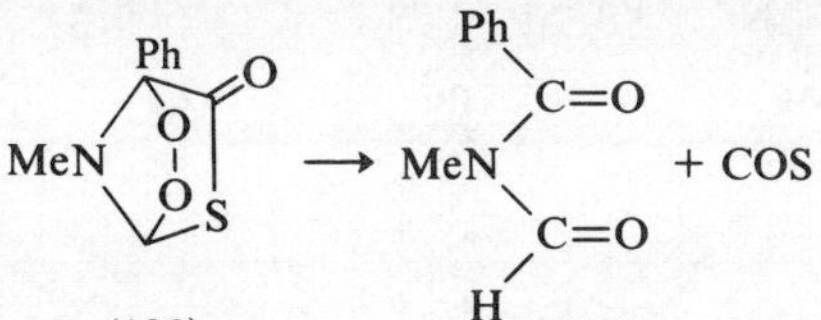

(190) (189)

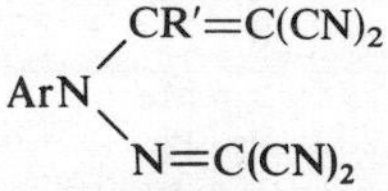

(191)

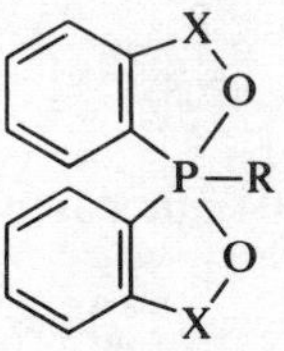

(192) $X = CO, R = OH$
(193) $X = CMe_2, R = OH$
(194) $X = CMe_2, R = H$

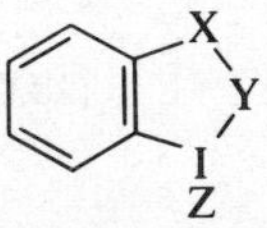

(195) $X = CO, Y = NCH_2CH_2Cl, Z = Cl$
(196) $X = CMe_2, Y = O, Z = Br$
(197) $X = CMe_2, Y = O, Z = OR_F$

Martin, Granoth, and co-workers have continued their studies on spiro-phosphoranes, and three notes record the synthesis of (192),[198] the pK_{R^+} of (193),[199] and the spectral effects of deprotonation of (194)[200] [*cf. Ann. Reports* (*B*), 1978, **75**, 264]. The same group have also turned their attention to hypervalent iodine compounds: an *X*-ray structure determination of a benziodazole (195) has been made,[201] and the *I*-bromobenziodoxole (196) was prepared.[202] The iodoxole (197) exchanges its perfluoroalkoxy-group very rapidly with a perfluoroalkoxide anion. In these compounds the iodine has approximately trigonal-bipyramidal geometry, with two equatorial electron pairs; the exchange intermediate probably has a square-planar arrangement of groups about the iodine atom.[202]

[195] N. H. Toubro, B. Hansen, N. Harrit, A. Holm, and K. T. Potts, *Tetrahedron*, 1979, **35**, 229.
[196] C. Bjerre, C. Christophersen, B. Hansen, N. Harrit, F. M. Nicolaisen, and A. Holm, *Tetrahedron*, 1979, **35**, 409.
[197] H. C. Berk and J. E. Franz, *J. Org. Chem.*, 1979, **44**, 2395.
[198] Y. Segall and I. Granoth, *J. Amer. Chem. Soc.*, 1979, **101**, 3687.
[199] I. Granoth and J. C. Martin, *J. Amer. Chem. Soc.*, 1979, **101**, 4618.
[200] I. Granoth and J. C. Martin, *J. Amer. Chem. Soc.*, 1979, **101**, 4623.
[201] T. M. Balthazor, D. E. Godar, and B. R. Stults, *J. Org. Chem.*, 1979, **44**, 1447.
[202] R. L. Amey and J. C. Martin, *J. Org. Chem.*, 1979, **44**, 1779.

Dico-ordinate phosphorus has also attracted much attention recently. Ketonic hydrazones have been known for some years to form diazaphospholes; *e.g.*, (198) from acetone methylhydrazone and PCl_3.[203] Now it has been found that the isomeric system (199) is also produced in this reaction. It is not known just how this happens, and the authors were not able to cause the isomers to interconvert.[204] Russian workers have shown that the diazaphosphole (200) reacts with diphenyl-diazomethane at room temperature in ether to give the fused phosphiran (201), which ring-opens in methanol to form the six-membered ring (202). These results were confirmed by *X*-ray crystallography.[205] The benzodiazaphospholes (203), prepared as indicated, exist as monomer, dimer, and trimer in solution, the proportion of monomer being *ca.* 30% in xylene at 140 °C and 100% in nitrobenzene. With BF_3 two species were observed, which the authors suggest are the adducts of the monomer in which one or the other nitrogen atom is co-ordinated to the boron.[206]

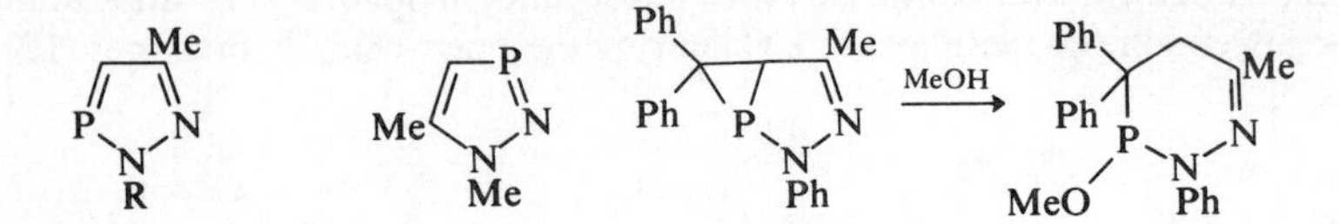

(198) R = Me
(200) R = Ph
(199)
(201)
(202)

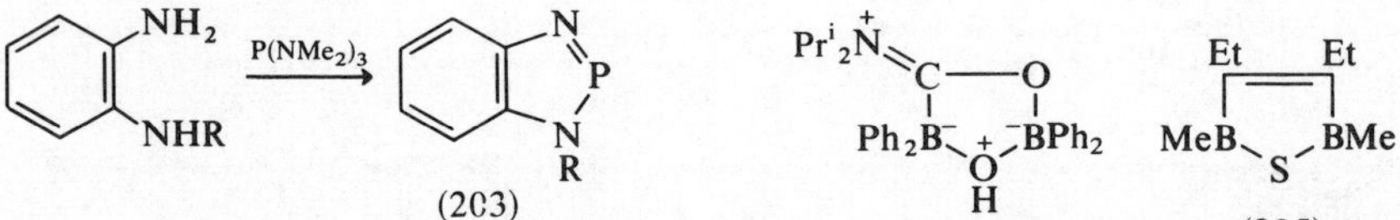

(203)
(204)
(205)

A novel compound containing boron is the very curious species (204), prepared from diphenylbromoborane and di-isopropylcarbamoyl-lithium.[207]

Metal sandwich compounds are becoming more of a mouthful every year.[208] Four molecules of the thiadiborolen (205) form the bread of a black centrosymmetric (by *X*-ray) 'tetra-decker' sandwich complex, containing two atoms of cobalt and one of iron between the rings.[209]

6 Six-membered Rings

A number of significant advances during the year have added to the heterocyclic chemist's already well-stocked armoury. 2-Chloro-3-substituted quinolines, and hence the 3-substituted quinolines (by reduction), are easily prepared from acyl-anilides under Vilsmeier formylation conditions. Quinoxalines are produced in the

[203] J. Luber and A. Schmidpeter, *Angew. Chem. Internat. Edn.*, 1976, **15**, 111.
[204] J. H. Weinmaier, J. Luber, A. Schmidpeter, and S. Pohl, *Angew. Chem. Internat. Edn.*, 1979, **18**, 412.
[205] B. A. Arbuzov, E. N. Dianova, and Yu. Yu. Samitov, *Doklady Akad. Nauk S.S.S.R.*, 1979, **244**, 1117 [*Proc. Acad. Sci.* (*U.S.S.R.*), 1979, **244**, 59].
[206] C. Malavaud, M. T. Boisdon, Y. Charbonnel, and J. Barrans, *Tetrahedron Letters*, 1979, 447.
[207] A. S. Fletcher, W. E. Paget, K. Smith, K. Swaminathan, J. H. Beynon, R. P. Morgan, M. Bozorgzadeh, and M. J. Haley, *J.C.S. Chem. Comm.*, 1979, 347.
[208] *cf.* W. Shakespeare, 'As You Like It', Act III, Scene 2, line 239.
[209] W. Siebert, W. Rothermel, C. Böhle, C. Krüger, and D. J. Brauer, *Angew. Chem. Internat. Edn.*, 1979, **18**, 949.

corresponding reaction using dimethylnitrosamine instead of the formamide.[210] Quinoline derivatives are also accessible by photochemical cyclization of thioamides (206); the synthesis is adaptable to a range of 2- and 3-substituents.[211] A general method for *c*-fused pyridines uses the thermal decomposition of appropriate *ortho*-methyl vinyl azides, *e.g.* (207) → (208),[212] and a synthesis of pyrimidin-4-ones has been described in which diphenylcyclopropenone reacts with amidoximes.[213] The versatility of β-imino-enamines as heterocyclic precursors is demonstrated by their reactions with ethyl chloroformate, carbon disulphide, aldehydes, and ketones, to give pyrimidin-2-ones and pyrimidine-2-thiones[214] and 1,2-dihydropyrimidines,[215] while with thionyl chloride they produce the thiadiazine oxides (209) at low temperatures; these lose SO to form the pyrazoles at high temperatures.[216] Two rhodium-catalysed reactions convert aniline into quinaldine; using ethylene under pressure,[217] and with acetaldehyde in the presence of nitrobenzene.[218]

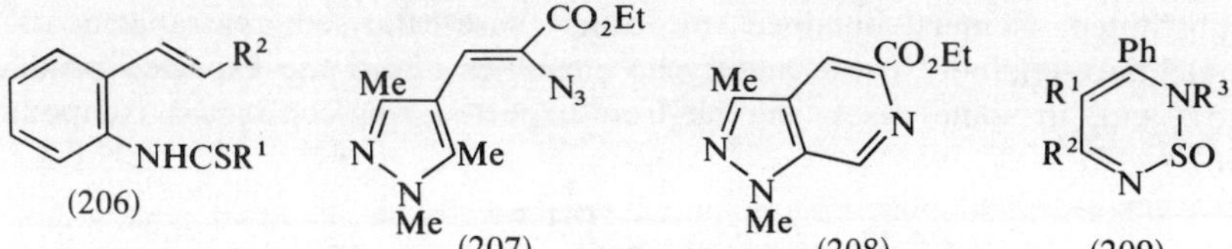

(206) (207) (208) (209)

A new method for the introduction of substituents at the 4-position of pyridine uses the sequence of Scheme 7. The precursor is prepared from pyridine in two steps. The *N*-substituent is designed to shield the 2-position and to activate the 4-position towards nucleophilic attack. The conversion is successful with a fairly wide range of

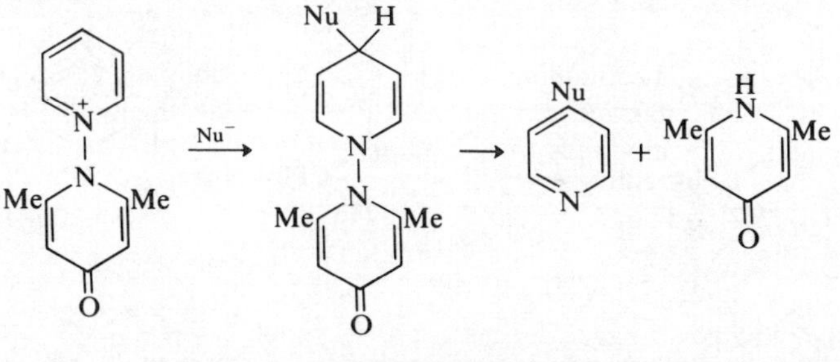

Scheme 7

[210] B. Narine and O. Meth-Cohn, *Tetrahedron Letters*, 1978, 2045; O. Meth-Cohn, B. Narine, and B. Tarnowski, *ibid.*, 1979, 3111; O. Meth-Cohn, S. Rhouati, and B. Tarnowski, *ibid.*, p. 4885.
[211] P. de Mayo, L. K. Sydnes, and G. Wenska, *J.C.S. Chem. Comm.*, 1979, 499.
[212] T. L. Gilchrist, C. W. Rees, and J. A. R. Rodrigues, *J.C.S. Chem. Comm.*, 1979, 627.
[213] M. Takahashi and S. Watanabe, *Chem. Letters*, 1979, 1213.
[214] J. Barluenga, M. Tomás, V. Rubio, and V. Gotor, *J.C.S. Chem. Comm.*, 1979, 675.
[215] J. Barluenga, M. Tomás, S. Fustero, and V. Gotor, *Synthesis*, 1979, 346.
[216] J. Barluenga, J. F. López-Ortiz, and V. Gotor, *J.C.S. Chem. Comm.*, 1979, 891.
[217] S. E. Diamond, A. Szalkiewicz, and F. Mares, *J. Amer. Chem. Soc.*, 1979, **101**, 4902.
[218] Y. Watanabe, M. Yamamoto, and S. C. Shim, *Chem. Letters*, 1979, 1025.

nucleophiles Nu⁻.[219] A more familiar route to 4-substituted pyridines is *via* nitration of the *N*-oxide. The corresponding bromination has seldom proved satisfactory, but in difficult cases the use of thallium(III) acetate and bromine may be effective.[220] Deoxygenation of nitropyridine oxides with phosphorus reagents can also be temperamental, and irradiation of a dilute solution of the *N*-oxide with trimethyl phosphite is an alternative to the thermal reaction.[221] Picoline 1-oxides and *N*-phenylbenzimidoyl chloride, with base, give side-chain acyl-aminated (210) and acylamino-arylated (211) products, by rearrangement of the anhydro-bases of type (212).[222] In a related series, photochemical rearrangement of (213) gives (214) together with the 5-substituted isomer. The corresponding 1-octyloxy-pyridone gives the 3-octyloxy-isomer, but the 1-phenethoxy-analogue gives the 3-benzyl derivative with loss of CH_2O.[223] Rearrangement to the 3-position of pyridine has also been observed in the reaction of *N*-(aryloxy)-pyridinium ions with azide (and some other) ions. The products, 3-(*o*-hydroxyphenyl)-pyridines, contrast with the 2-substituted isomers obtained by other base-catalysed rearrangements.[224] Pentachloropyridine 1-oxide and cyclic enamines afford the expected products (215), and, in some cases, the far-from-expected ring-contracted compounds (216).[225]

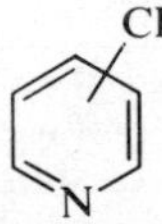

(210) R = N(Ph)COPh
(211) R = $C_6H_4N(H)COPh$

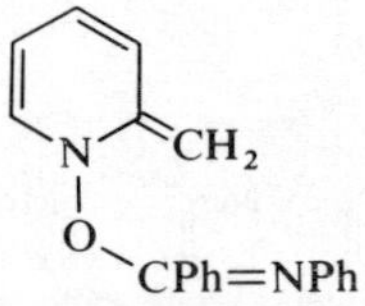

(212)

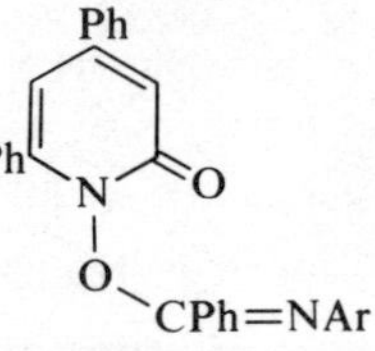

(213)

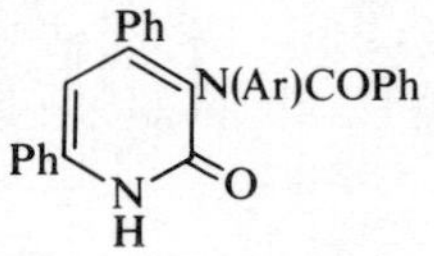

(214)

(Ar = *p*-MeC_6H_4)

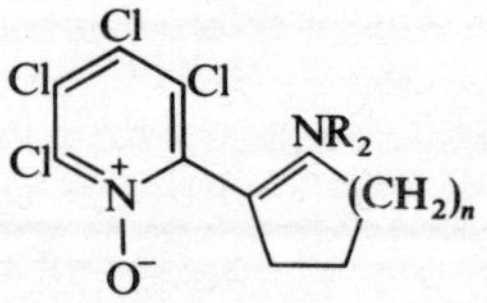

(215)

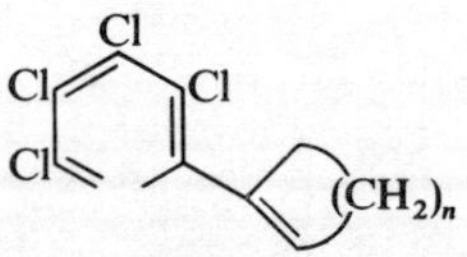

(216)

[219] A. R. Katritzky, H. Beltrami, J. G. Keay, D. N. Rogers, M. P. Sammes, C. W. F. Leung, and C. M. Lee, *Angew. Chem. Internat. Edn.*, 1979, **18**, 792; A. R. Katritzky, H. Beltrami, and M. P. Sammes, *J.C.S. Chem. Comm.*, 1979, 137.
[220] H. Saito and M. Hamana, *Heterocycles*, 1979, **12**, 475.
[221] C. Kaneko, A. Yamamoto, and M. Gomi, *Heterocycles*, 1979, **12**, 227.
[222] R. A. Abramovitch, D. A. Abramovitch, and P. Tomasik, *J.C.S. Chem. Comm.*, 1979, 956.
[223] A. R. Katritzky, A. V. Chapman, M. J. Cook, and G. H. Millet, *J.C.S. Chem. Comm.*, 1979, 395.
[224] R. A. Abramovitch, A. L. Miller, T. A. Radzikowska, and P. Tomasik, *J. Org. Chem.*, 1979, **44**, 464.
[225] H. Suschitzky, B. J. Wakefield, and J. P. Whitten, *J.C.S. Chem. Comm.*, 1979, 183.

Novel ring transformations include two conversions of pyrimidine-diones into pyridones. *i.e.* (217) → (218)[226] and (219) → (220).[227] In the former, the N—C—N unit is replaced by the C—C—N of an α-substituted amide, while the latter conversion involves a dianion, with rearrangement and extrusion of cyanate ion. Thermolysis of dihydro-triazines (221) and -pyrimidines (222) involves ring-opening, 1,7-hydrogen transfer, re-cyclization [*cf. Ann. Reports* (*B*), 1978, **75**, 267], and loss of ammonia, to form pyrimidines (223; X = N) and pyridines (223; X = CH) respectively.[228] With alkaline hydrogen peroxide the pyridinium betaine (224) gives (225) and (226), *via* a 2,3-dihydro-2,3-dihydroxypyridinium betaine intermediate. In contrast, 2,4,6-triaryl-pyridinium salts give β-amino- and β-aroylamino-chalcones.[229]

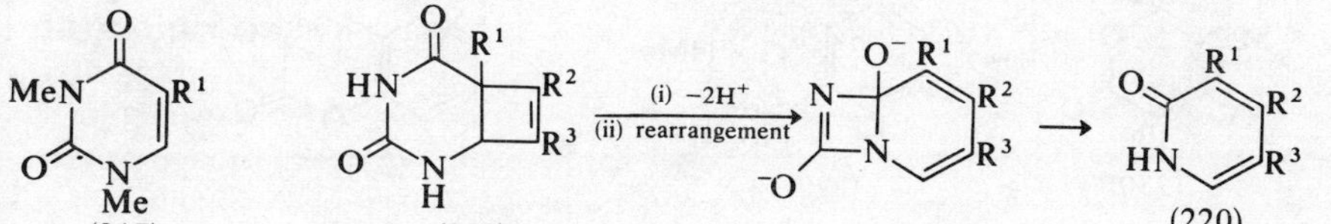

(217) (219) (220)

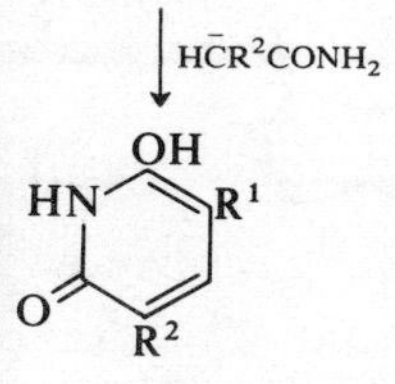

(218)

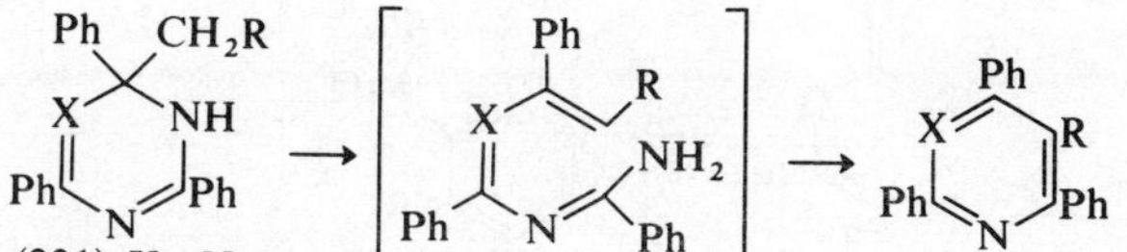

(221) X = N
(222) X = CH

(223)

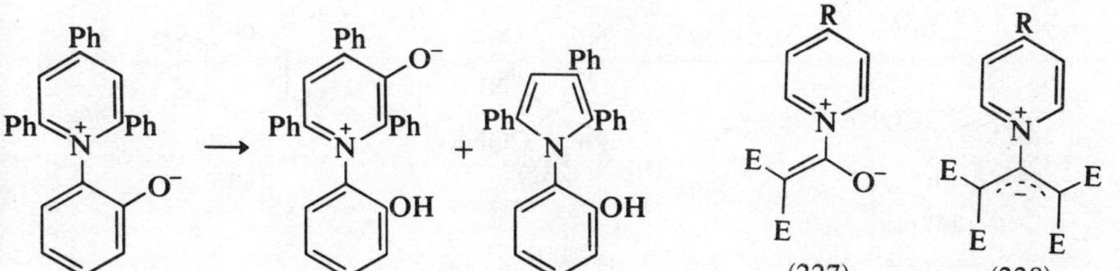

(227) (228)

(224) (225) (226) (E = CO_2Et)

Some interesting pyridinium betaines (227) and (228) are formed, reversibly, from the pyridines with diethoxycarbonylketen, tetraethoxycarbonylallene, or their precursors. The allene adducts (228) may be prepared by the reaction of (227) with the free keten.[230]

[226] K. Hirota, Y. Kitade, S. Senda, M. J. Halat, K. A. Watanabe, and J. J. Fox, *J. Amer. Chem. Soc.*, 1979, **101**, 4423.
[227] R. N. Comber, J. S. Swenton, and A. J. Wexler, *J. Amer. Chem. Soc.*, 1979, **101**, 5411.
[228] L. S. Cook and B. J. Wakefield, *Tetrahedron Letters*, 1979, 1241.
[229] A. R. Katritzky, C. A. Ramsden, Z. Zakaria, R. L. Harlow, and S. H. Simonsen, *J.C.S. Chem. Comm.*, 1979, 363.
[230] R. Gompper and U. Wolf, *Annalen*, 1979, 1388, 1406.

An unusual product (229) of reaction between the mesomeric betaine (230) and *o*-chloranil was verified by *X*-ray analysis.[231] Elimination of CO_2 from the intramolecular Diels–Alder reaction product of the coumarin esters (231) provides the benzo-phthalides (232).[232] Other intramolecular cycloaddition reactions to pyrimidine rings [*cf. Ann. Reports* (*B*), 1977, **74**, 272] have been described.[233]

Loss of CO_2 from the adduct formed between the oxazinone (233) and maleimide gives the pyridine (234). In contrast, the reaction of (233) with an ynamine forms (235) and (236), presumably by addition to the valence isomer (237), the former by a [4+2] cycloaddition, the latter by a [2+2] cycloaddition to give (238) as an intermediate.[234]

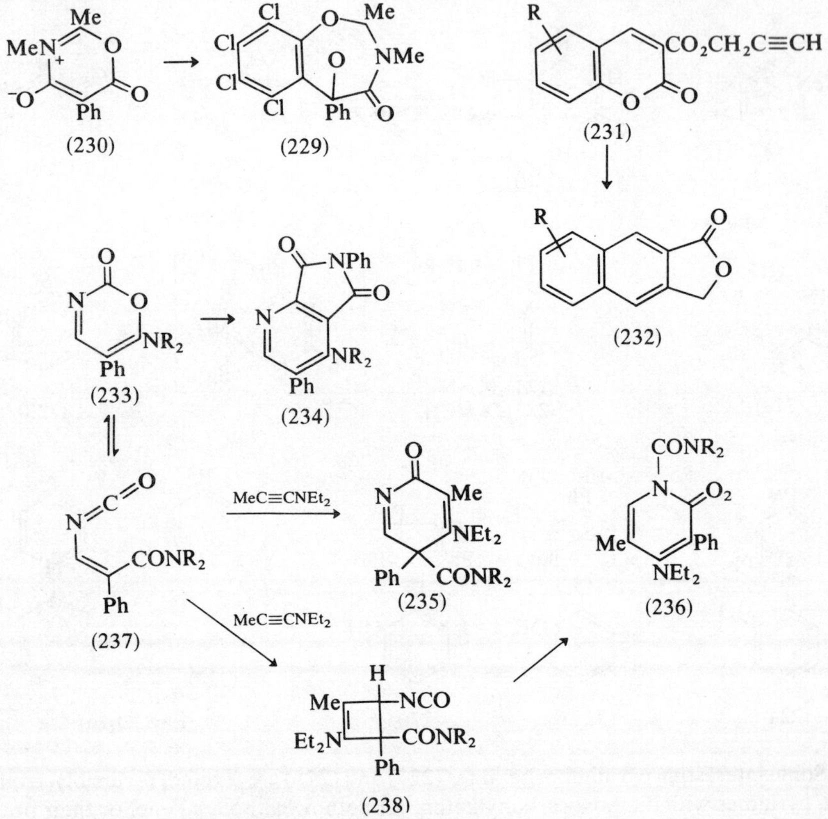

The Diels–Alder reaction of 4-cyano-1-methyl-2-pyridone with 2,3-dimethylbutadiene offers a novel entry into the isocarbostyril series,[235] and further investiga-

231 W. Friedrichsen, C. Krüger, E. Kujath, G. Liebezeit, and S. Mohr, *Tetrahedron Letters*, 1979, 237.
232 G. A. Kraus, J. O. Pezzanite, and H. Sugimoto, *Tetrahedron Letters*, 1979, 853.
233 T. Jojima, H. Takeshiba, and T. Kinoto, *Heterocycles*, 1979, **12**, 665.
234 A. E. Baydar, G. V. Boyd, P. F. Lindley, and F. Watson, *J.C.S. Chem. Comm.*, 1979, 178.
235 H. Kato, R. Fujita, H. Hongo, and H. Tomisawa, *Heterocycles*, 1979, **12**, 1.

tions of 2-pyridones in the role of dienes reveal that the reaction with dimethyl acetylenedicarboxylate is favoured by the presence of a 6-methyl group[236] and by high pressures.[237]

Different types of product arise from Diels–Alder reactions of the triazines (239) and (240). After loss of nitrogen, the intermediate (241; X = Cl) decomposes, *via* a 1,5-shift and loss of HCl, to a 2,6-dichloropyridine, but the fluoro-analogue undergoes a further [4 + 2] cycloaddition with another molecule of alkene.[238]

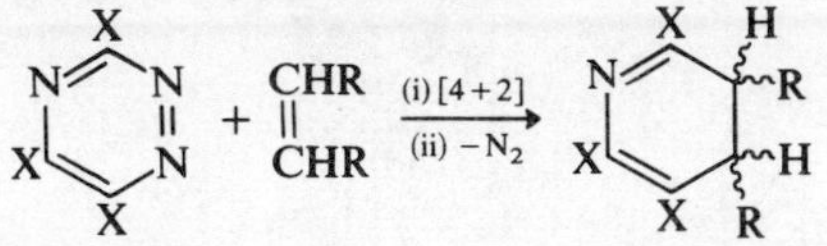

(239) X = Cl
(240) X = F
(241)

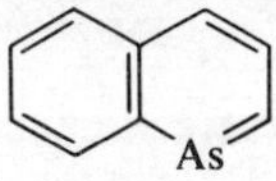

(242)

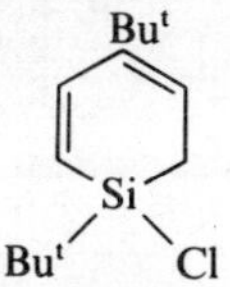

(243)

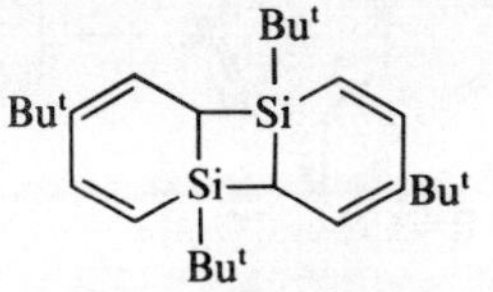

(244)

The hitherto unknown 1-arsanaphthalene (242) has been prepared by addition of benzyne to arsenin (arsabenzene) and then removing one of the etheno-bridges by standard tetrazine methods. It is very air-sensitive.[239] Dehydrochlorination of the silacyclohexadiene (243) affords the dimer (244), *via* the monomeric 1,4-di-t-butylsilabenzene, which could be trapped by dienes.[240]

7 Seven-membered and Larger Rings

Thiepins are notoriously unstable, extruding sulphur *via* a benzene episulphide intermediate. A remarkably stable example is (245), which has $t_{\frac{1}{2}}$ 7.1 h at 131 °C in [$^2H_8$]toluene. The t-butyl groups are believed to provide steric hindrance to the cyclization in the decomposition pathway, and their effectiveness is well shown by the contrast provided by the di-isopropyl analogue, which loses sulphur at −70 °C.[241]

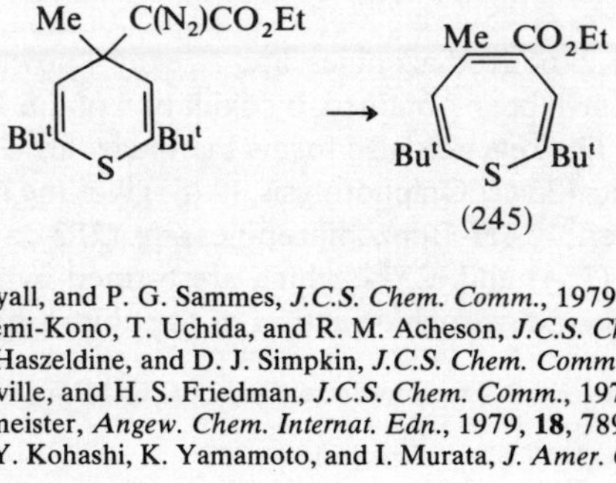

(245)

[236] G. P. Gisby, S. E. Royall, and P. G. Sammes, *J.C.S. Chem. Comm.*, 1979, 501.
[237] K. Matsumoto, Y. Ikemi-Kono, T. Uchida, and R. M. Acheson, *J.C.S. Chem. Comm.*, 1979, 1091.
[238] M. G. Barlow, R. N. Haszeldine, and D. J. Simpkin, *J.C.S. Chem. Comm.*, 1979, 658.
[239] A. J. Ashe, D. J. Bellville, and H. S. Friedman, *J.C.S. Chem. Comm.*, 1979, 880.
[240] G. Märkl and P. Hofmeister, *Angew. Chem. Internat. Edn.*, 1979, **18**, 789.
[241] K. Nishino, S. Yano, Y. Kohashi, K. Yamamoto, and I. Murata, *J. Amer. Chem. Soc.*, 1979, **101**, 5059.

The thiepin (246) could not be isolated, but a sulphur-containing species could be trapped, the thiol derivative (247) being amongst the products isolated.[242]

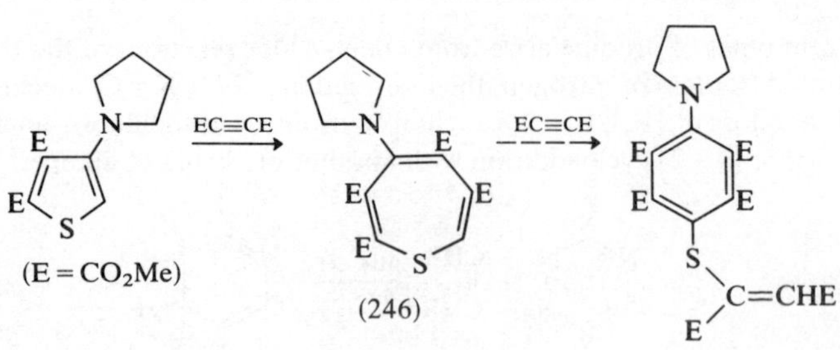

($E = CO_2Me$) (246) (247)

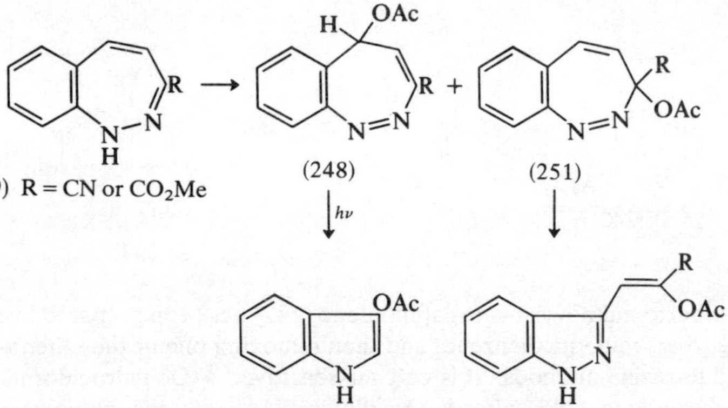

(249) R = CN or CO_2Me (248) (251) (252) (250)

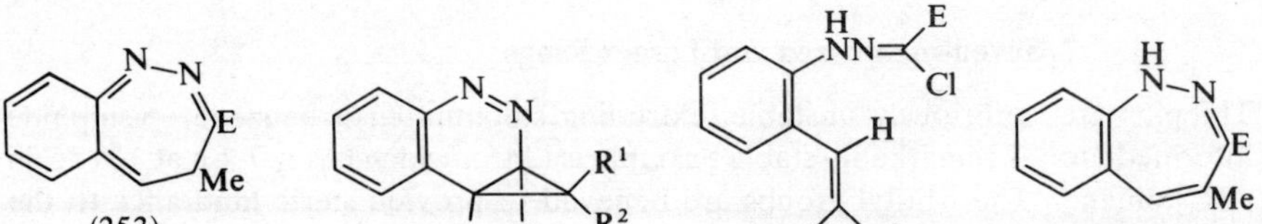

(253) ($E = CO_2Me$) (254) $R^1 = Me, R^2 = H$ (255) $R^1 = H, R^2 = Me$ (256) $R^1 = Me, R^2 = H$ (257) $R^1 = H, R^2 = Me$ (258)

The 1*H*- and 3*H*-1,2-benzodiazepines are already known, and now the first 5*H*-compounds (248) have been obtained, by oxidation of the 1*H*-derivatives (249) with lead(IV) acetate. The reaction also forms the indazoles (250), presumably *via* the unstable 3*H*-isomers (251). On photolysis, (248) gives the indole (252), through a tricyclic valence isomer.[243] 4*H*-Benzodiazepines, *e.g.* (253), are valence isomers of cyclopropa-cinnolines (254) and (255), which are formed by cyclization of styryl-nitrilimines, obtained by dehydrochlorination of the chloro-hydrazones (256) and

[242] D. N. Reinhoudt, G. Okay, W. P. Trompenaars, S. Harkema, D. M. W. van der Ham, and G. J. van Hummel, *Tetrahedron Letters*, 1979, 1529.
[243] T. Tsuchiya and J. Kurita, *J.C.S. Chem. Comm.*, 1979, 803.

(257). However, (253) is not an intermediate in their formation, since it has been found that the *cis*- and *trans*-isomers (256) and (257) form the *endo*- and *exo*-isomers (254) and (255) stereospecifically. Compound (253) is undoubtedly an intermediate in the thermal interconversion of (254) and (255), which takes place spontaneously in solution, and also in their slower transformation into the 1*H*-isomers (258).[244]

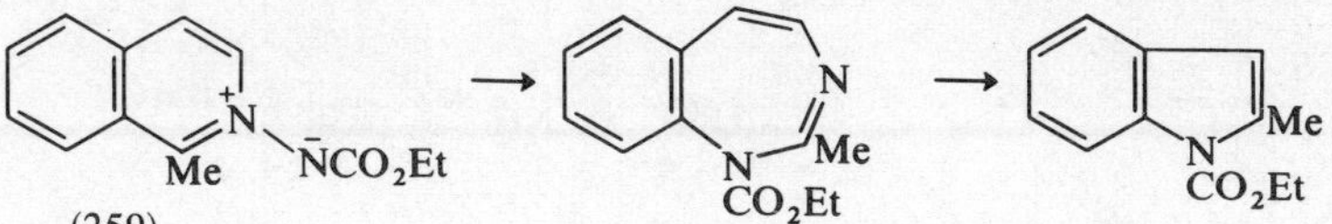

(259) (260) (261)

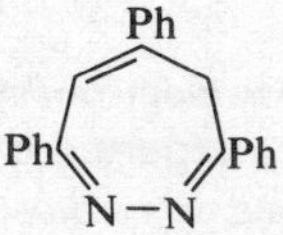

(262)

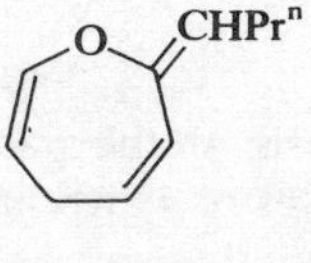

(265)

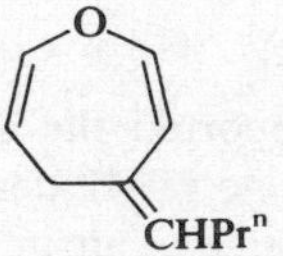

(266)

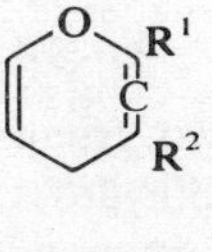

(268)

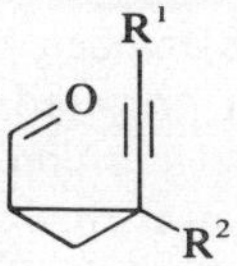

(263) $R^1 = Bu^n, R^2 = H$
(264) $R^1 = H, R^2 = Bu^n$
(267) $R^1 = H, R^2 = Ph$

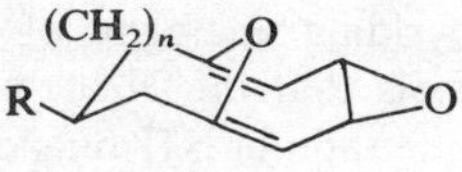

(269) $R = OMe, n = 1$
(270) $R = H, n = 1$
(271) $R = H, n = 3$

Photolysis of the *N*-imide (259) leads to the 1*H*-1,3-benzodiazepine (260), in a reaction which finds analogies in the *N*-oxide series. Further irradiation of the product gives the indole (261);[245] *cf.* (248) → (252). In the monocyclic diazepine series, it is found that the anions generated by deprotonation of (262) can be used to prepare 4-substituted derivatives of this compound.[246]

Thermolysis of the cyclopropanes (263) and (264) gives the (*E*)- and (*Z*)-isomers of the dihydro-oxepins (265) and (266) respectively. In contrast, (267) gives 4-phenylphenol, *via* the unstable compound 4-phenyloxepin. The cyclic allenes (268) have been postulated as intermediates in the two pathways[247] [*cf. Ann. Reports* (*B*), 1977, **74**, 254]. The crystal structure of the bicyclic dihydro-oxepin (269) has been reported: the double bonds, though having normal bond lengths, are twisted, and the bridgehead carbons approach pyramidal geometry.[248] The parent (270) is found not to undergo the Cope rearrangement, unlike its homologue (271). The lack of reactivity of (270) has been attributed to its inability to be converted from the structure with transoid oxygen atoms (shown here) into the cisoid isomer.[249]

[244] A. Padwa and S. Nahm, *J. Org. Chem.*, 1979, **44**, 4746.
[245] T. Tsuchiya, M. Enkaku, J. Kurita, and H. Sawanishi, *J.C.S. Chem. Comm.*, 1979, 534.
[246] L. Bemi, M. T. Thomas, and V. Snieckus, *Synthesis*, 1979, 130.
[247] F. Bourelle-Wargnier, M. Vincent, and J. Chuche, *J.C.S. Chem. Comm.*, 1979, 584.
[248] W. H. Rastetter, T. J. Richard, J. Bordner, and G. L. A. Hennessee, *J. Org. Chem.*, 1979, **44**, 999.
[249] W. H. Rastetter and T. J. Richard, *J. Amer. Chem. Soc.*, 1979, **101**, 3893.

In the sulphur series, the *cis*- and *trans*-2,3-divinylthiirans have been thermolysed. The *cis*-isomer gives the product of Cope rearrangement but the *trans*-isomer forms (272) and (273), presumably *via* initial C—S bond cleavage and subsequent reactions of the biradical.[250]

(272)

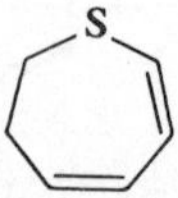

(273)

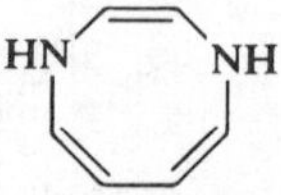

(274)

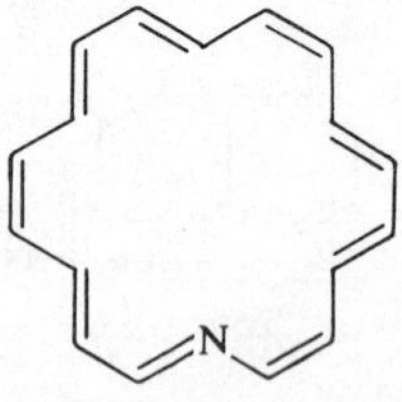

(275)

Two groups have reported the synthesis of the parent 10π-electron system 1,4-dihydro-1,4-diazocine (274), derivatives of which have been reported earlier. The compound is sensitive to air in solution but can be sublimed *in vacuo* and is stable to acid and basic media. In contrast to derivatives bearing electron-withdrawing groups at nitrogen, the ring is practically planar, and *X*-ray and spectral data are indicative of aromatic character.[251] The green aza[18]annulene (275), the first monocyclic pyridine analogue of this type, has been prepared, and its n.m.r. spectrum reveals that the N atom occupies an internal position, as shown. On protonation, the ratio of NH-outside to NH-inside cations is *ca.* 4 : 1.[252]

A correction should be noted to last year's Report on polycyclic medium rings. 3,11-Dimethyl-1,5,9,13-tetra-azatricyclo[9.5.1.1^{3,9}]octadecane, shown as the (chiral) *cis*-isomer [*Ann. Reports* (*B*), 1978, **75**, 274; structure (274)] is in fact the *meso-trans*-isomer.[253] The diaza-bicyclotetradecane (276) is mono- ($pK_a = 6.5$) and di-protonated ($pK_a = -3.25$) from the outside, but the inside-protonated ion is formed on standing in strong acid for a week. However, it appears that this does not arise by simple proton transfer, but *via* an aminium cation radical. Support for this comes from its accelerated production in the presence of a one-electron oxidant.[254]

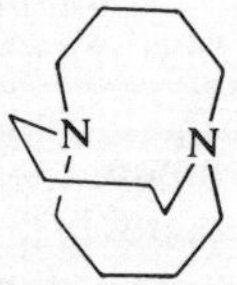

(276)

There are now so many apparently well-funded groups working on crowns and cryptands that it is inevitable that many more interesting and significant papers in this

[250] M. P. Schneider and M. Schnaithmann, *J. Amer. Chem. Soc.*, 1979, **101**, 255.
[251] H.-J. Altenbach, H. Stegelmeier, M. Wilhelm, B. Voss, J. Lex, and E. Vogel, *Angew. Chem. Internat. Edn.*, 1979, **18**, 962; M. Breuninger, B. Gallenkamp, K.-H. Müller, H. Fritz, H. Prinzbach, J. J. Daly, and P. Schönholzer, *ibid.*, p. 964.
[252] W. Gilb and G. Schröder, *Angew. Chem. Internat. Edn.*, 1979, **18**, 312.
[253] D. S. Kemp, R. V. Punzar, and J. C. Chabala, *Tetrahedron Letters*, 1979, 4240.
[254] R. W. Alder and R. B. Sessions, *J. Amer. Chem. Soc.*, 1979, **101**, 3651; R. W. Alder, A. Casson, and R. B. Sessions, *ibid.*, p. 3652.

area of feverish publication have to be omitted than can be chosen for mention here. Fortunately, detailed reviews of the subject have begun to appear: see the following section. The photochemical reduction of the metal in N-crown-complexed silver(I) to the zerovalent state[255] and the complexation of alkali metals[256] as (M^+–crown) M^- are two less usual aspects of their chemistry. Multi-looped polyether systems have been developed, both spiro-fused[257,258] and with other arrangements,[259] and an amusing exploitation of bullvalene isomerization has led to the production of 'breathing crown ethers', *i.e.* crowns which, within limits, can adapt their ring size according to wearer. Thus, the ring of (277) varies from eleven to thirteen members, and similarly the bullvaleno-[20—22]-crown-6 can change within the limits indicated.[260] Bis-monoaza-crown ethers (278) form 1 : 1 complexes with the bis-salt (279), each ring acting as host to one cationic moiety,[261] and a [27]-crown-9-hexacarboxylate forms complexes with guanidinium and imidazolium ions which are stable and also selective even in aqueous solution.[262]

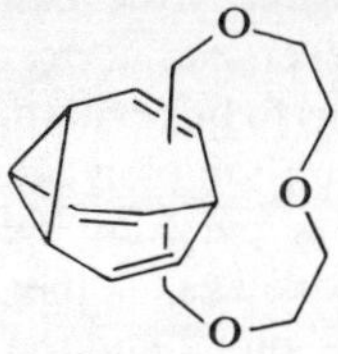

(277)

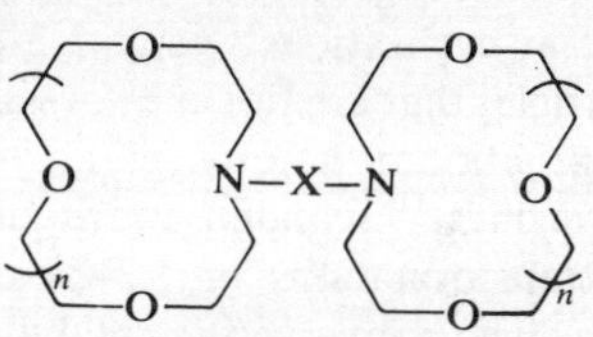

(278) $n = 2$, X = $(CH_2)_4$ or $(CH_2)_5$

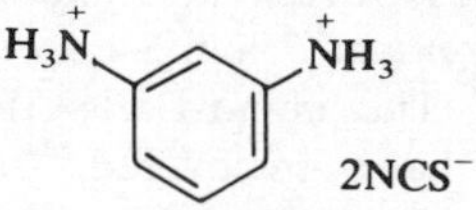

(279)

8 Monographs and Reviews

In 'The Chemistry of Heterocyclic Compounds' (Weissberger–Taylor series), three volumes on a major ring system (thiazole),[263] a third on another (indole),[264] and one covering the condensed pyrazines[265] have been published. A seven-volume monograph on the porphyrins has appeared,[266] and B. D. Tilak's 60th birthday is

[255] R. Humphry-Baker, M. Grätzel, P. Tundo, and E. Pelizzetti, *Angew. Chem. Internat. Edn.*, 1979, **18**, 630.

[256] J. L. Dye, *Angew. Chem. Internat. Edn.*, 1979, **18**, 587.

[257] E. Weber, *Angew. Chem. Internat. Edn.*, 1979, **18**, 219.

[258] V. Prelog and D. Bedekovic, *Helv. Chim. Acta*, 1979, **62**, 2285.

[259] R. C. Helgeson, T. L. Tarnowski, and D. J. Cram, *J. Org. Chem.*, 1979, **44**, 2538.

[260] G. Schröder and W. Witt, *Angew. Chem. Internat. Edn.*, 1979, **18**, 311.

[261] M. R. Johnson, I. O. Sutherland, and R. J. Newton, *J.C.S. Chem. Comm.*, 1979, 306.

[262] J. M. Lehn, P. Vierling, and R. C. Hayward, *J.C.S. Chem. Comm.*, 1979, 296.

[263] 'Thiazole and its Derivatives', ed. J. V. Metzger (Weissberger and Taylor's 'The Chemistry of Heterocyclic Compounds'), Wiley–Interscience, New York, 1979, Vol. 34, Parts 1—3.

[264] 'Chemistry of Indoles', ed. W. J. Houlihan (Weissberger and Taylor's 'The Chemistry of Heterocyclic Compounds'), Wiley–Interscience, New York, 1979, Vol. 25, Part 3.

[265] 'Condensed Pyrazines', ed. G. W. H. Cheeseman and R. F. Cookson (Weissberger and Taylor's 'The Chemistry of Heterocyclic Compounds'), Wiley–Interscience, New York, 1979, Vol. 35.

[266] 'The Porphyrins', Vols. I and II, 'Structure and Synthesis'; Vols. III–V, 'Physical Chemistry'; Vols. VI and VII, 'Biochemistry', ed. D. Dolphin, Academic Press, New York, 1978–9.

commemorated in a volume of reviews[267] on a variety of interesting short topics: 1-hydroxy-indoles,[a] naphtho-indolizines,[b] 3,5-disubstituted pyridines and related pyridinophanes,[c] phospholes and benzophospholes ('phosphindoles'),[d] 3-oxido-pyridinium betaines,[e] dibenzoxazepines,[f] and benzazetes;[g] also covered are some aspects of cycloaddition to azoles,[h] azirine chemistry[i] and aziridine stereochemistry,[j] pyrylium[k] and thiiranium salts,[l] the reaction of azides with indoles,[m] synthesis of benzo[*b*]thiophens, the use of thiophens in the synthesis of optically active quaternary hydrocarbons,[o] the photochemical construction of heterocyclic compounds,[p] and quantum-chemical aspects of electrophilic substitution in five-membered rings.[q]

Crowns, cryptands, and other multidentate macrocycles are the subject of a monograph,[268] and three other reviews deal with certain aspects of this area: one[256] describes the stabilization of derivatives of alkali-metal anions, in which crown ethers play an important role; another[269] covers cyclic di- and tetra-esters of polyether systems, and a third,[270] entitled 'From Carbohydrates to Enzyme Analogues', covers a lot of ground (including a section extolling the virtues of dreaming[271]).

1,5-Dipolar cyclizations[272] and 1,3-dipolar cyclo-reversions[273] are discussed in two long articles that are full of interest. Other reactions to be reviewed include the photo-oxygenation of heterocycles,[274] the cyclotrimerization of cyano-compounds to 1,3,5-triazines,[275] condensations of alkyl groups in pyrylium salts, [276] acid-catalysed transformations of 1,3-dioxans and 1,3-dioxolans,[277] the reactions of heterocycles which involve the catalytic use of cyanide ion,[278] and the reactions of nitrones with ketens (including an interesting account of Staudinger's 'nitrenes').[279] More modern nitrenes are also reviewed,[280] and other areas of interest to heterocyclic synthesis are the diazenium[281] and 3-chloro-2-propeniminium salts,[282] and hexamethylenetetramine.[283] The 'imino-Diels–Alder' synthesis of *N*-heterocycles (particularly 3-piperideines) has been covered,[284] and also the synthesis of hetero-

[267] (*a*) R. M.. Acheson; (*b*) N. R. Ayyangar and A. G. Lugade; (*c*) K. Deuchert and S. Hünig; (*d*) A. N. Hughes; (*e*) A. R. Katritzky and N. Dennis; (*f*) K. Nagaragan; (*g*) C. W. Rees; (*h*) R. B. Mitra, G. H. Kulkarni, G. S. Shirwaikar, and R. S. Jagtap; (*i*) A. Hassner and V. Alexanian; (*j*) P. Tarburton, C. A. Kingsbury, and N. H. Cromwell; (*k*) A. T. Balaban; (*l*) V. N. Gogte and H. M. Modak; (*m*) J. M. Peach and A. S. Bailey; (*n*) B. Iddon; (*o*) L. A. Hulshof and H. Wynberg; (*p*) A. Padwa, P. H. J. Carlsen, and N. Kamigata; (*q*) I. A. Abronin, L. I. Balen'kii, and Ya. L. Gol'dfarb, in 'New Trends in Heterocyclic Chemistry', ed. R. B. Mitra, N. R. Ayyangar, V. N. Gogte, R. M. Acheson, and N. H. Cromwell, Elsevier, Amsterdam, Oxford, and New York, 1979.

[268] 'Synthetic Multidentate Macrocyclic Compounds', ed. R. M. Izatt and J. J. Christensen, Academic Press, New York, 1978.

[269] J. S. Bradshaw, G. E. Maas, R. M. Izatt, and J. J. Christensen, *Chem. Rev.*, 1979, **79**, 37.

[270] J. F. Stoddart, *Chem. Soc. Rev.*, 1979, **8**, 85.

[271] *cf.* W. Shakespeare, 'Julius Caesar', Act I, Scene 2, line 24. For more on dreams and their place in chemistry, see A. Kekulé, quoted by K. Hafner, *Angew. Chem. Internat. Edn.*, 1979, **18**, 641.

[272] E. C. Taylor and I. J. Turchi, *Chem. Rev.*, 1979, **79**, 181.

[273] G. Bianchi, C. De Micheli, and R. Gandolfi, *Angew. Chem. Internat. Edn.*, 1979, **18**, 721.

[274] M. V. George and V. Bhat, *Chem. Rev.*, 1979, **79**, 447.

[275] D. Martin, M. Bauer, and V. A. Pankratov, *Russ. Chem. Rev.*, 1978, **47**, 975.

[276] V. V. Mezheritskii, A. L. Wasserman, and G. N. Dorofeenko, *Heterocycles*, 1979, **12**, 51.

[277] D. L. Rakhmankulov, E. A. Kantor, and R. A. Karakanov, *Heterocycles*, 1979, **12**, 1039.

[278] E. Hayashi and T. Higashino, *Heterocycles*, 1979, **12**, 837.

[279] M. A. Abou-Gharbia and M. M. Joullié, *Heterocycles*, 1979, **12**, 819.

[280] B. Iddon, O. Meth-Cohn, E. F. V. Scriven, H. Suschitzky, and P. T. Gallagher, *Angew. Chem. Internat. Edn.*, 1979, **18**, 900.

[281] M. A. Kuznetsov, *Russ. Chem. Rev.*, 1979, **48**, 563.

[282] L. Liebscher and H. Hartmann, *Synthesis*, 1979, 241.

[283] N. Blazevic, D. Kolbah, B. Berlin, V. Sunjic, and F. Kajfez, *Synthesis*, 1979, 161.

[284] S. M. Weinreb and J. I. Levin, *Heterocycles*, 1979, **12**, 949.

cycles using heterocumulenes.[285] The uses of isoxazoles in synthesis have been collected.[286]

Specific ring systems which have been covered during the year include, besides those mentioned above, the azetidines,[287] benzo[*c*]cinnolines,[288] pyridazines,[289] pyrrolo[3,2-*c*]quinolines;[290] pyrrolizidines,[291] 1,10-phenanthrolines[292] and other polyaza-phenanthrenes,[293] quinazolines,[294] 1,4-thiazines,[295] and thiirenium ions.[296] Other reviews of special areas deal with Reissert compounds,[297] 1,2-azoles (isoxazoles, pyrazoles, and isothiazoles),[298] selenium–nitrogen heterocycles,[299] three-membered rings with two heteroatoms,[300] physicochemical aspects of purines,[301] the ^{1}H n.m.r.[302] and u.v.[303] spectroscopy of phytoxanthones, and the stereochemistry of quinolizidines, indolizidines, and pyrrolizidines.[304] The application of photoelectron spectroscopy to conformational analysis is illustrated with many heterocyclic examples,[305] and a review by Kauffmann on 'areno-analogy', while stopping short of a clear definition of the term, is certainly packed with plenty of interesting heterocyclic chemistry.[306]

The latest volume of the Specialist Periodical Reports on sulphur, selenium, and tellurium seems to give excellent value: more than half of the chapters deal with heterocyclic ring systems, and there is also a section on the β-lactam antibiotics.[307]

285 O. Tsuge, *Heterocycles*, 1979, **12**, 1067.
286 C. Kashima, *Heterocycles*, 1979, **12**, 1343.
287 N. H. Cromwell and B. Phillips, *Chem. Rev.*, 1979, **79**, 331.
288 J. W. Barton, *Adv. Heterocyclic Chem.*, 1979, **24**, 151.
289 M. Tišler and B. Stanovnik, *Adv. Heterocyclic Chem.*, 1979, **24**, 363.
290 M. A. Khan and J. F. da Rocha, *Heterocycles*, 1979, **12**, 857.
291 D. J. Robins, *Adv. Heterocyclic Chem.*, 1979, **24**, 247.
292 W. Sliwa, *Heterocycles*, 1979, **12**, 1207.
293 W. Sliwa and H. Zamarlik, *Heterocycles*, 1979, **12**, 529.
294 W. L. F. Armarego, *Adv. Heterocyclic Chem.*, 1979, **24**, 1.
295 R. J. Stoodley, *Adv. Heterocyclic Chem.*, 1979, **24**, 293.
296 G. Capozzi, V. Lucchini, and G. Modena, *Rev. Chem. Intermediates*, 1979, **2**, 347.
297 F. D. Popp, *Adv. Heterocyclic Chem.*, 1979, **24**, 187.
298 S. D. Sokolov, *Russ. Chem. Rev.*, 1979, **48**, 289.
299 I. Lalezari, A. Shafiee, and M. Yalpani, *Adv. Heterocyclic Chem.*, 1979, **24**, 109.
300 E. Schmitz, *Adv. Heterocyclic Chem.*, 1979, **24**, 63.
301 J. H. Lister, *Adv. Heterocyclic Chem.*, 1979, **24**, 215.
302 M. Afzal and J. M. Al-Hassan, *Heterocycles*, 1979, **12**, 421.
303 M. Afzal, J. M. Al-Hassan, and F. N. Al-Masad, *Heterocycles*, 1979, **12**, 269.
304 I. M. Skvortsov, *Russ. Chem. Rev.*, 1979, **48**, 262.
305 M. Klessinger and P. Rademacher, *Angew. Chem. Internat. Edn.*, 1979, **18**, 826.
306 T. Kauffmann, *Angew. Chem. Internat. Edn.*, 1979, **18**, 1.
307 'Organic Compounds of Sulphur, Selenium, and Tellurium', ed. D. R. Hogg (Specialist Periodical Reports) The Chemical Society, London, 1979, Vol. 5.

12 Organometallic Chemistry
Part (i) The Transition Elements

By H. M. COLQUHOUN and J. HOLTON
ICI Corporate Laboratory, The Heath, Runcorn, Cheshire WA7 4QE
and
M. V. TWIGG
ICI Agricultural Division, Billingham, Cleveland TS23 1LD

1 Introduction

In this Report the emphasis is on the application of organo-transition-metal species in organic synthesis, and during 1979 the high level of interest in this area continued.

Although there has been considerable discussion on mechanistic aspects of catalytic olefin metathesis, no major practical advances have been reported, and accordingly, the section dealing with this topic has been omitted. New sections deal with carboxylation and synthetic procedures involving phase-transfer catalysis – areas in which useful developments are being made.

Relevant reviews include those dealing with palladium-catalysed reactions of olefins with aryl halides,[1] syntheses involving polybromo-ketones/iron carbonyl reactions,[2] and heterogenized catalysts.[3] Reduction of organic compounds with low-valent transition-metal species,[4] cobalt-catalysed reactions of hydrosilanes and carbon monoxide with unsaturated organic compounds,[5] and catalytic hydrogenation of arenes[6] have also been reviewed. A book concerned with Ziegler–Natta catalysts and polymerizations[7] has been published, and a volume of *Advances in Organometallic Chemistry* was devoted to the synthesis of organic compounds.[8] The growing acceptance of transition-metal reagents in synthesis is reflected by the appearance of several books concerned with this topic.[9]

1 R. F. Heck, *Accounts Chem. Res.*, 1979, **12**, 146.
2 R. Noyori, *Accounts Chem. Res.*, 1979, **12**, 61.
3 A. Y. Yuffa and G. V. Lisichkin, *Russ. Chem. Rev.*, 1978, **47**, 751.
4 T. L. Ho, *Synthesis*, 1971, 1.
5 S. Murai and N. Sonoda, *Angew. Chem. Internat. Edn.*, 1979, **18**, 837.
6 E. L. Muetterties and J. R. Bleeke, *Accounts Chem. Res.*, 1979, **12**, 324.
7 J. Boor, 'Ziegler–Natta Catalysts and Polymerisations', Academic Press, London, 1978.
8 'Advances in Organometallic Chemistry', Vol. 17, ed. F. G. A. Stone and R. West, Academic Press, London, 1979.
9 R. P. Houghton, 'Metal Complexes in Organic Chemistry', Cambridge University Press, 1979; M. Freidfelder, 'Catalytic Hydrogenation in Organic Synthesis: Procedures and Commentary', John Wiley, New York, 1979; J. K. Kochi, 'Organometallic Mechanisms and Catalysis', Academic Press, London, 1978; 'Transition Metal Organics in Organic Synthesis', Vol. II, ed. H. Alper, Academic Press, London, 1978; 'Fundamental Research in Homogeneous Catalysis II', ed. Y. Ishii and M. Tsutsui, Plenum Press, London, 1979.

2 Hydrogenation

The scope of reductions by sodium borohydride has been extended by several reports this year of transition-metal-promoted reductions by sodium borohydride. The reduction of a ketone in the presence of an aldehyde usually necessitates a three-step process, *viz.* protection of the aldehyde, reduction of the ketone, and finally liberation of the aldehyde. This lengthy procedure is often poorly selective, and leads to separation problems and low yields. A simple one-step route has now been devised, using $NaBH_4$ and $CeCl_3{\cdot}6H_2O$ (in molar ratio 1.5 : 1).[10] It is known that, in aqueous solution, non-conjugated aldehydes form hydrates to a greater extent than ketones, while conjugated ones are not hydrated. Protection of a non-conjugated aldehyde is achieved by formation of a *gem*-diol, which is stabilized by its co-ordination to a cerium(III) ion. The presence of Ce^{III} is sufficient to ensure adequate protection during reduction but does not hamper recovery of the aldehyde during work-up. Thus in an equimolar mixture of hexanal and cyclohexanone only 2% of the aldehyde is reduced, compared to 100% of the ketone; similarly, (1) is obtained in 75% isolated yield, as shown in Scheme 1. Selective reduction cannot be achieved with conjugated aldehydes (*e.g.* benzaldehyde), which are readily reduced. This, however, allows selective reduction of conjugated aldehydes, in the presence of non-conjugated aldehydes, so that, for example, benzaldehyde is reduced to the alcohol (85%) while cyclohexanecarboxaldehyde is not reduced.

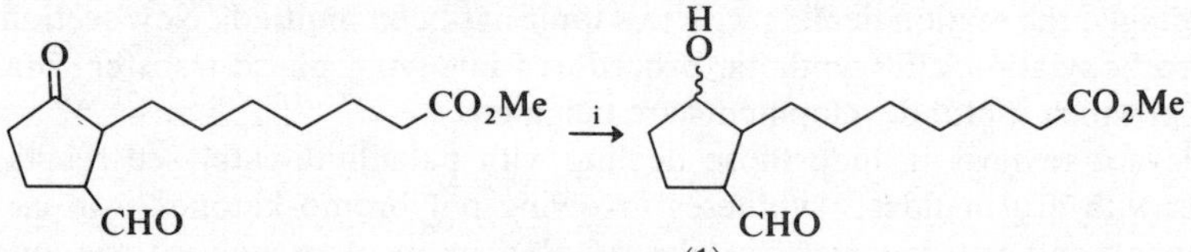

(1)

Reagents: i, $NaBH_4$, $CeCl_3{\cdot}6H_2O$

Scheme 1

The complex formed by a cobalt(II) salt and sodium borohydride reduces alkenes, in high yield, with high steric selectivity (mono- > di- > tri- and tetra-substituted olefins).[11] In limonene, only the disubstituted exocyclic double bond is reduced, and not the trisubstituted endocyclic double bond. Alkynes undergo facile reduction to alkanes, but partial reduction to alkenes is not controllable.

Aryl bromides are not readily reduced by sodium borohydride, but, in the presence of a catalytic amount of $[Ni(PPh_3)_3]$, substantial debromination occurs in *NN*-dimethylformamide (DMF) at 70 °C over 3—20 hours.[12] Aryl chlorides also react, but more slowly. Reduction of certain functional groups, *e.g.* nitro- or cyano-groups, occurs in preference to dehalogenation, and hydroxy and amino functions hinder the reaction.

[10] J. Luche and A. L. Gemal, *J. Amer. Chem. Soc.*,1979, **101**, 5848.
[11] S. Chung, *J. Org. Chem.*, 1979, **44**, 1014.
[12] S. Lin and J. A. Roth, *J. Org. Chem.*, 1979, **44**, 309.

Aromatic nitro-compounds are not reduced by sodium borohydride alone. However, in the presence of copper(II) acetylacetonate,[13] reduction to the corresponding amine occurs.

The chromium(II)–amine complex, formed by rapid addition of an aqueous solution of chromium(II) perchlorate to a cold solution of ethylenediamine (or triethylamine) in DMF, reduces terminal and alkyl-phenylacetylenes to the corresponding olefin with a high selectivity for the *cis*-isomer.[14] Dialkyl-acetylenes are not reduced by this reagent. Propargyl alcohols undergo facile reduction, but with low stereoselectivity, the [Cr^{II}–ethylenediamine] complex favouring *trans*-addition whereas [$Cr^{II}(NEt_3)$] favours *cis*-addition. *trans*-Addition is the exclusive pathway in the absence of amine. In the presence of amines, reduction of (2) occurs quantitatively, with *cis* : *trans* selectivities of 55:45 and 10:90 for ethylenediamine and triethylamine respectively. Thus the relative proportions of *cis*- and *trans*-isomers depend on both the nature of the acetylene and the amine.

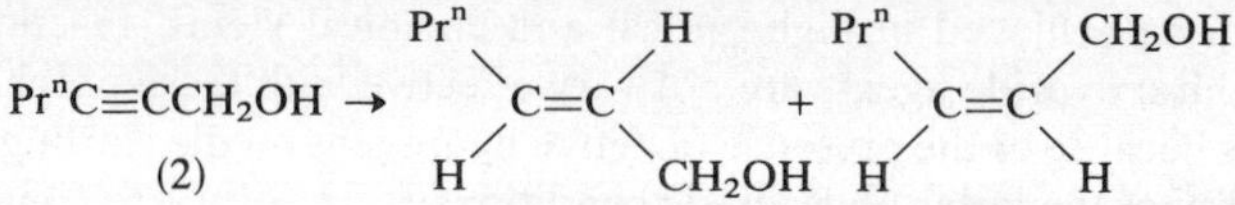

(2) *trans* *cis*

The rhodium hydride [$RhH(PPr^i_3)_3$] and, to a lesser extent, [$Rh_2H_2(\mu$-$N_2)$-$\{P(cyclohexyl)_3\}_4$] are active homogeneous catalysts for the hydrogenation of nitriles.[15] Aliphatic and aromatic nitriles are hydrogenated under ambient conditions in the presence of these hydrides to the corresponding primary amine. Using unsaturated nitriles, hydrogenation of the olefinic group proceeds more readily than that of the nitrile function. It is interesting to compare [$RhH(PPr^i_3)_3$] with the well-known hydrogenation catalyst [$RhCl(PPh_3)_3$], which reduces only the olefinic group under similar conditions.

The selective hydrogenation of a nitro-group in the presence of an acetylenic function may be achieved, using ruthenium supported on carbon or alumina.[16] Results with various acetylenic compounds show the reactivity sequence RC≡CH > ArC≡CH > ArC≡CR. Although selective reduction occurs with monosubstituted acetylenes, catalyst turnover is poor, owing to strong absorption of the terminal acetylene onto the catalyst. However, the use of a blocking group, such as dimethylcarbinol, to increase steric crowding around the triple bond maintains a high rate of hydrogenation of a nitro-group by protecting the catalyst from deactivation. Hence 2-methyl-4-(3-nitrophenyl)-3-butyn-2-ol is selectively reduced to the corresponding amine in 100% yield (50 °C, 50—60 psi of H_2). The blocking group is easily removed by a catalytic amount of sodium hydroxide.

The bidentate ligand anthranilic acid has been anchored to polystyrene beads to form (3); its reaction with [$RhCl_3 \cdot 3H_2O$], followed by reduction with sodium borohydride, produces a polymer-supported rhodium(I) species.[17] This is a useful hydrogenation catalyst of high activity, with long-term stability and a considerable

[13] K. Hanaya, T. Muramatsu, H. Kudo, and Y. L. Chow, *J.C.S. Perkin I*, 1979, 2409.
[14] J. K. Crandall and W. R. Heitmann, *J. Org. Chem.*, 1979, **44**, 3471.
[15] T. Yoshida, T. Okano, and S. Otsuka, *J.C.S. Chem. Comm.*, 1979, 870.
[16] A. Onopchenko, E. T. Sabourin, and C. M. Selwitz, *J. Org. Chem.*, 1979, **44**, 1233.
[17] N. L. Holy, *J. Org. Chem.*, 1979, **44**, 239.

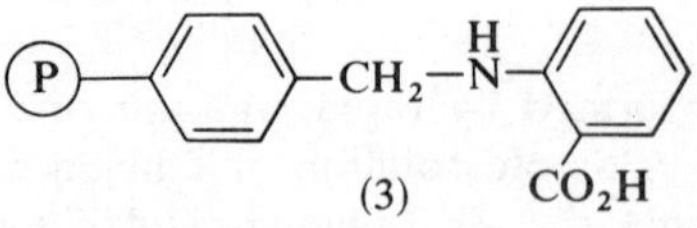

(3)

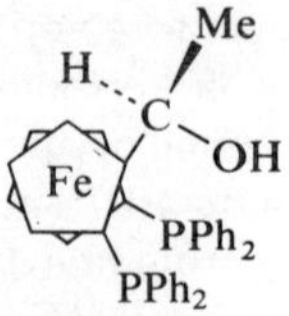

[(*R*)-(*S*)-BPPFOH]
(4)

tolerance to poisons (*e.g.* air). Using this catalyst, a variety of olefinic and aromatic hydrocarbons may be reduced, as well as carbonyl, nitrile, and nitro functional groups.

A new asymmetric synthesis of 2-amino-1-aryl-ethanols has been developed, based on the reduction of prochiral ketones.[18] Using a chiral rhodium complex of the hydroxyalkyl-ferrocenylphosphine (4), catalytic hydrogenation of aminomethyl aryl ketones can be achieved in high optical and chemical yields [reaction(1)]. Conventional chiral hydride agents are not very effective with these prochiral carbonyl compounds because of the presence of active hydrogens on the starting ketones and the instability of the latter under basic conditions.

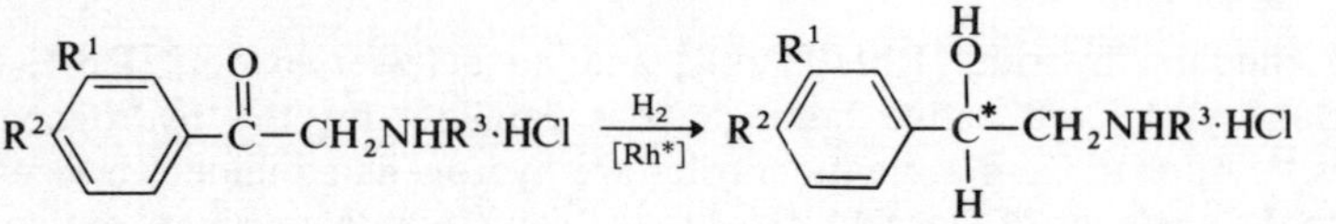

(1)

A remarkable effect of pressure on stereoselectivity has been described for the asymmetric hydrogenation of α-acylaminocinnamic acids.[19] Using neutral or cationic rhodium complexes with, for instance, DIOP to catalyse the hydrogenation of (*Z*)-α-benzamidocinnamic acid, it was found that (*R*)-*N*-benzoylphenylalanine was the preferred product at atmospheric pressure. However, the (*S*)-isomer predominated at high pressure (50—100 atm of H_2). This result may be explained in terms of two competing mechanisms; at low pressure, co-ordination of the olefin is followed by addition of hydrogen, whereas at high pressure the addition of hydrogen precedes complexation of the olefin. The stereochemistries of [Rh*] and [Rh*H_2] dictate the orientation of the olefin upon co-ordination, and in this example lead to inversion in the configuration of the product. Triethylamine is often added to such rhodium catalysts to improve optical yields. It is of interest to note that here the base supresses the pressure effect, *i.e.* prevents pre-co-ordination of H_2 at high pressures, as well as improving optical yields.

3 Dimerization, Oligomerization, and Polymerization

Interest in oligomerization and telomerization of butadiene has continued. The reaction of butadiene with an aldehyde (acetaldehyde, benzaldehyde, or acrolein) in the presence of [Ni(cod)$_2$] and PPh_3 gives[20] predominantly (>70%) the linear 2:1

[18] T. Hayashi, A. Katsumura, M. Konishi, and M. Kumada, *Tetrahedron Letters*, 1979, 425.
[19] I. Ojima, T. Kogure, and N. Yoda, *Chem. Letters*, 1979, 495.
[20] R. Baker and M. J. Crimmin, *J.C.S. Perkin I*, 1979, 1264.

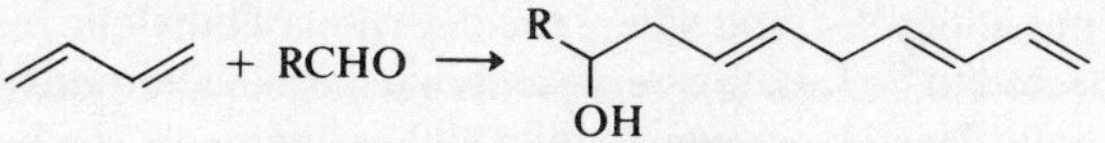

(2)

R = Me, Ph, or CH_2=$CHCH_2$

telomer, *i.e.* 1-substituted 3,6,8-nonatrien-1-ol [see reaction (2)]. With tricyclohexylphosphine the linear product is no longer preferred, and up to 85% of branched product is obtained. A similar reaction between isoprene and acetaldehyde does not occur in the presence of [$Ni(cod)_2$], but, using nickel, cyclododecatriene, and triphenylphosphine, the 2 : 1 adduct is obtained, in which isoprene is dimerized in a head-to-tail manner. The two major products are 3,7-dimethyl-3-vinyloct-7-en-2-ol and its isomer 3,7-dimethyl-3-vinyloct-6-en-2-ol.

The effect of changing the phosphorus ligand on the distribution of products[21] has been studied by Heimbach *et al.*, who have also examined the effect of ligand concentration.[22] A study of the system nickel(O)–triphenylphosphine–butadiene has shown that, at low nickel to phosphorus ratios, the formation of cyclo-trimers predominates, but at higher ratios cyclo-dimers are favoured. 'Partial control' graphs of product yield against ligand : metal ratio were drawn, and they allowed the optimization of the catalysed process, as well as a better analysis of the mechanism.

The linear dimerization of butadiene produces octatrienes, but, to obtain octadienes, a hydrogen-donating source (such as a Grignard reagent, alcohol, or formic acid) is required. It has been shown that nickel(O) complexes, together with morpholine and formaldehyde, catalytically convert butadiene into octadienes.[23] The predominant products are the dihydro-dimers *trans,trans*-octa-2,6-diene and *trans*-octa-1,6-diene. *N*-Hydroxymethyl-morpholine, which is the product of the reaction between morpholine and formaldehyde, is an efficient hydrogen-donor molecule for this catalytic hydrodimerization. The reaction is thought to proceed in one step, and not through the intermediate formation of an octatriene.

The selective dimerization of ethylene to 1-butene by [$Ta(CH_2CMe_3)_3(CHCMe_3)$] and PMe_3 has prompted the proposal of a new mechanism for the oligomerization of ethylene.[24] It has been suggested that the C_4 chain does not form by insertion of ethylene into a metal–ethyl bond but *via* a metallocyclopentane intermediate (see Scheme 2). For metals in a low oxidation state (the metal is formally oxidized upon

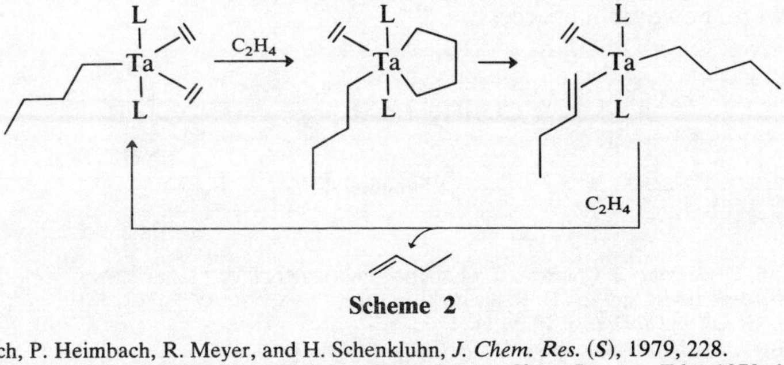

Scheme 2

[21] A. Busch, P. Heimbach, R. Meyer, and H. Schenkluhn, *J. Chem. Res.* (*S*), 1979, 228.
[22] F. Brille, P. Heimbach, J. Kluth, and H. Schenkluhn, *Angew. Chem. Internat. Edn.*, 1979, **18**, 400.
[23] J. Thivolle-Cazat and I. Tkatchenko, *J.C.S. Chem. Comm.*, 1979, 377.
[24] J. D. Fellmann, G. A. Rupprecht, and R. R. Schrock, *J. Amer. Chem. Soc.*, 1979, **101**, 5099.

formation of a metallocycle), and where the insertion of ethylene into a metal–alkyl bond is not expected to be fast, this represents a plausible alternative mechanism. If but-1-ene is not displaced by ethylene, then higher oligomers can be formed and, in the extreme, polyethylene.

The proposal of a novel mechanism for olefin polymerization, *via* α-elimination to form a transient carbene species, has led workers to re-examine the established views on Ziegler–Natta catalysis[25] (*i.e.* insertion into a metal–alkyl bond). Model systems have shown that the α-elimination step is possible,[26] and during the year the proposed mechanism has undergone subtle changes,[27] making it simpler and therefore perhaps more plausible. On the other hand, evidence for the Cosse mechanism has been found in the cobalt system $[Co(\eta^5\text{-}C_5H_5)(PPh_3)Me_2]$, where deuterium-labelling studies demonstrate that ethylene inserts into the metal–methyl bond.[28] The question of the mechanism has yet to be resolved, indeed, each mechanism may operate in different systems, but lively interest and debate has been aroused in the field of Ziegler–Natta catalysis, and further reports are likely.

4 Carbon Monoxide Chemistry

Carbonylation.—Cobalt-catalysed carbonylation of aldehydes in the presence of trialkyl-silanes affords moderate yields (*ca.* 50%) of α-siloxy-aldehydes (Scheme 3), which are useful intermediates in organic synthesis and are otherwise difficult to prepare.[29] A silyl-cobalt carbonyl $[R_3SiCo(CO)_3L]$ ($L = CO$ or PPh_3) is suggested to be the active species (see also Section 6).

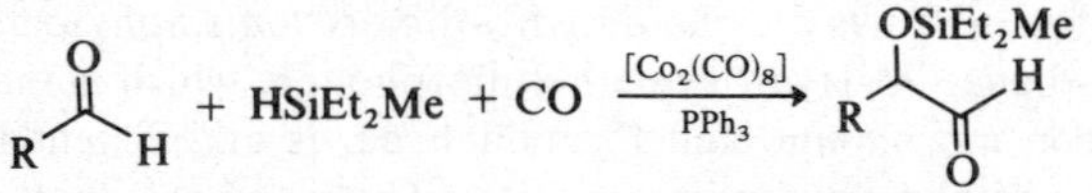

$R = n\text{-}C_3H_7$, $n\text{-}C_6H_{13}$, or cyclo-C_6H_{11}

Scheme 3

A study of the Wakamatsu synthesis, in which an aldehyde is catalytically carbonylated in the presence of an amide, giving *N*-acyl-amino-acids (Scheme 4), shows that the reaction proceeds with high selectivity, in virtually quantitative yield. Attempts to obtain asymmetric induction, by incorporating chiral ligands into the system, were, however, unsuccessful.[30]

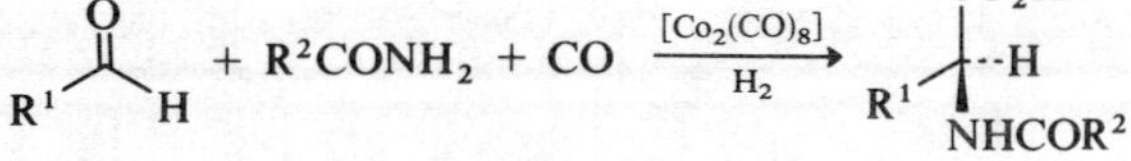

Scheme 4

[25] G. Ghiotti, E. Garrone, S. Coluccia, C. Monterra, and A. Zecchina, *J.C.S. Chem. Comm.*, 1979, 1032.
[26] C. D. Wood, S. J. McLain, and R. R. Schrock, *J. Amer. Chem. Soc.*, 1979, **101**, 3210; N. J. Cooper and M. L. H. Green, *J.C.S. Dalton*, 1979, 1121.
[27] R. J. Al-essa and R. J. Puddephatt, *J.C.S. Chem. Comm.*, 1980, 45.
[28] E. R. Evitt and R. G. Bergman, *J. Amer. Chem. Soc.*, 1979, **101**, 3973.
[29] S. Murai, T. Kato, N. Sonoda, Y. Seki, and K. Kawamoto, *Angew. Chem. Internat. Edn.*, 1979, **18**, 393.
[30] J. J. Parnaud, G. Campari, and P. Pino, *J. Mol. Catalysis*, 1979, **6**, 341.

A novel catalyst, which allows carbonylation of aryl halides at atmospheric pressure, is prepared by addition of cobalt(II) acetate to a suspension of sodium hydride and sodium alkoxide in THF under carbon monoxide.[31] Carbonylation of iodobenzene in the presence of sodium neopentoxide, for example, gave a 70% yield of benzoic acid and neopentyl benzoate, in the ratio 9 : 1. The mechanism of this reaction is unknown, but the formation of 2-aryl-tetrahydrofurans as by-products (up to 10%) suggests that free radicals may well be involved.

The rhodium-induced carbonylation of 2-aryl-azirines under ambient conditions affords highly reactive vinyl isocyanates in good yield (Scheme 5). A somewhat speculative mechanism proposes that carbon monoxide adds to an intermediate vinylnitrene–rhodium complex, followed by decomplexation of the resulting isocyanate.[32]

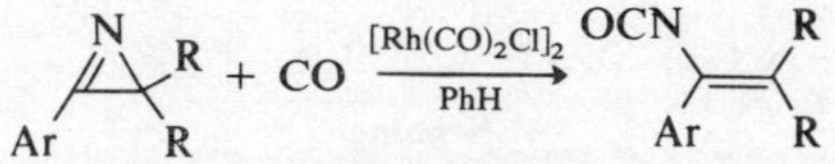

Scheme 5

The complex obtained by the reaction of bis(cyclo-octa-1,5-diene)nickel with methacrylamide in the presence of tricyclohexylphosphine reacts instantly with carbon monoxide at room temperature,[33] to give a quantitative yield of 3-methylsuccinimide (Scheme 6). It is suggested that an analogous cobaltacyclic amide may be an intermediate in the known *catalytic* synthesis of 3-methylsuccinimide from methacrylamide and carbon monoxide.

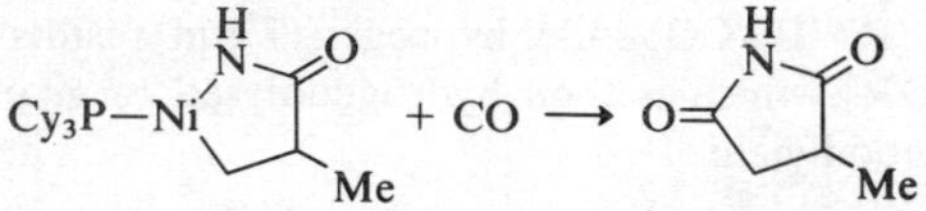

Scheme 6

Chemistry of Synthesis Gas.—The first reported Fischer–Tropsch-type chain growth on a soluble mononuclear catalyst results in formation of alkyl-benzenes $Ph(CH_2)_nH$ ($n = 1$—5) when a solution of $[W(CO)_6]$ and $AlCl_3$ in benzene is treated with carbon monoxide and hydrogen at 200 °C (20—100 atm). Turnover numbers >200 (CO : W) may be achieved in this reaction,[34] which appears to involve an initial arylation of tungsten followed by successive insertion/reduction steps, as proposed for the corresponding heterogeneous reaction.

The high-pressure Union Carbide synthesis of ethane-1,2-diol from carbon monoxide and hydrogen continues to provide a stimulus for research into possible mechanisms whereby two molecules of carbon monoxide may be reductively coupled. Wood and Schrock[35] have described the rapid reaction between $[Ta(\eta^5\text{-}C_5Me_5)Me_4]$ and carbon monoxide, to give an η^2-acetone complex, which

[31] J. J. Brunet, C. Sidot, B. Loubinoux, and P. Caubère, *J. Org. Chem.*, 1979, **44**, 2199.
[32] T. Sakakibara and H. Alper, *J.C.S. Chem. Comm.*, 1979, 458.
[33] T. Yamamoto, K. Igarishi, J. Ishizu, and A. Yamamoto, *J.C.S. Chem. Comm.*, 1979, 554.
[34] G. Henrici-Olivé and S. Olivé, *Angew. Chem. Internat. Edn.*, 1979, **18**, 77.
[35] C. D. Wood and R. R. Schrock, *J. Amer. Chem. Soc.*, 1979, **101**, 5421.

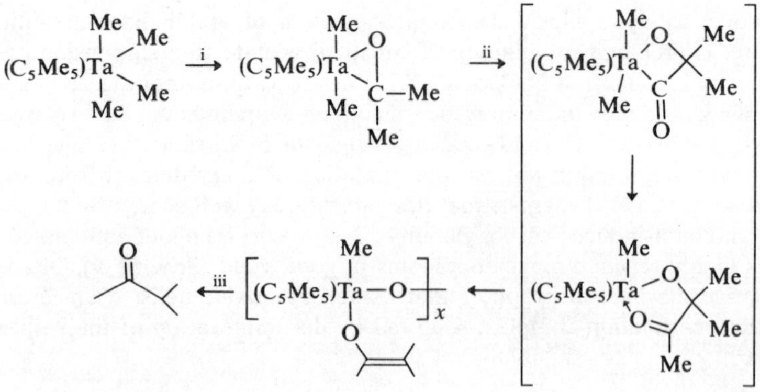

Reagents: i, CO, Et_2O, at −78 °C; ii, CO, at 25 °C; iii, H_2O

Scheme 7

then takes up a second molecule of carbon monoxide more slowly. Hydrolysis of the final product yields one equivalent of methyl isopropyl ketone (Scheme 7).

A more conventional scheme for coupling of carbon monoxide envisages the reduction of co-ordinated carbon monoxide to a hydroxymethyl ligand, insertion of carbon monoxide into the metal–carbon σ-bond, and hydrogenolysis of the resulting hydroxyacetyl complex. This scheme is now supported by isolation and characterization of the first stable low-valent transition-metal hydroxymethyl complex[36] (Scheme 8). Moreover, thermolysis of the acyloxymethyl complex $[Mn(CO)_5CH_2OR]$ ($R = Bu^tCO$) under hydrogen (7 atm) results in rearrangement to $[Mn(CO)_4COCH_2OR]$, which is then hydrogenolysed to give the mono-ester of ethane-1,2-diol in good yield.[37]

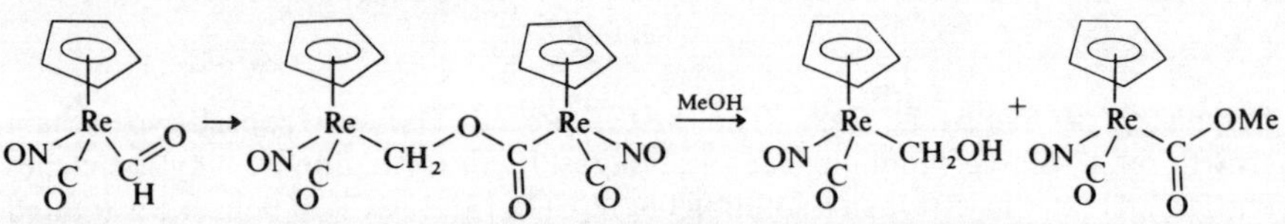

Scheme 8

Yet a third route to coupled products of carbon monoxide has been described by Bercaw and co-workers,[38] who found that $[(\eta^5\text{-}C_5Me_5)_2ZrH_2]$ reacts with $[(\eta^5\text{-}C_5Me_5)Zr(CO)_2]$ under hydrogen to give a quantitative yield of complex (5), which contains a bridging ethene-1,2-diolate ligand. When $[(\eta^5\text{-}C_5H_5)_2WCO]$ replaces the zirconium carbonyl in this reaction, however, a 'zirconoxy-carbene' complex of tungsten (6) is obtained.[39]

[36] C. P. Casey, M. A. Andrews, and D. R. McAllister, *J. Amer. Chem. Soc.*, 1979, **101**, 3371.
[37] B. D. Dombek, *J. Amer. Chem. Soc.*, 1979, **101**, 6466.
[38] J. M. Manriquez, D. R. McAllister, R. D. Sanner, and J. E. Bercaw, *J. Amer. Chem. Soc.*, 1978, **100**, 2716.
[39] P. T. Wolczanski, R. S. Threlkel, and J. E. Bercaw, *J. Amer. Chem. Soc.*, 1979, **101**, 218.

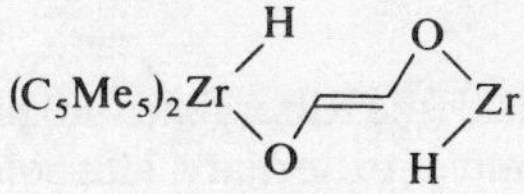

(5)

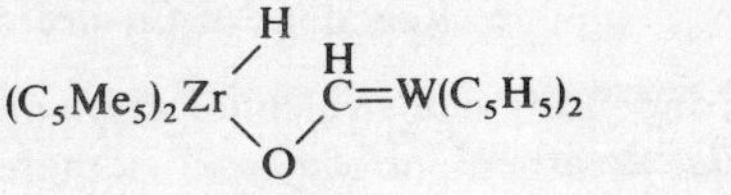

(6)

5 Carboxylation

The palladium(O)-catalysed reaction of carbon dioxide with isopropylidenecyclopropane gives a mixture of two-five-membered lactones, in proportions which depend on the nature of the catalyst (Scheme 9). Thus, using PPh_3 and [Pd(dibenzylideneacetone)$_2$], a 69% yield of (7) was obtained with only 8% of (8) whereas with [$Pd(Ph_2PCH_2CH_2PPh_2)_2$] the lactone (8) was formed in 48% yield, with only a trace of (7). Other methylene-substituted methylenecyclopropanes reacted similarly, but when ring-substituted analogues were used, only starting materials could be recovered. A mechanism involving addition of carbon dioxide to a trimethylenemethane–palladium complex was proposed.[40]

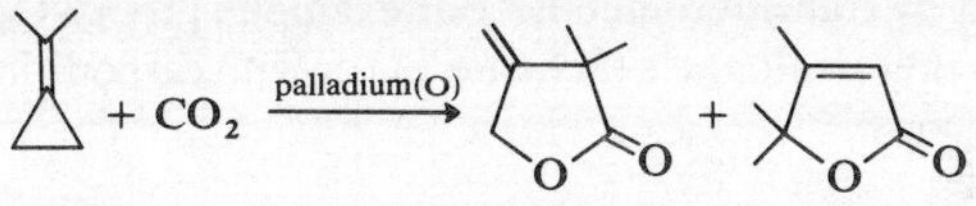

(7) (8)

Scheme 9

The addition of carbon dioxide to the cyanomethyl complex [$Fe(H)CH_2CN(Ph_2PCH_2CH_2PPh_2)_2$] (formed from the reaction of acetonitrile with the corresponding naphthyl iron hydride) gives a cyanoacetate complex which liberates cyanoacetic acid on heating in acetonitrile.[41] It is not clear from this paper whether the original cyanomethyl complex is re-formed, but if so this would seem to provide a simple catalytic route to cyanoacetic acid.

The paramagnetic titanium aryl [Ti(η^5-C_5H_5)(C_6H_4-*o*-CH_2NMe_2)$_2$] eliminates *NN*-dimethylbenzylamine on heating in toluene, and the resulting species (possibly a benzyne complex) incorporates carbon dioxide to give a terdentate amino-aryl-carboxylate ligand (Scheme 10).[42]

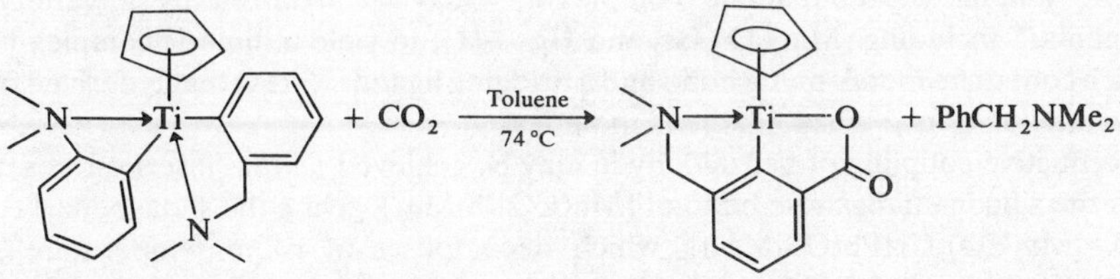

Scheme 10

[40] Y. Inoue, T. Hibi, M. Satake, and H. Hashimoto, *J.C.S. Chem. Comm.*, 1979, 982.
[41] S. D. Ittel, C. A. Tolman, A. D. English, and J. P. Jesson, *J. Amer. Chem. Soc.*, 1978, **100**, 7577.
[42] L. E. Manzer, *J. Amer. Chem. Soc.*, 1978, **100**, 8068.

6 Reactions of Co-ordinated Ligands

Condensation of an alcohol with a nitrile in strong acid (the Ritter reaction) proceeds *via* a carbenium ion, and is therefore often limited to tertiary alcohols. This limitation may, however, be overcome in some instances by utilizing the enhanced stability of secondary and even primary carbenium ions adjacent to transition metals. Thus, generation of the complexed benzyl cation $[Cr(\eta^6\text{-}PhCH_2)(CO)_3]^+$ from the corresponding alcohol by treatment with sulphuric acid, followed by addition of excess nitrile, allows the ready preparation of amide complexes $[Cr(\eta^6\text{-}PhCH_2NHCOR)(CO)_3]$ (R = Me, Ph, Pr^n, *etc.*) in very high yield. The alkyne–cobalt complex $[Co_2(HC{\equiv}CCH_2OH)(CO)_6]$ behaves similarly, and acid-catalysed rearrangements, characteristic of the uncomplexed alcohol, are avoided.[43]

A potentially valuable methylene-transfer reagent, *i.e.* $[Fe(\eta^5\text{-}C_5H_5)(CO)_2\text{-}CH_2SMe_2]^+$ BF_4^-, is obtained by the methylation of $[Fe(\eta^5\text{-}C_5H_5)\text{-}(CO)_2CH_2SMe]$ with Me_3O^+ BF_4^-. This reagent is an air-stable crystalline solid which, when refluxed with olefins in dioxan,[44] affords cyclopropanes in high yield.

Considerable interest is currently being shown in the coupling of unsaturated ligands, as induced by transition metals. For example, $[Ti(\eta^5\text{-}C_5H_5)_2(CO)_2]$ reacts readily with dialkyl ketomalonates (Scheme 11) or with carbodi-imides (Scheme 12) to give coupled ligands.[45]

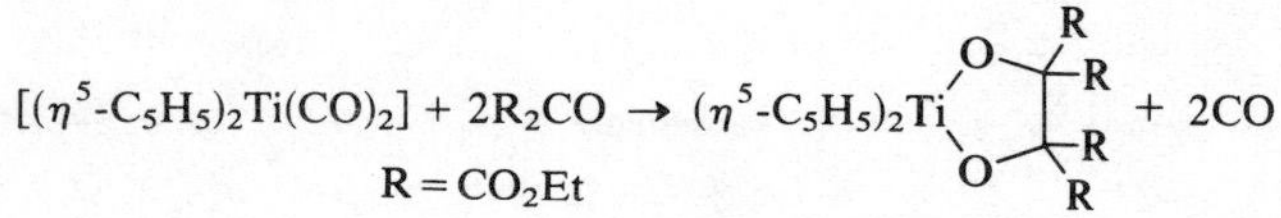

Scheme 11

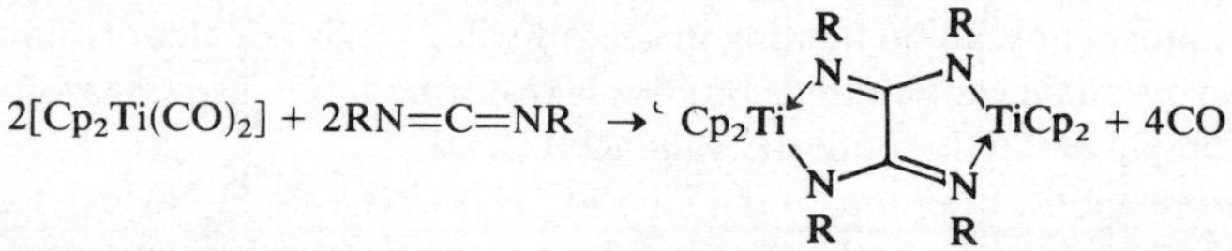

Scheme 12

The anionic diazabutadiene complex (9) may be oxidized by a variety of reagents,[46] including $[Mn(CO)_5Br]$ and Hg^{2+}–H^+, to yield a dimeric complex (10), which contains a metal–metal bond and a bridging ligand (*X*-ray) that is derived from the coupling of two diazabutadiene moieties.

Reductive coupling of benzaldehyde may be achieved *via* an 'insertion' reaction into the silicon–manganese bond of $[Mn(CO)_5SiMe_3]$, giving the siloxybenzyl complex $[Mn(CO)_5CHPh(OSiMe_3)]$, which decomposes at room temperature[47] to $[PhCH(OSiMe_3)]_2$ and $[Mn_2(CO)_{10}]$. Insertion of an aldehyde into a silicon–trans-

[43] S. Top and G. Jaouen, *J.C.S. Chem. Comm.*, 1979, 224.
[44] S. Brandt and P. Helquist, *J. Amer. Chem. Soc.*, 1979, **101**, 6473.
[45] M. Pasquali, C. Floriani, A. Chiesi-Villa, and C. Guastini, *J. Amer. Chem. Soc.*, 1979, **101**, 4740.
[46] L. H. Staal, A. Oskam, K. Vrieze, E. Roosendaal, and H. Schenk, *Inorg Chem.*, 1979, **18**, 1634.
[47] D. L. Johnson and J. A. Gladysz, *J. Amer. Chem. Soc.*, 1979, **101**, 6433.

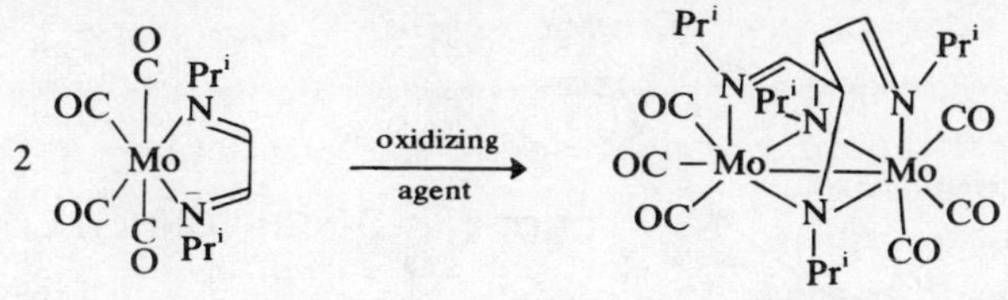

(9) (10)

ition-metal bond has been proposed as a key step in Murai's carbonylation/hydrosilylation synthesis,[5] but this has not been previously demonstrated.

The spiro[4.5]decane derivative (11) may be obtained *via* a tricarbonyldienylium iron complex (Scheme 13), and the corresponding spiro[5.5]decane system is accessible by a similar route.[48]

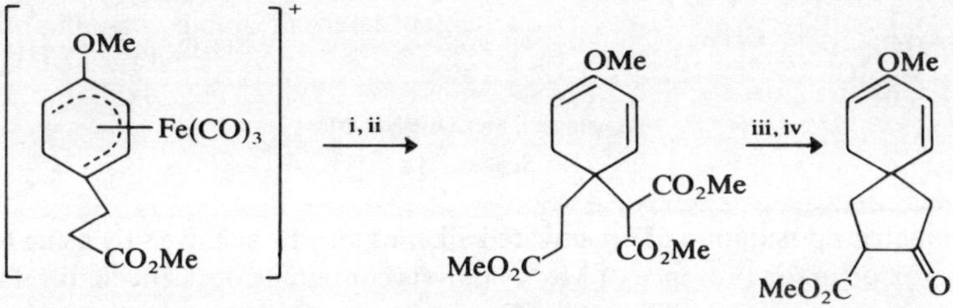

(11)

Reagents: i, $\bar{C}H(CO_2Me)_2$; ii, Me_3NO; iii, NH_4OAc; iv, NaH, THF

Scheme 13

7 Asymmetric Synthesis

Catalytic hydrocyanation of norbornene in the presence of [Pd{(+)-DIOP}]$_n$ gives 2-*exo*-cyanonorbornene (80% yield). The (1*S*,2*S*,4*R*)-(+)-enantiomer is present in *ca.* 30% enantiomeric excess, and the reaction of norbornadiene under similar conditions gives 2-*exo*-cyanonorborn-5-ene with *ca.* 17% enantiomeric excess of the (1*R*,2*R*,4*S*)-enantiomer. 7,7-Dimethylnorbornene does not react, however, suggesting that the reaction is very susceptible to steric hindrance.[49]

1-(*NN*-Dimethylaminomethyl)-2-formylcymantrene (AFCMT), when resolved into its enantiomers, is a useful reagent for the asymmetric synthesis of amino-acids.[50] For example, condensation of diglycine with (+)-AFCMT in the presence of a copper(II) salt yields an optically active complex. Hydroxyethylation of this complex with acetaldehyde, followed by acid hydrolysis, gives (*R*)-threonine (Thr) and (*R*)-allothreonine (αThr) in 92—98% enantiomeric excess (Scheme 14).

Hydroformylation using polymer-supported, asymmetric, diphosphine–rhodium catalysts[51] is reported to give unique selectivity, and optical yields (*ca.* 30%) comparable to those of homogeneous catalysts. The supported catalysts may be re-used after filtration, without loss of activity or selectivity.

[48] A. J. Pearson, *J.C.S. Perkin I*, 1979, 1255.
[49] P. S. Elmes and W. R. Jackson, *J. Amer. Chem. Soc.*, 1979, **101**, 6128.
[50] Yu. N. Belokon, I. E. Zel'tzer, N. M. Loim, V. A. Tsiryapkin, Z. N. Parnes, D. N. Kursanov, and V. M. Belikov, *J. C. S. Chem. Comm.*, 1979, 789.
[51] S. J. Fritschel, J. J. H. Ackerman, T. Keyser, and J. K. Stille, *J. Org. Chem.*, 1979, **44**, 3152.

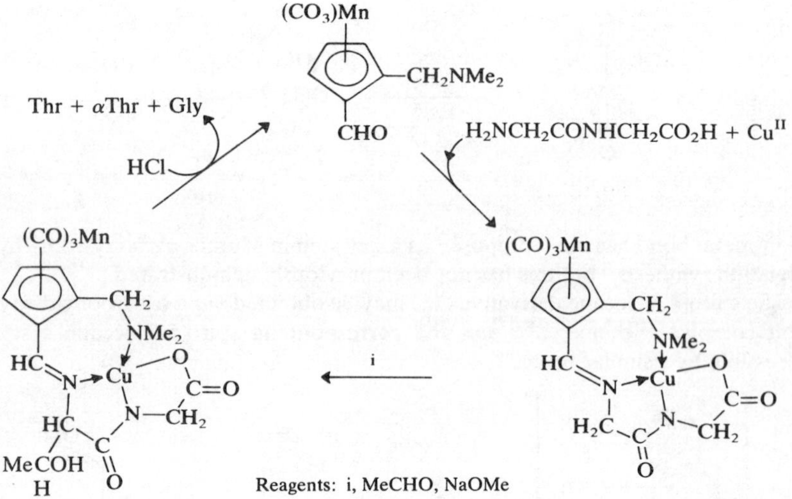

Reagents: i, MeCHO, NaOMe

Scheme 14

Asymmetric epoxidation of unactivated alkenes may be achieved by using t-butyl hydroperoxide in the presence of Mo^{VI} catalysts containing optically active diols,[52a] but optical yields are poor (7—14%). Development of this technique may be slow, owing to the difficulty of designing suitable chiral ligands that are stable to oxidation. In view of this limitation, it is of interest that asymmetric epoxidation has been achieved indirectly, *via* the cyclization of propylene chlorohydrins, in the presence of an optically active cobalt complex.[52b] Optical yields of up to 35% were achieved, by a mechanism that apparently involves hydroxyalkyl-cobalt intermediates.

8 Isomerization

Examples of rhodium-catalysed migration of a double bond include the clean, rapid conversion of (12) into the more substituted isomer (13) by refluxing it in 95%

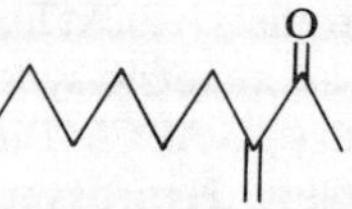

(12)

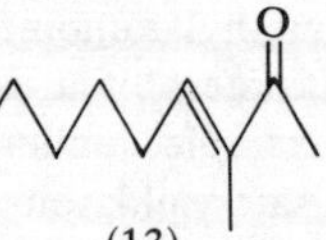

(13)

ethanol with a small quantity of rhodium trichloride.[53] That nickel acetylacetonate (1 mole %) at 165 °C catalyses the isomerization of the substituted 2,5-dienone (14) to the fully conjugated 2,4-isomer (15) with high selectivity is of interest since it has economic attractions.[54] Moreover, the catalyst may be recycled without loss of activity.

[52] (*a*) K. Tani, M. Hanafusa, and S. Otsuka, *Tetrahedron Letters*, 1979, 3017; (*b*) T. Takeichi, M. Ishimori, and T. Tsuruta, *Bull. Chem. Soc. Japan*, 1979, **52**, 2614.

[53] D. F. Taber, and B. P. Gunn, *J. Org. Chem.*, 1979, **44**, 450.

[54] T. Onishi, Y. Fujita, and T. Nishida, *Chem. Letters*, 1979, 765.

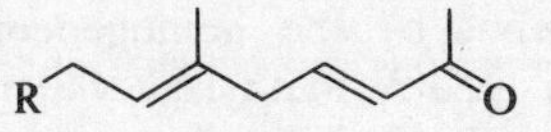

(14)

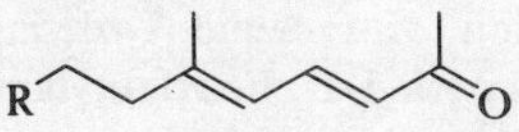

(15)

In contrast to the previous use of iron pentacarbonyl for the conversion of *trans*-steroidal dienes into *cis*-isomers, chromium hexacarbonyl brings about highly regioselective isomerization of steroidal *cis*-dienes to the *trans*-isomers. For instance, the smooth conversion of ergosterol acetate (16) into ergosteryl B_2 acetate (17) is accomplished by refluxing with chromium hexacarbonyl in n-octane. Unlike the *cis*-diene, the *trans*-diene, once formed, presumably does not function as a four-electron donor to the chromium centre, and so, under appropriate conditions, the isomerization is effectively irreversible.[55]

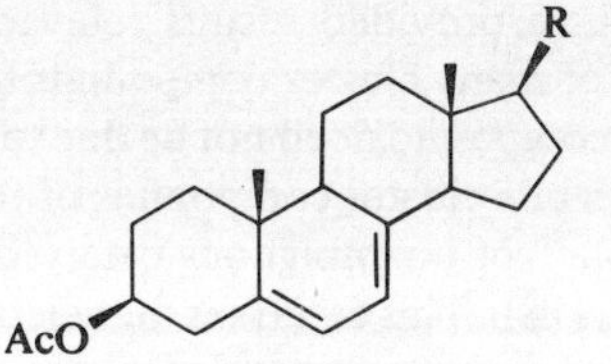

(16)

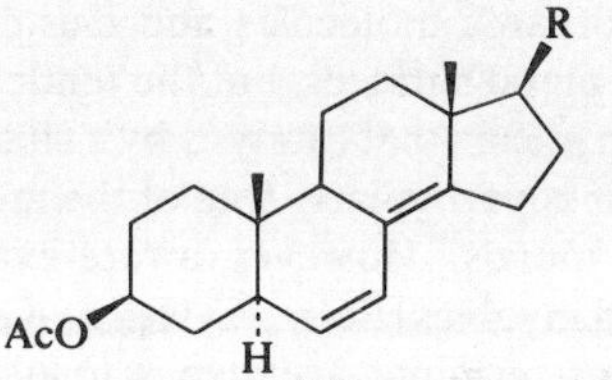

(17)

The reaction of allylic acetates with catalytic quantities of $[Pd(PPh_3)_4]$ in refluxing THF can result in rapid positional and stereochemical isomerization. The proposed mechanism involves co-ordination of freed acetate ion to the metal centre followed by stereo-controlled migration to carbon (Scheme 15). Prolonged reaction affords complete conversion into a 1,3-diene, with no aromatization. This reaction thus provides a route to 1,3-dienes from allylic acetates.[56]

In principle, two enantiomers could have different reactivities towards chiral isomerization catalysts, resulting in selective isomerization of one enantiomer. The

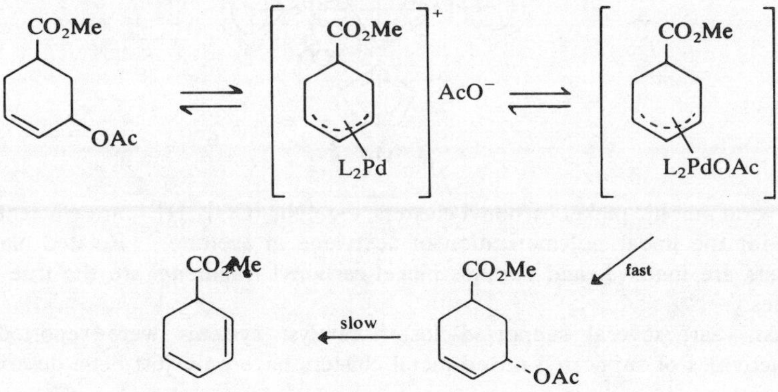

Scheme 15

[55] D. H. R. Barton, S. G. Davies, and W. B. Motherwell, *Synthesis*, 1979, 265.
[56] B. M. Trost, T. R. Verhoeven, and J. M. Fortunak, *Tetrahedron Letters*, 1979, 2301.

isomerization of racemic 1-olefins to 2-olefins by the homogeneous catalyst (*R*)-*NN*-dimethyl-1-phenylethylamine–$AlBu^i_3$–[Ni(*N*-methylsalicylideneamine)$_2$] is reported[57] to constitute just such a reaction. Both the unchanged 1-olefin (*e.g.* 4-methylhex-1-ene) and the product isomeric (*E*)-2-olefin are optically active, demonstrating that isomerization in the presence of the chiral amine is stereoselective. Although optical purities are low, this reaction is an interesting example of the kinetic resolution of a racemic mixture.

9 Metal Clusters In Catalysis

There has been much interest in the use of soluble transition-metal cluster compounds as potential models for catalysis on metal surfaces, and as possible selective catalysts for organic reactions. *X*-Ray and n.m.r. studies of the bonding between organic molecules and clusters have provided results relevant to interactions at metal surfaces, but the tendency of many cluster compounds to fragment in solution means that catalysis by a cluster compound need not be due to the cluster itself. This is particularly true of the more labile cluster compounds of the first-row transition metals. However, a rare example[58] of unambiguous catalysis by such a compound involves [Fe(η^5-C_5H_5)(μ_3-CO)]$_4$ (18). This catalyses the selective hydrogenation of terminal acetylenes to olefins (in the presence of olefin or internal acetylene) and of aryl-nitro-compounds to anilines, and the hydrogenation of activated terminal olefins. The bridging carbonyl groups in (18) are thought to be responsible for its marked reluctance to fragment. Under reaction conditions, the related dimer [Fe(η^5-C_5H_5)(CO)$_2$]$_2$ is converted into inactive [Fe(η^5-C_5H_5)(CO)$_2$H], so the cluster (18) itself must be responsible for the observed catalysis.

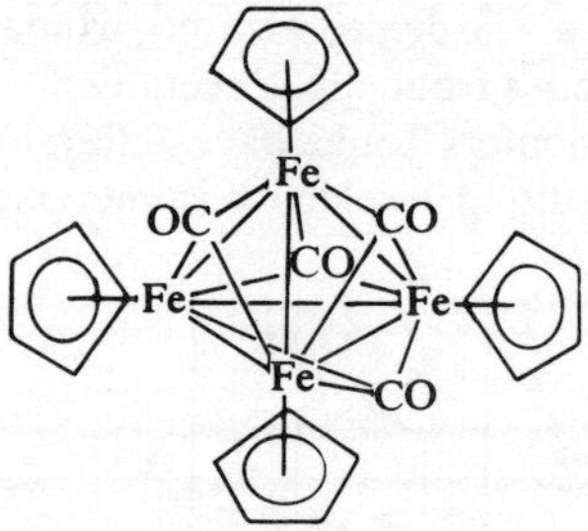

(18)

Several anionic nickel carbonyl clusters, *e.g.* $[Ni_{12}(CO)_{21}H]^{3-}$, are efficient catalysts for the linear polymerization of acetylene in acetone.[59] Related platinum clusters are inactive, and perhaps nickel-carbonyl fragments are the true active species.

Last year, several supported-cluster catalyst systems were reported, but the activities of supported mixed-metal clusters have only just been described.[60]

[57] G. Giacomelli, L. Lardicci, R. Menicagli, and L. Bertero, *J.C.S. Chem. Comm.*, 1979, 633.
[58] C. U. Pittman, R. C. Ryan, J. McGee, and J. P. O'Connor, *J. Organometallic Chem.*, 1979, **178**, C48.
[59] S. Ceriotti, G. Longoni, and P. Chini, *J. Organometallic Chem.*, 1979, **174**, C27.
[60] R. Pierantozzi, K. J. McQuade, B. C. Gates, M. Wolf, H. Knözinger, and W. Ruhmann, *J. Amer. Chem. Soc.*, 1979, **101**, 5436.

Phosphinated poly(styrene–divinylbenzene) supports containing $[Fe_2Pt(CO)_8(Ph_2P—Ⓟ)_2]$ or $[RuPt_2(CO)_5(Ph_2P—Ⓟ)_3]$ are long-lived catalysts for hydrogenation of ethylene. Silica-supported $[H_2Os_3(CO)_9(Ph_2P\text{-sil})]$ and $[HAuOs_3(CO)_{10}(Ph_2P\text{-sil})]$, where sil is silica, are inactive for hydrogenation of propylene, but they do catalyse the isomerization of but-1-ene. No doubt, further reports on supported mixed-metal cluster catalysts will appear.

10 Phase-transfer Catalysis

Phase-transfer catalysis, using a quaternary ammonium salt or a crown ether as transfer agent, is now a much used technique in organic chemistry.[61] Recently, stoicheiometric and catalytic phase-transfer methods have been successfully extended to reactions of organometallic compounds. Examples include an exceptionally facile synthesis of ferrocene derivatives,[62a] as well as preparative procedures for π-allylcobalt tricarbonyl,[62b] alkylidynetricobalt nonacarbonyl,[62b] and acyl-cobalt tetracarbonyls.[62c] Phase transfer of hydroxide ion catalyses carbonyl-substitution reactions of metal carbonyls,[62d] and facilitates the preparation of hydroplatinum compounds.[62e] More importantly, phase-transfer methods allow convenient generation of reactive organometallic species *in situ* in the synthesis of organic compounds.

Fulvenes are obtained by the novel desulphurization and coupling (Scheme 16) of thiobenzophenones with $[Fe(\eta^5\text{-}C_5H_5)(CO)_2]^-$. The reactive carbonyl may be generated *in situ*, *via* phase transfer of hydroxide ion: a thioketone is treated with $[Fe(\eta^5\text{-}C_5H_5)(CO)_2]_2$, aqueous sodium hydroxide, benzene, and a phase-transfer catalyst. After stirring the mixture overnight at room temperature, followed by simple work-up, better yields of fulvenes are obtained than when the carbonyl anion is produced by conventional methods.[63]

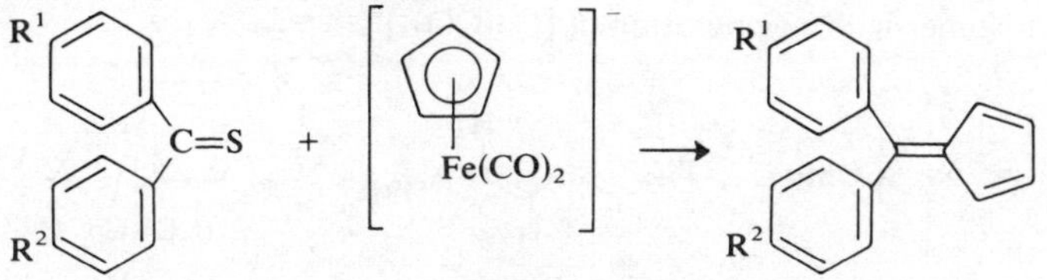

Scheme 16

Tetracarbonylferrates are versatile reagents in organic chemistry, but their preparation from iron pentacarbonyl is time-consuming, and care is needed in their manipulation. They can, however, be conveniently generated in an organic solvent directly from iron pentacarbonyl and aqueous sodium hydroxide, under phase-

[61] C. M. Starks and C. Liotta, 'Phase Transfer Catalysis – Principles and Techniques', Academic Press, London, 1978; W. E. Keller, 'Compendium of Phase Transfer Reactions and Related Synthetic Methods', Fluka AG, CH-9470 Buchs, Switzerland, 1979; W. P. Weber and G. W. Gokel, 'Phase Transfer in Organic Synthesis', Springer-Verlag, Berlin, 1977.

[62] (*a*) M. Salisova and H. Alper, *Angew. Chem. Internat. Edn.*, 1979, **18**, 792; (*b*) H. Alper, H. des Abbayes, and D. des Roches, *J. Organometallic Chem.*, 1976, **121**, C31; H. des Abbayes and A. Buloup, *ibid.*, 1979, **179**, C21; (*d*) K.-Y. Hui and B. L. Shaw, *ibid.*, 1977, **124**, 262; (*e*) M. E. Fakley and A. Pidcock, *J.C.S. Dalton*, 1977, 1444.

[63] H. Alper and H.-N. Paik, *J. Amer. Chem. Soc.*, 1978, **100**, 508.

transfer conditions. This forms the basis of an efficient synthesis of symmetrical and unsymmetrical ketones from halides.[64] Stirring a mixture of benzyl bromide, iron pentacarbonyl, aqueous sodium hydroxide, benzene, and transfer agent (tetra-butylammonium bromide) for three hours, at room temperatures, under an inert atmosphere, affords dibenzyl ketone in 94% yield. Similar results were obtained with several substituted benzyl bromides, but not with less reactive halides such as benzyl chloride or alkyl bromides. These react with $[Fe(CO)_5]$ to form the intermediate alkyl–iron complex; further reaction with a second molecule of the halide does not take place. The intermediate will, however, react with a more reactive halide to give an unsymmetrical ketone. Practically, this sequential reaction of two halides is an attractive route to unsymmetrical ketones. Catalytic carbonylation involving $[Co(CO)_4]^-$ is the basis of several synthetically useful reactions that are normally carried out using relatively high pressures of carbon monoxide. It has been demonstrated that $[Co(CO)_4]^-$ can be generated from $[Co_2(CO)_8]$ and hydroxide ion under phase-transfer conditions, allowing the smooth conversion of benzyl halides into carboxylic acids, in good yield, at atmospheric pressure and room temperature.[65] A variety of conjugated dienes and trienes undergo regiospecific acylation[66] with methyl iodide, carbon monoxide, and a catalytic amount of $[Co(CO)_4]^-$ (generated *via* phase transfer). Analogous reactions with alkynes afford hydroxybutenolides.[67] For instance, 1-vinylcyclohexene is acylated only at the least substituted carbon of the diene, to give the (*E*)(*trans*)-isomer (19) in 60% isolated yield; 1-phenylbuta-1,3-diene is acylated at the 4-position to give the (*E*)-product (20) (86%). It is thought that this reaction proceeds *via* 1,2-addition of $[MeCOCo(CO)_4]$ to the least substituted double-bond of the diene (or triene). Loss of carbon monoxide from the metal then leads to formation of a β-keto π-allyl complex, *e.g.* (21). Subsequent deprotonation of the acidic methylene, and removal of the cobalt carbonyl fragment in the presence of carbon monoxide, gives the thermodynamically stable (*E*)-isomer with regeneration of $[Co(CO)_4]^-$.

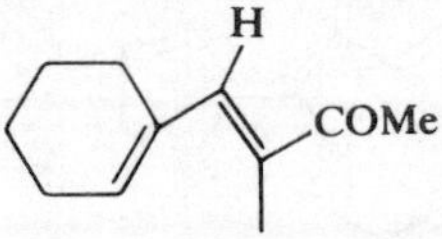

(19)

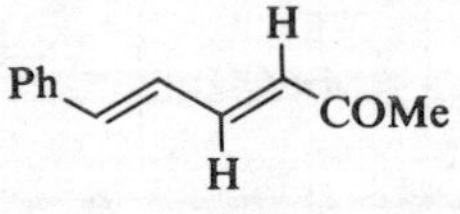

(20)

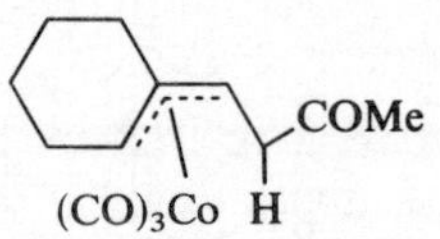

(21)

The inexpensive hydrogenation catalyst $K_3[Co(CN)_4H]$ has not found widespread application, owing to its insolubility in organic solvents. However, the use of $K_3[Co(CN)_5H]$ under phase-transfer conditions allows a variety of conjugated carbon–carbon bonds to be selectively hydrogenated.[68] Only *trans*-pent-2-ene (80% yield) was obtained from penta-1,3-diene, and hydrogenation of 2,3-dimethylbuta-1,3-diene afforded a mixture of 2,3-dimethylbut-2-ene (80%) and 2,3-dimethylbut-1-ene (20%). $\alpha\beta$-Unsaturated ketones react less rapidly, but with

[64] Y. Kimura, Y. Tomita, S. Nakanishi, and Y. Otsuji, *Chem. Letters*, 1979, 321.

[65] L. Cassar and M. Foa, *J. Organometallic Chem.*, 1977, **134**, C15; H. Alper and H. des Abbayes, *ibid.*, 1977, **134**, C11.

[66] H. Alper and J. K. Currie, *Tetrahedron Letters*, 1979, 2665.

[67] H. Alper, J. K. Currie, and H. des Abbayes, *J.C.S. Chem. Comm.*, 1978, 311.

[68] D. L. Reger, M. M. Habib, and D. J. Fauth, *Tetrahedron Letters*, 1979, 115.

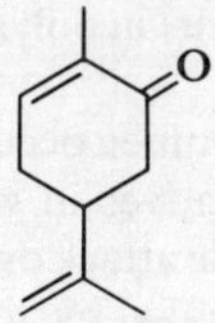

(22)

high selectivity. The conjugated carbon–carbon double bond in *l*-carvone (22) is hydrogenated almost quantitatively.

Miscellaneous uses of phase-transfer catalysis in synthesis *via* organometallics include the nickel(O)-catalysed cyanation of aromatic halides.[69] It is claimed that phase-transfer conditions are more convenient and produce better yields than conventional methods.

It may be expected that more examples of this technique will be reported in the future, and that continued progress (*e.g.* in the selection of an appropriate organic solvent and transfer catalyst to optimize yields) will be made in this growth area.

11 Miscellaneous

A novel ketone synthesis, involving the formal addition of an aldehyde C—H bond to an alkene, has been described by Suggs.[70] The aldehyde is first converted into an aldimine, which adds readily to $[Rh(PPh_3)_3Cl]$, yielding an iminoacyl-rhodium(III) hydride complex. The reaction of the latter with an alkene, followed by hydrolysis, affords a ketone in up to 80% yield. The reaction can be catalytic in rhodium (Scheme 17).

Cycloaddition of butadiene and dialkyl-ynamines is catalysed by the iron(O) complex $[Fe(cyclo\text{-}octatetraene)_2]$, and affords cyclohexa-1,4-dienamines in up to 80% yield. Other catalysts, such as CuCl or $PdCl_2$, cause dimerization of the

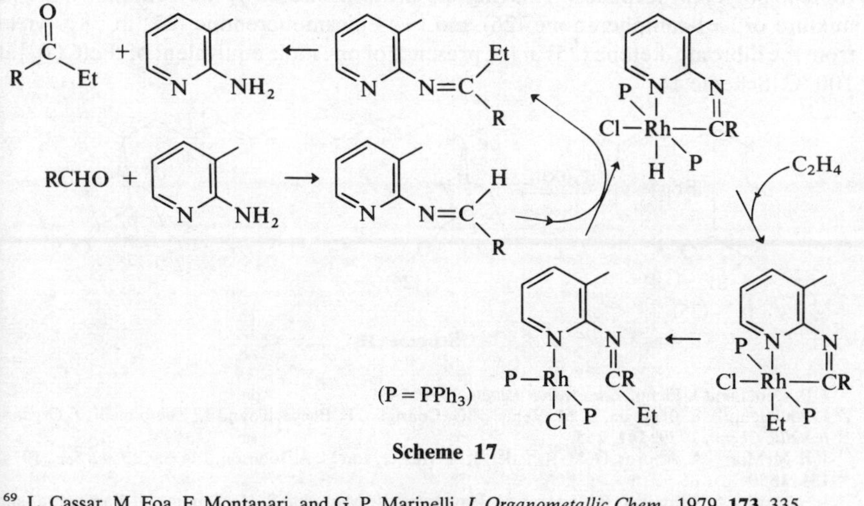

Scheme 17

[69] L. Cassar, M. Foa, F. Montanari, and G. P. Marinelli, *J. Organometallic Chem.*, 1979, **173**, 335.
[70] J. W. Suggs, *J. Amer. Chem. Soc.*, 1979, **101**, 489.

ynamine, while $[RuH_2(PPh_3)_3]$ results in polymerization of the latter under the same conditions.[71]

The cyclization of pent-4-enylamines occurs regiospecifically in the presence of $[PtCl_4]^{2-}$, giving pyrrolidine derivatives in good yield. The proposed mechanism involves intramolecular nucleophilic attack on a co-ordinated double bond, followed by cleavage of the resulting metal alkyl by acid.[72]

Synthesis involving Iron Carbonyls.—Iron carbonyl species have many potential uses in organic synthesis, and, being readily available, at moderate cost, they can be attractive reagents. Two reports illustrate their practical utility. The final stage of one approach[73] to the synthesis of aphidicolin (a diterpenoid tetraol with strong activity against Herpes virus) was accomplished by direct carbonylation/ring expansion of the tosylate derivative (23) with $[Na_2Fe(CO)_4]$. This procedure afforded the desired cyclohexanone (24), which had previously been transformed into aphidicolin. The trapping of reactive 2-oxyallyliron(II) intermediates (derived from bromo-ketones and an iron carbonyl) with monoenes or 1,3-dienes is an efficient route to cyclopentanones and cycloheptenones respectively (for a review see ref. 2).

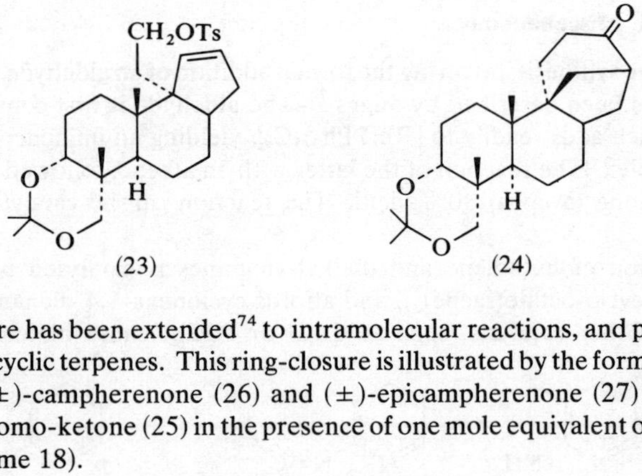

(23) (24)

This procedure has been extended[74] to intramolecular reactions, and provides a new route to polycyclic terpenes. This ring-closure is illustrated by the formation of a 2 : 1 mixture of (±)-campherenone (26) and (±)-epicampherenone (27) in 58% yield from the dibromo-ketone (25) in the presence of one mole equivalent of $[Fe(CO)_5]$ at 100 °C (Scheme 18).

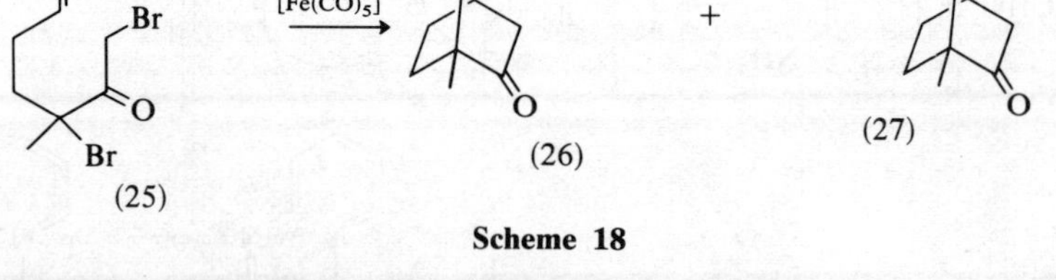

(25) (26) (27)

Scheme 18

[71] J. P. Genet and J. Ficini, *Tetrahedron Letters*, 1979, 1499.
[72] J. Ambuehl, P. S. Pregosin, L. M. Venanzi, G. Consiglio, F. Bachechi, and L. Zambonelli, *J. Organometallic Chem.*, 1979, **181**, 255.
[73] J. E. McMurry, A. Andrus, G. M. Ksander, J. H. Musser, and M. A. Johnson, *J. Amer. Chem. Soc.*, 1979, **101**, 1330.
[74] R. Noyori, M. Nishizawa, F. Shimizu, Y. Hayakawa, K. Maruoka, S. Hashimoto, H. Yamamoto, and H. Nozaki, *J. Amer. Chem. Soc.*, 1979, **101**, 220.

This intramolecular [3+4] reaction (*i.e.* employing a dibromo-keto-diene) also permits direct synthesis of the oxidoperhydroazulene skeleton. Although yields from these intramolecular reactions are not very high, it may be expected that in some instances the approach will be of value.

The cluster hydride $[Et_4N]^+[HFe_3(CO)_{11}]^-$ is readily prepared[75] from iron pentacarbonyl, but, unlike tetracarbonylferrate salts, it has the advantage of being stable in air, and can be stored indefinitely at room temperature. Treatment of a range of aromatic nitro-compounds with this reagent produces anilines in excellent yield; aldehydes are not reduced and ring dechlorination is not important. Double bonds in $\alpha\beta$-unsaturated carbonyl compounds are selectively reduced.

Palladium Chemistry.—Palladium complexes are versatile reagents in organic synthesis, and this is clearly demonstrated in a simple synthesis of (±)-pyrethrolone,[76] which is a component of naturally occurring pyrethroids. Three different palladium-catalysed reactions (see Scheme 19) are used to produce the key intermediate octa-5,7-dien-2-one (28) from butadiene and phenol. The mixture of the isomers of (28) is then readily converted into (*E*)- and (*Z*)-pyrethrolone, which can easily be separated.

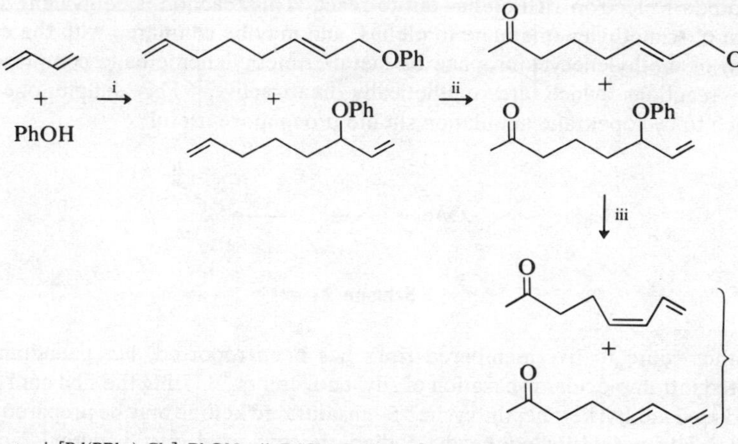

(28)

Reagents: i, $[Pd(PPh_3)_2Cl_2]$, PhONa; ii, $PdCl_2$, CuCl; iii, $[Pd(OAc)_2]$, PPh_3

Scheme 19

Simple organo-palladium chemistry (oxidative addition and insertion of carbon monoxide) has been used to advantage in developing a new route to butenolides from acetylenic alcohols.[77] The acetylene alcohol is first converted into the corresponding vinyl iodo-alcohol (29), which is then catalytically converted into the butenolide (25 °C; 3 atm; 2 days); see Scheme 20. Good overall yields are generally achieved, and, because of the mild conditions, a wide variety of functional groups can be tolerated.

[75] G. P. Boldrini, A. Umani-Ronchi, and M. Panunzio, *J. Organometallic Chem.*, 1979, **171**, 85.
[76] J. Tsuji, T. Yamakawa, and T. Mandai, *Tetrahedron Letters*, 1979, 3741.
[77] A. Cowell and J. K. Stille, *Tetrahedron Letters*, 1979, 133.

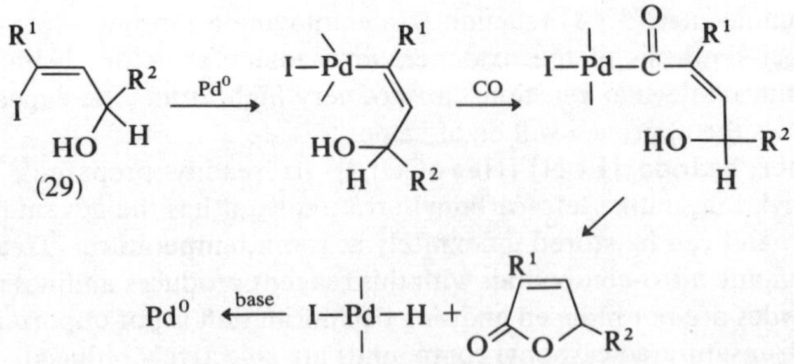

The synthon 2-acetoxymethyl-3-trimethylsilylprop-1-ene reacts with olefins bearing electron-withdrawing groups,[78] in the presence of a palladium(O) catalyst, to produce cyclopentane systems (Scheme 21). Yields of 50–85% are achieved with olefins bearing ester, nitrile, ketone, or sulphone substituents, but simple alkyl-substituted or electron-rich olefins fail to react. This reaction is equivalent to the addition of trimethylenemethane to olefins, and may be compared with the cycloaddition of methylenecyclopropane or metal–trimethylenemethane complexes to olefins – reactions which are synthetically unattractive. This simple one-step approach to cyclopentane annulation should prove more useful.

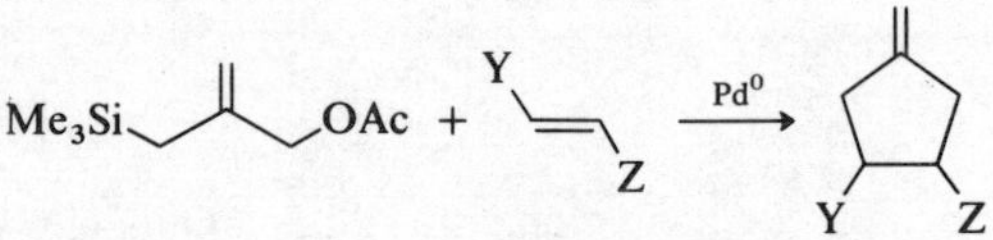

Scheme 21

Another route to five-membered rings has been reported, *via* palladium(II)-promoted intramolecular cyclization of silyl enol ethers.[79] Using the silyl enol ether of an alkenyl methyl ketone, the cyclic $\alpha\beta$-unsaturated ketone may be prepared. For example, 2-trimethylsilyloxy-hexa-1,5-diene is converted into 3-methyl-2-cyclopentenone (87%) by an equimolar amount of [Pd(OAc)$_2$] in acetonitrile. A variety of silyl enol ethers were successfully used, but attempts to prepare six-membered rings or larger gave poor yields.

[78] B. M. Trost and D. M. T. Chan, *J. Amer. Chem. Soc.*, 1979, **101**, 6429.
[79] Y. Ito, H. Aoyama, T. Hirao, A. Mochizuki, and T. Saegusa, *J. Amer. Chem. Soc.*, 1979, **101**, 494.

12 Organometallic Chemistry

Part (ii) Main-Group Elements

By J. L. WARDELL

Department of Chemistry, University of Aberdeen
Meston Walk, Old Aberdeen AB9 2UE

1 Introduction

An authoritative review of the synthetic uses of α-metallated isocyanides has appeared.[1] A comprehensive survey of the preparation of aryl- and heteroaryl-trimethylsilanes was made: topics covered the direct C—Si coupling, cycloaddition, modification of silyl-aromatics by incorporation of further functional groups, and conversion of existing substituents.[2a] Another useful review on organosilanes was concerned with electrophilic substitution.[2b]

Among the points covered in a review on the stereochemistry of organo-arsenic compounds were configurational stability and the energy barriers to inversion of asymmetric amines.[3]

Several methods of direct synthesis were also reported, with direct electrochemical synthesis being reviewed.[4] This method has been successfully used for Cd, Zn, Hg, Al, In, and Sn derivatives. The low-temperature co-deposition of metal vapours with solvents in high excess, followed by warming and subsequent partial re-clustering of metal atoms, allows preparations of very active metal atoms, in the form of slurries, to be made; organometallic derivatives of Cd, Zn, Al, In, Sn, and Pb were readily obtained from these slurries and organic halides.[5] Another synthesis is the co-condensation of reactive free radicals [*e.g.* $CH_3\cdot$ and $CF_3\cdot$; produced from C_2H_6 and C_2F_6 by radio-frequency glow discharge] with metals at −196 °C; derivatives of Cd, Hg, Bi, Ge, and Sn were so prepared.[6]

2 Group 1

Ab initio M.O. calculations (STO-3G) were made on pentaco-ordinate cations $CH_3M_2^+$(M = Li and Na as well as BeH and MgH). The structures (1), of symmetry C_s, can be envisaged as models for electrophilic substitutions occurring with retention, while the D_{3h} structure (2) is a model for processes that occur with inversion. The D_{3h} form was calculated to be the more stable for $CH_3Li_2^+$, a species that is

[1] U. Schöllkopf, *Pure Appl. Chem.*, 1979, **51**, 1347.
[2] (*a*) D. Habich and F. Effenberger, *Synthesis*, 1979, 841; (*b*) T. H. Chan and I. Fleming, *ibid.*, p. 761.
[3] F. D. Yambushev and V. I. Savin, *Russ. Chem. Rev.*, 1979, **48**, 582.
[4] D. G. Tuck, *Pure Appl. Chem.*, 1979, **51**, 2005.
[5] K. J. Klabunde and T. O. Murdock, *J. Org. Chem.*, 1979, **44**, 3901.
[6] T. J. Juhlke, R. W. Braun, T. R. Bierschenk, and R. J. Lagow, *J. Amer. Chem. Soc.*, 1979, **101**, 3229.

known in the gas phase. The results suggest (i) that three-centre two-electron bonds may favour linear over cyclic arrangements, and (ii) electrophilic aliphatic substitution may proceed either with retention or inversion of configuration, depending on the circumstances.[7a] It was also pointed out that a second-row substituent, *e.g.* M = Na, stabilizes methyl cations CH_2M^+ more effectively than does its first-row counterpart.[7b]

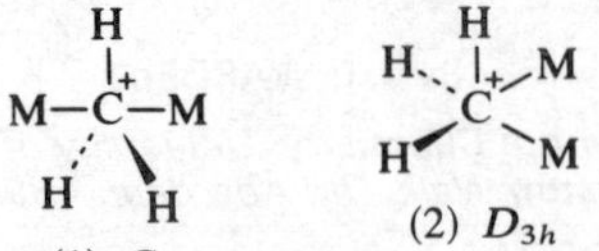

(1) C_s (2) D_{3h}

The stabilizing influence of Li substituents on strained or unsaturated molecules such as ethylene, cyclopropane, acetylene, and various bi- and poly-cyclic species was calculated (STO-3G basis sets) to be significant; *e.g.*, the strain in tetralithiotetrahedrane (3) was calculated to be less than one-fourth of that in tetrahedrane itself.[8] The compound Li_4C_4 (3), reported last year[9] to be produced by photolysis of Li_2C_2, reacts with MeI to give a species having all the properties expected of tetramethyltetrahedrane.[10]

von Schleyer[11] also looked at the structures, energies, and bonding of lithium-substituted allenes, propynes, and cyclopropenes by *ab initio* methods. The most stable arrangements involve acetylidic and bridging lithium atoms; *e.g.*, allenyllithium has a bent carbon skeleton which accommodates simultaneous bonding of Li to C-1 and to C-3, as shown in (4), but not to the nearest atom, C-2; for C_3Li_4 (5)

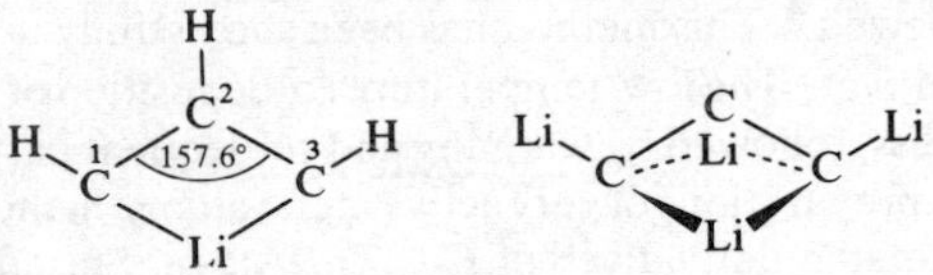

(4) (5)

there are two such bridging lithium atoms and two acetylide-type lithium atoms. The carbenoid CHF_2Li is suggested (from *ab initio* calculations) to exist in three isomeric forms (i) a CHF_2^-, Li^+ ion-pair, (ii) a $CHFLi^+$, F^- ion-pair (6), and (iii) (the least stable) a CHF.LiF complex; *i.e.*, none has a classical structure that contains tetrahedral carbon.[12] The ^{13}C n.m.r. spectra of carbenoids (7), CBr_3Li, and $MeCBr_2Li$, all labelled with ^{13}C at the carbenoid centre, were recorded at −100 °C in THF solution. From the CBr_3Li spectrum, three compounds could be detected;[13] these

[7] (*a*) E. D. Jemmis, J. Chandrasekhar, and P. von R. Schleyer, *J. Amer. Chem. Soc.*, 1979, **101**, 527; (*b*) T. Clark and P. von. R. Schleyer, *Tetrahedron Letters*, 1979, 4641.

[8] J. D. Dill, A. Greenberg, and J. F. Liebman, *J. Amer. Chem. Soc.*, 1979, **101**, 6814.

[9] M. G. Hutchins, *Ann. Reports (B)*, 1978, **75**, 120.

[10] N. S. Zefirov, V. N. Kirin, N. M. Yur'eva, A. S. Koz'min, N. S. Kulikov, and Yu. N. Luzikov, *Tetrahedron Letters*, 1979, 1925.

[11] E. D. Jemmis, J. Chandrasekhar, and P. von R. Schleyer, *J. Amer. Chem. Soc.*, 1979, **101**, 2848.

[12] T. Clark and P. von R. Schleyer, *Tetrahedron Letters*, 1979, 4963.

[13] D. Seebach, H. Siegel, K. Müllen, and K. Hiltbrunner, *Angew. Chem. Internat. Edn.*, 1979, **18**, 784; H. Siegel, K. Hiltbrunner, and D. Seebach, ibid., p. 785.

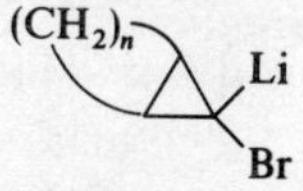

(7) $n = 4$ or 6

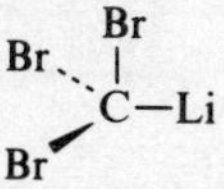

(8)

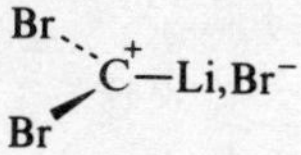

(9)

were the carbene $:CBr_2$ and two CBr_3Li species, one having tetrahedral carbon and the other (more stable) either like (8) or (9), *i.e.* analogous to (6). All carbenoids exhibit large downfield shifts of the carbenoid carbons.

Previously, a preparation of primary alkyl-lithiums from olefins (Scheme 1) was reported; now a complement to it – the preparation of secondary or tertiary alkyl-lithium from the same olefins (Scheme 2) – has been published.[14]

$$R^1R^2C{=}CH_2 + PhSH \xrightarrow{i} R^1R^2CHCH_2SPh \xrightarrow{ii} R^1R^2CHCH_2Li$$

Reagents: i, Radical initiator; ii, Li, THF

Scheme 1

$$R^1R^2C{=}CH_2 + PhSH \xrightarrow{i} R^1R^2C(SPh)CH_2 \xrightarrow{ii} R^1R^2CL_iCH_2$$

Reagents: i, $HClO_4$; ii, Li, THF

Scheme 2

The relative *ortho*-directing ability of groups Y in aromatic lithiations has been further studied,[15] both intramolecularly, using $YC_6H_4CONEt_2$ (10), and intermolecularly, involving competition between phenyloxazoline and PhY for BuLi. After suitable trapping, the metallation of (10) by Bu^sLi and tetramethylethylenediamine in THF at $-100\,°C$ indicated $CONEt_2$ to be superior to SO_2NEt_2, oxazoline, MeO, Me_2NCH_2, Cl, CO_2H, and Me in directing lithiation to an *ortho*-position. In the intermolecular reactions, a sequence of *ortho*-directing ability was established as $Me_2NSO_2 > CONEt_2 >$ oxazoline $> CH_2NMe_2$; anion crossover was detected, *e.g.* between the compound o-$LiC_6H_4\overline{C{=}NCMe_2CH_2O}$ and $PhSO_2NMe_2$.

Metallation of the acetal (H-2) protons in 1,3-dioxolans,1,3-dioxans, and open-chain acetals is only possible if the proton can occupy an equatorial-like conformation; *i.e.*, a preferred equatorial deprotonation, even in the case of a carbanion α to oxygen.[16] A hydroxyl substituent activates cyclopropanes towards metallation by Pr^iLi; the major product of reaction of cyclo-$C_3H_5CHR(OH)$, after carboxylation,[17] was *cis*-1-RCH(OH)-2-HO_2C-cyclo-C_3H_4, the other products being the *trans*-isomer and 1-RCH(OH)-1-HO_2C-cyclo-C_3H_4. Lithiation of (*E*)-$R^1SCH{=}CHCH_2XR^2$ (11; X = O or S) by BuLi occurs[18] at the sp^3 carbon; in contrast, lithiation of (11; $R^1 = Ph$, $XR^2 = NMe_2$) happens as shown in Scheme 3.

[14] C. G. Screttas and M. Micha-Screttas, *J. Org. Chem.*, 1979, **44**, 713.
[15] (*a*) P. Beak and R. A. Brown, *J. Org. Chem.*, 1979, **44**, 4463; (*b*) A. I. Meyers and K. Lutomski, *ibid.*, p. 4464.
[16] A. I. Meyers, A. L. Campbell, A. G. Abatjoglou, and E. L. Eliel, *Tetrahedron Letters*, 1979, 4159.
[17] G. W. Klumpp, M. Kool, M. Schakel, R. F. Schmitz, and C. Boutkan, *J. Amer. Chem. Soc.*, 1979, **101**, 7065.
[18] J. J. Fitt and H. W. Gschwend, *J. Org. Chem.*, 1979, **44**, 303.

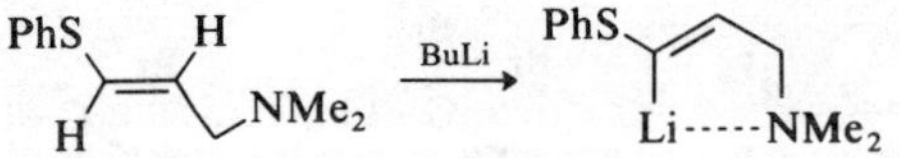

Scheme 3

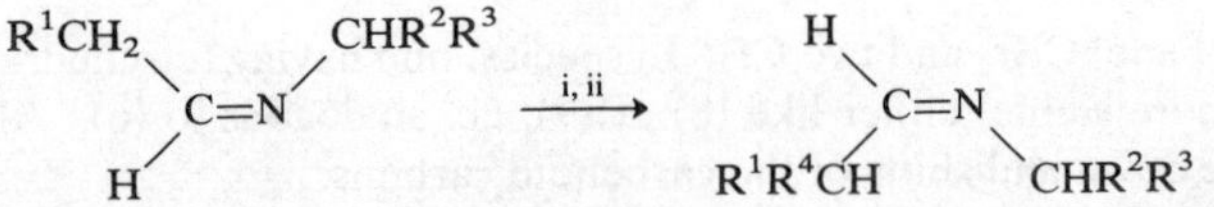

Reagents: i, $LiNPr^i_2$; ii, R^4X

Scheme 4

The lithiation and alkylation (Scheme 4) of aliphatic aldimines has been shown to give a ratio of *syn*- to *anti*-products of 96:4. This is interpreted as representing a minimum value for the *syn* : *anti* ratio of the lithiated aldimine intermediates.[19] The factor responsible for the preferential *syn* stabilization, though not yet identified, is determined to be >18 kJ mol^{-1}. Complete regiochemical control[20] pertains in the generation of alkenyl-lithiums from arenesulphonylhydrazones (*e.g.* Scheme 5).

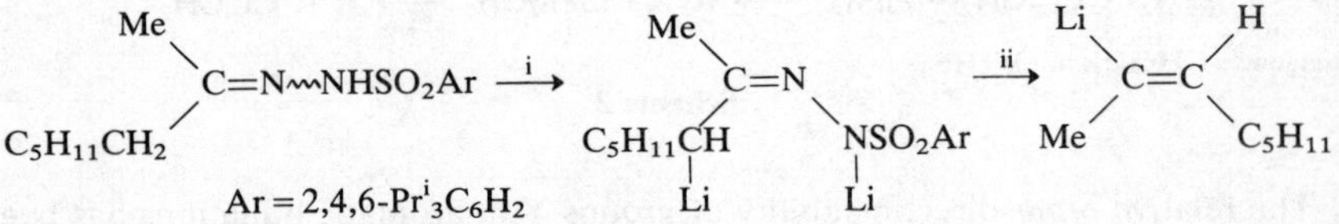

$Ar = 2,4,6\text{-}Pr^i_3C_6H_2$

Reagents: i, Bu^sLi, at −78 °C, THF; ii, 0 °C

Scheme 5

Transmetallation of allyl-tin and -lead compounds by organolithiums continues to be a most useful source of unsymmetrically substituted allyl-lithiums; the utility of the method is enhanced by development of new allylic-tin species *via* Wittig-style reactions.[21] Another source of allyl-lithiums, *i.e.* the conrotatory ring-opening of cyclopropyl-lithiums (*e.g.*, see Scheme 6), has also attracted further attention. The ring-opening of (12; X = CN) to *endo,endo*- and *exo,exo*-(13) proceeds at a slower rate than the isomerization of the latter to the thermodynamically more stable isomer *exo, endo*-(13). Percentages of equilibrium conformations and rotational barriers in

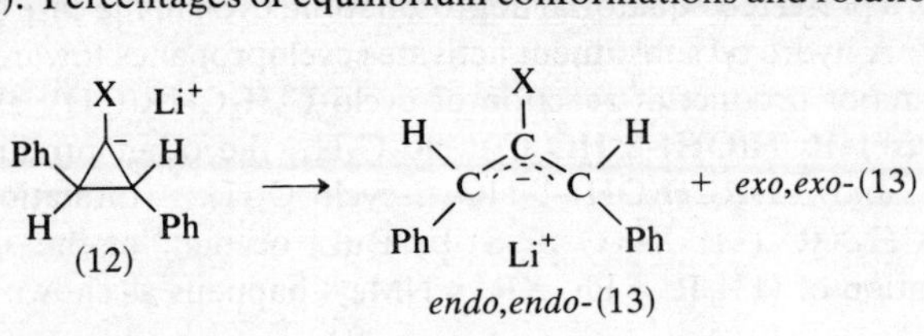

X = H, alkyl, aryl, or CN

Scheme 6

[19] R. R. Fraser and J. Banville, *J.C.S. Chem. Comm.*, 1979, 47.
[20] A. R. Chamberlin and F. T. Bond, *Synthesis*, 1979, 44.
[21] D. Seyferth and R. E. Mammarella, *J. Organometallic Chem.*, 1979, **177**, 53; D. Seyferth and K. R. Wursthorn, *ibid.*, 1979, **182**, 455; B. Mauze, *ibid.*, 1979, **170**, 265.

these and other allyl–alkali-metal compounds were determined. The potential steric congestion in the *endo,endo*-isomers is not realized, as the rings adopt a cyclophane-like conformation; in addition, an expansion of the angles at sp^2 carbon atoms of allyl occurs.[22]

The dynamic behaviour and structure of propyl-lithium, enriched with ^{13}C and ^{6}Li, was studied by n.m.r. spectroscopy. An advantage of using ^{6}Li is that its quadrupole relaxation is too slow (unlike that of ^{7}Li) to perturb the spectra. The hexamer undergoes fast intra-aggregate C—Li bond exchange to at least −80 °C and inter-aggregate exchange at higher temperatures. Evidence was also gained for different species at lower temperatures.[23]

3 Group 2

Interest[24] in (cyclopentadienyl)$_2$beryllium remains high. New electron-diffraction data at 120° were interpreted as being compatible with a C_{5v} symmetry, and in particular with a slipped sandwich (or *pentahapto*, *trihapto*) model, *e.g.* (14) – a

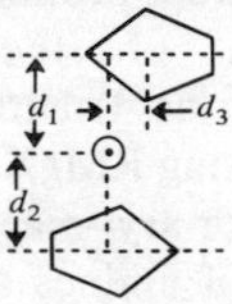

$d_1 = 1.86(2)$ Å
$d_2 = 1.468(10)$ Å
$d_3 = 0.79(10)$ Å
angle between C_5H_5 planes = −4(3)°

(14)

structure similar to that obtained by *X*-ray diffraction, in which d_3 was shown to be 1.2 Å but not compatible with D_{5d} symmetry or [π-C_5H_5, σ-C_5H_5] structures. The potential energy of the molecule was found not to alter much as d_3 changes between 0 and 1.2 Å. *Ab initio* calculations with a double-ζ basis were also carried out. Lower energies are obtained for the [π-C_5H_5, σ-C_5H_5] and D_{5d} models, in keeping with various earlier calculations, but clearly in conflict with the electron-diffraction data. Ionization potentials, obtained from the He(I) photoelectron spectrum, were compared with the calculated orbital energies, satisfactory agreement being found for the slipped sandwich structure. The Raman spectra of solid (at −100 and 25 °C) and liquid [$(C_5H_5)_2Be$](at 65 °C) also suggested the presence of π-bonded rings – one *pentahapto* and the other *polyhapto*. Allylic magnesium compounds and benzyne, generated *in situ*, undergo three competing reactions: (*a*) nucleophilic addition, (*b*) ($\pi^2 + \pi^2$) and (*C*) ($\pi^4 + \pi^2$) cycloadditions (Scheme 7). With cyclohexyne, only nucleophilic additions result.[25] While the addition of R^1MgX (R^1 = alkyl or aryl) to [$BrMgC{\equiv}CCHR^2OMgBr$] (R^2 = H or Me) in the presence of Cu^IX provides

[22] G. Boche, K. Buckl, D. Martens, D. R. Schneider, and H. U. Wagner, *Chem. Ber.*, 1979, **112**, 2961; G. Boche, K. Buckl, D. Martens, and D. R. Schneider, *Tetrahedron Letters*, 1979, 4967; T. B. Thompson and W. T. Ford, *J. Amer. Chem. Soc.*, 1979, **101**, 5459.

[23] G. Fraenkel, A. M. Fraenkel, M. J. Geckle, and F. Schloss, *J. Amer. Chem. Soc.*, 1979, **101**, 4745.

[24] A. Almenningen, A. Haaland, and J. Lusztyk, *J. Organometallic Chem.*, 1979, **170**, 271; R. Gleiter, M. C. Böhm, A. Haaland, R. Johansen, and J. Lusztyk, *ibid.*, p. 285; J. Lusztyk and K. B. Starowieyski, *ibid.*, p. 293.

[25] J. G. Duboudin, B. Jousseaume, and M. Pinet-Vallier, *J. Organometallic Chem.*, 1979, **172**, 1.

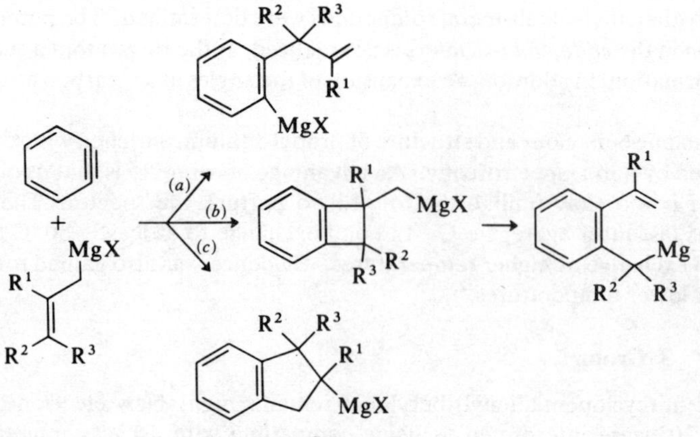

Scheme 7

$[(BrMg)_2C{=}CR^1CHR^2OMgBr]$, allylmagnesium bromide[26] produces the cyclized material (15). The carbomagnesiation of the unsaturated alcohols. $Ph_2(HO)C(CH_2)_nCH{=}CH_2$ (16; $n = 0$, 1, 2, or 4) and of cycloalkenols, *e.g.* (2-cyclohexenyl)diphenylcarbinol (17), occurs using RMgX (R = alkyl, $PhCH_2$, or Bu^t) and $[(allyl)_2Mg]$, but not with primary alkyl- or aryl-magnesiums.[27] The reaction is catalysed by Ni but is retarded in THF or if amines are present. The order of reactivity of compounds (16) was established as $n = 1 > 0 \gg 2 \gg 4$. The mechanism of the uncatalysed reaction of (17) with $[(CH_2{=}CHCH_2)MgX]$ is shown in Scheme 8, in

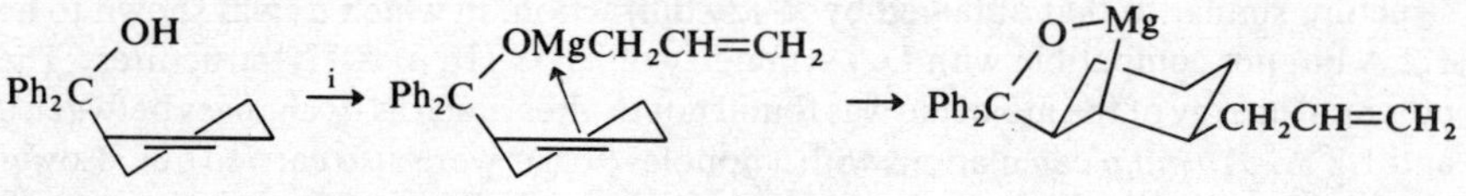

Reagents: i, $CH_2{=}CH_2MgX$.

Scheme 8

which a unimolecular *cis*-addition (with respect to the OH group) occurs.[27] The Grignard reagent (18) from $MeCHClCH_2CHMeCH{=}CH_2$ cyclizes reversibly in ethereal solutions to stereoisomers of (19); at 100 °C, in THF, the equilibrium constant $K (= [(19)]/[(18)])$ is 3.4, with a forward rate constant of 6.8×10^{-6} s; these values compare[28] with those for the unsubstituted cyclobutylmethylmagnesium

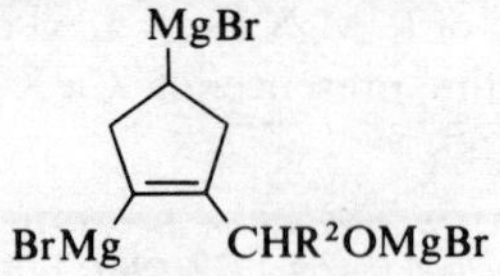

(15)

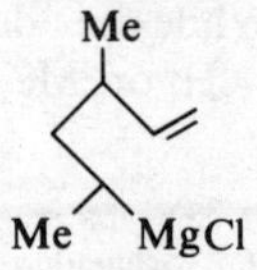

(18)

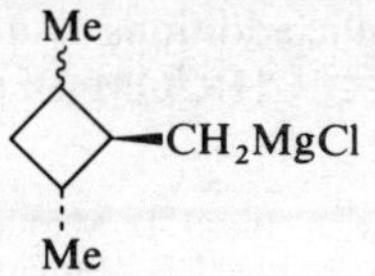

(19)

[26] J. G. Duboudin and B. Jousseaume, *Synth. Comm.*, 1979, **9**, 53.

[27] J. J. Eisch and J. H. Merkley, *J. Amer. Chem. Soc.*, 1979, **101**, 1148; J. J. Eisch, J. H. Merkley, and J. E. Galle, *J. Org. Chem.*, 1979, **44**, 587.

[28] E. A. Hill and M. M. Myers, *J. Organometallic Chem.*, 1979, **173**, 1.

system of 0.94×10^{-4} and 2×10^{-8} s. Compounds R^1MgOR^2 (20), as well as the zinc and aluminium analogues, decompose on heating to an alkane R^1H, an olefin R^2H, and the metal oxide,[29] in a process involving a cyclic six-centred transition state; *e.g.*, methylmagnesium-*threo*-1,2-diphenyl-1-propoxide produces exclusively (*Z*)-diphenylpropene. As (20) is produced from $R^1{}_2Mg$ and an alcohol, the process is, in essence, a useful method of dehydrating alcohols.[29]

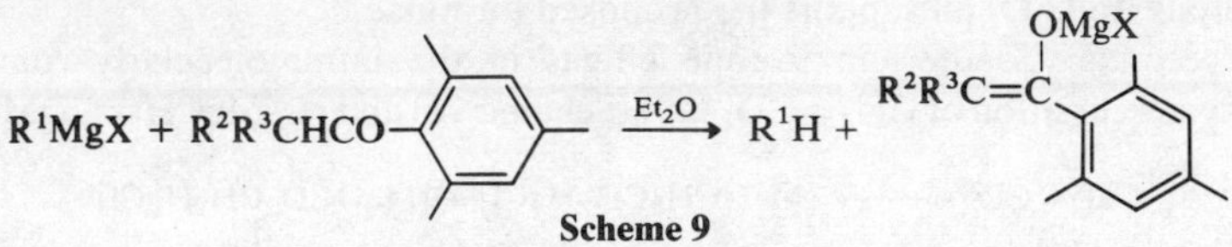

Scheme 9

The reactivity of R^1MgX in the enolization of ketones, *e.g.* as shown in Scheme 9, was established as $R^1 = Et > Pr^i > Me \gg Bu^t$, and is a consequence of both steric and electronic factors. The reaction involved co-ordination of R^1MgX to the ketone and subsequent rate-determining removal of hydrogen from the α-carbon in a six-membered transition state.[30] Results from a detailed study of the reaction of $PhCH_2MgCl$ with H_2CO indicate that it is not the paradox as often stated, but has many features in common with other aldehyde–benzyl-Grignard reactions; the products, in yields that are dependent on the conditions,[31] are *o*-$Me_2C_6H_4CH_2OH$, *o*-$HOCH_2CH_2C_6H_4CH_2OH$, and $PhCH_2CH_2OH$.

Several π-arene complexes of cadmium(II) and zinc(II), *i.e.* $[M^{2+}(ArH)]$, have been isolated from solutions of $[Cd(AsF_6)_2]$, $[Cd(SbF_6)_2]$, and $[Zn(SbF_6)_2]$ with methylbenzenes in liquid SO_2. Related species $[Hg^{2+}(ArH)_n]$ ($n = 1$ or 2) have also been prepared. Such complexes are sensitive to water and organic solvents. Stability constants for the complexes $[Cd(AsF_6)_2 \cdot ArH]$ are smaller than those of $[Hg_2(ArH)]^{2+}$ and $[Hg(ArH)]^{2+}$, and are in the range 0.5—2.1 l mol^{-1}. Localized bonding of the arene to the metal is indicated from ^1H and ^{13}C n.m.r. data; complexation generally produces deshielding of the ^{113}Cd resonance.[32]

The first well-characterized organo-mercury(I) derivatives (21) have been described[33] (Scheme 10). The corresponding Hg^{II} ketenides (22) are similarly

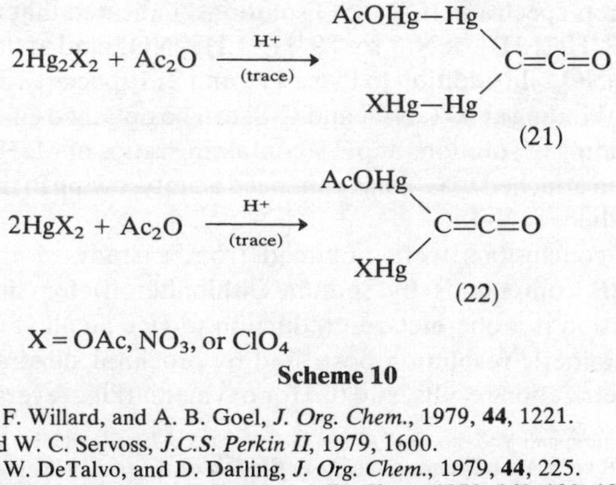

$X = OAc, NO_3$, or ClO_4

Scheme 10

[29] E. C. Ashby, G. F. Willard, and A. B. Goel, *J. Org. Chem.*, 1979, **44**, 1221.
[30] A. G. Pinkus and W. C. Servoss, *J.C.S. Perkin II*, 1979, 1600.
[31] R. A. Benkeser, W. DeTalvo, and D. Darling, *J. Org. Chem.*, 1979, **44**, 225.
[32] L. C. Damude and P. A. W. Dean, *J. Organometallic Chem.*, 1979, **168**, 123; 1979, **181**, 1.
[33] E. T. Blues, D. Bryce-Smith, and H. Karimpour, *J.C.S. Chem. Comm.*, 1979, 1043.

prepared (Scheme 10). Complexes (21) and (22) [i.r.: ν(CCO) *ca.* 2070(s), ν(HgC) 276(m), and ν(CCO) 620(w)] are probably polymeric, being involatile, infusible, and insoluble species, and (happily) non-explosive, except for $X = ClO_4$. Reactions with dilute hydrochloric acid [providing Hg^{n+} ($n = 1$ or 2) from (22) or (21)], with hydrogen chloride gas [producing keten and AcCl], and with bromine {giving $[(XHg_n)(AcOHg_n)CBrCOBr]$; $X = ClO_4$, $n = 1$ or 2 from (22) or (21)}, as well as thermolysis (to C_2O) all support the proposed formulae.

The secondary deuterium isotope effects in the intramolecularly competing methoxymercuriation of $CH_2{=}CD_2$ (23) (Scheme 11) and *cis*-MeCH=CDMe were

$$\underset{(23)}{CH_2{=}CD_2} \longrightarrow \underset{\alpha}{[MeOCH_2CD_2HgCl]} + \underset{\beta}{[MeOCD_2CH_2HgCl]}$$

Reagents: i, $[Hg(OAc)_2]$, MeOH; ii, NaCl

Scheme 11

determined from the isomer ratios $\alpha : \beta$ to be 1.12 and 1.06 respectively. These values indicate that the transition states do not have symmetrical structures, as do mercurinium ions, but have appreciable C—O bonding.[34] 4-Alkyl-cyclohexenes undergo oxymercuriation–demercuriation in a remarkedly stereoselective but non-regioselective manner; for 3-alkyl-cyclohexanes, both a stereo- and regio-selectivity results.[35]

Organomercury complexes of guanosine (24) ($\equiv LH_2$; $R = NH_2$) and related species, including inosine (24; R = H), were amongst those studied.[36] Infrared

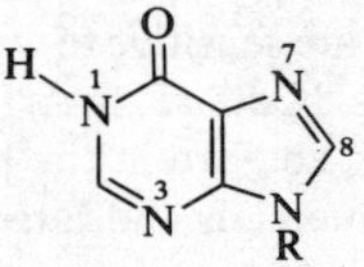

(24) R = ribosyl

spectra of solids, prepared from $MeHgNO_3$ or PhHgOH (R^1HgX) at an appropriate pH, as well as n.m.r. spectra in $(CD_3)_2SO$ solutions, indicated that the binding sites were at N-1 for $[R^1Hg(LH)]$, at N-7 for $[R^1Hg(LH_2)]NO_3$, and at both N-1 and N-7 for $[(R^1Hg)_2(LH)]NO_3$. In addition to these 1 : 1 and 2 : 1 species, a 3 : 1 complex, *i.e.* $[(MeHg)_3L]NO_3$, (binding at N-1, N-7, and C-8) can be obtained either on heating or on prolonged standing of solutions at pH 7 containing ratios of $MeHgNO_3$ to LH_2 of 3 : 1. The hydrogen attached to C-8 has enhanced acidity, owing to the co-ordination of MeHg at the adjacent N-7.

The following conclusions were obtained from a study of the reduction of organo-mercury(II) compounds by sodium dithionite: (i) for simple alkyls the predominant reaction is a one-electron reduction to give an alkyl radical, with the loss of any enantiomeric resolution possessed by prochiral substrates, (ii) for aryl substrates, symmetrization results, and (iii) for oxymercurials, reversion to an alkene

34 S. Shinoda, M. Isemura, and Y. Saito, *Bull. Chem. Soc. Japan*, 1979, **52**, 1855.
35 H. C. Brown, G. J. Lynch, W. J. Hammar, and L. C. Liu, *J. Org. Chem.*, 1979, **44**, 1910.
36 E. Buncel, A. R. Norris, W. J. Racz, and S. E. Taylor, *J.C.S. Chem. Comm.*, 1979, 562; A. J. Canty and R. S. Tobias, *Inorg. Chem.*, 1979, **18**, 413.

occurs.[37] The reaction of oxymercurials with $H_2C{=}CHR$ ($R = CN, CO_2Et$, or Ph) in the presence of $[NaBH(OMe)_3]$ leads[38] to the replacement of HgX by CH_2CH_2R (Scheme 12).

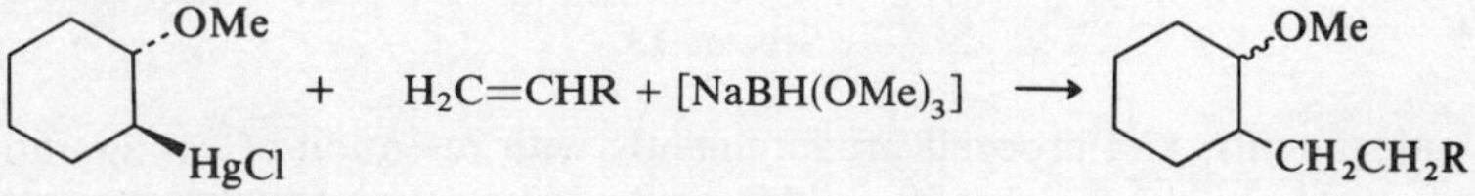

Scheme 12

4 Group 3

A theoretical study, using double-ζ basis sets, was made on the complex between an aluminium atom and acetylene, whose e.s.r. spectrum had previously been obtained at liquid-helium temperatures and interpreted as arising from a σ-bound complex. However, the most stable of the bound isomers was calculated to be the vinylidene structure (25) (binding energy *ca.* 20 kcal mol^{-1}) while the *cis*-(and *trans*-)-$AlCH{=}\dot{C}H$ species (26) had a binding energy of only *ca.* 8 kcal mol^{-1}. A resolution of this apparent conflict was suggested[39] to be that, at the temperature of liquid helium, there is insufficient energy to overcome the barrier of the 1,2-hydrogen shift required for conversion of (26) into (25). Matrix-isolation techniques were used to obtain the complex between an aluminium atom and benzene.[40] It was deduced that Al complexes with one of the C=C units of the ring.

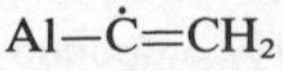

(25)

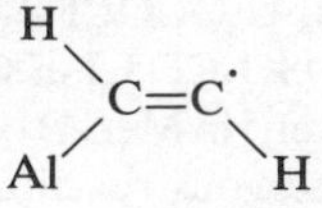

(26)

Hydroalumination of alkynes by $[HAl(NR_2)_2]$, in the presence of $[Cp_2TiCl_2]$ as catalyst, in benzene, occurs stereospecifically with *cis*-addition; *e.g.*, from $PhC{\equiv}CMe$, (*E*)-$[PhCH{=}CMeAl(NR_2)_2]$ (10%) and (*E*)-$[(R_2N)_2Al(Ph)C{=}CHMe]$ (90%) were obtained. The regiochemistry of the product(s) is determined by that of the intermediate alkynyl-titanium compound.[41]

Carbometallation of certain alkynols can be accomplished under mild conditions with transition-metal–organo-aluminium systems; *e.g.*, $[Et_2AlCl]$, $[Cp_2TiCl_2]$, and $HC{\equiv}CCH_2CH_2OH$, at 0 °C, provides, after hydrolysis, *trans*-$EtCH{=}CHCH_2CH_2OH$ and $CH_2{=}CEtCH_2CH_2OH$ in a 1:1 ratio. Changing the catalyst to the more bulky and less readily reduced $[(\eta^5\text{-}MeC_6H_4)_2TiCl_2]$ increases the overall yields, but without having much effect on the regioselectivity.[42]

Organo-zirconium(IV) complexes have been used[43] as precursors of organo-aluminiums (Scheme 13). The transmetallations occur more readily for alkenyl than

[37] L. M. Sayre and F. R. Jensen, *J. Org. Chem.*, 1979, **48**, 228.
[38] B. Giese and K. Heuck, *Chem. Ber.*, 1979, **112**, 3759, 3766.
[39] M. Trenary, M. E. Casida, B. R. Brooks, and H. F. Schaefer, *J. Amer. Chem. Soc.*, 1979, **101**, 1638.
[40] P. H. Kasai and D. McLeod, *J. Amer. Chem. Soc.*, 1979, **101**, 5860.
[41] E. C. Ashby and S. R. Noding, *J. Organometallic Chem.*, 1979, **177**, 117; see also *J. Org. Chem.*, 1979, **44**, 4364.
[42] D. C. Brown, S. A. Nicholas, A. B. Gilpin, and D. W. Thompson, *J. Org. Chem.*, 1979, **44**, 3457.
[43] D. B. Carr and J. Schwartz, *J. Amer. Chem. Soc.*, 1979, **101**, 3521.

$$[Cp_2Zr(H)Cl] \longrightarrow Cp_2Zr(Cl)(R) \xrightarrow{ii} [Cp_2ZrCl_2] + (RAlCl_2)_n$$

Reagents: i, alkene or alkyne; ii, $AlCl_3$

Scheme 13

alkyl substituents, and proceed predominantly with retention of configuration at carbon. Acyl groups can also be transferrred. Silica-supported titanium chloride and polystyrene-supported titanocene dichloride have been used[44] to catalyse the hydroalumination of alkanes and dienes by $LiAlH_4$. Racemic 1-olefins are isomerized by the homogeneous catalytic system (*R*)-*NN*-dimethyl-1-phenylethylamine–$AlBu^i_3$–nickel(*N*-methylsalicylideneamine)$_2$: both the unchanged 1-olefins and the (*E*)-2-olefins that are formed are optically active, indicating that the reaction is stereoselective; *e.g.*, $EtCPhHCH_2CH{=}CH_2$ is 35% isomerized to (−)-(*R*)-(*E*)-$EtCPhHCH{=}CHCH_3$ (0.24% optical purity), with the unchanged 1-olefin having a (+)-(*S*) configuration[45] (0.12% optical purity). Cyclopropylmethyl, allylic, and benzylic acetates are alkylated by trialkyl-aluminiums in CH_2Cl_2 solution at room temperature; carbo-cationic intermediates have been detected.[46]

Esters may be converted into amides by treatment with $R^1{}_2AlNR^2{}_2$; thus *m*-$NO_2C_6H_4CO_2Me$ and $[Et_2AlNH_2]$, from $[Et_3Al]$ and NH_3, produced *m*-$NO_2C_6H_4CONH_2$ in 91% yield.[47]

Thallium(III) tris(trifluoromethylsulphonate) is much more effective in aromatic thallations[48] than is $[Tl(OCOCF_3)_3]$; mono-thallation of polyfluoro-benzenes readily occurs, using $[Tl(OSO_3CF_3)_3]$ in CF_3CO_2H. Oxythallation of norbornene derivatives, using $[Tl(OAc)_3]$ in MeOH, provides *cis-exo*-acetoxythallated adducts (27), in contrast to the analogous reaction of the mercury compound, which leads to the methoxy-derivatives. Treatment of the adducts (27) with $NaBH_4$ provides olefins and some *exo*-alcohols.[49]

5 Group 4

The preparation and spectroscopic characterization of the elusive species dimethylsilylene, Me_2Si (28), has been achieved. Irradiation, at 254 nm, of $(Me_2Si)_6$ in rigid hydrocarbon glasses at 77 K produces singlet silylene (28) ($\nu = 1220\ cm^{-1}$, $\lambda_{max} = 453$ nm). Trapping of (28) by various agents, including hexene but not $Me_2Si(OMe)_2$, was possible. Irradiation of (28) with visible light led to the formation of 2-silapropene.[50] As well as the usual traps for silylenes, *e.g.* β-diketones and alkenes,[51] a silylene such as (28) can be trapped as its dimer, $Me_2Si{=}SiMe_2$, by anthracene;[52] thus, heating 2,3-dibenzo-7,7-dimethyl-1,4,5,6-tetraphenyl-7-

44 F. Sato, H. Ishikawa, Y. Takahashi, M. Miura, and M. Sato, *Tetrahedron Letters*, 1979, 3745.
45 G. Giacomelli, L. Lardicci, R. Menicagli, and L. Bertero, *J.C.S. Chem. Comm.*, 1979, 633.
46 A. Itoh, K. Oshima, S. Sasaki, H. Yamamoto, T. Hiyama, and H. Nozaki, *Tetrahedron Letters*, 1979, 4751.
47 J. L. Wood, N. A. Khatri, and S. M. Weinreb, *Tetrahedron Letters*, 1979, 4907.
48 G. B. Deacon and D. Tunaley, *Austral. J. Chem.*, 1979, **32**, 737.
49 S. Uemura, H. Miyoshi, M. Okano, I. Morishima, and T. Inubushi, *J. Organometallic Chem.*, 1979, **165**, 9.
50 T. J. Drahnak, J. Michl, and R. West, *J. Amer. Chem. Soc.*, 1979, **101**, 5427.
51 W. Ando and M. Ikeno, *J.C.S. Chem. Comm.*, 1979, 655; M. Ishikawa, K. I. Nakagawa, and M. Kumada, *J. Organometallic Chem.*, 1979, **178**, 105.
52 Y. Nakadaira, T. Kobayashi, T. Otsuka, and H. Sakurai, *J. Amer. Chem. Soc.*, 1979, **101**, 486.

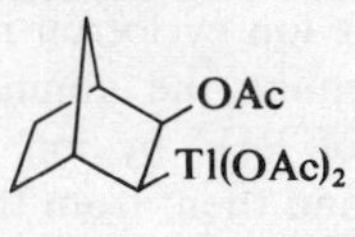

(27)

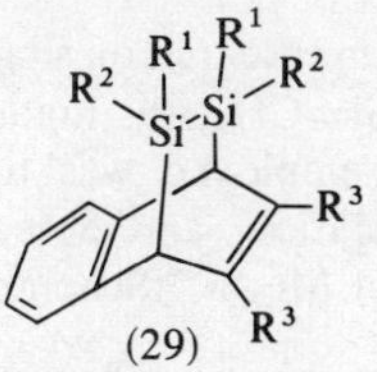

(29)

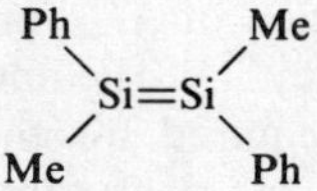

(30)

silanorbornadiene in the presence of anthracene, at 350 °C, provides (29; $R^1 = R^2 =$ Me, R^3–R^3 = —CH=CHCH=CH—). Geometric isomerism in Si=Si bonded species, *e.g.* (30), has been shown.[53] Thermolysis of (29; R^1 = Me, R^2 = Ph, R^3 = H) in the presence of anthracene at 300 °C provides (29; R^1 = Me, R^2 = Ph, R^3–R^3 = —CH=CHCH=CH), *via trans*-(30). In a similar fashion, the *cis*-isomer also reacts stereospecifically. The use of higher temperatures and less reactive enophiles than anthracene lead to a reduced stereoselectivity. These results clearly show the Si=Si to be a true double bond, with slow $(E) \rightleftharpoons (Z)$ isomerism at the temperatures used.[53]

During an *ab initio* study, using non-empirical pseudopotential methods, on various C_2H_4Si isomers it was calculated[54] that the stability sequence is silylacetylene < silacyclopropene (31) ≃ silacyclopropylidene (32) < 2-sila-allene < 2-silapropyne. Particularly noteworthy was the stability of singlet (32) and the non-aromatic character of (31). Photolysis of $HC{\equiv}CSiR^1_2SiR^2_3$ provides silylethylenes, silacyclopropene, and silapropadiene derivatives (*e.g.* as in Scheme 14). These

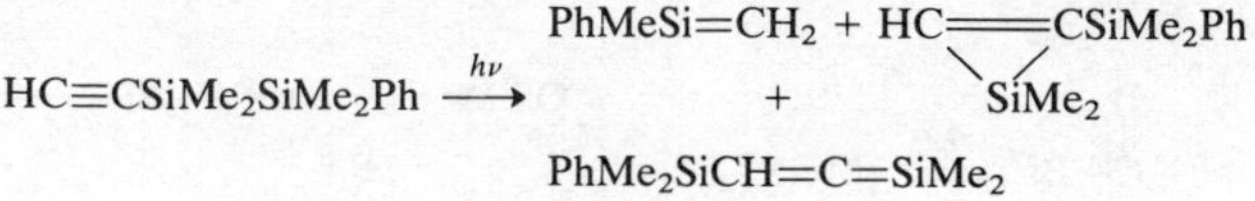

Scheme 14

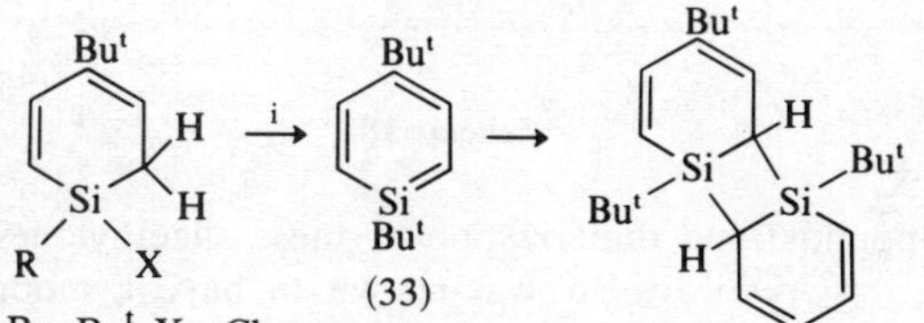

(34) R = Bu^t, X = Cl (33)

Reagents: i, $LiNPr^i_2$, C_5H_{12}, 0 °C

Scheme 15

were trapped[55] by MeOH. Successful attempts have been made[56] to generate and trap the silylbenzene derivative (33). Compound (34) was generated from 1,1-di-n-butyl-4-t-butyl-4-methoxy-1,4-dihydrostannin. To contrast with Scheme 15, only substitution at silicon occurs when the less hindered (34; R = Me, X = Cl) is treated with $LiNEt_2$, to give initially (34; R = Me, X = NEt_2). Trapping of (33) by conjugated dienes provides $(\pi^4 + \pi^2)$ cyclo-adducts.

[53] H. Sakurai, Y. Nakadaira, and T. Kobayashi, *J. Amer. Chem. Soc.*, 1979, **101**, 487.
[54] J. C. Barthelat, G. Trinquier, and G. Bertrand, *J. Amer. Chem. Soc.*, 1979, **101**, 3785.
[55] M. Ishikawa, H. Sugisawa, K. Yamamoto, and M. Kumada, *J. Organometallic Chem.*, 1979, **179**, 377.
[56] G. Märkl and P. Hofmeister, *Angew. Chem. Internat. Edn.*, 1979, **18**, 789.

Interest has been maintained in research on silaethylenes. A direct experimental determination of ΔH_f for $Me_2Si{=}CH_2$ was made, using ion cyclotron resonance spectroscopy.[57a] The method employed was to determine the minimum base strength required for the formation, and detection, of BH^+ by ion cyclotron resonance, *via* deprotonation of Me_3Si^+ (Scheme 16), and then, from the known

$$B + Me_3SiCl \xrightarrow{e^-} Me_3Si^+ \xrightarrow{B} Me_2Si{=}CH_2 + BH^+$$

Scheme 16

enthalpy of proton transfer to the base B, a value for ΔH_f (*ca.* 20.5 kcal mol^{-1}) was obtained. A π-bond energy of 34 kcal mol^{-1} was also calculated. Lower values were reported for these properties from calculations based on the thermodynamic cycle involving the thermal decomposition and dissociative ionization processes for 1,1-dimethyl-1-silacyclobutane.[57b]

Photolysis of acyl-polysilanes[58] at 360 nm gives rise to carbenes and/or silaethylenes of enhanced stability (Schemes 17 and 18). While the trapping (by alcohols and

$$R^1_3SiSiR^2_2C(O)R^3 \xrightarrow{h\nu} R^1_3SiSiR^2_2{-}O{-}\ddot{C}R^2 + R^2_2Si{=}C(OSiR^1_3)R^3$$

Scheme 17

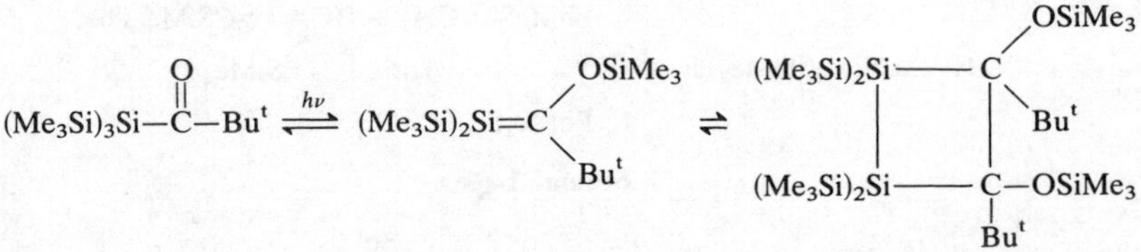

(35)

Scheme 18

unsaturated compounds) and dimerization of these silaethylenes were generally studied, $(Me_3Si)_2Si{=}C(OSiMe_3)Bu^t$ was shown to have a moderate lifetime at ambient temperature, and to be in equilibrium with its head-to-head dimer (35). Spectral characterizations were made by u.v. and by ^1H, ^{13}C, and ^{29}Si n.m.r. The enhanced stability was attributed to both steric and electronic factors, particularly of the Me_3Si groups. The *X*-ray structure of (35) and another head-to-head dimer of a silylethylene, *i.e.* $(Me_3Si)_2C{=}C{=}SiPh_2$, were reported.[59] Other silicon–carbon double-bonded intermediates were formed in the photolysis of 1- and 2-naphthyl-SiR_2SiR_3. From the 1-naphthyl derivatives,[60] even in the presence of a trapping

[57] (*a*) W. J. Pietro, S. K. Pollack, and W. J. Hehre, *J. Amer. Chem. Soc.*, 1979, **101**, 7126; (*b*) L. E. Gusel'nikov and N. S. Nametkin, *J. Organometallic Chem.*, 1979, **169**, 155.

[58] A. G. Brook, J. W. Harris, J. Lennon, and M. El-Sheikh, *J. Amer. Chem. Soc.*, 1979, **101**, 83; A. G. Brook, S. C. Nyburg, W. F. Reynolds, Y. C. Poon, Y. M. Chang, J. S. Lee, and J. P. Picard, *ibid.*, p. 6750.

[59] M. Ishikawa, T. Fuchikami, M. Kumada, T. Higuchi, and S. Miyamoto, *J. Amer. Chem. Soc.*, 1979, **101**, 1348.

[60] M. Ishikawa, M. Oda, N. Miyoshi, L. Fabry, M. Kumada, Y. Yamabe, K. Akagi, and K. Fukui, *J. Amer. Chem. Soc.*, 1979, **101**, 4612.

agent, 1-HR_2Si-2-R_3Si-naphthalenes were obtained *via* 1,3-hydrogen shifts in the intermediate Si=C-bonded species.

Claisen sigmatropic rearrangements of $PhOSiMe_2CH{=}CH_2$ (36) and related species were reported;[61] (36) gave, on heating at 350 °C, 1-oxa-5,6-benzo-2,2-dimethyl-2-silacyclohexane. Photolysis of $Me_3SiC({=}N_2)CO_2R^1$ in R^2OH yields products (i) $Me_3SiCHOR^2CO_2R^1$, from insertion of the corresponding carbene into the O—H bond of R^2OH, (ii) $Me_3SiC(OR^3)HCO_2R^2$ ($R^3 = R^1$ or R^2), by Wolff rearrangement, and (iii) $Me_2SiOR^2CMeCO_2R^1$ (37), apparently from the silene $Me_2Si{=}CMeCO_2R^1$. However in the gas phase,[62] the precursor to (37) was shown to be the keten $Me_2Si(OR^1)C(Me){=}C{=}O$. A method of preparation of tri-substituted ethylenes stereoselectively is based on the reaction of $R^1COCR^2HSiR^3{}_3$ with R^4Li and on the subsequent highly discriminative elimination of $R^3{}_3Si$ and the oxide group from the adducts. *syn*-Elimination occurs under basic conditions whereas *anti*-elimination proceeds with acids.[63] Two preparations of that most useful reagent Me_3SiI *in situ* have been published by Olah;[64] these are the reactions of $(Me_3Si)_2$ with I_2 and of Me_3SiCl with NaI in MeCN. The addition of Me_3SiI to $\alpha\beta$-unsaturated ketones is a new use of this reagent.[65]

Gielen continued his work[66] on chiral tin compounds with details of the synthesis, optical stability, and stereo-reactions of hydrides, such as $[(PhMe_2CCH_2)R(Ph)SnH]$ (R = Me or Bu^t). The chiral hydrides, obtained by asymmetric reduction of the corresponding racemic halides by chiral $[HAl(OC_6H_3Me_2\text{-}2,6)\text{-}(OCHPhCHMeNMe_2)]^- Li^+$, only slowly racemize in benzene solution at room temperature ($t_{\frac{1}{2}}$ 17 days). The H–D exchange with $[Ph_3SnD]$ occurs with retention of configuration at tin.

In contrast to the structure of $[Ph_3SnOSnPh_3]$, those of $O[(PhCH_2)_3M]_2$ (M = Ge or Sn) have linear M—O—M fragments.[67] The synthesis and structure of the *nido*-cluster (38) was reported;[68] compound (38) is sensitive to air and to moisture. As well as the fast intramolecular exchanges in $[C_5H_5SnMe_3]$ and [(cyclo-nonatetraenyl)$SnMe_3$] (by 1,9-sigmatropic shifts) in solvents such as THF or DME,

$[(Me_5C_5)_2Sn^{II}] + HBF_4 \longrightarrow$

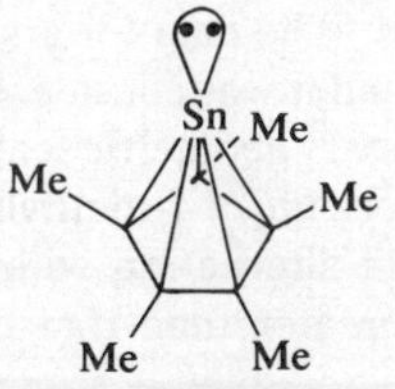

(38)

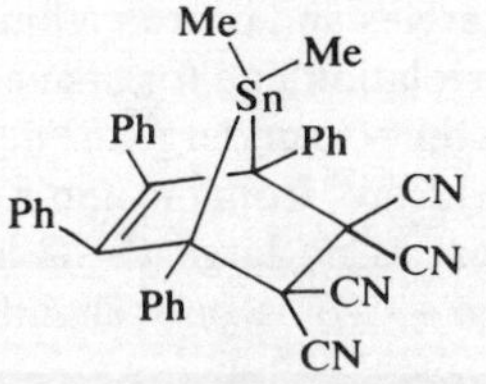

(39)

[61] J. Ancelle, G. Bertrand, M. Joanny, and P. Mazerolles, *Tetrahedron Letters*, 1979, 3153.
[62] W. Ando, A. Sekiguchi, T. Hagiwara, T. Migata, V. Chowdhry, F. H. Westheimer, S. L. Kammula, M. Green, and M. Jones, *J. Amer. Chem. Soc.*, 1979, **101**, 6393.
[63] M. Obayashi, K. Utimoto, and H. Nozaki, *Bull. Chem. Soc. Japan*, 1979, **52**, 1760.
[64] G. A. Olah, S. C. Narang, B. G. B. Gupta, and R. Malhotra, *J. Org. Chem.*, 1979, **44**, 1247; *Angew. Chem. Internat. Edn.*, 1979, **18**, 612.
[65] R. D. Miller and D. R. McKean, *Tetrahedron Letters*, 1979, 2305.
[66] M. Gielen and Y. Tondeur, *J. Organometallic Chem.*, 1979, **169**, 265.
[67] C. Glidewell and D. C. Liles, *J.C.S. Chem. Comm.*, 1979, 93; *Acta Cryst.*, 1979, **35B**, 1689.
[68] P. Jutzi, F. Kohl, and C. Krüger, *Angew. Chem. Internat. Edn.*, 1979, **18**, 59.

intermolecular exchanges *via* ion-pairs in the presence of HMPT have been recognized.[69] The 7-stannanorbornane derivative (39), from 1,1-dimethyl-2,3,4,5-tetraphenyl-stannacyclopentadiene and $(CN)_2C{=}C(CN)_2$, is stable below −30 °C; at higher temperatures (39) decomposes (at −10 °C, $t_{\frac{1}{2}} = 17$ min) to 5,5,6,6-tetracyano-1,2,3,4-tetraphenyl-1,3-cyclohexadiene and Me_2Sn, which can be trapped or allowed to polymerize.[70] The insertion of $(CN)_2C{=}C(CN)_2$ into C—Sn bonds proceeds *via* charge-transfer complexes under both photochemical and thermal activation. The activation step is transfer of an electron from the charge-transfer complex, the same paramagnetic and intermediate species being formed in both the thermal and photochemical reactions.[71] A similar study was concerned[72] with the insertion of $(CH)_2C{=}C(CN)_2$ into H—M bonds (M = Si, Ge, or Sn). The sequence of events is given in Scheme 19, in which tcne is tetracyanoethylene, (40) is a charge-transfer complex, and $k_{e.t.}$ is the rate constant for electron transfer.

$$R_3MH + \text{tcne} \rightarrow \underset{(40)}{[HMR_3\text{—tcne}]} \xrightarrow{k_{e.t.}} [HMR_3^{+\cdot}\ \text{tcne}^{-\cdot}]$$

$$\downarrow$$

$$[(CN)_2HC(CN)_2CMR_3] \leftarrow [H^{\cdot}\ MR_3^{+}\ \text{tcne}^{-\cdot}]$$

Scheme 19

Palladium complexes, in particular $[PhCH_2Pd(PPh_3)_2Cl]$, catalyse the coupling of R_4Sn with $PhCH_2X$ and ArX. Various functional groups are tolerated in this high-yielding synthesis. α-Deuteriobenzyl bromide and Me_4Sn provide the inverted PhCHDMe product.[73] Compounds $[ArPb(OCOR)_3]$, prepared by plumbylation and by transmetallation of mercury and silicon derivatives, are useful phenylating reagents towards ArH, phenols, and β-diketones.[74]

6 Group 5

It has been variously established that correlations exist between core-ionization energies and proton affinities. The data for arsabenzene, however, fall well off the correlation line for arsines, and it was considered that the low basicity is due to the inability to undergo the necessary geometric rearrangement on protonation.[75] It was deduced[76] from the n.m.r. spectrum of 4-methylstibabenzene in a nematic phase that there is no large C—C bond alternation within the ring, and that the length of C-α—C-β is only slightly greater than that of C-β—C-γ. The thermally labile

[69] (*a*) G. Boche, F. Heidenhain, and B. Staudig, *Angew. Chem. Internat. Edn.*, 1979, **18**, 218; (*b*) A. Bonny and S. R. Stobart, *J.C.S. Dalton*, 1979, 786.

[70] C. Grugel, W. P. Neumann, and M. Schriewer, *Angew. Chem. Internat. Edn.*, 1979, **18**, 543.

[71] S. Fukuzumi, K. Mochida, and J. K. Kochi, *J. Amer. Chem. Soc.*, 1979, **101**, 5961; see also O. A. Reutov, V. I. Rozenberg, G. V. Gavrilova, and V. A. Nikanorov, *J. Organometallic Chem.*, 1979, **177**, 101.

[72] R. J. Klinger, K. Mochida, and J. K. Kochi, *J. Amer. Chem. Soc.*, 1979, **101**, 6626.

[73] D. Milstein and J. K. Stille, *J. Amer. Chem. Soc.*, 1979, **101**, 4992.

[74] H. C. Bell, J. R. Kalman, J. T. Pinhey, and S. Sternhell, *Austral. J. Chem.*, 1979, **32**, 1521; H. C. Bell, J. R. Kalman, G. L. May, J. T. May, J. T. Pinhey, and S. Sternhell, *ibid.*, p. 1531; H. C. Bell, J. T. Pinhey, and S. Sternhell, *ibid.*, p. 1551; J. T. Pinhey and B. A. Rowe, *ibid.*, p. 1561.

[75] A. J. Ashe, M. K. Bahl, K. D. Bomben, W. T. Chan, J. K. Gimzewski, P. G. Sitton, and T. D. Thomas, *J. Amer. Chem. Soc.*, 1979, **101**, 1764.

[76] T. C. Wong, M. G. Ferguson, and A. J. Ashe, *J. Mol. Structure*, 1979, **52**, 231.

arsanaphthalene (41), previously unknown, was prepared[77] as shown in Scheme 20. The ^{13}C n.m.r. spectrum of (41) was obtained despite its slow decomposition in solution. Treatment of (41) with $CF_3C{\equiv}CCF_3$ provides (42; R = CF_3).

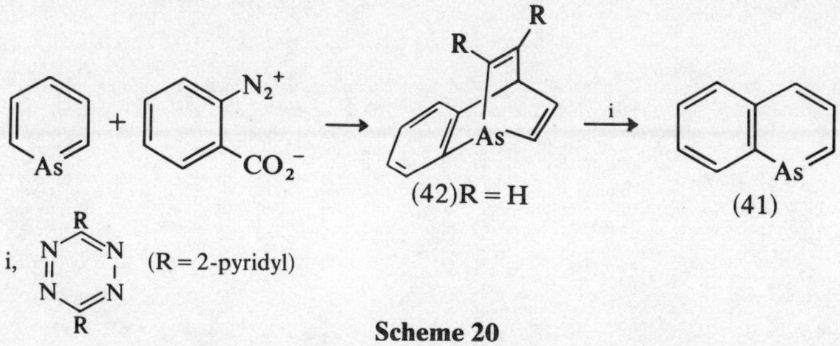

Reagents: i, (R = 2-pyridyl)

Scheme 20

Mesityl(diphenylmethylene)arsane, [2,4,6-$Me_3C_6H_2As{=}CPh_2$], which is a compound with an isolated double bond, is stable at ambient temperature, in the absence of air, and shows little tendency to oxidize.[78] It was synthesized from [2,4,6-$Me_3C_6H_2As(Cl)CHPh_2$] and diazabicycloundecene.

[77] A. J. Ashe, D. J. Bellville, and H. S. Friedman, *J.C.S. Chem. Comm.*, 1979, 880.
[78] T. C. Klebach, H. van Dongen, and F. Bickelhaupt, *Angew. Chem. Internat. Edn.*, 1979, **18**, 395.

13 Organoboron Chemistry

By K. SMITH and W. E. PAGET

*Department of Chemistry, University College of Swansea
Singleton Park, Swansea SA2 8PP*

This Report is concerned with major advances in organoboron chemistry over the past five years. It is perhaps appropriate that in 1979 the importance of the field and the unparallelled contributions of H. C. Brown were recognized by the award of a Nobel Prize. H. C. Brown has himself recently reflected upon forty years of hydride reductions.[1]

There have been many recent reviews covering individual aspects of organoboron chemistry, but we cite only one book[2a] and a review[2b] because together they contain key references to all areas of organoboron chemistry that are likely to be of interest to organic chemists.

1 Reductions using Organoboron Compounds

Alkali-metal trialkylhydridoborates (trialkylborohydrides) are increasingly being recognized as useful reducing agents, and detailed studies of their preparations from trialkylboranes and a variety of simple and complex hydrides have appeared.[3] The sterically hindered compounds are particularly useful for stereoselective reduction of carbonyl compounds (*e.g.* Scheme 1).[4] Some trialkylborohydrides are now commercially available.

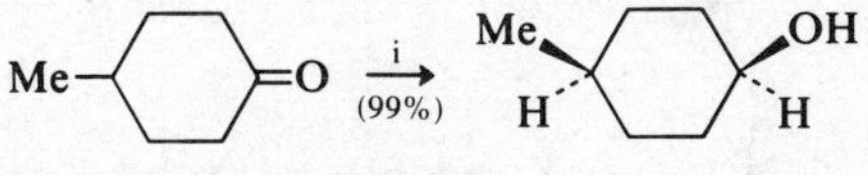

Reagents: i, $Li[Me_2CHCH(Me)]_3BH$, at −78 °C

Scheme 1

$\alpha\beta$-Enones which are not encumbered at the β-position generally undergo conjugate reduction to give enolates, which can be directly alkylated *in situ* if required (Scheme 2).[5] On the other hand, branching at the β-position generally leads to direct reduction of the carbonyl group.[5,6]

[1] H. C. Brown and S. Krishnamurthy, *Tetrahedron*, 1979, **35**, 567.

[2] (*a*) H. C. Brown, 'Organic Synthesis *via* Boranes,' Wiley–Interscience, New York, 1975; (*b*) A. Pelter and K. Smith, in 'Comprehensive Organic Chemistry', ed. D. H. R. Barton and W. D. Ollis, Pergamon, Oxford, 1978, Vol. 3, pp. 689–940.

[3] (*a*) C. A. Brown and S. Krishnamurthy, *J. Organometal. Chem.*, 1978, **156**, 111; (*b*) C. A. Brown and J. L. Hubbard, *J. Am. Chem. Soc.*, 1979, **101**, 3964; (*c*) H. C. Brown. S. Krishnamurthy, and J. L. Hubbard, *ibid.*, 1978, **100**, 3343; *J. Organometal. Chem.*, 1979, **166**, 271.

[4] S. Krishnamurthy and H. C. Brown, *J. Am. Chem. Soc.*, 1976, **98**, 3383.

[5] J. M. Fortunato and B. Ganem, *J. Org. Chem.*, 1976, **41**, 2194.

[6] W. G. Dauben and J. W. Ashmore, *Tetrahedron Lett.*, 1978, 4487.

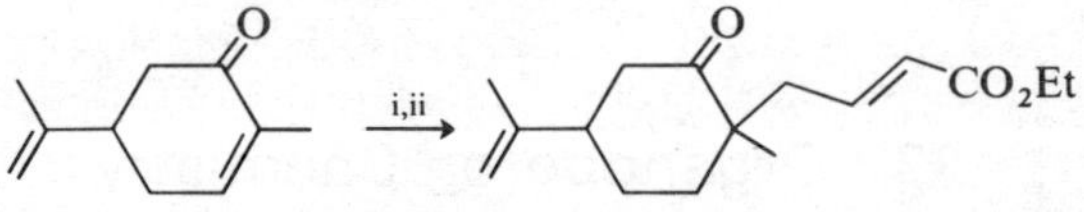

Reagents: i, $LiBu^s{}_3BH$; ii, I CO_2Et

Scheme 2

Trialkylborohydrides are extremely powerful reducing agents, and as such are useful for the deoxygenation of alcohols by reduction of their tosylates[7] or oxyphosphonium salts (applied particularly to glycosyl derivatives),[8] and for the demethylation of quaternary ammonium salts.[9] They also react with elemental sulphur and selenium to give active, anhydrous lithium sulphides (Li_2S or Li_2S_2, depending upon the stoicheiometry) and selenides, which are useful reagents for introduction of sulphur and selenium into organic molecules.[10] Metal carbonyls may be converted into metal formyl complexes,[11] and dimeric carbonyl compounds cleaved to metal carbonyl anions,[12] by the action of lithium triethylborohydride.

9-Alkyl-9-borabicyclo[3.3.1]nonanes (9-alkyl-9-BBN's) and their *B*-alkyl ate-complexes are interesting new types of reducing agents. The boranes (9-alkyl-9-BBN's) eliminate alkene on treatment with an aldehyde, and there is concomitant reduction of the aldehyde. *B*-Siamyl-9-BBN (siamyl = 1,2-dimethylpropyl) appears to be particularly promising for selective reduction of aldehydes in the presence of ketones,[13] whilst optically active *B*-isopinocampheyl-9-BBN (formed by hydroboration of α-pinene, using 9-BBN) is capable of producing primary alcohols RCHDOH of very high optical purity (>90%) from deuterio-aldehydes RCDO.[14]

Lithium 9,9-dibutyl-9-boratabicyclo[3.3.1]nonane donates a hydride ion from one of the bridging positions and itself undergoes rearrangement (Scheme 3).[15] However, it shows considerable discrimination in the facility with which it gives up its hydride. Thus, for example, it reacts with 4-methylcyclohexanone (*cf.* Scheme 1) to give the *cis*-alcohol (84%) in the presence of methanol or the *trans*-isomer (90%) in the presence of lithium methoxide.[15b] It can also be used for selective

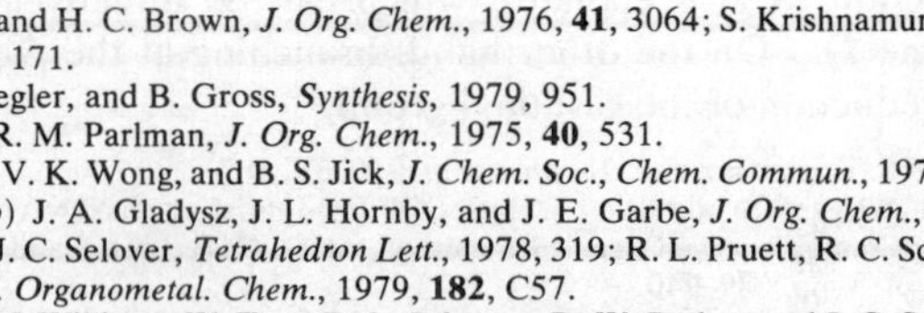

Scheme 3

[7] S. Krishnamurthy and H. C. Brown, *J. Org. Chem.*, 1976, **41**, 3064; S. Krishnamurthy, *J. Organometal. Chem.*, 1978, **156**, 171.
[8] P. Simon, J.-C. Ziegler, and B. Gross, *Synthesis*, 1979, 951.
[9] M. P. Cooke and R. M. Parlman, *J. Org. Chem.*, 1975, **40**, 531.
[10] (*a*) J. A. Gladysz, V. K. Wong, and B. S. Jick, *J. Chem. Soc., Chem. Commun.*, 1978, 838; *Tetrahedron*, 1979, **35**, 2329; (*b*) J. A. Gladysz, J. L. Hornby, and J. E. Garbe, *J. Org. Chem.*, 1978, **43**, 1204.
[11] J. A. Gladysz and J. C. Selover, *Tetrahedron Lett.*, 1978, 319; R. L. Pruett, R. C. Schoening, J. L. Vidal, and R. A. Fiato, *J. Organometal. Chem.*, 1979, **182**, C57.
[12] J. A. Gladysz, G. M. Williams, W. Tam, D. L. Johnson, D. W. Parker, and J. C. Selover, *Inorg. Chem.*, 1979, **18**, 553.
[13] M. M. Midland and A. Tramontano, *J. Org. Chem.*, 1978, **43**, 1470.
[14] M. M. Midland, S. Greer, A. Tramontano, and S. A. Zderic, *J. Am. Chem. Soc.*, 1979, **101**, 2352.
[15] (*a*) G. W. Kramer and H. C. Brown, *J. Am. Chem. Soc.*, 1976, **98**, 1964; (*b*) Y. Yamamoto, H. Toi, A. Sonoda, and S.-I. Murahashi, *ibid.*, p. 1965.

dehalogenation of reactive (tertiary alkyl, allyl, or benzyl) halides,[16] or for regio-selective cleavage of epoxides in a manner opposite to that usually encountered.[17]

Acyloxy-borohydrides have also been recommended as novel types of reducing agents,[18] whilst new applications of the established reagents 9-BBN (*e.g.* reduction of enones to allylic alcohols)[19] and catecholborane (unsaturated tosylhydrazones to alkenes or allenes)[20] have also been reported.

2 Hydroboration

Details of the hydroboration of alkenes and alkynes and of the cyclic hydroboration of dienes, using monochloroborane diethyl etherate[21] or the more stable and convenient adducts of dimethyl sulphide with monohalogeno-boranes,[22] have appeared. The dimethyl sulphide adducts, in particular, promise to be amongst the more important hydroborating agents in future years. 1,3,2-Dithiaborolan[23] is another new hydroborating agent.

In some respects, 9-BBN is an unusual hydroborating reagent. For example, the rate of its hydroboration of many alkenes is independent of the nature or concentration of the alkene because the rate-limiting step is dissociation of the 9-BBN dimer into monomer.[24] Also, *B*-cyclo-octyl-9-BBN derivatives do not undergo the isomerization of other *B*-cyclo-octyl compounds,[25] and 9-BBN, unlike other dialkyl-boranes, hydroborates most alkenes faster than the corresponding alkynes.[26] Details of the general applications of 9-BBN have appeared,[27] and the preparation of a *B*-deuterio-derivative has also been reported.[28]

Much effort has been devoted to the development of a convenient procedure for preparation of monoisopinocampheylborane, a relatively unhindered asymmetric hydroborating agent.[29] One method involves displacement of α-pinene from di-isopinocampheylborane, using tetramethylethylenediamine (TMEDA), and results in a product of very high optical purity, much higher than that of the original α-pinene used in the preparation of the di-isopinocampheylborane.[29b] The TMEDA complex of thexylborane (1,1,2-trimethylpropylborane) reacts with

[16] Y. Yamamoto, H. Toi, S.-I. Murahashi, and I. Moritani, *J. Am. Chem. Soc.*, 1975, **97**, 2558.
[17] Y. Yamamoto, H. Toi, A. Sonoda, and S.-I. Murahashi, *J. Chem. Soc., Chem. Commun.*, 1976, 672.
[18] N. Umino, T. Iwakuma, and N. Itoh, *Tetrahedron Lett.*, 1976, 763, 2875; G. W. Gribble, W. J. Kelly, and S. E. Emery, *Synthesis*, 1978, 763.
[19] S. Krishnamurthy and H. C. Brown, *J. Org. Chem.*, 1975, **40**, 1864; H. C. Brown, S. Krishnamurthy, and N. M. Yoon, *ibid.*, 1976, **41**, 1778.
[20] G. W. Kabalka, D. T. C. Yang, and J. D. Baker, *J. Org. Chem.*, 1976, **41**, 574; G. W. Kabalka, R. J. Newton, J. H. Chandler, and D. T. C. Yang, *J. Chem. Soc., Chem. Commun.*, 1978, 726.
[21] H. C. Brown and N. Ravindran, *J.Am. Chem. Soc.*, 1976, **98**, 1785, 1798, H. C. Brown and M. Zaidlewicz, *ibid.*, p. 4917.
[22] H. C. Brown, N. Ravindran, and S. U. Kulkarni, *J. Org. Chem.*, 1979, **44**, 2417, H. C. Brown and S. U. Kulkarni, *ibid.*, p. 2422.
[23] S. Thaisrivongs and J. D. Wuest, *J. Org. Chem.*, 1977, **42**, 3243.
[24] H. C. Brown, C. G. Scouten, and K. K. Wang, *J. Org. Chem.*, 1979, **44**, 2589.
[25] H. Taniguchi, L. Brener, and H. C. Brown, *J. Am. Chem. Soc.*, 1976, **98**, 7107.
[26] C. A. Brown and R. A. Coleman, *J. Org. Chem.*, 1979, **44**, 2328.
[27] H. C. Brown, R. Liotta, and G. W. Kramer, *J. Org. Chem.*, 1978, **43**, 1058; H. C. Brown, C. G. Scouten, and R. Liotta, *J. Am. Chem. Soc.*, 1979, **101**, 96; J. C.-S. Chen, *J. Organometal. Chem.*, 1978, **156**, 213.
[28] M. M. Midland and S. Greer, *Synthesis*, 1978, 845.
[29] (*a*) H. C. Brown and A. K. Mandal, *Synthesis*, 1978, 146; (*b*) H. C. Brown, J. R. Schwier, and B. Singaram, *J. Org. Chem.*, 1978, **43**, 4395; (*c*) A. Pelter, D. J. Ryder, J. H. Sheppard, C. Subrahmanyam, H. C. Brown, and A. K. Mandal, *Tetrahedron Lett.*, 1979, 4777.

alkenes to give complexes of monoalkyl-boranes, from which the free borane can easily be liberated.[30] Diphenylborane, another potentially useful reagent for preparation of 'mixed' organoboranes, has also been prepared in high yield.[31]

Hydroboration of allenes[32] is very dependent upon the nature of the hydroborating agent, and reaction with 9-BBN is especially useful for the preparation of allyl-boranes.[32a,b] Hydroboration–oxidation of an allene was used as a key step in a total synthesis of isocaryophyllene (Scheme 4).[33]

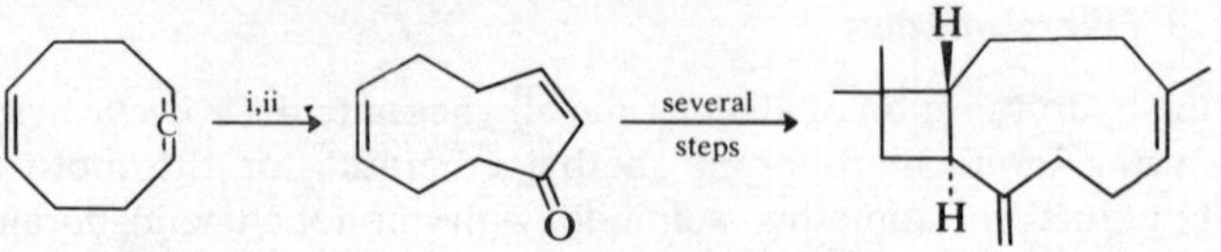

Reagents: i, BH_3, THF, at 160 °C, for 3 h; ii, CrO_3

Scheme 4

Vinylsilanes are hydroborated regioselectively with dialkyl-boranes to give over 95% of the β-dialkylboryl-silane, whereas borane–THF favours the formation of the α-isomer.[34] Borane converts β-enamino-esters into the corresponding $\alpha\beta$-unsaturated esters,[35] and vinyl-aziridines into allyl-amines.[36] Hydroboration of 1-bromo-alkynes with a dialkyl-borane, followed by treatment with t-butyl-lithium or with lithium triethylborohydride, provides a method of synthesizing *cis*-alkenyl-boranes, *i.e.* alkenyl-boranes with stereochemistry opposite to that obtained by simple hydroboration of a terminal alkyne.[37] It has finally been proved that hydroboration occurs in a *cis* manner, even in acyclic cases.[38]

3 Synthetic Applications of Organoboranes

Potassium tri-isopropoxyborohydride is a superior reagent for hydride-induced carbonylation of organoboranes, a reaction which gives an aldehyde after oxidative work-up.[39] The reaction occurs only under conditions which allow the existence of a small amount of free trialkylborane;[40] if acid is added subsequent to the carbonylation step, a second rearrangement occurs, giving rise to a secondary alcohol upon work-up.[41] The carbonylation reaction has been applied in the synthesis of benzaldehyde from triphenylborane,[42] of ^{14}C-labelled aldehydes, ketones, and tertiary alcohols,[43] and of juvabione and its epimer (Scheme 5).[44]

[30] H. C. Brown, J. R. Schwier, and B. Singaram, *J. Org. Chem.*, 1979, **44**, 465.
[31] P. Jacob, *J. Organometal. Chem.*, 1978, **156**, 101.
[32] (*a*) L. Chevolot, J. Soulié, and P. Cadiot, *Tetrahedron Lett.*, 1974, 3435; (*b*) H. C. Brown, R. Liotta, and G. W. Kramer, *J. Am. Chem. Soc.*, 1979, **101**, 2966; (*c*) V. V. Ramana Rao, S. K. Agarwal, D. Devaprabhakara, and S. Chandrasekaran, *Synth. Commun.*, 1979, **9**, 437.
[33] A. Kumar, A. Singh, and D. Devaprabhakara, *Tetrahedron Lett.*, 1976, 2177.
[34] J. A. Soderquist and A. Hassner, *J. Organometal. Chem.*, 1978, **156**, C12.
[35] J. Froburg, G. Magnusson, and S. Thorén, *Tetrahedron Lett.*, 1975, 1621.
[36] R. Chaabouni, A. Laurent, and B. Marquet, *Tetrahedron Lett.*, 1976, 757.
[37] E. Negishi, R. M. Williams, G. Lew, and T. Yoshida, *J. Organometal. Chem.*, 1975, **92**, C4; J. B. Campbell and G. A. Molander, *ibid.*, 1978, **156**, 71.
[38] G. W. Gabalka, R. J. Newton, and J. Jacobus, *J. Org. Chem.*, 1978, **43**, 1567.
[39] H. C. Brown, J. L. Hubbard, and K. Smith, *Synthesis*, 1979, 701.
[40] H. C. Brown and J. L. Hubbard, *J. Org. Chem.*, 1979, **44**, 467.
[41] J. L. Hubbard and H. C. Brown, *Synthesis*, 1978, 676.
[42] G. W. Kabalka and J. W. Fernell, *Synth. Commun.*, 1979, **9**, 443.
[43] G. W. Kabalka, E. E. Gooch, C. J. Collins, and V. F. Raaen, *J. Chem. Soc., Chem. Commun.*, 1979, 607.
[44] E. Negishi, M. Sabanski, J.-J. Katz, and H. C. Brown, *Tetrahedron*, 1976, **32**, 925.

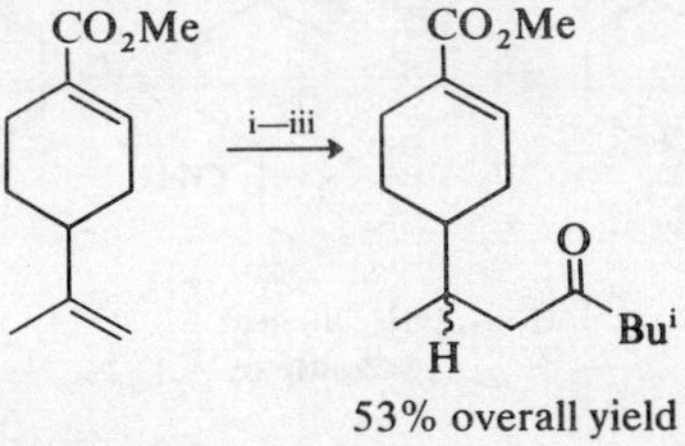

Reagents: i, $(Me_2CHCMe_2)BHBu^i$; ii, CO, 70 atm, 50 °C; iii, [O]

Scheme 5

A number of new reagents for oxidation of organoboranes have been reported, of which the most notable are trimethylamine oxide dihydrate[45] and pyridinium chlorochromate.[46] The latter oxidizes organoboranes and alkoxyboranes directly to aldehydes and ketones. The reagent $PhI(OAc)_2$ converts alkyl- and vinyl-boron compounds into the corresponding alkyl or vinyl acetates.[47]

The first quantitative conversion of trialkylboranes into chloro-alkanes utilizes nitrogen trichloride as the reagent.[48] Methoxide ion is a superior base to hydroxide ion for the base-induced iodinolysis of organoboranes, allowing all three alkyl groups of a primary trialkylborane to be converted into iodo-alkane (two for secondary alkyl groups).[49] Base-induced iodinolysis and brominolysis of organoboranes both involve *inversion* of stereochemistry at the displaced centre.[50] Iodinolysis of a vinyl-catecholborane has been applied in the synthesis of some substituted prostaglandins.[51]

Iron(III) salts have been used to convert organoboranes into alkyl selenocyanates[52] and azides,[53] whilst Chloramine-T reacts with trialkylboranes to give the corresponding alkyl toluenesulphonamide.[54] The direct transfer of two alkyl groups from a trialkylborane to a single nitrogen atom ('nitrogen stitching') has been accomplished in one case, as part of a synthesis of the perhydroazaphenalene ring system, which is found in alkaloids of the Coccinellidae (Scheme 6).[55]

Trialkylboranes react with the Grignard reagent derived from 1,5-dibromopentane to give a spiroborate complex and the alkyl Grignard reagent.[56] Alkenyl groups of alkenyl-boranes may be hydrolysed under the influence of palladium(II) acetate in THF,[57] or coupled to give (*E*,*E*)-dienes, using a reagent composed of

45 G. W. Kabalka and H. C. Hedgecock, *J. Org. Chem.*, 1975, **40**, 1776.
46 V. V. Ramana Rao, D. Devaprabhakara, and S. Chandrasekaran, *J. Organometal. Chem.*, 1978, **162**, C9; C. Gundu Rao, S. U. Kulkarni, and H. C. Brown, *ibid.*, 1979, **172**, C20; H. C. Brown, S. U. Kulkarni, and C. Gundu Rao, *Synthesis*, 1979, 702, 704.
47 Y. Masuda and A. Arase, *Bull. Chem. Soc. Jpn.*, 1978, **31**, 901; Y. Masuda, A. Arase, and A. Suzuki, *Chem. Lett.*, 1978, 665.
48 H. C. Brown and N. R. De Lue, *J. Organometal. Chem.*, 1977, **135**, C57.
49 N. R. De Lue and H. C. Brown, *Synthesis*, 1976, 114.
50 H. C. Brown, N. R. De Lue, G. W. Kabalka, and H. C. Hedgecock, *J. Am. Chem. Soc.*, 1976, **98**, 1290; D. E. Bergbreiter and D. P. Rainville, *J. Organometal. Chem.*, 1976, **121**, 19.
51 P. W. Collins, E. Z. Dajani, M. S. Bruhn, C. H. Brown, J. R. Palmer, and R. Pappo, *Tetrahedron Lett.*, 1975, 4217.
52 A. Arase and Y. Masuda, *Chem. Lett.*, 1976, 785.
53 A. Suzuki, M. Ishidoya, and M. Tabata, *Synthesis*, 1976, 687.
54 V. B. Jigajinni, A. Pelter, and K. Smith, *Tetrahedron Lett.*, 1978, 181.
55 R. H. Mueller, *Tetrahedron Lett.*, 1976, 2925.
56 K. Kondo and S.-I. Murahashi, *Tetrahedron Lett.*, 1979, 1237.
57 H. Yatagai, Y. Yamamoto, and K. Maruyama, *J. Chem. Soc., Chem. Commun.*, 1978, 702.

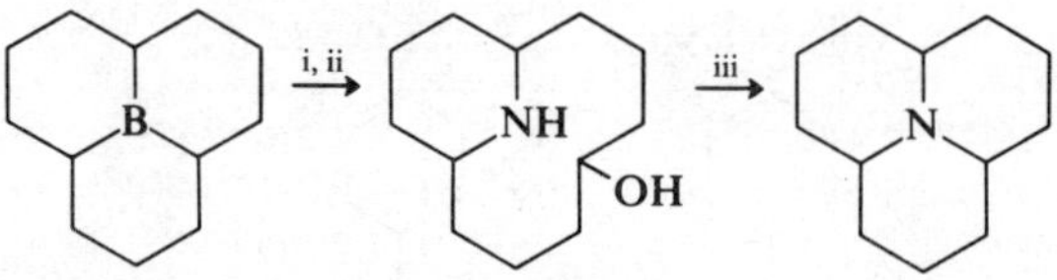

Reagents: i, $[ClNHOC_6H_3(NO_2)_2]$; ii, H_2O_2, OH^-; iii, heat

Scheme 6

palladium(II) chloride, lithium chloride, and triethylamine.[58] On the other hand, they can be coupled with aryl or vinyl halides under the influence of palladium(O) and a base.[59] Diboryl-alkanes react with silver nitrate to give substantial amounts of cycloalkanes[60] or (*E*)-alkenes (from 1,2-diboryl-alkanes).[61]

The previously discovered oxygen-induced conjugate addition of trialkylboranes to enones and similar compounds has been extended to 1-acyl-2-vinyl-cyclopropanes, giving $\gamma\delta$-unsaturated ketones (Scheme 7).[62] *B*-Alkenyl-9-BBN derivatives undergo conjugate addition of the alkenyl-boron unit to enones, apparently by a non-radical mechanism, providing another approach to $\gamma\delta$-unsaturated ketones.[63] Electrolysis of trialkylboranes in the presence of 2-alkylacrylates is an alternative way of inducing conjugate addition.[64]

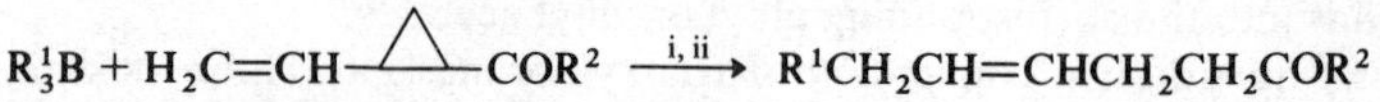

Reagents: i, O_2; ii, H_2O

Scheme 7

Diethylborylpivalate (or triethylborane in the presence of derivatives of pivalic acid) reacts with hydroxylic compounds relatively readily, liberating ethane. Such systems have been extensively utilized for the selective protection of sugars and other polyhydroxy-compounds.[65]

Inversion occurs at the migration terminus during base-induced rearrangement of α-halogenoalkyl-boranes and during the formation of cyclopropanes from γ-chloroalkyl-boranes.[66] Kinetic and competition experiments provide evidence for a dehydroboration–rehydroboration process in the alkene–alkyl exchange reactions of *B*-alkyl-9-BBN compounds,[67] and, by implication, also in the isomerization of alkyl-boranes. A potentially useful procedure for selective elimination of the secondary alkyl groups present in a typical tri-n-alkylborane prepared by hydroboration of a terminal alkene involves heating the borane first with anisole and then with a small amount of DMSO.[68]

58 V. V. Ramana Rao, C. V. Kumar, and D. Devaprabhakara, *J. Organometal. Chem.*, 1979, **179**, C7.
59 N. Miyaura and A. Suzuki, *J. Chem. Soc., Chem. Commun.*, 1979, 866, N. Miyaura, K. Yamada, and A. Suzuki, *Tetrahedron Lett.*, 1979, 3437.
60 R. Murphy and R. H. Prager, *Tetrahedron Lett.*, 1976, 463.
61 K. Avasthi, S. S. Ghosh, and D. Devaprabhakara, *Tetranedron Lett.*, 1976, 4871.
62 N. Miyaura, M.Itoh, N. Sasaki, and A. Suzuki, *Synthesis*, 1975, 317.
63 P. Jacob and H. C. Brown, *J. Am. Chem. Soc.*, 1976, **98**, 7832.
64 Y. Takahashi, K. Yuasa, M. Tokuda, M. Itoh, and A. Suzuki, *Bull. Chem. Soc. Jpn.*, 1978, **51**, 339.
65 W. V. Dahloff and R. Köster, *J. Org. Chem.*, 1976, **41**, 2316; and references cited therein.
66 M. M. Midland, A. R. Zolopa, and R. L. Halterman, *J. Am. Chem. Soc.*, 1979, **101**, 248; H. L. Goering and S. L. Trenbeath, *ibid.*, 1976, **98**, 5016.
67 M. M. Midland, J. E. Petre, and S. A. Zderic, *J. Organometal. Chem.*, 1979, **182**, C53.
68 Y. Masuda, M. Hoshi, and A. Arase, *Bull Chem. Soc. Jpn.*, 1979, **52**, 271.

4 Synthetic Applications of Organoborates

Probably the most significant development in organoborane chemistry over the past decade has been the realization of the synthetic importance of the reactions of four-co-ordinate borate salts with electrophiles. Full details of the reactions of cyanoborates with electrophiles, the first reactions to demonstrate the synthetic importance of organoborates, were published at the start of the current review period.[69] The most useful processes involve acylation of the cyanoborates with reagents such as trifluoroacetic anhydride, leading ultimately to ketones or tertiary alcohols (Scheme 8). The adducts of the cyclic intermediates with water have been characterized by *X*-ray crystallography,[70] and the reaction has been applied in the annulation of terpenoid polyenes.[71]

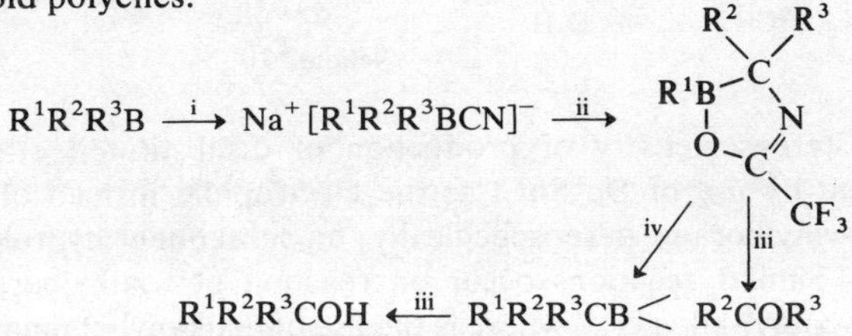

Reagents: i, NaCN; ii, $(CF_3CO)_2O$; iii, H_2O_2, OH^-; iv, excess $(CF_3CO)_2O$.

Scheme 8

The formation of a tertiary alcohol from *cis,cis*-perhydro-9b-boraphenalene is interesting in that it gives the *trans,trans,trans*-alcohol, in contrast to the *cis,cis,cis*-alcohol obtained by carbonylation (see Scheme 9).[72]

Alkynylborates react with acids or alkylating agents to give intermediate vinylboranes, which may be oxidized to ketones or hydrolysed to alkenes.[73] The reactions with α-nitro-alkenes are Michael-type addition processes.[73f] With simple alkylating

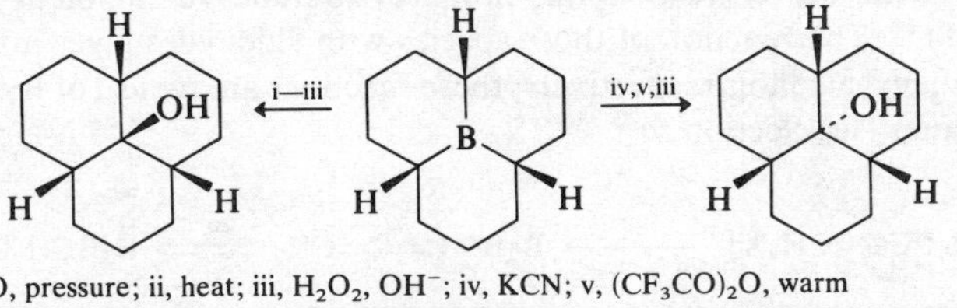

Reagents: i, CO, pressure; ii, heat; iii, H_2O_2, OH^-; iv, KCN; v, $(CF_3CO)_2O$, warm

Scheme 9

[69] A. Pelter, K. Smith, M. G. Hutchings, and K. Rowe, *J. Chem. Soc.*, *Perkin Trans. 1*, 1975, 129, 138; A. Pelter, M. G. Hutchings, and K. Smith, *ibid.*, p. 142; A. Pelter, M. G. Hutchings, K. Smith, and D. J. Williams, *ibid.*, p. 145.

[70] P. R. Mallinson, D. N. J. White, A. Pelter, K. Rowe, and K. Smith, *J. Chem. Res.* 1978, (*S*) 234, (*M*) 3101.

[71] R. Murphy and R. H. Prager, *Aust. J. Chem.*, 1976, **29**, 617; *J. Organometal. Chem.*, 1978, **156**, 133.

[72] A. Pelter, P. J. Maddocks, and K. Smith, *J. Chem. Soc., Chem. Commun.*, 1978, 805.

[73] (*a*) H. C. Brown, A. B. Levy, and M. M. Midland, *J. Am. Chem. Soc.*, 1975, **97**, 5017; (*b*) M. M. Midland and H. C. Brown, *J. Org. Chem.*, 1975, **40**, 2845; (*c*) A. Pelter,, C. R. Harrison, C. Subrahamanyam, and D. Kirkpatrick, *J. Chem. Soc.*, *Perkin Trans. 1*, 1976, 2435; (*d*) A. Pelter, T. W. Bentley, C. R. Harrison, C. Subrahmanyam, and R. J. Laub, *ibid.*, p. 2419; (*e*) A. Pelter, C. Subrahmanyam, R. J. Laub, K. J. Gould, and C. R. Harrison, *Tetrahedron Lett.*, 1975, 1633; (*f*) A. Pelter and L. Hughes, *J. Chem. Soc., Chem. Commun.*, 1977, 913.

agents, mixtures of the (*E*)- and (*Z*)-isomers of the vinylboranes are obtained, but with more complex ones (*e.g.* propargyl halides, α-halogenocarbonyl compounds, and the like) the reaction is stereospecific, giving rise to (*Z*)-alkenyl derivatives on hydrolysis (Scheme 10).[74] The intermediate vinylboranes can also be worked up in other ways; for example, by application of the Zweifel alkene synthesis, using iodine and base.[75]

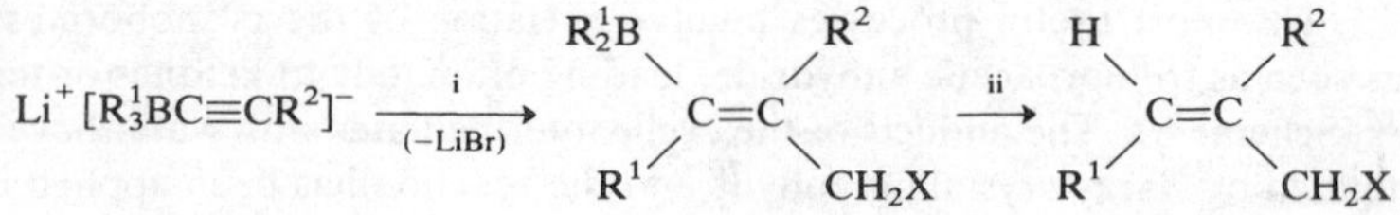

Reagents: i, $BrCH_2X$; ii, $MeCO_2H$

Scheme 10

The stereoselectivity of production of disubstituted ethenes can be greatly improved by use of Bu^n_3SnCl as the electrophile instead of acid.[76] This gives a stannyl-vinylborane stereospecifically; on subsequent hydrolysis, this gives a (*Z*)-alkene. Similar reactions occur on reaction of trialkylboranes directly with 1-trialkylstannyl-alkynes,[77] whereas the use of dialkynyl-stannanes gives rise to bora-stanna-heterocycles.[78]

Other electrophiles which give rise to useful products include iodine (gives alkynes)[79] and ethylene oxide (gives alkenylethanol derivatives).[80] Substituents in the original alkyne also affect the course of the reaction. Thus, (trimethylsilyl)ethynyl-borates give overwhelmingly one geometrical isomer of the vinylborane, even on reaction with simple alkylating agents.[81] The lithiated derivatives of lithium ethynyltrialkylborates can be alkylated by alkyl halide, and then treated with iodine to give unsymmetrical internal alkynes.[82] Lithiated propargyl chlorides and acetates react with trialkylboranes to give allenyl-boranes, which equilibrate with the corresponding propargyl-boranes at ambient temperature (Scheme 11).[83] The reaction of these species with aldehydes gives homopropargyl and homoallenyl alcohols respectively; these reactions are typical of βγ-unsaturated organoboranes (see Section 5).

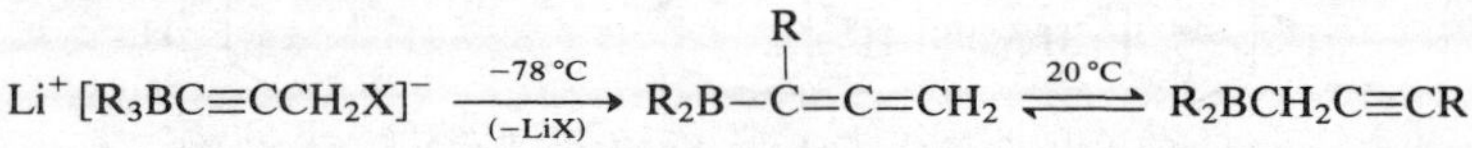

Scheme 11

[74] A. Pelter, K. J. Gould, and C. R. Harrison, *J. Chem. Soc., Perkin Trans. 1*, 1976, 2428.
[75] G. Zweifel and R. P. Fisher, *Synthesis*, 1975, 376.
[76] J. Hooz and R. Mortimer, *Tetrahedron Lett.*, 1976, 805.
[77] B. Wrackmeyer and R. Zentgraf, *J. Chem. Soc., Chem. Commun.*, 1978, 402.
[78] L. Killian and B. Wrackmeyer, *J. Organometal. Chem.*, 1978, **153**, 153; and references cited therein.
[79] A. Suzuki, N. Miyaura, S. Abiko, M. Itoh, H. C. Brown, J. A. Sinclair, and M. M. Midland, *J. Am. Chem. Soc.*, 1973, **95**, 3080.
[80] K. Utimoto, T. Furubayashi, and H. Nozaki, *Chem. Lett.*, 1975, 397.
[81] K. Utimoto, M. Kitai, M. Naruse, and H. Nozaki, *Tetrahedron Lett.*, 1975, 4233; R. Köster and L. A. Hagelee, *Synthesis*, 1976, 118.
[82] K. Utimoto, Y. Yabuki, K. Okada, and H. Nozaki, *Tetrahedron Lett.*, 1976, 3969.
[83] (*a*) G. Zweifel, S. J. Backlund, and T. Leung, *J. Am. Chem. Soc.*, 1978, **100**, 5561; (*b*) M. M. Midland and D. C. McDowell, *J. Organometal. Chem.* 1978, **156**, C5.

Dialkyl(alkenyl)(alkynyl)borates undergo reactions with electrophiles in the same way as trialkyl(alkynyl)borates, except that the products are conjugated dienes rather than mono-enes, and so on.[84] Similarly, dialkyl(dialkynyl)borates react with iodine to give conjugated di-ynes.[85] If the method of preparation of the dialkynyl-borate is chosen so as to allow the synthesis of derivatives with two different alkynyl groups on a single boron atom, the conjugated di-ynes are also unsymmetrical, thus providing a more convenient alternative to the Cadiot–Chodkiewicz reaction.[85b,c]

Alkenyl(trialkyl)borates undergo the same types of reactions as alkynyl-borates, for example with iodine to give alkenes,[86] but in general they have been less widely studied. The alkene synthesis works well on alkenyl(alkyl)dimethoxyborates, allowing better utilization of alkyl groups, and this approach has been applied in the synthesis of prostaglandin model compounds.[87] Alkenyl-boranes can be hydrolysed under basic conditions if they are first converted into the corresponding borates by treatment with butyl-lithium.[88] Alkenyl-borates are also intermediates in the synthetically useful reactions of trialkylboranes with methoxyvinyl-lithium (Scheme 12).[89]

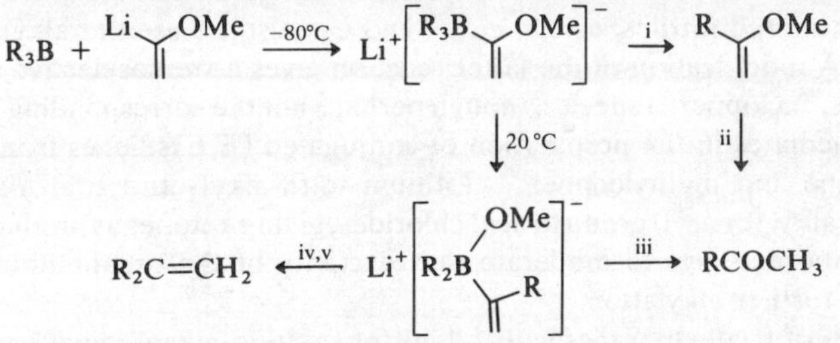

Reagents: i, I_2, at −80 °C; ii, H_3O^+; iii, H_2O_2, HO^-; iv, $BF_3 \cdot Et_2O$; v, I_2, HO^-

Scheme 12

Allyl-borates are formed with considerable regioselectivity, and then react very selectively with electrophiles, which attack at the position γ- to boron. This allows head-to-tail coupling of allylic borates with allyl halides, for example.[90] Propargyl-borates are intermediates in the formation of alkynyl- and allenyl-silanes from lithiated 3-phenoxy-1-trimethylsilylpropyne and trialkylboranes.[91]

Ate-complexes formed from trialkylboranes and 2-lithiated furan, thiophen, *N*-methylpyrrole, and *N*-methylindole undergo reaction with iodine to give the 2-alkylated heterocycles.[92] Other electrophiles can also be used to induce

[84] G. Zweifel and S. J. Backlund, *J. Organometal. Chem.*, 1978, **156**, 159.
[85] (*a*) A. Pelter, K. Smith, and M. Tabata, *J. Chem. Soc., Chem. Commun.*, 1975, 857; (*b*) J. A. Sinclair and H. C. Brown, *J. Org. Chem.*, 1976, **41**, 1078; (*c*) A. Pelter, R. J. Hughes, K. Smith, and M. Tabata, *Tetrahedron Lett.*, 1976, 4385.
[86] N. J. Lalima and A. B. Levy, *J. Org. Chem.*, 1978, **43**, 1279.
[87] D. A. Evans, T. C. Crawford, R. C. Thomas, and J. A. Walker, *J. Org. Chem.*, 1976, **41**, 3947.
[88] E. Negishi and K.-W. Chiu, *J. Org. Chem.*, 1976, **41**, 3484.
[89] A. B. Levy, S. J. Schwartz, N. Wilson, and B. Christie, *J. Organometal. Chem.*, 1978, **156**, 123.
[90] Y. Yamamoto and K. Maruyama, *J. Am. Chem. Soc.*, 1978, **100**, 6282; Y. Yamamoto, H. Yatagai, and K. Maruyama, *J. Chem. Soc., Chem. Commun.*, 1979, 157; *Chem. Lett.*, 1979, 385.
[91] T. Yogo, J. Koshino, and A. Suzuki, *Tetrahedron Lett.*, 1979, 1781.
[92] I. Akimoto and A. Suzuki, *Synthesis*, 1979, 146; T. Sotoyama, S. Hara, and A. Suzuki, *Bull. Chem. Soc. Jpn.*, 1979, **52**, 1865; E. R. Marinelli and A. B. Levy, *Tetrahedron Lett.*, 1979, 2313; A. B. Levy, *J. Org. Chem.*, 1978, **43**, 4684.

rearrangement, and in the case of *N*-methylindole this has been used to prepare 2,3-disubstituted indoles (Scheme 13; EX = *e.g.* ICH_2CN, ICH_2CONH_2, $CH_2{=}CHCH_2Br$, *etc.*).[93] Two heteroaromatic units can be coupled by the reaction of the cyclic ethanolamine-borates of di(heteroaryl)borinic acids with bromine or *N*-bromosuccinimide,[94] whilst the presence of the bromine substituent in 6-bromo-2-lithiopyridine causes rearrangement of its initially formed trialkylborate salts, leading ultimately to unsaturated nitriles.[95]

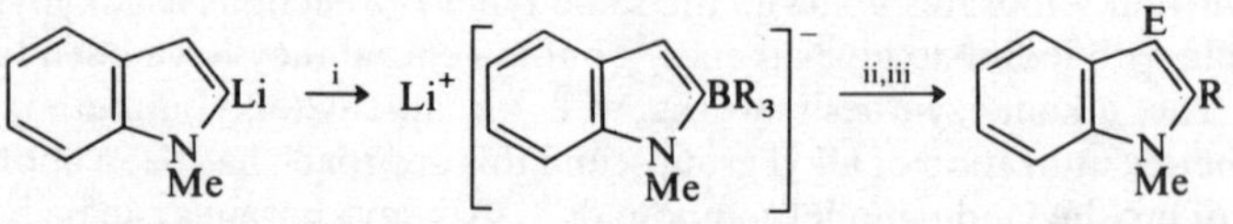

Reagents: i, R_3B; ii, EX; iii, [O]

Scheme 13

Addition of CuI to lithium trialkylmethylborates gives uncharacterized species, assumed to be $Cu^+[R_3BMe]^-$, which undergo conjugate addition of the alkyl group to $\alpha\beta$-unsaturated nitriles, esters, *etc.*,[96] and can also be used to alkylate reactive halides.[97] A modification of the latter reaction gives a stereoselective synthesis of vinylsilanes.[98] Copper(I) species, though perhaps not the corresponding borates, are also intermediates in the preparation of conjugated (*E*,*E*)-dienes from dialkenylchloroborane and methylcopper.[99] Lithium tetra-alkyl- and trialkylaryl-borates transfer an alkyl (or aryl) group to acid chlorides, giving ketones as products.[100] Thus, the trialkylborane acts to moderate the reactivity of the organolithium reagent, preventing further alkylation.

Reactions of trialkylboranes with 1,1-bis(phenylthio)alkanes give borates, which spontaneously rearrange to give compounds that are oxidizable to ketones. Addition of an electrophile such as $HgCl_2$ prior to oxidation results in the formation of tertiary alcohols (Scheme 14).[101] If 2-alkylbenzo-1,3-dithioles are used instead of 1,1-bis(phenylthio)alkanes, this allows the reactions to be applied to hindered organoboranes.[102] The reaction of the ate-complex from a trialkylborane and methylthiomethyl-lithium with methyl iodide gives the homologated organoborane R_2BCH_2R.[103]

[93] A. B. Levy, *Tetrahedron Lett.*, 1979, 4021.
[94] G. M. Davies, P. S. Davies, W. E. Paget, and J. M. Wardleworth, *Tetrahedron Lett.*, 1976, 795.
[95] K. Utimoto, N. Sakai, M. Obayashi, and H. Nozaki, *Tetrahedron*, 1976, **32**, 769.
[96] N. Miyaura, M. Itoh, and A. Suzuki, *Tetrahedron Lett.*, 1976, 255; K. Yamada, N. Miyaura, M. Itoh, and A. Suzuki, *Bull. Chem. Soc. Jpn.*, 1977, **50**, 3431; Y. Yamamoto, H. Yatagai, and K. Maruyama, *J. Org. Chem.*, 1979, **44**, 1744.
[97] K. Yamada, T. Yano, N. Miyaura, and A. Suzuki, *Bull. Chem. Soc. Jpn.*, 1979, **52**, 275.
[98] K. Uchida, K. Utimoto, and H. Nozaki, *J. Org. Chem.*, 1976, **41**, 2941.
[99] Y. Yamamoto, H. Yatagai, and I. Moritani, *J. Am. Chem. Soc.*, 1975, **97**, 5606; Y. Yamamoto, H. Yatagai, K. Maruyama, A. Sonoda, and S. Murahashi, *Bull. Chem. Soc., Jpn.*, 1977, **50**, 3427.
[100] E. Negishi, K.-W. Chiu, and T. Yosida, *J. Org. Chem.*, 1975, **40**, 1676; E. Negishi, A. Abramovitch, and R. E. Merrill, *J. Chem. Soc., Chem. Commun.*, 1975, 138.
[101] R. J. Hughes, S. Ncube, A. Pelter, K. Smith, E. Negishi, and T. Yoshida, *J. Chem. Soc., Perkin Trans. 1*, 1977, 1172.
[102] S. Ncube, A. Pelter, and K. Smith, *Tetrahedron Lett.*, 1979, 1893, 1895.
[103] E. Negishi, T. Yoshida, A. Silveira, and B. L. Chiou, *J. Org. Chem.*, 1975, **40**, 814.

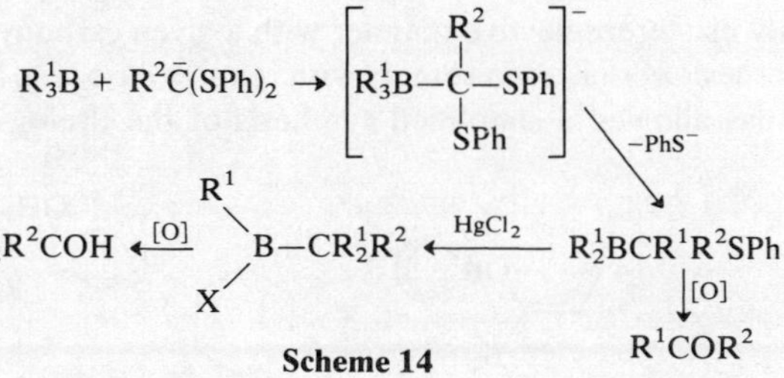

Scheme 14

The reactions of lithium aldimines with dialkyl-chloroboranes, followed by treatment with electrophiles, offer an alternative method for the synthesis of ketones and tertiary alcohols.[104]

5 Applications of $\beta\gamma$-Unsaturated Organoboron Compounds

Allyl-boranes have been extensively studied by Mikhailov's group for many years. In particular, their reactions with alkynes, allenes, and reactive alkenes, resulting in addition of the allyl-boron unit (with rearrangement of the allyl group) across the multiple bond, have led to interesting, novel boron-containing systems.[105] Reactions of allyl-boron compounds with aldehydes and ketones also involve rearrangement, and provide syntheses of homoallylic alcohols,[106] sometimes with considerable diastereoselectivity.[106c,d] Propargyl- and allenyl-boron compounds react with carbonyl compounds to give homoallenyl and homopropargyl alcohols respectively.[83a,107]

Following Mukaiyama's demonstration of crossed aldol reactions between boron enolates (vinyloxy-boranes) and carbonyl compounds,[108] much effort has been devoted to studying such reactions. Methods for the synthesis of vinyloxy-boranes include addition of boron compounds to $\alpha\beta$-unsaturated carbonyl compounds[108,109] or ketens,[110] reactions of trialkylboranes with α-diazocarbonyl compounds,[111] and boron-compound-promoted enolysis of ketones.[112] By careful choice of method it is often possible to obtain either the (*Z*)- or the (*E*)-isomer of the vinylborane with considerable selectivity,[113,114] and it has been shown that a given geometrical isomer

[104] Y. Yamamoto, K. Konda, and I. Moritani, *Bull. Chem. Soc. Jpn.*, 1975, **48**, 3682; *J. Org. Chem.*, 1975, **40**, 3644.

[105] B. M. Mikhailov, T. K. Kozminskaya, and B. I. Bryantsev, *Zh. Obshch. Khim.*, 1976, **46**, 87; Yu. N. Bubnov, B. A. Kazanskii, O. A. Nesmeyanova, T. Yu. Rudashevskaya, and B. M. Mikhailov, *Izv. Akad. Nauk SSSR, Ser. Khim.*, 1977, 2545.

[106] (*a*) G. W. Kramer and H. C. Brown, *J. Org. Chem.*, 1977, **42**, 2292; (*b*) B. M. Mikhailov, Yu. N. Bubnov, and A. V. Tsyban, *J. Organometal. Chem.*, 1978, **154**, 113; (*c*) T. Herold and R. W. Hoffmann, *Angew. Chem., Int. Ed. Engl.*, 1978, **17**, 768; (*d*) R. W. Hoffmann and H.-J. Zeiss, *ibid.*, 1979, **18**, 306.

[107] É. Favre and M. Gaudemar, *J. Organometal. Chem.*, 1975, **92**, 17.

[108] M. Muraki, K. Inomata, and T. Mukaiyama, *Bull. Chem. Soc. Jpn.*, 1975, **48**, 3200; T. Mukaiyama and T. Inoue, *Chem. Lett.*, 1976, 559.

[109] W. Fenzl, R. Köster, and H. J. Zimmermann, *Justus Liebigs Ann. Chem.*, 1975, 2201; 1976, 1116.

[110] P. Paetzold and M. Lasch, *Chem. Ber.*, 1979, **112**, 663.

[111] S. Masamune, S. Mori, D. Van Horn, and D. W. Brooks, *Tetrahedron Lett.*, 1979, 1665.

[112] W. Fenzl and R. Köster, *Justus Liebigs Ann. Chem.*, 1975, 1322; W. Fenzl, H. Kosfeld and R. Köster, *ibid.*, 1976, 1370.

[113] (*a*) M. Hirama and S. Masamune, *Tetrahedron Lett.*, 1979, 2225; (*b*) D. E. Van Horn and S. Masamune, *ibid.*, p. 2229; (*c*) M. Hirama, D. S. Garvey, L. D.-L. Lu, and S. Masamune, *ibid.*, p. 3937.

[114] D. A. Evans, E. Vogel, and J. V. Nelson, *J. Am. Chem. Soc.*, 1979, **101**, 6120.

reacts in a highly diastereoselective manner with a given carbonyl compound, thus allowing the synthesis of diastereoisomers with relative ease (*e.g.* Scheme 15).[113,114] This approach has allowed a simplified synthesis of the Prelog–Djerassi lactonic acid.[113c]

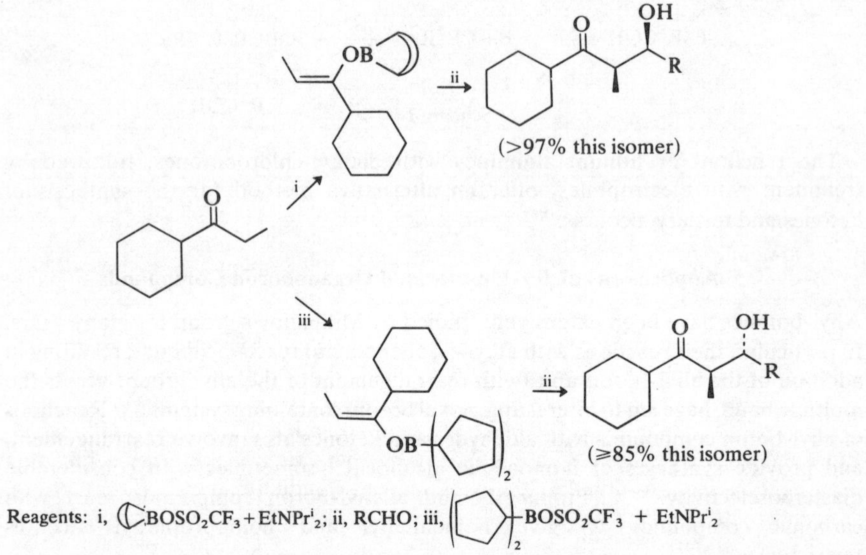

Scheme 15

Potassium trialkyl(vinyloxy)borates, formed by addition of potassium enolates to trialkylboranes, react more selectively with alkyl halides than do the potassium enolates themselves.[115] Vinylamino-boranes also undergo aldol condensation reactions,[116] whilst the related reactions of phenylamino-boron compounds allow specific *ortho*-substitution of aniline derivatives.[117] Particularly useful is the formation of *ortho*-formyl derivatives from the reactions with isocyanides (Scheme 16).[118] Related reactions have been used to prepare indoles[119] and acridines.[120]

6 Novel Types of Organoboron Compound

Attempts to synthesize highly hindered tri-t-butylborane have now been successful.[121] A number of bora-adamantane compounds have been prepared, including 1-bora-adamantane,[122] 2-bora-adamantane,[123] the dibora-compound (1),[124] and the

[115] E. Negishi, M. J. Idacavage, F. DiPasquale, and A. Silveira, *Tetrahedron Lett.*, 1979, 845.
[116] T. Sugasawa, T. Toyoda, and K. Sasakura, *Synth. Commun.*, 1979, **9**, 515; T. Sugasawa and T. Toyoda, *ibid.*, p. 553, *Tetrahedron Lett.*, 1979, 1423.
[117] T. Sugasawa, T. Toyoda, M. Adachi, and K. Sasakura, *J. Am. Chem. Soc.*, 1978, **100**, 4842.
[118] T. Sugasawa, H. Hamana, T. Toyoda, and M. Adachi, *Synthesis*, 1979, 99.
[119] T. Sugasawa, M. Adachi, K. Sasakura, and A. Kitagawa, *J. Org. Chem.*, 1979, **44**, 578.
[120] B. M. Mikhailov, V. A. Dorokhov, and O. G. Boldyreva, *Izv. Akad. Nauk SSSR, Ser. Khim.*, 1978, 2574.
[121] H. Noth and T. Taeger, *J. Organometal. Chem.*, 1977, **142**, 281.
[122] B. M. Mikhailov and T. K. Baryshnikova, *Dokl. Akad. Nauk SSSR*, 1978, **243**, 929.
[123] B. M. Mikhailov, T. A. Shchegoleva, E. M. Shashkova, and V. G. Kiselev, *Izv. Akad. Nauk SSSR, Ser. Khim.*, 1977, 894.
[124] S. U. Kulkarni and H. C. Brown, *J. Org. Chem.*, 1979, **44**, 1747.

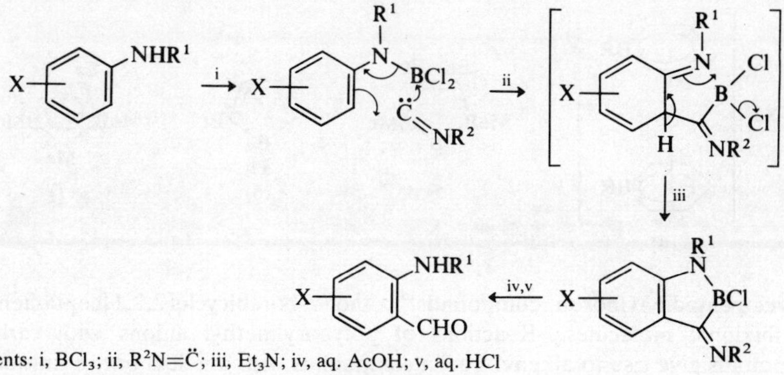

Reagents: i, BCl_3; ii, $R^2N{=}\ddot{C}$; iii, Et_3N; iv, aq. AcOH; v, aq. HCl

Scheme 16

hexabora-adamantane (2).[125] Such compounds form extremely strong complexes with donor molecules.[122,124] A bora-adamantane with a betaine structure has also been prepared.[126] The reactions of dialkyl-chloroboranes with di-iso-propylcarbamoyl-lithium give rise to other novel types of organoboron betaines with interesting properties.[127] For example, compound (3) apparently forms stable addition compounds with simple amides.[127a]

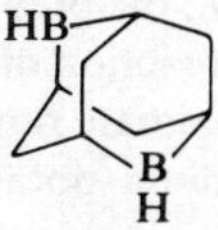

(1)

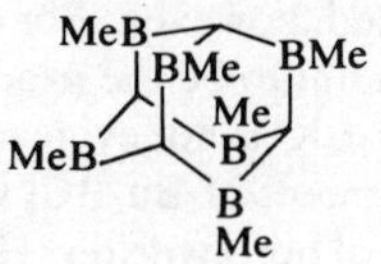

(2)

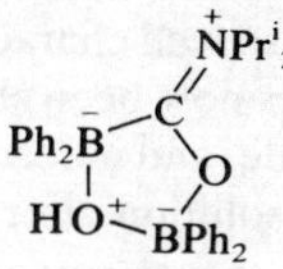

(3)

A plethora of transition-metal complexes of boron compounds has been synthesized. Of particular interest are complexes of borabenzene derivatives,[128] including a new class of stable twenty-electron sandwich complexes (4),[129] and multi-decker sandwiches in which cyclic boron compounds such as (5),[130] (6),[131] and (7)[132] are the ligands. The current multi-decker record holder, rivalling the 'Big Mac®' in terms of number of layers, is a trinuclear tetra-decker sandwich of (5) with consecutive Co, Fe, and Co atoms as the nuclei.[133]

[125] M. P. Brown, A. K. Holliday, and G. M. Way, *J. Chem. Soc., Dalton Trans.*, 1975, 148; I. Rayment and H. M. M. Shearer, *ibid.*, 1977, 136.

[126] W. Kliegel and E. Ahlenstiel, *Chem. Ber.*, 1976, **109**, 3547.

[127] (*a*) A. S. Fletcher, W. E. Paget, K. Smith, K. Swaminathan, J. H. Beynon, R. P. Morgan, M. Bozorgzadeh, and M. J. Haley, *J. Chem. Soc., Chem. Commun.*, 1979, 347; (*b*) A. S. Fletcher, W. E. Paget, K. Smith, K. Swaminathan, and M. J. Haley, *ibid.*, p. 573.

[128] G. E. Herberich, H. J. Becker, and G. Engelke, *J. Organometal. Chem.*, 1978, **153**, 265; and references cited therein; G. E. Herberich, G. Engelke, and W. Pahlmann, *Chem. Ber.*, 1979, **112**, 607; A. J. Ashe, W. Butler, and H. F. Sandford, *J. Amer. Chem. Soc.*, 1979, **101**, 7066.

[129] G. E. Herberich, W. Koch, and H. Lueken, *J. Organometal. Chem.*, 1978, **160**, 17.

[130] W. Siebert and W. Rothermel, *Angew. Chem., Int. Ed. Engl.*, 1977, **16**, 333.

[131] G. E. Herberich, J. Hengesbach, U. Kölle, G. Huttner, and A. Frank, *Angew. Chem., Int. Ed. Engl.*, 1976, **15**, 433.

[132] W. Siebert and M.Bochmann, *Angew. Chem., Int. Ed. Engl.*, 1977, **16**, 857; W. Siebert, J. Edwin, and M. Bochmann, *ibid.*, 1978, **17**, 868.

[133] W. Siebert, W. Rothermel, C. Bohle, C. Krüger, and D. J. Brauer, *Angew. Chem., Int. Ed. Engl.*, 1979, **18**, 949.

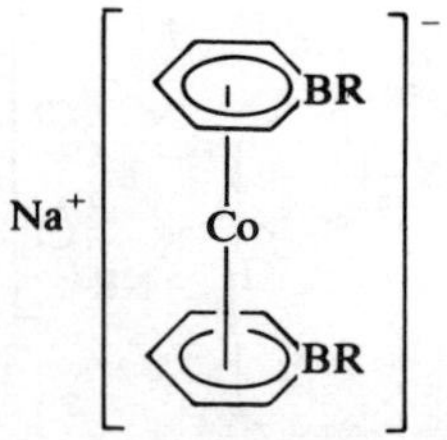

(4)

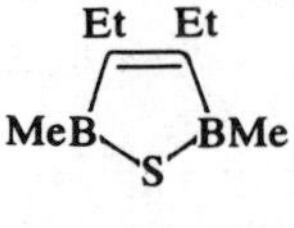

(5)

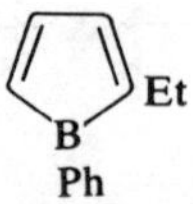

(6)

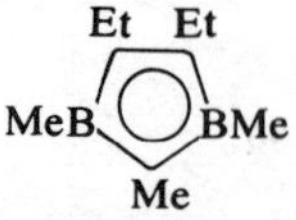

(7)

Cyclopentadienylboron compounds[134] and 7-borabicyclo[2.2.1]heptadienes[135] are fluxional molecules. Reactions of polyborylmethyl anions with carbonyl compounds give rise to alkenyl-boron compounds with one less boron unit on the *B*-attached carbon atom.[136] This type of approach has been applied in the synthesis of homologated aldehydes,[136a] methyl ketones,[136b] borazaroquinolines,[136c] and keten dithioacetals.[136d] In the last case, the reagent is an α-phenylthio-boron compound, other examples of which have been used in the synthesis of acetals and monothioacetals[137] and of iodomethaneboronic esters.[138] Several other (α-halogeno-alkyl)boronic acid derivatives have been prepared by various procedures.[139]

Amine adducts of cyano-, carbamoyl-, and carboxy-boranes, *e.g.* (8), are interesting as boron-containing analogues of α-amino-acids.[140] An isocyanide (9) has also been obtained.[141] A number of other potentially interesting new compound types have not been so well characterized, however. For example, the previously claimed Bu_2BCOPh has now been shown *not* to be the product of reaction of 'Bu_2BK' with benzoyl chloride, and indeed 'Bu_2BK' is not even an approximate representation of the species in solution after treatment of Bu_2BCl with sodium–potassium alloy.[142] The product is, in fact, some type of borohydride. However, highly interesting cyclic products such as (10) and (11) have been claimed, without substantial evidence, to be produced from the reactions of cyclohexene with the $MeBCl_2$–potassium system[143] and with the photolysis products of tri-1-naphthylborane.[144] The danger of making assumptions about these latter reactions has been pointed out.[145]

[134] P. Jutzi and A. Seufert, *J. Organometal. Chem.*, 1979, **169**, 327, 357, 373; *Chem. Ber.*, 1979, **112**, 2481; H. D. Johnson, T. W. Hartford, and C. W. Spangler, *J. Chem. Soc., Chem. Commun.*, 1978, 242.

[135] J. J. Eisch and J. E. Galle, *J. Organometal. Chem.*, 1977, **127**, C9.

[136] (*a*) D. S. Matteson, R. J. Moody, and P. K. Jesthi, *J. Am. Chem. Soc.*, 1975, **97**, 5608; (*b*) R. J. Moody and D. S. Matteson, *J. Organometal. Chem.*, 1978, **152**, 265, (*c*) D. S. Matteson, M. S. Biernbaum, R. A. Bechtold, J. D. Campbell, and R. J. Wilcsek, *J. Org. Chem.*, 1978, **43**, 950; (*d*) A. Mendoza and D. S. Matteson, *ibid.*, 1979, **44**, 1352.

[137] A. Mendoza and D. S. Matteson, *J. Chem. Soc., Chem. Commun.*, 1978, 357; *J. Organometal. Chem.*, 1978, **156**, 149.

[138] D. S. Matteson and D. Majumdar, *J. Organometal. Chem.*, 1979, **170**, 259.

[139] M. W. Rathke, E. Chao, and G.Wu, *J. Organometal. Chem.*, 1976, **122**, 145; D. S. Matteson and P. K. Jesthi, *ibid.*, 1976, **114**, 1; H. C. Brown, N. R. De Lue, Y. Yamamoto, and K. Maruyama, *J. Org. Chem.*, 1977, **42**, 3252.

[140] B. F. Spielvogel, F. Harchelroad, and P. Wisian-Nielson, *J. Inorg. Nucl. Chem.*, 1979, **41**, 1223; and references cited therein.

[141] J. L. Vidal and G. E. Ryschkewitsch, *J. Chem. Soc., Chem. Commun.*, 1976, 192.

[142] K. Smith and K. Swaminathan, *J. Chem Soc., Dalton Trans.*, 1976, 2297.

[143] S. M. van der Kerk, J. Boersma, and G. J. M. van der Kerk, *Tetrahedron Lett.*, 1976, 4765.

[144] B. G. Ramsey and D. M. Anjo, *J. Am. Chem. Soc.*, 1977, **99**, 3182.

[145] J. J. Eisch and H. P. Becker, *J. Organometal. Chem.*, 1979, **171**, 141; see also K. Smith, *Ann. Reports* (*B*), 1976, **73**, 122.

$Me_3\overset{+}{N}-\overset{-}{B}H_2CO_2H$ (8)

$Me_3\overset{+}{N}-\overset{-}{B}H_2NC$ (9)

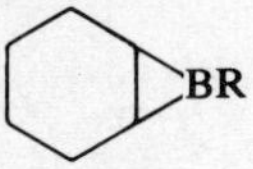

(10)

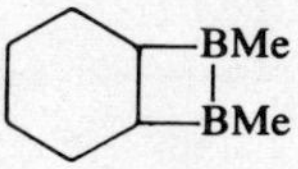

(11)

7 Miscellaneous Organoboron Chemistry

Tris(organylthio)boranes, $(RS)_3B$, react with carboxylic acids[146] or esters[147] to give thioesters, and their established application for the synthesis of thioketals has been extended.[148] They are also useful for synthesis of 1,3-bis(organylthio)alkenes from enones,[149] and for the formation of 2-hydroxyalkyl sulphides by ring-opening of epoxides.[150] The corresponding selenoboron compounds are applicable to selenoketal synthesis.[151]

Dialkyl(methylthio)boranes, R^1_2BSMe, give dialkyl-bromoboranes on reaction with bromine,[152] and dialkyl-bromoboranes can be converted into 'mixed' trialkylboranes, $R^1_2R^2B$, by treatment with sodium hydride in the presence of an alkene.[153] 9-Bromo-9-BBN is a convenient and selective reagent for cleavage of ethers,[154] and catecholborane has been used in the formation of amides and macrocyclic lactones from carboxylic acids.[155]

146 A. Pelter, T. E. Levitt, K. Smith, and A. Jones, *J. Chem. Soc., Perkin Trans. 1*, 1977, 1672.
147 T. Cohen and R. E. Gapinski, *Tetrahedron Lett.*, 1978, 4319.
148 D. R. Morton and S. J. Hobbs, *J. Org. Chem.*, 1979, **44**, 656; H. A. Klein, *Z. Naturforsch., Teil B*, 1979, **34**, 999.
149 T. Cohen, D. A. Bennett, and A. J. Mura, *J. Org. Chem.*, 1976, **41**, 2506.
150 H. A. Klein, *Chem. Ber.*, 1979, **112**, 3037.
151 D. L. J. Clive and S. M. Menchen, *J. Chem. Soc., Chem. Commun.*, 1978, 356; *J. Org. Chem.*, 1979, **44**, 1883.
152 A. Pelter, K. Rowe, D. N. Sharrocks, K. Smith, and C. Subrahmanyam, *J. Chem. Soc., Dalton Trans.*, 1976, 2087.
153 A. Pelter, K. Rowe, and K. Smith, *J. Chem. Soc., Chem. Commun.*, 1975, 532.
154 M. V. Bhatt, *J. Organometal. Chem.*, 1978, **156**, 221.
155 D. B. Collum, S.-C. Chen, and B. Ganem, *J. Org. Chem.*, 1978, **43**, 4393.

14 Organophosphorus Chemistry

By C. D. HALL
Department of Chemistry, King's College, (University of London)
Strand, London WC2R 2LS

During the six years since the last review,[1] more than ten thousand papers have been published in the field of organophosphorus chemistry; in consequence, it is inevitable that this report will be highly selective. In particular, lack of space has led to the omission of a section on phosphorus biochemistry. To a large extent, the period has been one of consolidation, with the major advances arising from the study of penta- and hexa-co-ordinate phosphorus compounds, many of which are now available as stable, isolable compounds. Interest in highly reactive phosphorus species (phosphorus radicals, methylenephosphines, and metaphosphates) has also flourished, and, as synthetic and instrumental techniques improve, one may envisage further growth in these areas.

The period has seen the publication of all seven volumes of the invaluable book by Kosolapoff,[2] and several useful texts or monographs have appeared.[3] In addition, the Specialist Periodical Reports,[4] now up to Volume 10, continue to provide excellent abstract coverage annually.

1 Pentaco-ordinate Phosphorus Compounds

Structure.—Phosphoranes vary in structure from trigonal-bipyramidal (tbp), *e.g.* (1)[5] and (2), to square-pyramidal (sqp), *e.g.* (3; X = O, Y = H, Z = Me), with the vast

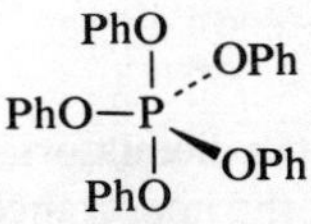

(1)

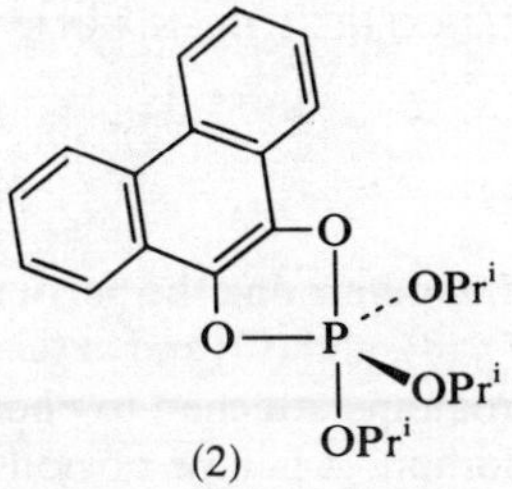

(2)

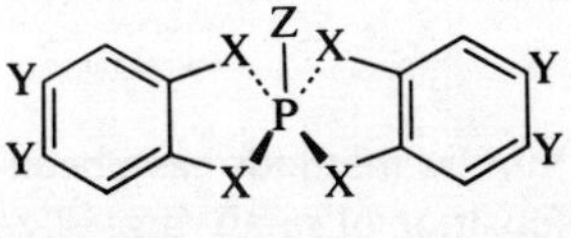

(3)

[1] S. Trippett, *Ann. Reports (B)*, 1973, **70**, Ch. 8, p. 268.
[2] 'Organic Phosphorus Compounds', ed. G. M. Kosolapoff and L. Maier, Wiley–Interscience, New York, 1973, Vols. 1–7.
[3] J. W. Emsley and C. D. Hall, 'The Chemistry of Phosphorus', Harper and Row, London, 1976; D. E. C. Corbridge, 'Phosphorus – An Outline of its Chemistry, Biochemistry and Technology', Elsevier, Amsterdam, 1978; R. Luckenbach, 'Dynamic Stereochemistry of Pentaco-ordinated Phosphorus and Related Elements', G. Thieme, Stuttgart, 1973.
[4] 'Organophosphorus Chemistry', ed. S. Trippett and D. W. Hutchinson (Specialist Periodical Reports), The Chemical Society, London 1970—78, Vols. 1—10.
[5] R. Sarma, F. Ramirez, B. McKeever, J. F. Maracek, and S. Lee, *J. Amer. Chem. Soc.*, 1976, **98**, 581.

majority of spirophosphoranes falling into the category of distorted tbp, *e.g.* (3; X = O, Y = H, Z = F).[6] Analysis of the distortions from ideal tbp in compounds of type (3) shows that these are towards sqp, *i.e.* along the pathway followed in a Berry pseudorotation,[7] and calculations reveal that the sqp is favoured by enhanced electronegativity in X and Y and an enhanced electropositive character for Z.[8]

Ligand Reorganization. —Interest in ligand reorganization ('permutational isomerization' or 'polytopal rearrangement') within phosphoranes has continued at a high level, especially with regard to the energy barriers for such rearrangements.[9] Dynamic n.m.r. spectroscopy remains the principal technique in this area, and studies of acyclic phosphoranes containing trifluoromethyl groups, *e.g.* (4), by ^{19}F n.m.r. and ^{13}C n.m.r. have revealed apicophilicities that decrease in the order X = F > Cl > OMe > SMe > NMe_2 > Me.[10] Likewise, ^{1}H and ^{19}F n.m.r. studies of monocyclic, *e.g.* (5), and bicyclic phosphoranes, *e.g.* (6), have extended the range

$(CF_3)_3P(CH_3)X$

(4)

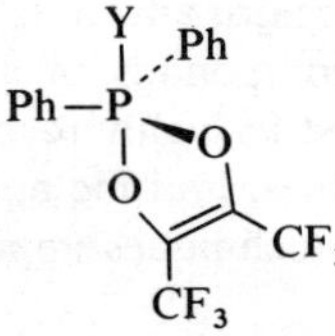

(5)

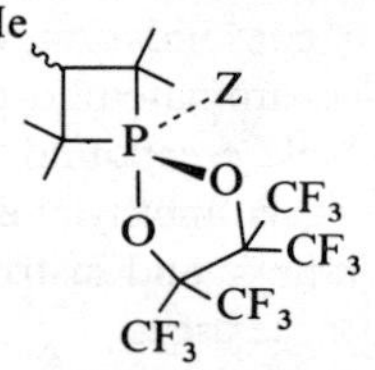

(6)

shown above and placed it on a more quantitative basis.[9,11] A combination of the available data results in a range of semi-quantitative, relative values of apicophilicity (Table 1). These energy differences must be treated with caution, but may be a useful guide to assessing the energetics of displacement reactions at phosphorus *via* pentaco-ordinate intermediates.

Table 1 *Relative apicophilicities for a variety of groups*

Group	F	CN	PhO,PhS	Cl,NCO,NCS	H	N_3	OMe	SMe	NMe_2	Me	Pr^i	Ph
$\Delta(\Delta G^{+})$/kJ mol^{-1}	0	1	4	5	~8*	11	12	20	27	31	34	50

* Values range from 4 to 12.

Similar methods have been used to determine the barriers facing the diequatorial disposition of small rings [*e.g.* in (7), (8), and (9)] and in this context, the importance of maintaining lone pairs on heteroatoms attached to phosphorus *in the equatorial plane* has been evaluated.[9] In addition, it is now obvious that one cannot ignore steric factors,[9,12] and the small size of the hydrogen atom may well account for its

[6] H. Wunderlich, D. Mootz, R. Schmutzler, and M. Wieber, *Z. Naturforsch.*, 1974, **29b**, 32.
[7] R. R. Holmes and J. A. Deiters, *J. Amer. Chem. Soc.*, 1977, **99**, 3318.
[8] R. R. Holmes, *J. Amer. Chem. Soc.*, 1975, **97**, 5379.
[9] S. Trippett, *Phosphorus Sulfur*, 1976, **1**, 89.
[10] K. I. The and R. G. Cavell, *Inorg. Chem.*, 1977, **16**, 2887; *ibid.*, 1976, **15**, 2518; R. G. Cavell, J. A. Gibson, and K. I. The, *ibid.*, 1978, **17**, 2880.
[11] R. K. Oram and S. Trippett, *J.C.S. Perkin I*, 1973, 1300; J. Brierley, J. I. Dickstein, and S. Trippett, *Phosphorus Sulfur*, 1979, **7**, 167.
[12] I. Szele, S. J. Kubisen jnr., and F. H. Westheimer, *J. Amer. Chem. Soc.*, 1976, **98**, 3533.

unexpectedly high apicophilicity. On the theoretical side, Holmes has developed a model which predicts the free energies of activation for Berry pseudorotations; the agreement with experiment over a wide range of phosphoranes is good.[13]

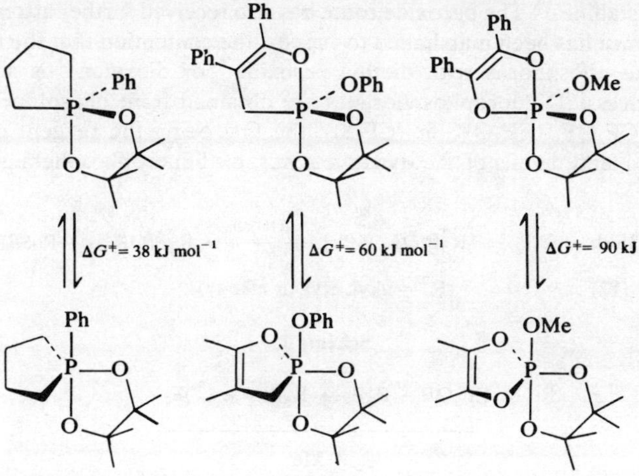

(7) (8) (9)

The Berry process is now widely accepted as the most likely *mechanism* for ligand reorganization. *Ab initio* calculations on PH_5 reveal $\Delta G^{\ddagger} \simeq 8$ kJ mol^{-1} (for Berry) *versus* 42 kJ mol^{-1} for the turnstile (TR) mechanism,[14] and compounds such as (10), with every incentive to adopt a 0° (2+3) conformation, turn out to be distorted tbp, but apparently still capable of undergoing rapid reorganization of ligands.[15] Furthermore, compounds such as (11) do not undergo ligand reorganization when, according to the TR mechanism, they should.[16]

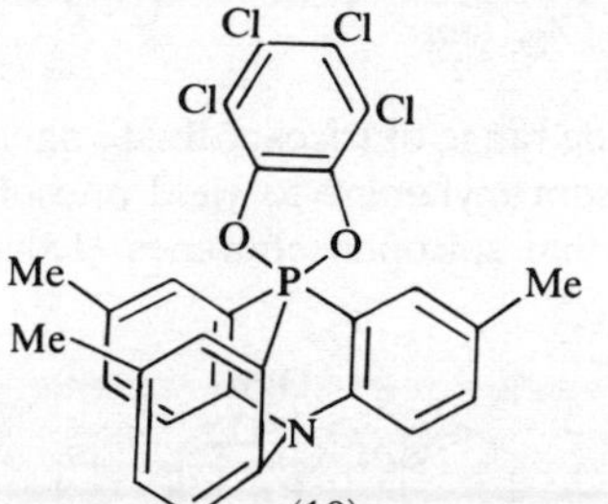

(10)

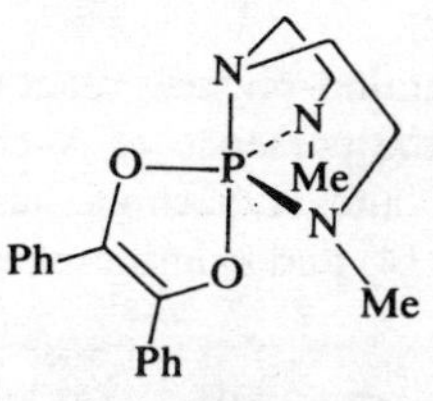

(11)

Preparation and Chemistry of Phosphoranes. —Reviews have appeared on the synthetic applications of oxyphosphoranes[17] and on the penta-alkyl derivatives of

[13] R. R. Holmes, *J. Amer. Chem. Soc.*, 1978, **100**, 433.
[14] J. A. Altmann, K. Yates, and I. G. Csizmadia, *J. Amer. Chem. Soc.*, 1976, **98**, 1450.
[15] D. Hellwinkel, W. Blaicher, W. Krapp, and W. S. Sheldrick, *Chem. Ber.*, 1980, **113**, 1406.
[16] D. B. Denney, personal communication.
[17] F. Ramirez and I. Ugi, *Bull. Soc. chim. France*, 1974, 453.
[18] H. Schmidbaur, *Adv. Organometallic Chem.*, 1976, **14**, 205.
[19] L. L. Chang, D. B. Denney, D. Z. Denney, and R. J. Kazior, *J. Amer. Chem. Soc.*, 1977, **99**, 2293; D. A. Bowman, D. B. Denney, and D. Z. Denney, *Phosphorus Sulfur*, 1978, **4**, 229.
[20] N. J. De'ath and D. B. Denney, *Phosphorus Sulfur*, 1977, **3**, 51.

elements of Group 5.[18] The reaction of sulphenate esters (12) with trico-ordinate phosphorus compounds (see Scheme 1) is now the method of choice for the preparation of acyclic alkoxyphosphoranes, which in some cases may be distilled or obtained crystalline.[19] The peroxide route has also received further attention,[20] and kinetic evidence has been marshalled to support the contention that the reaction of trico-ordinate phosphorus with diethyl peroxide[21] or dioxetan[22] is a concerted (biphilic) process. Difluorophosphoranes are obtained from phosphines or phosphites with CF_3OF, $(CF_3O)_2$, or $(CF_3S)_2$, the first being the reagent of choice[23] [reaction (1)], and details of the dynamic n.m.r. of difluorophosphetans (13) have appeared.[24]

$$R^1OSPh + R^2_3P \rightarrow [R^2_3P(OR^1)SPh] \xrightarrow{R^1OSPh} R^2_3P(OR^1)_2 + PhSSPh$$

(12) (R^2 = alkyl, aryl, or alkoxy)

Scheme 1

$$CF_3OF + R_3P \rightarrow R_3PF_2 + COF_2 \quad (1)$$

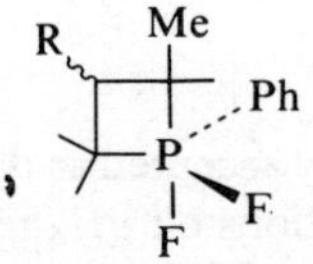

(13) a; R = H
b; R = Me

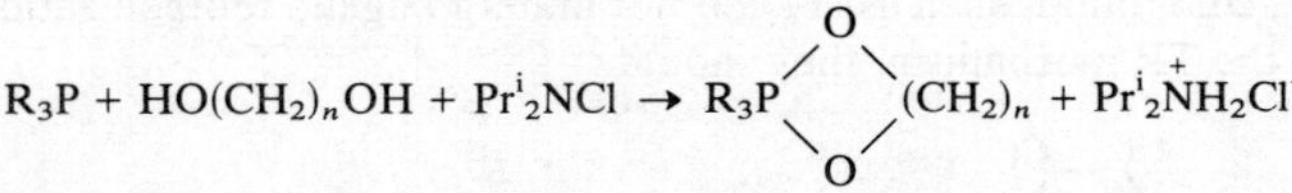

(2)

Diols and *o*-amino-phenols react with a wide range of trico-ordinate phosphorus compounds in the presence of *N*-chlorodi-isopropylamine to yield phosphoranes [reaction (2)],[25] and Kabachnik has shown that spirophosphoranes (15) may be obtained from (14) and *o*-aminophenol.[26]

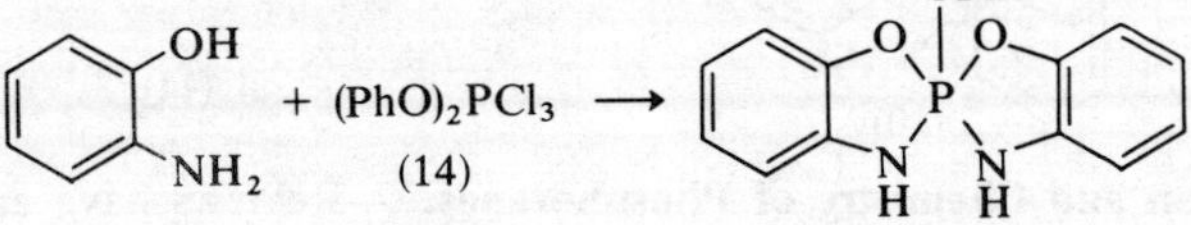

(15)

[21] G. Scott, P. J. Hammond, C. D. Hall, and J. D. Bramblett, *J.C.S. Perkin II*, 1977, 882.
[22] P. D. Bartlett, A. L. Baumstark, M. E. Landis, and C. L. Lerman, *J. Amer. Chem. Soc.*, 1974, **96**, 5267.
[23] N. J. De'ath, D. Z. Denney, D. B. Denney, and C. D. Hall, *Phosphorus*, 1974, **3**, 205.
[24] N. J. De'ath, D. B. Denney, D. Z. Denney, and Y. F. Hsu, *J. Amer. Chem. Soc.*, 1976, **98**, 768.
[25] S. Antczak, S. A. Bone, J. Brierley, and S. Trippett, *J.C.S. Perkin I*, 1977, 278.
[26] N. A. Tikhonina, V. A. Gilyarov, and M. P. Kabachnik, *Izvest. Akad. Nauk S.S.S.R., Ser. khim.*, 1976, 2442.

The condensations of trico-ordinate phosphorus compounds with dienes (16a),[27] $\alpha\beta$-unsaturated carbonyl compounds (16b),[28] and α-dicarbonyls or monothiocarbonyls (16c)[29] continue to offer useful routes to a variety of mono- and bi-cyclic phosphoranes [reaction(3)]. Numerous stable phosphoranes containing the P—H bond have been prepared,[30] and the positions of equilibria with the corresponding tervalent phosphorus compounds [reaction(4)] have been shown to depend to a large

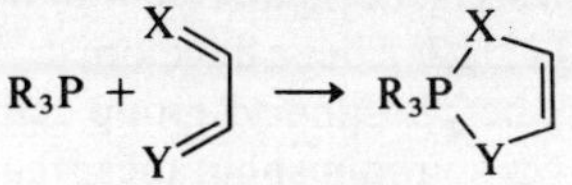

(3)

(16) a; X, Y = CH_2
b; X = O; Y = CH_2
c; X = O or S; Y = O

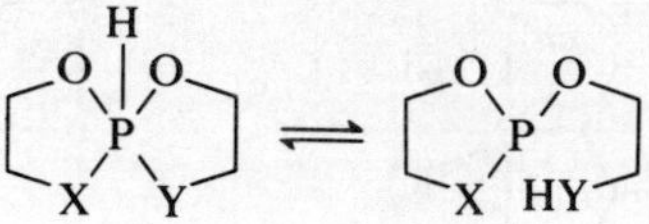

(4)

(X, Y = O or NMe)

extent on symmetry.[31] Furthermore, the first, stable hydroxyphosphorane (18a) was obtained recently by the action of dry HCl in CH_2Cl_2 on its trimethylsilyl ester (17).[32] Equilibrium to the phosphate (18b) was slow, on the n.m.r. time-scale, below 10 °C, in MeCN.

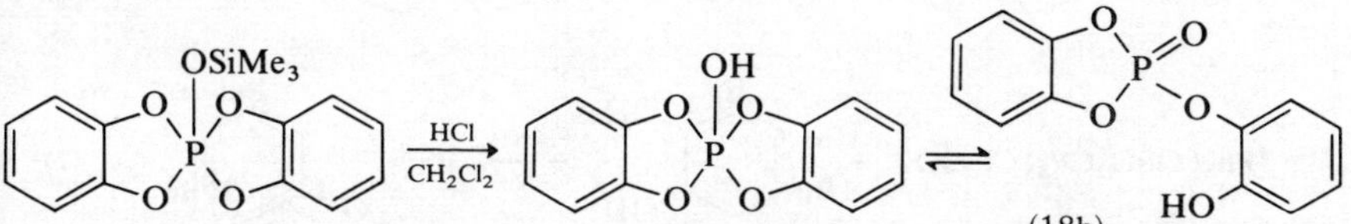

(17) (18a) (18b)

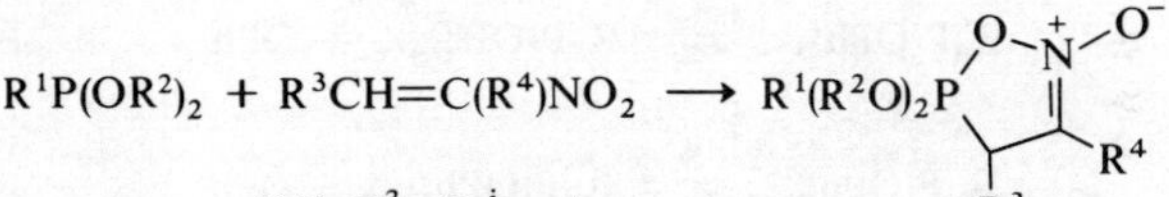

(19) R^3 = Pr^i or Ar
R^4 = H or Ph

(20)

High yields of monocyclic phosphoranes (20) are obtainable from phosphonites and nitro-olefins (19),[33] and activated olefins (21) have been shown to give good

[27] N. A. Razumova, N. A. Kurshakova, Z. L. Evtikhov, and A. A. Petrov, *J. Gen. Chem.* (*U.S.S.R.*), 1974, **44**, 1834.
[28] V. V. Vasil'ev, N. A. Razumova, and L. V. Dogadaeva, *J. Gen. Chem.* (*U.S.S.R.*), 1976, **46**, 461.
[29] B. A. Arbuzov, N. A. Polezhaeva, V. V. Smirnov, and A. A. Musina, *Bull. Acad. Sci. U.S.S.R.*, 1975, **24**, 1548; D. Bernard and R. Burgada, *Tetrahedron*, 1975, **31**, 797.
[30] D. Houalla, M. Sanchez, D. Gonbeau, and G. Pfister-Guillouzo, *Nouveau. J. Chem.*, 1979, **3**, 507; M. Koenig, A. Munoz, B. Garrigues, and R. Wolf, *Phosphorus Sulfur*, 1979, **6**, 435.
[31] R. Burgada and C. Laurenco, *J. Organometallic Chem.*, 1974, **66**, 255.
[32] F. Ramirez, M. Nowakowski, and J. F. Marecek, *J. Amer. Chem. Soc.*, 1977, **99**, 4515.
[33] R. D. Gareev, G. M. Loginova, and A. N. Pudovik, *J. Gen. Chem.* (*U.S.S.R.*), 1976, **46**, 1843; J. I. G. Cadogan, R. A. North, and A. G. Rowley, *J. Chem. Res.*(*S*), 1978, 1.

yields of acyclic, monocyclic, and bicyclic phosphoranes with phosphinites or phosphonites in the presence of alcohols, diols, or catechols[34] [reaction(5)].

$$Ph_2POEt + CH_2{=}CHX + EtOH \rightarrow \underset{(21)}{Ph_2P(OEt)_2CH_2CH_2X} \qquad (5)$$

$$(X = CN \text{ or } CO_2Et)$$

Some bicyclic phosphoranes, *e.g.* (22), are remarkably stable,[35] and general routes to analogous phosphoranes (23) have been developed.[36] Finally, although the routes from phosphonium salts or the phosphoryl group remain relative rarities, careful control of the conditions gives spirophosphoranes [reaction(6)],[37] and Trippett has developed a method from alkylated phosphine oxides or sulphides by treatment with catechols in the presence of di-isopropylamine [reaction(7)].[38]

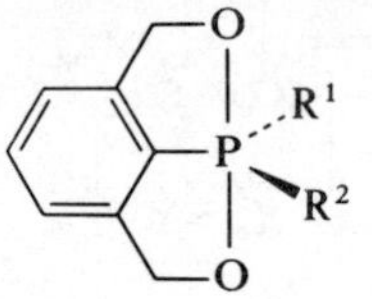

(22)

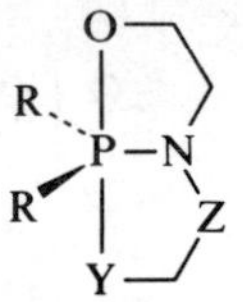

(23) $Y = CH_2$, NH, or O
$Z = CH_2$ or NH

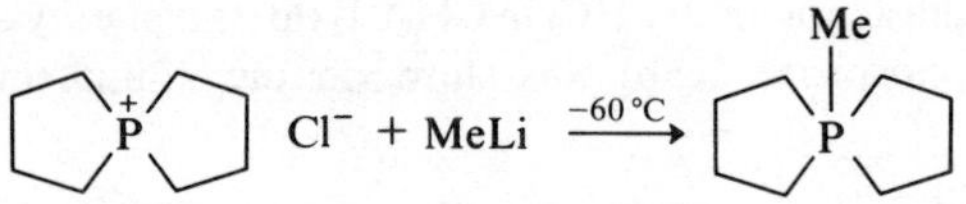

(6)

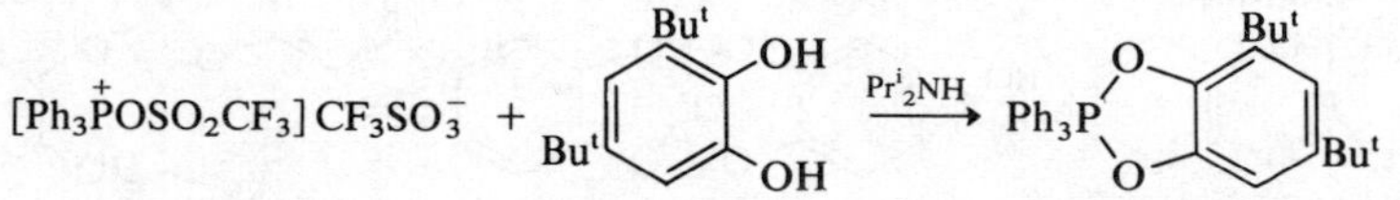

(7)

$$R_nP(OPh)_{5-n} \overset{K_1}{\rightleftharpoons} R_n\overset{+}{P}(OPh)_{4-n} + \bar{O}Ph \qquad (8)$$

$$R_n\bar{P}(OPh)_{6-n} \overset{K_2}{\rightleftharpoons} R_nP(OPh)_{5-n} + \bar{O}Ph \qquad (9)$$

Phosphoranes have been detected by low-temperature n.m.r. as intermediates in the Arbusov reaction,[39] and conductivity measurements in acetonitrile have allowed the calculation of equilibrium constants for the dissociation of both pentaco-ordinate and hexaco-ordinate phosphorus [reactions (8) and (9)].[40] On the other hand, a kinetic study of the hydrolysis of phosphoranes has shown that the reaction proceeds through a hexaco-ordinate intermediate (or transition state) rather than through a

[34] P. D. Beer, R. C. Edwards, C. D. Hall, J. R. Jennings, and R. J. Cozens, *J.C.S. Chem. Comm.*, 1980, 351.
[35] D. Hellwinkel and W. Krapp, *Chem. Ber.*, 1978, **111**, 13.
[36] A. Schmidpeter, J. H. Weinmaier, and E. Glaser, *Angew. Chem. Internat. Edn.*, 1977, **16**, 549, 865.
[37] H. Schmidbaur, P. Holl, and F. H. Köhler, *Angew. Chem. Internat. Edn.*, 1977, **16**, 722.
[38] S. Antczak and S. Trippett, *J.C.S. Perkin I*, 1978, 1326.
[39] J. Michalski, J. Mikolajczak, M. Pakulski, and A. Skowronska, *Phosphorus Sulfur*, 1978, **4**, 233; D. B. Denney, D. Z. Denney, and G. Di Miele, *ibid.*, p. 125.
[40] I. S. Sigal and F. H. Westheimer, *J. Amer. Chem. Soc.*, 1979, **101**, 5329, 5334.

dissociative mechanism.[41] By analogy with orthoesters, alkoxyphosphoranes are powerful alkylating agents,[42] and those containing a P—H bond will add to aldehydes or imines to form phosphoranes [reaction (10)].[43]

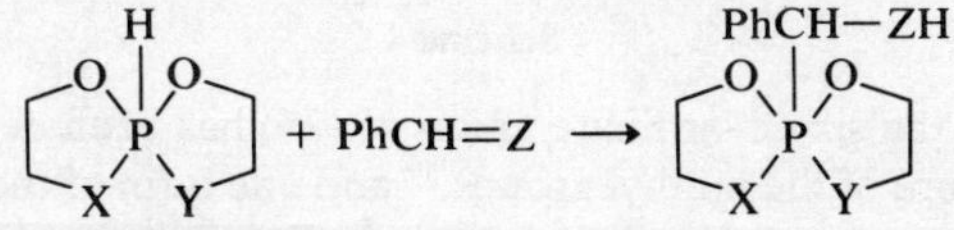

(X, Y = O or NMe) (Z = O, NMe, or NPh) (10)

A trioxaphosphetan (24) has been recognized as a convenient source of singlet oxygen,[44] and pure isomers of 1,3,2-oxazaphospholidines, *e.g.* (25), have been prepared by asymmetric induction and their isomerization followed polarimetrically.[45]

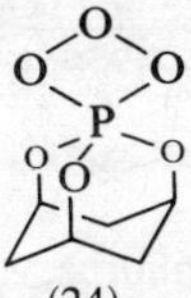

(24)

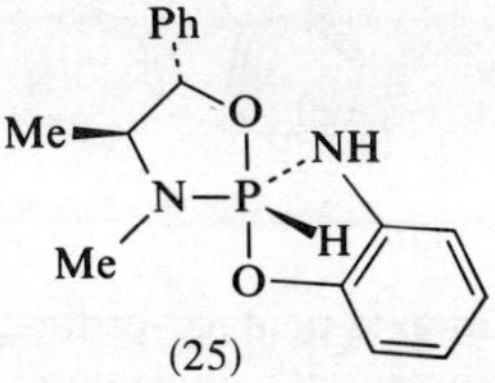

(25)

2 Hexaco-ordinate Phosphorus Compounds

Interest in the preparation of hexaco-ordinate phosphorus compounds has expanded from the cyclic phosphonite method (Scheme 2)[46] through a facile synthesis from

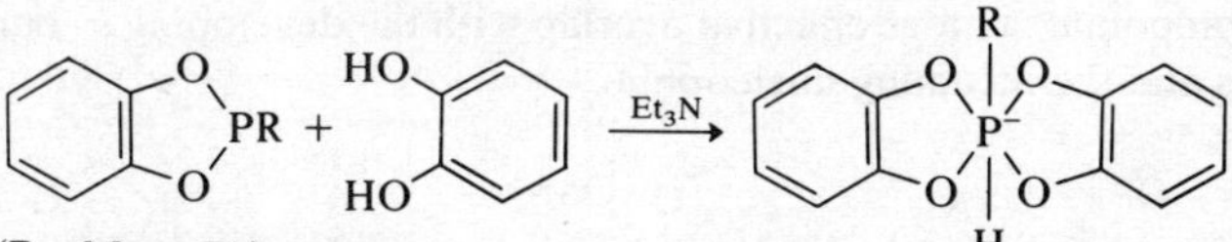

(R = Me or Ph)

Scheme 2

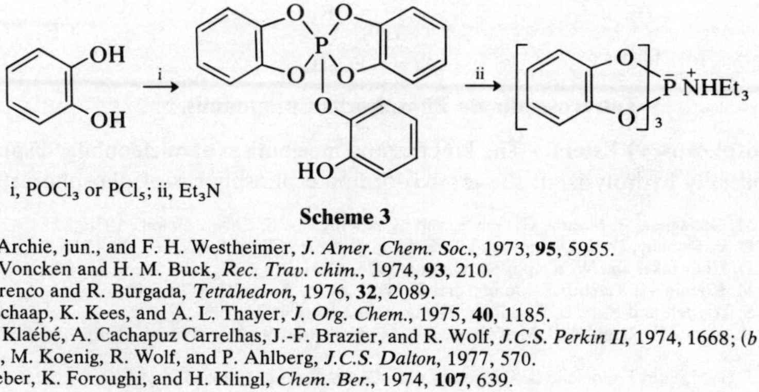

Reagents: i, $POCl_3$ or PCl_5; ii, Et_3N

Scheme 3

[41] W. C. Archie, jun., and F. H. Westheimer, *J. Amer. Chem. Soc.*, 1973, **95**, 5955.
[42] W. G. Voncken and H. M. Buck, *Rec. Trav. chim.*, 1974, **93**, 210.
[43] C. Laurenco and R. Burgada, *Tetrahedron*, 1976, **32**, 2089.
[44] A. P. Schaap, K. Kees, and A. L. Thayer, *J. Org. Chem.*, 1975, **40**, 1185.
[45] (*a*) A. Klaébé, A. Cachapuz Carrelhas, J.-F. Brazier, and R. Wolf, *J.C.S. Perkin II*, 1974, 1668; (*b*) A. Klaébé, M. Koenig, R. Wolf, and P. Ahlberg, *J.C.S. Dalton*, 1977, 570.
[46] M. Wieber, K. Foroughi, and H. Klingl, *Chem. Ber.*, 1974, **107**, 639.

phosphorus oxychloride (Scheme 3)[47] to a simple preparation of hexamethoxyphosphate anion (Scheme 4).[48]

$$(MeO)_5P + MeOK\,(excess) \rightleftharpoons (MeO)_6\bar{P}\,K^+$$

Scheme 4

The chirality of the six-co-ordinate anion of (26) has been demonstrated by the diastereotopic nature of the methyl groups,[49] and one form of the salt (27) has been obtained pure, again by asymmetric induction.[50] Equilibration of the diastereomers can be followed, as with the pentaco-ordinate systems, by an n.m.r. or a polarimetric temperature jump method.[45b]

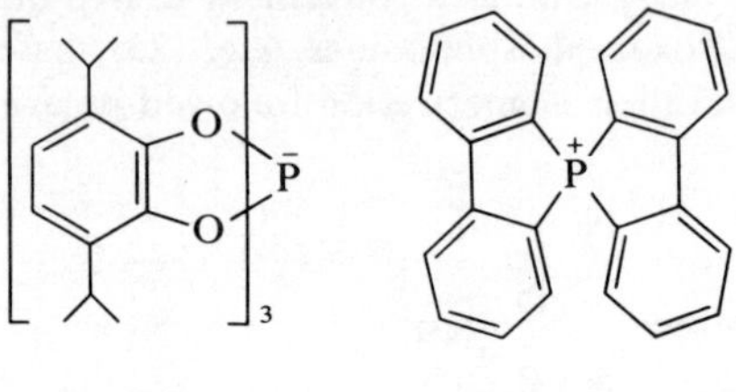

(26)

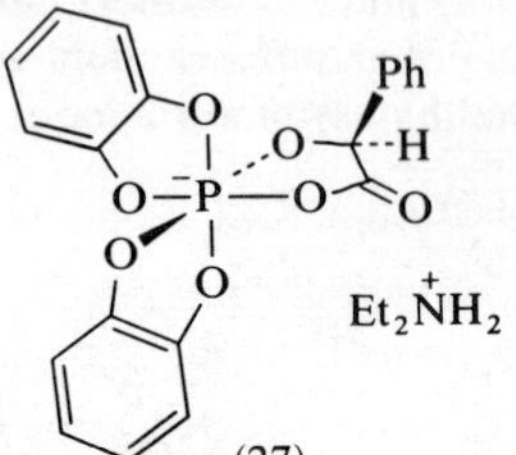

(27)

Nucleophilic substitution at pentaco-ordinate phosphorus, *e.g.* in (28), occurs *via* hexaco-ordinate structures with inversion and/or retention at phosphorus.[51] Although the stable configuration of spiro-bicyclic, six-co-ordinate phosphorus is apparently *cis*,[52] it appears that, under certain conditions, nucleophilic attack at P^V is under kinetic control and gives, initially, the less stable *trans* structure (29) before isomerization to the *cis* form (30).[53] This is probably an example of stereoelectronic control by lone pairs on heteroatoms within the substrate, which give, in the case of the *trans* compound, a more effective overlap with the developing σ-bond between phosphorus and the incoming nucleophile.

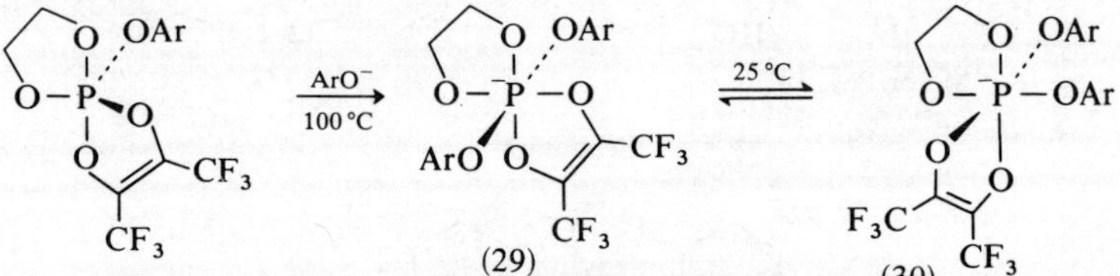

(28) (29) (30)

Ar = p-FC_6H_4

3 Tetraco-ordinate Phosphorus Compounds

Phosphorus(V) Esters.—The kinetics and mechanism of nucleophilic displacement, especially hydrolysis, at the tetraco-ordinate phosphorus of phosphorus(V) esters

47 M. Gallagher, A. Munoz, G. Gence, and M. Koenig, *J.C.S. Chem. Comm.*, 1976, 321.
48 D. B. Denney, D. Z. Denney, and C.-F. Ling, *J. Amer. Chem. Soc.*, 1976, **98**, 6755.
49 D. Hellwinkel and W. Krapp, *Phosphorus*, 1976, **6**, 91.
50 M. Koenig, A. Klaébé, A. Munoz, and R. Wolf, *J.C.S. Perkin II*, 1976, 955.
51 S. Trippett and R. E. L. Waddling, *Tetrahedron Letters*, 1979, 193.
52 R. Sarma, F. Ramirez, B. McKeever, J. F. Maracek, and V. A. V. Prasad, *Phosphorus Sulfur*, 1979, **5**, 323.
53 J. H. M. Font Freide and S. Trippett, *J.C.S. Chem. Comm.*, 1980, 157.

have continued to receive considerable attention,[54] and among the more unusual aspects are reports of micellar catalysis,[55] an α-effect with hydroperoxide as nucleophile,[56] and the effects of heterocyclic groups on the alkaline hydrolysis of phosphinates, *e.g.* (31), and phosphonates (32).[57] A large proportion of

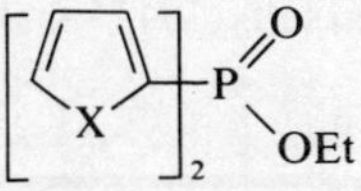

(31) X = O or S

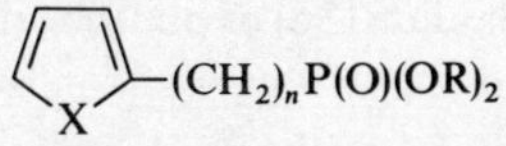

(32) X = O, S, or NMe

the work, however, has concentrated on the stereochemistry of nucleophilic displacement, and much has been documented on the synthesis, resolution, and determination of enantiomeric purity of tetraco-ordinate organophosphorus esters[58] and acids.[59] In this respect, the star prize must go to the synthesis of the chiral (1*R*)-[^{16}O,^{17}O,^{18}O]phospho-(*S*)-propane-1,2-diol (34) by the reaction of (33), which originates from ephedrine, with the *O*-benzyl ether of (*S*)-propane-1,2-diol[60] (see Scheme 5).

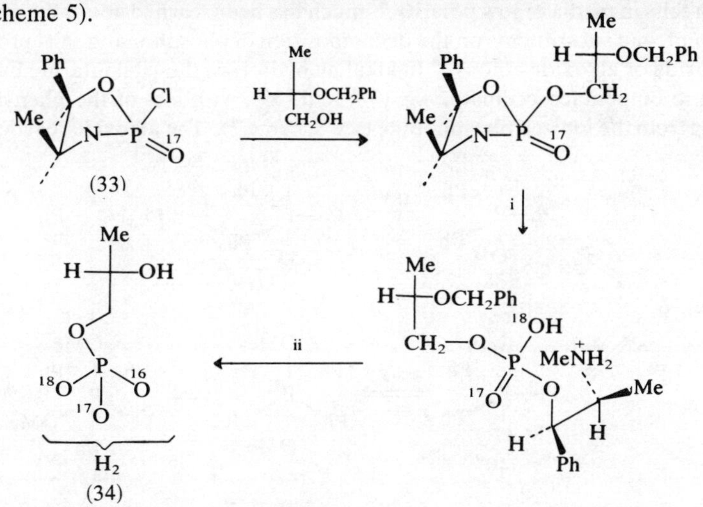

Reagents: i, $H_3{}^{18}O^+$, CF_3CO_2H; ii, H_2, Pd/C

Scheme 5

Derivatives of the anti-cancer drug cyclophosphamide continue to attract attention,[61] and optically active forms of the drug have been reported.[62] Within a welter

[54] V. E. Belśkii, *Russ. Chem. Rev.*, 1977, **46**, 1578; F. H. Westheimer, *Pure Appl. Chem.*, 1977, **49**, 1059.
[55] C. A. Bunton, S. Diaz, G. M. Van Fleteren, and C. Park, *J. Org. Chem.*, 1978, **43**, 258.
[56] L. Horner and A. Parg, *Annalen*, 1977, 61.
[57] D. W. Allen, B. G. Hutley, and M. T. J. Mellor, *J.C.S. Perkin II*, 1977, 1705, 789.
[58] C. R. Hall and T. D. Inch, *Phosphorus Sulfur*, 1979, **7**, 171.
[59] M. Mikolajczyk and M. Leitloff, *Russ. Chem. Rev.*, 1975, 44, 670.
[60] S. J. Abbott, S. R. Jones, S. A. Weinman, F. M. Bockhoff, F. W. McLafferty, and J. R. Knowles, *J. Amer. Chem. Soc.*, 1979, **101**, 4323.
[61] H. J. Hohurst, G. Peter, and R. F. Struck, *Cancer Res.*, 1976, **36**, 2278.
[62] Ger. Offen. 2 644 905, 1977 (*Chem. Abs.*, 1977, **87**, 53 401); T. Kawashima, R. D. Kroshefsky, R. A. Kok, and J. G. Verkade, *J. Org. Chem.*, 1978, **43**, 1111.

of good reviews, those on phosphorus–selenium compounds,[63] phosphorus esters from oxyphosphoranes,[64] and phosphorylation[65] are perhaps of widest appeal.

Phosphonium Salts.—Among the many syntheses of phosphonium salts, one of the most useful is the acid-catalysed cyclization of hydroxy-phosphines (35) to give good yields of phospholanium (36) or phosphorinanium salts (37),[66] as shown in Scheme 6.

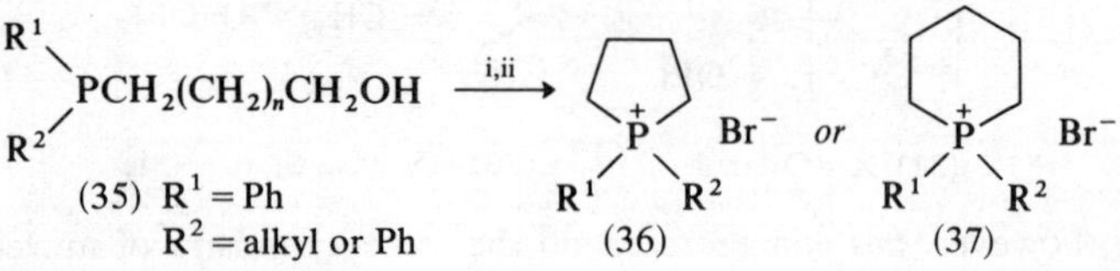

(35) R^1 = Ph
R^2 = alkyl or Ph
n = 1 or 2

(36) (37)

Reagents: i, HBr; ii, $NaHCO_3$

Scheme 6

Since the early reports of enormously enhanced rates of hydrolysis of phosphonium salts in media of low polarity,[67] much has been learned about the effects of salt, solvent, and substituents on the decomposition of phosphonium salts promoted by hydroxide or alkoxide anions.[68] In alkaline hydrolysis the vital intermediates are believed to be pentaco-ordinate, *e.g.* (38) and (39), with loss of the phenyl anion occurring from the ionized phosphorane (see Scheme 7). The alkoxide reactions are,

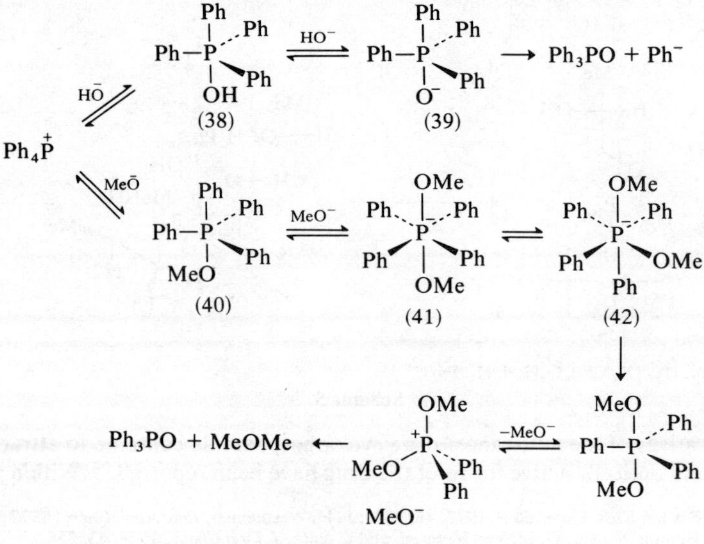

Scheme 7

[63] J. Michalski, *Chemica Scripta* (*A*), 1975, **8**, 58.
[64] F. Ramirez and I. Ugi, *Phosphorus Sulfur*, 1976, **1**, 231.
[65] L. A. Slotin, *Synthesis*, 1977, 737.
[66] W. R. Purdum and K. D. Berlin, *J. Org. Chem.*, 1975, **40**, 2801.
[67] A. Schnell, J. G. Dawber, and J. C. Tebby, *J.C.S. Perkin II*, 1976, 633; F. Y. Khalil and G. Aksnes, *Z. phys. Chem.* (*Frankfurt*), 1975, **97**, 179.
[68] G. Aksnes, *Phosphorus Sulfur*, 1977, **3**, 227; G. Aksnes, F. Y. Khalil, and P. J. Majewski, *ibid.*, p. 157.

however, thought to proceed *via* hexaco-ordinate intermediates (41) and (42), thus avoiding a route involving de-alkylation of a methoxyphosphorane (40).[68] Notice the suggestion of *trans* attack on (40) to form (41) followed by rearrangement to *cis* before loss of the phenyl anion (*cf.* ref. 53). Not unexpectedly, the hydrolysis of optically active phosphonium salts in media of low polarity gives *racemic* oxide, presumably due to the extended lifetimes of the pentaco-ordinate intermediates.[69] A thorough investigation confirms that the presence of bulky substituents at phosphorus [*e.g.* R = Bu^t in (43)] leads to alkaline hydrolysis with retention of configuration at phosphorus whereas smaller groups (*e.g.* R = Pr^n or Pr^i) give rise to almost complete inversion.[70] Allen has added to his study of the hydrolysis of heteroarylphosphonium salts (44),[71] and alkoxyphosphonium salts (45), which are intermediates in the Arbusov reaction, have been isolated and characterized.[72]

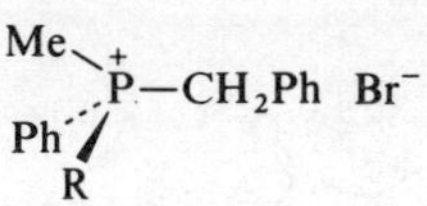
(43)

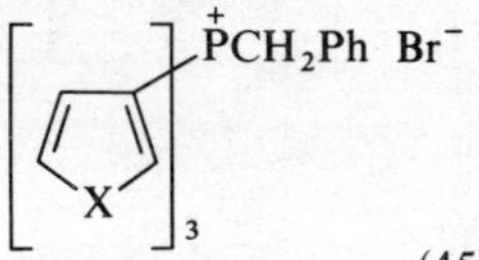
(44) X = O or S

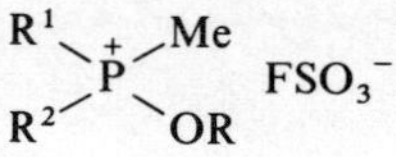
(45) R^1, R^2 = Et, Ph, OMe, or OPh

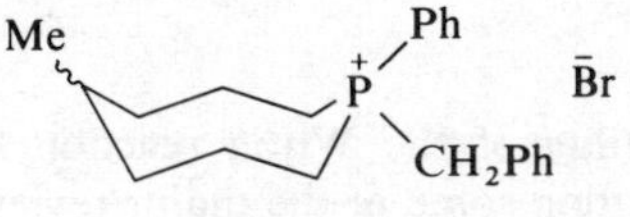
(46)

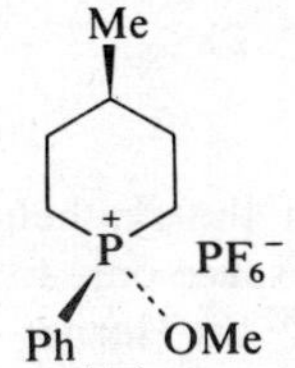
(47)

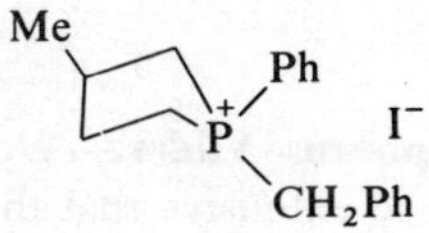
(48)

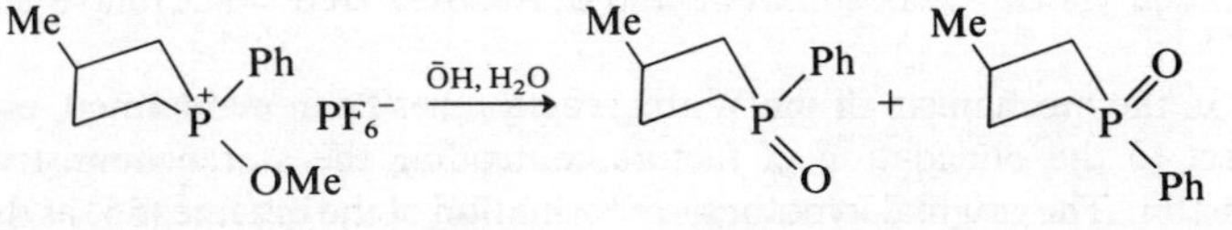
(49) 51% 49%

The excellent work with cyclic phosphonium salts has been continued by Marsi with the observation of inversion in phosphocanium salts (46)[73] and alkoxyphosphorinanium salts (47),[74] retention in a totally resolved phospholanium salt (48),[75] but a non-stereospecific reaction with an alkoxyphospholanium salt (49),[74] all of

[69] A. Schnell and J. C. Tebby, *J.C.S. Perkin I*, 1977, 1883.
[70] R. Luckenbach, *Z. Naturforsch.*, 1976, **31b**, 1127.
[71] D. W. Allen and B. G. Hutley, *J.C.S. Perkin I*, 1978, 675.
[72] K. S. Colle and E. S. Lewis, *J. Org. Chem.*, 1978, **43**, 571.
[73] K. L. Marsi and F. B. Burns, *Phosphorus*, 1974, **4**, 211.
[74] K. L. Marsi and J. L. Jasperse, *J. Org. Chem.*, 1978, **43**, 760; K. L. Marsi, *ibid.*, 1975, **40**, 1779.
[75] K. L. Marsi and A. Tuinstra, *J. Org. Chem.*, 1975, **40**, 1843.

which can be rationalized in terms of ring strain *versus* apicophilicity. The stereochemical properties of the 1-chlorophospholenium chloride (50) and its hydrolysis to a mixture of oxides have been discussed[76] and the isomerization of *cis*-phospholen (52) to the *trans*-compound (53) in the presence of traces of phosgene (present in $CDCl_3$) has been noted;[77] the mechanism of isomerization probably involves (51). Finally, a useful review has appeared on the synthesis of heterocyclic compounds from vinylphosphonium salts,[78] and it may be of interest to note that phosphonium salts of type (54) have been shown to possess anti-leukaemic activity.[79]

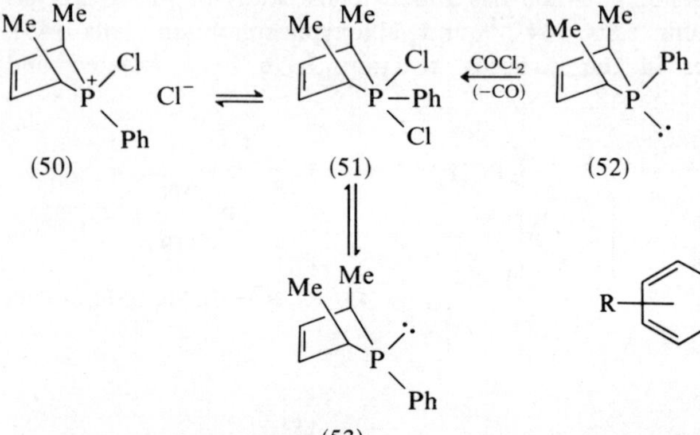

(50) (51) (52) (53)

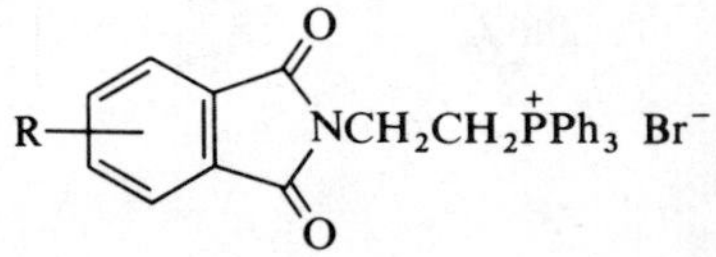

(54)

Phosphorus Ylides.—Work on the synthetic utility of the Wittig reaction has been so extensive that there is space only to mention some of the major reviews; these include historical aspects,[80] the Horner–Wadsworth–Emmons modification,[81] carotenoids,[82] insect pheromones,[83] heterocyclic compounds,[78] non-benzenoid aromatics,[84] cyclophanes,[85] and syntheses employing phosphacumulenes and phospha-allene ylides.[86] Phosphazene and derivatives ($R_3P{=}NX$) have also been reviewed.[87]

Interest in the mechanism of the Wittig reaction has been maintained, especially with respect to the elucidation of factors controlling the stereochemistry of the resultant olefin. The original hypothesis of formation of the betaine (55) as the *initial* step followed by collapse to an oxaphosphetan (56), which subsequently rearranges to products,[2,3] has been superseded by the idea that (56) is formed directly in the

[76] L. D. Quin and R. C. Stocks, *Phosphorus Sulfur*, 1977, **3**, 151.
[77] P. J. Hammond and C. D. Hall, *Phosphorus Sulfur*, 1977, **3**, 351.
[78] E. Zbiral, *Synthesis*, 1974, 775.
[79] R. J. Dubois, C.-C. Lin, and J. A. Beisler, *J. Medicin. Chem.*, 1978, **21**, 303.
[80] G. Wittig, *Accounts Chem. Res.*, 1974, **7**, 6.
[81] J. Boutagy and R. Thomas, *Chem. Rev.*, 1974, **74**, 87.
[82] H. Pommer, *Angew. Chem. Internat. Edn.*, 1977, **16**, 423; B. C. L. Weedon, *Pure Appl. Chem.*, 1976, **47**, 161.
[83] R. Rossi, *Synthesis*, 1977, 817.
[84] K. P. C. Vollhardt, *Synthesis*, 1975, 765.
[85] B. Thulin, O. Wennerstrom and I. Somfai, *Acta. Chem. Scand. (B)*, 1978, **32**, 109.
[86] H. J. Bestmann, *Angew. Chem. Internat. Edn.*, 1977, **16**, 349.
[87] R. A. Shaw, *Phosphorus Sulfur*, 1979, **5**, 363.

rate-limiting step[88,89] and then decomposes by cleavage of the phosphorus–carbon bond to form (57), which undergoes a rapid β-elimination of phosphine oxide to generate the olefin (see Scheme 8). Bestmann[89a] suggests that (56) is formed with the *cis*-configuration (58) (see Scheme 9) and that when R is electron-donating (*i.e.* the ylide is 'reactive') the carbanion (59) eliminates phosphine oxide very rapidly, with retention of stereochemistry, to form the *cis*-(Z)-olefin (60). If R is electron-withdrawing, however, the carbanion is stable enough for rotation about the C—C bond to compete with the elimination rate, and the *trans*-(E)-olefin (62) results. Incidentally, it may be necessary for (58) to pseudorotate to (61) before the phosphorus–carbon bond can cleave from an *apical* position. In contrast, a

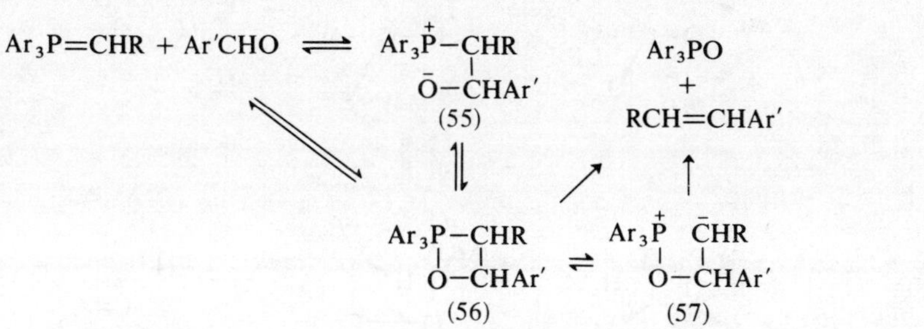

Scheme 8

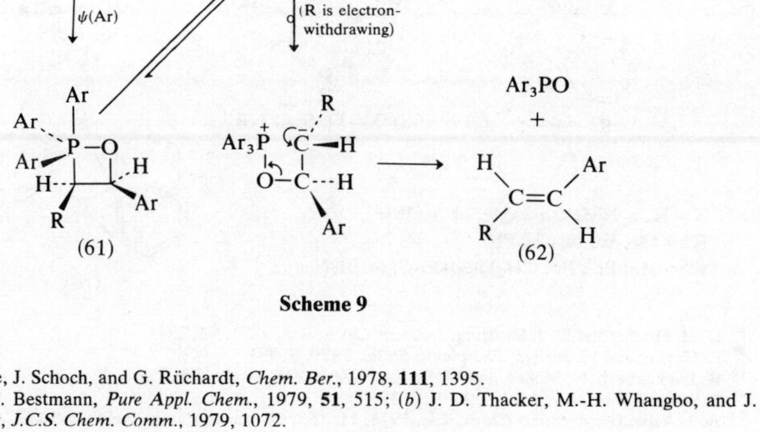

Scheme 9

[88] B. Giese, J. Schoch, and G. Rüchardt, *Chem. Ber.*, 1978, **111**, 1395.
[89] (*a*) H. J. Bestmann, *Pure Appl. Chem.*, 1979, **51**, 515; (*b*) J. D. Thacker, M.-H. Whangbo, and J. Bordner, *J.C.S. Chem. Comm.*, 1979, 1072.

subsequent view[89b] suggests that there is initial formation of the *trans*-oxaphosphetan (63), which ionizes to (64) and, *if R is electron-donating, in the presence of salt,* then gives *syn*-elimination to form *trans*-olefin (*cf.* certain *E*1cB eliminations).[90] In salt-free media, however, (64) is alleged to rotate about the C—C bond to give (65), which allows *anti*-elimination, to form the *cis*-olefin, as shown in Scheme 10. If R is electron-withdrawing (*e.g.* CO_2Et) the carbanion would be planar, *e.g.* (66), and would tend to form the *trans*-olefin.

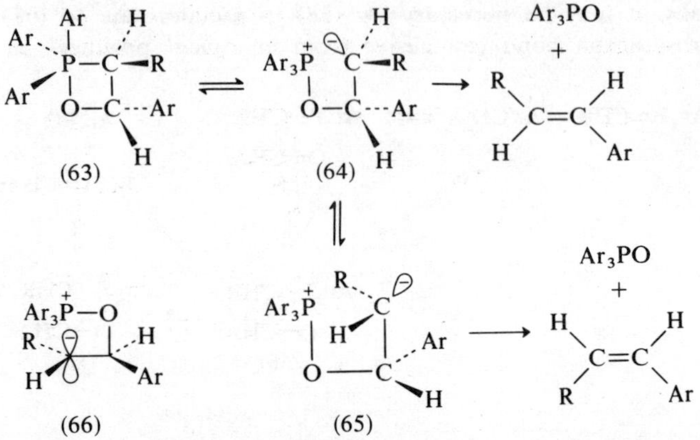

Scheme 10

4 Trico-ordinate Phosphorus Compounds

Phosphines.—Considerable emphasis has been placed on the synthesis and resolution of chiral phosphines, *e.g.* (67),[91] and their use in stereochemical reaction cycles has been reviewed.[92] The synthesis, configuration, and conformation of phosphorus heterocycles such as (68) has been described[93] and the chemistry of λ^3-phosphorins [*i.e.* phosphabenzenes, (69)] has been discussed.[94] One of the most remarkable

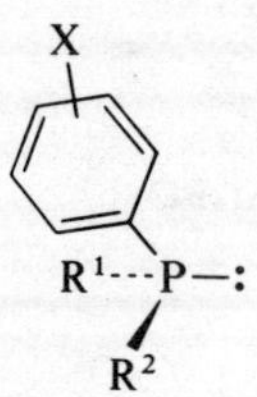

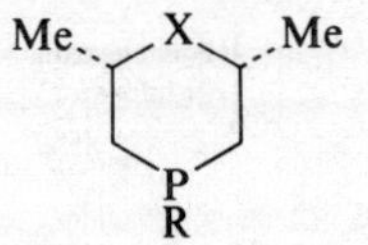

(68) X = O, S, or NR_2

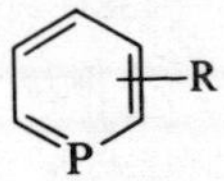

(69)

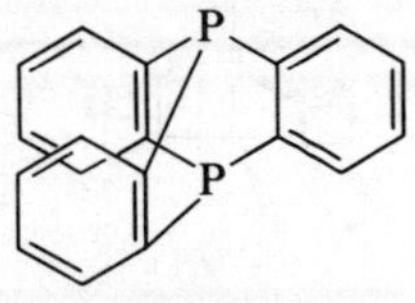

(67) X = H, *o*-NMe_2, *o*-OMe, or *p*-OMe
R^1 = Me, Et, or CH_2Ph
R^2 = Me, Pr^n, Pr^i, CH_2Ph, *o*-tolyl, or Et_2N

(70)

[90] D. H. Hunter and D. J. Shearing, *J. Amer. Chem. Soc.*, 1971, **93**, 2348.
[91] L. Horner and M. Jordan, *Phosphorus Sulfur*, 1979, **6**, 491.
[92] R. Luckenbach, N. Müller, and W. Endres, *Chem.-Ztg.*, 1976, **100**, 320.
[93] S. Samaan, *Phosphorus Sulfur*, 1979, **7**, 89.
[94] A. J. Ashe, III, *Accounts Chem. Res.*, 1978, **11**, 153; G. Märkl, *Phosphorus Sulfur*, 1977, **3**, 77.

reports in this area is of the synthesis of the diphosphine (70) from white phosphorus and *o*-dichlorobenzene.[95]

The quaternization of phosphines with alkyl halides proceeds through a transition state which is reactant-like,[96] and the rate data correlate with the ionization potentials of the lone pairs on phosphoranes.[97] The reaction of triphenylphosphine with CCl_4 may be controlled to give any one of three products (71)—(73), and (71) is a convenient source of the dichloro-ylide (74).[98]

$$Ph_3P + CCl_4 \rightarrow \underset{(71)}{Ph_3\overset{+}{P}CCl_3\ Cl^-} \ \textit{or}\ \underset{(72)}{Ph_3\overset{+}{P}CHCl_2\ Cl^-} \ \textit{or}\ \underset{(73)}{Ph_3\overset{+}{P}CH_2Cl\ Cl^-}$$

$$(71) \xrightarrow{(Me_2N)_3P} \underset{(74)}{Ph_3P{=}CCl_2} + (Me_2N)_3PCl_2$$

The synthesis and use of optically active phosphines as co-catalysts for hydrogenation is still actively pursued,[99] and the co-ordination properties of acyclic and constrained (bicyclic) amino-phosphines have been discussed.[100] Nucleophilic displacement at acyclic phosphines, *e.g.* (75), proceeds to form (76) or (77) with inversion.[101] Finally, one should mention a useful conversion of acid chlorides into aromatic aldehydes using Δ^2-phospholens (78).[102]

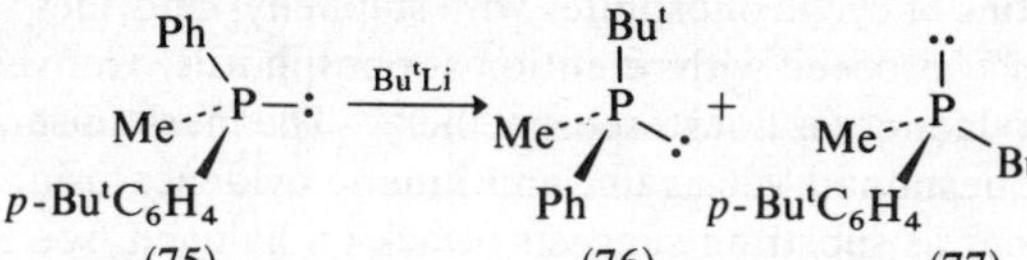

(75) (76) (77)

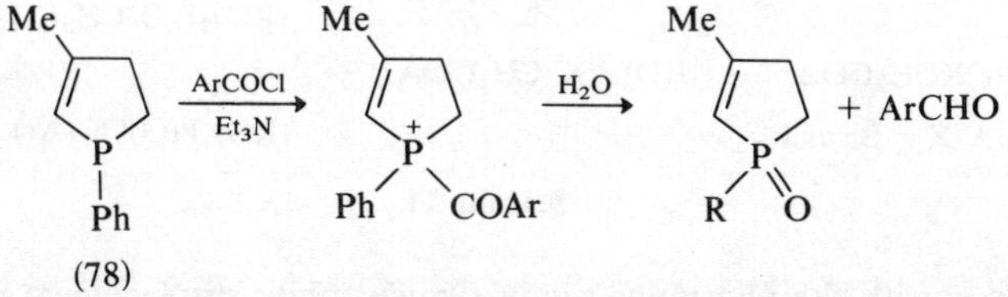

(78)

Phosphorus(III) Esters.—Trialkyl phosphites are now available from white phosphorus, carbon tetrachloride, and alkoxide ion,[103] and the first chiral trico-ordinate

[95] K. G. Weinberg, *J. Org. Chem.*, 1975, **40**, 3586.
[96] W. E. McEwan, J. E. Fountaine, D. N. Schulz, and W.-I. Shiau, *J. Org. Chem.*, 1976, **41**, 1684; T. Thorstenson and J. Songstad, *Acta. Chem. Scand.* (*A*), 1976, **30**, 781, 724.
[97] O. Dahl and L. Henriksen, *Acta. Chem. Scand.* (*B*), 1977, **31**, 427.
[98] R. Appel and W. Morbach, *Synthesis*, 1977, 699.
[99] L. Horner and B. Schlotthauer, *Phosphorus Sulfur*, 1978, **4**, 155.
[100] R. D. Kroshefsky, J. G. Verkade, and J. R. Pipal, *Phosphorus Sulfur*, 1979, **7**, 377.
[101] E. P. Kyba, *J. Amer. Chem. Soc.*, 1976, **98**, 4805.
[102] C. A. Scott, D. G. Smith, and D. J. H. Smith, *Synth. Comm.*, 1976, **6**, 135.
[103] C. Brown, R. F. Hudson, G. A. Wartew, and H. Coates, *J.C.S. Chem. Comm.*, 1978, 7.

ester (79), albeit in low optical purity, has been prepared by Mikolajczyk.[104] A study of the synthesis and stereochemistry of a range of 2-substituted-4-methyl-1,3,2-dioxaphospholans (80) revealed that, in contrast to the 1,3-dioxans, the *trans*-isomer was favoured.[105] The pure *trans*-isomer of (81) was obtained by fractional distillation from a mixture of (−)-ephedrine and PCl_3.[106]

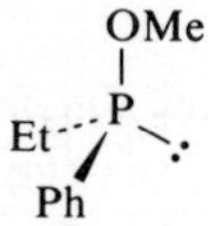

(79)

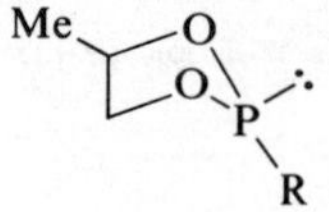

(80)

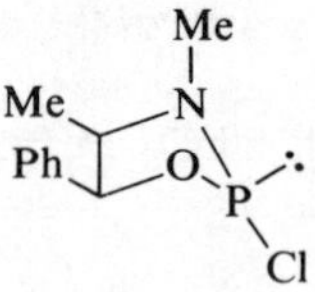

(81)

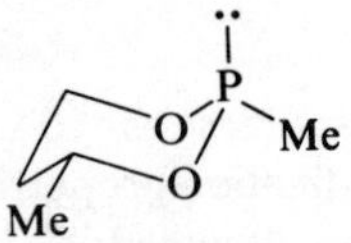

(82)

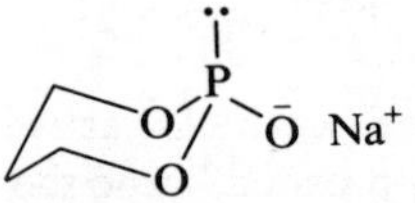

(83)

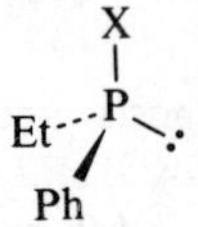

(84)

The reaction of cyclic phosphorus esters, *e.g.* (82), or cyclic phosphate anions (83) with methyl iodide is stereospecific, and occurs with retention at phosphorus.[107] Likewise, the reactions of cyclic phosphites with sulphenyl chlorides[108] and oxygen, sulphur, or selenium[109] proceed with retention at phosphorus. Conversely, reaction with *N*-chloro-dialkylamines is not stereospecific.[110] The mechanism of the Perkow reaction has been questioned yet again, and kinetic evidence using α-bromo- or α-iodo-acetophenones as substrate suggests attack on halogen (see Scheme 11) as the initial, common step for both the Perkow and the Arbusov reaction.[111]

$$(RO)_3P + XCH_2COAr \rightarrow [(RO)_3\overset{+}{P}X\ \overset{-}{C}H_2COAr] \rightarrow \begin{array}{l}(RO)_2P(O)CH_2COAr \\ \textit{or} \\ (RO)_2P(O)OC(Ar){=}CH_2\end{array}$$

(X = Br or I

Scheme 11

In harmony with the phosphine picture, nucleophilic displacement at tervalent phosphorus esters (84) occurs with predominant inversion.[112] The interesting displacement by oxime nucleophiles is followed by a rearrangement which has been shown by CIDNP techniques to proceed through a radical-cage mechanism (see

[104] M. Mikolajczk, J. Drabowicz, J. Omelanczuk, and E. Fluck, *J.C.S. Chem. Comm.*, 1975, 382.
[105] W. G. Bentrude and H. W. Tun, *J. Amer. Chem. Soc.*, 1976, **98**, 1850.
[106] C. R. Hall and T. D. Inch, *Tetrahedron Letters*, 1976, 3645.
[107] R. A. Adamcik, L. L. Chang, and D. B. Denney, *J.C.S. Chem. Comm.*, 1974, 986; K. Lesiak, B. Uznanski, and W. Stec, *Phosphorus*, 1975, **6**, 65.
[108] D. B. Denney and M. Moskal, *Phosphorus*, 1974, **4**, 77.
[109] W. J. Stec, *Khim. Primen. Fosfororg Soedin., Tr. Konf., 5th*, 1972, 351 (*Chem. Abs.*, 1976, **84**, 17 295).
[110] L. L. Chang and D. B. Denney, *J. Org. Chem.*, 1977, **42**, 782.
[111] L. Toke, I. Petriehazy, and G. Szakal, *J. Chem. Res.* (*S*), 1978, 155.
[112] J. Omelanczuk and M. Mikolajczyk, *J.C.S. Chem. Comm.*, 1976, 1025.

Scheme 12).[113] Optically active alcohols can be converted into optically active halides, using the well-known reaction with PCl_3, but an initial temperature of −25 °C, followed by prolonged stirring at 4 °C, is crucial.[114]

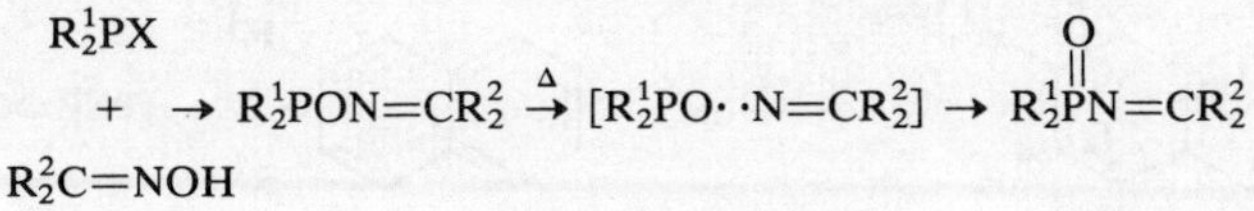

Scheme 12

It has been shown that protonation or alkylation of the bicyclic phosphite (85) leads to a tricyclic structure (86) with a P—N bond;[115] bicyclic phosphites such as (87) are now known to be very toxic.[116]

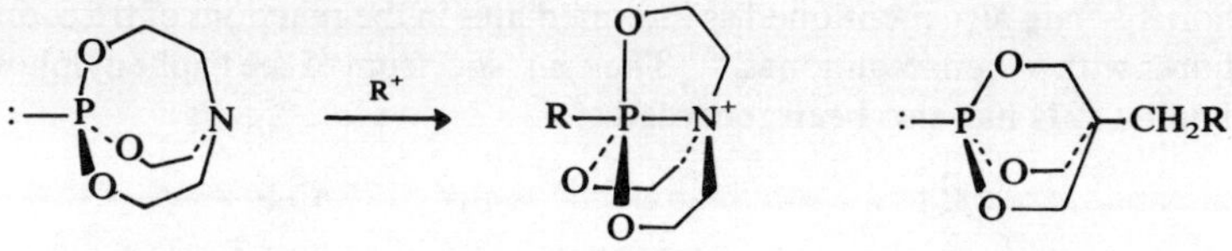

(85) (86) (87)

5 Dico-ordinate Phosphorus

For many years the existence of methylenephosphines (>C=P—) was thought to be extremely unlikely, because of 'poor 2*p*–3*p* overlap'. The synthesis of a stable methylenephosphine (88)[117] by the sequence shown in Scheme 13 has effectively undermined this idea, and this must be regarded as one of the major advances of the period.

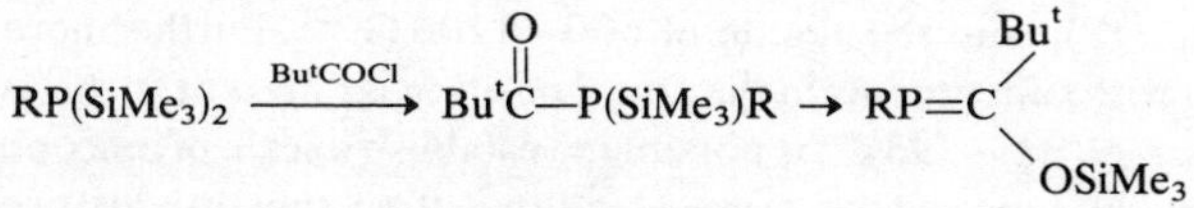

(88)

($R = Me_3Si$, Me, Bu^t, cyclohexyl, or Ph)

Scheme 13

6 Phosphorus Radicals and Reactive Species

Mono- and Di-co-ordinate Phosphorus Radicals.—A comprehensive review covering phosphinidenes ($R\ddot{P}$:) and phosphinidene oxides/sulphides RP=X (X = O or S) has appeared.[118] Typical reactions in which phosphinidene oxides are

[113] C. Brown, R. V. Hudson, A. Maron, and K. A. F. Record, *J.C.S. Chem. Comm.*, 1976, 663.
[114] R. O. Hutchins, D. Masilamani, and C. A. Maryanoff, *J. Org. Chem.*, 1976, **41**, 1071.
[115] D. S. Milbrath and J. G. Verkade, *J. Amer. Chem. Soc.*, 1977, **99**, 6607.
[116] D. S. Milbrath, J. P. Springer, J. C. Clardy, and J. G. Verkade, *J. Amer. Chem. Soc.*, 1976, **98**, 5493.
[117] G. Becker, *Z. anorg. Chem.*, 1976, **423**, 242; *ibid.*, 1977, **430**, 66.
[118] U. Schmidt, *Angew. Chem. Internat. Edn.*, 1975, **14**, 523.

postulated as intermediates are shown in reactions (11) and (12),[119] and the e.s.r. spectrum of the dico-ordinate phosphorus radical cation (89) has been reported.[120]

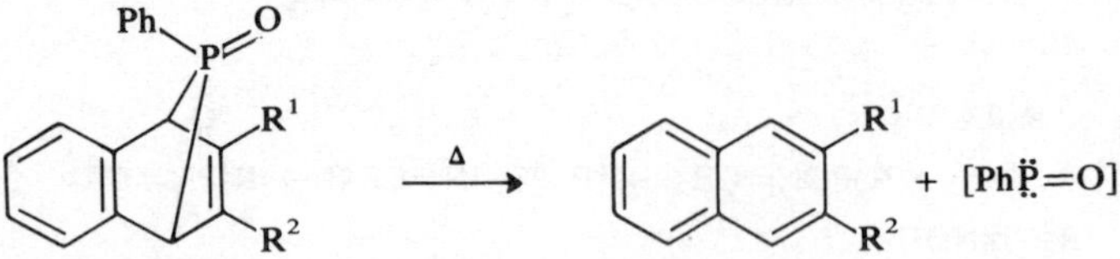

(11)

$$2(MeO)P(O)H_2 \xrightarrow{\text{base}} [MeO\ddot{P}=O] + H_3PO + MeOH \qquad (12)$$

Trico-ordinate Phosphorus Radicals.—Phosphinium radical cations (90), which may be produced by γ-irradiation of phosphines[121] or anodic oxidation of trico-ordinate phosphorus,[122] are also mentioned as intermediates in the reactions of trico-ordinate phosphorus with *p*-benzoquinones.[123] The e.s.r. spectrum of the triphenylphosphine radical anion (91) has also been recorded.[124]

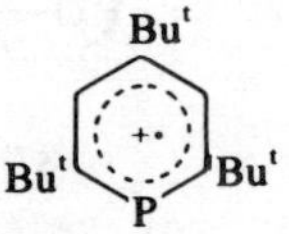

(89)

$Ph\overset{+\cdot}{P}R_2$ (90)

$Ph_3\overset{-\cdot}{P}$ (91)

Tetraco-ordinate Phosphorus Radicals.—A review on phosphoranyl radicals, with particular reference to the interpretation of their e.s.r. spectra, covers the literature to mid-1979.[125] The majority of phosphoranyl radicals have structures which approximate to tbp (92), with the unpaired electron (shown in an equatorial position) distributed between phosphorus and the apical ligands. This is revealed by theoretical calculations and by e.s.r. spectra in which the hyperfine splitting constant to phosphorus, a (P), is in the region of 600—1200 G.[125] Furthermore, since hyperfine splitting with substituents in the apical position is different to that with equatorial substituents, *e.g.* (93)—(95), it is possible to establish a scale of apicophilicity by e.s.r. spectroscopy; in general, the more electronegative substituents prefer the apical

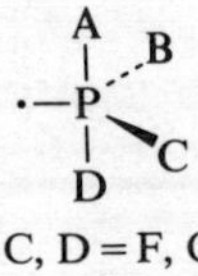

(92) A, B, C, D = F, Cl, CF_3O, RO, alkyl, or H

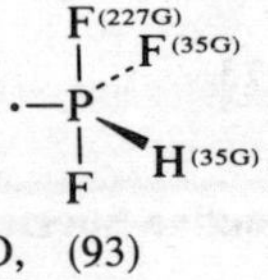

(93)

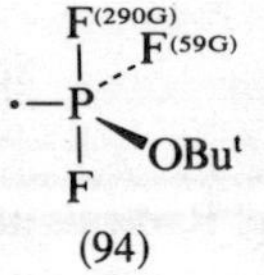

(94)

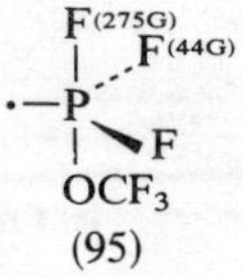

(95)

[119] T. H. Chan and K. T. Nwe, *Tetrahedron*, 1975, **31**, 2537; C. J. R. Fookes and M. J. Gallagher, *J.C.S. Perkin I*, 1975, 1876.
[120] D. Griller, K. Dimroth, T. M. Fyles, and K. U. Ingold, *J. Amer. Chem. Soc.*, 1975, **97**, 5526.
[121] G. W. Eastland and M. C. R. Symons, *J.C.S. Perkin II*, 1977, 833.
[122] H. Ohmori, S. Nakai, and M. Masui, *J.C.S. Perkin I*, 1979, 2023.
[123] G. Boekestein and H. M. Buck, *Phosphorus Sulfur*, 1978, **5**, 61.
[124] R. Nasirov, S. P. Solodovnikov, and M. I. Kabachnik, *Bull. Acad. Sci. U.S.S.R.*, 1976, 2230.
[125] B. P. Roberts, in 'Advances in Free Radical Chemistry' ed. G. H. Williams, Heyden and Sons, London, 1980, Vol. 6, Ch. 5.

position. Thus the apicophilicity in $X\dot{P}F_3$ may be shown to increase in the order $X = H < Bu^tO < F < OCF_3$.

Recent work has revealed, however, that, if one ligand bond provides a much deeper 'electron trap' than the others, the phosphoranyl radical is better described as a distorted tetrahedral structure, *e.g.* (96), with local C_{3v} symmetry at phosphorus and with the unpaired electron confined to a σ^* antibonding P—Cl orbital, as shown.[126,127] Furthermore, low values of a(P) (10—20 G) demonstrate that some phosphoranyl radicals, *e.g.* (97), have almost all the unpaired electron density in the phenyl ring, and are essentially tetrahedral.[128]

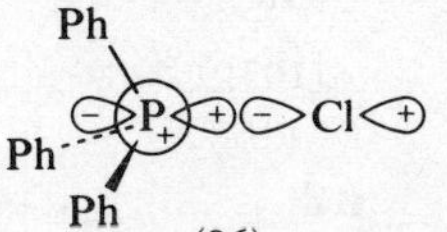

(96)

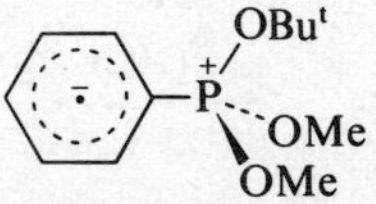

(97)

Ligand exchange in tetra-alkoxyphosphoranyl radicals is now known to be rapid,[129,130] and a clear distinction has been made between the *mode* of ligand exchange in phosphoranyl radicals and the *mechanism* of the process.[125] There is good evidence to suggest that the latter involves a discrete σ^* phosphoranyl intermediate (99), which avoids the 'pairwise' (Berry) mode and may apply to acyclic and monocyclic phosphoranyl radicals, *e.g.* (101)[130]

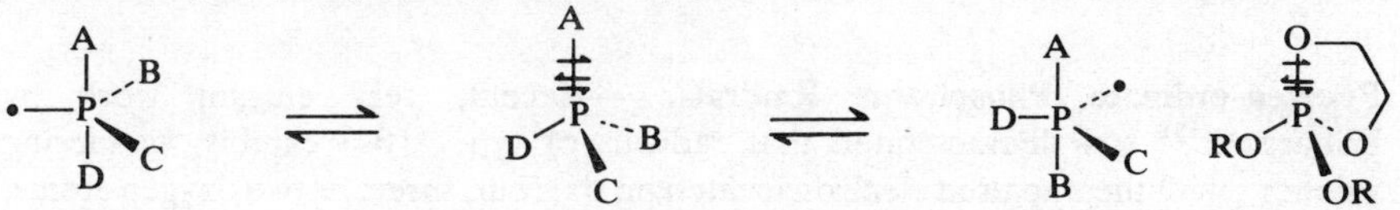

(98) (99) (100) (101)

(A, B, C, D = F, Cl, CF_3O, RO, alkyl, or H)

Phosphoranyl radicals may undergo α- or β-scission [reactions (13a) and (13b)], and the evidence points strongly to α-scission occurring from the apical position†

$Bu^tO\dot{P}Bu^n_3$ —(α-scission, 80%)→ $Bu^n\cdot + Bu^n_2POBu^t$ (13a)

$Bu^tO\dot{P}Bu^n_3$ —(β-scission, 20%)→ $Bu^n_3PO + Bu^t\cdot$ (13b)

† Hence radical attack on phosphorus(III) compounds must give entry into an apical position.

126 T. Berclaz, M. Geoffroy, and E. A. C. Lucken, *Chem. Phys. Letters*, 1975, **36**, 677.
127 G. W. Eastland and M. C. R. Symons, *J.C.S. Perkin II*, 1977, 833.
128 G. Boekestein, E. H. J. M. Jansen, and H. M. Buck, *J.C.S. Chem. Comm.*, 1974, 118.
129 A. Nakanishi, K. Mishibida, and W. G. Bentrude, *J. Amer. Chem. Soc.*, 1978, **100**, 6403.
130 R. S. Hay and B. P. Roberts, *J.C.S. Perkin II*, 1978, 770.

and (less strongly) to β-scission occurring preferentially from an equatorial position.[125] There is no evidence, however, for β-scission occurring *before* ligand exchange in alkoxyphosphoranyl radicals, and no evidence for the existence of a 'memory effect', *i.e.* the ability of a phosphoranyl radical to retain a 'memory' of its origin in terms of the relative concentrations of (102) and (103).[129,130]

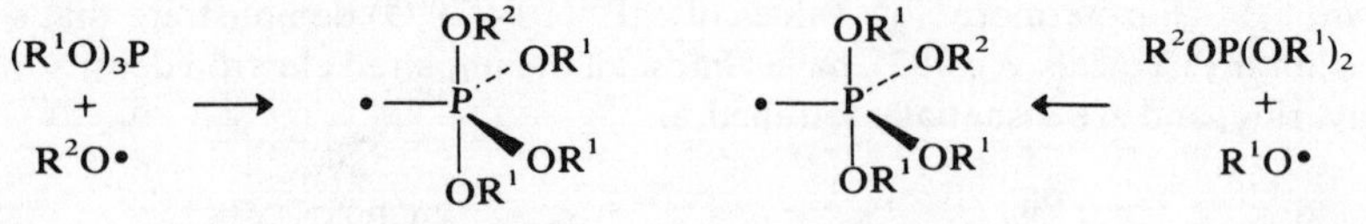

(102) (103)

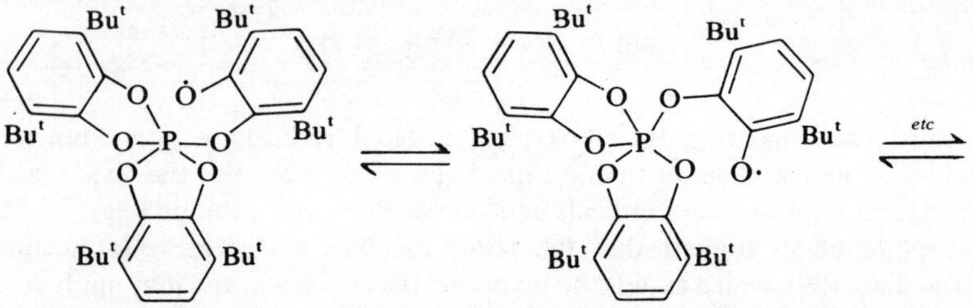

(104)

Pentaco-ordinate Phosphorus Radicals. —Recent, very elegant work by Kabachnik[131] has demonstrated that radicals of type (104) exhibit 'wandering valency', with the unpaired electron located on six, four, three, or two oxygen atoms, depending on the substituents in the phenyl rings. In the totally symmetrical structure shown, migration occurred over all six oxygen atoms, but migration ceased on decreasing the temperature, and the electron was localized on a single ligand.

131 A. I. Prokof'ev, A. A. Khodak, M. A. Malysheva, P. V. Petrovsky, N. N. Bubnov, S. P. Solodovnikov, and M. I. Kabachnik, *Doklady Akad. Nauk. S.S.S.R.*, 1978, **240**, 92.

15 Synthetic Methods

By R. BRETTLE
Department of Chemistry, The University, Sheffield S3 7HF

1 Introduction

Certain broad aspects of synthesis, by their nature, cannot be satisfactorily covered in a Report of this type, but continue to receive a lot of attention. For example, the latest of several recent reviews on asymmetric synthesis appeared at the end of the year.[1] The concept of reactivity umpolung is now widely taught and understood, and its principles lie behind many of the synthetic methods reported below. Seebach has contributed an important summarizing review article;[2] it is too early to say whether a new nomenclature for reagents which he introduces will be widely adopted, but the concepts which they codify will remain of the greatest importance. Two other areas, now rather losing their original novelty, are concerned with the modification of well-known synthetic methods by the use of either phase-transfer catalysts or supported reagents; the second of these approaches has been comprehensively reviewed.[3]

2 Alkanes

Further methods for the deamination of primary amines, which complement other recent procedures, have been developed. In particular, the reduction of isocyanides, which can easily be obtained by formylation–dehydration, using tributylin hydride in the presence of a radical initiator, has been thoroughly investigated.[4,5] Tertiary, secondary, and primary isocyanides demand progressively higher temperatures.[4] In the case of 6α-alkyl-6β-isocyano-penicillanates, the reaction proceeds with inversion to give 6β-alkyl-penicillanates.[5] A modified work-up procedure for reductions employing tributyltin hydride has been recommended.[6] Primary amines can also be deaminated by first converting them into pyridinium salts by reaction with the 2,3,4,5-tetraphenylpyrylium ion; reduction with borohydride and pyrolysis of the resultant dihydropyridines then gives the hydrocarbons.[7] This is one of a series (see below) of reactions developed by Katritzky's group.

[1] J. W. ApSimon and R. P. Seguin, *Tetrahedron*, 1979, **35**, 2797.
[2] D. Seebach, *Angew. Chem. Internat. Edn.*, 1979, **18**, 239.
[3] A. McKillop and D. W. Young, *Synthesis*, 1979, 401, 481.
[4] D. H. R. Barton, G. Bringmann, G. Lamotte, R. S. Hay Motherwell, and W. B. Motherwell, *Tetrahedron Letters*, 1979, 2291.
[5] D. I. John, E. J. Thomas, and N. D. Tyrrell, *J.C.S. Chem. Comm.*, 1979, 345.
[6] J. M. Berge and S. M. Roberts, *Synthesis*, 1979, 471
[7] A. R. Katritzky, K. Horvath, and B. Plau, *J.C.S. Chem. Comm.*, 1979, 300.

The deoxygenation of non-hindered primary and secondary alcohols can conveniently be accomplished by reduction of the derived thiocarbamates in t-butylamine with potassium that is solubilized by 18-crown-6.[8] The deoxygenation of aldehydes and ketones, including aryl alkyl ketones, by the reduction of the derived tosylhydrazones results in an improved yield if carried out in chloroform containing tetrabutylammonium acetate.[9]

Selective olefin reductions are possible using borohydride in the presence of cobalt(II) chloride, a simpler procedure than the known alternatives.[10] Limonene (1) gives (2).

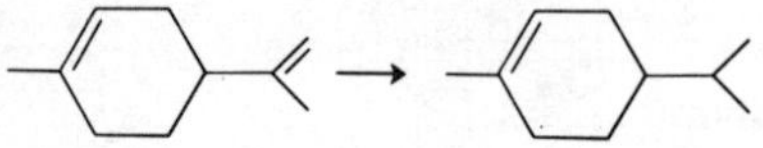

(1) (2)

3 Alkenes

Adamantene has been simply prepared[11] by the debromination of 1,2-dibromoadamantane with bis(trimethylsilyl)mercury, an almost forgotten reagent for this type of reaction. The gas-phase pyrolysis of bridgehead acetates or chlorides can be used to make bridgehead olefins so long as the simple procedures for the removal of the accompanying acid, described in a recent paper,[12] are adhered to.

Selenoxide eliminations continue to be used, and it has been shown that ozone can be used to oxidize a phenylselenenyl group during the simultaneous ozonization of a double bond[13] (*e.g.* see Scheme 1). The phenylselenenic acid produced in such

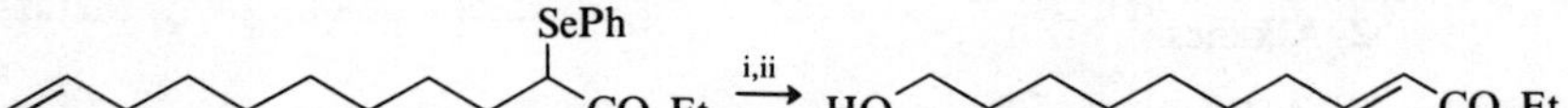

Reagents: i, O_3; ii, $NaBH_4$

Scheme 1

eliminations can be troublesome, and it has therefore, for instance, been trapped by a secondary amine. However, in a case where the product was an $\alpha\beta$-olefinic ketone, capable of acting as a Michael acceptor of such an amine, the elimination was performed in acetic acid containing potassium acetate (giving the β-acetoxy-ketone) and the phenylselenenic acid was trapped by the addition of vinyl acetate or ethyl vinyl ether.[14] Alternatively, base-catalysed elimination of a selenonium salt can be a useful alternative to selenoxide elimination.[15]

[8] A. G. M. Barrett, P. A. Prokopiou, and D. H. R. Barton, *J.C.S. Chem. Comm.*, 1979, 1175.
[9] G. W. Kabalka and J. H. Chandler, *Synth. Comm.*, 1979, **9**, 275.
[10] S.-K. Chung, *J. Org. Chem.*, 1979, **44**, 1015.
[11] J. I. G. Cadogan and R. Leardini, *J.C.S. Chem. Chem.*, 1979, 783.
[12] K. B. Becker and R. W. Pfluger, *Tetrahedron Letters*, 1979, 3713.
[13] T. A. Hase and R. Kwikari, *Acta Chem. Scand.* (*B*), 1979, **33**, 589.
[14] P. E. Eaton, G. D. Andrews, E.-P. Krebs, and A. Kunai, *J. Org. Chem.*, 1979, **44**, 2824.
[15] S. Halazy and A. Krief, *Tetrahedron Letters*, 1979, 4233.

Many examples of the alkylation of (*Z*)-dialk-1-enylcuprates (3), derivable from alkynes and dialkylcuprates, with a wide variety of halides (including those containing compatible functionality, *e.g.* CO_2Et) have been given.[16] Both alkenyl groups in the cuprate are converted into the product, and the (*Z*) configuration is retained.

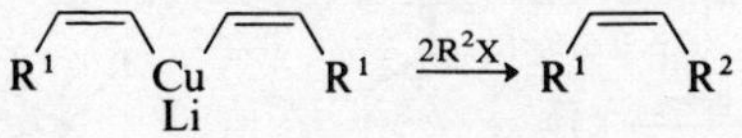

(3)

Enol ethers and enol thioethers are converted into alkenes by reaction with methylmagnesium bromide and other Grignard reagents lacking a β-hydrogen atom, in the presence of a nickel catalyst[17] (*e.g.* see Scheme 2); the configuration at

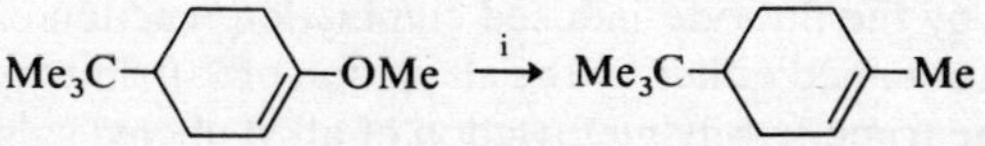

Reagents: i, MeMgBr, $[(PPh_3)_2NiCl_2]$

Scheme 2

the double bond is retained. Allylic sulphides[18] and sulphones[19] undergo a similar, but faster, cross-coupling with Grignard reagents, catalysed by nickel or copper complexes. Usually coupling occurs at both the α- and γ-positions, but $\gamma\gamma$-disubstituted sulphones condense only at the α-position, and the geometry of the olefin is conserved. Allylic sulphones are regioselectively desulphonylated by organotin hydrides; the reaction[20] involves a double migration of the double bond, and should be applicable to electrophiles other than H^+ (Scheme 3). Allylic acetates

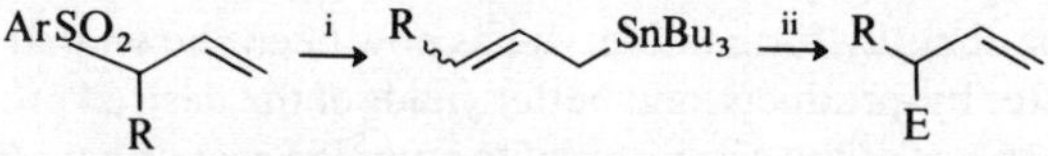

Reagents: i, Bu_3SnH; ii, E^+(E = H or D)

Scheme 3

and phenyl ethers are deoxygenated by catalytic reduction with formate ion [see *Ann. Reports* (*B*), 1978, **75**, 115]; the major product is the terminal alkene (4).[21]

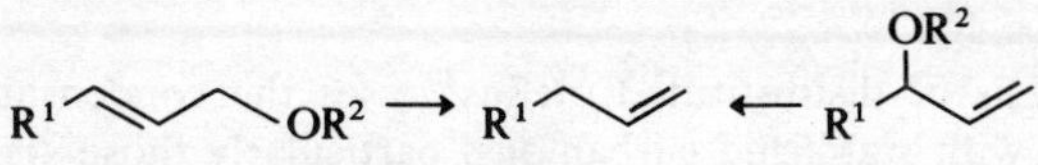

(4)

[16] A. Alexakis, G. Cahiez, and J. F. Normant, *Synthesis*, 1979, 826.
[17] E. Wenkert, E. L. Michelotti, and C. S. Swindell, *J. Amer. Chem. Soc.*, 1979, **101**, 2246; E. Wenkert, T. W. Ferreira, and E. L. Michelotti, *J.C.S. Chem. Comm.*, 1979, 637; H. Okamura, M. Miura, and H. Takei, *Tetrahedron Letters*, 1979, 43.
[18] H. Okamura and H. Takei, *Tetrahedron Letters*, 1979, 3425.
[19] M. Julia, A. Righini, and J.-N. Verpeaux, *Tetrahedron Letters*, 1979, 2393.
[20] Y. Ueno, S. Aoki, and M. Okaware, *J. Amer. Chem. Soc.*, 1979, **101**, 5416.
[21] J. Tsuji and T. Yamakawa, *Tetrahedron Letters* 1979, 613.

Allylic carbamates are converted into alkenes, *e.g.* (5), by lithium dimethylcuprate in a regio- and stereo-specific process.[22] Allylsilanes can conveniently be prepared by a Wittig reaction, thereby adding a two-carbon unit to a carbonyl compound. A wide

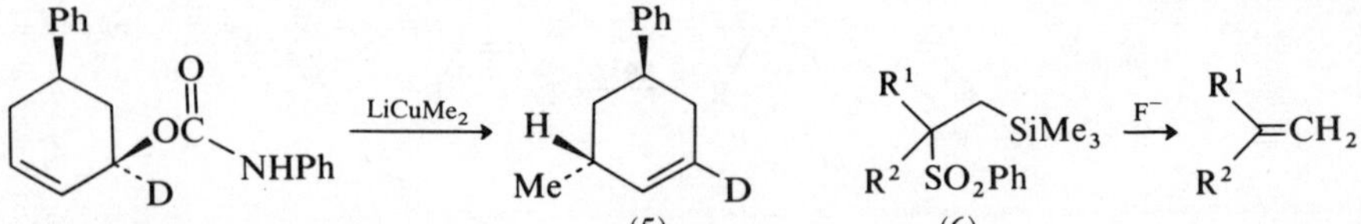

(5) (6)

variety of substituted alkenes can then be obtained by the use of the appropriate electrophile (Scheme 4); protiodesilylation gives the parent alkene.[23] Alkenes can also be prepared by the fluoride-induced elimination reaction of β-silyl sulphones (6), which can be obtained either by the alkylation of 2-(trimethylsilyl)ethyl phenyl sulphone or by the trimethylsilylmethylation of alkyl phenyl sulphones.[24]

Reagents: i, $Ph_3P{=}CHCH_2SiMe_3$; ii, BF_3, $MeCO_2H$ (E = H) or BF_3, Bu^tCl (E = Bu^t)

Scheme 4

Two other eliminative routes start from β-hydroxy-acids. It was reported in 1975 that their decarboxylative dehydration could be brought about by treatment with the dimethyl acetal of dimethylformamide. It has now been shown that the dineopentyl acetal gives no ester by-products, and better yields of the desired products,[25] and this and many other aspects of the chemistry of formamide acetals have been reviewed.[26] A highly stereoselective synthesis of alkenes is possible if the pure *threo*-isomer of the β-hydroxy-acid is used, and it is known that this isomer can be obtained by the action of aldehydes on metallated carboxylate salts; the (Z)-alkene is formed by reaction with triphenylphosphine and diethyl azodicarboxylate, and the (E)-alkene by conversion into the β-lactone, which then loses carbon dioxide on distillation.[27] The method is particularly recommended for the stereocontrolled synthesis of enol ethers.

A new synthesis of disubstituted olefins[28] uses the condensation of aldehyde tosylhydrazones with stabilized carbanions, particularly those stabilized by a sul-

[22] C. Gallina and P. G. Ciattini, *J. Amer. Chem. Soc.*, 1979, **101**, 1035.
[23] I. Fleming and I. Paterson, *Synthesis*, 1979, 446.
[24] P. J. Kocienski, *Tetrahedron Letters*, 1979, 2649.
[25] J. Mulzer and G. Brüntrup, *Tetrahedron Letters*, 1979, 1909.
[26] R. F. Abdulla and R. S. Brinkmeyer, *Tetrahedron*, 1979, **35**, 1675.
[27] J. Mulzer, A. Pointner, A. Chucholowski, and G. Brüntrup, *J.C.S. Chem. Comm.*, 1979, 52.
[28] E. Vedejs, J. M. Dolphin, and W. T. Stolle, *J. Amer. Chem. Soc.*, 1979, **101**, 249.

phone group, but it fails for the hydrazones of sterically hindered or $\alpha\beta$-olefinic aldehydes (Scheme 5).

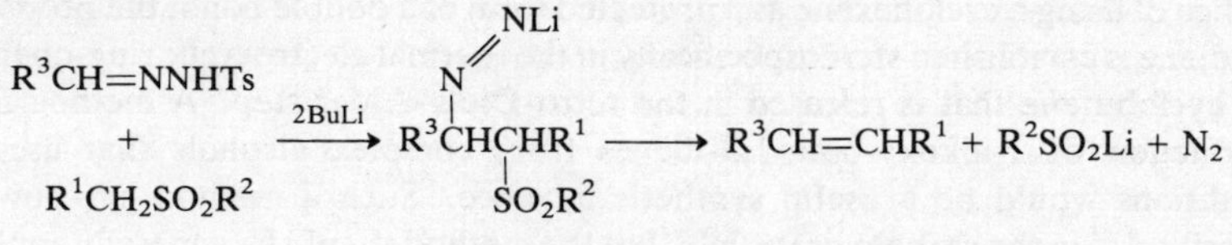

Scheme 5

Symmetrical (*E,E*)-1,3-dienes (7) can be obtained[29] under mild conditions, with very high stereochemical purity, by the silver(I)-ion-promoted coupling of (*E*)-alkenyl-pentafluorosilicates (8), which are available from alkynes and are air-stable. Unsymmetrical (*E,E*)-1,3-dienes can be obtained by the cross-coupling

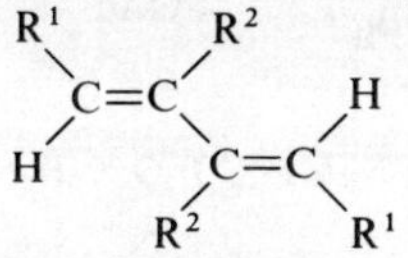

(7)

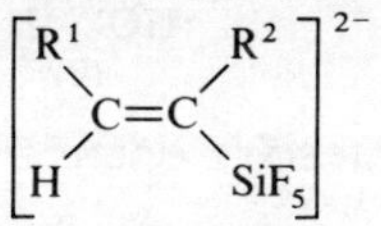

(8)

of alk-1-enylboranes with (*E*)-alk-1-enyl halides, using tetrakis(triphenyl-phosphine)palladium as the catalyst, in the presence of base; in the absence of base, the coupling fails.[30] Coupling with alk-1-ynyl halides gives internal (*E*)-alkenynes.[30] Internal (*Z*)-alkenynes can be prepared[31] by the copper-catalysed reaction of Grignard reagents with 3-acetoxyalk-1-en-4-ynes (Scheme 6).

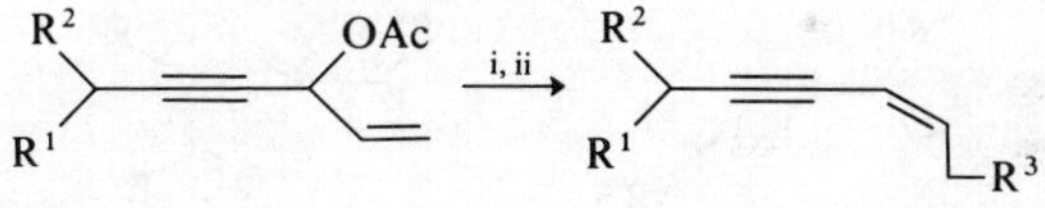

Reagents: i, R^3MgX, CuI; ii, H_3O^+

Scheme 6

The desire to extend the range of the Diels–Alder reaction continues to encourage new syntheses of heterosubstituted 1,3-dienes. Three recent examples are discussed here. Trost has shown that the acyloin (9), readily obtained from the Diels–Alder adduct from cyclopentadiene and maleic anhydride, does have the stereochemistry

[29] K. Tamao, H. Matsumoto, T. Kakui, and M. Kumada, *Tetrahedron Letters*, 1979, 1137.
[30] N. Miyaura, K. Yamada, and A. Suzuki, *Tetrahedron Letters*, 1979, 3437.
[31] G. Cassani, P. Massardo, and P. Piccardi, *Tetrahedron Letters*, 1979, 633.

shown, and has used it to prepare 1,4-diheterosubstituted butadienes. Scheme 7 illustrates a simple example.[32] The success of the method depends on the common artifice of using a cyclohexene as a protected form of a double bond; the geometry of the diene is established stereospecifically in the thermal electrocyclic ring-opening of the cyclobutene that is released in the retro-Diels–Alder step. A method for the production of 1-alkoxy-buta-1,3-dienes from complex alcohols that uses mild conditions would be a useful synthetic advance. Such a method has now been described,[33] in the carbohydrate field, but the method should be generally applicable (Scheme 8). *trans-N*-Acyl-*N*-alkyl-1-amino-1,3-dienes can be prepared from dienamines by the very mild procedure[34] shown in Scheme 9.

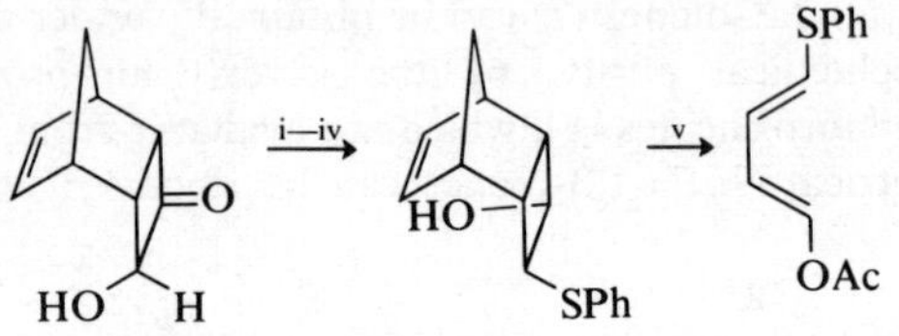

(9)

Reagents: i, $C_7H_7SO_2Cl$, C_5H_5N; ii, $PhS^- Na^+$; iii, $NaBH_4$; iv, $(CH_3CO)_2O$, C_5H_5N; v, Flash vacuum pyrolysis

Scheme 7

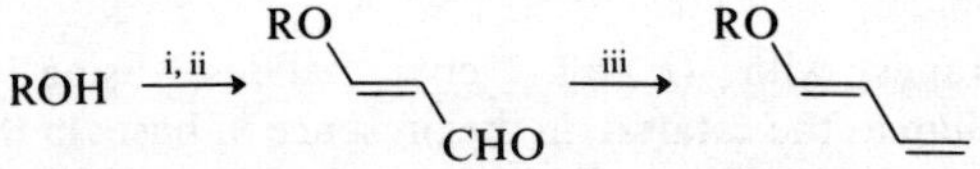

Reagents: i, NaH; ii, 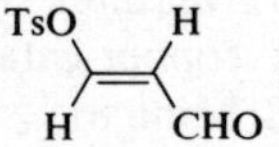

; iii, $Ph_3P{=}CH_2$

Scheme 8

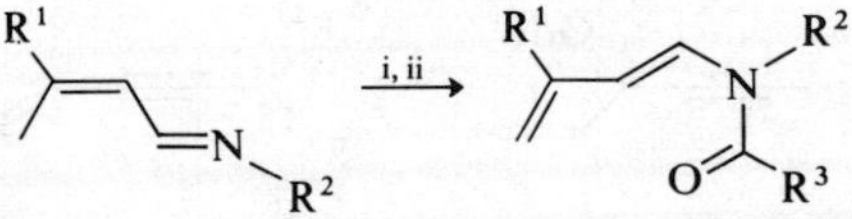

Reagents: i, R^3COCl, 25 °C; ii, $PhNMe_2$, 25 °C

Scheme 9

4 Alkynes

Three novel methods of forming a carbon–carbon triple bond have been described during the year. Perhaps the most interesting is the one illustrated in Scheme 10, in which the key step is the surprisingly ready condensation of trichloroethene with an enolate ion.[35] Non-conjugated ketones and simple esters do not react under the

32 B. M. Trost, S. A. Godleski, and J. Ippen, *J. Org. Chem.*, 1978, **43**, 4559.
33 S. David and J. Eustache, *J.C.S. Perkin I*, 1979, 2521.
34 W. Oppolzer, L. Bieber, and E. Francotte, *Tetrahedron Letters*, 1979, 981.
35 A. S. Kende, M. Benechie, D. P. Curran, and P. Fludzinski, *Tetrahedron Letters*, 1979, 4513.

conditions given in Scheme 10, but active-methylene compounds react under more vigorous conditions. Another new method,[36] which would be more useful if the starting nitrimines were more readily available, is shown in Scheme 11; ketones with both α- and α′-hydrogen atoms give some allene as well. The third method is a modification of the well-known Eschenmoser fragmentation of the tosylhydrazones of αβ-epoxy-ketones, which is now even used on an industrial scale. The original method suffers from the disadvantage that sterically crowded enones are not easily epoxidized, but in the modification[37] the tosylhydrazone of the parent enone is used. The way in which the fragmentation is brought about is shown in Scheme 12.

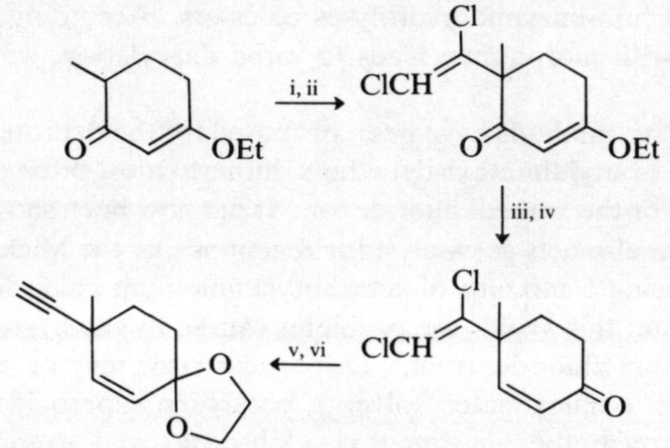

Reagents: i, LDA, THF, at −78 °C; ii, $ClCH{=}CCl_2$, HMPT, −78 to +15 °C; iii, DIBAL; iv, H_3O^+; v, $HOCH_2CH_2OH$, TsOH; vi, BuLi

Scheme 10

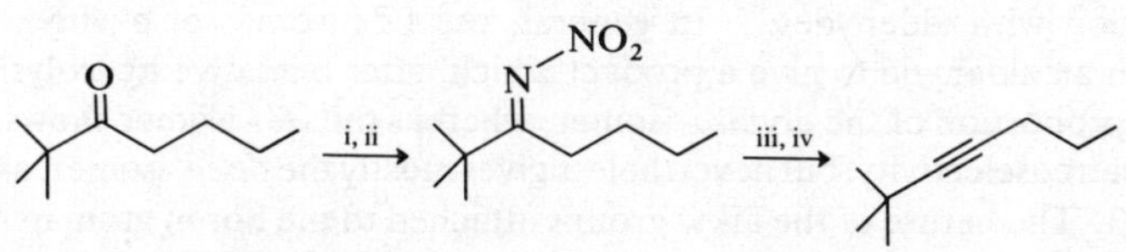

Reagents: i, NH_2OH; ii, $NaNO_2$, H_2SO_4; iii, $(CH_3CO)_2O$, C_5H_5N; iv, reflux

Scheme 11

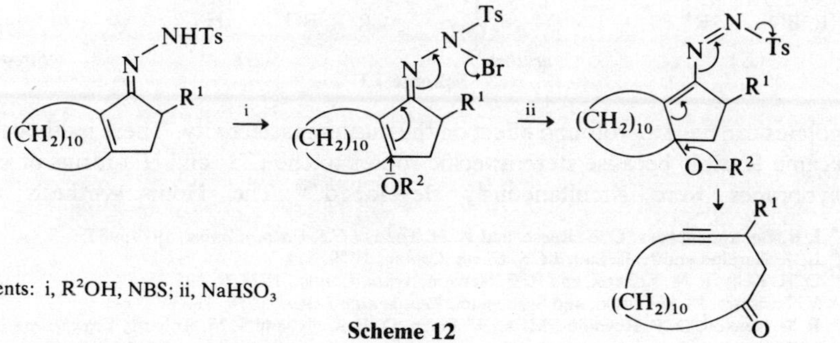

Reagents: i, R^2OH, NBS; ii, $NaHSO_3$

Scheme 12

[36] G. Büchi and H. Wüest, *J. Org. Chem.*, 1979, **44**, 4116.
[37] C. Fehr, G. Ohloff, and G. Büchi, *Helv. Chim. Acta*, 1979, **62**, 2655.

5 Alcohols

The need to protect hydroxy-groups during synthetic operations continues to produce new ideas and modifications to earlier methods. This year Reese has introduced the concept of the protected protecting group.[38] Alcohols are converted into their 2-dibromomethylbenzoyl derivatives, which are comparatively stable to both acid and alkali. The deprotection takes place in two stages. In the first, treatment with silver perchlorate in a slightly wet solvent, in the presence of a base such as collidine (which keeps the reaction medium virtually neutral) gives the formyl ester. It has been known for some time that the hydrolysis of a 2-formylbenzoate is one of the fastest non-enzymic hydrolyses of esters. Accordingly, in the second stage, treatment with morpholine leads to rapid deacylation, with release of the alcohol.

Several alternative methods have been proposed for the deprotection of alcohols protected as their t-butyldimethylsilyl ethers, hitherto most often deblocked by the use of dilute acids or the 'naked' fluoride ion. It has now been shown[39] that 'naked' fluoride ion (which also acts as catalyst for reactions like the Michael reaction) can be obtained by using a mixture of tetrabutylammonium chloride and potassium difluoride dihydrate; this avoids the problems caused by the hygroscopic nature of tetrabutylammonium fluoride itself. Three new reagents are boron trifluoride etherate[40] (used in a halogenated solvent), potassium superoxide[41] (used in conjunction with a crown ether, in dimethyl sulphoxide), and acetonitrile containing some aqueous hydrogen fluoride.[42] The last of these does not cause simultaneous dehydration of β-ketols. Further significant advances have been made in controlling the diastereoselectivity of condensations of the aldol type. Two groups have demonstrated that vinyloxyboranes show a very high stereoslectivity in their condensation with aldehydes.[43] In general, the (*Z*)-isomer of a vinyloxyborane reacts with an aldehyde to give a product which, after oxidative hydrolysis, gives a very high proportion of the *erythro*-isomer, whereas the (*E*)-isomer shows a slightly lower diastereoselectivity, but nevertheless gives mostly the *threo*-isomer, as shown in Scheme 13. The nature of the alkyl groups attached to the boron atom in the boron

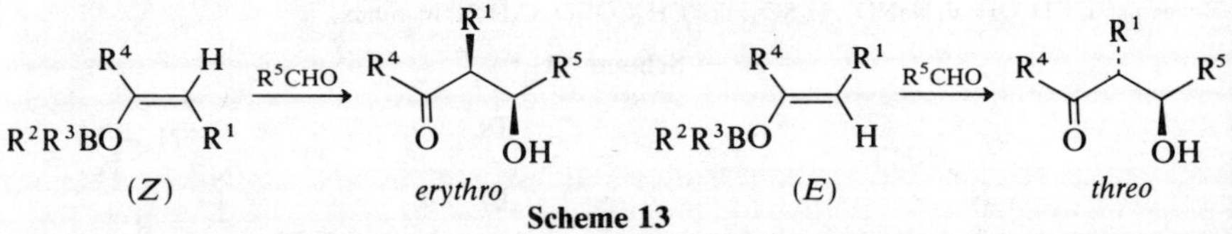

Scheme 13

enolates can have a profound effect on the diastereoselectivity. These methods only became feasible because stereospecific routes to the (*Z*)- and (*E*)-forms of vinyloxyboranes were simultaneously developed.[43] The Hooz synthesis from

38 J. B. Chattopadhyaya, C. B. Reese, and A. H. Todd, *J.C.S. Chem. Comm.*, 1979, 987.
39 L. A. Carpino and A. C. Sau, *J.C.S. Chem. Comm.*, 1979, 514.
40 D. R. Kelly, S. M. Roberts, and R. F. Newton, *Synth. Comm.*, 1979, **9**, 295.
41 Y. Torisawa, M. Shibasaki, and S. Ikegami, *Tetrahedron Letters*, 1979, 1865.
42 R. F. Newton, D. P. Reynolds, M. A. W. Finch, D. R. Kelly, and S. M. Roberts, *Tetrahedron Letters*, 1979, 3981.
43 S. Masamune, S. Mori, D. Van Horn, and D. W. Brooks, *Tetrahedron Letters*, 1979, 1665; D. A. Evans, E. Vogel, and J. V. Nelson, *J. Amer. Chem. Soc.*, 1979, **101**, 6120.

diazoketones and trialkylboranes has now been shown to lead almost exclusively to (*E*)-vinyloxyboranes, but equilibration with lithium phenoxide or pyridine gives the (*Z*)-isomer, so that the other non-stereospecific syntheses, such as the conjugate addition of trialkylboranes to $\alpha\beta$-olefinic ketones, can also be used to prepare the pure (*Z*)-isomer. Vinyloxyboranes can also be prepared from ketones by the action of trifluoromethanesulphonyloxyboranes in the presence of base.

Similar diastereoselectivity has also been observed[44] in the synthesis of the thiol-esters of β-hydroxy-α-methyl-carboxylic acids, again based on the use of vinyloxyboranes (Scheme 14). Very high diastereoselectivity with respect to the

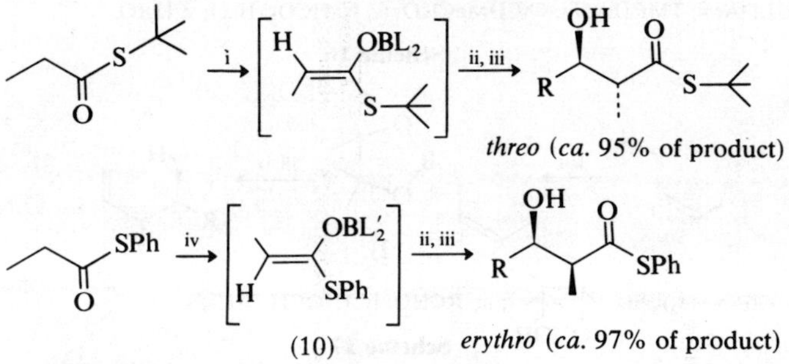

Reagents: i, $(C_5H_9)_2BOSO_2CF_3$, $Pr^i{}_2NEt$; ii, RCHO; iii, H_2O_2, H_2O, MeOH; iv, 9-borabicyclo[3.3.1]non-9-yl trifluoromethanesulphonate, $Pr^i{}_2NEt$

Scheme 14

positions α- and β- to the carbonyl group was also observed when a condensation of the type illustrated in Scheme 14 was carried out with an aldehyde that was so constituted that a further chiral centre was introduced at the γ-position, but a rather low diastereoselectivity was observed with respect to the centres at the β- and γ-positions (Scheme 15). However, much greater diastereoselectivity with respect

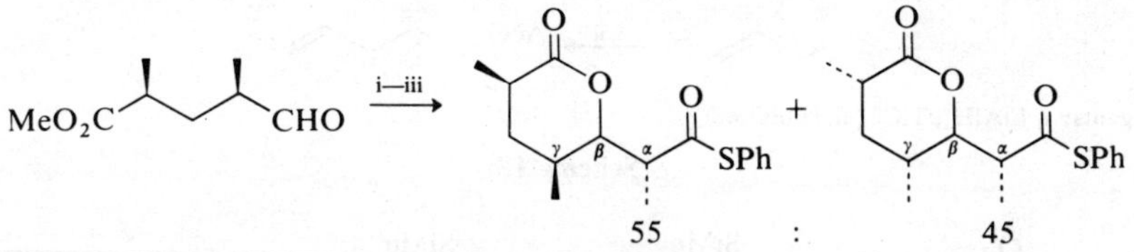

Reagents: i, (10); ii, H_2O_2, H_2O, MeOH; iii, CF_3CO_2H

Scheme 15

to the β- and γ-positions has been observed in another synthesis of β-hydroxy-γ-methyl-carboxylic acids. The reagent (11) was selected on the basis of prior considerations[45] which led to the conclusion that it would show very high *erythro*-selectivity with respect to the α- and β-positions and because the aldol product could

[44] M. Hirama and S. Masamune, *Tetrahedron Letters*, 1979, 2225; M. Hirama, D. S. Garvey, L. D.-L. Lu, and S. Masamune, *ibid.*, p. 3937.
[45] C. H. Heathcock and C. T. White, *J. Amer. Chem. Soc.*, 1979, **101**, 7076.

be readily converted into the desired carboxylic acid. The condensation of (11) with α-phenylpropanal led to a single product, with the stereochemistry shown[46] (Scheme 16). β-Methylhomoallylic alcohols can be prepared[47] in the *erythro*-form from (*Z*)-but-2-enylboronates (12), which can be obtained by the route shown in Scheme 17.

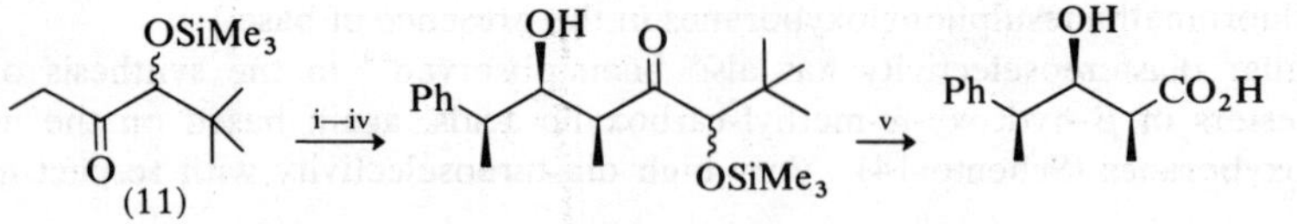

(11)

Reagents: i, LDA; ii, TMEDA; iii, PhCHMeCHO; iv, $NaHCO_3$, H_2O; v, H_5IO_6

Scheme 16

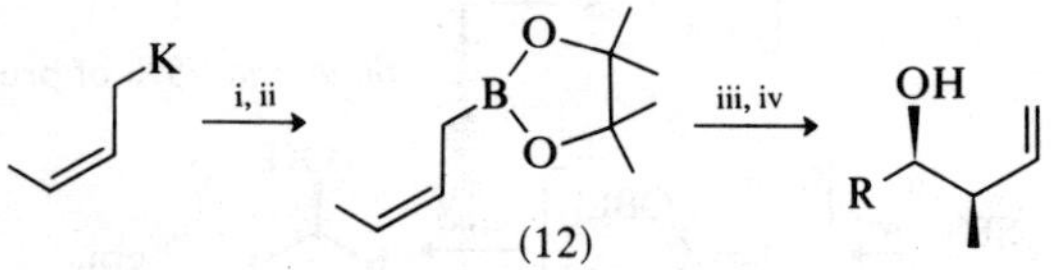

(12)

Reagents: i, $ClB(NMe_2)_2$; ii,  ; iii, RCHO; iv, $(HOCH_2CH_2)_3N$

Scheme 17

The unexpectedly sluggish reaction of 9-borabicyclo[3.3.1]nonane with triple bonds permits the selective hydroboronation of primary olefins containing a non-conjugated acetylenic group, and hence provides a route to acetylenic alcohols.[48] The selective monohydroboronation of the analogous non-conjugated primary and secondary dienes cannot be achieved, but the development of methods for selective hydroalumination makes possible the synthesis of the olefinic primary acetate (Scheme 18).[49] The photosensitized reaction of vinylsilanes with singlet oxygen leads to α-silylated allyl alcohols, and in conjunction with known reactions (Scheme 19) this gives a new 1,2-oxygen transposition, leading from a ketone to an allylic

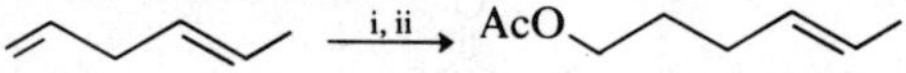

Reagents: i, $LiAlH_4$, $TiCl_4$; ii, $[Pb(OAc)_4]$

Scheme 18

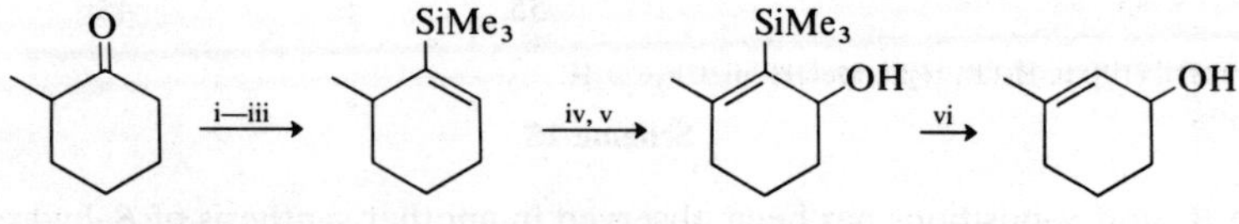

Reagents: i, $PhSO_2NHNH_2$; ii, BuLi; iii, Me_3SiCl; iv, $^1O_2$; v, $NaBH_4$; vi, $Bu_4N^+ F^-$

Scheme 19

[46] C. M. Heathcock, M. C. Pirrung, C. T. Buse, J. P. Hagan, S. D. Young, and J. E. Sohn, *J. Amer. Chem. Soc.*, 1979, **101**, 7077.
[47] R. W. Hoffmann and H.-J. Zeiss, *Angew. Chem. Internat. Edn.*, 1979, **18**, 306.
[48] C. A. Brown and R. A. Coleman, *J. Org. Chem.*, 1979, **44**, 2328.
[49] F. Sato, Y. Mori, and M. Sato, *Tetrahedron Letters*, 1979, 1405.

alcohol.[50] Highly alkylated oxirans are converted into allylic alcohols[51] under very mild conditions by using the sequence shown in Scheme 20. In the case of trisubstituted oxirans the ring opens at the more highly substituted carbon atom: this is the opposite of the earlier selenium-based methods of Sharpless.

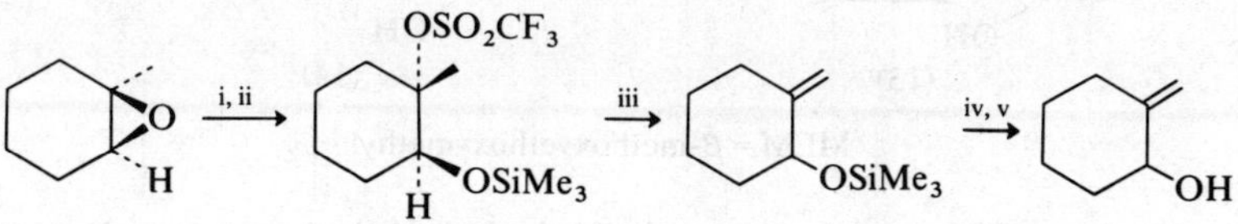

Reagents: i, $CF_3SO_2OSiMe_3$; ii, 2,6-lutidine; iii, DBU; iv, KF; v, H_3O^+ Cl^-

Scheme 20

A few new reductive routes to alcohols have been investigated during the year, and H. C. Brown (with Krishnamurthy) has surveyed 'Forty Years of Hydride Reductions'; a fitting report[52] in the year in which Brown shared the Nobel Prize for Chemistry with Wittig. Both names have appeared very frequently over the years in this section of *Annual Reports*, as a result of their outstanding contributions to organic synthesis. 1-Succinimidyl esters, commonly used as activated esters in peptide synthesis, are reduced to primary alcohols by sodium borohydride,[53] and ketones can be reduced in the presence of aldehydes by using sodium borohydride in

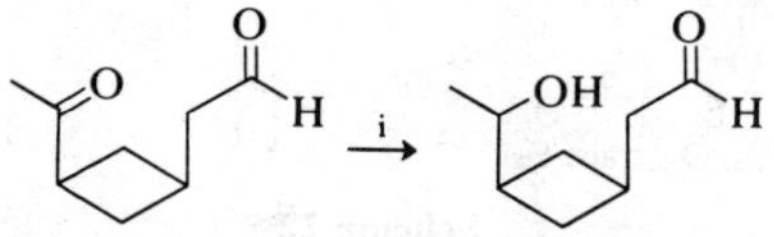

Reagents: i, $NaBH_4$, $CeCl_3{\cdot}6H_2O$

Scheme 21

the presence of lanthanoids.[54] An example is shown in Scheme 21. (±)-Di-isobornyloxyaluminium isopropoxide proved to be a very much better reagent than aluminium isopropoxide for the clean reduction of a prostaglandin-type enone to the corresponding allylic alcohol,[55] and sodium bis-(2-methoxyethoxy)aluminium hydride in the presence of DABCO, at −20 °C, was successfully used[56] to reduce the

[50] W. E. Fristad, T. R. Bailey, L. A. Paquette, R. Gleiter, and M. C. Böhm, *J. Amer. Chem. Soc.*, 1979, **101**, 4420.
[51] S. Murata, M. Suzuki, and R. Noyori, *J. Amer. Chem. Soc.*, 1979, **101**, 2738.
[52] H. C. Brown and S. Krishnamurthy, *Tetrahedron*, 1979, **35**, 567.
[53] J.-I. Nikawa and T. Shiba, *Chem. Letters*, 1979, 981.
[54] J. L. Luche and A. L. Gemal, *J. Amer. Chem. Soc.*, 1979, **101**, 5848.
[55] J. Hutton, *Synth. Comm.*, 1979, **9**, 483.
[56] E. J. Corey and J. G. Smith, *J. Amer. Chem. Soc.*, 1979, **101**, 1038.

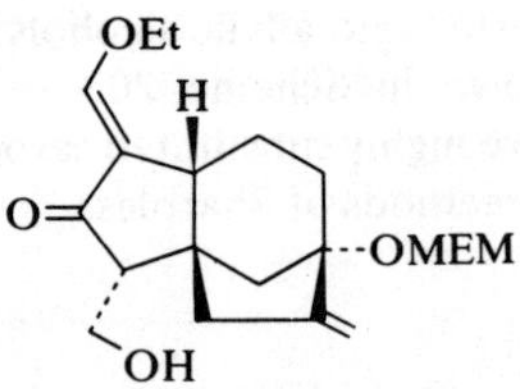

(13)

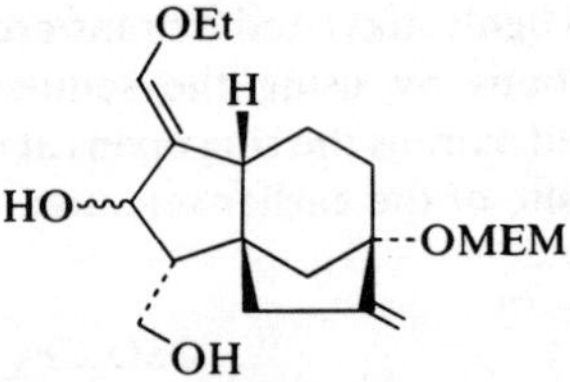

(14)

MEM = β-methoxyethoxymethyl

β-ethoxy-enone (13) to the β-ethoxy-allylic alcohols (14) when many other reagents had failed to work on related model systems.

There have been further reports of regio- and stereo-selective oxygenations at remote unactivated sites. $1\alpha,3\alpha$-Diacetoxycholestane was prepared in 45% yield from 3α-cholestanyl xanthate *via* a transient iron–oxygen complex, the iron itself being complexed by a sulphur atom in the xanthate group[57] (Scheme 22). 6-Methylhexan-2-ol was converted into its 6-hydroxy-derivative in 88% yield on a scale of five grams by the reaction of the derived alkyl hydrogen succinate, adsorbed on silica gel, with a solution of ozone in a Freon.[58]

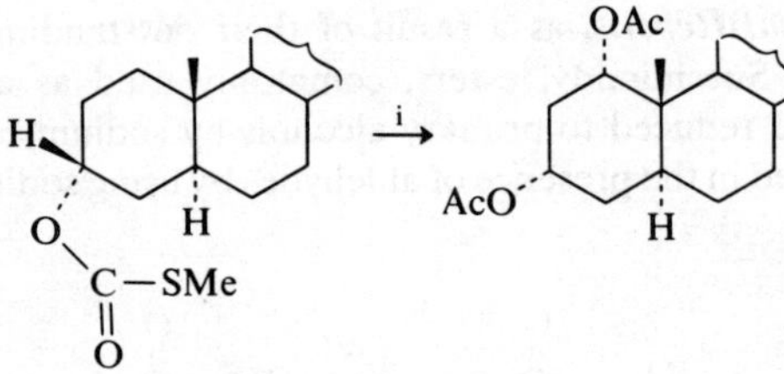

Reagents: i, AcOH, $Fe(ClO_4)_2$, O_2, trace Fe^{3+}

Scheme 22

6 Ethers

Methods for the epoxidation of alkenes continue to be discovered. A development of the Corey sulphur ylide method which can be applied on a large scale involves the use of solid trimethylsulphonium chloride (where the use of the hard counter-ion is crucial) and solid sodium hydroxide in dimethyl sulphide containing a phase-transfer catalyst. This system permitted a virtually quantitative conversion of the rather unreactive ketone β-ionone (15) into the epoxide (16) and hence, by rearrangement,

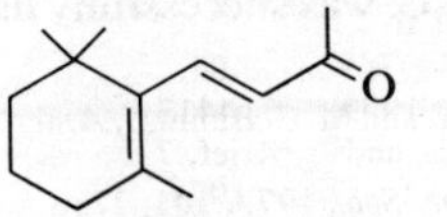

(15)

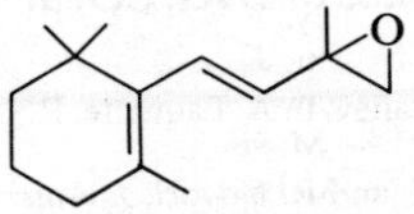

(16)

[57] H. Patin and G. Mignani, *J.C.S. Chem. Comm.*, 1979, 685.
[58] A. L. J. Beckwith and T. Duong, *J.C.S. Chem. Comm.*, 1979, 690.

into the homologous aldehyde.[59] Another successful method used a polymeric sulphonium salt and a phase-transfer catalyst in a water–dichloromethane system.[60] New epoxidizing agents include triphenylsilyl hydroperoxide[61] and 2-hydroperoxyhexafluoropropan-2-ol;[62] the latter gives excellent yields, even with acid-senstive products, owing to the very low acidity of the hexafluoropropane-2,2-diol which is the other product, and it is exceptionally stereoselective; for example, in the epoxidation of 3-hydroxycyclohex-1-ene.

Enol thioethers have hitherto found only limited use in organic synthesis, but suggestions in a recent paper[63] by Trost could alter this. Oxidation with lead tetra-acetate leads to bis-acetoxylation at the double bond, and the product can then be converted into the substituted allylic acetate. Amongst the synthetic operations possible with these are conversion into the thioenol ether of the 1,2-diketone (Scheme 23), which might form the basis for a 1,2-carbonyl transposition, and coupling with cuprates to give the enol thioethers of substituted ketones.

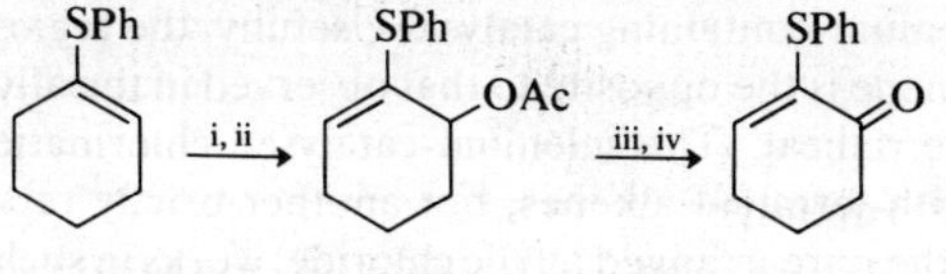

Reagents: i, $[Pb(OAc)_4]$; ii, $BF_3{\cdot}Et_2O$; iii, K_2CO_3, MeOH, H_2O, iv, MnO_2

Scheme 23

7 Halides

Diphosphorus tetraiodide is an excellent reagent for the synthesis of alkyl iodides from alcohols. Primary, secondary, and tertiary alcohols, including neopentyl-type alcohols, are transformed, with inversion, but without rearrangement, into the iodides in a slow reaction in carbon disulphide solution at room temperature.[64] The reagent also converts aldoximes and primary nitroalkanes into nitriles, and deoxygenates epoxides, sulphoxides, and selenoxides.[65]

Primary aliphatic amines can be converted into alkyl chlorides by reaction with 2,4,6-triphenylpyrylium chloride, with azeotropic removal of water, and subsequent thermolysis of the resulting *N*-alkyl-pyridinium chlorides[66] (*cf.* Section 2). Aliphatic, aromatic, and heteroaromatic iodides can similarly be prepared, generally in high yields, from 2,4,6-triphenylpyrylium iodide.[67] Analogous procedures permit the conversion of primary amines into alkyl thiocyanates, thiocarbonates, acetates, and benzoates.[68]

[59] M. Rosenberger, P. McDougal, G. Saucy, and J. Bahr, *Pure Appl. Chem.*, 1979, **51**, 871.
[60] M. J. Farrall, T. Durst, and J. M. J. Frechet, *Tetrahedron Letters*, 1979, 203.
[61] J. Rebek and R. McCready, *Tetrahedron Letters*, 1979, 4337.
[62] R. P. Heggs and B. Ganem, *J. Amer. Chem. Soc.*, 1979, **101**, 2484.
[63] B. M. Trost and Y. Tanigawa, *J. Amer. Chem. Soc.*, 1979, **101**, 4413.
[64] M. Lauwers, B. Regnier, M. van Eenoo, J. N. Denis, and A. Krief, *Tetrahedron Letters*, 1979, 1801.
[65] H. Suzuki, T. Fuchita, A. Iwasa, and T. Mishina, *Synthesis*, 1978, 905; J. N. Denis and A. Krief, *Tetrahedron Letters*, 1979, 3995.
[66] A. R. Katritzky, K. Horvath, and B. Plau, *Synthesis*, 1979, 437.
[67] A. R. Katritzky, N. F. Eweiss, and P.-L. Nie, *J.C.S. Perkin I*, 1979, 433.
[68] A. R. Katritzky, U. Gruntz, N. Mongelli, and M. C. Rezende, *J.C.S. Perkin I*, 1979, 1953; A. R. Katritzky, U. Gruntz, D. H. Kenny, M. C. Rezende, and H. Sheikh, *ibid.*, p. 430.

Some new methods for the ionic chlorination of alkenes to give allylic chlorides have been developed.[69] The reaction of the alkene (17) with *N*-chlorosuccinimide in the presence of diphenyl diselenide is very selective, and gives a good yield of the rearranged chloride (18); the addition of some pyridine suppresses the formation of isomeric allylic chlorides. β-Pinene (19) is exceptional in that the product is almost

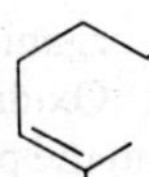

(17)

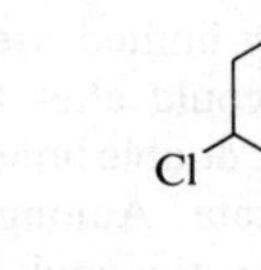

(18)

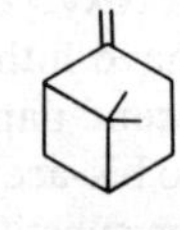

(19)

exclusively the thermodynamically less stable unrearranged allylic chloride; the different regioselectivity in this case has been attributed to the formation of a different active selenium-containing catalyst. Usefully, the regioselectivity which is observed with β-pinene is the opposite to that observed in the allylic chlorination by t-butyl hypochlorite radical. The selenium-catalysed chlorination with *N*-chlorosuccinimide fails with terminal alkenes, but another briefly reported new system, which largely gives the unrearranged allylic chloride, works in such cases; the catalyst in this system is *N*-sulphinyltoluene-*p*-sulphonamide.

8 Nitriles

The cycloaddition of α-nitroso-acrylic esters or of α-nitrosoacrolein, prepared *in situ*, to alkenes gives access to 5,6-dihydro-4*H*-1,2-oxazine-3-carboxylic acids, which on thermal decomposition give γ-hydroxy-nitriles, corresponding to the *syn*-addition of $HOCH_2CN$ to the alkene[70] (Scheme 24). The trimethylsilyl ethers of

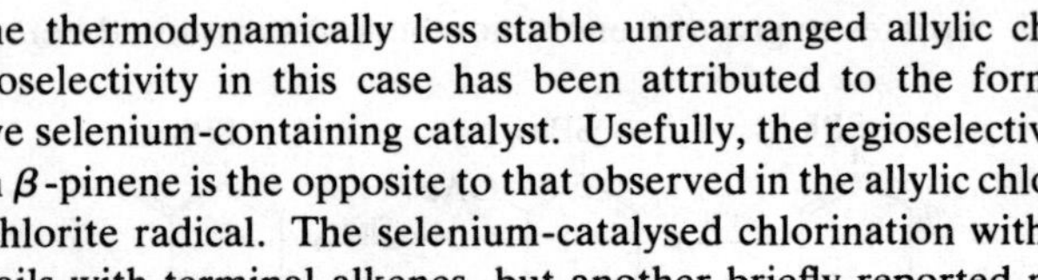

Reagents: i, $BrCH_2C(=NOH)CO_2Et$, Na_2CO_3; ii, NaOH; iii, $ClCH_2C(=NOH)CHO$; iv, Ag_2O; v, heat to 150 °C

Scheme 24

γ-hydroxy-nitriles can be prepared by the reaction of an epoxide with the anion from 2-trimethylsilylacetonitrile at low temperatures, followed by quenching with water; a more convenient route to trimethylsilylacetonitrile was also reported.[71] The *anti*-cyanohydroxylation of alkenes can, of course, be achieved by opening of the derived epoxide with cyanide ion. A new procedure for *syn*-cyanohydroxylation,

[69] T. Hari and K. B. Sharpless, *J. Org. Chem.*, 1979, **44**, 4204, 4208.
[70] T. L. Gilchrist and T. G. Roberts, *J.C.S. Chem. Comm.*, 1979, 1090.
[71] I. Matsuda, S. Murata, and Y. Ishii, *J.C.S. Perkin I*, 1979, 26.

which is more acceptable than earlier methods using cyanogen *N*-oxide or fulminic acid, involves the 1,3-dipolar addition of benzenesulphonylnitrile oxide, prepared *in situ*, to the alkene, followed by reductive cleavage of the benzenesulphonyl-isoxazoline[72] (Scheme 25).

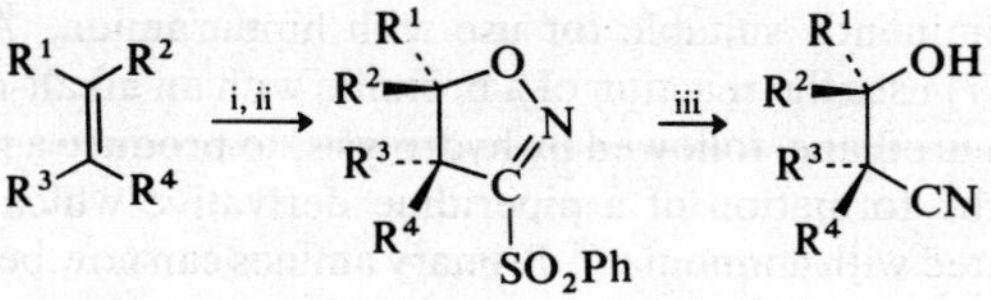

Reagents: i, $PhSO_2C(Br)=NOH$; ii, K_2CO_3; iii, 2% Na—Hg alloy

Scheme 25

The reductive hydrocyanation of alkynes, to give alkyl cyanides [see *Ann. Reports* (*B*), 1978, **75**, 318], can in fact be carried out without using either hydrogen or hydrogen cyanide by means of the tetracyanonickelate ion in conjunction with sodium borohydride.[73] An old reaction, originally applied to camphor by Lapworth, has been revived to prepare an α-cyano-ketone from a hydroxymethylene ketone.[74] The cyano-ketone undergoes a sort of Michael reaction with 1,1-diethoxyprop-2-ene (Scheme 26). This reaction has further synthetic potential.

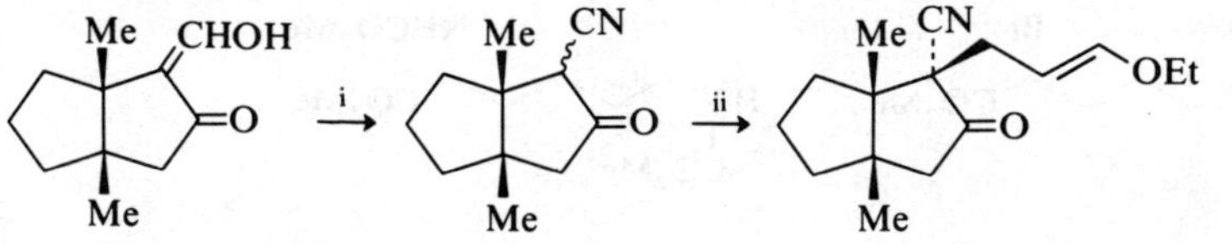

Reagents: i, NH_2OH (5 equiv.), NaOMe, MeOH, reflux; ii, $CH_2=CHCH(OEt)_2$, benzene, 80 °C

Scheme 26

9 Nitro-compounds

The uses of nitroacetic acid and its esters in organic synthesis have been reviewed.[75] Two general methods for the *C*-alkylation of pre-formed nitronate anions have at last been discovered.[76] In one the alkylating agent is an *N*-alkyl-pyridinium salt (see Sections 2 and 7) and in the other, which is a photochemical radical chain reaction, an alkyl-mercury halide.

[72] P. A. Wade and H. R. Hinney, *J. Amer. Chem. Soc.*, 1979, **101**, 1319.
[73] T. Funabiki and Y. Yamazaki, *J.C.S. Chem. Comm.*, 1979, 1110.
[74] R. M. Coates, S. K. Shah, and R. W. Mason, *J. Amer. Chem. Soc.*, 1979, **101**, 6765.
[75] M. T. Shipchandler, *Synthesis*, 1979, 666.
[76] A. R. Katritzky, G. de Ville, and R. C. Patel, *J.C.S. Chem. Comm.*, 1979, 602; G. A. Russell, J. Hershberger, and K. Owens, *J. Amer. Chem. Soc.*, 1979, **101**, 1312.

10 Amines

The use of bis(trifluoroacetoxy)iodobenzene at room temperature in aqueous acetonitrile to convert simple amides into the corresponding primary amines represents a mild alternative to the classical Hofmann degradation.[77] It cannot be used with aromatic amides, owing to the further oxidation of the product by the reagent, but is eminently suitable for use with hippuramide. A new synthesis of lysine (Scheme 27) uses the reaction of a bromide with an alkali-metal cyanate in an alcohol to form a urethane, followed by hydrolysis, to produce a primary amine, and thereby avoids the formation of a piperidine derivative which occurs if the 1,5-dibromide is treated with ammonia.[78] Primary amines can now be α-alkylated by the procedure[79] shown in Scheme 28. In the α-alkylation of secondary amines through the nitrosamines, Seebach, the originator of the method, has shown that a two-stage reduction of the alkylated nitrosamine by lithium aluminium hydride (which gives the hydrazine) and then Raney nickel is a safe method insofar as it avoids any contamination of the product by carcinogenic nitrosamines.[80] A transition-metal-catalysed transformation of tertiary amines into secondary amines and sulphides with thiolate anions has synthetic potential; the order of preference in the cleavage of the C—N bonds is tertiary > secondary > primary.[81] An application of this reaction in the alkaloid field is illustrated in Scheme 29.

The rearrangement sequence shown in Scheme 30 provides an efficient route to *N*-allyl-toluene-*p*-sulphonamides.[82] Allylamines can be prepared through the transition-metal-catalysed process[83] shown in Scheme 31. In a cyclic chiral system, both configurations are formed at the site of the new C—N bond.

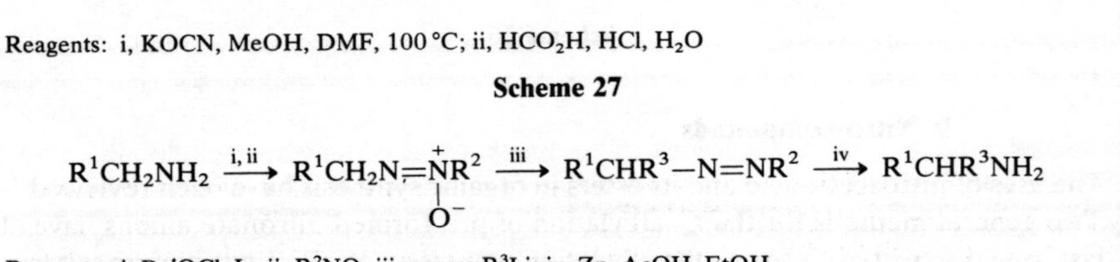

Reagents: i, KOCN, MeOH, DMF, 100 °C; ii, HCO_2H, HCl, H_2O

Scheme 27

$$R^1CH_2NH_2 \xrightarrow{\text{i, ii}} R^1CH_2N{=}\overset{+}{N}(O^-)R^2 \xrightarrow{\text{iii}} R^1CHR^3{-}N{=}NR^2 \xrightarrow{\text{iv}} R^1CHR^3NH_2$$

Reagents: i, Bu^tOCl, I_2; ii, R^2NO; iii, excess R^3Li; iv, Zn, AcOH, EtOH

Scheme 28

77 A. S. Radhakrishna, M. E. Parham, R. M. Riggs, and G. M. London, *J. Org. Chem.*, 1979, **44**, 1746.
78 F. Effenberger and K. Drauz, *Angew. Chem. Internat. Edn.*, 1979, **18**, 474.
79 D. H. R. Barton, G. Lamotte, W. B. Motherwell, and S. C. Narang, *J.C.S. Perkin I*, 1979, 2030.
80 D. Seebach and W. Wykypiel, *Synthesis*, 1979, 423.
81 S.-I. Murashi and T. Yano, *J.C.S. Chem. Comm.*, 1979, 270.
82 M. Kakimoto, T. Yamamoto, and M. Okawa, *Tetrahedron Letters*, 1979, 623.
83 B. M. Trost and E. Keman, *J. Org. Chem.*, 1979, **44**, 3451.

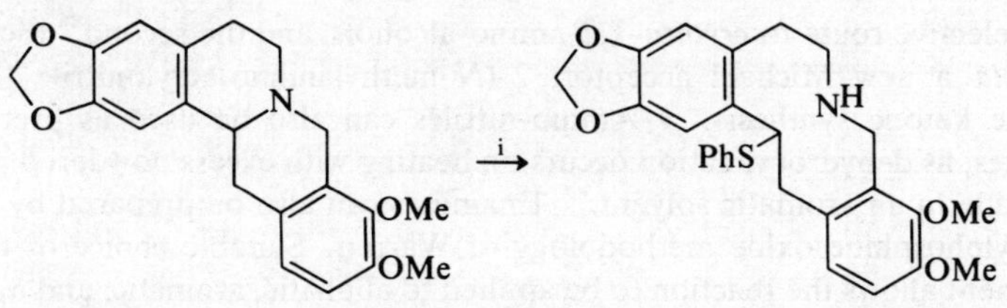

Reagents: i, $PhS^- Na^+$, $RuCl_3$

Scheme 29

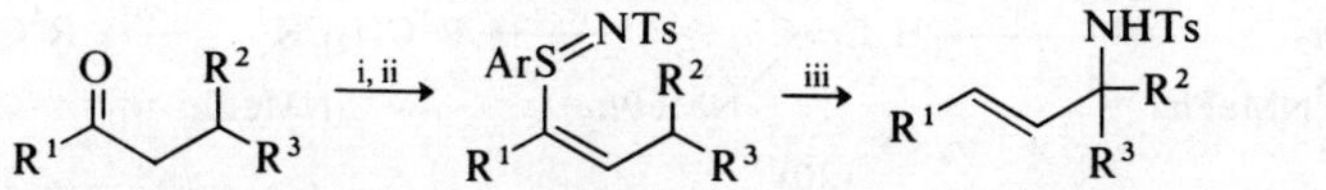

Reagents: i, ArSH, TsOH, C_6H_6; ii, NaClNTs; iii, NaOEt, EtOH, 50 °C

Scheme 30

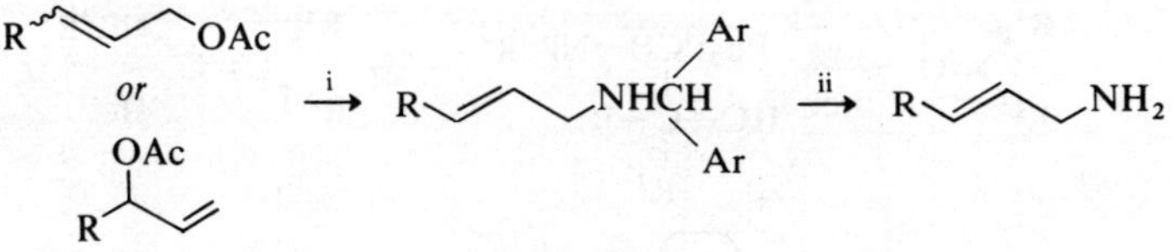

Ar = *p*-$MeOC_6H_4$

Reagents: i, Ar_2CHNH_2, $[(PPh_3)_4Pd]$; ii, >88% HCO_2H, 80 °C

Scheme 31

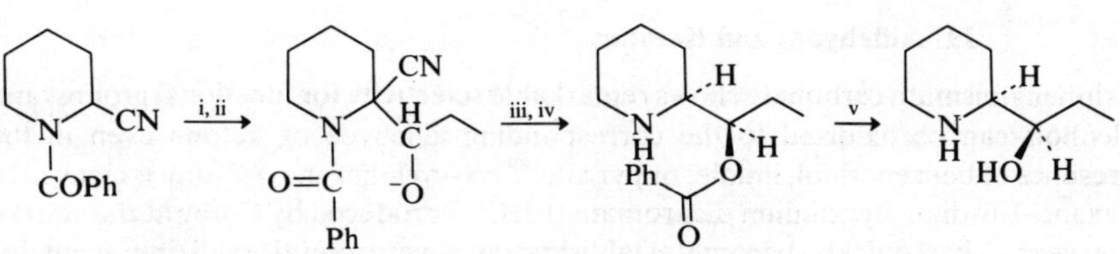

Reagents: i, LDA; ii, EtCHO, at −78 °C; iii, warm to 0 °C; iv, $NaBH_4$; v, MeOH

Scheme 32

Diethyl phosphorocyanidate can be used with an amine in a non-aqueous solvent for the preparation of α-amino-nitriles from aldehydes and ketones – the Strecker synthesis.[84] The use of α-amino-nitriles as acyl equivalents has been reported before [see *Ann. Reports* (*B*), 1978, **75**, 318], as has their reduction by borohydride to give the corresponding amines. A variety of electrophiles can be used in the reaction with α-amino-nitrile anions, and two useful synthetic sequences based on this methodology have appeared. The first[85] (Scheme 32) provides a highly

[84] S. Harusawa, Y. Hamada, and T. Shioiri, *Tetrahedron Letters*, 1979, 4663.
[85] G. Stork, R. M. Jacobson, and R. Levitz, *Tetrahedron Letters*, 1979, 771.

stereoselective route to *erythro*-1,2-amino-alcohols, and the second[86] (Scheme 33) leads *via* a new Michael acceptor, 2-(*N*-methylanilino)acrylonitrile (20), to a versatile ketone synthesis. α-Amino-nitriles can also be used as precursors of enamines, as dehydrocyanation occurs on heating with excess powdered potassium t-butoxide in an aromatic solvent.[87] Enamines can also be prepared by using the diphenylphosphine oxide methodology of Warren. Suitable choice of the amine component allows the reaction to be applied to aliphatic, aromatic, and $\alpha\beta$-olefinic aliphatic aldehydes, and to ketones[88] (Scheme 34).

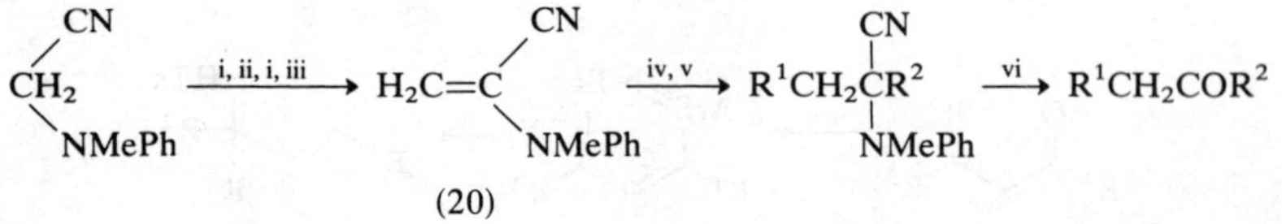

Reagents: i, LDA; ii, Me_3SiCl; iii, CH_2O; iv, R^1Li; v, R^2X; vi, H_3O^+ Cl^-

Scheme 33

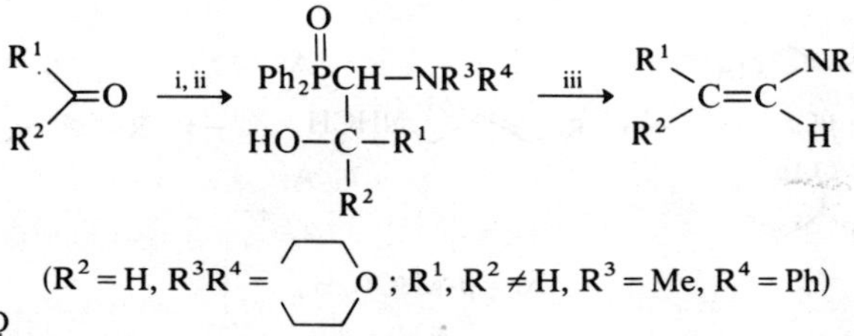

Reagents: i, $Ph_2P(O)CHLi{-}NR^3R^4$; ii, NH_4Cl; iii, KH

Scheme 34

11 Aldehydes and Ketones

Triphenylbismuth carbonate shows remarkable selectivity for functional groups, and alcohols can be oxidized to the corresponding aldehyde or ketone even in the presence of benzenethiol, indole, or pyrrole;[89] *cis*-cyclohexane-1,2-diol is cleaved to hexane-1,6-dial. Pyridinium dichromate (PDC), introduced by Corey at the start of the year,[90] has quickly become established as a very useful oxidizing agent for alcohols. It is an alternative to Collin's reagent, which has to be used in large excess, and can be used in more nearly neutral conditions than pyridinium chlorochromate (PCC). PDC can easily be prepared and stored, and is used either as a solution in DMF, when it oxidizes primary (but not allylic) alcohols to the carboxylic acids, or more generally as a suspension in dichloromethane, when all types of primary alcohols are oxidized to the corresponding aldehydes, and secondary alcohols give ketones. A complex secondary alcohol containing a thioacetal group was selectively oxidized at the alcohol group by PDC.[90] A similarly functionalized molecule was

86 H. Albrecht and K. Pfaff, *Synthesis*, 1978, 897.
87 H. Albrecht, W. Raab, and C. Vonderheid, *Synthesis*, 1979, 127.
88 N. L. J. M. Broekhof, F. L. Jonkers, and A. van der Gen, *Tetrahedron Letters*, 1979, 2433.
89 D. H. R. Barton, D. J. Lester, W. B. Motherwell, and M. T. B. Papoula, *J.C.S. Chem. Comm.*, 1979, 705.
90 E. J. Corey and G. Schmidt, *Tetrahedron Letters*, 1979, 399.

selectively oxidized to the ketone, retaining the thioacetal group, by DMSO in the presence of DCC and a minimal amount of phosphoric acid (the original Pfitzner–Moffatt reagent).[91]

Minor, but significant, modifications to the two-phase dichromate method for the oxidation of primary alcohols to aldehydes reported last year [see *Ann. Reports* (*B*), 1978, **75**, 311] have appeared.[92] A suspension of potassium permanganate and hydrated copper sulphate in benzene smoothly oxidizes secondary alcohols in benzene.[93] Despite the warning given last year [see *Ann. Reports* (*B*), 1978, **75**, 311] about the instability of tetrabutylammonium permanganate, the use of quaternary ammonium permanganates continues to be explored. The benzyltriethylammonium salt has been put forward as a reagent for the selective oxidation of olefins to dialdehydes under homogeneous non-aqueous conditions,[94] and yet the solid reagent, originally introduced[95] for the oxidation of hydrocarbons and ethers, has since been shown (following an explosion) to be thermally unstable and shock-sensitive.[96] Surely such reagents would be better avoided.

Alcohols can be efficiently oxidized to carbonyl compounds by t-butyl hydroperoxide or chloramine T when the reaction is catalysed by diaryl diselenides. Olefinic bonds and selenides are not oxidized, and the cyclization that accompanies the oxidation of an alcohol such as citronellol can be suppressed by the addition of a little of a secondary amine.[97]

Trialkylborates can be oxidized to carbonyl compounds by pyridinium chlorochromate (PCC). This is of no advantage if the borate has been prepared from the alcohol, but there can be advantages if the borate has been generated directly as a reaction intermediate. Scheme 35 shows a good synthesis of aldehydes from acids, based on this.[98]

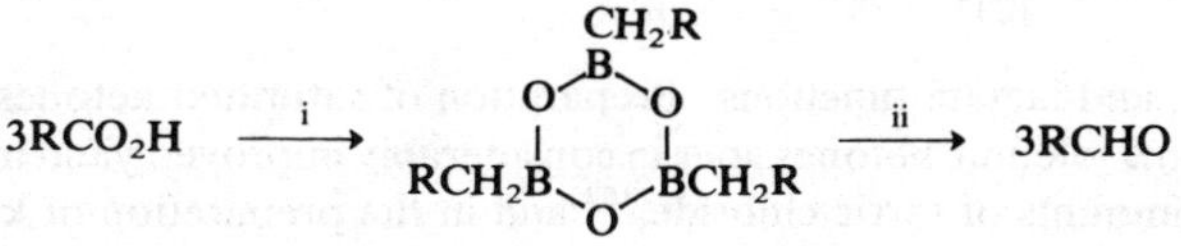

Reagents: i, $BH_3{\cdot}Me_2S$; ii, PCC

Scheme 35

The synthesis of aldehydes and ketones (and also of carboxylic acids) from lower carbonyl compounds by carbon–carbon coupling reactions has been reviewed. The syntheses are classified in terms of the number of carbon atoms separating the new carbonyl group from the original carbonyl carbon atom.[99] The enamine synthesis

91 I. Dyong, R. Hermann, and G. von Kiedrowski, *Synthesis*, 1979, 527.
92 D. Pletcher and S. J. D. Tait, *J.C.S. Perkin I*, 1979, 788; D. Landini, F. Montanari, and F. Rolla, *Synthesis*, 1979, 134.
93 F. M. Menger and C. Lee, *J. Org. Chem.*, 1979, **44**, 3446.
94 T. Ogino and K. Mochizuba, *Chem. Letters*, 1979, 443.
95 H.-J. Schmidt and H. J. Schäfer, *Angew. Chem. Internat. Edn.*, 1979, **18**, 68, 69.
96 H. Jäger, J. Lütolf, and M. W. Meyer, *Angew. Chem. Internat. Edn.*, 1979, **18**, 786; H.-J. Schmidt and H. J. Schäfer, *ibid.*, p. 787.
97 M. Shimizu and I. Kuwajima, *Tetrahedron Letters*, 1979, 2801.
98 H. C. Brown, S. U. Kulkarni, and C. Gundu Rao, *Synthesis*, 1979, 702; H. C. Brown, C. Gundu Rao, and S. U. Kulkarni, *ibid.*, p. 704.
99 S. F. Martin, *Synthesis*, 1979, 633.

shown in Scheme 34 provides a recent example of the synthesis of an aldehyde (generated in this case by hydrolysis of the enamine) with one more carbon atom than the starting aldehyde or ketone, and a second recent example is provided by the route illustrated in Scheme 36.[100]

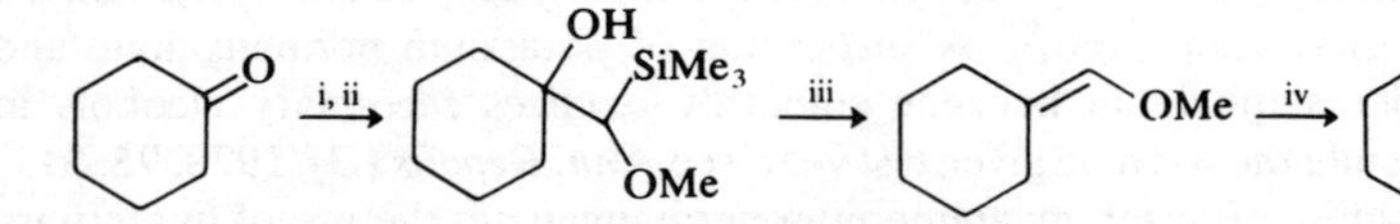

Reagents: i, Me_3SiCH_2OMe; ii, Bu^sLi; iii, KH; iv, H_2O, HCO_2H

Scheme 36

It is in fact possible to prepare ketones in excellent yield by the action of a Grignard reagent on an acid chloride,[101] although to avoid the formation of the tertiary alcohol the reaction must be carried out in THF, at −78 °C. Similarly, ketones can be prepared from esters by the action of a Grignard reagent and triethylamine at −35 °C.[102] The reaction was successfully performed on the non-conjugated ester group of compound (21), despite the fact that the molecule also contains conjugated

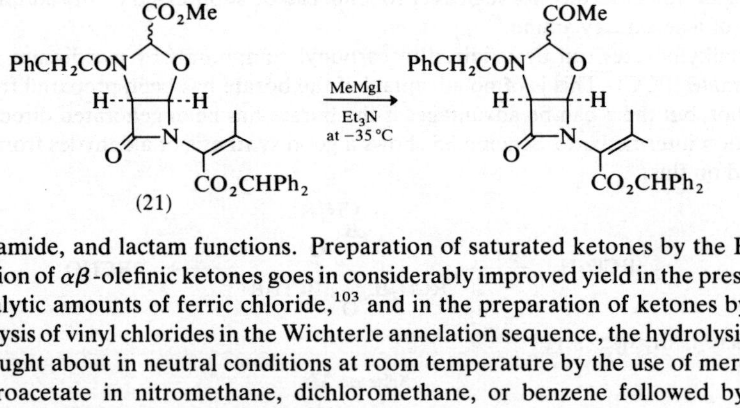

(21)

ester, amide, and lactam functions. Preparation of saturated ketones by the Birch reduction of $\alpha\beta$-olefinic ketones goes in considerably improved yield in the presence of catalytic amounts of ferric chloride,[103] and in the preparation of ketones by the hydrolysis of vinyl chlorides in the Wichterle annelation sequence, the hydrolysis can be brought about in neutral conditions at room temperature by the use of mercuric trifluoroacetate in nitromethane, dichloromethane, or benzene followed by the addition of dilute hydrochloric acid.[104] An as yet unexplained transposition of the oxygen occurs if methanol is used as the solvent. An efficient method for carrying out the transformation (22) → (23) has been worked out, and used to construct a key intermediate in the Woodward reserpine synthesis. The method,[105] which uses a

(22)

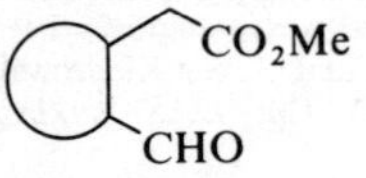

(23)

[100] P. Magnus and G. Roy, *J.C.S. Chem. Comm.*, 1979, 822.
[101] F. Sato, M. Inoue, K. Oguro, and M. Sato, *Tetrahedron Letters*, 1979, 4303.
[102] M. Yoshioka, I. Kikkawa, T. Tsuji, Y. Nishitani, S. Mori, K. Okada, M. Murakami, F. Matsubara, M. Yamaguchi, and W. Nagata, *Tetrahedron Letters*, 1979, 4287.
[103] G. S. R. Subba Rao and N. S. Sundar, *J. Chem. Res. (S)*, 1979, 282.
[104] H. Yoshioka, T. Takasabi, M. Kobayashi, and T. Matsumoto, *Tetrahedron Letters*, 1979, 3489.
[105] B. A. Pearlman, *J. Amer. Chem. Soc.*, 1979, **101**, 6398, 6404.

de Mayo photochemical cyclization to develop a β-hydroxycarbonyl system, which then undergoes retro-aldol fission, is outlined in Scheme 37.

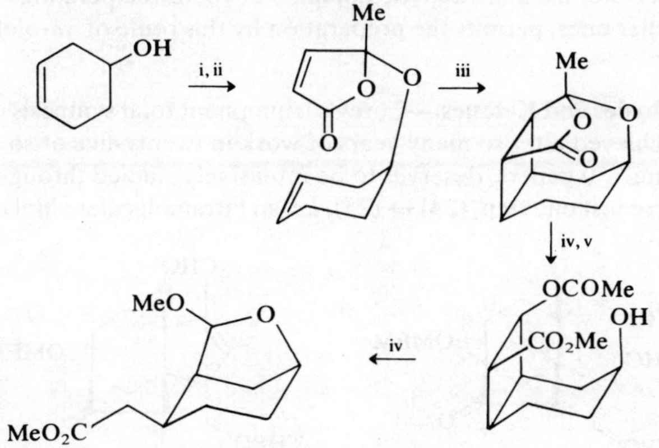

Reagents: 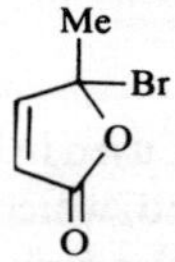

; ii, Ag_2O, $CaSO_4$; iii, $h\nu$; iv, H_2SO_4, MeOH; v, CF_3CO_3H

Scheme 37

Synthetic methods based on the Shapiro approach often call for the regeneration of ketones from their arylsulphonylhydrazones. Many reagents have been suggested for this purpose in recent years. The latest ones include hydrogen peroxide–potassium carbonate,[106] aqueous bromine,[107] and aqueous trifluoroacetic acid–sodium nitrite.[108] The last of these is an additional reagent for the regeneration of aldehydes and ketones from their thioacetals.[109]

β-Keto-esters.—Three useful syntheses of β-keto-esters have been published. The most significant of these, from Masamune's group, which clearly closely parallels the acylation mechanism operating in the biosynthesis of fatty acids, allows β-keto-esters to be formed under virtually neutral conditions. The acylating agent is an acyl-imidazole, which is condensed with the *neutral* magnesium salt of a malonic half-ester,[110] as shown in Scheme 38.

A closely related process employs the condensation of an acyl chloride with the dianion from a malonic half-ester, and, like the first method, gives nearly

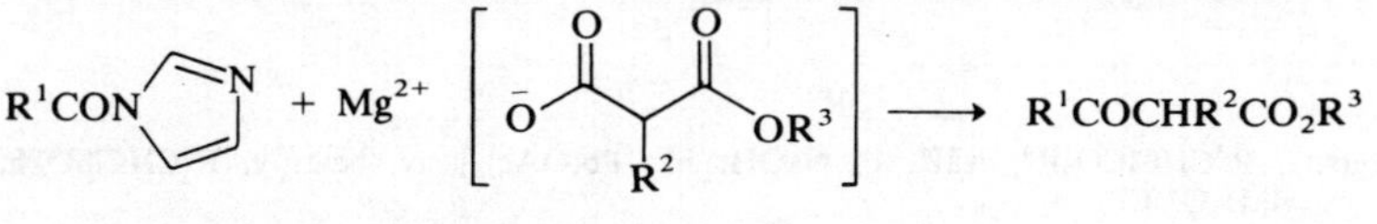

Scheme 38

[106] J. Jirieny, D. M. Orere, and C. B. Reese, *Synthesis*, 1978, 919.
[107] G. A. Olah, Y. D. Vankov, and G. K. S. Prakash, *Synthesis*, 1979, 113.
[108] L. Caglioti, F. Gasparrini, D. Misiti, and G. Palmieri, *Synthesis*, 1979, 207.
[109] G. A. Olah, S. C. Narang, G. F. Salem, and B. G. B. Gupta, *Synthesis*, 1979, 273.
[110] D. W. Brooks, L. D.-L. Lu, and S. Masamune, *Angew. Chem. Internat. Edn.*, 1979, **18**, 72.

quantitative yields.[111] In the third method (a modification of earlier ones), α-diazo-β-hydroxy-esters are converted into β-keto-esters by treatment with a new reagent, rhodium(II) acetate, used in catalytic amounts at room temperature; this reagent, unlike the earlier ones, permits the preparation by this route of $\gamma\delta$-olefinic-β-keto-esters.[112]

Olefinic Aldehydes and Ketones.—Corey's triumphant total synthesis of gibberellic acid, finally achieved after so many years of work in twenty-five or so laboratories, leading to some 150 papers, deserves to be intensively studied through the original papers.[113] Here just one step, (24) → (25), *i.e.* an intramolecular aldol condensation

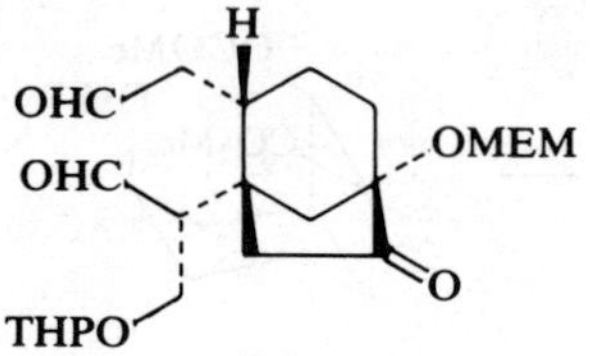

(24)

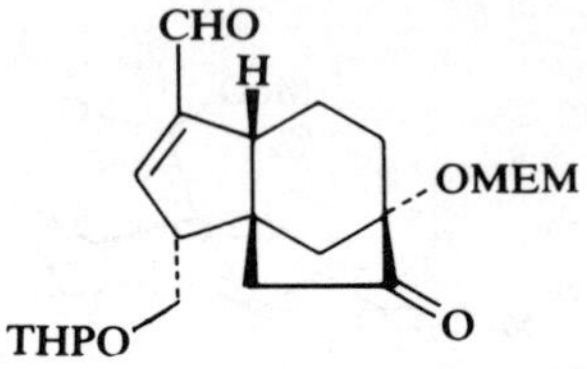

(25)

leading to an $\alpha\beta$-olefinic aldehyde, will be considered. The reagent, used in benzene at 50 °C, was dibenzylammonium trifluoroacetate, a crystalline solid, which proved to be an outstanding selective reagent for this transformation. It was arrived at by systematic variation of the amine and acid components, in the belief that to activate the methylene group adjacent to the less hindered of the two aldehyde groups as an enamine would require a not-too-basic, sterically discriminating secondary amine under almost neutral, aprotic conditions.

Several syntheses of $\alpha\beta$-olefinic ketones have appeared. Two of them (see Scheme 39) build the acyl group onto the original carbon skeleton by means of an

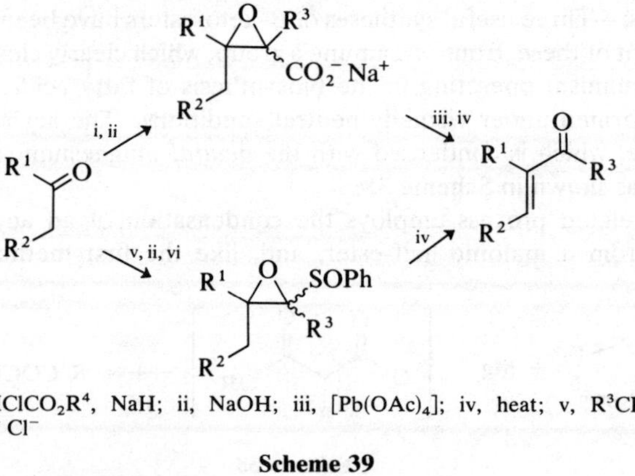

Reagents: i, $R^3CHClCO_2R^4$, NaH; ii, NaOH; iii, $[Pb(OAc)_4]$; iv, heat; v, $R^3CHClSOPh$, LDA; vi, H_3O^+ Cl^-

Scheme 39

[111] W. Werenga and H. I. Skulnick, *J. Org. Chem.*, 1979, **44**, 310.
[112] R. Pellicciari, R. Fringuelli, P. Ceccherelli, and E. Sisani, *J.C.S. Chem. Comm.*, 1979, 959.
[113] E. J. Corey, R. L. Danheiser, S. Chandrasekaran, P. Siret, G. E. Keek, and J.-L. Gras, *J. Amer. Chem. Soc.*, 1978, **100**, 8031; E. J. Corey, R. L. Danheiser, S. Chandrasekaran, G. E. Keck, B. Gopalan, S. D. Larsen, P. Siret, and J.-L. Gras, *ibid.*, p. 8034.

epoxide.[114] Another[115] uses a connective method (Scheme 40), rather than the existing methods based on interchanges of functional groups, to prepare phenylselenomethyl ketones, which can be converted into $\alpha\beta$-olefinic ketones by known methods that involve alkylation followed by selenoxide elimination.

Reagents: i, PhSeBr, EtOH; ii, H_3O^+; iii, RMgX, H_3O^+; iv, 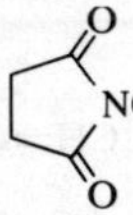 NCl, Me_2S, Et_3N

Scheme 40

A new approach to ketone synthesis[116] is based on the enolonium equivalents (26) and (27), and it can be put into practice through the palladium-catalysed reactions of

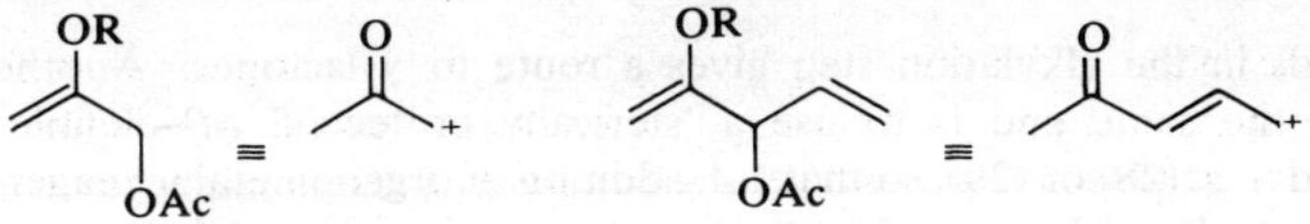

(26) (27)

allylic acetates, although special conditions are required because of the presence of the electron-donating alkoxy-substituent. A rather simple application, leading to an $\alpha\beta$-olefinic γ-keto-ester, is shown in Scheme 41.

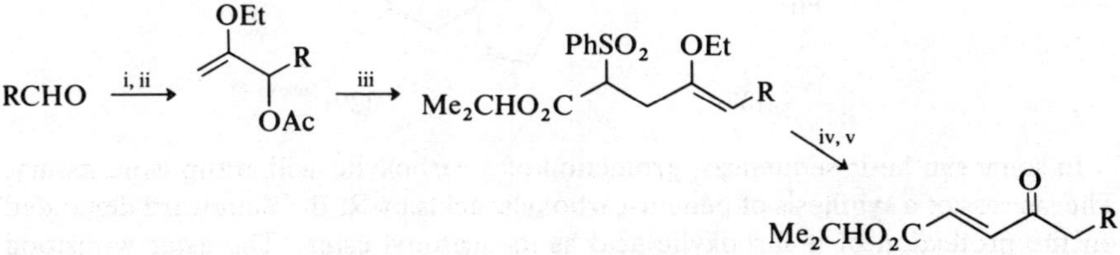

Reagents: i, 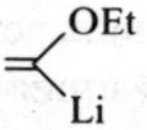 ; ii, Ac_2O, C_5H_5N; iii, $PhSO_2CH_2CO_2CHMe_2$, $[Pd(Ph_3P)_4]$, DBU; iv, 0.02M-HCl, THF; v, Et_3N

Scheme 41

Reduction of Mannich base methiodides, followed by exchange of the anion, gives γ-hydroxy-quaternary ammonium hydroxides, which on thermolysis fragment to give an olefin and a carbonyl group.[117] The procedure can be used to prepare olefinic aldehydes, where the better-known base-catalysed fragmentation of 1,3-diol monotosylates cannot be used.

[114] V. Reutrakul, S. Nimgirawath, S. Panichanum, and Y. Srikirin, *Tetrahedron Letters*, 1979, 1321; D. F. Taber and B. P. Gunn, *J. Org. Chem.*, 1979, **44**, 450.
[115] R. Baudat and M. Petrzilka, *Helv. Chim. Acta*, 1979, **62**, 1406.
[116] B. M. Trost and F. W. Gowland, *J. Org. Chem.*, 1979, **44**, 3448.
[117] L. H. F. Tietze, G. Kinast, and H. C. Uzar, *Angew. Chem. Internat. Edn.*, 1979, **18**, 541.

12 Carboxylic Acids and Derivatives

Benzyltrialkylammonium permanganates have been suggested as reagents for the oxidation of aldehydes to carboxylic acids[118] or of ethers to esters,[95] but, in view of the extremely hazardous nature of the solid oxidants[95] (see Section 11), such reagents cannot be recommended here.

Two groups have developed related syntheses[119] of saturated β-alkyl-substituted carboxylic esters from $\alpha\beta$-olefinic aldehydes (Scheme 42). The use of carbonyl

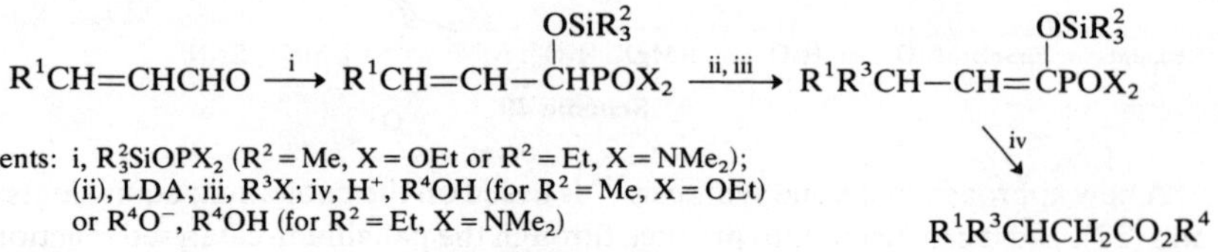

Reagents: i, $R^2_3SiOPX_2$ ($R^2 = Me$, $X = OEt$ or $R^2 = Et$, $X = NMe_2$); (ii), LDA; iii, R^3X; iv, H^+, R^4OH (for $R^2 = Me$, $X = OEt$) or R^4O^-, R^4OH (for $R^2 = Et$, $X = NMe_2$)

Scheme 42

compounds in the alkylation step gives a route to γ-lactones. Another way of achieving the same end is to use a 'sterically protected' $\alpha\beta$-olefinic carbonyl compound, *e.g.* (28) or (29), so that 1,4-addition of organometallic reagents occurs. In these cases the addition product can be converted into the β-alkyl-substituted acid by the Haller–Bauer cleavage of the ketone, or by hydrolysis, coupled with a retro-Mannich cleavage, of the amide.[120]

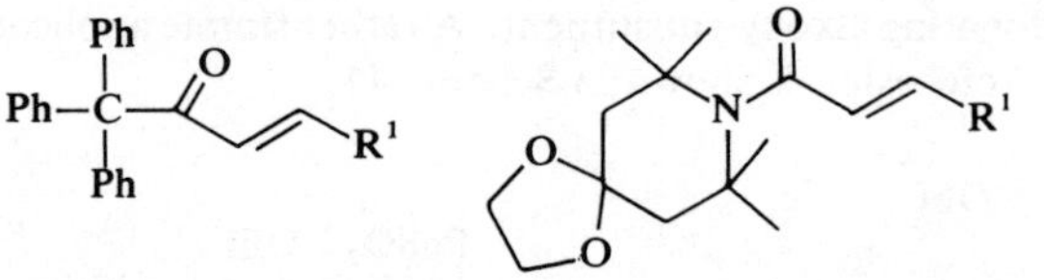

(28) (29)

In some synthetic sequences, protection of a carboxylic acid group is necessary. The success of a synthesis of penem-carboxylic acids by R. B. Woodward depended on the protection of a carboxylic acid as its acetonyl ester. The ester withstood procedures involving trifluoroacetic acid and sodium bicarbonate, but it was efficiently cleaved by hydrolysis with a stoicheiometric amount of aqueous sodium hydroxide at 0 °C.[121] Methylthiomethyl (MTM) esters can be prepared quantitatively under very mild Pummerer-like conditions by treating the sodium carboxylate with the adduct from DMSO and t-butyl bromide,[122] and can be cleaved under basic conditions, if required, by molybdate-catalysed oxidation to the sulphone followed by alkaline hydrolysis.[123]

Activation of acids continues to attract attention. Acyl chlorides can be made at room or lower temperatures, without the formation of hydrogen chloride, by the

118 D. Scholz, *Monatsh.*, 1979, **110**, 1471.
119 D. A. Evans, J. M. Takacs, and K. M. Hurst, *J. Amer. Chem. Soc.*, 1979, **101**, 371; T. Hata, M. Nakajima, and M. Sekine, *Tetrahedron Letters*, 1979, 2047.
120 D. Seebach and R. Locher, *Angew. Chem. Internat. Edn.*, 1979, **18**, 957.
121 H. R. Pfaendler, J. Gosteli, and R. B. Woodward, *J. Amer. Chem. Soc.*, 1979, **101**, 6306.
122 A. Dossena, R. Marchelli, and G. Casnati, *J.C.S. Chem. Comm.*, 1979, 370.
123 J. M. Gerdes and L. G. Wade, jr., *Tetrahedron Letters*, 1979, 689.

treatment of acids with the enamine (30).[124] Much of the work is directed towards the synthesis of macrolides, and a very full account by Mukaiyama of the activation of carboxy- and hydroxy-groups, based on the onium salts of aza-arenes, has appeared.[125] A better method for preparing pyridine-2-thiol carboxylic esters has appeared,[126] but if the 3-cyano-4,6-dimethylpyridinethiol esters, now reported,[127] are used they react more rapidly with alcohols, and a silver ion catalyst is no longer required.[127] t-Butylthiol esters[128] and benzeneselenol esters,[129] which are all very powerful acyl-transfer agents, have also been investigated. Esterification reactions employing isoureas, some of which [*e.g.* (31), which is an alternative to the use of

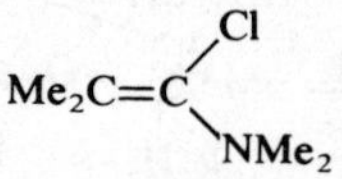

(30)

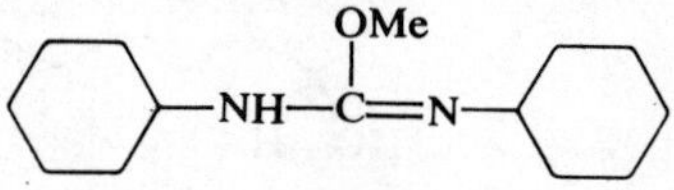

(31)

diazomethane] are commercially available, have been reviewed.[130] They avoid the *N*-acyl-ureas that are persistently formed with carbodi-imides; this drawback can also be overcome, whilst still using carbodi-imides, if the esterification is carried out in pyridine containing a little toluene-*p*-sulphonic acid.[131]

New syntheses of $\alpha\beta$-olefinic esters include the use of the Claisen orthoester rearrangement with trimethyl 3-phenylseleno-orthopropionate followed by selenoxide elimination[132] (Scheme 43) and the reaction of primary cuprates with the enol-phosphates of β-keto-esters,[133] an example of which is given in Scheme 44. A

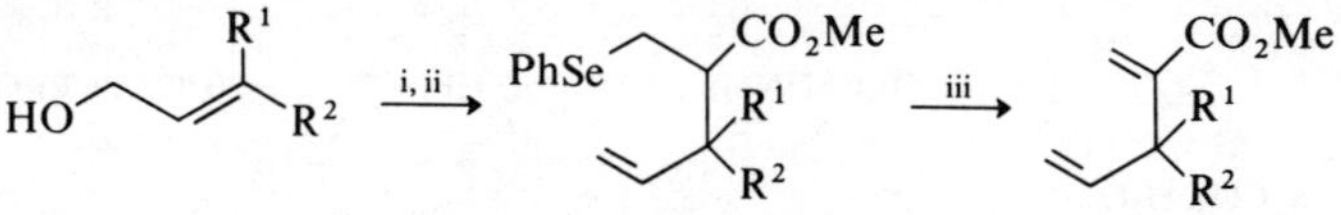

Reagents: i, $PhSeCH_2CH_2C(OMe)_3$; ii, Me_3CCO_2H, 170 °C; iii, H_2O_2

Scheme 43

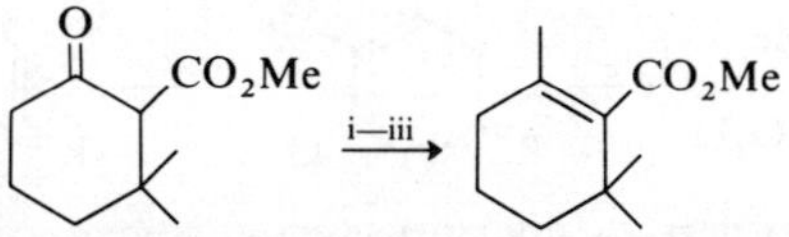

Reagents: i, NaH; ii, $ClPO(OEt)_2$; iii, $LiMe_2Cu$

Scheme 44

[124] A. Devos, J. Remion, A.-M. Frisque-Hesbian, A. Colens, and L. Ghosez, *J.C.S. Chem. Comm.*, 1979, 1180.
[125] T. Mukaiyama, *Angew. Chem. Internat. Edn.*, 1979, **18**, 707.
[126] E. J. Corey and D. A. Clark, *Tetrahedron Letters*, 1979, 2875.
[127] U. Schmidt and D. Heermann, *Angew. Chem. Internat. Edn.*, 1979, **18**, 308.
[128] D. N. Harpp, T. Aida, and T. H. Chan, *Tetrahedron Letters*, 1979, 2853.
[129] T. G. Back and S. Collins, *Tetrahedron Letters*, 1979, 2661.
[130] L. J. Mathias, *Synthesis*, 1979, 561.
[131] K. Holmberg and B. Hansen, *Acta Chem. Scand.* (*B*), 1979, **33**, 410.
[132] S. Raucher, K.-J. Hwang, and J. E. Macdonald, *Tetrahedron Letters*, 1979, 3057.
[133] F. W. Sum and L. Weiler, *Canad. J. Chem.*, 1979, **57**, 1431; F. W. Sum and L. Weiler, *J. Amer. Chem. Soc.*, 1979, **101**, 4401.

new connective synthesis of $\alpha\beta$-olefinic acids is illustrated in Scheme 45.[134] It is notable for, amongst other things, the use of a t-butyl ketone as a masked carboxylic acid, the acid finally being generated from the cyanide which results from the Beckmann type-II cleavage of the ketoxime.

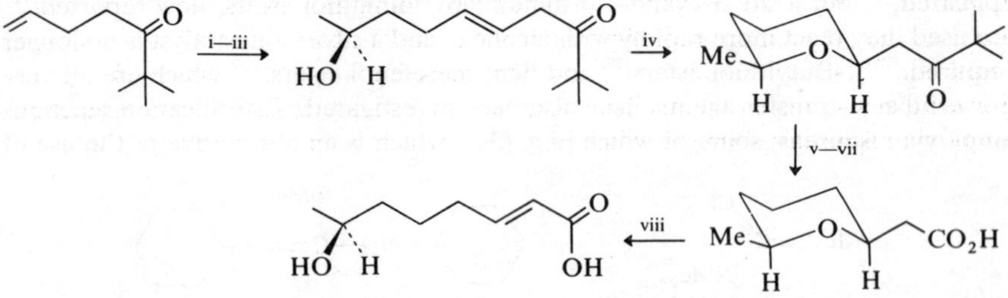

Reagents: i, KH, 25 °C; ii, Bu^sLi, at −78 °C; iii, 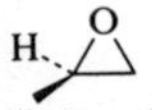 , 0 °C; iv, MeONa, 4 days; v, NH_2OH; vi, PCl_5; vii, KOH, H_2O, $(HOCH_2)_2$, 150 °C; viii, 2 equiv. LDA, quench

Scheme 45

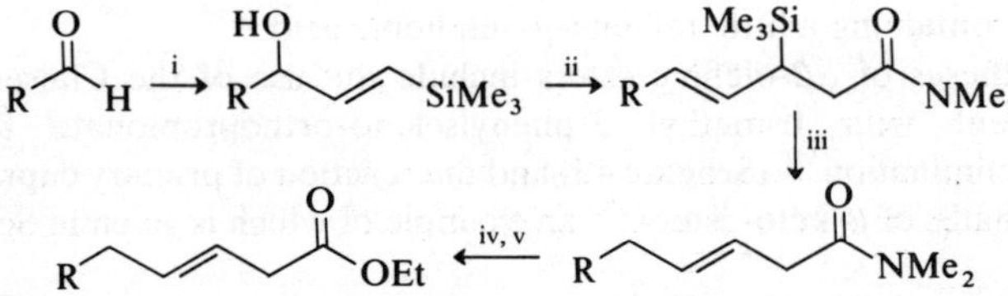

Reagents: i, 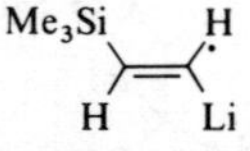 ; ii, $(MeO)_2CMeNMe_2$, 80 °C; iii, HF, −75 to −20 °C; iv, $Et_3O^+ BF_4^-$; v, K_2CO_3, H_2O

Scheme 46

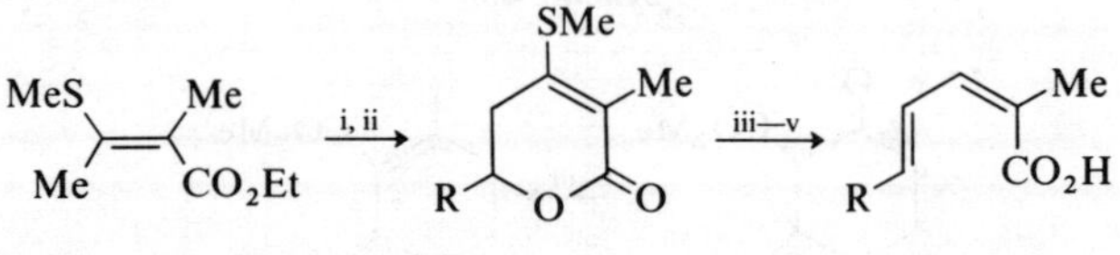

Reagents: i, LDA; ii, RCHO; iii, Raney nickel W-2; iv, 2.1 equiv. $K^+ Bu^tO^-$; v, NH_4Cl, H_2O

Scheme 47

The Claisen amide acetal rearrangement as applied to 3-(trimethylsilyl)allyl alcohols, which can be prepared from aldehydes and trimethylsilylvinyl-lithium, provides a route to the (E)-forms of $\beta\gamma$-olefinic amides and esters (Scheme 46).[135] A method for the stereospecific synthesis of the type of α-methyl-substituted dienoic acid fragment present in rifamycin has been described,[136] and is illustrated in Scheme 47.

[134] D. Seebach and M. Pohmakotr, *Helv. Chim. Acta*, 1979, **62**, 843.
[135] P. R. Jenkins, R. Gut, H. Wetter, and A. Eschenmoser, *Helv. Chim. Acta*, 1979, **62**, 1922.
[136] E. J. Corey and G. Schmidt, *Tetrahedron Letters*, 1979, 2317.

Lactones.—The synthesis and the synthetic uses of halogeno-lactones, particularly their reduction to the halogen-free lactones and their dehydrohalogenation to olefinic lactones, have been reviewed.[137]

N-Bromosuccinimide in completely dry dimethyl sulphoxide oxidizes cyclic enol ethers, *e.g.* (32), to α-bromo-lactones.[138] Benzeneseleninic anhydride is an efficient

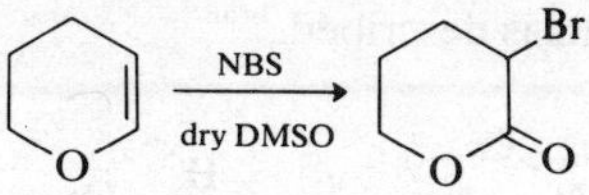

(32)

reagent for the $\alpha\beta$-dehydrogenation of lactones, but is inefficient with acyclic esters.[139] A new butenolide synthesis[140] uses the condensation of an aldehyde or ketone with a metallated β-bromo-acrylamide, and is of some generality, since β-bromo-acraldehydes are readily available from ketones through the Vilsmeier reaction. The overall sequence is shown in Scheme 48. Another route to butenolides,[141] based on known reactions not hitherto linked together to achieve this result, starts from γ-iodoallyl alcohols; the carbonyl group is introduced by a palladium-catalysed carbonylation under mild conditions.

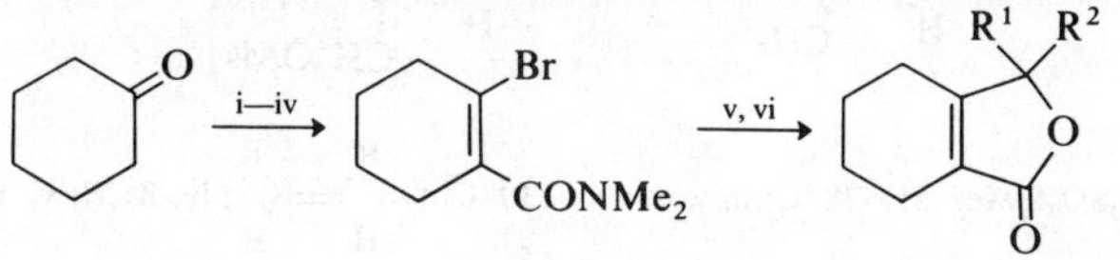

Reagents: i, $Me_2\overset{+}{N}{=}CHBr\ Br^-$; ii, Ag_2O; iii, $SOCl_2$; iv, Me_2NH; v, 2.1 equiv. Bu^tLi; vi, R^1R^2CO

Scheme 48

Remarkable, though not unexpected, stereospecificity is displayed in the ferric-chloride-induced cyclization of the ethylenetricarboxylate (33), which gives the single stereoisomer (34). Compound (33) decomposed at temperatures where acyclic allylic dimethyl ethylenetricarboxylates undergo the thermal ene reaction to give β-alkenyl-lactones.[142]

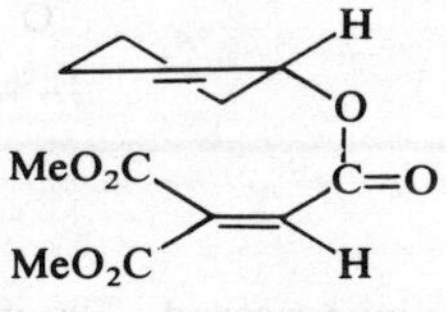

(33)

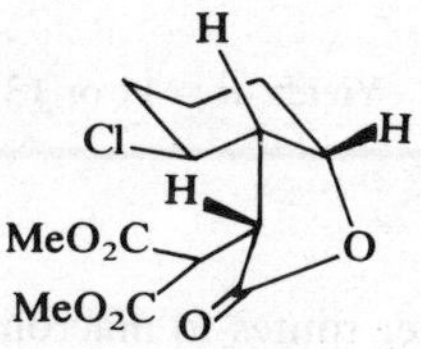

(34)

[137] M. D. Dowle and D. I. Davies, *Chem. Soc. Rev.*, 1979, **8**, 171.
[138] W. F. Berkowitz, I. Sasson, P. S. Sampathkumar, J. Hrabe, S. Choudhry, and D. Pierce, *Tetrahedron Letters*, 1979, 1641.
[139] D. H. R. Barton, R. A. H. F. Hui, D. J. Lester, and S. V. Ley, *Tetrahedron Letters*, 1979, 3331.
[140] W. R. Baker and R. M. Coates, *J. Org. Chem.*, 1979, **44**, 1022.
[141] A. Cowell and J. K. Stille, *Tetrahedron Letters*, 1979, 133.
[142] B. B. Snider and D. M. Roush, *J. Org. Chem.*, 1979, **44**, 4229.

A new α-methylenelactone synthesis related to Eschenmoser's chloro-nitrone route [see *Ann. Reports* (*B*), 1973, **70**, 688] has appeared.[143] The new method is illustrated in Scheme 49. Earlier work on phenylselenolactonization has been published in full,[144] and it has been shown that the procedure can be extended to the formation of phenylselenyl-substituted macrolides (Scheme 50) if the phenylselenation is carried out with *N*-phenylseleno-phthalimide or -succinimide; the preparation of these reagents is described.[145]

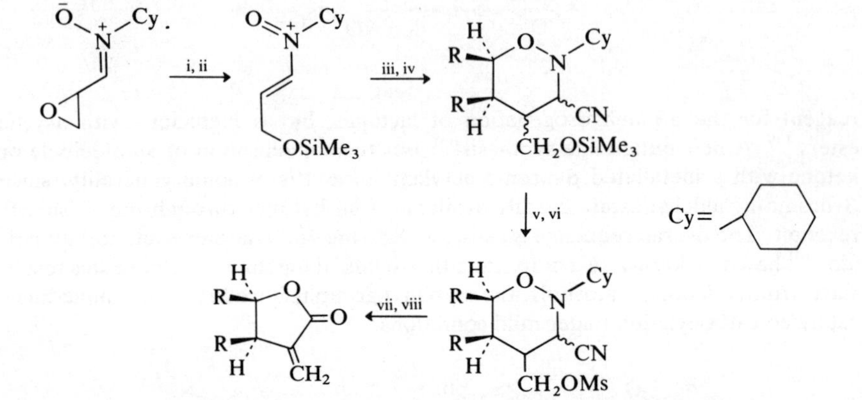

Reagents: i, $CF_3SO_3SiMe_3$, at −78 °C; ii, warm to −30 °C; iii, 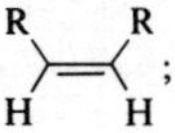; iv, KCN; v, 2M-HCl, MeOH, CH_2Cl_2; vi, $MeSO_2Cl$, C_5H_5N; vii, K^+ Bu^tO^-; viii, H_2SO_4

Scheme 49

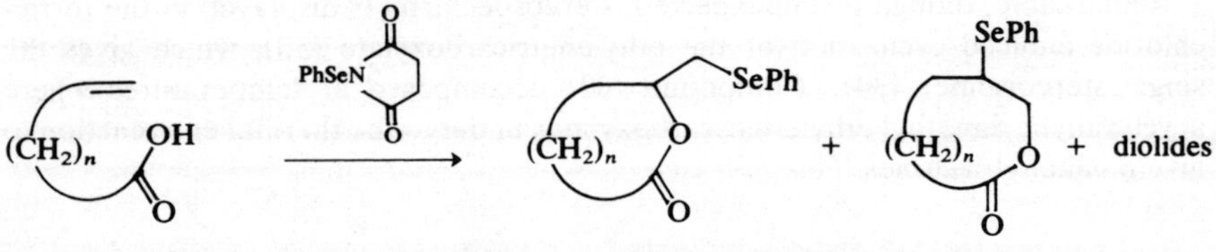

Yields (n = 11 or 13) 50% 15% 10%

Scheme 50

Several other routes to macrolides have been investigated. The intramolecular cyclization of an ω-hydroxy-acid can be brought about by activating the acid group, as its mixed anhydride with 2,4,6-trichlorobenzoic acid, and then adding the hydroxy-anhydride very slowly to refluxing benzene containing 4-dimethylamino-

[143] M. Riediker and W. Graf, *Helv. Chim. Acta*, 1979, **62**, 205, 1586.
[144] K. C. Nicolaou, S. P. Seitz, W. J. Sipic, and J. F. Bount, *J. Amer. Chem. Soc.*, 1979, **101**, 3884.
[145] K. C. Nicolaou, D. A. Claremon, W. E. Barnette, and S. P. Seitz, *J. Amer. Chem. Soc.*, 1979, **101**, 3704.

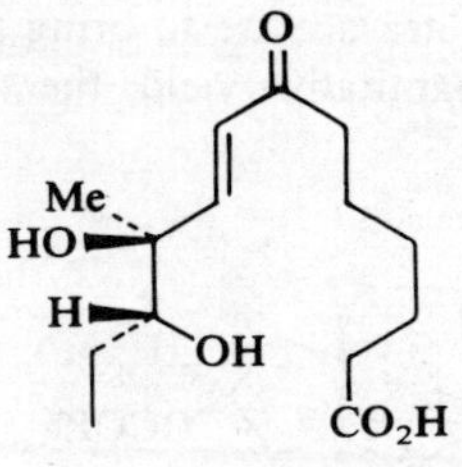

(35)

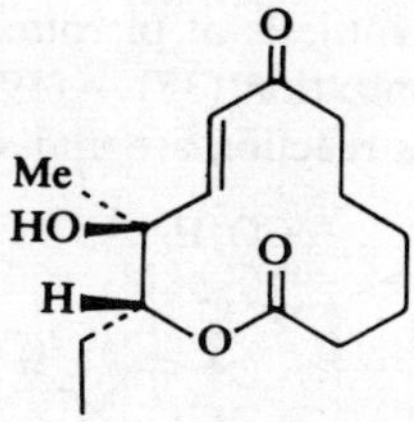

(36)

pyridine.[146] The cyclization of (35) to (36) was achieved by this method, in 46% yield. In macrolide syntheses based on ring closure at the site of the ester function it has normally been the acyl–oxygen bond which has been formed, as in this case. A new procedure[147] is based on the formation of the alkyl–oxygen bond. Caesium salts of ω-iodo-acids cyclize at 40 °C, in DMF, without the use of high-dilution or slow-addition techniques: the reaction was successful with a secondary iodo-acid. Two groups have looked at the use of intramolecular keto-phosphonate cyclizations in macrolide syntheses.[148] The method is particularly useful for the construction of *trans*-$\alpha\beta$-olefinic macrolides, using phosphonate esters (Scheme 51), although in

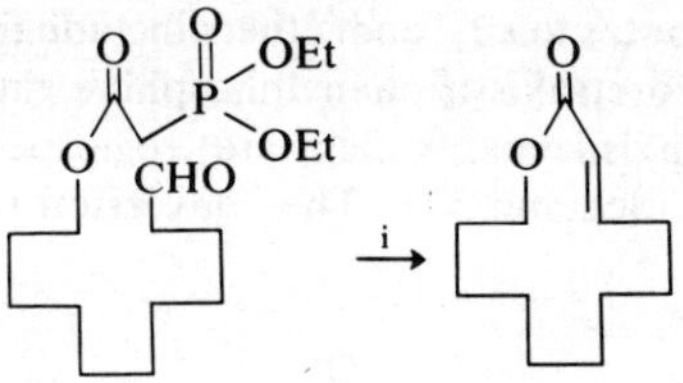

60—70%

Reagents: i, Li^+ Pr^iO^-, 1% HMPT, THF, high-dilution conditions

Scheme 51

some cases both geometries at the double bond are formed. Eschenmoser has used a decarboxylative double fragmentation to generate a doubly unsaturated macrolide (37) in very high yield.[149]

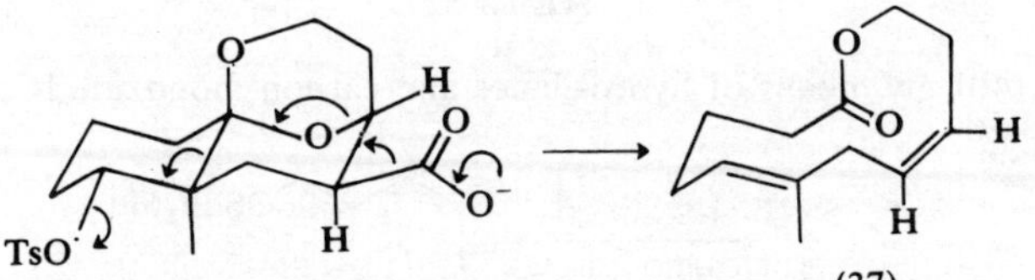

(37)

[146] J. Inanaga, K. Hirata, H. Saeki, T. Katsuki, and M. Yamaguchi, *Bull. Chem. Soc. Japan*, 1979, **52**, 1989; J. Inanaga, T. Katsuki, S. Takimoto, S. Ouchida, K. Inoue, A. Nakano, N. Okukado, and M. Yamaguchi, *Chem. Letters*, 1979, 1021.

[147] W. H. Kruitzinga and R. M. Kellogg, *J.C.S. Chem. Comm.*, 1979, 285.

[148] G. Stork and E. Nakamura, *J. Org. Chem.*, 1979, **44**, 4010; K. C. Nicolaou, S. P. Switz, M. R. Pavia, and N. A. Petasis, *J. Org. Chem.*, 1979, **44**, 4011.

[149] D. Sternbach, M. Shibuya, F. Jaisli, M. Bonett, and A. Eschenmoser, *Angew. Chem. Internat. Edn.*, 1979, **18**, 634.

[150] E. J. Corey and H. L. Pearce, *J. Amer. Chem. Soc.*, 1979, **101**, 5841.

Corey's total synthesis of picrotoxin used lead tetra-acetate to bring about the unusual bis-lactonization (38) → (39) in almost quantitative yield: the scope and mechanism of this reaction are under investigation.[150]

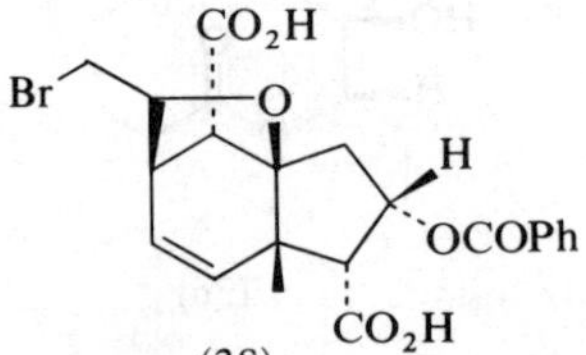

(38)

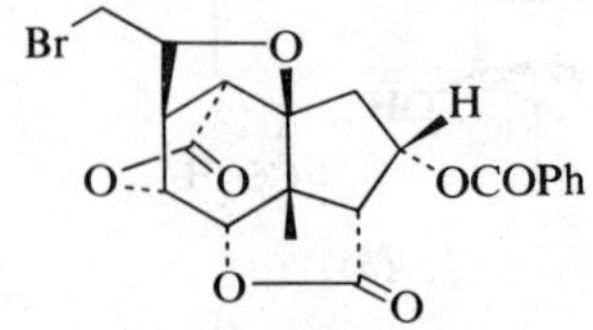

(39)

13 Alkylation

There has been a lot of interest in the reactions of trimethylsilyl enol ethers, and in methods for their preparation. Several useful modifications to the standard methods for the preparation of kinetically controlled enol ethers have been published.[151] Surprisingly, it has been found that treatment of an aldehyde, ketone, or enone with trimethylsilyl iodide (for which a convenient new synthesis has appeared[152]) and hexamethyldisilazane at room temperature or below, in tetrachloromethane, leads to the thermodynamic equilibrium mixture of silyl enol ethers within a few minutes.[153] Other good routes to silyl enol ethers include the rearrangement of allyl silyl ethers with dihydrotetrakis(triphenylphosphine)ruthenium[154] and several routes from acyl-trimethylsilanes, which are regiospecific, and in one case stereoselective as well[155] (Scheme 52). The conversion of alkenes into enol silyl

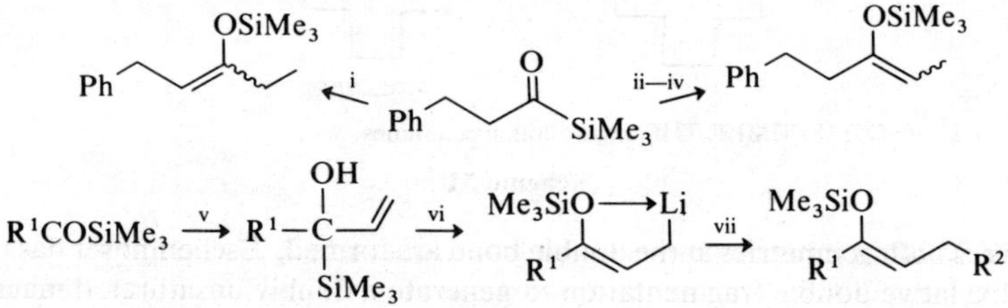

Reagents: i, $PhSO_2CHLiMe$; ii, LDA; iii, $PhSO_2SPh$; iv, EtLi; v, $H_2C{=}CHMgBr$; vi, BuLi; vii, R^2X

Scheme 52

ethers, *e.g.* (40), by means of hydrosilanes and carbon monoxide is covered by a review.[156]

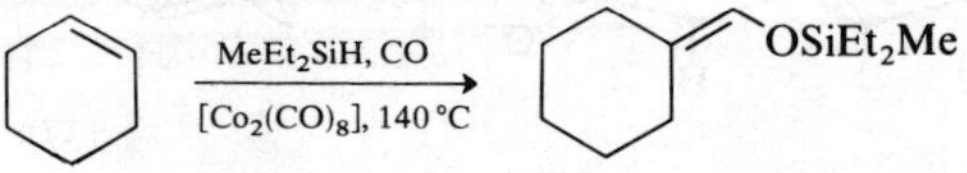

(40)

[151] I. Fleming and I. Paterson, *Synthesis*, 1979, 736; H. Vorbrüggen and K. Krolikiewicz, *ibid.*, p. 34.
[152] H. Sakurai, A. Shirahata, K. Sasaki, and A. Hosomi, *Synthesis*, 1979, 740.
[153] R. D. Miller and D. R. McKean, *Synthesis*, 1979, 730.
[154] H. Suzuki, Y. Koyama, Y. Moro-oka, and T. Ikawa, *Tetrahedron Letters*, 1979, 1415.
[155] H. J. Reich, J. J. Rusek, and R. E. Olson, *J. Amer. Chem. Soc.*, 1979, **101**, 2225; I. Kuwajima and M. Kato, *J.C.S. Chem. Comm.*, 1979, 708.
[156] S. Murai and N. Sonoda, *Angew. Chem. Internat. Edn.*, 1979, **18**, 837.

Silyl enol ethers do not react with alkyl halides in the absence of a catalyst, but they do react with a variety of halides in the presence of Lewis-acid catalysts, *e.g.* titanium and tin tetrachlorides and zinc bromide. The reactions succeed with the enol ethers of ketones, aldehydes, and esters, provided that the appropriate Lewis-acid catalyst is selected. Phenylthioalkylation, followed by desulphurization with Raney nickel or sulphoxide elimination, then leads to the alkylated and alkylidenated products respectively.[157] The use of 1-chloro-1-phenylthio-2,2-dimethylpropane allows a neopentyl group to be introduced. The choice of kinetic or thermodynamic enol ethers permits regiospecific substitution, and the dienol ethers usually lead to a high proportion of the γ-alkylated products. In a similar way, the use of phenylsulphenyl chloride converts an $\alpha\beta$-olefinic ketone into the γ-phenylthio-derivative, and elimination of sulphoxide then leads to the conjugated dienone.[158] Alkylation with certain alkyl halides has been reported; cation rearrangement does not occur.[159] With 2-halogeno-propanes, elimination takes place, but isopropylation is still possible, using a ferric-chloride-catalysed reaction with 2,2-bis(ethylthio)propane, followed by desulphurization. Both enol silyl ethers and enol acetates react with methoxymethyl chloride in the presence of active zinc catalysts.[160] The bis-silyl ether (41), a protected form of methyl acetoacetate, similarly reacts with the cyclic bis-acetal (42) to give the bicyclic product (43).[161] One enol derivative that does

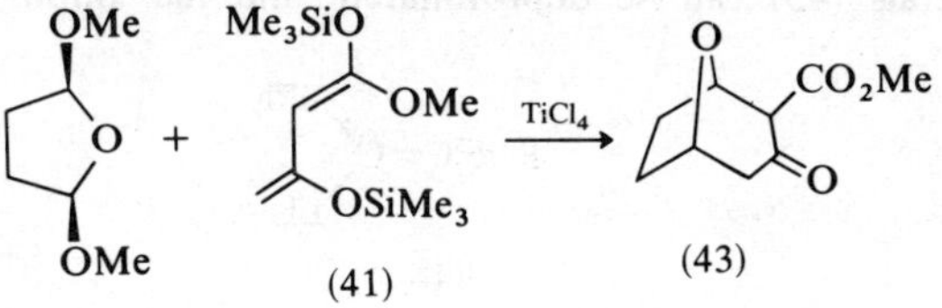

(42) (41) (43)

undergo an uncatalysed reaction with alkyl halides is the potassium enoxytriethylborate (see Scheme 53).[162]

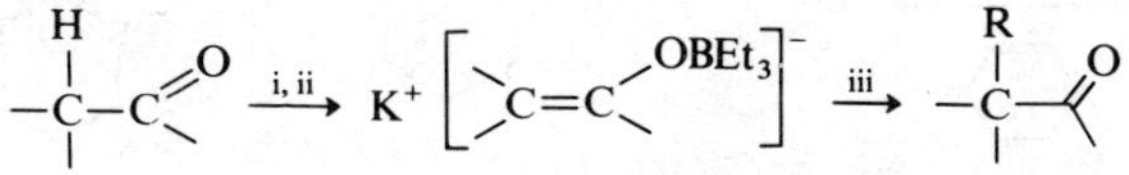

Reagents: i, KH; ii, Et$_3$B; iii, RX, ≤25 °C

Scheme 53

There has been further interest in the synthetic applications of keten thioacetals. There are advantages in the use of open-chain thioacetals, and these can be prepared

[157] I. Paterson and I. Fleming, *Tetrahedron Letters*, 1979, 993, 995, and 2179; I. Fleming, J. Goldhill, and I. Paterson, *ibid.*, p. 3209.
[158] I. Fleming, J. Goldhill, and I. Paterson, *Tetrahedron Letters*, 1979, 3205.
[159] M. T. Reetz, S. Hüttenhain, P. Walz, and U. Löwe, *Tetrahedron Letters*, 1979, 4971; M. T. Reetz, I. Chatziiosifidis, U. Löwe, and W. K. Maier, *ibid.*, p. 1427; I. Paterson, *ibid.*, 1979, 1519.
[160] T. Shono, I. Nishiguchi, T. Komamura, and M. Sasaki, *J. Amer. Chem. Soc.*, 1979, **101**, 984.
[161] P. Brownbridge and T.-H. Chan, *Tetrahedron Letters*, 1979, 4437; T.-H. Chan and P. Brownbridge, *J.C.S. Chem. Comm.*, 1979, 578.
[162] E.-I. Negishi, M. J. Idacavage, F. DiPasquale, and A. Silveira, jr., *Tetrahedron Letters*, 1979, 845.

by heating carboxylic acids or esters with aluminium thiophenoxide,[163] from aldehydes and ketones by reaction with the lithium derivatives of cyclic esters of bis(phenylthio)methaneboronic acid,[164] and by the reaction of dimethyl disulphide with lithio-*NN*-dimethyl-3-(phenylthio)-prop-2-enylamine, which gives a keten thioacetal containing a basic centre.[165] In the last reaction, metallation (and hence electrophilic attack) occurs at the sp^2 carbon atom next to sulphur (Scheme 54), whereas the related phenylthio-substituted allylic ethers and thioethers undergo metallation at the sp^3 carbon atom, to give allylic carbanions.

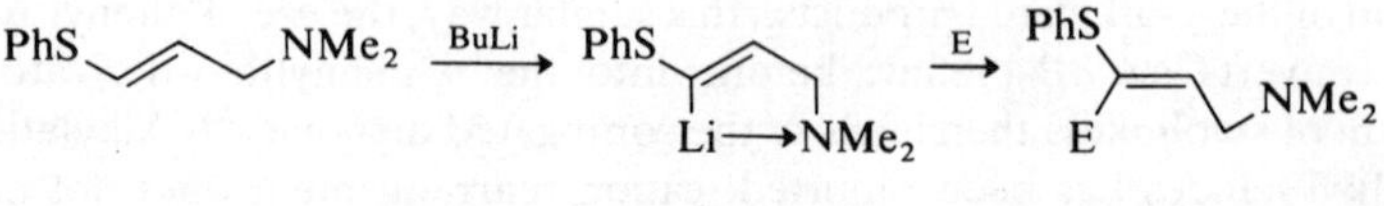

Scheme 54

New methods for the preparation of cyclic keten thioacetals, derived from 1,3-dithian, include a route from thioamides[166] and a dehydrogenation of 2-alkyl-1,3-dithians.[167] Keten thioacetals, on reductive lithiation with lithium naphthalenide at −70 °C, give metallated vinyl sulphides (44), known to be useful acyl-anion equivalents, which now become readily available from acids and esters.[168] Keten thioacetals (45) can be deprotonated, and the anions will then add to

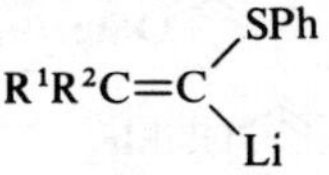

(44)

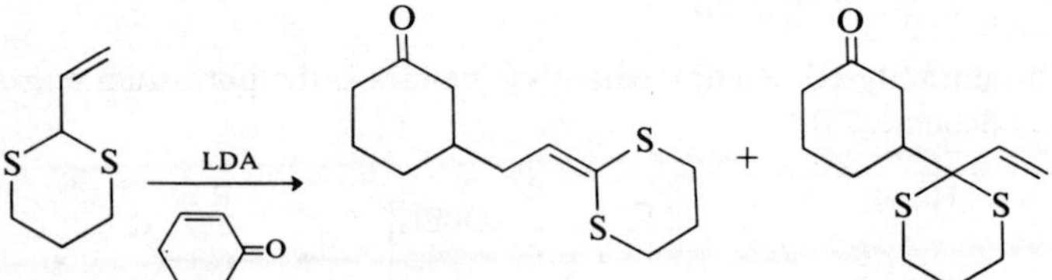

(45)

Product ratio:	71 parts	20 parts
in presence of CuI and $(MeO)_3P$:	2 parts	98 parts

$\alpha\beta$-olefinic ketones.[169] Most of the addition occurs in a conjugate manner, and the regiospecificity with respect to the allyl anion can be largely controlled by the reaction conditions.

[163] T. Cohen, R. E. Gapinski, and R. R. Hutchins, *J. Org. Chem.*, 1979, **44**, 3598.
[164] A. Mendoza and D. S. Matten, *J. Org. Chem.*, 1979, **44**, 1352.
[165] J. J. Fitt and H. W. Gschwend, *J. Org. Chem.*, 1979, **44**, 303.
[166] T. Harada, Y. Tamaru, and Z. Yoshida, *Tetrahedron Letters*, 1979, 3525.
[167] Y. Nagao, K. Seno, and E. Fujita, *Tetrahedron Letters*, 1979, 4403.
[168] T. Cohen and R. B. Weisenfeld, *J. Org. Chem.*, 1979, **44**, 3601.
[169] F. E. Ziegler and C. C. Tam, *Tetrahedron Letters*, 1979, 4717.

Conjugate addition of the anions from *saturated* thioacetals to $\alpha\beta$-olefinic ketones takes place provided that hexamethylphosphoric triamide is present.[170] The conjugated thioacetal (46) can be prepared as shown in Scheme 55. Alkylation of the derived anion gives a thioacetal, which on hydrolysis gives a conjugated dienone.[171]

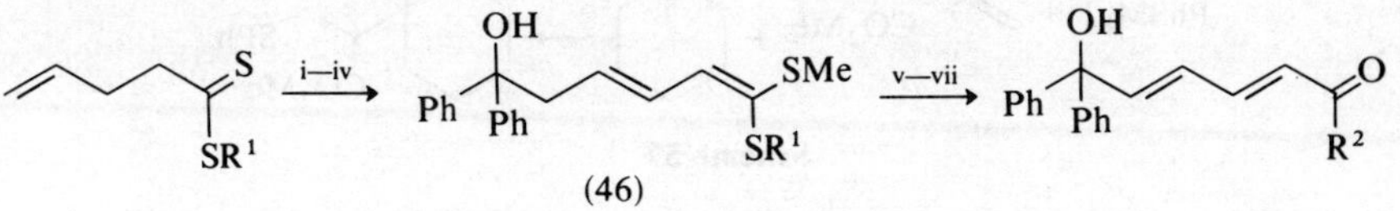

Reagents: i, KH; ii, Bu^sLi; iii, Ph_2CO; iv, MeI; v, LDA, HMPT; vi, R^2X; vii, MeI, H_2O, THF, MeCN, $CaCO_3$

Scheme 55

Many other anion equivalents have been prepared and their reactions with electrophiles studied; several have already been described in earlier sections of this Report. Two further representative examples will be discussed here. The acyl-anion equivalent lithio-4,4-dimethyl-1,3-oxathiolan *S,S*-dioxide (47) reacts with many

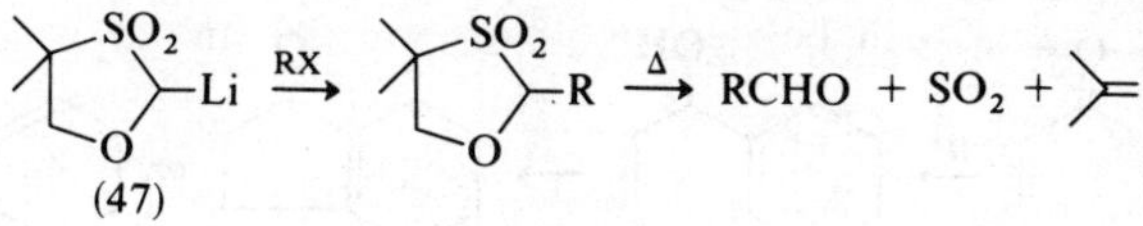

electrophiles. Pyrolysis of the alkylated material gives an aldehyde, and the only other products are gaseous.[172] A homoenolate ion equivalent is produced by the deprotonation of the phosphonamide derivative (48), which is readily prepared from the $\alpha\beta$-olefinic aldehyde. A simple example[173] of the use of this reagent is shown in Scheme 56.

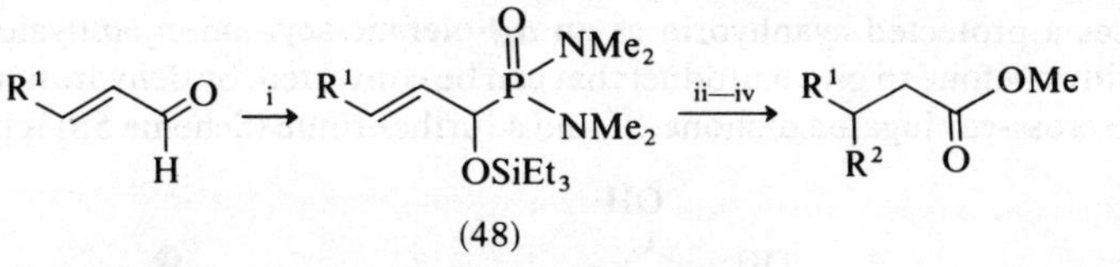

Reagents: i, $Et_3SiOP(NMe_2)_2$; ii, BuLi; iii, R^2X; iv, MeOH, MeO^- Na^+

Scheme 56

Shono has developed his reaction involving the conjugate addition of an organometallic reagent to an active olefin, with trapping of the intermediate by a carbonyl

[170] C. A. Brown and A. Yamamaichi, *J.C.S. Chem. Comm.*, 1979, 100; J. Lucchetti, W. Dumont, and A. Krief, *Tetrahedron Letters*, 1979, 2695.
[171] M. Pohmakotr and D. Seebach, *Tetrahedron Letters*, 1979, 2271.
[172] G. W. Gokel, H. M. Gerdes, D. E. Miles, J. M. Hufnal, and G. A. Zerby, *Tetrahedron Letters*, 1979, 3375.
[173] D. A. Evans, J. M. Takacs, and K. M. Hurst, *J. Amer. Chem. Soc.*, 1979, **101**, 371.

compound, by substituting (phenylthio)magnesium iodide for the organometallic reagent, as in the example shown in Scheme 57.[174] The presence of a phenylthio-group in the products adds to their utility as synthetic intermediates.

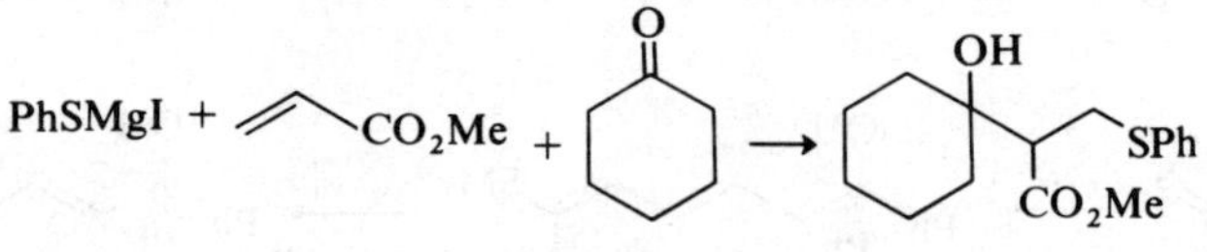

Scheme 57

The generation of alkenyl-lithiums from alkyl methyl ketone aryl-sulphonylhydrazones, described previously [*cf. Ann. Reports* (*B*), 1978, **75**, 185], normally gave predominantly an alk-1-enyl-lithium. A variation has now made it possible to generate the (*Z*)-alk-2-enyl-lithium, which reacts with electrophiles in the normal way, with retention of the geometry of the olefin.[175]

14 Ring Synthesis

Considerable ingenuity has been displayed in devising routes to cycloalkenyl vinyl ketones (49) which can be cyclized by the Nazarov cyclization, thus providing

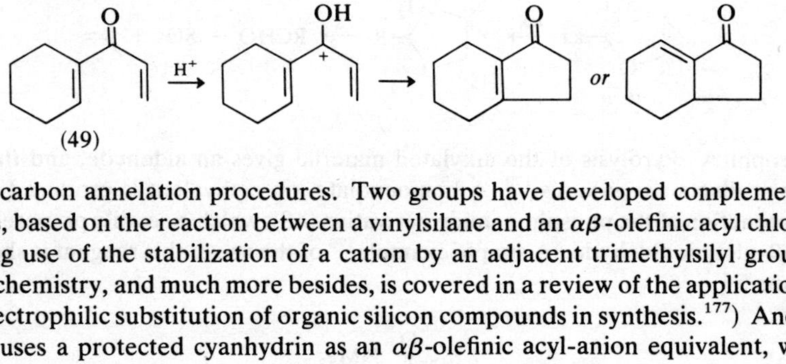

three-carbon annelation procedures. Two groups have developed complementary routes, based on the reaction between a vinylsilane and an $\alpha\beta$-olefinic acyl chloride, making use of the stabilization of a cation by an adjacent trimethylsilyl group.[176] (This chemistry, and much more besides, is covered in a review of the applications of the electrophilic substitution of organic silicon compounds in synthesis.[177]) Another route uses a protected cyanhydrin as an $\alpha\beta$-olefinic acyl-anion equivalent, which reacts with a ketone to give a product that can be converted, by dehydration, into the requisite cross-conjugated dienone,[178] and a further route (Scheme 58) is based on a

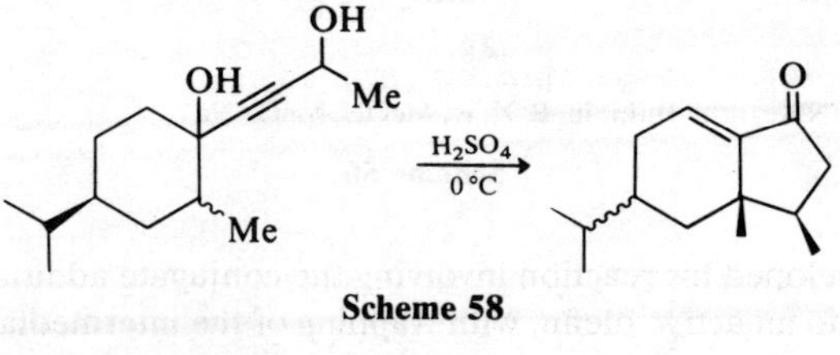

Scheme 58

[174] T. Shono, Y. Matsumura, S. Kashimura, and K. Hatanaka, *J. Amer. Chem. Soc.*, 1979, **101**, 4752.
[175] A. R. Chamberlin and F. T. Bond, *Synthesis*, 1979, 44.
[176] F. Cooke, J. Schwindeman, and P. Magnus, *Tetrahedron Letters*, 1979, 1995; W. E. Fristad, D. S. Dime, T. R. Bailey, and L. A. Paquette, *ibid.*, p. 1999.
[177] T. H. Chan and I. Fleming, *Synthesis*, 1979, 761.
[178] R. M. Jacobson and G. P. Lahm, *J. Org. Chem.*, 1979, **44**, 462.

little-known reaction of α,α'-dihydroxy-acetylenes.[179] Another route (Scheme 59) that is based on the acylation of a silane,[180] in this case a trimethylsilyl-alkyne, also leads to cyclopentenones, though not by a Nazarov cyclization.

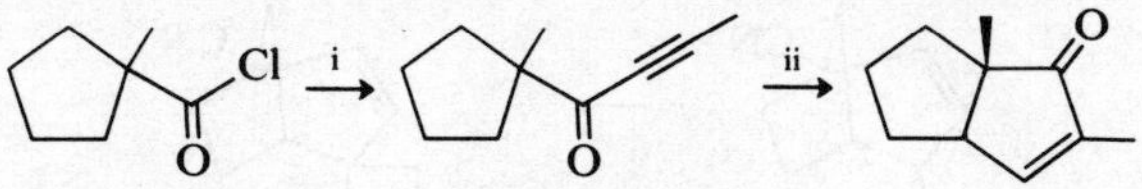

Reagents: i, $Me_3SiC{\equiv}CMe$; ii, 650 °C

Scheme 59

Other noteworthy syntheses of cyclopentenones have appeared during the year; attention has already been drawn to an example of a regiospecific intramolecular aldol condensation [(24) → (25)]. Pentannelations based on intramolecular Horner–Emmons[181] and Wittig[182] reactions used respectively the phosphonate (50) and the

$$BrCH_2\overset{\displaystyle OEt}{\overset{|}{C}}{=}CH\overset{\displaystyle O}{\overset{\|}{P}}(OMe)_2$$

(50)

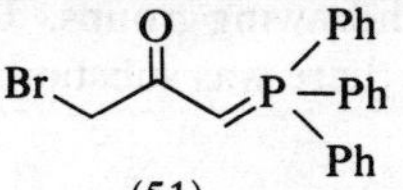

(51)

ylide (51) as three-carbon units to alkylate ketone enolates. Two other cyclopentenone syntheses[183,184] are illustrated in Schemes 60 and 61; the former[183] occurs under very mild conditions.

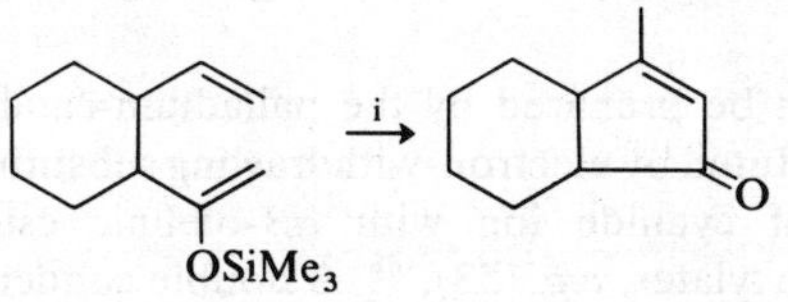

Reagent: i, $Pd(OAc)_2$, 20 °C

Scheme 60

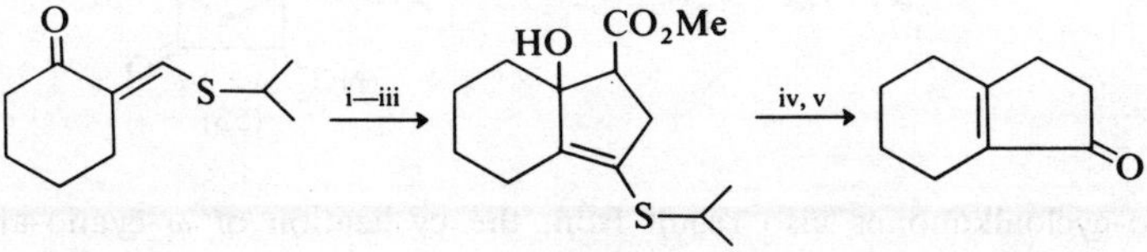

Reagents: i, LiN; ii, CO_2Me; iii, NH_4Cl, H_2O; iv, HCl, H_2O, THF; v, Me_3SiI

Scheme 61

179 T. Hiyama, M. Shinoda, and H. Nozaki, *J. Amer. Chem. Soc.*, 1979, **101**, 1599.
180 M. Karpf and A. S. Dreiding, *Helv. Chim. Acta*, 1979, **62**, 852.
181 E. Piers, B. Abeysekera, and J. R. Scheffer, *Tetrahedron Letters*, 1979, 3279.
182 J.-J. Altenbach, *Angew. Chem. Internat. Edn.*, 1979, **18**, 940.
183 Y. Ito, H. Aoyama, T. Hirao, A. Mochizuki, and T. Saegusa, *J. Amer. Chem. Soc.*, 1979, **101**, 494.
184 J. P. Marino and L. C. Katterman, *J.C.S. Chem. Comm.*, 1979, 946.

An interesting cyclopentanone synthesis[185] employed Stetter's thiazolium reagent to bring about the intramolecular Michael condensation of an aldehyde, converted *in situ* into an acyl-anion equivalent, with an $\alpha\beta$-olefinic ester (Scheme 62), and a

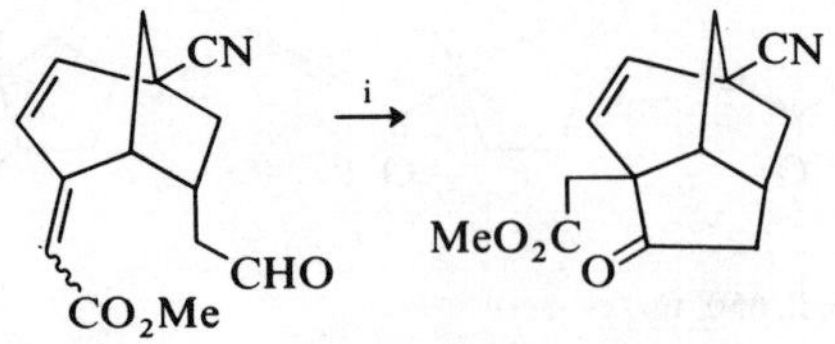

Reagents: i, 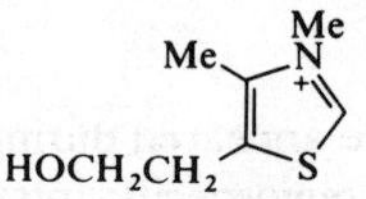 I^-, Et_3N (excess), refluxing Me_2CHOH

Scheme 62

synthesis[186] of methylenecyclopentanes used the palladium-catalysed condensation of (2-acetoxymethylallyl)trimethylsilane (52) with alkenes carrying one or two electron-withdrawing groups. The (*E*)-alkenes gave the *trans*-products, but with the (*Z*)-alkenes there was substantial crossover.

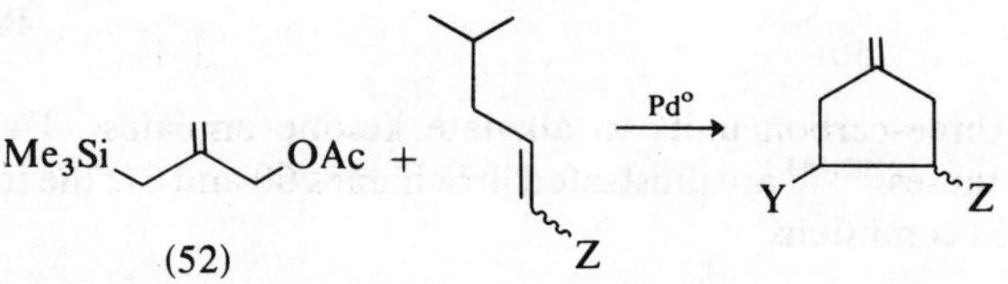

Cyclopentenes can be prepared by the palladium-catalysed cyclization of 1,6-dienes that are substituted by electron-withdrawing substituents at the 4-position,[187] and the reaction of cyanide ion with $\alpha\beta$-olefinic esters gives alkyl cyano-cyclopentanonecarboxylates, *e.g.* (53).[188] A double condensation is shown.

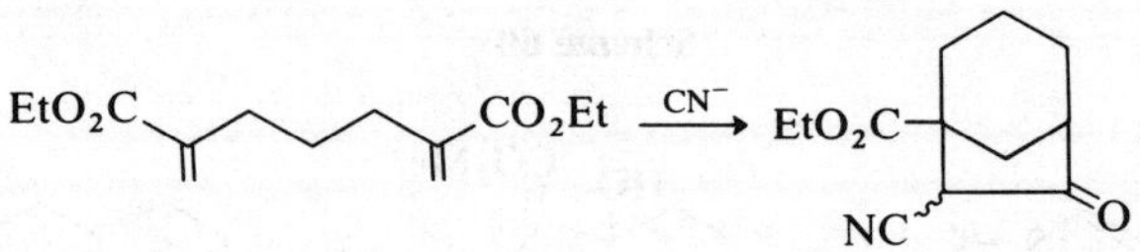

(53)

Cyano-cycloalkanones also result from the cyclization of ω-cyano-amides, a reaction which has advantages over the Thorpe–Ziegler $\alpha\omega$-dicyanide cyclization because the regiospecificity is controlled in unsymmetrically substituted systems.[189] Two aspects of the stereochemistry of formation of five-membered rings have been investigated. Cyclization of $\gamma\delta$-olefinic aldehydes (54) to cyclopentanones is

[185] B. M. Trost, C. D. Shiney, F. DiNinno, jr., and S. M. McElvain, *J. Amer. Chem. Soc.*, 1979, **101**, 1284.
[186] B. M. Trost and D. M. T. Chan, *J. Amer. Chem. Soc.*, 1979, **101**, 6429.
[187] R. Grigg, T. R. B. Mitchell, and A. Ramasubba, *J.C.S. Chem. Comm.*, 1979, 669.
[188] H. Stetter and H. Kuhlmann, *Annalen*, 1979, 303, 944, and 1122.
[189] M. Larcheveque and P. Mulot, *Canad. J. Chem.*, 1979, **57**, 17.

stereospecific,[190] and ring-closure of (55) by an S_CN' process leads to transfer of chirality in the sense shown, with an *anti* relationship between the incoming and departing groups.[191]

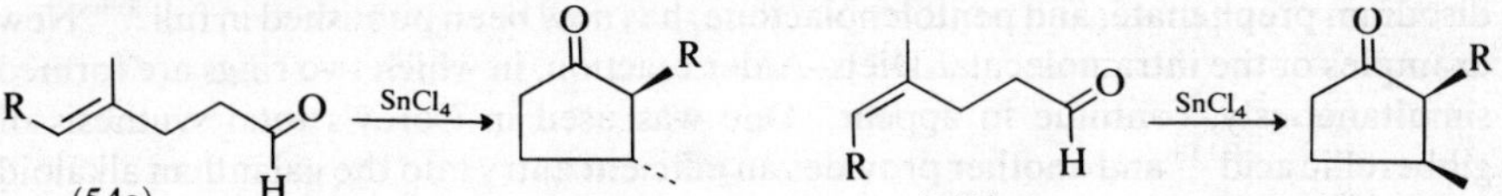

(54a) (54b)

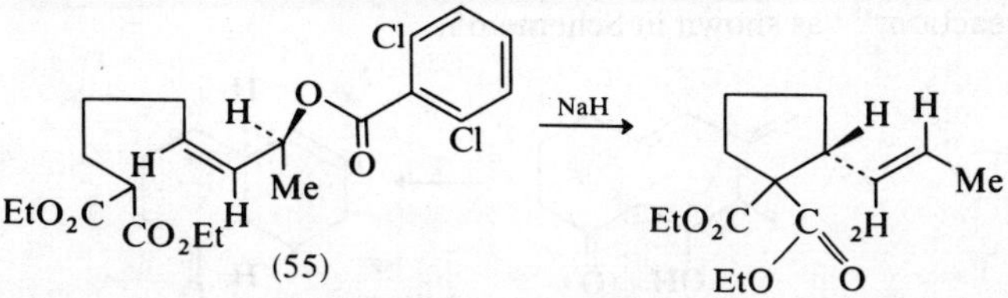

(55)

A new ring-closure reaction (Scheme 63) depends on participation by a phenylselenyl group and at the same time gains by the synthetic potential offered by a phenylselenyl group in the product.[192]

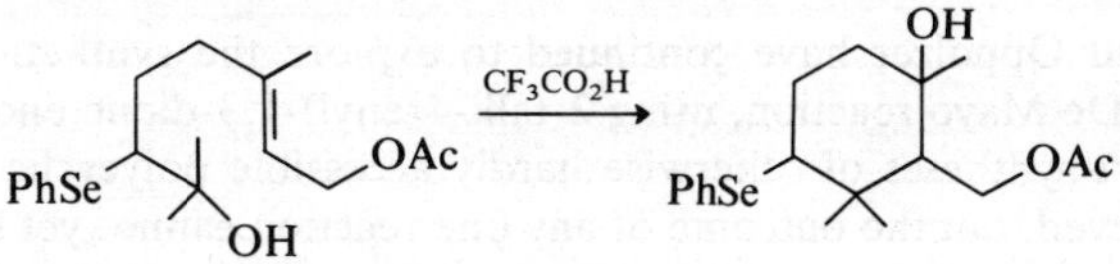

Scheme 63

The synthesis of five- and seven-membered rings by the reaction of polybromoketones with iron carbonyl has been reviewed[193] and also used intramolecularly in a terpene synthesis which converts geraniol into camphor, dihydrocarvone, and other minor products.[194] The use of disodium tetracarbonylferrate to cyclize a pent-4-enyl tosylate to a cyclohexanone (*e.g.* Scheme 64) has now been used intramolecularly in a total synthesis of aphidicolin.[195]

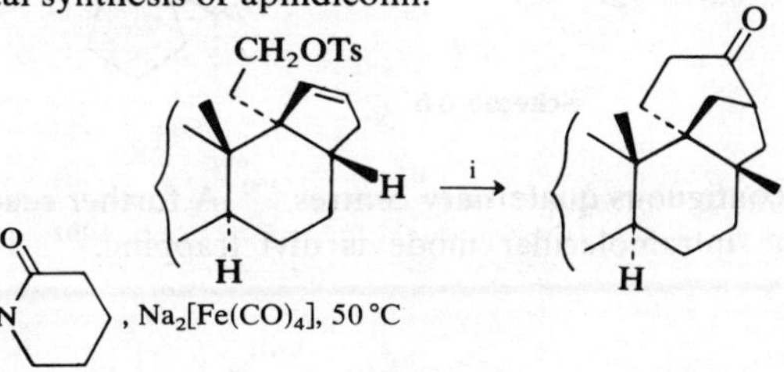

Reagents: i, MeN(piperidinone), $Na_2[Fe(CO)_4]$, 50 °C

Scheme 64

[190] R. C. Cookson and S. A. Smith, *J.C.S. Chem. Comm.*, 1979, 145.
[191] G. Stork and A. R. Schoofs, *J. Amer. Chem. Soc.*, 1979, **101**, 5081.
[192] T. Kametani, K. Suzuki, H. Kurobe, and H. Nemoto, *J.C.S. Chem. Comm.*, 1979, 1128.
[193] R. Noyari, *Acc. Chem. Res.*, 1979, **12**, 61.
[194] R. Noyari, M. Nishizawa, F. Shimizu, Y. Hayakawa, K. Maruoka, S. Hashimoto, H. Yamamoto, and H. Hozaki, *J. Amer. Chem. Soc.*, 1979, **101**, 220.
[195] J. E. McMurry, A. Andrus, G. M. Ksander, J. H. Musser, and M. A. Johnson, *J. Amer. Chem. Soc.*, 1979, **101**, 1330.

Some of the finest examples of the use of the classical intermolecular Diels–Alder reaction are from the work of Danishefsky, using methoxy- and trimethylsilyloxy-substituted dienes. The work, which culminated in elegant syntheses of griseofulvin, disodium prephenate, and pentolenolactone, has now been published in full.[196] New examples of the intramolecular Diels–Alder reaction, in which two rings are formed simultaneously, continue to appear. One was used in Corey's total synthesis of gibberellic acid[113] and another provides an efficient entry into the galanthan alkaloid series.[197] *cis*-Decalones can also be obtained from acyclic precursors through this type of reaction,[198] as shown in Scheme 65.

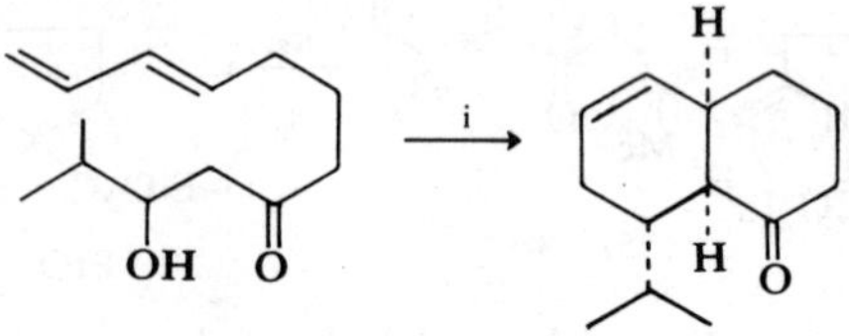

Reagents: i, $MeSO_2Cl$, Et_3N

Scheme 65

Pattenden and Oppolzer have continued to explore the synthetic uses of the intramolecular De Mayo reaction, using 2-(alk-4-enyl)-1,3-dione enol acetates.[199] Some very short syntheses of otherwise hardly accessible polycyclic ring systems have been achieved, but the outcome of any one reaction cannot yet be predicted; some are not regioselective, or take unexpected courses, and in some cases the photo-Fries rearrangement intrudes. Another, related [2+2] intramolecular photocyclization (Scheme 66) is notable for giving a single stereoisomer, and that in a

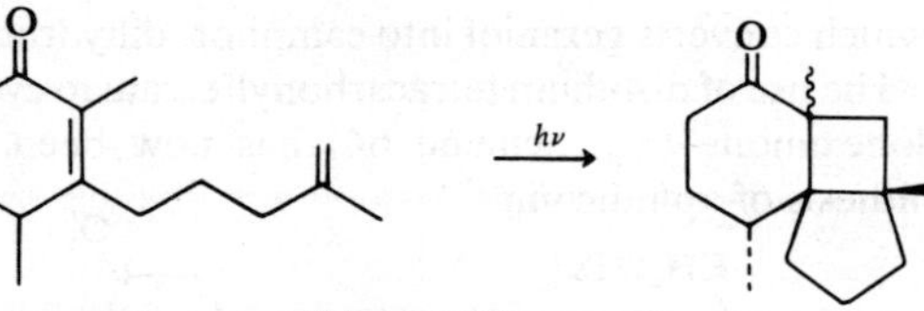

Scheme 66

molecule with three contiguous quaternary centres.[200] A further reaction which is more effective in the intramolecular mode is diyl trapping.[201] The very high

[196] S. Danishefsky, T. Kitahara, C. F. Yan, and J. Morris, *J. Amer. Chem. Soc.*, 1979, **101**, 6996; S. Danishefsky, C. F. Yan, R. K. Singh, R. B. Gammill, P. McCurry, N. Fritsch, and C. J. Clardy, *ibid.*, p. 7001; S. Danishefsky, T. Harayama, and R. K. Singh, *ibid.*, p. 7008; S. Danishefsky, M. Hirama, N. Fritsch, and C. J. Clardy, *ibid.*, p. 7013; S. Danishefsky and F. J. Walker, *ibid.*, p. 7018; S. Danishefsky, M. Hirama, K. Gombatz, T. Harayama, E. Berman, and P. F. Schuda, *ibid.*, p. 7020.

[197] D. J. Morgans, jr. and G. Stork, *Tetrahedron Letters*, 1979, 1959.

[198] J.-L. Gras and M. Bertrand, *Tetrahedron Letters*, 1979, 4549.

[199] M. J. Begley, M. Mellor, and G. Pattenden, *J.C.S. Chem. Comm.*, 1979, 235; W. Oppolzer and T. G. C. Bird, *Helv. Chim. Acta*, 1979, **62**, 1199.

[200] M. C. Pirrung, *J. Amer. Chem. Soc.*, 1979, **101**, 7130.

[201] R. D. Little and G. W. Muller, *J. Amer. Chem. Soc.*, 1979, **101**, 7129.

stereoselectivity observed (Scheme 67) was rationalized in terms of bonding interactions of secondary orbitals.

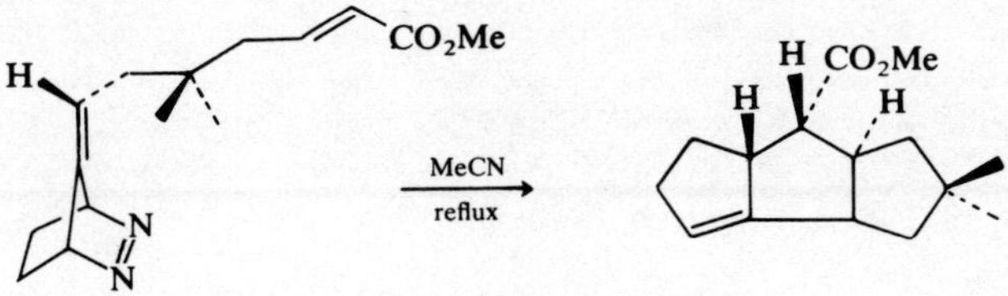

Scheme 67

Finally, Robinson's famous tropinone synthesis, 62 years on, has been applied in a spectacular synthesis of the ladybird defence alkaloids (Scheme 68), in a reaction that creates three rings and five chiral centres at one time, in an aqueous buffer, and from which a single stereoisomer crystallized out.[202]

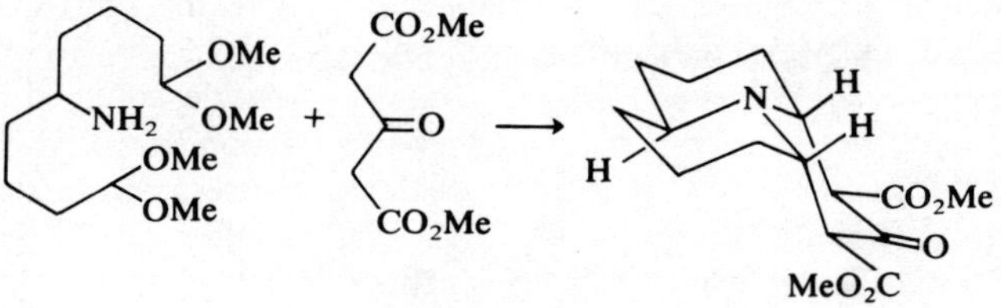

Scheme 68

[202] R. V. Stevens and A. W. M. Lee, *J. Amer. Chem. Soc.*, 1979, **101**, 7032.

16 Biological Chemistry
Part (i) Terpenoids and Steroids

By J. R. HANSON
School of Molecular Sciences, University of Sussex, Brighton BN1 9QJ

1 Introduction

The terpenoids and steroids were last reviewed in these Reports in 1973[1a] and 1975[1b] respectively. However, rather than attempt to cover the intervening years in the limited space available, attention has been concentrated on the past two years, with the omission of biosynthesis, which was covered last year.[1c] For detailed reviews the reader is referred to the Specialist Periodical Reports on 'Terpenoids and Steroids'[2] and to the relevant chapters in the companion volume on 'Biosynthesis'.[3]

The advent of high-field ^{1}H n.m.r. spectroscopy and the widespread use of ^{13}C n.m.r. in structural elucidation have made a significant impact on the number and variety of known terpenoids. Some useful reviews of the ^{13}C data have appeared.[4] An increasing number of structures without a heavy atom have been elucidated by *X*-ray analysis.

2 Monoterpenoids

Amongst the many novel terpenoids which have been isolated from marine organisms are a number of halogenated monoterpenoids, exemplified by chondrocolactone (1).[5] The number of known highly oxygenated iridoids has continued to grow, and these now include the *S*-methylthiocarbonate paedorside (2)[6] and xylomollin (3).[7] The latter has a relatively unusual *trans*-fused ring junction. Amongst the new irregular monoterpenoids are the epimeric santolinyl lactones, the santolinolides (4), which were isolated[8] from *Artemisia tridentata*.

Many monoterpenoids play an important role as insect pheromones.[9] Consequently, attention has been directed not only at their identification but also at their

[1] (*a*) B. A. Marples, *Ann. Reports* (*B*), 1973, **70**, 549; (*b*) D. N. Kirk, *ibid.*, 1975, **72**, 366; (*c*) J. R. Hanson, *ibid.*, 1978, **75**, 329.
[2] 'Terpenoids and Steroids,' ed. J. R. Hanson (Specialist Periodical Reports), The Chemical Society, London, 1979, Vol. 9 and preceding volumes.
[3] 'Biosynthesis,' ed. J. D. Bu'Lock (Specialist Periodical Reports), The Chemical Society, London, 1977, Vol. 5 and preceding volumes.
[4] F. W. Wehrli and T. Nishida, *Fortschr. Chem. org. Naturstoffe*, 1978, **36**, 1; W. B. Smith, *Ann. Rep. N.M.R. Spectroscopy*, 1978, **8**, 199.
[5] F. X. Woolard, R. E. Moore, D. van Engen, and J. Clardy, *Tetrahedron Letters*, 1978, 2367.
[6] G. J. Kapadia, Y. N. Shukla, A. K. Bose, H. Fujiwara, and H. A. Lloyd, *Tetrahedron Letters*, 1979, 1937.
[7] M. Nakane, C. R. Hutchinson, D. van Engen, and J. Clardy, *J. Amer. Chem. Soc.*, 1978, **100**, 7079.
[8] W. W. Epstein and L. A. Gaudioso, *J. Org. Chem.*, 1979, **44**, 3113.
[9] For a recent review see J. M. Brand, J. C. Young, and R. M. Silverstein, *Fortschr. Chem. org. Naturstoffe*, 1979, **37**, 1; see also R. Baker and D. A. Evans, *Ann. Reports* (*B*), 1977, **74**, 367.

synthesis. Ipsenol (5) and ipsdienol (6), aggregation pheromones of the bark beetle *Ips confusus*, have been the subject of several syntheses.[10] The use of methylcyclobutene (7) as an isoprene equivalent figured[11] in one such synthesis. The unusual ether lineatin (8), a pheromone of the beetle *Trypodendron lineatum*, which attacks the Douglas fir, has also been synthesized.[12] The success of the pyrethroids as insecticides[13] has stimulated a number of syntheses of chrysanthemic acid (9).[14]

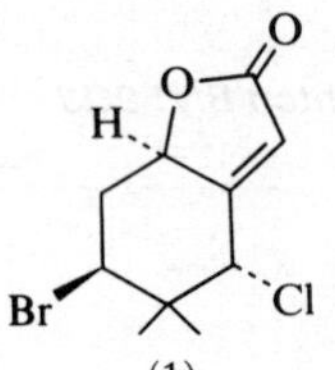

(1)

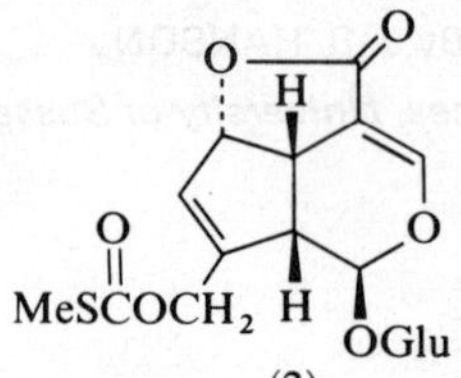

(2)

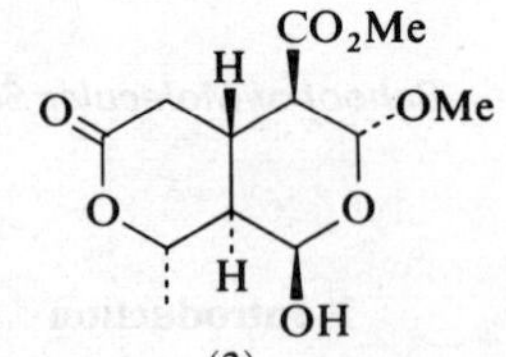

(3)

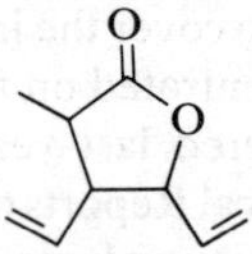

(4)

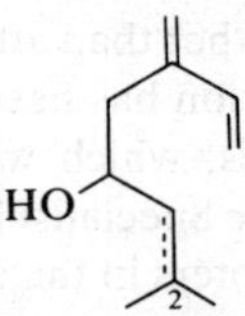

(5)

(6) Δ^2

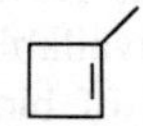
(7)

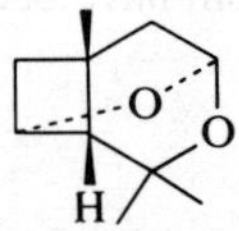

(8)

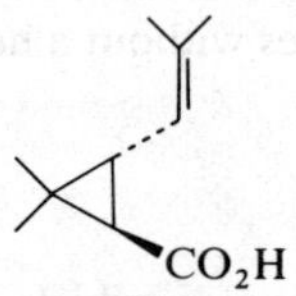

(9)

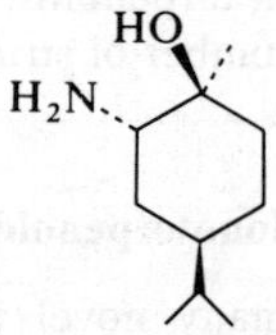

(10)

Monoterpenoids have provided the basis for a number of reagents employed for asymmetric induction in synthesis. Thus improvements have been described[15] in the preparation of mono- and di-isopinocamphenylborane, which are useful reagents for hydroboronation–oxidation, whilst some terpene alkanolamines, *e.g.* (10), have been recommended[16] for the asymmetric synthesis of 2-substituted aldehydes by alkylation of the corresponding metallo-enamines. The enantiomeric purity of C-2 chiral aldehydes has been determined,[17] using the position of the ^{1}H n.m.r. signal of the aldimine proton in their condensation products with 2-amino-1-methoxymenth-

[10] M. Bertrand and J. Viala, *Tetrahedron Letters*, 1978, 2575; A. Hosomi, M. Saito, and A. Sakurai, *ibid.*, 1979, 429; K. Mori, T. Takigawa, and T. Matsuo, *Tetrahedron*, 1979, **35**, 933.
[11] S. R. Wilson, L. R. Phillips, and K. J. Natalie, *J. Amer. Chem. Soc.*, 1979, **101**, 3340.
[12] K. Mori and M. Sasaki, *Tetrahedron Letters*, 1979, 1329.
[13] For a recent review see M. Elliott and N. F. James, *Chem. Soc. Rev.*, 1978, **7**, 473.
[14] See, for example, M. J. De Vos, J. N. Denis, and A. Krief, *Tetrahedron Letters*, 1978, 1847; M. J. De Vos and A. Krief, *ibid.*, 1979, 1511, 1515, and 1891.
[15] H. C. Brown, J. R. Schwier, and B. Singaram, *J. Org. Chem.*, 1978, **43**, 4395; A. Pelter, D. J. Ryder, J. H. Sheppard, C. Subrahmanyam, H. C. Brown, and A. K. Mandal, *Tetrahedron Letters*, 1979, 4777.
[16] A. I. Meyers, Z. Brich, G. W. Erickson, and S. G. Traynor, *J.C.S. Chem. Comm.*, 1979, 566.
[17] A. I. Meyers and Z. Brich, *J.C.S. Chem. Comm.*, 1979, 567.

8-enes. Monoterpenoid ketones have been used[18] in the synthesis of optically active α-deuterio-ketones in which the optical activity of the final product arises solely from the centre bearing deuterium. A useful review of *X*-ray structure analyses of monoterpenoids has appeared.[19]

Monoterpenoids often provide suitable model substrates for examining the scope of novel reactions. Thus a 1,2-carbonyl transposition based on the epoxidation of enol silyl ethers, reduction with lithium aluminium hydride, and oxidation with chromic acid has been examined[20] with menthone. The formation of allylic alcohols by the reduction[21] of $\alpha\beta$-unsaturated ketones with sodium borohydride–cerium(III) chloride and the *cis*-hydroxylation of alkenes using tetraethylammonium acetate and t-butyl peroxide in acetone, catalysed by osmium tetroxide,[22] are useful procedures which have been examined in the monoterpene series.

Further studies have been reported[23] on the hydrolysis of geranyl chloride, phosphate, and pyrophosphate, which give acyclic primary and tertiary substitution products, whilst neryl derivatives give, with participation, cyclization products. Biomimetic cyclizations of geraniol and nerol in aqueous citric acid[24] and under more vigorous acidic conditions (85% phosphoric acid)[25] have also been described. The *syn*-addition of DCl to the less-hindered face of α- and β-pinenes (11) has been demonstrated,[26] using a combination of ^{13}C and ^{2}H n.m.r. spectroscopy. The initial tertiary chloride that is formed rapidly isomerizes to bornyl chloride (12). The mechanism of the acid-catalysed opening of the cyclopropane ring of the thujols (13) to form a methylcyclopentenium ion (14) in FSO_3H–SbF_5–SO_2 at $-80\,°C$ has been studied by deuterium labelling.[27]

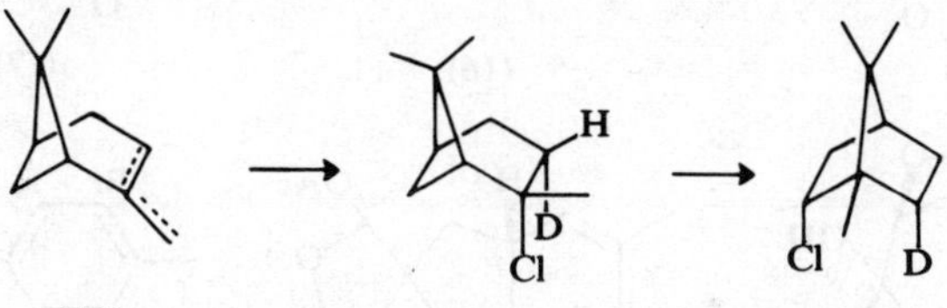

(11) (12)

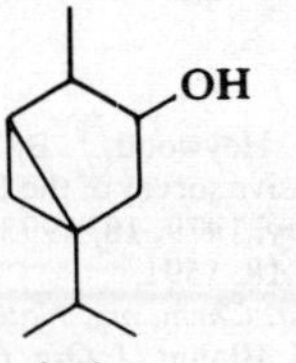

(13)

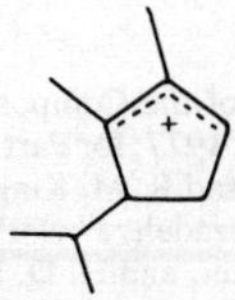

(14)

[18] S. G. Levine and B. Gopalakrishnan, *Tetrahedron Letters*, 1979, 699.
[19] R. B. Yeats in ref. 2 p. 5.
[20] W. A. Fristad, T. R. Bailey, and L. A. Paquette, *J. Org. Chem.*, 1978, **43**, 1620.
[21] J.-L. Luche, L. Rodriguez-Hahn, and P. Crabbé, *J.C.S. Chem. Comm.*, 1978, 601.
[22] K. Akashi, R. E. Palermo, and K. B. Sharpless, *J. Org. Chem.*, 1978, **43**, 2063.
[23] C. A. Bunton, O. Cori, D. Hachey, and J.-P. Leresche, *J. Org. Chem.*, 1979, **44**, 3238.
[24] D. L. Baxter, W. A. Laurie, and D. McHale, *Tetrahedron*, 1978, **34**, 2195.
[25] J. P. McCormick and D. L. Barton, *Tetrahedron*, 1978, **34**, 325.
[26] E. F. Weigand and H. J. Schneider, *Chem. Ber.*, 1979, **112**, 3031.
[27] J. C. Rees and D. Whittaker, *J.C.S. Chem. Comm.*, 1978, 1096.

3 Sesquiterpenoids

The number of known sesquiterpenoids has continued to increase rapidly, partly as a result of an extensive phytochemical survey of the Compositae as a whole[28] and of species of *Eupatorium*, *Helenium*, and related genera in particular in the search for tumour-inhibitory substances. The sesquiterpenoid lactones isolated from the genus *Artemisia* have been reviewed,[29] whilst a lengthy survey of the sesquiterpenoid lactones isolated from both plants and fungi has appeared.[30] Typical of the lactones are esters of the germacranolide (15)[31] and the guaianolide mikanocryptin (16), the *X*-ray analysis of which has been reported.[32] Novel microbial metabolites include pentalenolactone G (17), isolated from a Streptomycete,[33] and a carotane, aspterric acid (18), obtained[34] from an *Aspergillus* species. Marine organisms have yielded a number of unusual sesquiterpenoids, such as spirodysin (19)[35] and the iodinated compounds (20) and (21).[36] The *X*-ray analysis of the sulphide (22), obtained from mint, has been described.[37] The liverwort *Plagiochila semidecurrens* contains[38] some 2,3-seco-alloaromadendranes, such as (23) (ovalifoliene), whilst (−)-bicyclohumulenone (24) has been isolated[39] from another *Plagiochila* species.

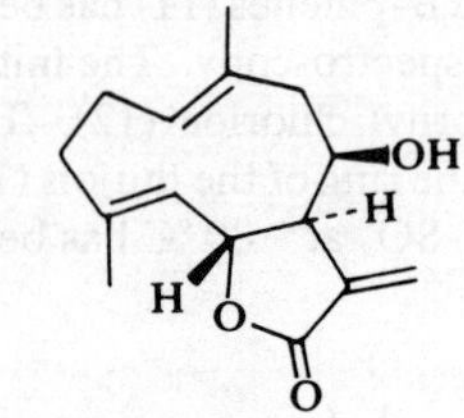

(15)

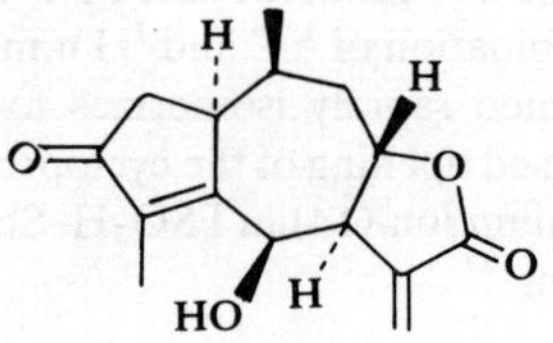

(16)

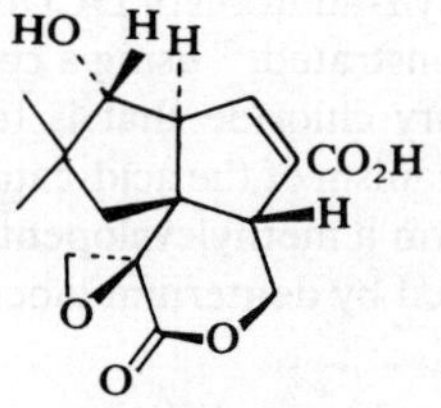

(17)

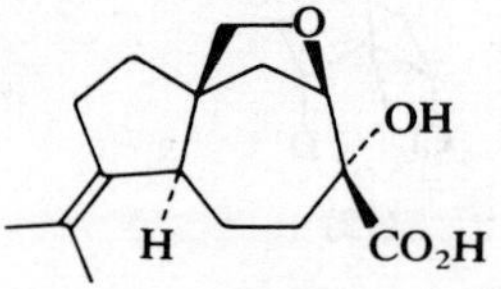

(18)

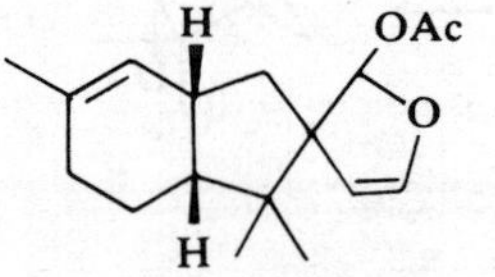

(19)

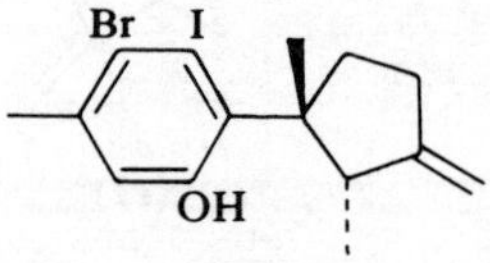

(20)

[28] 'Biology and Chemistry of the Compositae', ed. V. V. Heywood, J. B. Harborne, and B. L. Turner, Academic Press, London, 1977; for Part 231 of an extensive survey of the Compositae, see F. Bohlmann, C. Zdero, H. Robinson, and R. M. King, *Phytochemistry*, 1979, **18**, 1993, and refs therein.

[29] R. G. Kelsey and F. Shafizadeh, *Phytochemistry*, 1979, **18**, 1591.

[30] M. H. Fischer, E. J. Olivier, and H. D. Fischer, *Fortschr. Chem. org. Naturstoffe*, 1979, **38**, 47.

[31] W. Herz, R. de Groote, R. Murari, N. Kumar, and J. F. Blount, *J. Org. Chem.*, 1979, **44**, 2784.

[32] M. J. Bovill, M. H. P. Guy, G. A. Sim, D. N. J. White, and W. Herz, *J.C.S. Perkin II*, 1979, 53.

[33] H. Seto, T. Sasaki, J. Uzawa, S. Takeuchi, and H. Yonehara, *Tetrahedron Letters*, 1978, 923, 4411.

[34] Y. Tsuda, M. Kaneda, A. Tada, K. Nitta, Y. Yamamoto, and Y. Iitaka, *J.C.S. Chem. Comm.*, 1978, 160.

[35] R. Kazlauskas, P. T. Murphy, and R. J. Wells, *Tetrahedron Letters*, 1978, 4949; for a review, see R. J. Wells, *Pure Appl. Chem.*, 1979, **51**, 1829.

[36] R. R. Izac and J. J. Sims, *J. Amer. Chem. Soc.*, 1979, **101**, 6136

[37] T. Yoshida, S. Muraki, K. Takahashi, T. Kato, C. Kabuto, T. Suzuki, T. Uyehara, and T. Ohnuma, *J.C.S. Chem. Comm.*, 1979, 512.

[38] A. Matsuo, K. Atsumi, M. Nakayama, S. Hayashi, and K. Kuriyama, *J.C.S. Chem. Comm.*, 1979, 1010; see also Y. Asakawa, M. Toyota, T. Takemoto, and C. Suire, *Phytochemistry*, 1979, **18**, 1355.

[39] A. Matsuo, H. Nozaki, M. Nakayama, Y. Kushi, S. Hayashi, T. Komori, and N. Kamijo, *J.C.S. Chem. Comm.*, 1979, 174.

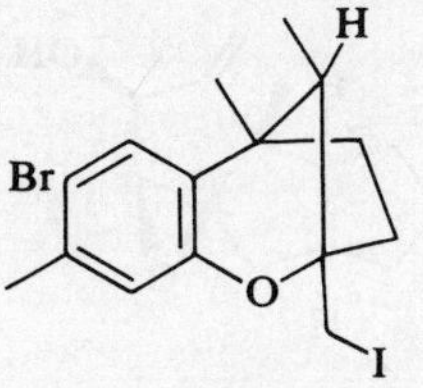

(21)

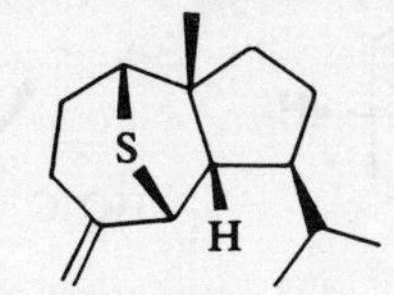

(22)

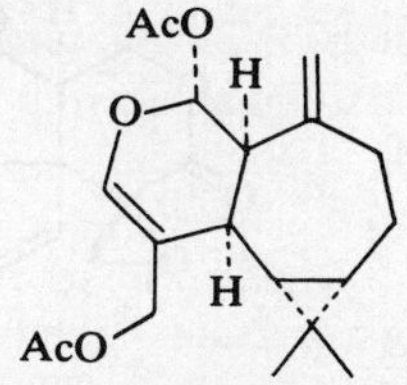

(23)

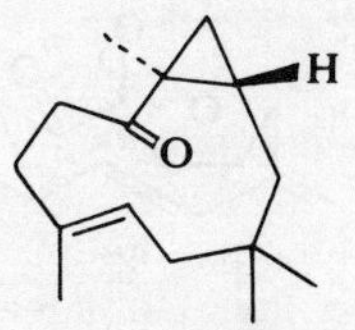

(24)

The biological activity of various sesquiterpenoids has provided the stimulus for many syntheses. A number of new strategies have been developed[40] for the construction of the α-methylene lactone moiety found in many of the cytotoxic substances. Further syntheses of the tumour-inhibitory bis-methylene lactone vernolepin (25) have been reported.[41] The unsaturated ketone (26) has formed a key

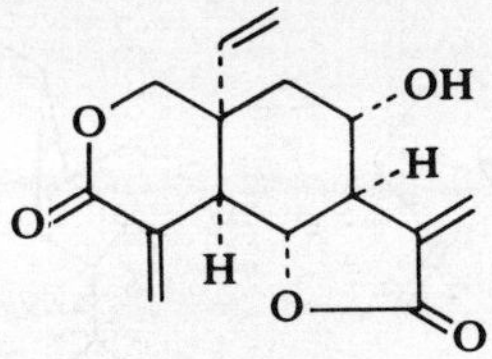

(25)

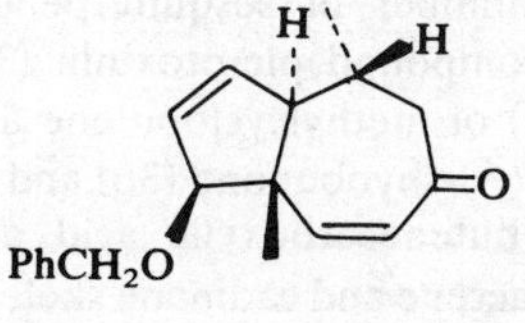

(26)

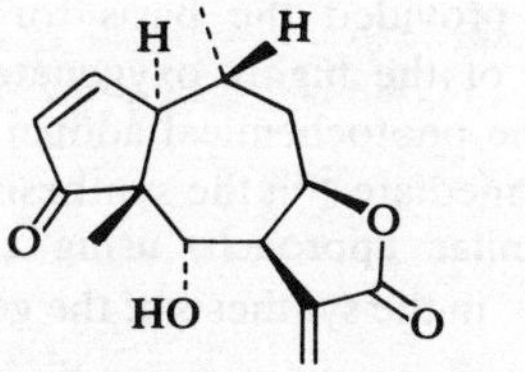

(27)

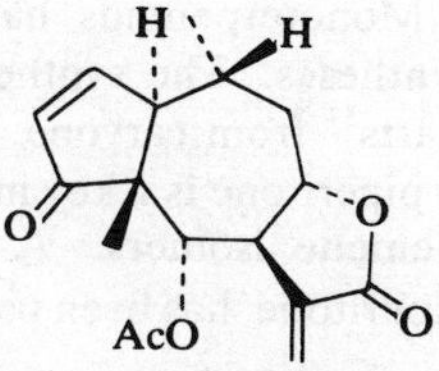

(28)

intermediate in the synthesis of both the potent cytotoxic substance helenalin (27)[42] and of the pseudoguaianolide bigelovin (28).[43] The sex-attractant pheromone of the American cockroach (*Periplaneta americana*), *i.e.* periplanone B (29), hitherto available in microgram quantities, has been synthesized[44] and its absolute stereochemistry determined[45] by the exciton chirality method. The variety of ring systems found amongst the sesquiterpenoids have also provided a number of synthetic

[40] R. M. Adlington and A. G. M. Barrett, *J.C.S. Chem. Comm.*, 1978, 1071.

[41] M. Iio, M. Isobe, and T. Goto, *J. Amer. Chem. Soc.*, 1979, **101**, 6076; F. Zutterman, H. de Wilde, R. Mijnheer, P. de Clerq, and M. Vandewalle, *Tetrahedron*, 1979, **35**, 2389; *cf.* P. A. Grieco, M. Nishizawa, T. Oguri, S. D. Burke, and N. Marinovic, *J. Amer. Chem. Soc.*, 1977, **99**, 5773; S. Danishefsky, P. F. Schuda, T. Kitahara, and S. J. Etheredge, *ibid.*, p. 6066.

[42] Y. Ohfune, P. A. Grieco, C.-L. J. Wang, and G. Majetich, *J. Amer. Chem. Soc.*, 1978, **100**, 5946.

[43] P. A. Grieco, Y. Ohfune, and G. Majetich, *J. Org. Chem.*, 1979, **44**, 3092.

[44] W. C. Still, *J. Amer. Chem. Soc.*, 1979, **101**, 2493.

[45] M. A. Adams, K. Nakanishi, W. C. Still, E. V. Arnold, J. Clardy, and C. J. Persoons, *J. Amer. Chem. Soc.*, 1979, **101**, 2495.

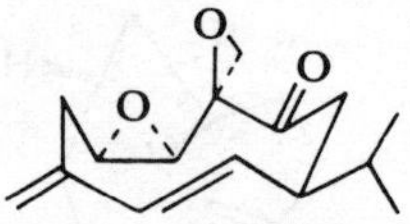

(29)

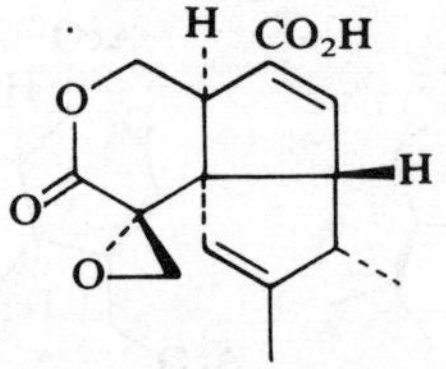

(30)

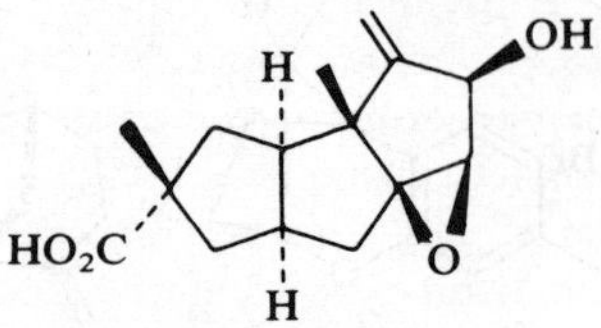

(31)

challenges. Thus the tumour-inhibitory antibiotic pentalenolactone (30),[46] which possesses an unusual pentalene ring system and a spiro-epoxymethylene lactone, and another antibiotic, hirsutic acid (31),[47] have both been the subject of stereocontrolled syntheses. The parent hydrocarbon, hirsutene, has also been synthesized.[48] Amongst the bridged structures which have attracted interest are the tricyclic compound gymnomitrol (32)[49] and the isocyanopupukeananes, *e.g.* (33).[50]

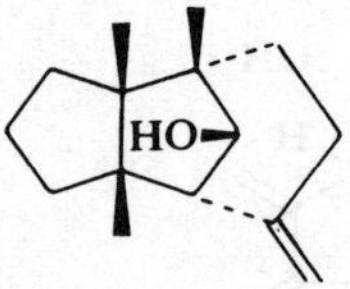

(32)

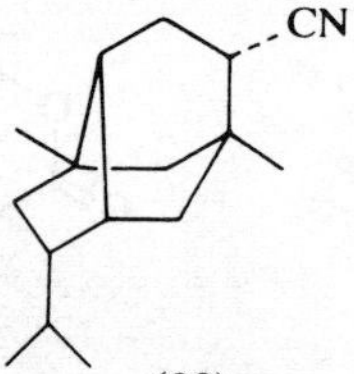

(33)

Monoterpenoids have provided the basis for a number of sesquiterpenoid syntheses. The synthesis of the highly oxygenated compound picrotoxinin (34) starts[51] from carvone. The photochemical adduct (35) of methylcyclobutene and L-piperitone is a key intermediate[52] in the synthesis of (−)-shyobunone (36) and its elemene isomers. A similar approach, using cyclobutenecarboxylic acid and piperitone, has been used[53] in the synthesis of the germacrene and cadinane skeleta.

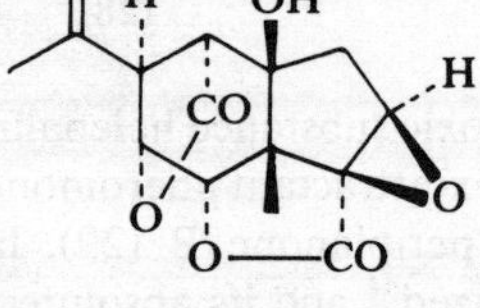

(34)

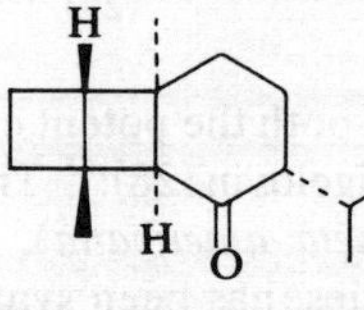

(35)

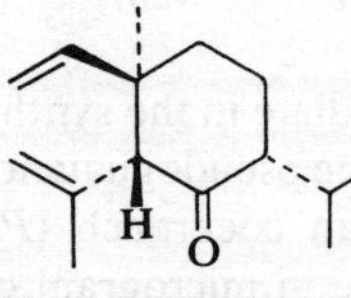

(36)

[46] S. Danishefsky, M. Hirama, K. Gombatz, T. Harayama, E. Berman, and P. F. Schuda, *J. Amer. Chem. Soc.*, 1978, **100**, 6536; **101**, 7020.

[47] B. M. Trost, C. D. Shuey, F. DiNinno, jr., and S. S. McElvain, *J. Amer. Chem. Soc.*, 1979, **101**, 1284.

[48] K. Hayano, Y. Ohfune, H. Shirahama, and T. Matsumoto, *Tetrahedron Letters*, 1978, 1991; K. Tatsuta, K. Akimoto, and M. Kinoshita, *J. Amer. Chem. Soc.*, 1979, **101**, 6116.

[49] R. M. Coates, S. K. Shah, and R. W. Mason, *J. Amer. Chem. Soc.*, 1979, **101**, 6765; G. Büchi and P.-S. Chu, *ibid.*, p. 6767; S. C. Welch and S. Chayabunjonglerd, *ibid.*, p. 6768.

[50] E. J. Corey and M. Ishiguro, *Tetrahedron Letters*, 1979, 2745; E. J. Corey, M. Behforouz, and M. Ishiguro, *J. Amer. Chem. Soc.*, 1979, **101**, 1608; H. Yamamoto and H. L. Shan, *ibid.*, p. 1609.

[51] E. J. Corey and H. L. Pearce, *J. Amer. Chem. Soc.*, 1979, **101**, 5841.

[52] J. R. Williams and J. F. Callahan, *J.C.S. Chem. Comm.*, 1979, 404.

[53] G. L. Lange and F. C. McCarthy, *Tetrahedron Letters*, 1978, 4749.

A number of biomimetic studies have focused on humulene (37) as a possible precursor of polycyclic sesquiterpenoids. The relationship between the conformation of humulene and its cyclization products has been examined.[54] The conversion of humulene into the proto-illudane skeleton has been reported.[55] Epoxides have been considered as possible initiators of cyclization. The biomimetic acid-catalysed cyclization of humulene 4,5-epoxide to compounds of the africanol type (38) and of epoxygermacrene-D (39) to compounds of the selinane (40) and oppositol (41) skeleta has been examined[56] in this context.

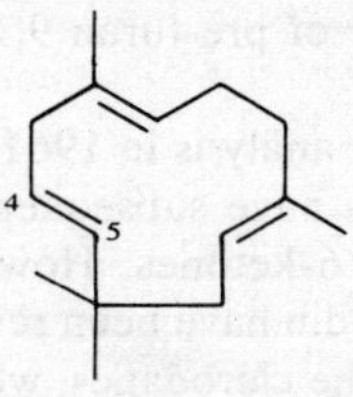

(37)

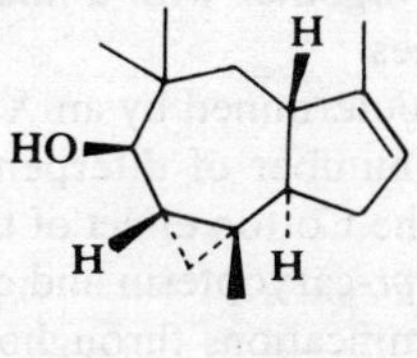

(38)

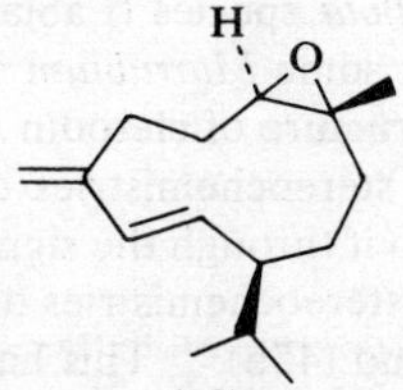

(39)

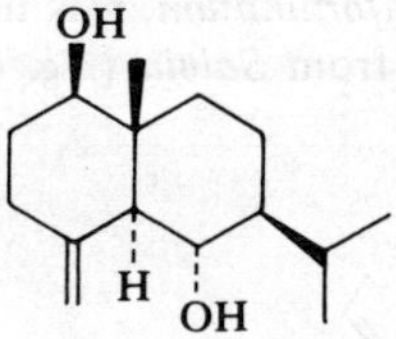

(40)

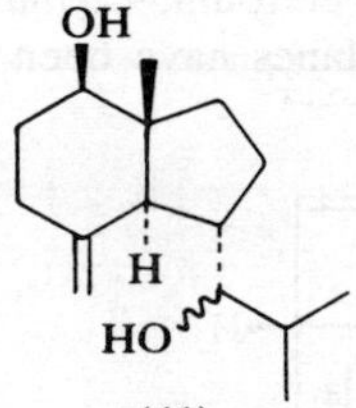

(41)

4 Diterpenoids

The diversity of diterpenoid skeleta now parallels that of the sesquiterpenoids. An interesting cyclic ether, zoapatanol (42), has been obtained[57] from a Mexican plant, *Montanoa tomentosa*, which is used as an abortifacient. Phytochemical surveys of both the Compositae and the Labiatae have also yielded many new diterpenoids. Thus a series of compounds, *e.g.* (43),[58] related to marrubiin have been obtained

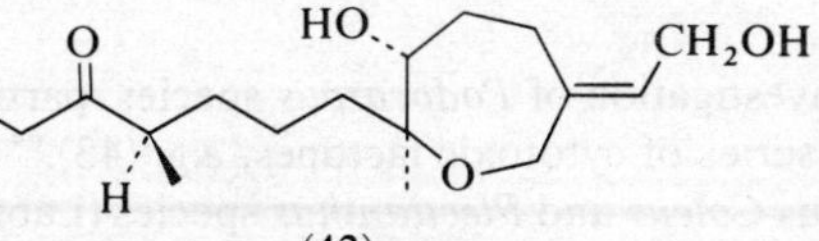

(42)

[54] H. Shirahama, E. Osawa, and T. Matsumoto, *Tetrahedron Letters*, 1978, 1987.

[55] S. Misumi, T. Ohtsuka, Y. Ohfune, K. Sugita, H. Shirahama, and T. Matsumoto, *Tetrahedron Letters*, 1979, 31; S. Misumi, T. Ohtsuka, H. Hashimoto, Y. Ohfune, H. Shirahama, and T. Matsumoto, *ibid.*, p. 35; also see W. Renold, G. Ohloff, and T. Norin, *Helv. Chim. Acta*, 1979, **62**, 985.

[56] M. Niwa, M. Iguchi, and S. Yamamura, *Tetrahedron Letters*, 1978, 4043; *ibid.*, 1979, 4291; J. A. Mlotkiewicz, J. Murray-Rust, P. Murray-Rust, W. Parker, F. G. Riddell, J. S. Roberts, and A. Sattar, *ibid.*, 1979, 3887.

[57] S. D. Levine, R. E. Adams, R. Chen. M. L. Cotter, A. F. Hirsch, V. V. Kane, R. M. Kanojia, C. Shaw, M. P. Wachter, E. Chin, R. Huettemann, P. Ostrowski, J. L. Mateos, L. Noriega, A. Guzman, A. Mijarez, and L. Tovar, *J. Amer. Chem. Soc.*, 1979, **101**, 3404.

[58] G. Savona, F. Piozzi, J. R. Hanson, and M. Siverns, *J.C.S. Perkin I*, 1978, 1271, and refs therein.

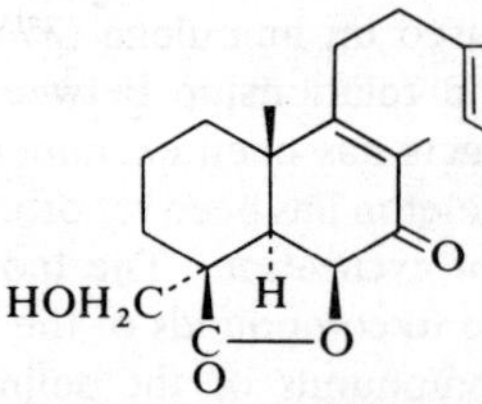

(43)

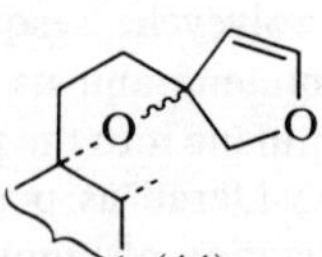

(44)

from *Ballota* species (Labiatae) together with a number of pre-furan 9,13-ethers (44)[59] for some *Marrubium* species.

The structure of clerodin was determined by an *X*-ray analysis in 1961, and the absolute stereochemistries of a number of diterpenoids have subsequently been related to it through the sign of the Cotton effect of their 6-ketones. However, the absolute stereochemistries of 3-*epi*-caryopterin and clerodin have been reversed to (45a)[60] and (45b).[61] This has ramifications throughout the clerodanes, which, like many diterpenoids, have members in both the '*ent*' and 'normal' series. The original source of clerodin, *Clerodendron infortunatum*, was indeed aptly named. A range of clerodanes have been isolated from *Salvia* [*e.g.* (46)] and *Teucrium* species [*e.g.* (47)].[62,63]

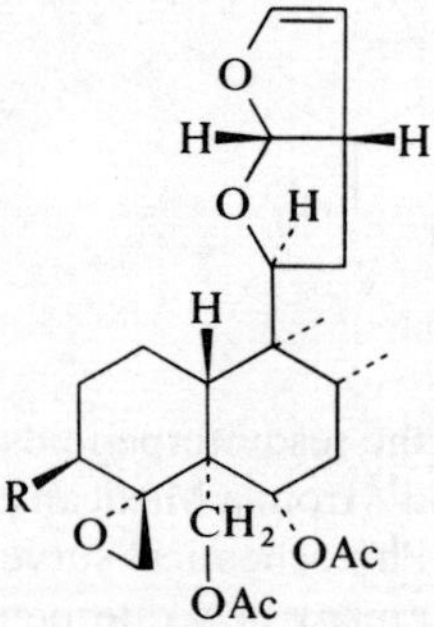

(45) a; R = OAc
b; R = H

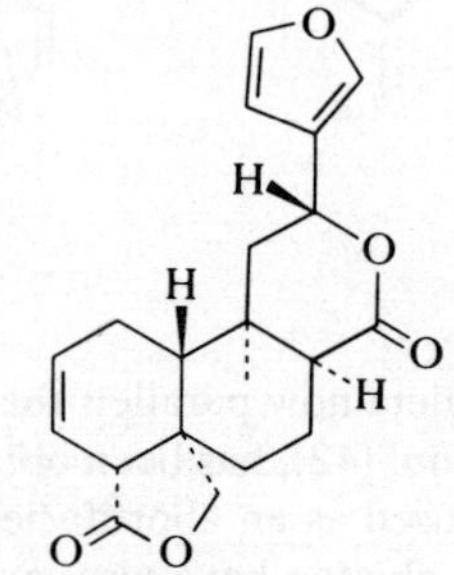

(46)

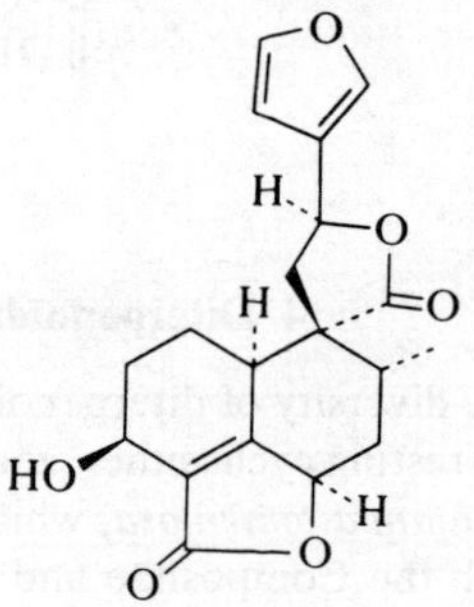

(47)

An extensive investigation of *Podocarpus* species, particularly *P. nagi*, has led to the isolation of a series of cytotoxic lactones, *e.g.* (43).[64] The leaf-gland colouring materials of various *Coleus* and *Plectranthus* species (Labiatae) contain a number of highly oxygenated diterpenoids known as the coleons. Recent examples include

[59] G. Savona, F. Piozzi, L. M. Aranguez, and B. Rodriguez, *Phytochemistry*, 1979, **18**, 859.
[60] N. Harada and H. Uda, *J. Amer. Chem. Soc.*, 1978, **100**, 8022.
[61] D. Rogers, G. G. Unal, D. J. Williams, S. V. Ley, G. A. Sim, B. S. Joshi, and K. R. Ravindranath, *J.C.S. Chem. Comm.*, 1979, 97.
[62] G. Savona, M. P. Paternostro, F. Piozzi, J. R. Hanson, P. B. Hitchcock, and S. A. Thomas, *J.C.S. Perkin I*, 1978, 643; *ibid.*, 1979, 1915; G. Savona, M. P. Paternostro, F. Piozzi, and J. R. Hanson, *ibid.*, 1979, 533.
[63] E. Gacs-Baitz, L. Radics, G. B. Oganessian, and V. A. Mnatsakanian, *Phytochemistry*, 1978, **17**, 1967.
[64] Y. Hayashi, T. Matsumoto, and T. Sakan, *Heterocycles*, 1978, **10**, 123 (and refs. therein); see also Y. Hayashi, T. Matsumoto, T. Hyono, N. Nishikawa, M. Uemura, M. Nishizawa, M. Togami, and T. Sakan, *Tetrahedron Letters*, 1979, 3311.

coleon U and 3β-acetoxyfuerstion (49).[65] *Solenostemon* species (Labiatae) are the source[66] of some unusual rearranged diterpenoids in which one of the substituents at C-4 has migrated to C-3. The compounds, *e.g.* (50), also contain a spirocyclopropane ring, as found in barbatusin.

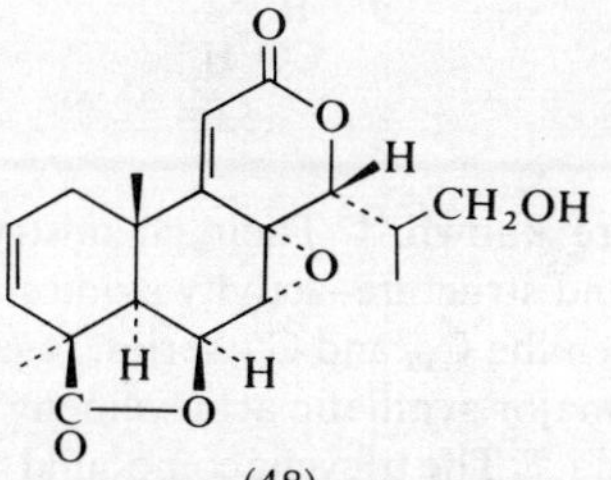

(48)

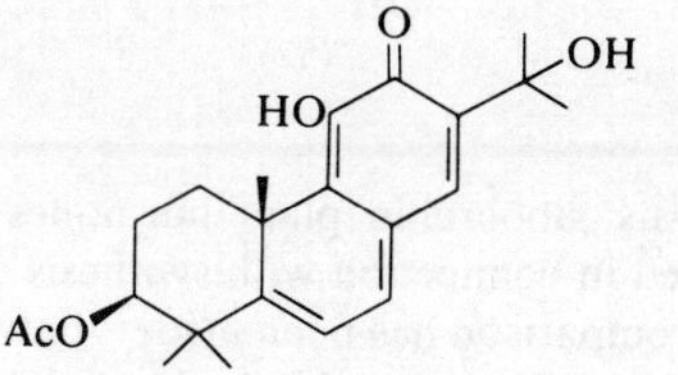

(49)

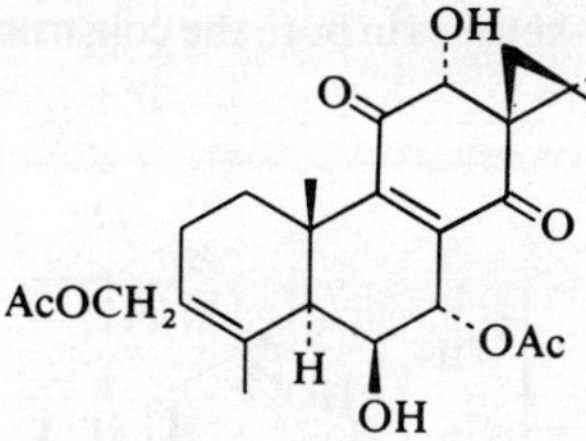

(50)

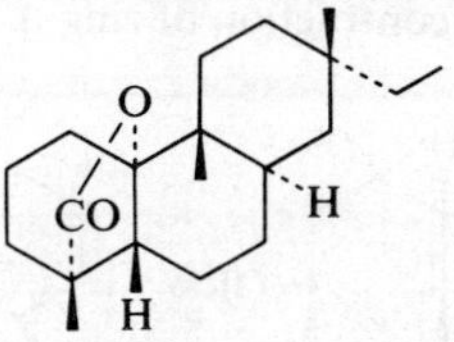

(51)

After some controversy, the stereochemistry of the γ-lactone (51), obtained by treatment of dihydroisopimaric acid with acid, has been established[67] by an *X*-ray analysis. A variety of synthetic strategies have been developed, directed at the tricyclic diterpenoids. One route[68] involves the Wittig reaction between α-cyclocitral (ring A) and various benzylphosphonium salts (ring C) followed by partial reduction and cyclization. This has afforded ferruginol, totarol, and some more highly oxidized relatives, such as the royleanones.

A large number of polyhydroxylated *ent*-kaur-15- and -16-enes bearing various oxygen functions at C-3, -7, -15, -18, and -19 have been isolated from *Sideritis* and *Isodon* species (Labiatae). Some of the latter, *e.g.* (52),[69] show tumour-inhibitory properties. Enmein, obtained from *Isodon* species, has formed the starting material for a number of diterpenoid partial syntheses.[70] The unusual skeleton (53), reminiscent of aphidicolin, has been assigned to helifulvanolic acid.[71]

[65] P. Ruedi and C. H. Eugster, *Helv. Chim. Acta*, 1978, **61**, 709; G. Buchbauer, P. Ruedi, and C. H. Eugster, *ibid.*, p. 1969.

[66] T. Miyase, P. Ruedi, and C. H. Eugster, *J.C.S. Chem. Comm.*, 1977, 859.

[67] W. Herz and J. F. Blount, *J. Org. Chem.*, 1979, **44**, 1172.

[68] T. Matsumoto and S. Usui, *Bull. Chem. Soc. Japan*, 1979, **52**, 213; T. Matsumoto and A. Suetsugu, *ibid.*, p. 1450; T. Matsumoto and S. Harada, *ibid.*, p. 1459; T. Matsumoto and S. Takeda, *Chem. Letters*, 1979, 409.

[69] I. Kubo, M. J. Pettei, K. Hirotsu, H. Tsuji, and T. Kubota, *J. Amer. Chem. Soc.*, 1978, **100**, 628

[70] T. Fujita, I. Masuda, S. Takao, and E. Fujita, *J.C.S. Perkin I*, 1979, 915.

[71] F. Bohlmann, C. Zdero, R. Zeisberg, and W. S. Sheldrick, *Phytochemistry*, 1979, **18**, 1359.

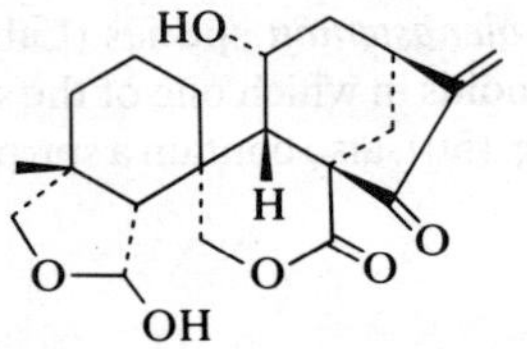
(52)

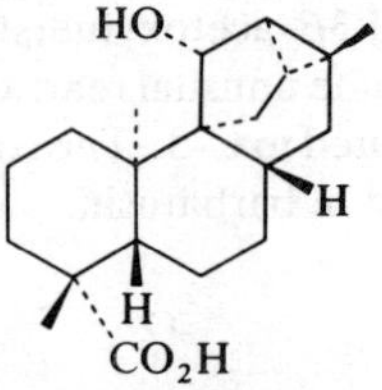
(53)

Fifty-six gibberellin plant hormones are known.[72] Their chemistry has been examined in connection with synthesis[73] and structure–activity studies,[74] whilst an *X*-ray comparison has been made[75] between the C_{19} and C_{20} series, revealing some interesting differences. Undoubtedly the major synthetic achievement in this area has been the total synthesis of gibberellic acid.[76] The tricyclic compound (54) formed a key intermediate, and this was converted into the homoannular diene (55) and thence into gibberellic acid (56). Another successful strategy[77] for gibberellin synthesis makes use of the reactions of diazo-ketones in both the construction of ring D and the contraction of ring B.

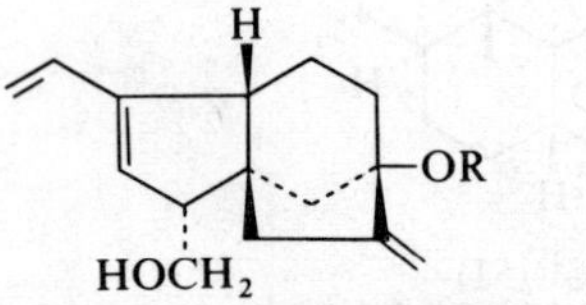
(54) $R = CH_2OCH_2CH_2OMe$

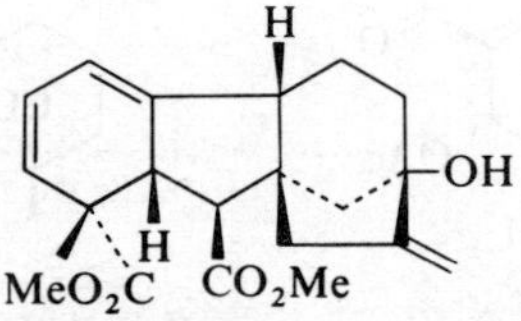
(55)

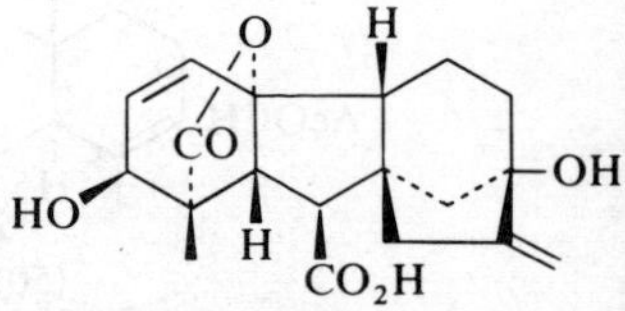
(56)

An increasing number of diterpenoids have been isolated with structures based on the macrocyclic cembrane skeleton.[78] These include many marine products isolated from various corals.[79] The *X*-ray analysis has been reported[80] of the tetra-acetate of ingol (57), a member of the lathyrane series, plausible cyclizations of which may lead to the tigliane, daphnane, and ingenane series of tumour-promoting esters of the Euphorbiaceae.[81]

[72] For recent reviews see J. E. Graebe and H. J. Ropers, in 'Phytohormones and Related Compounds', ed. D. S. Lethan, P. B. Goodwin, and J. J. V. Higgins, Elsevier, Amsterdam, 1978, Vol. 1, p. 107; J. MacMillan, *Pure Appl. Chem.*, 1978, **50**, 995; for gibberellins A_{54-56} see N. Murofushi, M. Sugimoto, K. Itoh, and N. Takahashi, *Agric. Biol. Chem.* (*Japan*), 1979, **43**, 2179.

[73] G. Stork and J. Singh, *J. Amer. Chem. Soc.*, 1979, **101**, 7109.

[74] B. E. Cross and A. Erasmuson, *J.C.S. Chem. Comm.*, 1978, 1013; K. Boulton and B. E. Cross, *J.C.S. Perkin I*, 1979, 1354.

[75] G. Ellames, J. R. Hanson, P. B. Hitchcock, and S. A. Thomas, *J.C.S. Perkin I*, 1979, 1922.

[76] E. J. Corey, R. L. Danheiser, S. Chandrasekaran, C. E. Keck, B. Gopalan, S. D. Larsen, P. Siret, and J. L. Gras, *J. Amer. Chem. Soc.*, 1978, **100**, 8034; E. J. Corey and J. Gorzynski-Smith, *ibid.*, 1979, **101**, 1038.

[77] L. N. Mander and S. G. Pyne, *J. Amer. Chem. Soc.*, 1979, **101**, 3373.

[78] For a review see A. J. Weinheimer, C. W. J. Chang, and J. A. Matson, *Fortschr. Chem. org. Naturstoffe*, 1979, **36**, 285.

[79] See, for example, B. N. Ravi and D. J. Faulkner, *J. Org. Chem.*, 1978, **43**, 2127; B. F. Bowden, J. C. Coll, S. J. Mitchell, and G. J. Stokie, *Austral. J. Chem.*, 1979, **32**, 653.

[80] H. Lotter, H. J. Opferkuch, and E. Hecker, *Tetrahedron Letters*, 1979, 77.

[81] For a review, see F. J. Evans and C. J. Soper, *Lloydia*, 1978, **41**, 193; E. Hecker, *Pure. Appl. Chem.*, 1977, **49**, 1423.

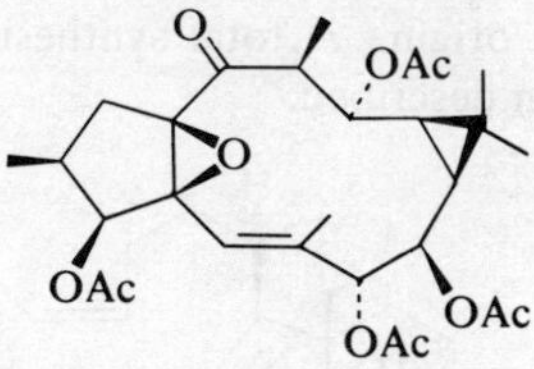

(57)

A number of diterpenoids have been obtained with structures which resemble prenylated sesquiterpenes. These include dictyodial (58), which has been isolated[82] from various marine *Dictyota* (algae) and *Aplysia* (sea hare) species. Amongst unusual diterpenoids are the trinervitene alcohols, *e.g.* (59),[83] obtained from termites, the fungal cyathins, *e.g.* (60),[84] and the verrucosanes, *e.g.* (61),[85] obtained from the liverwort *Mylia verrucosa*.

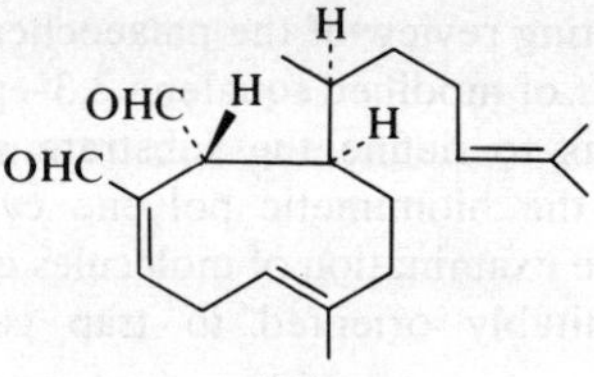

(58)

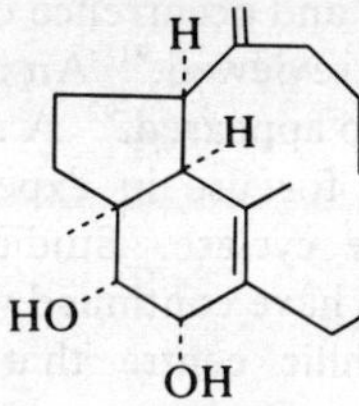

(59)

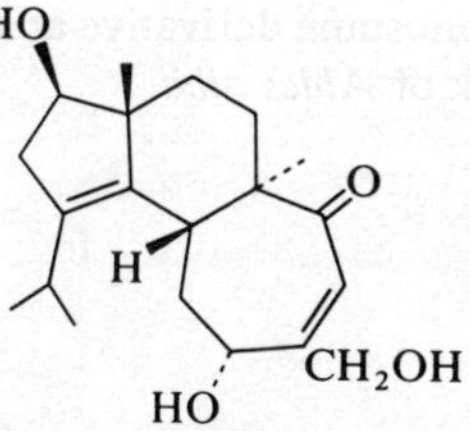

(60)

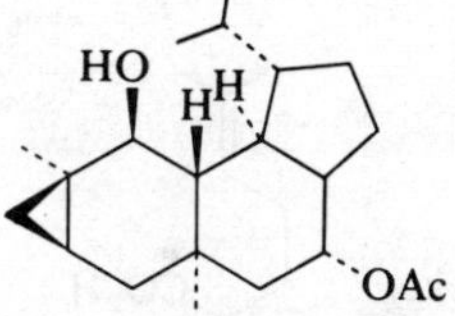

(61)

5 Sesterterpenoids

The number of known sesterterpenoids has continued to increase, particularly in isolates from marine organisms,[86] from fungi,[87] and from insects.[88] However, not all C_{25} fungal metabolites are sesterterpenes. Thus the andibenins[89] appear to be of

[82] J. Finar, J. Clardy, W. Fenical, L. Minale, R. Riccio, M. Kirkup, and R. E. Moore, *J. Org. Chem.*, 1979, **44**, 2044.
[83] J. Vroc, M. Budesinsky, and P. Sedmera, *Coll. Czech. Chem. Comm.*, 1978, **43**, 1125, 2478.
[84] W. A. Ayer, T. Yoshida, and D. M. J. van Schie, *Canad. J. Chem.*, 1978, **56**, 2113.
[85] S. Hayashi, A. Matsuo, H. Nozaki, M. Nakayama, D. Takaoka, and M. Hiroi, *Chem. Letters*, 1978, 953.
[86] See, for example, G. Cimino, S. de Stefano, L. Minale, R. Riccio, K. Hirtosu, and J. Clardy, *Tetrahedron Letters*, 1979, 3619.
[87] C. Rossi and L. Tuttobello, *Tetrahedron Letters*, 1978, 307.
[88] F. Miyamoto, H. Naoki, T. Takemoto, and Y. Naya, *Tetrahedron*, 1979, **35**, 1913.
[89] A. W. Dunn, R. A. W. Johnstone, T. J. King, L. Lessinger, and B. Sklarz, *J.C.S. Perkin I*, 1979, 2113; T. J. Simpson, *ibid.*, p. 2118.

mixed polyketide–terpenoid origin. A total synthesis and the *X*-ray analysis of gascardic acid (62) have been described.[90]

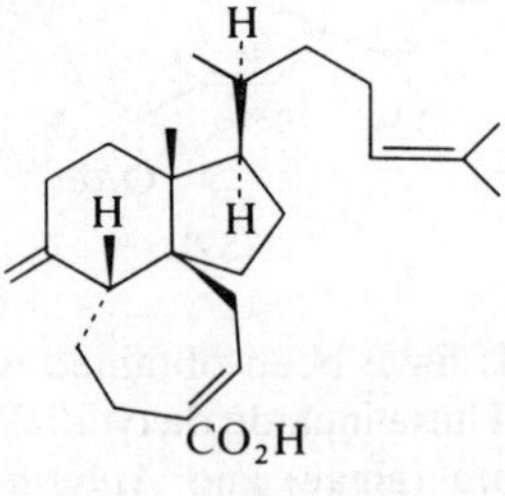

(62)

6 Triterpenoids

The identification and occurrence of triterpenoids, and particularly of the friedelin group, have been reviewed.[91] An interesting review of the palaeochemistry of the hopanoids has also appeared.[92] A number of modified squalene 2,3-epoxides have been prepared,[93] for use in experiments to define the substrate specificity of 2,3-oxidosqualene cyclase. Studies on the biomimetic polyene cyclizations of squalene relatives have continued with the examination of molecules containing an internal nucleophilic centre that is suitably oriented to trap carbo-cationic intermediates.[94]

A number of new fungal metabolites, *e.g.* (63), related to fusidic acid have been obtained[95] from *Fusidium coccineum*. The lanostane derivative abietospiran (64) is present[96] in quite large quantities in the bark of *Abies alba*.

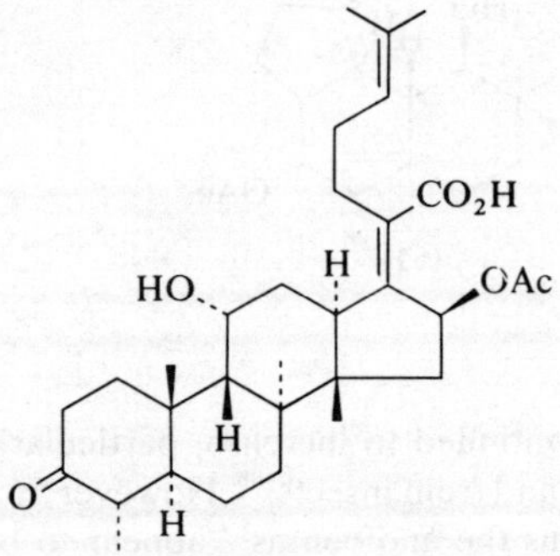

(63)

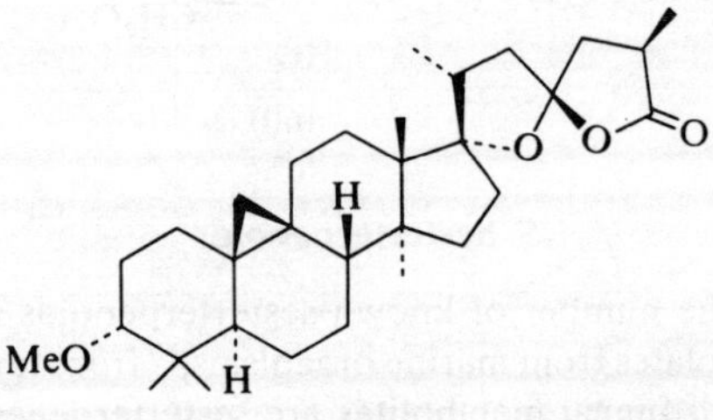

(64)

[90] R. K. Boeckman, D. M. Blum, and E. V. Arthur, *J. Amer. Chem. Soc.*, 1979, **101**, 5060; for *X*-ray analysis, see: R. K. Boeckman, D. M. Blum, E. V. Arthur, and J. Clardy, *Tetrahedron Letters*, 1979, 4609.
[91] P. Pant and R. P. Rastogi, *Phytochemistry*, 1979, **18**, 1095; R. F. Chandler and S. N. Hooper, *ibid.*, p. 711.
[92] G. Ourisson, P. Albrecht, and M. Rohmer, *Pure Appl. Chem.*, 1979, **51**, 709.
[93] E. E. van Tamelen, A. D. Pedlar, E. Li, and D. R. James, *J. Amer. Chem. Soc.*, 1977, **99**, 6678; M. Herin, P. Sandra, and A. Krief, *Tetrahedron Letters*, 1979, 3103.
[94] M. E. Garst, Y.-F. Cheung, and W. S. Johnson, *J. Amer. Chem. Soc.*, 1979, **101**, 1279, 4404.
[95] W. O. Godtfredsen, N. Rastrup-Andersen, S. Vangedal, and W. D. Ollis, *Tetrahedron*, 1979, **35**, 2419.
[96] W. Steglich, M. Klaar, L. Zechlin, and H. J. Hecht, *Angew. Chem. Internat. Edn.*, 1979, **18**, 698.

A short synthesis of cycloartenol from 11β-oxygenated lanostanes by photolysis of nitrite esters has been described.[97] The rearrangements of lanostane 9,11-epoxides have been examined, with a view to inducing the rearrangement of the methyl group at C-10 to C-9 and the formation of the cucurbitane skeleton.[98]

The tumour-inhibitory properties of the cucurbitacins have attracted some attention. Isocucurbitacin D (65) and its 3-epimer are new tumour-inhibitory compounds which have been isolated from the leaves of *Phormium tenax*.[99] A number of the limonoids also show tumour-inhibitory properties. Aphnastatin (66), from *Aphnamixis grandifolia*, was isolated[100] in this context. Some highly oxygenated limonoids, *e.g.* polystachin (67), have been obtained[101] from *A. polystacha*. The insect antifeedant activity of the tetranortriterpenoids has also attracted attention.[102] The quassinoids are another group of biologically active nortriterpenoids. The compound (68), isolated[103] from *Simarouba amara*, has antineoplastic properties. An interesting variant amongst the nortriterpenoids is the compound clausenolide (69), whose structure was determined by an *X*-ray analysis.[104]

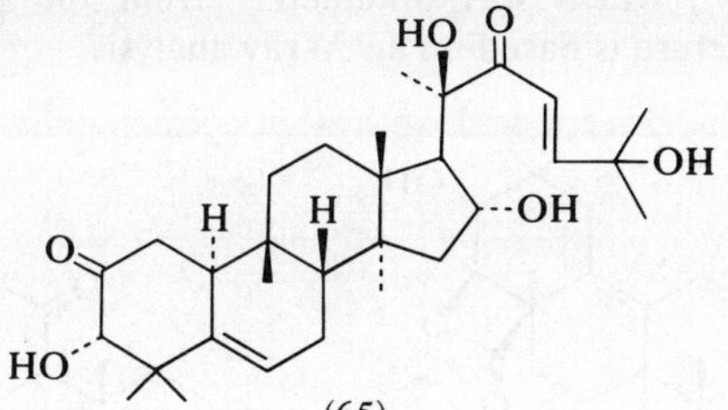

(65)

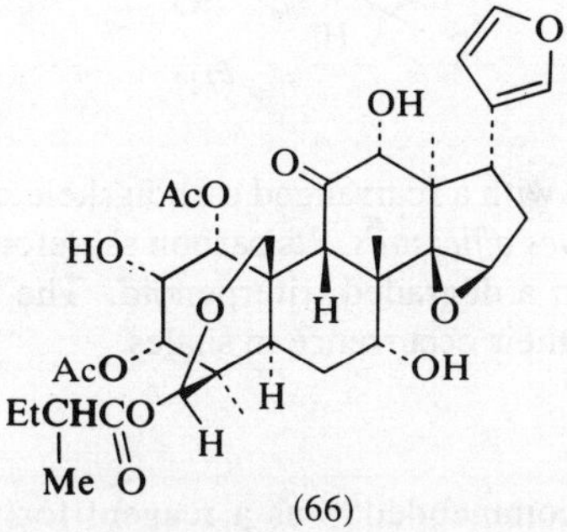

(66)

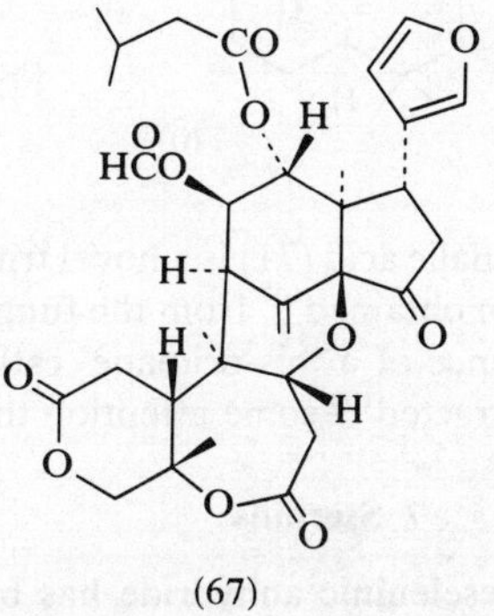

(67)

[97] R. B. Boar and D. P. Copsey, *J.C.S. Perkin I*, 1979, 563.

[98] G. V. Baddeley, H. J. Samaan, J. J. H. Simes, and T. H. Ai, *J.C.S. Perkin I*, 1979, 7; Z. Paryzek, *ibid.*, p. 1222.

[99] S. M. Kupchan, H. Meshulam, and A. T. Sneden, *Phytochemistry*, 1978, **17**, 767.

[100] J. Polonsky, Z. Varon, B. Arnoux, C. Pascard, G. R. Pettit, J. H. Schmidt, and L. M. Lange, *J. Amer. Chem. Soc.*, 1978, **100**, 2575.

[101] D. A. Mulholland and D. A. H. Taylor, *J. Chem. Res.*, 1979, (*S*)294, (*M*)3101; J. D. Connolly, C. Labbe, D. S. Rycroft, D. A. Okorie, and D. A. H. Taylor, *J. Chem. Res.*, 1979, (*S*)256, (*M*)2858.

[102] See, for example, W. Kraus, W. Grimminger, and G. Sawitski, *Angew. Chem. Internat. Edn.*, 1978, **17**, 476.

[103] J. Polonsky, Z. Varon, H. Jacquemin, and G. R. Pettit, *Experientia*, 1978, **34**, 1122; *cf.* M. E. Wall and M. C. Wani, *J. Medicin. Chem.*, 1978, **21**, 1186.

[104] D. P. Chakraborty, P. Bhattacharyya, S. P. Bhattacharyya, J. Bordner, G. L. A. Hennessee, and B. Weinstein, *J.C.S. Chem. Comm.*, 1979, 246.

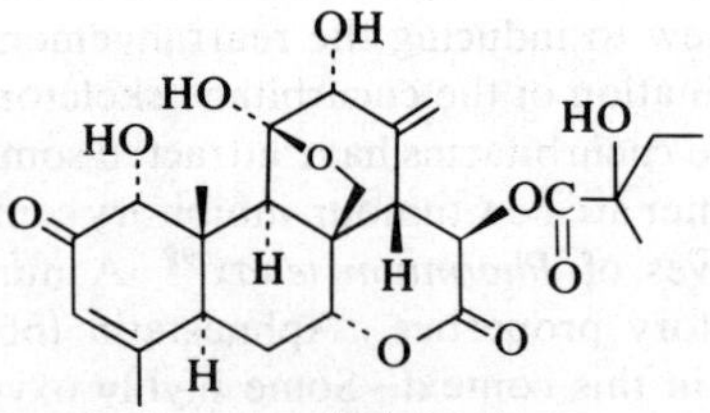
(68)

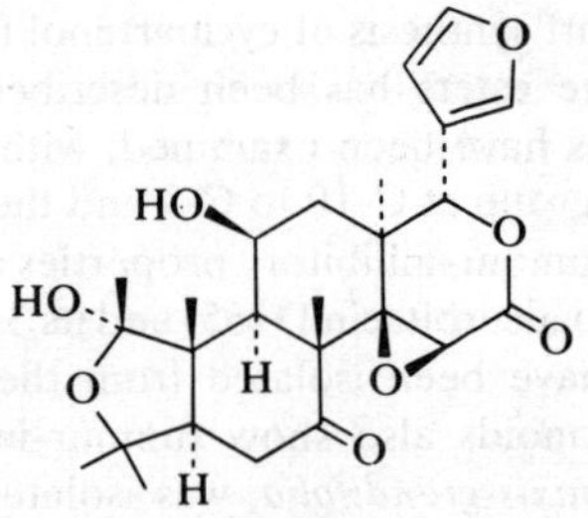
(69)

A simple stereoselective route for the preparation of pentacyclic intermediates used in triterpene synthesis has been described.[105] It is based, as are a number of steroid syntheses, on the cycloaddition reaction of *o*-quinodimethanes derived by the thermolysis of benzocyclobutanes. An unusual oleanane is the β-lactone papyriogenin G (70), which was obtained[106] from the leaves of *Tetrapanax papyriferum*. Its structure is based on an *X*-ray analysis.

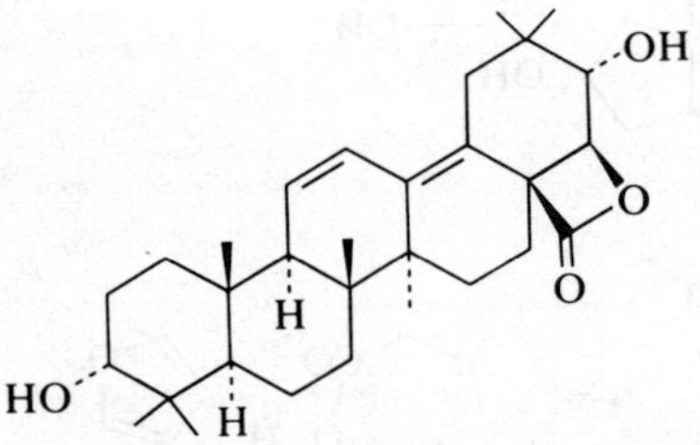
(70)

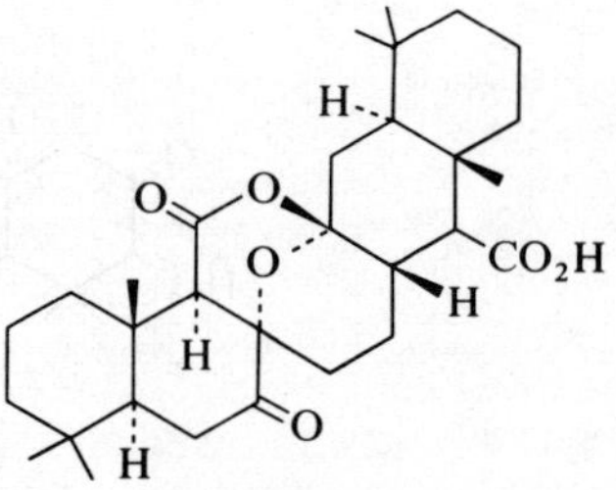
(71)

Officinalic acid (71) is a novel triterpene with a rearranged carbon skeleton which has been obtained[107] from the fungus *Fomes officinalis*. Its carbon skeleton has the appearance of a 'bis-drimane' rather than a degraded triterpenoid. The hopanes have attracted[108] some attention through their occurrence in shales.

7 Steroids

Benzeneseleninic anhydride has been recommended[109] as a reagent for the preparation of the biologically active steroidal 1,4-dien-3-ones. The corresponding thiones have also been prepared.[110] The scope of diethylaminosulphurtrifluoride as a reagent for introducing fluorine into the steroid skeleton has been examined

[105] T. Kametani, Y. Hirai, Y. Shiratori, K. Fukumoto, and F. Satoh, *J. Amer. Chem. Soc.*, 1978, **100**, 554.
[106] Y. Ogihara, M. Asada, and Y. Iitaka, *J.C.S. Chem. Comm.*, 1978, 364; *cf.* S. Amagaya, T. Takeda, Y. Ogihara, and Y. Yamisaki, *J.C.S. Perkin I*, 1979, 2044.
[107] W. W. Epstein, F. W. Sweat, G. Vanhear, F. M. Lovell, and E. J. Gabe, *J. Amer. Chem. Soc.*, 1979, **101**, 2748.
[108] W. K. Siefert, J. M. Moldowan, G. W. Smith, and E. V. Whitehead, *Nature*, 1978, **271**, 436.
[109] D. H. R. Barton, A. G. Brewster, R. A. H. F. Hui, D. J. Lester, S. V. Ley, and T. G. Back, *J.C.S. Chem. Comm.*, 1978, 952.
[110] D. H. R. Barton, L. L. Choi, R. M. Hesse, M. M. Pechet, and C. Wilshire, *J.C.S. Perkin I*, 1979, 1166.

further.[111] Neighbouring-group participation has been noted[112] in the formation and hydrolysis of steroidal 3-hydroxy-1,2-epoxides. The stereochemistry of various addition reactions of steroidal Δ^2-enes has continued to attract attention.[113] The participation of a 19-oxygen function in the addition of hypobromous acid to a Δ^2-ene is manifest[114] in the formation of 19–2 ether and lactone rings. Some curious results have been noted[115] in the unimolecular solvolysis of 3-trifluoroacetoxy-cholest-4-enes which imply conformational differences between the carbo-cations arising from the 3-epimers. The stereochemistry of the formation of A-homo-steroids in the acid-catalysed cleavage of 2,3-dihalogenocyclopropyl-3-acetoxy-steroids has been examined,[116] as has the contraction of the ring during the cleavage of cyclo-steroids with hydrogen chloride gas.[117a] Another ring-contraction of ring A which possibly involves the formation of cyclo-steroid intermediates is the base-catalysed rearrangement of 5β,6β-epoxy-4-ketones (72).[117b] The stereochemistry of some reactions of ring A in the 9-methyl-5α/5β,9β,10α-oestran-3-one series (73) has been examined.[118]

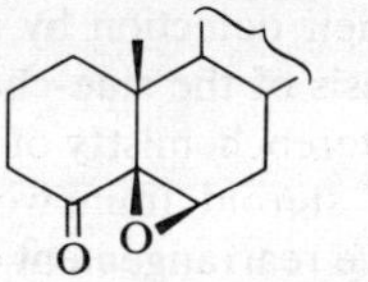
(72)

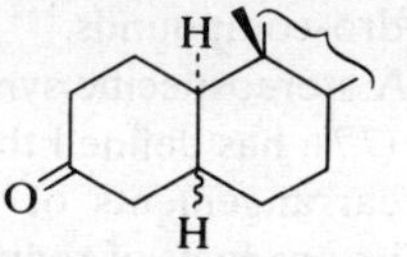

(73)

The equilibration of 6α- and 6β-nitrocholest-4-enes reveals[119] the influence of the homoallylic participation of the olefin. The product of solvolysis of 3β-tosyloxy-5β-hydroxycholestan-6-one (74) has been shown to be the A-homo-B-nor-keto-oxide (75), which is formed by participation of the carbonyl oxygen in the displacement reaction.[120]

The 4-phenyl-1,2,4-triazoline-3,5-dione adducts of steroidal 5,7-dienes have been used to protect this system in the partial synthesis of derivatives of vitamin D.

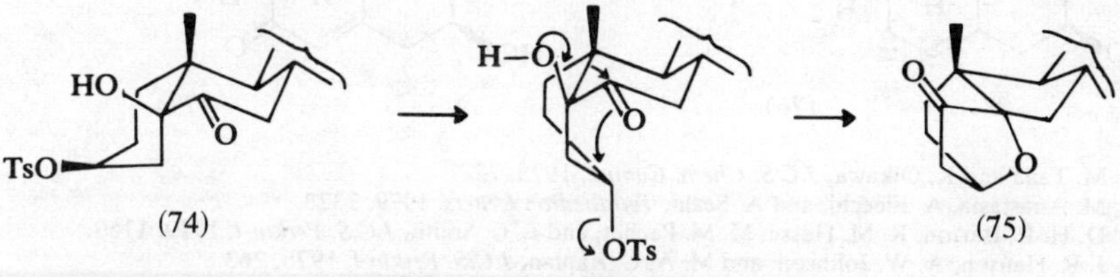

(74) (75)

[111] S. Rozen, Y. Faust, and H. Ben-Yakov, *Tetrahedron Letters*, 1979, 1823.
[112] E. Glotter and P. Krinsky, *J.C.S. Perkin I*, 1978, 408, 413; D. Baldwin, J. R. Hanson, and D. Raines, *ibid.*, 1979, 344.
[113] R. C. Cambie, H. H. Lee, P. S. Rutledge, and P. D. Woodgate, *J.C.S. Perkin I*, 1979, 757.
[114] P. Kocovsky and V. Czerny, *Coll. Czech. Chem. Comm.*, 1978, **43**, 327, 1924.
[115] G. Ortar, M. P. Paradisi, E. Morera, and A. Romeo, *J.C.S. Perkin I*, 1978, 5.
[116] P. Crabbé, J.-L. Luche, J.-C. Damiano, M.-J. Luche, and A. Cruz, *J. Org. Chem.*, 1979, **44**, 2929.
[117] E. J. Brunke, *Chem. Ber.*, 1979, **112**, 1606; J. R. Hanson, D. Raines, and H. Wadsworth, *J.C.S. Perkin I*, 1978, 743.
[118] J. C. A. Boeyens, J. R. Bull, J. Floor, and A. Tuinnan, *J.C.S. Perkin I*, 1978, 808.
[119] J. T. Pinhey and G. C. Smith, *Austral. J. Chem.*, 1978, **31**, 2563.
[120] V. Dave and E. W. Warnhoff, *J. Org. Chem.*, 1978, **43**, 4622; R. W. G. Foster and B. A. Marples, *Tetrahedron Letters*, 1979, 2071.

A simplified method of deprotection has been described.[121] The oxidation of the 5,7-dienes with potassium permanganate has been the subject of some controversy.[122] Further studies have been made on the mechanism of the Barton reaction in efforts[123] to identify the origin and hence suppress the formation of various by-products. The formation of Δ^5-7-ones in the reaction of 5β,6β-epoxy-7-hydroxy-androstanes with thionyl chloride containing sulphur dioxide has been observed.[124]

The formation of D-homo-C-nor-steroids has been reviewed.[125] Further studies of the acid-catalysed D-homoannulation of 17α-hydroxy-pregnan-20-ones have been published.[126] The reaction of thallium(I) acetate and iodine with enol-acetates has been examined[127] in the context of the formation of 16-iodo-17-ketones.

The majority of sterols, including the moulting hormones and metabolites of vitamin D, possess the *R* configuration at C-20. A number of synthetic efforts have been made[128,129] to generate this system. Marine organisms have yielded several sterols with unusual side-chains,[130] exemplified by xestosterol (76)[131] and various 23-methyl-22-dehydro-compounds.[132] Their detection by mass spectrometry has been reviewed.[133] A stereospecific synthesis of the side-chain of the *Achlya* sex-hormone, oogoniol (77), has defined the stereochemistry of this substance.[134]

The molecular rearrangements of the steroid framework have continued to attract attention. The products of reductive rearrangement of androst-4-ene-3,17-dione in HF–SbF_5 in the presence of a hydrogen donor are 6α- and 6β-methyl-14β-

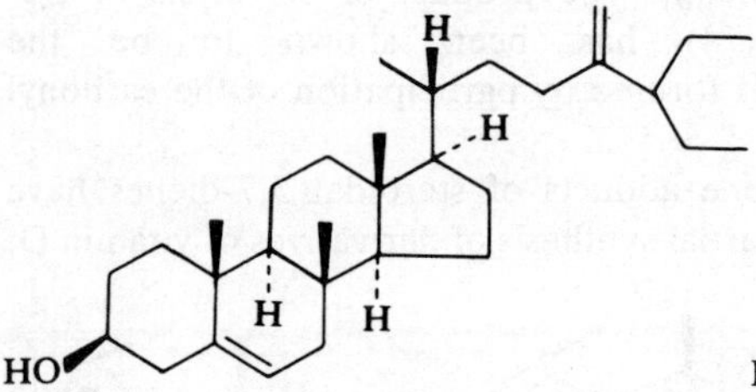

(76)

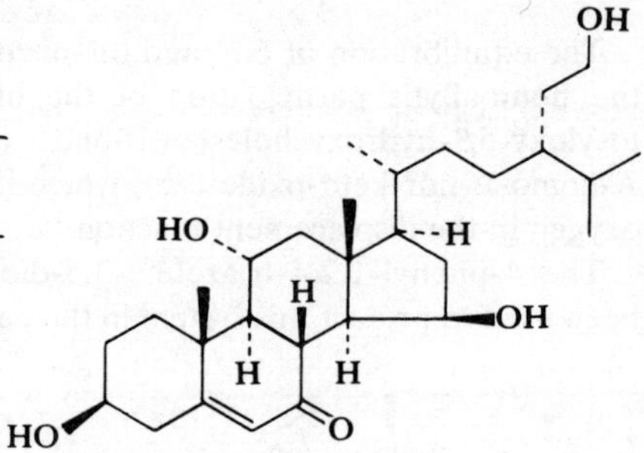

(77)

[121] M. Tada and K. Oikawa, *J.C.S. Chem. Comm.*, 1978, 727.
[122] M. Anastasia, A. Fiecchi, and A. Scala, *Tetrahedron Letters*, 1979, 3323.
[123] D. H. R. Barton, R. M. Hesse, M. M. Pechet, and L. C. Smith, *J.C.S. Perkin I*, 1979, 1159.
[124] J. R. Hanson, A. W. Johnson, and M. A. C. Kaplan, *J.C.S. Perkin I*, 1978, 263.
[125] E. Brown and M. Ragault, *Tetrahedron*, 1979, **35**, 911.
[126] D. N. Kirk and C. R. McHugh, *J.C.S. Perkin I*, 1978, 173.
[127] R. C. Cambie, R. C. Hayward, J. L. Jurlina, P. S. Rutledge, and P. D. Woodgate, *J.C.S. Perkin I*, 1978, 126.
[128] For a review, see D. M. Piatak and J. Wicha, *Chem. Rev.*, 1978, **78**, 199.
[129] B. M. Trost, P. R. Bernstein, and P. C. Funfschilling, *J. Amer. Chem. Soc.*, 1979, **101**, 4378; P. A. Grieco, T. Takigawa, and D. R. Moore, *ibid.*, p. 4380.
[130] W. Hofheinz and G. Oesterhelt, *Helv. Chim. Acta*, 1979, **62**, 1307; W. C. Kokke, M. C. Pak, W. Fenical, and C. Djerassi, *ibid.*, p. 1310.
[131] W. C. Kokke, C. Tarchini, D. B. Stierle, and C. Djerassi, *J. Org. Chem.*, 1979, **44**, 3385.
[132] W. C. Kokke, I. J. Massey, N. W. Withers W. Fenical, and C. Djerassi, *Tetrahedron Letters*, 1979, 3601.
[133] C. Djerassi, *Pure Appl. Chem.*, 1978, **50**, 171.
[134] J. R. Wiersig, N. Waespe-Sarcevic, and C. Djerassi, *J. Org. Chem.*, 1979, **44**, 3374; M. W. Preus and T. C. McMorris, *J. Amer. Chem. Soc.*, 1979, **101**, 3066.

oestranediones.[135] The aromatization of steroid rings has continued to be of interest. The aromatization of ring C by rearrangement of some $\Delta^{9(11)}$-17-hydroxy-steroids[136] and of 9,11-epoxy-17-hydroxy-steroids[137] has been described. The mechanism of the rearrangement 2α-hydroxycholest-4-en-3-one to 2-methoxy-4-methyl-19-norcholesta-1,3,5(10)-triene has been examined.[138] The chromogenic reactions used in the assay of testosterone have been examined[139] and C-14 carbo-cations proposed as the chromophores.

A series of papers have appeared that give details of the synthesis of various derivatives of vitamin D. These include syntheses of various intermediates, 1α-hydroxy-vitamins D_3 and D_4, 22,23-epoxy-vitamin D_2, and 24-fluoro-25-hydroxy-vitamin D_3.[140,141]

A growing number of pregnanes have been isolated from marine organisms, including the unusual 18-acetoxypregna-1,4,20-trien-3-one.[142] A number of esters of lucibufagenin (78) form the defensive steroids of the firefly *Photinus pyralis*.[143]

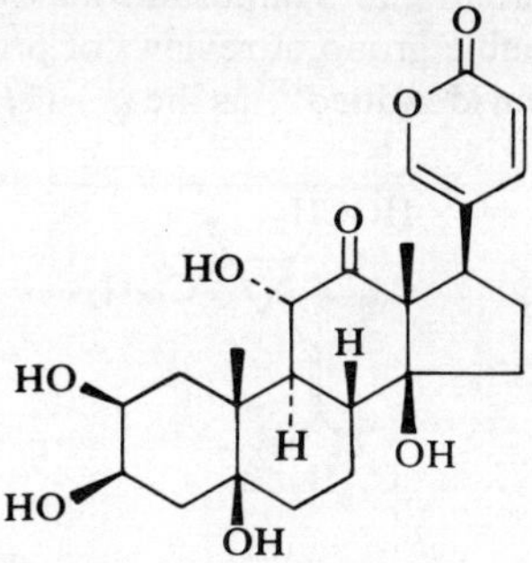

(78)

Further steroidal alkaloids of the Solanaceae and of the C-nor-D-homocevanine series have been isolated.[144]

A number of novel syntheses of steroids have been described, particularly those based on *o*-quinodimethane cycloadditions.[145] Thus thermolysis of the benzocyclobutene (79) affords an enantioselective route to 3-methoxy-1,3,5(10)-

[135] R. Jacquesy and C. Narbonne, *Bull. Soc. chim. France, Part II*, 1978, 163.
[136] A. B. Turner, *J.C.S. Perkin I*, 1979, 1333.
[137] H. T. A. Cheung, R. G. McQueen, A. Vadasz, and T. R. Watson, *J.C.S. Perkin I*, 1979, 1048.
[138] B. R. Davis, G. W. Rewcastle, and P. D. Woodgate, *J.C.S. Perkin I*, 1978, 735.
[139] T. Miura, H. Takagi, K. Harita, and M. Kimura, *Chem. and Pharm. Bull. (Japan)*, 1979, **27**, 452; T. Miura, H. Takagi, and M. Kimura, *ibid.*, p. 783.
[140] B. Lythgoe, R. Manwaring, J. R. Milner, T. A. Moran, M. E. N. Nambudiry, and J. Tideswell, *J.C.S. Perkin I*, 1978, 387; B. Lythgoe, T. A. Moran, M. E. N. Nambudiry, J. Tideswell, and P. W. Wright, *ibid.*, p. 590; P. J. Kocienski, B. Lythgoe, and D. A. Roberts, *ibid.*, p. 834; P. J. Kocienski, B. Lythgoe, and S. Ruston, *ibid.*, 1979, 1290; P. J. Kocienski, B. Lythgoe, and I. Waterhouse, *Tetrahedron Letters*, 1979, 4419; D. W. Guest and D. H. Williams, *J.C.S. Perkin I*, 1979, 1695.
[141] Y. Kobayashi, T. Taguchi, T. Terada, J.-I. Oshida, M. Morisaki, and N. Ikekawa, *Tetrahedron Letters*, 1979, 2023; K. Ochi, I. Matsunaga, M. Shindom, and C. Kaneko, *J.C.S. Perkin I*, 1979, 161; M. Tada and A. Kawa, *ibid.*, p. 1858.
[142] R. A. Ross and P. J. Scheur, *Tetrahedron Letters*, 1979, 4701.
[143] J. Meinwald, D. F. Wiemer, and T. Eisner, *J. Amer. Chem. Soc.*, 1979, **101**, 3055.
[144] K. Kaneko, M. Tanaka, K. Haruki, M. Naruse, and H. Mitsuhashi, *Tetrahedron Letters*, 1979, 3737.
[145] W. Oppolzer, *Synthesis*, 1978, 793.

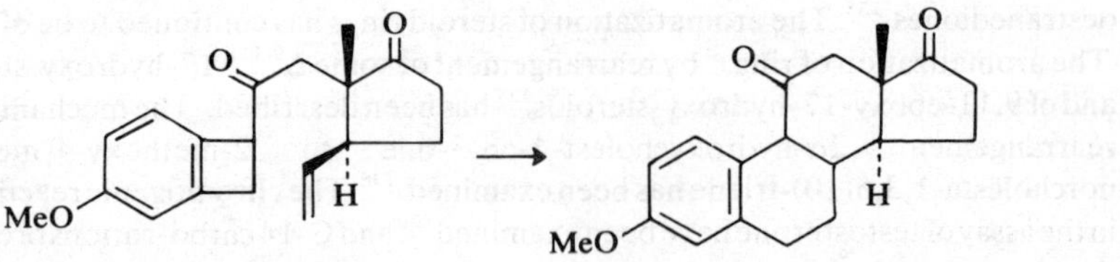

(79) (80)

oestratriene-11,17-dione (80).[146] Similar routes have been developed in other laboratories,[147] and the generation of the *o*-quinodimethane by the chelotropic elimination of sulphur dioxide has recently been described.[148]

8 Carotenoids

Lectures from the 5th International Symposium on Carotenoids have been published[149] and provide a valuable group of reviews of progress in this area. Natural prephytoene alcohol has been identified[150] as the (+)-(1*R*,2*R*,3*R*)-isomer (81). The

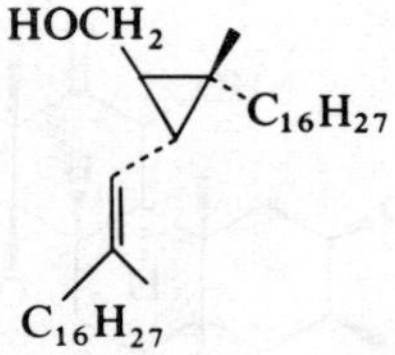

(81)

structure 7,9,7′,9′-tetra-*cis*-ψ,ψ-carotene has been assigned[151] to the poly(*cis*)carotenoid prolycopene. The *cis*-pentaene chromophore of phytofluene has been synthesized.[152] Amongst novel more highly oxidized carotenoids are a number of marine products, whilst flavoxanthin (82) and its epimer, chrysanthemaxanthin, have been obtained[153] from flowers of *Taraxacum officinale*. A number of degraded

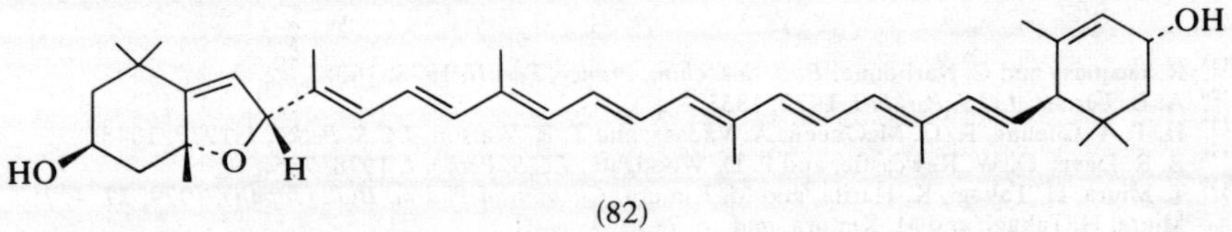

(82)

[146] W. Oppolzer, K. Battig, and M. Petrzilka, *Helv. Chim. Acta*, 1978, **61**, 1945.
[147] R. L. Funk and K. P. C. Vollhardt, *J. Amer. Chem. Soc.*, 1979, **101**, 215; T. Kametani, H. Matsumoto, H. Nemoto, and K. Fukumoto, *ibid.*, 1978, **100**, 6218.
[148] K.-C. Nicolaou and W. E. Barnette, *J.C.S. Chem. Comm.*, 1979, 1119.
[149] *Pure Appl. Chem.*, 1979, **51**, 435—675.
[150] L. J. Altman, R. C. Kowerski, H. C. Rilling, and D. R. Laungani, *J. Amer. Chem. Soc.*, 1978, **100**, 6174.
[151] G. Englert, B. O. Brown, G. P. Moss, B. C. L. Weedon, G. Britton, T. W. Goodwin, K. L. Simpson, and R. J. H. Williams, *J.C.S. Chem. Comm.*, 1979, 545; J. M. Clough and G. Pattenden, *ibid.*, p. 616.
[152] J. M. Clough and G. Pattenden, *Tetrahedron Letters*, 1979, 5043.
[153] H. Cadosch, U. Vogeli, P. Ruedi, and C. H. Eugster, *Helv. Chim. Acta*, 1978, **61**, 1511.

carotenoids have been isolated, particularly from tobacco.[154] The chemistry and biology of abscisic acid has been reviewed.[155]

There has been considerable synthetic activity in the carotenoid area, particularly in the preparation of compounds with unusual end-groups and partially degraded structures.[156] A novel synthesis of crocetin (84) used the cleavage of the dienone (83) that was obtained by alkylation of 2,6-dimethylphenol.[157] Novel syntheses of abscisic acid and its relatives have also been reported.[158] Electrochemical reduction has been used[159] to prepare astaxanthin from astecene.

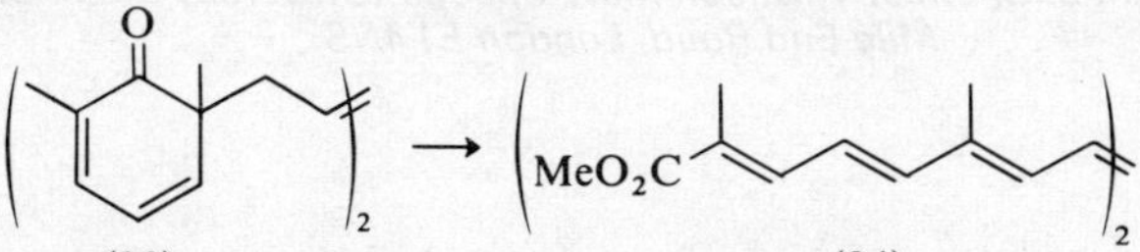

(83) (84)

The chromophoric unit of the visual pigments is known to consist of 11-*cis*-retinal covalently bound, as a protonated Schiff's base, to the ϵ-amino-group of lysine in opsin. An external point-charge model, based on an electrostatic interaction with a charged group on the rhodopsin, has been proposed[160] to account for the shift in absorption maximum between protonated Schiff's bases of retinal and the visual pigment bovine rhodopsin. Some simple compounds have been synthesized which show this effect.

[154] See, for example, E. Demole and P. Enngist, *Helv. Chim. Acta*, 1978, **61**, 2318; R. Uegaki, T. Fujimori, H. Kaneko, K. Kato, and M. Noguchi, *Agric. Biol. Chem.* (*Japan*), 1979, **43**, 1149.

[155] B. V. Milborow, in 'Phytohormones and Related Compounds', ed. D. S. Letham, P. B. Goodwin, and J. J. V. Higgins, Elsevier, Amsterdam, 1978, Vol. 1, p. 295.

[156] See, for example, M. Akhtar, A. E. Faruk, C. J. Harris, G. P. Moss, S. W. Russell, and B. C. L. Weedon, *J.C.S. Perkin I*, 1978, 1511.

[157] G. Quinkert, K. R. Schmieder, G. Durner, K. Hache, A. Stegk, and D. H. R. Barton, *Chem. Ber.*, 1977, **110**, 3582.

[158] F. Kienzle, H. Mayer, R. E. Minder, and H. Thommen, *Helv. Chim. Acta*, 1978, **61**, 2616.

[159] E. A. H. Hall, G. P. Moss, J. H. P. Utley, and B. C. L. Weedon, *J.C.S. Chem. Comm.*, 1978, 387.

[160] B. Honig, U. Dinur, K. Nakanishi, V. Balogh-Nair, M. A. Gawinowicz, M. Arnaboldi, and M. G. Motto, *J. Amer. Chem. Soc.*, 1979, **101**, 7084; M. Sheves, K. Nakanishi, and B. Honig, *ibid.*, p. 7086.

16 Biological Chemistry
Part (ii) Alkaloids

By D. G. BUCKLEY
Department of Chemistry, Queen Mary College (University of London) Mile End Road, London E1 4NS

1 Introduction

In the three years since alkaloids were last reviewed in *Annual Reports*[1] a large amount of new work has been published, and for a comprehensive survey the reader is referred to Volumes 8 and 9 of the Specialist Periodical Reports on the alkaloids,[2,3] which cover the period July 1976 to June 1978. Biosynthetic aspects are also treated in the companion volumes[4,5] on biosynthesis, and these include a tabular survey of all tracer incorporations into alkaloids reported in the years 1975–78. The period covered by this short review is 1977–79 inclusive.

2 Pyridine and Piperidine Alkaloids

Crombie and Whiting have reported[6—9] the isolation and characterization of eleven new celastraceous alkaloids of mainly Ethiopian and Kenyan origin, having molecular weights of *ca.* 600—*ca.* 1200 and a sesquiterpene core. They have been classified in three groups:[6] (*a*) cathedulins E2 and E8, which are esters of a new pentahydroxysesquiterpene ether (1) (reported previously[10] as cathedulins 2 and 8); (*b*) cathedulins K1, K2, K6, and K15, which are esters of the known nonahydroxysesquiterpene ether euonyminol (2), to which is attached a single dilactone bridge; and (*c*) cathedulins E3, E4, E5, E6, and K12, which are more complex esters of euonyminol containing two dilactone bridges. In each case the basic functionality is provided by one or more esterified nicotinic acid derivatives. The structure of

[1] D. G. Buckley, *Ann. Reports (B)*, 1976, **73**, 397.
[2] 'The Alkaloids', ed. M. F. Grundon (Specialist Periodical Reports), The Chemical Society, London, 1978, Vol. 8.
[3] 'The Alkaloids', ed. M. F. Grundon (Specialist Periodical Reports), The Chemical Society, London, 1979, Vol. 9.
[4] 'Biosynthesis', ed. J. D. Bu'Lock (Specialist Periodical Reports), The Chemical Society, London, 1977, Vol. 5.
[5] 'Biosynthesis', ed. J. D. Bu'Lock (Specialist Periodical Reports), The Chemical Society, London, 1980, Vol. 6.
[6] R. L. Baxter, L. Crombie, D. J. Simmonds, D. A. Whiting, O. J. Braenden, and K. Szendrei, *J.C.S. Perkin I*, 1979, 2965.
[7] R. L. Baxter, L. Crombie, D. J. Simmonds, and D. A. Whiting, *J.C.S. Perkin I*, 1979, 2972.
[8] L. Crombie, W. M. L. Crombie, D. A. Whiting, and K. Szendrei, *J.C.S. Perkin I*, 1979, 2976.
[9] R. L. Baxter, W. M. L. Crombie, L. Crombie, D. J. Simmonds, D. A. Whiting, and K. Szendrei, *J.C.S. Perkin I*, 1979, 2982.
[10] R. L. Baxter, L. Crombie, D. J. Simmonds, and D. A. Whiting, *J.C.S. Chem. Comm.*, 1976, 465.

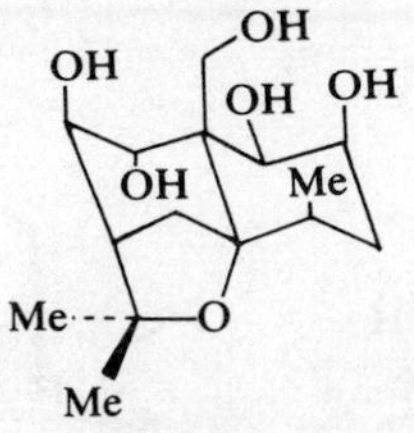

(1)

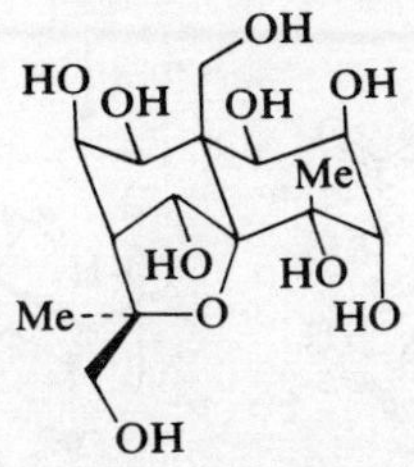

(2)

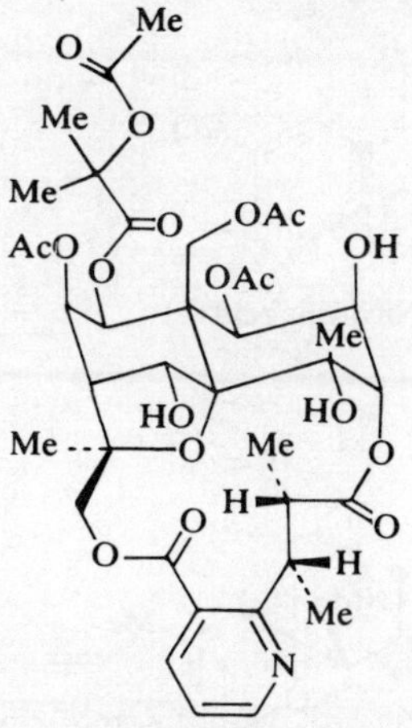

(3)

cathedulin K2 (3), the most abundant alkaloid isolated from Kenyan-grown plants, is illustrative of the complex structures of the cathedulins.

The biosynthesis of anatabine (4) is unusual[1,11,12] in that its dehydropiperidine ring derives from nicotinic acid (5) rather than lysine, which is the source of the similar fragment in anabasine (6).[13] Most interestingly, it has been shown[14] that the piperidine fragment of dioscorine (7) also derives from nicotinic acid (5): [2-^{14}C]- and [5,6-$^{13}C_2$,$^{14}C_2$]-nicotinic acids were tested as precursors, and in the latter case the ^{13}C-labelling of the dioscorine was shown by ^{13}C n.m.r. spectroscopy to be at C-1 and C-7. Thus 3,6-dihydronicotinic acid (8), which is probably an important intermediate in the biosynthesis of the pyridine ring of nicotine and anabasine[13] and of both rings in anatabine,[1,11] also seems to be involved in the biosynthesis of dioscorine (7) (Scheme 1). Details of the introduction of the remaining fragments (acetate-derived[15]) are not yet available, but clearly such details are well worth pursuing.

The alkaloids represented by tylophlorine (9) and cryptopleurine (10) exhibit various interesting biological properties, including anti-cancer action.[16] Exploitation of these properties in the synthesis of analogues requires efficient elaboration of these and related ring systems and, based on biogenetic considerations, Herbert has developed[17] an economical synthesis of septicine (11); this, on treatment with thallium(III) trifluoroacetate, gives tylophlorine (9) in high yield (Scheme 2).[18] Extension of this route to the synthesis of cryptopleurine (10) did not prove

[11] E. Leete and S. A. Slattery, *J. Amer. Chem. Soc.*, 1976, **98**, 6326.
[12] E. Leete, *J.C.S. Chem. Comm.*, 1978, 610.
[13] E. Leete and Y.-Y. Lin, *Phytochemistry*, 1973, **12**, 593; R. B. Herbert, in 'The Alkaloids', ed. J. E. Saxton (Specialist Periodical Reports), The Chemical Society, London, 1973, Vol. 3, p. 7.
[14] E. Leete and S. A. Slattery, *J. Amer. Chem. Soc.*, 1977, **99**, 648; *Phytochemistry*, 1977, **16**, 1705.
[15] R. B. Herbert, in ref. 2, p. 1.
[16] J. L. Hartwell and B. J. Abbott, *Adv. Pharmacol. Chemother.*, 1969, **7**, 117; N. R. Farnsworth, N. K. Hart, S. R. Johns, J. A. Lamberton, and W. Messmer, *Austral. J. Chem.*, 1969, **22**, 1805.
[17] R. B. Herbert, F. B. Jackson, and I. T. Nicholson, *J.C.S. Chem. Comm.*, 1976, 450.
[18] R. B. Herbert, *J.C.S. Chem. Comm.*, 1978, 794.

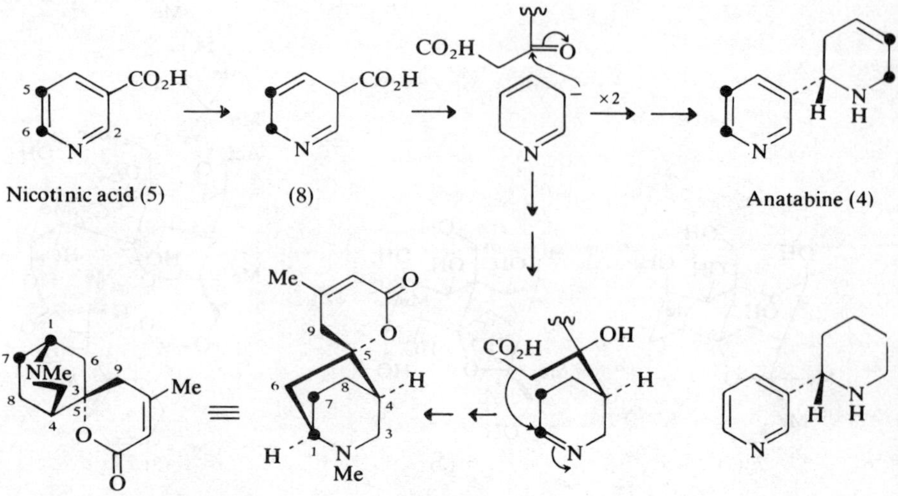

Nicotinic acid (5)
(8)
Anatabine (4)
Dioscorine (7)
Anabasine (6)

Scheme 1

straightforward, but has now been completed in fair overall yield.[18] The key intermediate (12), itself an alkaloid, was easily synthesized by condensation of 3,4-dimethoxybenzoylacetic acid (13) with Δ^1-piperidine, generated *in situ* from cadaverine by pea-seedling diamine oxidase. The required enamine–ketone condensation occurred readily in the case of (14),[17] but a Lewis acid such as titanium(IV) chloride was required to effect the ring-closure of (15) to give the known alkaloid julandine (16).[18] Transformation of (16) into cryptopleurine (10) was achieved in 69% yield by using thallium(III) trifluoroacetate, even though the linkage to one aryl ring of (16) is *meta* to the methoxy-substituent; as in the synthesis of tylophlorine (9), only one of the possible coupling products was detected.

3 Pyrrolizidine Alkaloids

Significant advances are now being made in the synthesis of various pyrrolizidine alkaloids.[19]

Danishefsky and co-workers have developed a novel route to the saturated pyrrolizidine alcohols,[20] outlined in Scheme 3. The lithium salt of propargyl alcohol tetrahydropyranyl ether (17) was converted into the (Z)-diazomalonate (18) in a number of steps. Intramolecular *syn*-addition of the carbenoid from (18) to give (19) was carried out by boiling with copper bronze in toluene for 22 h. Treatment of the activated cyclopropane (19) with an excess of hydrazine released the free amine, which underwent intramolecular homoconjugate addition with complete inversion of stereochemistry, and subsequent lactam formation, to give the pyrrolizidine hydrazide (20). Hydrolysis with decarboxylation produced a salt containing a

[19] D. J. Robins, in ref. 2, Ch. 3; in ref. 3, Ch. 4.
[20] S. Danishefsky, R. McKee, and R. K. Singh, *J. Amer. Chem. Soc.*, 1977, **99**, 4783.

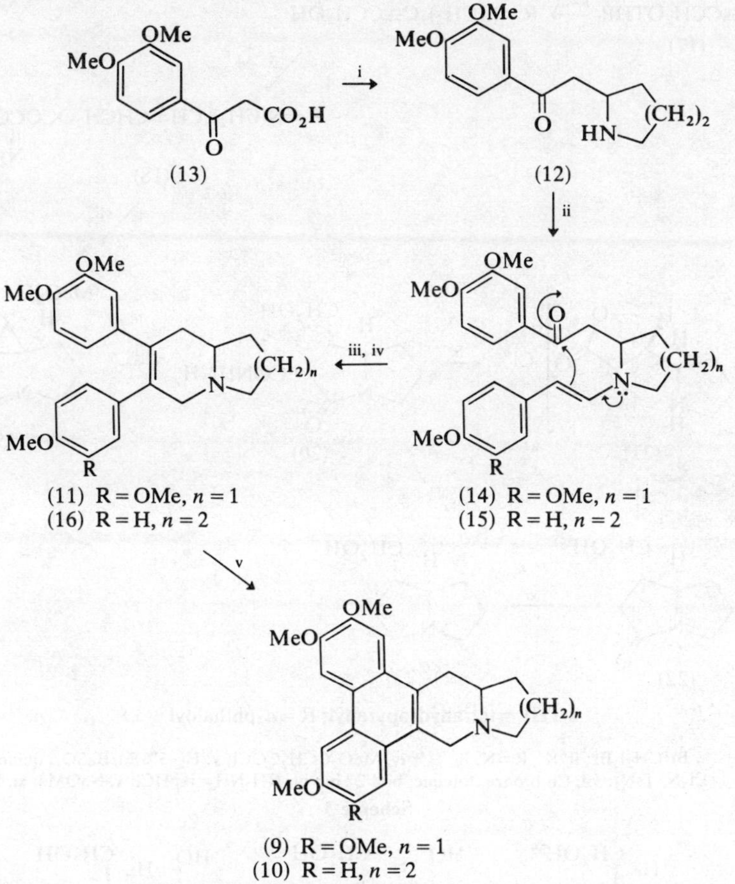

Reagents [for (13) → (10)]: i, $H_2N(CH_2)_5NH_2$, pea-seedling diamine oxidase, pH 7; ii, *p*-$MeOC_6H_4CH_2CHO$, C_6H_6; iii, $TiCl_4$; iv, $NaBH_4$, Pr^iOH; v, $Tl(OCOCF_3)_3$, CF_3CO_2H, room temperature

Scheme 2

γ-lactone function which was assigned the structure (21). The pyrrolizidine system (22) was regenerated using sodium methoxide, and a final reduction step yielded (±)-trachelanthamidine (23). Careful analysis of the pyrrolizidine derivative (22) by ^{1}H n.m.r. spectroscopy at 250 MHz showed that there was no contamination with the C-1 epimer. Analogous treatment of the (*E*)-isomer of (18) afforded this C-1 epimer, (±)-isoretronecanol (24).

Danishefsky *et al.*[21] have extended this strategy to synthesize necine bases with hydroxyl functions at C-7, namely (±)-hastanecine (25) and its epimer (±)-dihydroxyheliotridane (26).

[21] S. Danishefsky, R. McKee, and R. K. Singh, *J. Amer. Chem. Soc.*, 1977, **99**, 7711.

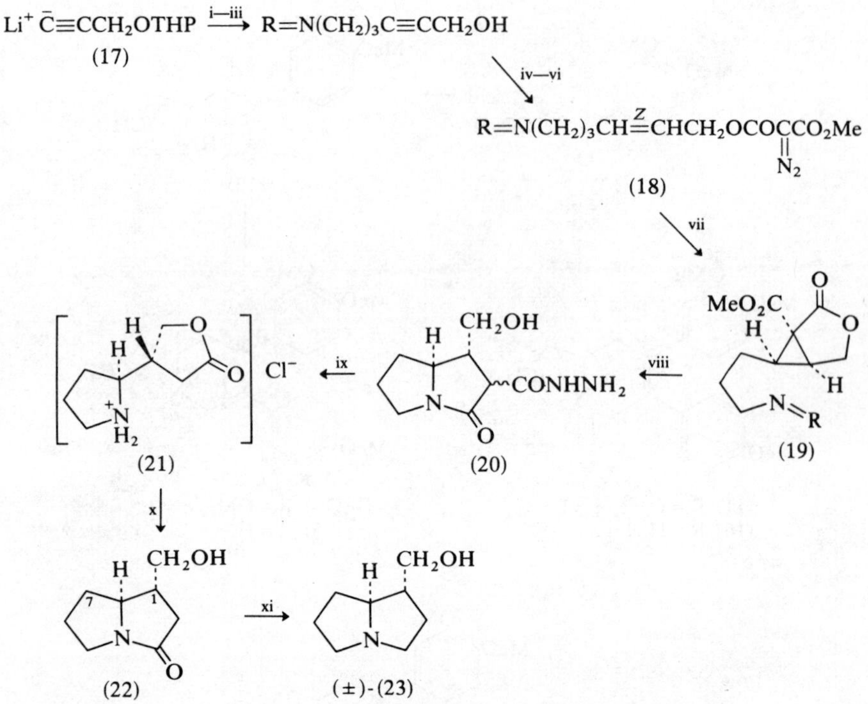

THP = tetrahydropyranyl; R = *o*-phthaloyl

Reagents: i, $Br(CH_2)_3Br$; ii, K^+ $R{=}\bar{N}$; iii, H^+; iv, MeO_2CCH_2COCl; v, H_2–5% $Pd/BaSO_4$, quinoline; vi, Et_3N, TsN_3; vii, Cu bronze, toluene, boil 22 h; viii, NH_2NH_2; ix, HCl; x, NaOMe; xi, $LiAlH_4$

Scheme 3

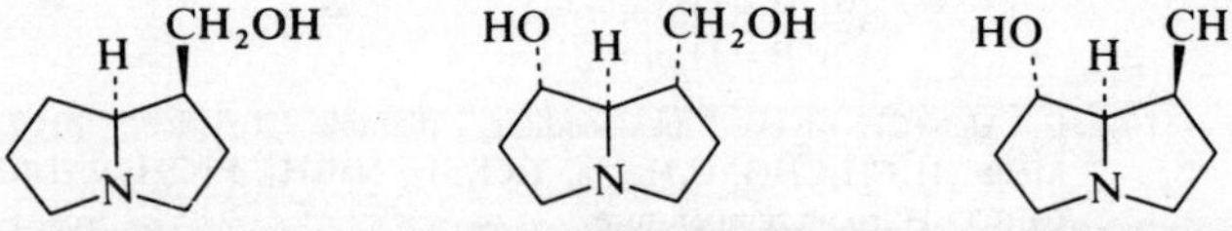

(±)-(24) (±)-(25) (±)-(26)

Robins' group has developed useful syntheses of both the fully saturated necine bases and the physiologically active 1,2-unsaturated pyrrolizidine derivatives, *e.g.* supinidine (27).[22,23] Their strategy involves the conversion of saturated 1-substituted pyrrolizidines into the required 1,2-didehydro-compounds by thermal elimination of a phenylseleno-group. This approach is exemplified by the synthesis of (±)-supinidine (27) (Scheme 4).[22]

The thermodynamically less stable racemate of ethyl pyrrolizidine-1-*endo*-carboxylate (28) was prepared in 80% overall yield using the method of Pizzorno and Albonico.[24] This involves regiospecific 1,3-dipolar cycloaddition of ethyl propiolate to the postulated azomethine ylide[24] that is formed by heating *N*-formyl-L-

[22] D. J. Robins and S. Sakdarat, *J.C.S. Perkin I*, 1979, 1734.
[23] D. J. Robins and S. Sakdarat, *J.C.S. Chem. Comm.*, 1979, 1181.
[24] M. T. Pizzorno and S. M. Albonico, *J. Org. Chem.*, 1974, **39**, 731.

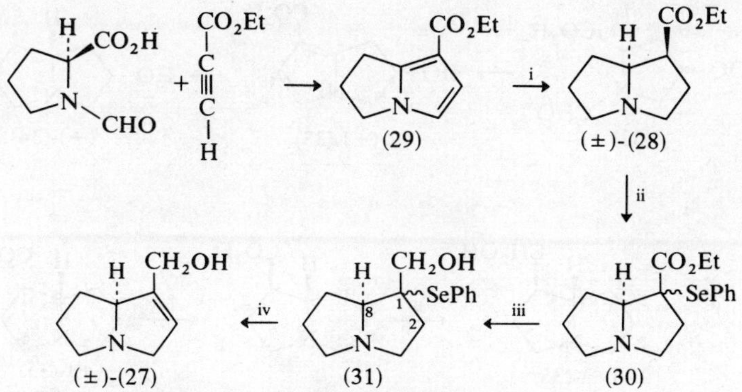

Reagents: i, H_2–10% Pd/C; ii, $Pr^i_2\bar{N}$ Li^+, PhSeCl; iii, $LiAlH_4$; iv, H_2O_2

Scheme 4

proline in acetic anhydride, to give the dihydropyrrolizine ester (29), followed by catalytic hydrogenation. Selenylation of the lithium enolate derived from the ester (28) was readily accomplished with phenylselenium chloride, to give the ester (30). Reduction with lithium aluminium hydride gave the derived primary alcohol (31); on treatment with hydrogen peroxide, this gave (±)-supinidine (27). The stereochemistry of the intermediate phenylselenides (Scheme 4) is uncertain, but the authors were able to show that elimination of the phenylselenyl group from (31) gave only the required 1,2-double bond, and that no 1,8-didehydropyrrolizidine was produced.

In an elegant extension of this procedure (Scheme 5), Robins *et al.*[23] have used the *NO*-diformyl derivative (32) to produce the chiral alcohol (33). Catalytic hydrogenation of (33) gave the crystalline pyrrolizidine alcohol (34), formed by stereospecific *cis*-addition of hydrogen to the less hindered β-face of (33). Replacement of the 6α-hydroxy-group of (34) by chlorine followed by catalytic hydrogenolysis gave the key optically active pyrrolizidine ester (35). Reduction of this ester with lithium aluminium hydride gave (+)-isoretronecanol (24), the overall yield of optically active product being 45% from the readily available (−)-4-hydroxy-L-proline. Epimerization at C-1 of the thermodynamically less stable ester (35) was achieved with sodium ethoxide in ethanol, and reduction of the product gave (+)-laburnine (36)* in 64% yield from (35). The ester (35) was also converted (in 21% overall yield) into (+)-supinidine (27), using the method outlined in Scheme 4; a higher overall yield was obtained when purification of intermediates was omitted.

Although a natural (−)-4-hydroxy-L-proline can be converted into its enantiomer by epimerization of both chiral centres,[25] thus allowing the synthesis of the enantiomers of (+)-(24), (+)-(36), and (+)-(27), the authors devised[23] a simpler procedure which involved efficient isomerization of the chiral intermediate (33). The hydroxy-ester (+)-(33) was first converted into its tosylate derivative in high yield.

[25] D. S. Robinson and J. P. Greenstein, *J. Biol. Chem.*, 1952, **195**, 383.

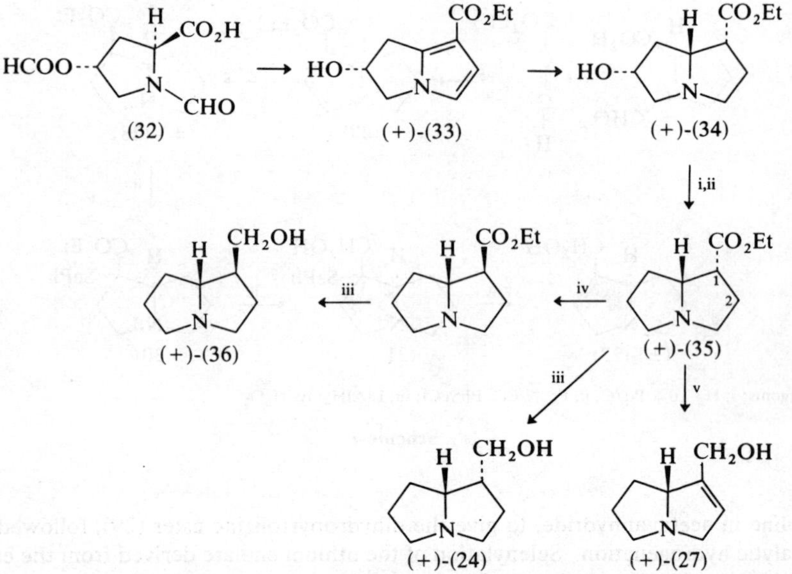

Reagents: i, $SOCl_2$; ii, H_2–Raney nickel; iii, $LiAlH_4$; iv, NaOEt, EtOH; v, steps ii—iv of Scheme 4

Scheme 5

Inversion of stereochemistry was then achieved by S_N2 displacement with formate anion to give, after hydrolysis, the enantiomer of (33) required for conversion, as outlined above, into the 8α-pyrrolizidine bases (−)-isoretronecanol [as (24)], (−)-trachelanthamidine [as (23)]*, and (−)-supinidine [as (27)].

Ornithine, putrescine, and arginine have been shown[26] to be specific precursors of retronecine (37), the most common necine-base portion of pyrrolizidine alkaloids, *e.g.* retrorsine (38). Degradations of retronecine (37), derived biosynthetically from [5-^{14}C]ornithine (39), [2-^{14}C]ornithine [as (39)], or [1,4-$^{14}C_2$]putrescine (40), demonstrated that, in each case, one quarter of the total radioactivity of the necine base was located at C-9, suggesting that C-2 and C-5 of the molecule of ornithine used to form ring B of retronecine (37) become equivalent (see Scheme 6). Further degradations of the retronecine have been necessary to establish whether the second molecule of ornithine involved in the biosynthesis also passes through a symmetrical intermediate. Robins's group has now reported[27] the first degradation designed to examine ring A. Their use of improved feeding techniques in these studies has led to much higher incorporations and to the identification of two new polyamine precursors.

* (+)-Laburnine (36)[23] is the enantiomer of (−)-trachelanthamidine [as (23)];[20,23] the racemate is named (±)-trachelanthamidine (23).[20]

[26] E. Nowacki and R. U. Byerrum, *Life Sciences*, 1962, **1**, 157; W. Bottomley and T. A. Geissman, *Phytochemistry*, 1964, **3**, 357; C. A. Hughes, R. Letcher, and F. L. Warren, *J. Chem. Soc.*, 1964, 4974; N. M. Bale and D. H. G. Crout, *Phytochemistry*, 1975, **14**, 2617.

[27] D. J. Robins and J. R. Sweeney, *J.C.S. Chem. Comm.*, 1979, 120.

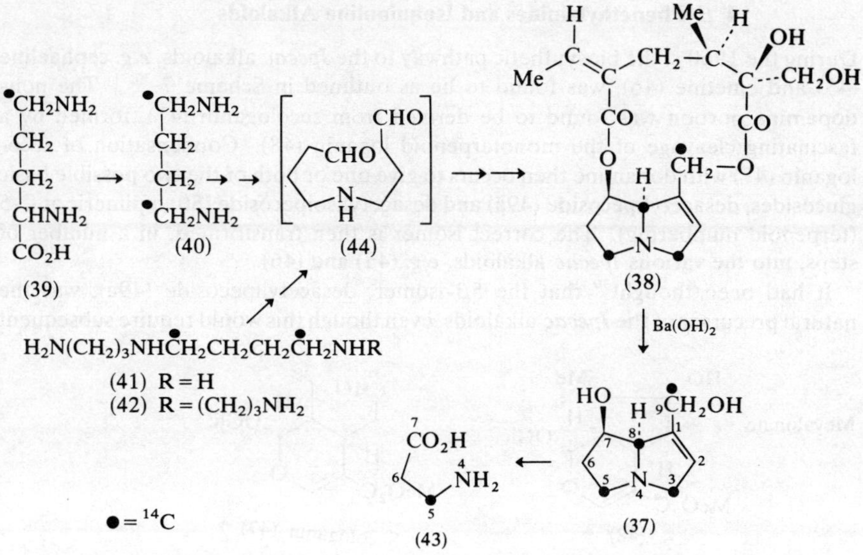

Scheme 6

Good incorporations of precursors into retrorsine (38) were achieved[27] by direct absorption through stem punctures in *Senecio isatideus*. These studies indicate that the best precursors for retronecine biosynthesis examined so far are putrescine (40) and the previously untested polyamines spermidine (41) and spermine (42). Putrescine was found to be a much more efficient precursor than ornithine (39), thus supporting the theory[28] that putrescine follows ornithine on the biosynthetic pathway.

The necine base retronecine (37), obtained by basic hydrolysis of retrorsine (38), was degraded in two ways for experiments involving each of the precursors (39), (40), (41), and (42). One set of degradations showed that 25, 24, 23, and 22% of the total ^{14}C activity was located at C-9 of retronecine (37) after incorporations of (39), (40), (41), and (42), respectively, in agreement with previous results.[26] Modified Kuhn–Roth oxidation[29] of (37) gave β-alanine (43), isolated as its 2,4-dinitrophenyl derivative. This fragment corresponds to (C-5 + C-6 + C-7) of retronecine (37) (Scheme 6). In experiments involving the same precursors, *viz*. (39), (40), (41), and (42), it was found that 22, 22, 23, and 24%, respectively, of the total ^{14}C activity of retronecine (37) was located in this fragment. Robins *et al.* suggest[27] that these results, taken together with earlier findings,[26,28] point to the involvement of a later symmetrical intermediate in the biosynthesis of retronecine (37), such as (44), derived from two molecules of putrescine (40).

[28] T. A. Geissman and D. H. G. Crout, 'Organic Chemistry of Secondary Plant Metabolism', Freeman, Cooper and Co., San Francisco, 1969, p. 448.
[29] Y. K. Ho, R. N. Gupta, D. B. MacLean, and I. D. Spencer, *Canad. J. Chem.*, 1971, **49**, 3352.

4 β-Phenethylamines and Isoquinoline Alkaloids

During the 1960's, the biosynthetic pathway to the *Ipecac* alkaloids, *e.g.* cephaeline (45) and emetine (46), was found to be as outlined in Scheme 7.[30,31] The non-dopamine portion was found to be derived from secologanin (47), formed by a fascinating cleavage of the monoterpenoid loganin (48). Condensation of secologanin (47) with dopamine then occurs to give one or both of the two possible basic glucosides, desacetylipecoside (49a) and desacetylisoipecoside (50), epimeric at C-5 (terpenoid numbering). The correct isomer is then transformed, in a number of steps, into the various *Ipecac* alkaloids, *e.g.* (45) and (46).

It had been thought[30] that the 5β-isomer, desacetylipecoside (49a), was the natural precursor of the *Ipecac* alkaloids, even though this would require subsequent

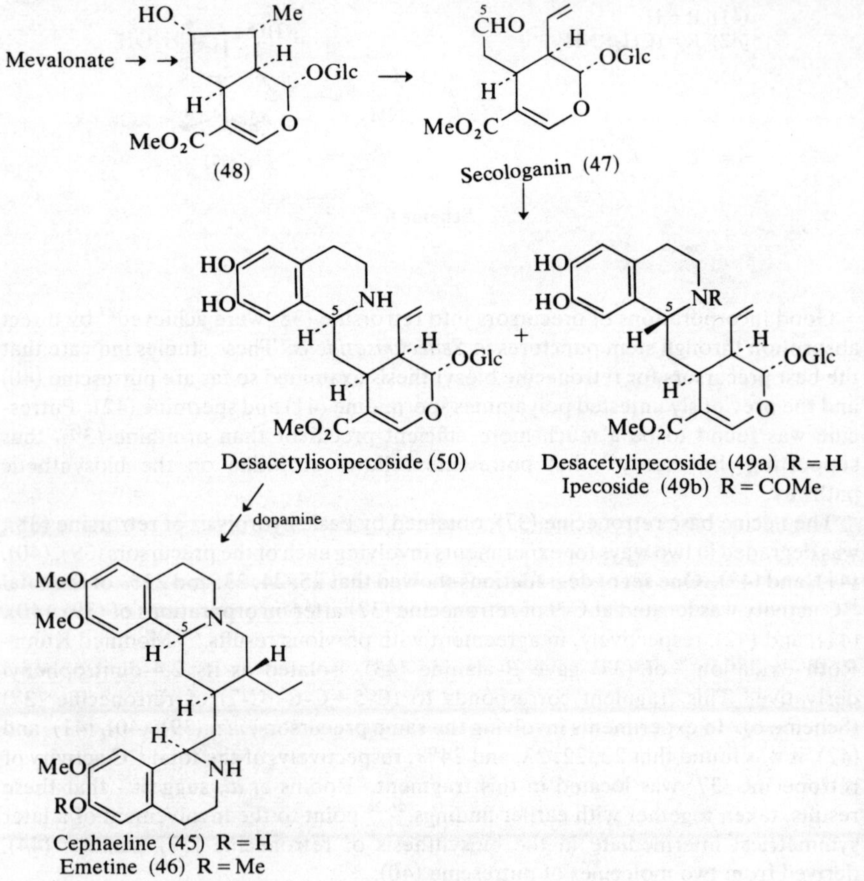

Scheme 7

[30] A. R. Battersby and R. J. Parry, *Chem. Comm.*, 1971, 901.
[31] J. Staunton, in 'The Alkaloids', ed. J. E. Saxton, (Specialist Periodical Reports), The Chemical Society, London, 1972, Vol. 2, p. 6.

inversion of configuration at C-5. [The (5*S*) stereochemistry of desacetylipecoside (49a) follows unambiguously from an *X*-ray analysis of its OO-dimethyl derivative.[32]] Recent work[33] has now shown that the 5α-isomer, desacetylisoipecoside (50), is the correct precursor of cephaeline (45) and emetine (46) in *Cephaelis ipecacuanha*, not the 5β-isomer (49a); the 5β-isomer (49a) is the precursor of the 5β-basic glucoside ipecoside (49b). Since the 5α configuration of desacetylisoipecoside (50) matches that of the corresponding centre in cephaeline (45) and emetine (46), it is now clear that this part of the biosynthetic pathway is stereochemically unexceptional.

As part of his Manchester group's biomimetic alkaloid syntheses, R. T. Brown has developed[34] a useful route to various *Ipecac* alkaloids, *e.g.* cephaeline (45) and emetine (46). The *Ipecac* alkaloids are formed *in vivo* (in the main) by condensation of a dopamine or tryptamine residue with protoemetine (51), which itself is comprised of a dopamine residue and secologanin (47) (see Schemes 7 and 8); identical absolute configurations at C-2 and C-7 in (51) and (47) reflect this biosynthetic relationship. Brown's synthetic strategy (Scheme 8) involves the conservation of stereochemistry at C-2 and C-7 of the natural precursor (47), an approach which avoids the problems of stereochemical control encountered in earlier syntheses[35] of the clinically important compound emetine (46).

Methyl 3,4-dihydrosecoxyloganin tetra-acetate (52) was prepared from secologanin (47) in 63% isolated yield.[36] Zemplen deacetylation and enzymic removal of the sugar afforded the aglycone (53), which was treated with 2-(3,4-dimethoxyphenyl)ethylamine and sodium cyanoborohydride to give the tetrahydropyridine (54a) in 42% overall isolated yield.[34] The tetrahydropyridine (54a) had been obtained previously, using a 'one-pot' procedure,[37] albeit in lower yield. Careful analysis and chemical investigation established that the above procedure did not result in any significant inversion at C-2 to give the more stable *trans*-isomer (54b), itself available by prior equilibration of the aglycone (53) under basic conditions. Selective hydrolysis and decarboxylation of (54a) gave (55), which was converted into the protected pyridone aldehyde (56) in 85% yield with an excess of *NN*-dimethylhydrazine. Although this key intermediate (56) was used in two different ways to produce various types of related alkaloids,[34] only the route to cephaeline (45) and emetine (46) is given in Scheme 8. Bischler–Napieralski cyclization, subsequent reduction, and hydrolysis of the *NN*-dimethylhydrazone afforded the unstable proto-emetine (51), which has the illustrated natural (*S*) configuration at C-5 (terpenoid numbering). Pictet–Spengler condensation of (51) gave cephaeline (45) together with a smaller amount of isocephaeline (57). Subsequent treatment of (45) with diazomethane completed the synthesis of emetine (46).

[32] O. Kennard, P. J. Roberts, N. W. Isaacs, F. H. Allen, W. D. S. Motherwell, K. H. Gibson, and A. R. Battersby, *Chem. Comm.*, 1971, 899.
[33] A. R. Battersby, N. G. Lewis, and J. M. Tippett, *Tetrahedron Letters*, 1978, 4849.
[34] R. T. Brown, A. G. Lashford, and S. B. Pratt, *J.C.S. Chem. Comm.*, 1979, 367.
[35] Reviewed by A. Brossi, S. Teitel, and G. V. Parry, in 'The Alkaloids', ed. R. H. F. Manske, Academic Press, New York, 1971, Vol. XIII, Ch. 3.
[36] R. T. Brown, C. L. Chapple, D. M. Duckworth, and R. Platt, *J.C.S. Perkin I*, 1976, 160.
[37] R. T. Brown and J. Leonard, *J.C.S. Chem. Comm.*, 1978, 726.

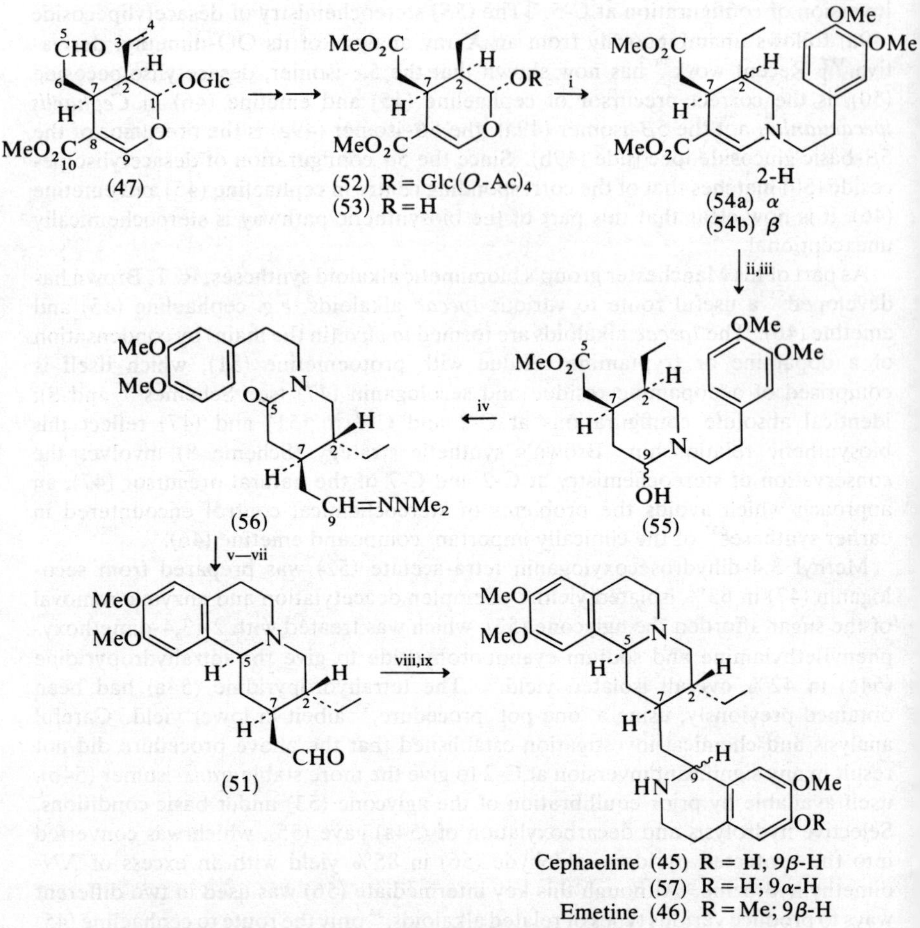

Reagents: i, 3,4-$(OMe)_2C_6H_3CH_2CH_2NH_2$, $NaCNBH_3$, MeOH, 15 min; ii, 1% HCl—MeOH, boil, 1 h; iii, Et_3N; iv, H_2NNMe_2; v, $POCl_3$, C_6H_5Me; vi, $NaBH_4$; vii, $Cu(OAc)_2$, H_2O; viii, 3-OH,4-OMe-$C_6H_3CH_2CH_2NH_2$, HOAc—H_2O; ix [for (45) → (46)], CH_2N_2

Scheme 8

The unravelling of the biosynthesis of protostephanine (58) and of hasubanonine (59) has proved to be a long and difficult task.[38] Previous investigations led to the conclusion that these alkaloids of *Stephania japonica* are both constructed from two tyrosine-derived C_6–C_2 units. The unit which is the source of ring C and the attached ethanamine side-chain in (58) and (59) combines as (60) with the other unit.[39]

Sixteen possible benzylisoquinolines, synthesized formally from (60) and the acids (64), have been examined[40] as precursors for (58) and (59). Of these, the bases with

[38] D. G. Buckley, in ref. 1, p. 402; R. B. Herbert, in ref. 3, pp. 8—10.
[39] A. R. Battersby, R. C. F. Jones, R. Kazlauskas, C. Poupat, C. W. Thornber, S. Ruchirawat, and J. Staunton, *J.C.S. Chem. Comm.*, 1974, 773.
[40] A. R. Battersby, A. Minta, A. P. Ottridge, and J. Staunton, *Tetrahedron Letters*, 1977, 1321.

O-methyl groups at C-8 and/or C-3′ [see (63)] were not incorporated, but (63) as well as (61) and (62) were, and specifically so where examined; the results also show that the timing of *N*-methylation is not critical. The biosynthesis indicated by these results is illustrated in Scheme 9. It is noteworthy that phenol oxidative coupling must occur on (63) itself, and not an *O*-methylated derivative of it, and thus this pathway is unique in that two hydroxy-groups must be present on one ring for coupling to occur. The important conclusion is that protostephanine (58) and hasubanonine (59) are members of the large family of modified benzylisoquinoline alkaloids.

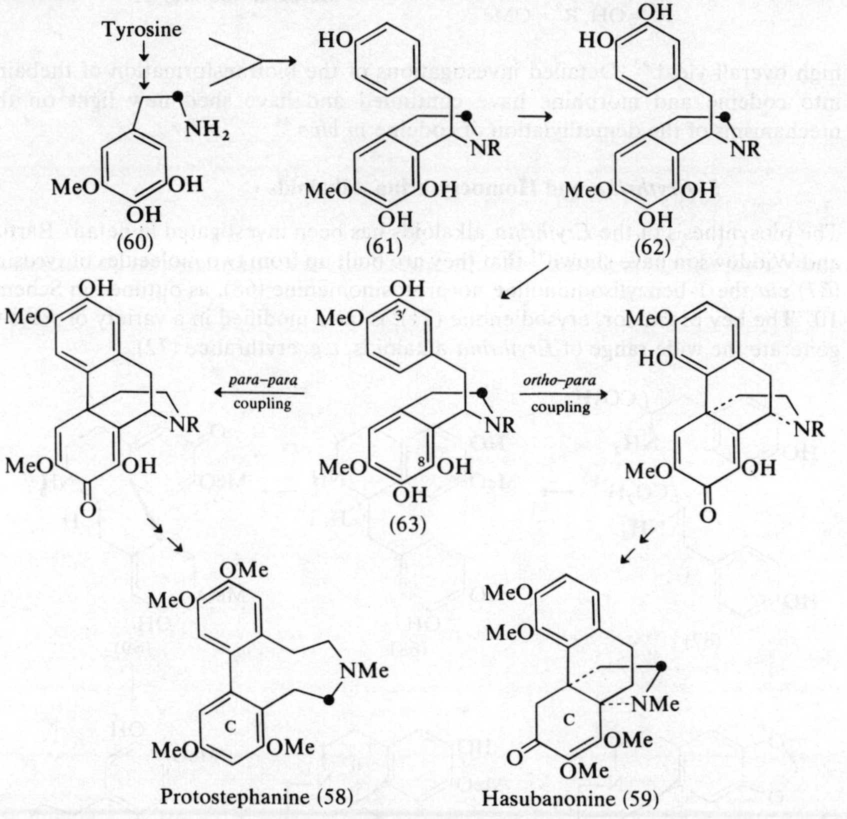

● = ^{14}C; R = H or Me

Scheme 9

New methods for the demethylation of codeine (65) to morphine (66), previously a capricious reaction, have been reported, the product being obtained in good yield in each case. Demethylation by boron tribromide in chloroform gives 90%[41] and by potassium t-butoxide in propanethiol gives 80% of morphine.[42] Various methods have been developed to convert thebaine into codeinone, and thence into codeine, in

[41] K. C. Rice, *J. Medicin. Chem.*, 1977, **20**, 164.
[42] J. A. Lawson and J. I. DeGraw, *J. Medicin. Chem.*, 1977, **20**, 165.

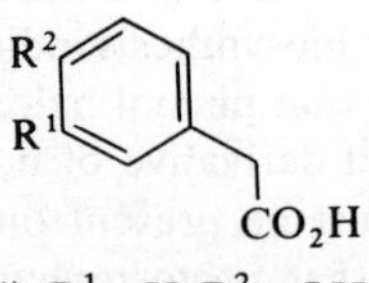

(64) $R^1 = H, R^2 = OH$
$R^1 = R^2 = OH$
$R^1 = OMe, R^2 = OH$
$R^1 = OH, R^2 = OMe$

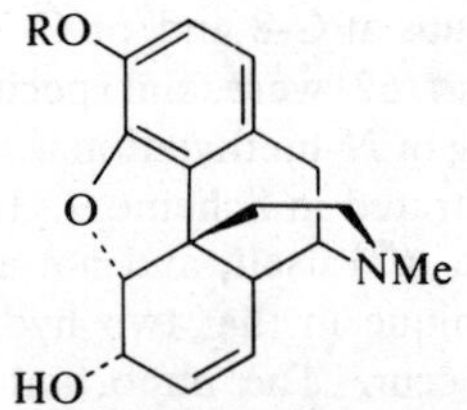

Codeine (65) R = Me
Morphine (66) R = H

high overall yield.[43] Detailed investigations of the biotransformation of thebaine into codeine and morphine have continued and have shed new light on the mechanisms of the demethylation of codeine *in vivo.*[44]

5 *Erythrina* and Homoerythrina Alkaloids

The biosynthesis of the *Erythrina* alkaloids has been investigated in detail. Barton and Widdowson have shown[45] that they are built up from two molecules of tyrosine (67) *via* the 1-benzylisoquinoline norprotosinomenine (68), as outlined in Scheme 10. The key precursor, erysodienone (71), is later modified in a variety of ways to generate the wide range of *Erythrina* alkaloids, *e.g.* erythraline (72).

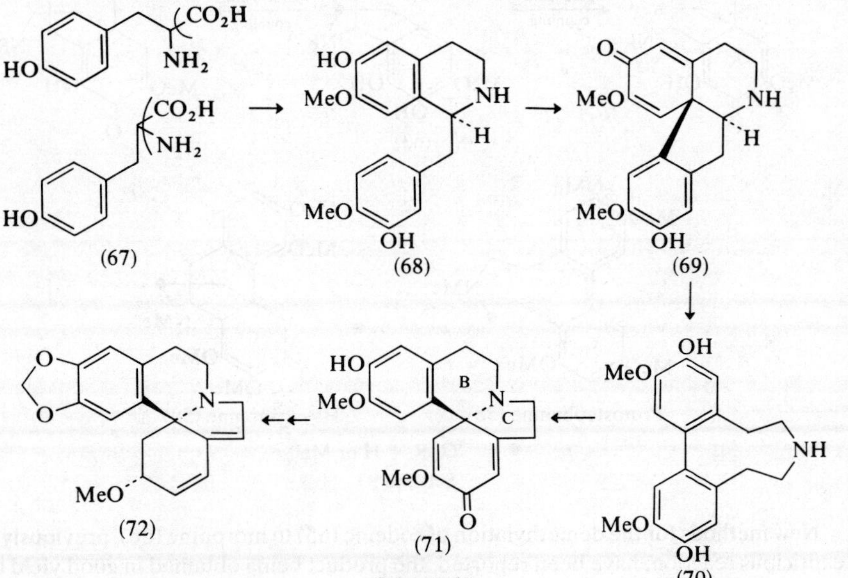

Scheme 10

[43] Reviewed by K. W. Bentley in ref. 2, pp. 112—114.
[44] Reviewed by R. B. Herbert in ref. 3, pp. 8—10.
[45] D. H. R. Barton, R. B. Boar, and D. A. Widdowson, *J. Chem. Soc. (C)*, 1970, 1213; see also D. H. R. Barton, R. D. Bracho, C. J. Potter, and D. A. Widdowson, *J.C.S. Perkin I*, 1974, 2278, and refs. cited therein.

Synthetic studies have shown that erysodienone (71) can be prepared by phenol oxidative coupling of the amine (73).[46] In a detailed investigation on model compounds, Barton *et al.* later showed[47] that this oxidation *in vitro* proceeded *via* an initial C–C coupling to give the biologically significant dibenzazonine (70). Subsequent oxidation to the diphenoquinone (74) and cyclization gave erysodienone (71) in 30% overall yield (Scheme 11). Thus, unexpectedly, formation of the nine-membered ring (70) must occur readily. Further model studies to investigate the structural requirements for formation of the intermediate dibenzazonine (70) have now been reported[48] by Barton's group, and their results support the mechanism outlined above. Their approach involved synthesis of the amide (75), ketone (76), and pentamethylene derivative (77) to investigate their reactions under various conditions designed to promote phenol oxidative coupling.

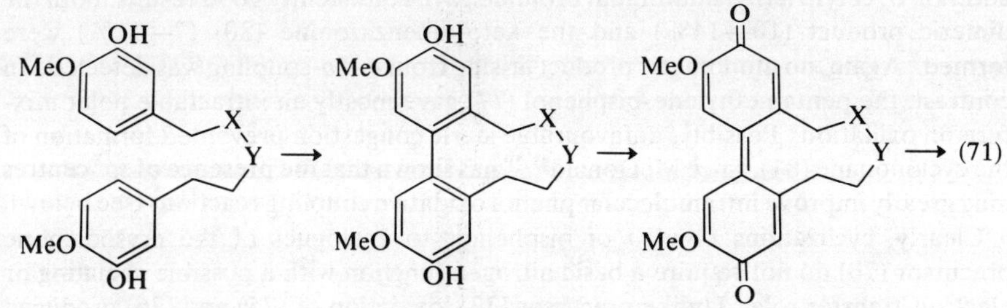

(73) $X = CH_2, Y = NH$
(75) $X = CO, Y = NH$
(76) $X = CO, Y = CH_2$
(77) $X = Y = CH_2$

(70) $X = CH_2, Y = NH$
(78) $X = CO, Y = NH$
(80) $X = CO, Y = CH_2$
(81) $X = Y = CH_2$

(74) $X = CH_2, Y = NH$
(82) $X = CO, Y = NH$
(83) $X = CO, Y = CH_2$

Scheme 11

The amide (75) was prepared[48] by mostly standard methods. Oxidation with alkaline potassium hexacyanoferrate(III) in chloroform and water gave the azoninone derivative (78) in 10—12% yield (allowing for recovered starting material); various modifications of the reaction conditions did not alter the yield significantly. Oxidation of (75) with vanadium oxychloride gave the product (78) in 16% yield. Oxidations with manganese dioxide (with or without silica as diluent) or the iron(III) chloride–DMF complex $[Fe(DMF)_3Cl_2]^+$ $[FeCl_4]^-$ gave no dibenzazonine (78). In each case, the structure of the product was shown to be that due to *para–para* phenol oxidative coupling, and no product arising from *ortho*-coupling could be detected. However, a dimeric product was also formed, but in much lower yield, and its structure was not determined.

Having established that the dibenzazonine (78) was readily obtained from the bisphenol (75), phenol oxidative coupling of the nitrogen-free analogues (76) and (77) was then examined. The required ketone (76) was prepared *via* alkylation of the isocyanide (79a) with the iodide (79b) and subsequent oxidative hydrolysis and

[46] J. E. Gervay, F. McCapra, T. Money, G. M. Sharma, and A. I. Scott, *J.C.S. Chem. Comm.*, 1966, 142; A. Mondon and M. Ehrhardt, *Tetrahedron Letters*, 1966, 2557.
[47] D. H. R. Barton, R. B. Boar, and D. A. Widdowson, *J. Chem. Soc. (C)*, 1970, 1208.
[48] A. G. M. Barrett, D. H. R. Barton, G. Franckowiak, D. Papioannou, and D. A. Widdowson, *J.C.S. Perkin I*, 1979, 662.

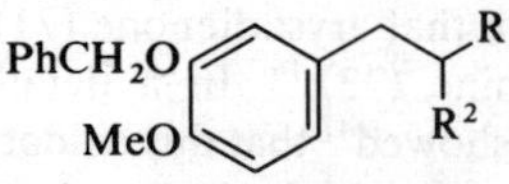

(79a) $R^1 = NC, R^2 = 4\text{-}MeC_6H_4SO_2$
(79b) $R^1 = H, R^2 = CH_2I$

hydrogenolysis.[48] Huang–Minlon reduction of the intermediate protected ketone followed by deprotection gave the corresponding alkane (77).

Phenol oxidative coupling of the keto-bisphenol (76), using alkaline potassium hexacyanoferrate(III) under various conditions, gave mostly polar material and a minor dimeric product. The use of 18-crown-6 as a phase-transfer catalyst increased the rate of reaction and traces of an additional compound were formed. However, addition of cetyltriethylammonium bromide gave consistently good results: both the dimeric product (10—14%) and the keto-dibenzazonine (80) (7—11%) were formed. Again, no monomeric product arising from *ortho*-coupling was detected. In contrast, the pentamethylene-bisphenol (77) gave mostly an intractable polar mixture on oxidation. Possibly, unfavourable steric congestion prevented formation of the cyclononane (81), since McDonald[49,50] has shown that the presence of sp^2 centres may greatly improve intramolecular phenol oxidative coupling reactions (see below).

Clearly, cyclizations *in vitro* of bisphenols to analogues of the erysodienone precursor (70) do not require a basic nitrogen function with a possible chelating or electron-transfer role. Unlike precursor (73), oxidation of (75) and (76) produced the cyclononane derivatives (78) and (80), respectively. These gave polar tars on further oxidation, presumably because neither amide nor ketone functions could trap the corresponding diphenoquinones (82) and (83) intramolecularly and thereby prevent competing intermolecular reactions (see Scheme 11). Thus these results provide further strong evidence in favour of the intermediacy of the dibenzazonine (70) in the formation of erysodienone (71) *in vitro*.

Recently, a series of B-homoerythrina alkaloids, *e.g.* schelhammeridine (84), has been found in plants of the genera *Schelhammera* and *Cephalotaxus*. It seems likely that these alkaloids are biosynthesized by a pathway analogous to that operating for the *Erythrina* alkaloids (Scheme 10), and in particular it is probable that the intermediates (85) and (86) are involved.[51] Furthermore, the dibenzazecine (85) might also serve as a precursor of the more unusual *Cephalotaxus* alkaloids, *e.g.* cephalotaxine (87)[52] (Scheme 12). No C-homoerythrina alkaloids have yet been reported, though, as McDonald has pointed out,[50] these might reasonably be expected from the dibenzazecine (85) *via* the 6;6-fused dienone (88)* (see Scheme 12).

* Closure of the ten-membered ring of (85) by oxidation and subsequent nucleophilic attack by nitrogen could lead to the formation of either two new six-membered rings, as in the 6;6-fused dienone (88), or new five- and seven-membered rings, as in the 5;7-fused dienone (86).

[49] E. McDonald and A. Suksamrarn, *J.C.S. Perkin I*, 1978, 434.
[50] E. McDonald and A. Suksamrarn, *J.C.S. Perkin I*, 1978, 440; preliminary communications, E. McDonald and A. Suksamrarn, *Tetrahedron Letters*, 1975, 4421, 4425.
[51] See A. R. Battersby, E. McDonald, J. A. Milner, S. R. Johns, J. A. Lamberton, and A. A. Sioumis, *Tetrahedron Letters*, 1975, 3419.
[52] See J. N. Schwab, M. N. T. Chang, and R. J. Parry, *J. Amer. Chem. Soc.*, 1977, **99**, 2368.

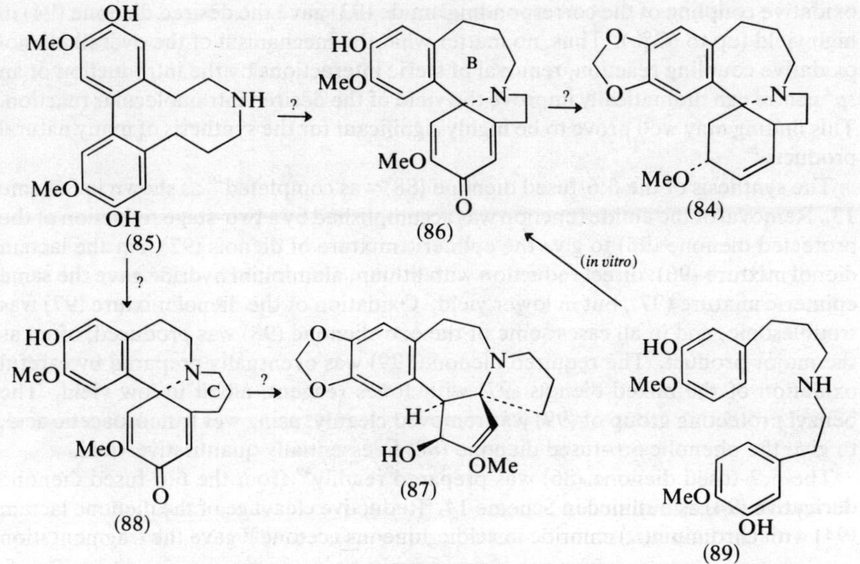

Scheme 12

Kametani had found that oxidation of the amino-bisphenol (89) with alkaline potassium hexacyanoferrate(III) gave, in 4% yield, a dienone, presumably formed *via* (85), to which he initially assigned[53] the 6;6-fused structure (88). After further investigation he changed the assignment in favour of the 5;7-fused dienone (86).[54]

McDonald *et al.* have now reported[49,50] the synthesis of the unknown C-homo-erysodienone (88). This synthesis,[49] outlined in Scheme 13, is of major significance since it has led to a better understanding of some of the factors involved in oxidative coupling reactions of phenols.[50]

The authors had originally envisaged that the best route to the required dienone (88) would be the reduction of (90), deprotection, and phenol oxidative coupling of the resulting bisphenol (91). However, inspection of molecular models revealed a set of three 1,3-diaxial interactions between the substituents at the marked atoms in the intermediate (92). Thus they concluded that the desired *intramolecular* phenol oxidative couplings would be unlikely, and that oxidation would favour *intermolecular* coupling, leading eventually to polymeric products. It was also evident that if one of the sp^3-hybridized carbon atoms C-9 and C-11 were replaced by an sp^2-hybridized atom, *two* of the three troublesome interactions would be removed.[50] The authors have now shown that these considerations are essentially correct.

The *N*-acetyltetrahydroquinoline (93) was prepared in the protected form (90), using standard procedures.[50] Reduction of (90) with lithium aluminium hydride gave the corresponding amine, which gave the amino-bisphenol (91) on hydrogenolysis. Oxidation of (91) under various conditions gave only complex mixtures, and none of the required dienone (88) could be detected. However, phenol

[53] T. Kametani and K. Fukumoto, *Chem. Comm.*, 1968, 26.
[54] T. Kametani and K. Fukumoto, *J. Chem. Soc.* (*C*), 1968, 2156.

oxidative coupling of the corresponding amide (93) gave the desired dienone (94) in high yield (up to 67%). Thus, no matter what the mechanism of the overall phenol oxidative coupling reaction, removal of steric interactions by the introduction of an sp^2 centre can dramatically improve the yield of the desired intramolecular reaction. This finding may well prove to be highly significant for the synthesis of many natural products.[50]

The synthesis of the 6;6-fused dienone (88) was completed[49] as shown in Scheme 13. Removal of the amide function was accomplished by a two-stage reduction of the protected dienone (95) to give the epimeric mixture of dienols (97), *via* the lactam dienol mixture (96); direct reduction with lithium aluminium hydride gave the same epimeric mixture (97), but in lower yield. Oxidation of the dienol mixture (97) was troublesome, and in all cases some of the oxo-dienone (98) was produced, often as the major product. The required dienone (99) was eventually prepared by careful oxidation of the mixed dienols (97) with Jones reagent, albeit in low yield. The benzyl protecting group of (99) was removed cleanly, using wet trifluoroacetic acid, to give the phenolic 6;6-fused dienone (88) in essentially quantitative yield.

The 5;7-fused dienone (86) was prepared readily[49] from the 6;6-fused dienone derivative (94) as outlined in Scheme 14. Reductive cleavage of the dienone lactam (94) with chromium(II) chloride in acidic aqueous acetone[55] gave the fragmentation

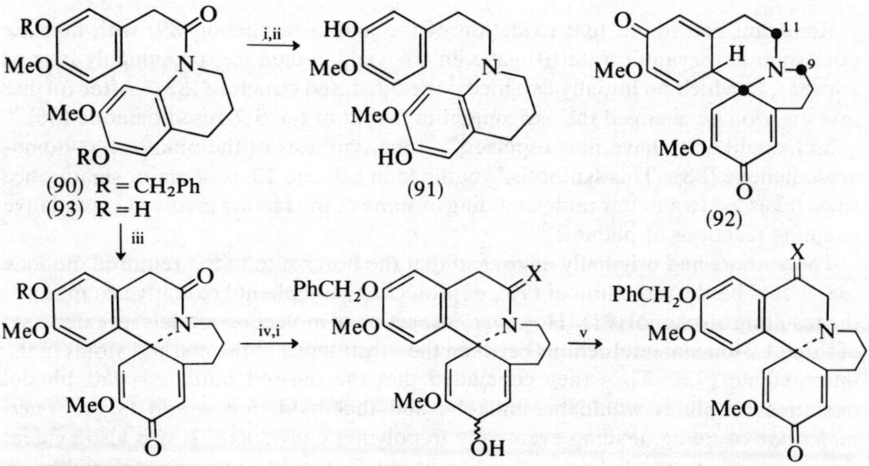

Reagents: i, $LiAlH_4$; ii, H_2–10% Pd/C; iii, 1:1 $CHCl_3$–5% $NaHCO_3$, $K_3[Fe(CN)_6]$; iv, $NaBH_4$; v, Jones oxidation, 0 °C, 45 s; vi, 3:1 CF_3CO_2H–H_2O

Scheme 13

[55] D. H. R. Barton, R. James, G. W. Kirby, D. W. Turner and D. A. Widdowson, *J. Chem. Soc.* (*C*), 1968, 1529.

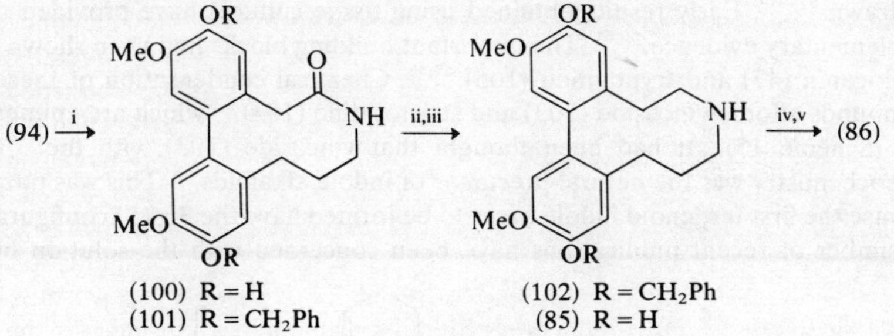

Reagents: i, $CrCl_2$ in 1:2 Me_2CO–3% HCl, N_2, 2 h; ii, K_2CO_3–MeOH, $PhCH_2Cl$; iii; $LiAlH_4$; iv, H_2–10% Pd/C, HCl–MeOH; v, 1:1 $CHCl_3$–5% $NaHCO_3$, $K_3[Fe(CN)_6]$

Scheme 14

product (100)* in 87% yield. Benzylation gave the lactam (101), which was reduced to the corresponding amine (102) with lithium aluminium hydride. Hydrogenolysis of (102) gave the diphenolic dibenzazecine (85), (a likely biosynthetic precursor of the *Schelhammera* alkaloids) in 44% overall yield from the 6;6-fused dienone (94). Two-phase oxidation of the bisphenol (85) with potassium hexacyanoferrate(III) gave the 5;7-fused dienone (86) in 61% yield, together with a trace of an unidentified product.

Marino and Samanen have reported[56] a different synthesis of the dibenzazecine (85). They obtained a 3:1 mixture of B-homoerysodienone (86) and C-homoerysodienone (88) on oxidation of the bisphenol (85), using a slightly different two-phase system.

Both of these studies *in vitro*,[49,50,56] together with Kametani's conversion of the open-chain bisphenol (89) intc B-homoerysodienone (86)[54] (see above), provide firm supporting evidence for the possible biosynthetic origin of the *Schelhammera* and *Cephalotaxus* alkaloids suggested in Scheme 12. Furthermore, oxidative cyclization of diphenolic dibenzazecines, *e.g.* (85), appears to provide a regioselective[49] and efficient synthetic entry to alkaloids of the *Schelhammera* type.

6 Terpenoid Indole Alkaloids

Recent publications concerned with the role of vincoside (103) and strictosidine (isovincoside) (104)† in the biosynthesis of terpenoid indole alkaloids are clearly most significant. However, this aspect is only a small part of the biosynthesis of these alkaloids, and must be seen in this perspective.

Extensive research on the elaboration of terpenoid indole alkaloids in whole plants by several research groups allowed a fairly clear picture of their biosynthesis to

* The correct structure of this fragmentation product is as formulated in (100) (Scheme 14), in agreement with the spectral data given in ref. 49 for this product and its dibenzyl ether (101). The lactam carbonyl group has been misplaced in the structures given for these two compounds in ref. 49 (and the names used reflect these incorrect structures); E. McDonald, personal communication.

† To simplify the literature, it is best that (104) should be known as strictosidine[62] and (103) as vincoside.

[56] J. P. Marino and J. H. Samanen, *J. Org. Chem.*, 1976, **41**, 179.

be drawn.[57—59] Early results obtained using tissue cultures have provided useful supplementary evidence.[60,61] The important building blocks had been shown to be secologanin (47) and tryptamine (105).[57,58] Chemical condensation of these two compounds affords vincoside (103) and strictosidine (104),[62] which are epimeric at C-3 (Scheme 15). It had been thought that vincoside (103) with the 3β (*R*) stereochemistry was the natural precursor of indole alkaloids.[63] This was puzzling, because the first terpenoid indole bases to be formed have the 3α (*S*) configuration. A number of recent publications have been concerned with the solution of this

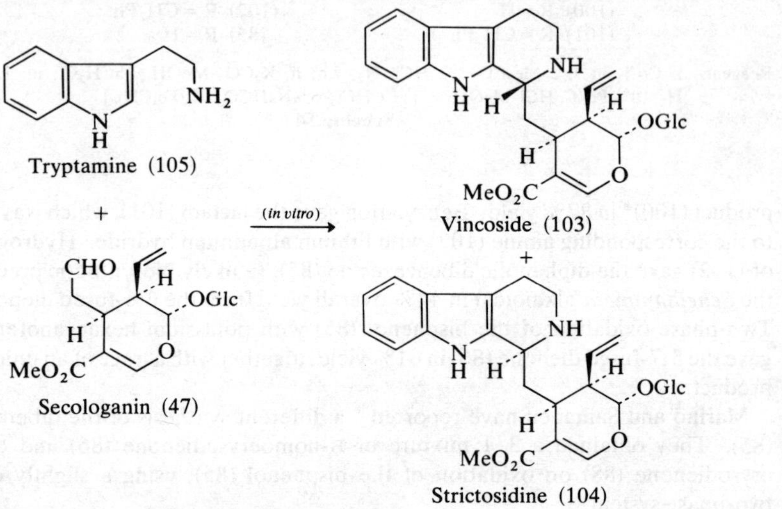

Scheme 15

When [2-^{14}C]tryptamine and secologanin (47) were incubated with an enzyme preparation from a *Catharanthus roseus* (syn. *Vinca rosea*) cell culture, in the presence of a β-glucosidase inhibitor or at low pH, strictosidine (104), and *not* vincoside (103), was formed; the reaction was clearly enzyme-dependent.[64,65] This crude enzyme preparation also converted labelled strictosidine (104) into ajmalicine (106), 19-*epi*-ajmalicine (107), and tetrahydroalstonine (108) in the presence of

57 A. R. Battersby, *Pure Appl. Chem.*, 1967, **14**, 117; A. R. Battersby, in 'The Alkaloids', ed. J. E. Saxton, (Specialist Periodical Reports), The Chemical Society, London, 1971, Vol. 1, p. 31.
58 For reviews, see (*a*) G. A. Cordell, *Lloydia*, 1974, **37**, 219; (*b*) R. J. Parry, in 'The *Catharanthus* Alkaloids', ed. W. I. Taylor and N. R. Farnsworth, Dekker, New York, 1975, p. 141; (*c*) R. B. Herbert, in 'Comprehensive Organic Chemistry', ed. D. H. R. Barton and W. D. Ollis, Pergamon, Oxford, 1978, Vol. 5, p. 1045. See also 'The Alkaloids', (Specialist Periodical Reports), The Chemical Society, London 1972–77, Vols. 2—7; R. B. Herbert, in ref. 2, p. 27.
59 R. B. Herbert in ref. 3, p. 18.
60 A. I. Scott and S.-L. Lee, *J. Amer. Chem. Soc.*, 1975, **97**, 6906; A. I. Scott, S.-L. Lee, and W. Wan, *Biochem. Biophys. Res. Comm.*, 1977, **75**, 1004.
61 J. Stöckigt, J. Treimer, and M. H. Zenk, *F.E.B.S. Letters*, 1976, **70**, 267; J. Stöckigt, H. P. Husson, C. Kan-Fan, and M. H. Zenk, *J.C.S. Chem. Comm.*, 1977, 164.
62 G. N. Smith, *Chem. Comm.*, 1968, 912; K. T. D. DeSilva, G. N. Smith, and K. E. Warren, *ibid.*, 1971, 905.
63 A. R. Battersby, A. R. Burnett, and P. G. Parsons, *J. Chem. Soc.* (*C*), 1969, 1193.
64 J. Stöckigt and M. H. Zenk, *F.E.B.S. Letters*, 1977, **79**, 233.
65 J. Stöckigt and M. H. Zenk, *J.C.S. Chem. Comm.*, 1977, 646.

NADPH. In the absence of NADPH, cathenamine (109) was produced, as noted previously[61] (see Scheme 16).

Cell-free preparations from cell-suspension cultures of various plants known to produce terpenoid indole alkaloids were also found to convert secologanin (47) and tryptamine (105) into strictosidine (104); the formation of vincoside (103) was never observed.[64,65] Thus it became clear that it must be strictosidine (104), and *not* vincoside (103), which is the natural precursor of the terpenoid indole alkaloids. This finding was also confirmed for whole-plant biosynthesis.[65] Doubly-labelled strictosidine (104) was incorporated, in *C. roseus*, into representatives of the three main groups of alkaloids, *i.e.* ajmalicine (106), serpentine (110), vindoline (111), and catharanthine (112), without significant changes of isotope ratio, indicating that there was intact incorporation. In parallel experiments, vincoside (103) was not utilized for alkaloid biosynthesis.[64,65]

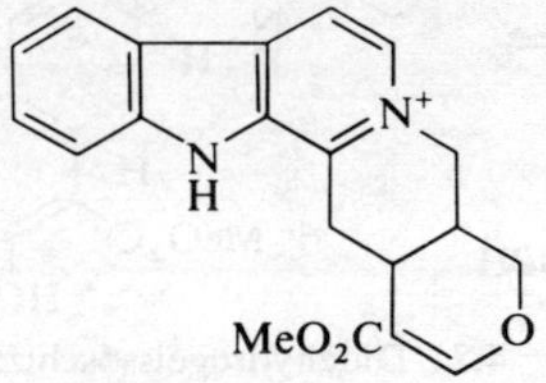

Serpentine (110)

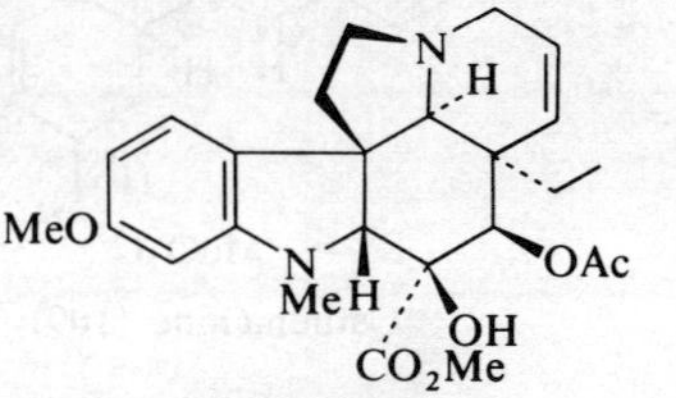

Vindoline (111)

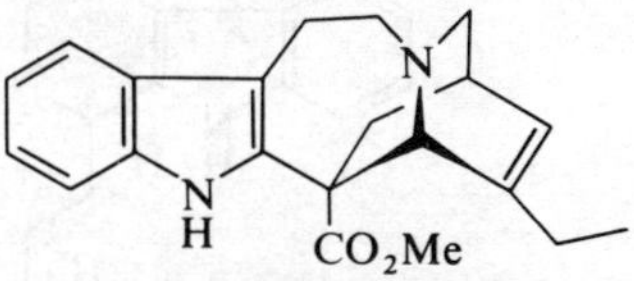

Catharanthine (112)

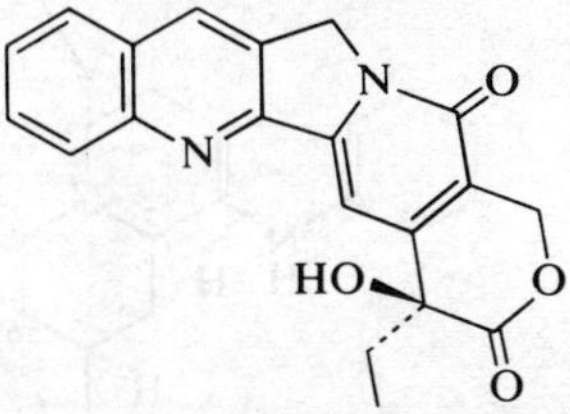

Camptothecin (113)

Many recent studies[33,59,66—69] have put the above finding beyond doubt. Furthermore, camptothecin (113), which had previously been deduced to arise from strictosidine lactam,[70] is now known to be derived initially from strictosidine (104).[71] Thus it is now clear that terpenoid indole alkaloids belonging to the *Corynanthe* (3α and 3β series), *Aspidosperma*, and *Iboga* types from various plant families, and related alkaloids, are all derived from strictosidine (104), the first terpenoid indole base on the biosynthetic pathway to all such alkaloids.[69]

Further advances in our understanding of the biosynthesis of the heteroyohimbine and related alkaloids have now been made as a result of detailed studies with enzyme preparations from *Catharanthus roseus* cell cultures.[72] Early work with such systems

[66] A. I. Scott, S. L. Lee, P. de Capite, M. G. Culver, and C. R. Hutchinson, *Heterocycles*, 1977, **7**, 979.
[67] M. Rueffer, N. Nagakura, and M. H. Zenk, *Tetrahedron Letters*, 1978, 1593.
[68] R. T. Brown, J. Leonard, and S. K. Sleigh, *Phytochemistry*, 1978, **17**, 899.
[69] N. Nagakura, M. Rüffer, and M. H. Zenk, *J.C.S. Perkin I*, 1979, 2308.
[70] C. R. Hutchinson, A. H. Heckendorf, P. E. Daddona, E. Hagaman, and E. Wenkert, *J. Amer. Chem. Soc.*, 1974, **96**, 5609.
[71] A. H. Heckendorf and C. R. Hutchinson, *Tetrahedron Letters*, 1977, 4253.
[72] M. Rueffer, C. Kan-Fan, H.-P. Husson, J. Stöckigt, and M. H. Zenk, *J.C.S. Chem. Comm.*, 1979, 1016.

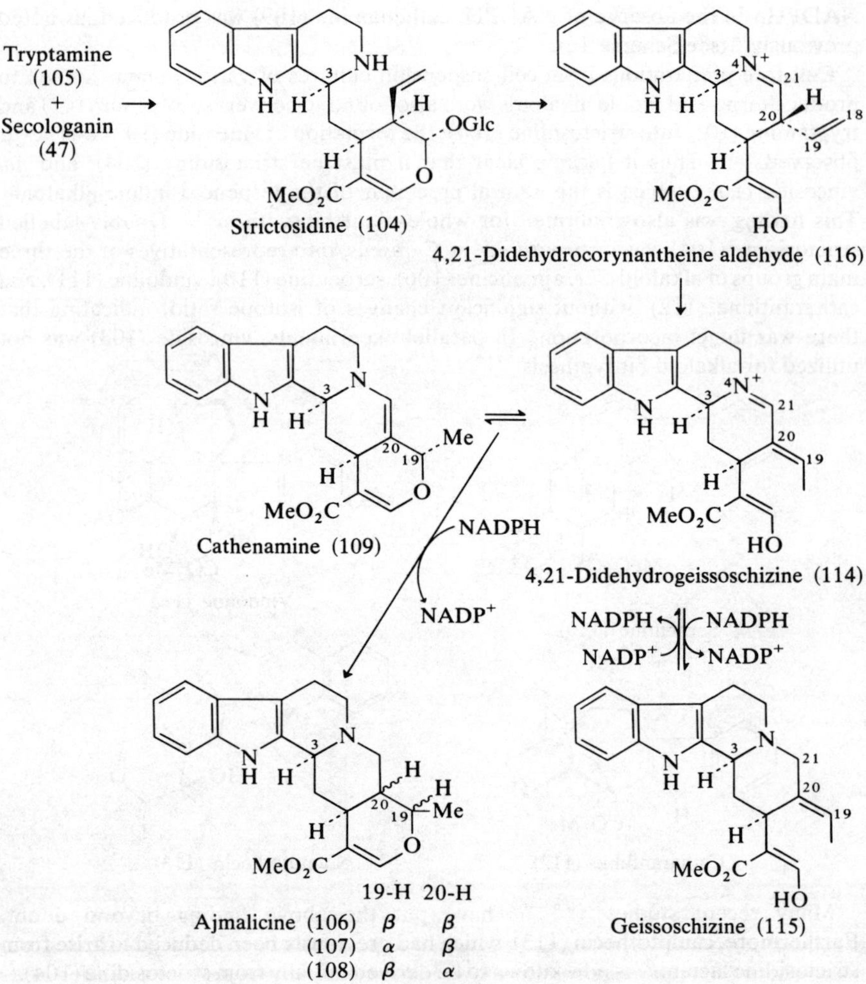

Scheme 16

had revealed[61] the intermediacy of cathenamine (109) in ajmalicine biosynthesis. The clue to another important intermediate came from a study of the plant *Guettarda eximia* (Rubiaceae), known[73] to be a rich source of strictosidine (104), from which 4,21-didehydrogeissoschizine (114) was isolated.[74] Chemical transformations[74] of (114) lead to either cathenamine (109) or geissoschizine (115). These biomimetic reactions reflect closely the postulated biosynthetic expectations of an intermediate located between 4,21-didehydrocorynantheine aldehyde (116) and cathenamine (109).

[73] H.-P. Husson, C. Kan-Fan, T. Sevenet, and J. P. Vidal, *Tetrahedron Letters*, 1977, 1889.
[74] C. Kan-Fan and H.-P. Husson, *J.C.S. Chem. Comm.*, 1979, 1015.

Rigorous studies[72] with enzyme preparations from *C. roseus* cell cultures established 4,21-didehydrogeissoschizine (114) as an obligatory intermediate in the biosynthetic pathway to the heteroyohimbine alkaloids (Scheme 16). Furthermore, it was shown that, as well as being the immediate precursor of (109), 4,21-didehydrogeissoschizine is transformed under NADPH-regenerating conditions into geissoschizine (115), which is known *not* to be a direct precursor of ajmalicine (106) and its isomers (107) and (108).[75] The authors concluded[72] that 4,21-didehydrogeissoschizine (114) occupies a crucial position at the branch point in the biosynthesis of the *Iboga*, *Aspidosperma*, and *Corynanthe* alkaloids.[58]

J. Stöckigt, *J.C.S. Chem. Comm.*, 1978, 1097.

16 Biological Chemistry
Part (iii) Insect Chemistry

By R. BAKER and J. W. S. BRADSHAW
Department of Chemistry, The University, Southampton SO9 5NH

1 Introduction

Since the last *Annual Report* on this subject,[1] the volume of literature on chemicals which mediate the behaviour and ecology of insects has increased, particularly during 1979, and while the scope of this review remains similar to that of the 1977 Report, it has proved necessary to limit coverage to the more detailed or important papers. A number of books and review articles have been published.[2]

2 Sex Attractants and Related Compounds

The discovery of attractants for male moths from a number of well-studied families has now become a routine procedure, and will not be reviewed here. The principle that the females of the majority of species produce more than one bioactive compound is well established, and the Table (see pages 406 and 407) lists structures from a number of detailed examinations.

The abdominal scent glands of female *Bombyx mori*, from which the first Lepidopteran sex pheromone, *i.e.* (*E*,*Z*)-hexadeca-10,12-dien-1-ol (bombykol), was isolated in 1959, have been found to contain the analogous aldehyde (bombykal)[3] and the (*E*,*E*)-isomer of the alcohol.[4] The first Lepidopteran potentially chiral female sex pheromone (1) has been reported from *Adoxophyes orana fasciata*.[5] No less than seven compounds, all but one of them aldehydes, have been isolated from

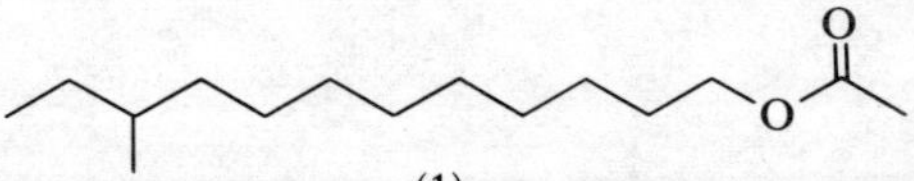

(1)

[1] R. Baker and D. A. Evans, *Ann. Reports* (*B*), 1977, **74**, 367.
[2] 'Biochemistry of Insects', ed. M. Rockstein, Academic Press, London, 1978; 'Arthropod Venoms', ed. S. Bettini, Springer, Berlin, 1978; E. Kramer, *Recept. Recognition, Ser. B.*, 1978, **5**, 231; J. Meinwald, G. D. Prestwich, K. Nakanishi, and I. Kubo, *Science*, 1978, **199**, 1167; R. Baker and D. A. Evans, in 'Aliphatic and Related Natural Product Chemistry', ed. F. D. Gunstone, (Specialist Periodical Reports). The Chemical Society, London 1979, Vol. 1, p. 102; 'Chemical Ecology: Odour Communications in Animals', ed. F. J. Ritter, Elsevier, Amsterdam, 1979; J. A. Pickett, *Educ. Chem.*, 1979, **16**, 44.
[3] G. Kasang, K. E. Kaissling, O. Vostrowsky, and H. J. Bestmann, *Angew. Chem. Internat. Edn.*, 1978, **17**, 60.
[4] G. Kasang, D. Schneider, and W. Schaefer, *Naturwiss.*, 1978, **65**, 337.
[5] Y. Tamaki, H. Noguchi, H. Sugie, R. Sato, and A. Kariya, *Appl. Entomol. Zool.*, 1979, **14**, 101.

the ovipositors of *Heliothis virescens*, and four from *Heliothis zea*.[6] A four-component pheromone has been demonstrated for a New World population of *Grapholitha molesta*,[7] and the role of each component in mediating behaviour of males has been investigated in detail.[8] One of these four components, dodecan-1-ol, could not be detected in a French population of the same species;[9] the role of ratios of (*E*)- and (*Z*)-dodec-8-en-1-yl acetates in maintaining sexual isolation in four species of the genus has been demonstrated.[10] Two forms of the European corn borer, *Ostrinia nubilalis*, which produce (and respond to) opposite (97:3 and 4:96) ratios of (*Z*)- and (*E*)-tetradec-11-en-1-yl acetates, have been investigated biochemically[11] and genetically,[12] to determine whether the differences in pheromones have led to reproductive isolation of the strains, and hence to the formation of two species. Geographical variations in compositions of sex pheromones within a species are probably more widespread than the current literature indicates.

Scent-scales on the wings of *Pieris* butterflies produce a mixture of terpenes (α-pinene, β-pinene, myrcene, *p*-cymene, limonene, neral, geranial, and linalool) and n-undecane; these compounds may be useful chemotaxonomic criteria.[13]

In the Coleoptera, four compounds (2)—(5), isolated from the frass of female boll weevils (*Anthonomus grandis*), attract males in the laboratory.[14] Females of the pecan weevil, *Curculio caryae*, also produce (3).[15] The sex pheromone of the drugstore beetle, *Stegobium paniceum*, is reported as (6),[16] while that of the cigarette beetle, *Lasioderma serricorne*, is (7).[17] Further investigations of the sex pheromones of the dermestid *Trogoderma granarium* have shown that the (−)-(*R*)-isomers of (8a)

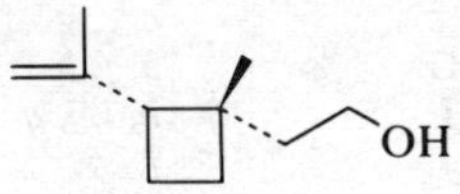

(2)

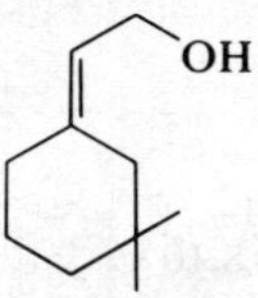

(3)

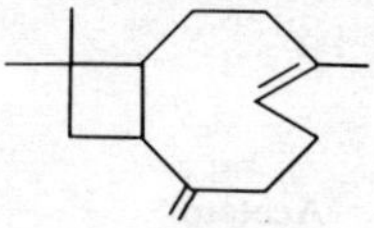

(4)

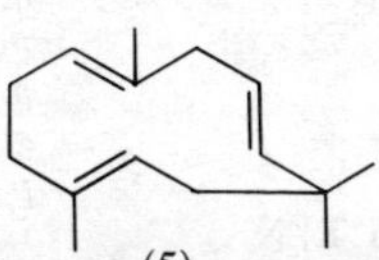

(5)

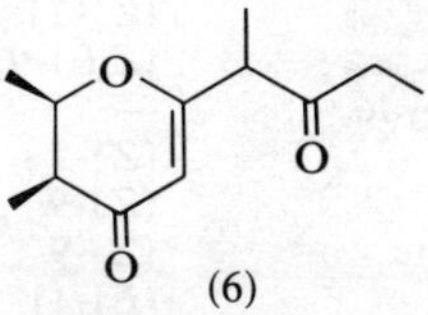

(6)

[6] J. A. Klun, J. R. Plimmer, B. A. Bierl-Leonhardt, A. N. Sparks, and O. L. Chapman, *Science*, 1979, **204**, 1328.
[7] A. M. Carde, T. C. Baker, and R. T. Carde, *J. Chem. Ecol.*, 1979, **5**, 423.
[8] T. C. Baker and R. T. Carde, *Environ. Entomol.*, 1979, **8**, 956.
[9] G. Biwer, C. Descoins, and M. Gallois, *Compt. rend.*, 1979, **288**, *D*, 413.
[10] G. Biwer and C. Descoins, *Compt. rend.*, 1978, **286**, *D*, 875.
[11] R. T. Carde, W. L. Roelofs, R. G. Harrison, A. T. Vawter, P. F. Brussard, A. Mutuura, and E. Munroe, *Science*, 1978, **199**, 555.
[12] J. A. Klun and S. Maini, *Environ. Entomol.*, 1979, **8**, 423.
[13] N. Hayashi, Y. Kuwahara, and H. Komae, *Experientia*, 1978, **34**, 684.
[14] P. A. Hedin, G. H. McKibben, E. B. Mitchell, and W. L. Johnson, *J. Chem. Ecol.*, 1979, **5**, 617.
[15] P. A. Hedin, J. A. Payne, T. L. Carpenter, and W. Neal, *Environ. Entomol.*, 1979, **8**, 521.
[16] Y. Kuwahara, H. Fukami, R. Howard, S. Ishii, F. Matsumura, and W. E. Burkholder, *Tetrahedron*, 1978, **34**, 1769.
[17] T. Chuman, M. Kohno, K. Kato, and M. Noguchi, *Tetrahedron Letters*, 1979, 2361.

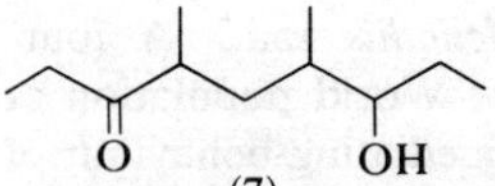

(7)

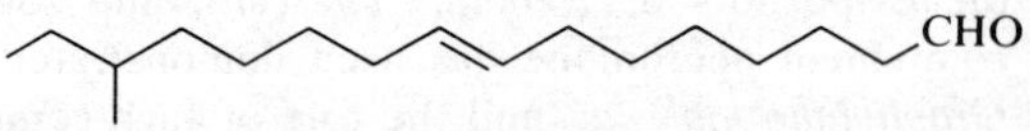

(8a) (8b has *Z* geometry)

and (8b) are much less effective in inducing attraction of unmated males than are the (+)-(*S*)-isomers.[18] A female-produced sex pheromone has been demonstrated in the Colorado beetle, *Leptinotarsa decemlineata.*[19]

The majority of the so-called sex pheromones of Diptera (true flies) operate as close-range or contact sex-recognition stimuli rather than as long-range attractants. The alkenes from the cuticle of the stable fly, *Stomoxys calcitrans*, have been described; that most active in inducing male copulatory behaviour is 13-methyl-

Table *Some female sex pheromones of Lepidoptera*

Parent chain length	*Terminal functional group*	*Unsaturation*	*Species*	*Refs.*
C_{10}	Acetate	(*Z*)-5	A, B	*a, b*
		(*Z*)-7	B	*b*
C_{12}	Alcohol	—	C, D	*c, d*
		(*Z*)-8	C	*c, e*
		(*E*)-9	E	*f*
	Acetate	—	D, F, G	*d, g, h*
		(*Z*)-7	J	*k*
		(*E*)-8	C, F(?)	*c, e, g*
		(*Z*)-8	C, F	*c, e, g*
		(*E*)-9	D, E, G	*d, f, h*
		(*Z*)-9	G, H, I	*h, i, j, q*
		(*Z*)-11	I	*j*
		(*E,E*)-8,10	F	*g*
	Acetate (10-methyl)	—	K	*l*
C_{14}	Alcohol	(*Z*)-9	L	*m*
		(*Z*)-11	M, N	*n, o*
		(*Z,E*)-9,12	L	*m*
	Aldehyde	—	O	*p*
		(*Z*)-7	P	*q*
		(*Z*)-9	O	*p*
	Acetate	(*Z*)-9	H, J, K, N	*i, k, l, o*
		(*E*)-11	K, M	*l, n*
		(*Z*)-11	I, K, M, N, Q	*j, l, n, o, r*
C_{16}	Alcohol	(*Z*)-11	O, R, S(?)	*p, s, t*
		(*E,E*)-10,12	T	*u*
	Aldehyde	—	O, U	*p*
		(*Z*)-7	O, U	*p*
		(*Z*)-9	O, U	*p*
		(*Z*)-11	O, R, S, U, V	*p, s, t, v*
		(*E, Z*)-10, 12	T, W	*w, x*
	Acetate	(*Z*)-11	V, X	*v, y*
C_{18}	Alcohol	(*Z,Z*)-3,13	Y	*z*
	Acetate	(*Z,Z*)-3,13	Y	*z*

[18] R. Rossi, P. A. Salvadori, A. Carpita, and A. Niccoli, *Naturwiss.*, 1979, **66**, 211.
[19] H. Z. Levinson, A. R. Levinson, and T. L. Jen, *Naturwiss.*, 1979, **66**, 472.

Species: A, *Agrotis segetum*; B, *A. fucosa*; C, *Grapholitha molesta*; D, *Homona coffearia*; E, *Rhyacionia subtropica*; F, *Hedya nubiferana*; G, *Cnephasia pumicana*; H, *Spodoptera frugiperda*; I, *Homona magnanima*; J, *Agrotis ipsilon*; K, *Adoxophyes orana*; L, *Homoeosoma electellum*; M, *Choristoneura rosaceana*; N, *Pandemis heperana*; O, *Heliothis* (*Helicoverpa*) *virescens*; P, *Prays oleae*; Q, *Tortrix viridana*; R, *Chilo partellus*; S, *Chrysoteuchia topiaria*; T, *Bombyx mori*; U, *Heliothis* (*Helicoverpa*) *zea*; V, *Plutella xylostella*; W, *Manduca sexta*; X, *Mamestra brassicae*; Y, *Aegeria tibialis.*

(*a*) H. J. Bestmann, O. Vostrowsky, K. H. Koschatzky, H. Platz, T. Brosche, I. Kantardjiew, M. Rheinwald, & W. Knauf *Angew. Chem. Internat. Edn.* 1978 **17** 768; (*b*) S. Wakamura *Appl. Entomol. Zool.* 1978 **13** 290; (*c*) A. M. Carde, T. C. Baker, & R. T. Carde *J. Chem. Ecol.* 1979 **5** 423; (*d*) J. P. Kochansky, W. L. Roelofs, & P. Sivapalen *ibid.* 1978 **4** 623; (*e*) G. Biwer, C. Descoins, & M. Gallois *Compt. rend.* 1979 **288** *D* 413; (*f*) W. L. Roelofs, A. S. Hill, C. W. Berisford, & J. F. Godbee *Environ. Entomol.* 1979 **8** 894; (*g*) B. Frerot, E. Priesner, & M. Gallois *Z. Naturforsch.* 1979 **34c** 1248; (*h*) G. Biwer, C. Descoins, M. Gallois, E. Priesner, J. P. Chambon, G. Genestier, & M. Martinez *Ann. Zool.-Ecol. Anim.* 1978 **10** 129; (*i*) R. L. Jones & A. N. Sparks *J. Chem. Ecol.* 1979 **5** 721; (*j*) H. Noguchi, Y. Tamaki, & T. Yushima *Appl. Entomol. Zool.* 1979 **14** 225; (*k*) A. S. Hill, R. W. Rings, S. R. Swier, & W. L. Roelofs *J. Chem. Ecol.* 1979 **5** 439; (*l*) Y. Tamaki, H. Noguchi, H. Sugie, R. Sato, & A. Kariya *Appl. Entomol. Zool.* 1979 **14** 101; (*m*) E. W. Underhill, A. P. Arthur, M. D. Chisholm, & W. F. Steck *Environ. Entomol.* 1979 **8** 740; (*n*) A. S. Hill & W. L. Roelofs *J. Chem. Ecol.* 1979 **5** 3; (*o*) B. Frerot, M. Gallois, & J. Einhorn *Compt. rend.* 1979 **288** *D* 1611; (*p*) J. A. Klun, J. R. Plimmer, B. A. Bierl-Leonhardt, A. N. Sparks, & O. L. Chapman *Science* 1979 **204** 1328; (*q*) D. G. Campion, L. J. McVeigh, J. Polyrakis, S. Michaelakis, G. N. Stravrakis, P. S. Beevor, D. R. Hall, & B. F. Nesbitt *Experientia* 1979 **35** 1146; M. Renon, C. Descoins, E. Priesner, M. Gallois, & M. Lettere *Compt. rend.* 1979 **288** *D* 1559; (*r*) H. Arn, E. Priesner, H. Bogenschutz, H. R. Buser, D. L. Struble, S. Rauscher, & S. Voerman *Z. Naturforsch.* 1979 **34c** 1281; (*s*) B. F. Nesbitt, P. S. Beevor, D. R. Hall, R. Lester, J. C. Davies, & K. V. S. Reddy *J. Chem. Ecol.* 1979 **5** 153; (*t*) L. M. McDonough & J. A. Kamm *ibid.* 1979 **5** 211; (*u*) G. Kasang, D. Schneider, & W. Schaefer *Naturwiss.* 1978 **65** 337; (*v*) Y. S. Chow, Y. M. Lin, & C. L. Hsu *Bull. Inst. Zool., Acad. Sin.* 1977 **16** 99; (*w*) G. Kasang, K. E. Kaissling, O. Vostrowsky, & H. J. Bestmann *Angew. Chem. Internat. Edn.* 1978 **17** 60; (*x*) A. N. Starratt, K. H. Dahm, N. Allen, J. G. Hildebrand, T. L. Payne, & H. Roeller *Z. Naturforsch.* 1979 **34c** 9; (*y*) H. J. Bestmann, O. Vostrowsky, K. H. Koschatzky, H. Platz & A. Szymanska *Tetrahedron Letters* 1978, 605; Y. Hirai, H. Kimura, K. Kawasaki, & Y. Tamaki *Appl. Entomol. Zool.* 1978 **13** 136; (*z*) E. W. Underhill, W. Steck, M. D. Chisholm, H. A. Worden, & J. A. G. Howe *Canad. Entomol.* 1978 **110** 495.

---

tritriacont-1-ene, and the activity is enhanced by adding (*Z*)-hentriacont-9-ene and the previously reported 11-methylhentriacontane.[20] The appearance, with increasing age, of sex-specific hydrocarbons in females of *Fannia* species, and their effects in stimulating male copulatory behaviour, have been reported.[21] The three most active compounds in the sex pheromone of the tsetse fly, *Glossina morsitans*, *i.e.* (9a), (9b), and (10),[22] may derive from a mixed isoprenoid and polyketide biosynthesis.

(9) a; $R^1 = C_{14}H_{29}$, $R^2 = C_{18}H_{37}$
b; $R^1 = R^2 = C_{16}H_{33}$

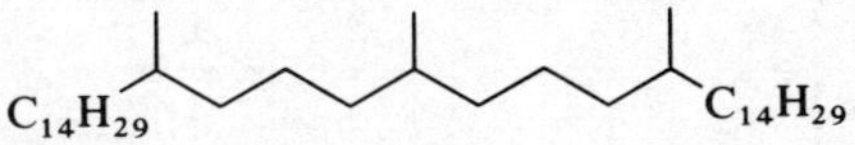

(10)

Chemoreception of these compounds in this species probably occurs *via* the feet (tarsi) rather than the antennae.[22] A contact mating pheromone has also been reported in the biting midge *Culicoides melleus*; the most active hydrocarbons from the cuticle are monomethyl-docosanes and -tricosanes.[23] Hexanal, isolated from whole extracts of *Sarcophaga bullata*, attracts females, but not males, of this species, and is thus a possible long-range sex pheromone.[24]

The structure of the major sex pheromone of the cockroach *Periplaneta americana* has been confirmed as (1*R*, 2*R*, 7*S*, 10*R*)-(5*E*)-1,2:10,14-diepoxygermacra-4(15),-

[20] P. E. Sonnet, E. C. Uebel, W. R. Lusby, M. Schwarz, and R. W. Miller, *J. Chem. Ecol.*, 1979, **5**, 353.

[21] E. C. Uebel, M. Schwarz, R. E. Menzer, and R. W. Miller, *J. Chem. Ecol.*, 1978, **4**, 73; E. C. Uebel, M. Schwarz, R. W. Miller, and R. E. Menzer, *ibid.*, p. 83.

[22] D. A. Carlson, P. A. Langley, and P. Huyton, *Science*, 1978, **201**, 750

[23] J. R. Linley and D. A. Carlson, *J. Insect Physiol.*, 1978, **24**, 423.

[24] J. E. Girard, F. J. Germino, J. P. Budris, R. A. Vita, and M. P. Garrity, *J. Chem. Ecol.*, 1979, **5**, 125.

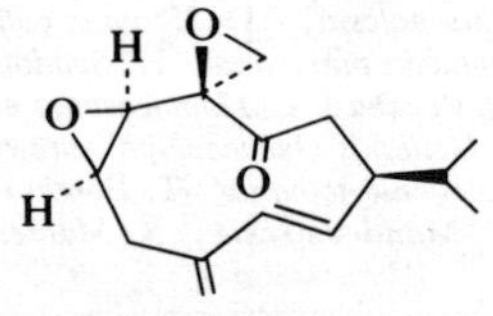

(11)

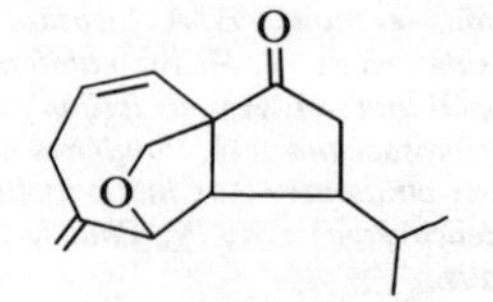

(12)

5-dien-9-one (11),[25] and the absolute configuration has been assigned by a c.d. study of the corresponding alcohol.[26] A second compound (12), isolated from the excreta of the same species, may be a natural degradation product.[27] One component of the sex pheromone of *Blatella germanica* has been shown to be the (*S,S*)-isomer of 3,11-dimethylnonacosan-2-one, but a mixture of epimers is reported to have the same behavioural effect as the single isomer produced by this coakroach.[28]

The female-produced sex pheromones of two species of scale insect have been shown to be related structures. Those of *Aonidiella aurantii*, the California red scale, are (13) and (14),[29] and (15) has been isolated from *A. citrina*, the yellow scale.[30]

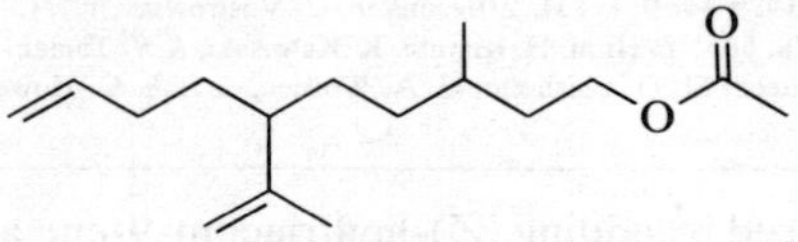

(13)

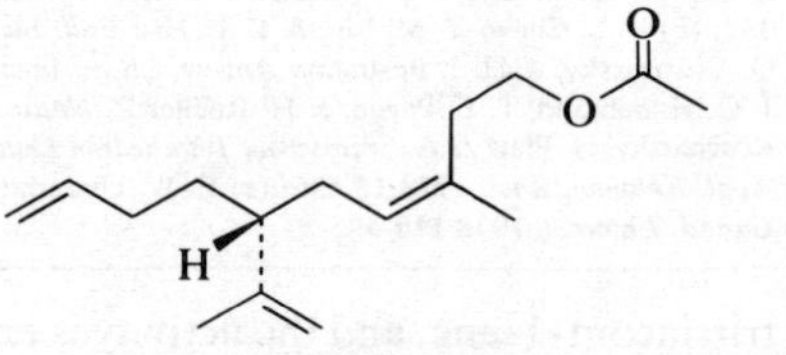

(14)

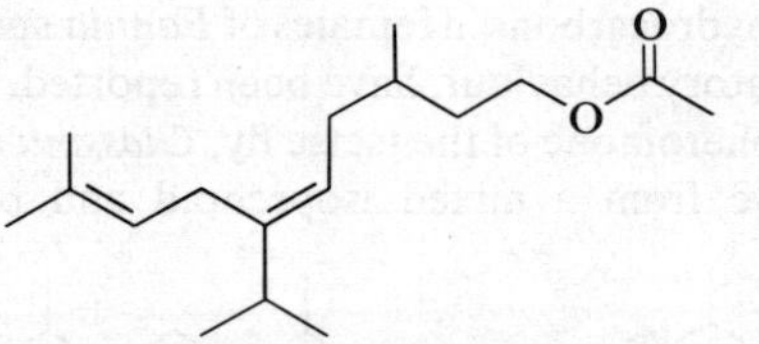

(15)

Compounds specific to males have been isolated from the abdominal glands of certain true bugs (Heteroptera). The dorsal abdominal glands of *Posidus maculiventris* contain representative compounds from aliphatic, terpenoid, and aromatic pathways.[31] A wide range of aromatics is found in the ventral abdominal gland of some *Leptoglossus* species, together with octan-1-ol.[32]

[25] C. J. Persoons, P. E. J. Verwiel, E. Talman, and F. J. Ritter, *J. Chem. Ecol.*, 1979, **5**, 221.
[26] M. A. Adams, K. Nakanishi, W. C. Still, E. V. Arnold, J. Clardy, and C. J. Persoons, *J. Amer. Chem. Soc.*, 1979, **101**, 2495.
[27] E. Talman, P. E. J. Verwiel, F . J. Ritter, and C. J. Persoons, *Israel J. Chem.*, 1978, **17**, 227.
[28] R. Nishida, Y. Kuwahara, H. Fumaki, and S. Ishii, *J. Chem. Ecol.*, 1979, **5**, 289.
[29] W. L. Roelofs, M. J. Gieselmann, A. Carde, H. Tashiro, D. S. Moreno, C. A. Henrick, and R. J. Anderson, *J. Chem. Ecol.*, 1978, **4**, 211; H. Tashiro, M. J. Gieselmann, and W. L. Roelofs, *Environ. Entomol.*, 1979, **8**, 931.
[30] M. J. Gieselmann, D. S. Moreno, J. Fargerlund, H. Tashiro, and W. L. Roelofs, *J. Chem. Ecol.*, 1979, **5**, 27.
[31] J. R. Aldrich, M. S. Blum, H. A. Lloyd, and H. M. Fales, *J. Chem. Ecol.*, 1978, **4**, 161.
[32] J. R. Aldrich, M. S. Blum, and H. M. Fales, *J. Chem. Ecol.*, 1979, **5**, 53.

3 Aggregation Pheromones and Population Attractants

Most of the studies in this area are concerned, as in previous years, with bark beetles (Scolytidae).[33] As in the Lepidoptera, population differences in pheromone chemistry have been established for a few species; in *Dendroctonus pseudotsugae*, populations from W. Oregon and Idaho produced different spectra of volatiles, including two compounds new to this group on insects, *i.e.* pent-3-en-1-ol and 6- (or 4-) methylhept-5-en-2-one. Cross-feeding tests indicated that the population differences were host-determined rather than genetic.[34] No neurophysiological differences were found between chemically distinct populations of *Ips pini*.[35] *cis*-Verbenol has been added to the aggregation pheromone blend of *I. acuminatus*. The (*S*)-isomers of *cis*-verbenol and ipsdienol attract this species, whereas the enantiomers suppress this response.[36] Females of *Dendroctonus jeffreyi* produce heptan-1-ol and heptan-2-ol, and these, together with heptane from the host, constitute the aggregation pheromone of this species.[37] Volatiles trapped from logs of *Ulmus procera* that were infested with *Scolytus scolytus* include (−)-*threo*- and (−)-*erythro*-4-methylheptan-3-ol, the former being produced only by the males.[38]

Males of the dermestid beetle *Dermestes maculatus* produce a unique blend of isopropyl esters in a subepidermal abdominal gland. The secretion of this gland attracts both males and females.[39] An aggregation pheromone, and possibly an alarm pheromone, are produced by certain stages of the subsocial bug *Oncopeltus fasciatus*.[40]

4 Chemistry of the Hymenoptera

The Hymenoptera exhibit the full range of sociality, from completely solitary species to the advanced social forms, such as certain bees, wasps, and the ants. From the sawflies (Symphyta), a relatively primitive group, adults of both sexes of *Neodiprion sertifer* contain (16);[41] this is produced separately from the sex pheromone. Macrocyclic lactones, such as 18-octadecanolide and the C_{20}, C_{22}, and C_{24} homologues, together with the corresponding ω-hydroxy-acids, have been reported from Dufour's gland in several genera of halictine[42] and *Colletes*[43] bees. In the latter genus at least, the brood cells are lined with a polyester, derived from these hydroxy-acids.[43] The mandibular glands of several species from the same genus contain linalool, neral, and geranial, which serve as aggregation pheromones.[44] A

33 M. C. Birch, *Amer. Sci.*, 1978, **66**, 409; K. Beck, *J. Chem. Educ.*, 1978, **55**, 567.
34 L. C. Ryker, L. M. Libbey, and J. A. Rudinsky, *Environ. Entomol.*, 1979, **8**, 789.
35 M. E. Angst and G. N. Lanier, *J. Chem. Ecol.*, 1979, **5**, 131.
36 A. Bakke, *Oikos*, 1978, **31**, 184.
37 J. A. A. Renwick and G. B. Pitman, *Environ. Entomol.*, 1979, **8**, 40.
38 M. M. Blight, L. J. Wadhams, and M. J. Wenham, *Insect Biochem.*, 1978, **8**, 135; *ibid.*, 1979, **9**, 525.
39 H. Z. Levinson, A. R. Levinson, T. L. Jen, J. L. D. Williams, G. Kahn, and W. Francke, *Naturwiss.*, 1978, **65**, 543; W. Francke, A. R. Levinson, T. L. Jen, and H. Z. Levinson, *Angew. Chem. Internat. Edn.*, 1979, **18**, 796.
40 T. Aller and R. L. Caldwell, *Physiol. Entomol.*, 1979, **4**, 287.
41 G. Ahlgren, G. Bergström, J. Lofqvist, A. Jansson, and T. Norin, *J. Chem. Ecol.*, 1979, **5**, 309.
42 A. Hefetz, M. S. Blum, G. C. Eickwort, and J. W. Wheeler, *Comp. Biochem. Physiol.*, 1978, **61B**, 129; G. Bergström and J. Tengö, *Acta Chem. Scand. (B)*, 1979, **33**, 390.
43 A. Hefetz, H. M. Fales, and S. W. T. Batra, *Science*, 1979, **204**, 415.
44 G. Bergström and J. Tengö, *J. Chem. Ecol.*, 1978, **4**, 437; A. Hefetz, S. W. T. Batra, and M. S. Blum, *Experientia*, 1979, **35**, 319.

cephalic secretion from the non-social sphecid wasp *Sceliphron caementarium* contains geranyl acetate and dec-2-en-1-ol, the latter a novel compound in arthropods.[45] Straight-chain alkanes and esters, and two sesquiterpene acetates, have been identified from various species of anthophorid bees;[46,47] the Dufour's gland of *Xylocarpa virginica* is used to mark flowers, and is a short-term repellent for conspecifics.[47]

The complex secretions of the mandibular gland of bumble bees, in this case several species of *Alpinobombus*, have received further attention.[48] Additional compounds have been reported from the stings of honeybees, *Apis mellifera*,[49] and the alarm pheromone from this source in *A. dorsata* and *A. florea* is dec-2-en-1-yl acetate.[50] Abdomens of the social wasp *Paravespula vulgaris* contin both (*E*)- and (*Z*)-isomers of the spiro-ketals (17) and (18), which may serve to protect individuals from attacks by their sister workers.[51] A brood-warming pheromone, (*Z*)-pentacos-9-ene, has been identified from pupae and brood cells of the hornet, *Vespa crabro*.[52]

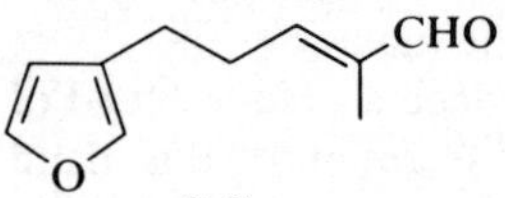

(16)

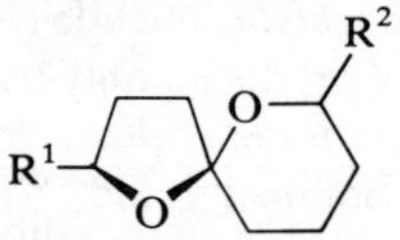

(17) $R^1 = H, R^2 = Me$
(18) $R^1 = Me, R^2 = H$

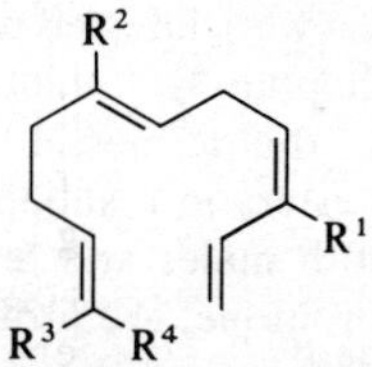

(19) a; $R^1 = R^2 = R^3 = R^4 = Me$
b; $R^1 + R^2 = Me + Et, R^3 = R^4 = Me$
c; $R^1 = R^2 = Et, R^3 + R^4 = Me + Et$

The ants (Formicidae) continue to be studied in detail. Reviews of chemistry, chemical communication, and chemosystematics have been published.[53] The species *Myrmica scabrinodis* has been investigated, and its pheromones have been compared to those of *M. rubra*.[54] The Dufour's gland of the former contains the terpenoids (19a), (19b), and (19c), which may be used in the marking of territories. Alkylpyrazines have been reported from a number of species. The leaf-cutting ant *Atta sexdens rubropilosa* produces (20) as a trail pheromone,[55] and related structures have been reported from ponerine ants[56] and an un-named species of *Calomyrmex*.[57] The latter produces a highly complex secretion from the mandibular glands, including

[45] A. Hefetz and S. W. T. Batra, *Experientia*, 1979, **35**, 1138.
[46] S. W. T. Batra and A. Hefetz, *Ann. Entomol. Soc. Amer.*, 1979, **72**, 514.
[47] S. B. Vinson, G. W. Frankie, M. S. Blum, and J. W. Wheeler, *J. Chem. Ecol.*, 1978, **4**, 315.
[48] B. G. Svensson and G. Bergström, *J. Chem. Ecol.*, 1979, **5**, 603.
[49] M. S. Blum, H. M. Fales, K. W. Tucker, and A. M. Collins, *J. Apicultural Res.*, 1978, **17**, 218.
[50] H. J. Veith, J. Weiss, and N. Koeniger, *Experientia*, 1978, **34**, 423.
[51] W. Francke, G. Hindorf, and W. Reith, *Angew. Chem. Internat. Edn.*, 1978, **17**, 862.
[52] H. J. Veith and N. Koeniger, *Naturwiss.*, 1978, **65**, 263.
[53] K. Parry and E. D. Morgan, *Physiol. Entomol.*, 1979, **4**, 161; B. Holldobler, *Adv. Study Behav.*, 1978, **8**, 75; J. M. Brand, *Biochem. Syst. Ecol.*, 1978, **6**, 337.
[54] E. D. Morgan, M. R. Inwood, and M. C. Cammaerts, *Physiol. Entomol.*, 1978, **3**, 107; E. D. Morgan, K. Parry, and R. C. Tyler, *Insect Biochem.*, 1979, **9**, 117; M. C. Cammaerts, M. R. Inwood, E. D. Morgan, K. Parry, and R. C. Tyler, *J. Insect Physiol.*, 1978, **24**, 207.
[55] J. H. Cross, R. C. Byler, U. Ravid, R. M. Silverstein, S. W. Robinson, P. M. Baker, J. Sabino de Oliveira, A. R. Jutsum, and J. M. Cherrett, *J. Chem. Ecol.*, 1979, **5**, 187.
[56] C. Longhurst, R. Baker, P. E. Howse, and W. Speed, *J. Insect. Physiol.*, 1978, **24**, 833.
[57] W. V. Brown and B. P. Moore, *Insect Biochem.*, 1979, **9**, 451.

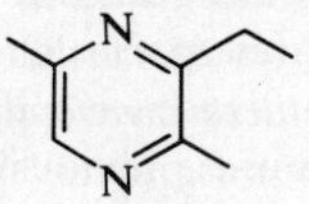

(20)

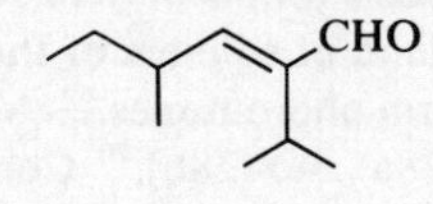

(21)

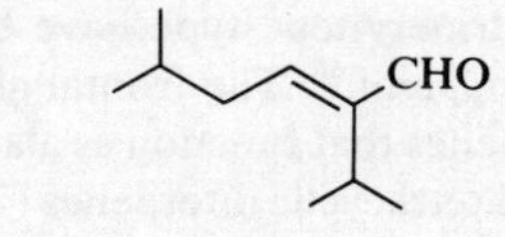

(22)

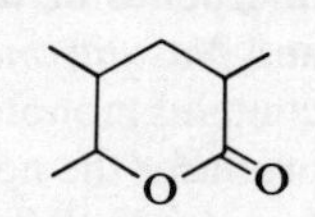

(23)

pyrazines, monoterpenes, and three compounds novel to insects, *i.e.* the aldehydes (21) and (22) and the lactone (23). The lactone is found in males only, and is probably a sex pheromone. A compound more typical of Lepidoptera, (*Z*)-hexadec-9-enal, is reported to be an aggregation pheromone of the Argentine ant, *Iridomyrmex humilis*.[58]

The chemical alarm communication systems of the African weaver ant, *Oecophylla longinoda*, have been studied in detail.[59—61] Major workers produce hexanal, hexan-1-ol, undecan-3-one, and (24) in their mandibular glands, and, after emission from the gland, these components form a complex sequential message in both space and time.[59] Variations in the composition of the contents of the manibular gland are apparent between major workers collected in different areas, between major and minor worker castes, and between the workers and the males; the latter produce acids, including (25a) and (25b).[60] The mandibular gland chemistry of a number of other species of West African ants has been investigated, and correlated with the ease of their detection by prey species of termite. This phenomenon has been termed chemical crypsis.[62] A specialized termite predator, *Megaponera foetens*, produces sulphides, including benzyl methyl sulphide, in its mandibular glands, and uses these with the poison-gland secretion to co-ordinate raids on termite colonies.[63]

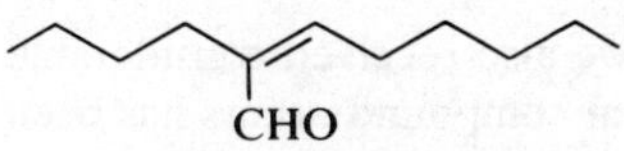

(24)

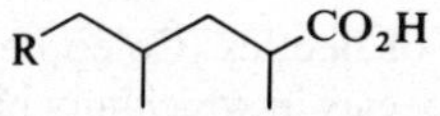

(25) a; R = Me
b; R = Et

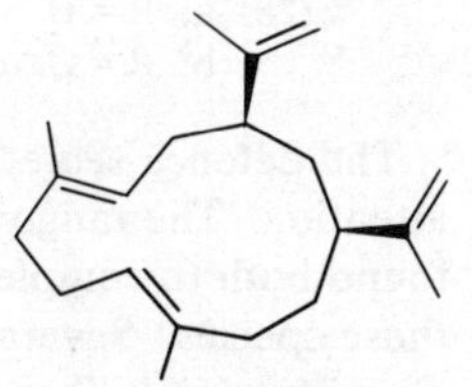

(26)

5 Defence Secretions

A large number of complex and diverse structures have been reported from the defence secretions of insects. Some of the most complex structures come from termites; chemical defence in this Order has been reviewed.[64] Three macrocyclic diterpenes, inluding cubitene (26)[65] and (3*Z*)-cembrene A (27),[66] have been isolated from the secretion of the frontal gland of soldiers of *Cubitermes umbratus*. Novel

[58] G. W. K. Cavill, P. L. Robertson, and N. W. Davies, *Experientia*, 1979, **35**, 989.
[59] J. W. S. Bradshaw, R. Baker, and P. E. Howse, *Physiol. Entomol.*, 1979, **4**, 15.
[60] J. W. S. Bradshaw, R. Baker, P. E. Howse, and M. D. Higgs, *Physiol. Entomol.*, 1979, **4**, 27
[61] J. W. S. Bradshaw, R. Baker, and P. E. Howse, *Physiol. Entomol.*, 1979, **4**, 39.
[62] C. Longhurst, R. Baker, and P. E. Howse, *Experientia*, 1979, **35**, 870.
[63] C. Longhurst, R. Baker, and P. E. Howse, *J. Chem. Ecol.*, 1979, **5**, 703.
[64] G. D. Prestwich, *J. Chem. Ecol.*, 1979, **5**, 459.
[65] G. D. Prestwich, D. F. Wiemer, J. Meinwald, and J. Clardy, *J. Amer. Chem. Soc.*, 1978, **100**, 2560.
[66] D. F. Wiemer, J. Meinwald, G. D. Prestwich, and I. Miura, *J. Org. Chem.*, 1979, **44**, 3950.

diterpenes of the trinervitene type have been found in *Trinervitermes gratiosus*[67] and *Nasutitermes rippertii*.[68] The frontal gland of soldiers of the latter species also contains monoterpenes that function as alarm pheromones.[69] *Nasutitermes octopilis* produces the novel tetracyclic diterpenes (28a) and (28b).[70] Compounds previously found in both *Trinervitermes* and *Nasutitermes* occur in *Grallatotermes africanus*, and confirm the taxonomic relationship between these three genera.[71] Novel sesquiterpenes have been elucidated in *Ancistrotermes cavithorax*; (29) is characteristic of the minor soldiers, while (30) is produced by major soldiers.[72] Four additional terpenes have been found in *Amitermes evuncifer*, including (31), (32a), and (32b).[73]

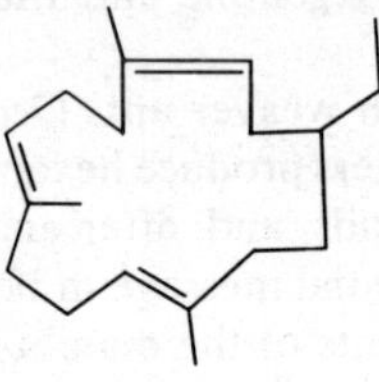

(27)

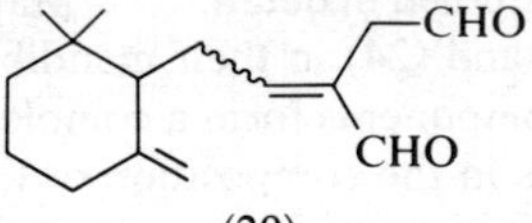

(29)

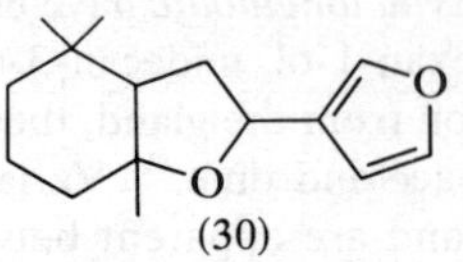

(30)

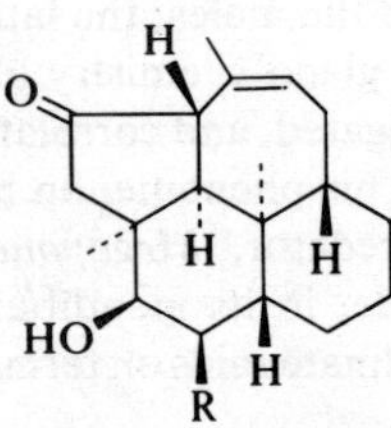

(28) a; R = H
b; R = OAc

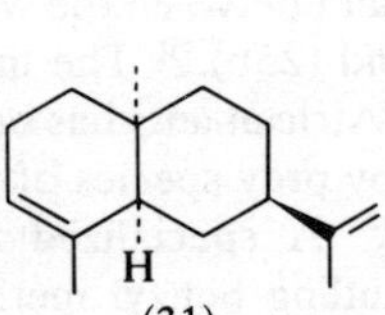

(31)

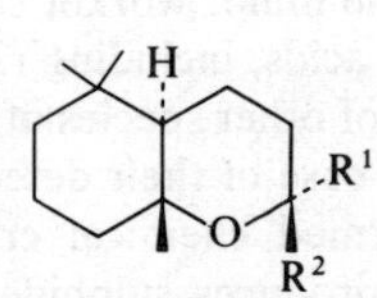

(32) a; $R^1 = Me, R^2 = C_2H_3$
b; $R^1 = C_2H_3, R^2 = Me$

The defence secretions of beetles (Coleoptera) have also received considerable attention. The range of species in which any particular compound occurs has been found both to complement[74] and disagree with[75] the conventional classification of those species. Several novel steroid pyrones have been isolated from the fireflies *Photinus ignitus*, *P. marginellus*, and *P. pyralis*, including (33). These compounds are responsible for the unpalatibility of these beetles to birds.[76] A chrysomelid beetle, *Chrysolina coerulans*, produces the cardenolides (34a) and (34b), which it does not derive directly from cardenolides in its food plants.[77] Chrysomelidial and its enol lactone have been reported from a second species of leaf beetle, *Gastrophysa*

[67] G. D. Prestwich, *Experientia*, 1978, **34**, 682.
[68] J. Vrkoc, M. Budesinsky, and P. Sedmera, *Coll. Czech. Chem. Comm.*, 1978, **43**, 1125.
[69] J. Vrkoc, J. Krecek, and I. Hrdy, *Acta Entomol. Bohemoslov.*, 1978, **75**, 1.
[70] G. D. Prestwich, J. W. Lauher, and M. S. Collins, *Tetrahedron Letters*, 1979, 3827.
[71] G. D. Prestwich, *Insect Biochem.*, 1979, **9**, 563.
[72] R. Baker, P. H. Briner, and D. A. Evans, *J.C.S. Chem. Comm.*, 1978, 410.
[73] R. Baker, D. A. Evans, and P. G. McDowell, *Tetrahedron Letters*, 1978, 4073.
[74] K. Dettner, *Biochem. Syst. Ecol.*, 1979, **7**, 129.
[75] B. P. Moore and W. V. Brown, *Insect Biochem.*, 1978, **8**, 393.
[76] T. Eisner, D. F. Wiemer, L. R. W. Haynes, and J. Meinwald, *Proc. Nat. Acad. Sci. U.S.A.*, 1978, **75**, 905; M. Goetz, D. F. Wiemer, L. R. W. Haynes, J. Meinwald, and T. Eisner, *Helv. Chim. Acta*, 1979, **62**, 1396; J. Meinwald, D. F. Wiemer, and T. Eisner, *J. Amer. Chem. Soc.*, 1979, **101**, 3055.
[77] D. Daloze and J. M. Pasteels, *J. Chem. Ecol.*, 1979, **5**, 63.

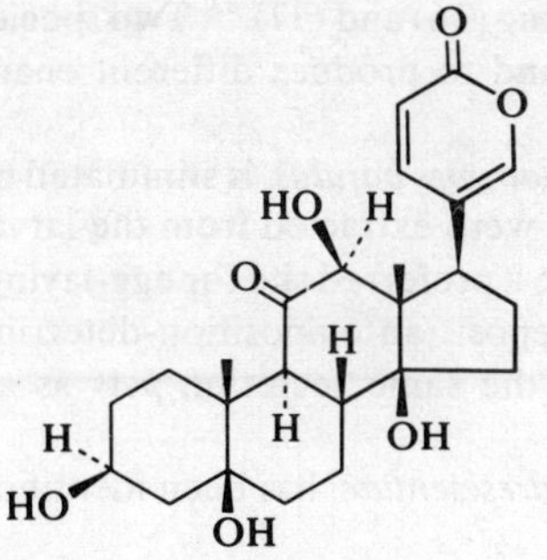

(33)

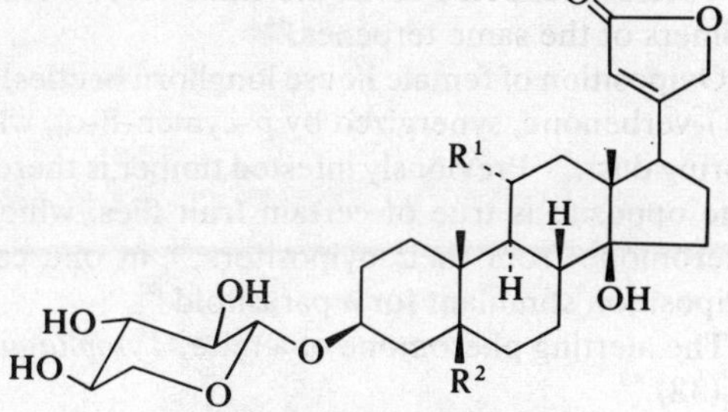

(34) a; $R^1 = OH, R^2 = H$
b; $R^1 = H, R^2 = OH$

cyanea.[78] A tenebrionid beetle, *Apsena pubescens*, uses, *inter alia*, 8-hydroxy-isocoumarin and 3,4-dihydro-8-hydroxycoumarin in a defensive role.[79]

p-Cresol and phenol constitute the defensive secretion of the Malayan cockroach.[80] Larvae of the moth *Schizura concinna* produce formic acid, decyl acetate, dodecyl acetate, and tridecan-2-one in their thoracic defensive glands;[81] such a blend of compounds is normally considered typical of the defensive secretion of formicine ants.

6 Miscellaneous Compounds

A number of compounds have been isolated from insects which are not readily categorized under the headings above. The unusual β-triketones (35a) and (35b) are found in the larval mandibular glands of the moth *Anagasta kuehniella*;[82] the secretion of these glands is a repellant to other larvae. The wax of coccids has proved

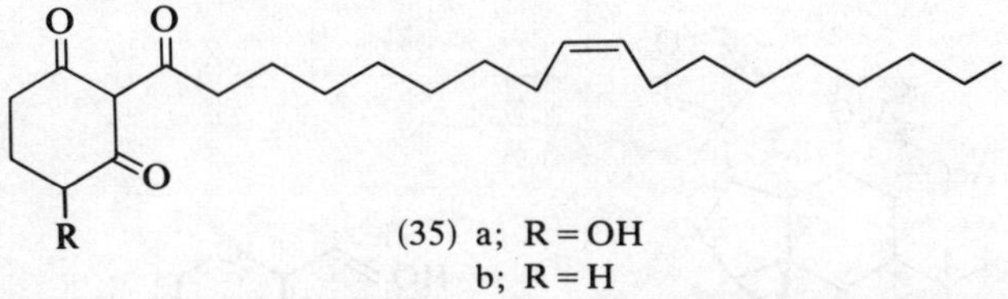

(35) a; R = OH
b; R = H

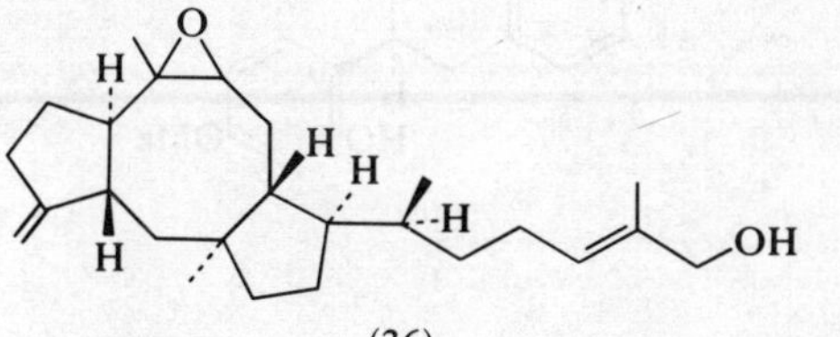

(36)

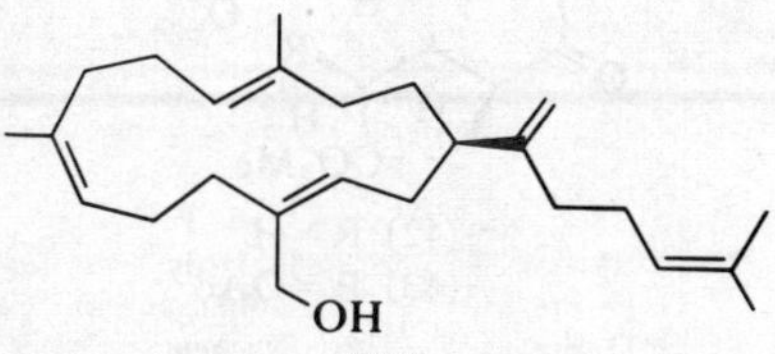

(37)

[78] M. S. Blum, J. B. Wallace, R. M. Duffield, J. M. Brand, H. M. Fales, and E. A. Sokoloski, *J. Chem. Ecol.*, 1978, **4**, 47.
[79] H. A. Lloyd, S. L. Evans, A. H. Kahn, W. R. Tschinkel, and M. S. Blum, *Insect Biochem.*, 1978, **8**, 333.
[80] U. Maschwitz and Y. P. Tho, *J. Chem. Ecol.*, 1978, **4**, 375.
[81] J. Weatherston, J. E. Percy, L. M. MacDonald, and J. A. MacDonald, *J. Chem. Ecol.*, 1979, **5**, 165.
[82] A. Mudd, *J.C.S. Chem. Comm.*, 1978, 1075.

to be a fruitful source of novel sesterterpenes, including (36) and (37).[83] Two species of coccids, collected from the same tree, were found to produce different enantiomers of the same terpenes.[84]

Oviposition of female house longhorn beetles, *Hylotrupes bajulus*, is stimulated by (−)-verbenone, synergized by *p*-cymen-8-ol, which were extracted from the larval boring dust.[85] Previously infested timber is therefore a preferred site for egg-laying. The opposite is true of certain fruit flies, which deposit an oviposition-deterring pheromone from their ovipositors;[86] in one case, the same secretion acts as an oviposition stimulant for a parasitoid.[87]

The alerting pheromone of a mite, *Tyrophagus putrescientiae*, has been identified as (38).[88]

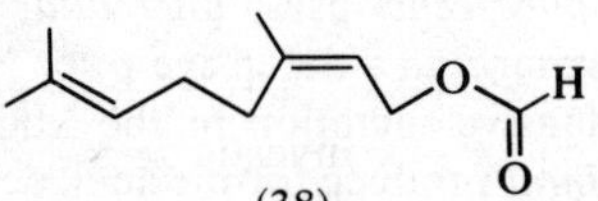

(38)

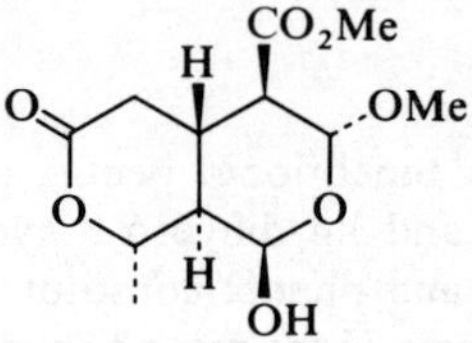

(39)

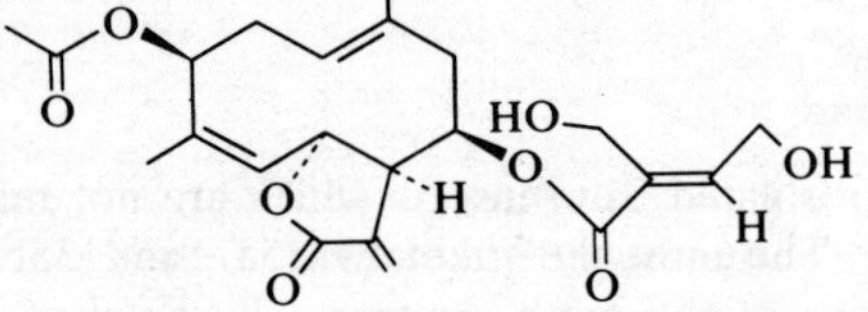

(40)

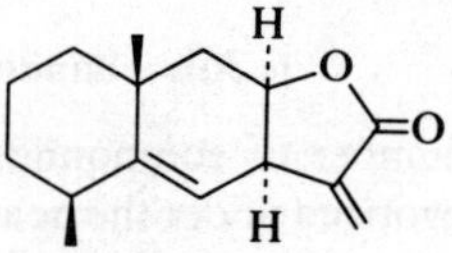

(41)

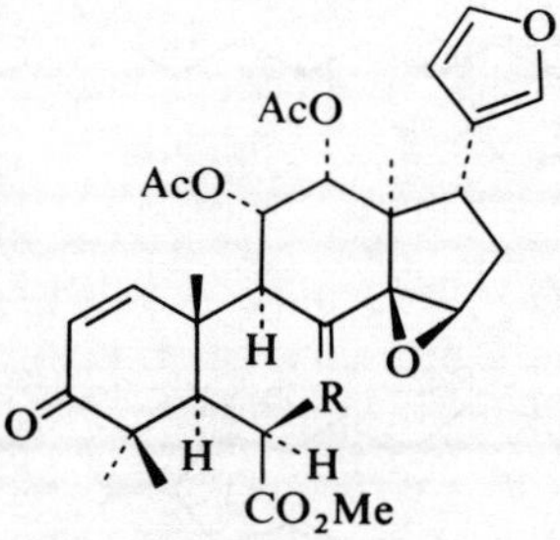

(42) R = H
(43) R = OAc

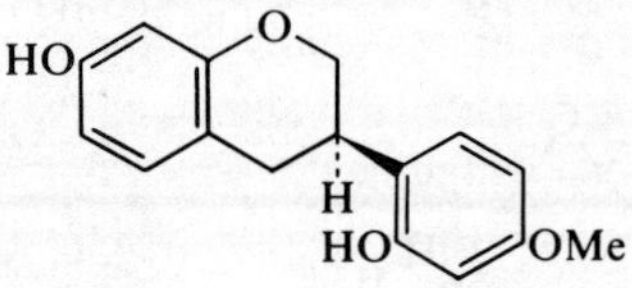

(44)

[83] J. S. Calderon, L. Quijano, and T. Rios, *Experientia*, 1978, **34**, 421; T. Kusimi, T. Kinoshita, K. Fujita, and H. Kakisawa, *Chem. Letters*, 1979, 1129; F. Miyamoto, H. Naoki, T. Takemoto, and Y. Naya, *Tetrahedron*, 1979, **35**, 1913.
[84] Y. Naya, F. Miyamoto, and T. Takemoto, *Experientia*, 1978, **34**, 984.
[85] M. D. Higgs and D. A. Evans, *Experientia*, 1978, **34**, 46.
[86] R. J. Prokopy, J. R. Ziegler, and T. T. Y. Wong, *J. Chem. Ecol.*, 1978, **4**, 55.
[87] R. J. Prokopy and R. P. Webster, *J. Chem. Ecol.*, 1978, **4**, 481.
[88] Y. Kuwahara, *Shokubutsu Boeki*, 1978, **32**, 62.

7 Antifeedants

Antifeedant, insecticidal, and hormonal effects have been reported from a number of plant constituents. The following are representative structures for which antifeedant properties have been demonstrated. Compounds active against *Spodoptera exempta* include xylomollin (39)[89] and schkuhrin (40).[90] Feeding of the flour beetle, *Tribolium confusum*, is reduced in a concentration-dependent way by alantolactone (41), found in many members of the Compositae.[91] Two novel 7,8-seco-tetranortriterpenoids (42) and (43) are antifeedants for the Mexican dried bean beetle, *Epilachna varivestis*.[92] The resistant pasture legume *Lotus pedunculatus* produces (−)-(3*R*)-vestitol (44), which is a feeding deterrent for the beetle *Costelytra zealandica*.[93]

8 Biosynthesis and Biotransformation

Elucidation of the pathways of synthesis for the terpenoid pheromones of bark beetle has continued. In a particularly elegant study, deuterium labelling was used to study the interconversions of ipsdienol, ipsenol, and ipsdienone in *Ips paraconfusus*.[94] The first two of the compounds mentioned have also been investigated in other species, where their production from myrcene is stimulated by analogues of juvenile hormone.[95] The oxidation of α-pinene in *Dendroctonus terebrans* is achieved by inducible microsomal enzymes.[96]

The sex pheromone of *Bombyx mori* is apparently present in the haemolymph of the female pupa and adult, in an inactive form, probably bound to a protein.[97] Similar phenomena may account for the low titre of free pheromone found in some other species. Incorporation of [^{14}C]acetate into all parts of (*Z*)-dodec-7-en-1-yl acetate in the moth *Trichoplusia ni* has been reported;[98] in contrast, labelled acetate is incorporated preferentially into the acetate moiety of decyl acetate in the ant *Formica schaufussi*.[99] Biosynthesis of formic acid in *Camponotus pennsylvanicus* occurs *via* tetrahydrofolate derivatives, from serine, glycine, or histidine.[100]

9 Chemoreception

Perception of pheromones and related compounds by insects is studied from several aspects;[101] only those directly concerned with insect chemistry will be considered here. Pheromone mimics have been shown to be sensitive probes of specialized

89 M. Nakane, C. R. Hutchinson, D. VanEngen, and J. Clardy, *J. Amer. Chem. Soc.*, 1978, **100**, 7079.
90 M. J. Pettei, I. Miura, I. Kubo, and K. Nakanishi, *Heterocycles*, 1978, **11**, 471.
91 A. K. Picman, R. H. Elliott, and G. H. N. Towers, *Biochem. Syst. Ecol.*, 1978, **6**, 333.
92 W. Kraus, W. Grimminger, and G. Sawitzki, *Angew. Chem. Internat. Edn.*, 1978, **17**, 452.
93 G. B. Russell, O. R. W. Sutherland, R. F. N. Hutchins, and P. E. Christmas, *J. Chem. Ecol.*, 1978, **4**, 571.
94 R. H. Fish, L. E. Browne, D. L. Wood, and L. B. Hendry, *Tetrahedron Letters*, 1979, 1465.
95 C. M. Harring, *Z. Angew. Entomol.*, 1978, **85**, 281; J. A. A. Renwick and J. C. Dickens, *Physiol. Entomol.*, 1979, **4**, 377.
96 R. A. White, R. T. Franklin, and M. Agosin, *Pestic. Biochem. Physiol.*, 1979, **10**, 233.
97 K. Hayashiya, M. Kitao, A. Yamazaki, M. Kumazawa, Y. Okeda, and J. Nishida, *Appl. Entomol. Zool.*, 1979, **23**, 28.
98 I. F. Jones and R. S. Berger, *Environ. Entomol.*, 1978, **7**, 666.
99 R. A. Graham, J. M. Brand, and A. J. Markovetz, *Insect Biochem.*, 1979, **9**, 331.
100 A. Hefetz and M. S. Blum, *Science*, 1978, **201**, 454; *Biochim. Biophys. Acta*, 1978, **543**, 484.
101 'Neurotoxicology of Insecticides and Pheromones', ed. T. Narahashi, Plenum, New York, 1978; W. D. Seabrook, *Ann. Rev. Entomol.*, 1978, **23**, 471; W. Roelofs, *Chemtech.*, 1979, **9**, 222.

chemoreceptors. A chiral analogue of the pheromone of the red-banded leaf-roller, *Argyrotaenia velutinana*, has been used to differentiate between stereochemically distinct receptor sites for the same (achiral) pheromone.[102] A photoaffinity-labelled pheromone mimic[103] and crown polyethers[104] have also been used to study perception mechanisms. The behavioural and neurophysiological effects of enantiomers of natural pheromones have been found to vary considerably between species.[105] A threshold hypothesis for pheromone perception has been advanced to account for some hitherto anomalous results from trapping Lepidoptera[106] when sex pheromones were used as the lure.

10 Techniques of Micro-scale Structure Elucidation

The trapping of insect-derived compounds from the vapour phase has been improved by two new methods, the one utilizing Tenax as the collecting medium[107] and the other a molecular sieve, followed by bromination of unsaturated compounds and analysis on a gas chromatograph fitted with an electron-capture detector.[108] By contrast, simple washings from the walls of flasks which had contained calling females gave good recovery of the sex pheromones of *Grapholitha molesta*.[7] Isomeric olefins, such as the sex pheromones of Lepidoptera, can be readily separated on a smectic liquid-crystal stationary phase by gas chromatography,[109] and ^{13}C n.m.r. spectroscopy has been used to distinguish (E)- and (Z)-isomers.[110] For less volatile and less stable molecules, a high-performance liquid chromatography technique has been developed,[111] particularly suited to the sesquiterpenoids found in water beetles.

11 Behaviour-modifying Chemicals in Pest Control

As part of a continuing expansion in the use of pheromones and related compounds in the field of pest control, investigations have been carried out into improved formulation methods,[112] both for population monitoring and for the disruption of mating. While the former application is now becoming routine for a variety of pest species, the latter, which relies on continuous release of relatively high levels of pheromone, has not always been entirely successful.[113]

102 O. L. Chapman, J. A. Klun, K. C. Mattes, R. S. Sheridan, and S. Maini, *Science*, 1978, **201**, 926; O. L. Chapman, K. C. Mattes, R. S. Sheridan, and J. A. Klun, *J. Amer. Chem. Soc.*, 1978, **100**, 4878.
103 I. Ganjian, M. J. Pettei, K. Nakanishi, and K. E. Kaissling, *Nature*, 1978, **271**, 157.
104 J. G. Kostelc, B. J. Garcia, G. W. Gokel, and L. B. Hendry, *J. Chem. Ecol.*, 1979, **5**, 179.
105 R. Rossi and A. Niccoli, *Naturwiss.*, 1978, **65**, 259; K. Mori, S. Tamada, and P. A. Hedin, *ibid.*, p. 653; R. Rossi, P. A. Salvadori, A. Carpita, and A. Niccoli, *ibid.*, 1979, **66**, 211; J. R. Miller and W. L. Roelofs, *Environ. Entomol.*, 1978, **7**, 42; M. Kraemer, H. C. Coppel, F. Matsumura, T. Kikukawa, and K. Mori, *ibid.*, 1979, **8**, 519.
106 W. L. Roelofs, *J. Chem. Ecol.*, 1978, **4**, 685.
107 M. Vanhaelen, R. Vanhaelen-Fastre, J. Geeraerts, and T. Wirthlin, *Microbios*, 1978, **23**, 199.
108 J. Caro, B. A. Bierl, H. P. Freeman, and P. E. Sonnet, *J. Agric. Food Chem.*, 1978, **26**, 461.
109 R. Lester, *J. Chromatogr.*, 1978, **156**, 55.
110 A. Barabas, A. A. Botar, A. Gocan, N. Popovici, and F. Hodosan, *Tetrahedron*, 1978, **34**, 2191.
111 A. T. Newhart and R. O. Mumma, *J. Chem. Ecol.*, 1978, **4**, 503.
112 D. G. Campion, R. Lester, and B. F. Nesbitt, *Pestic. Sci.*, 1978, **9**, 434; L. F. Wiener and J. L. Capinera, *Ann. Entomol. Soc. Amer.*, 1979, **72**, 369; W. F. Steck, B. K. Bailey, M. D. Chisholm, and E. W. Underhill, *Environ. Entomol.*, 1979, **8**, 732.
113 H. Kanno, S. Tatsuki, and K. Uchiumi, *Appl. Entomol. Zool.*, 1978, **13**, 321; E. F. Taschenberg and W. L. Roelofs, *Environ. Entomol.*, 1978, **7**, 103; R. J. Marks, B. F. Nesbitt, D. R. Hall, and R. Lester, *Bull. Entomol. Res.*, 1978, **68**, 11.

12 Synthetic Studies

Reviews on the synthesis of chiral components of insect pheromones[114] and applications of insoluble polymer supports in pheromone synthesis[115] have appeared. An extremely valuable appraisal of recent advances in the chemistry, biology, and application of insect phereomones has been published.[116]

Acyclic Derivatives.—A stereoselective synthesis of β-substituted $\alpha\beta$-unsaturated esters, using the reaction of dialkylcuprates with the enol phosphate of β-keto-esters, has been developed.[117] This method was employed in the preparation of (*E*,*E*)-10-hydroxy-3,7-dimethyldeca-2,6-dienoic acid (49), produced by the male Monarch butterfly (Scheme 1).[118] The dianion of methyl acetoacetate reacted with the

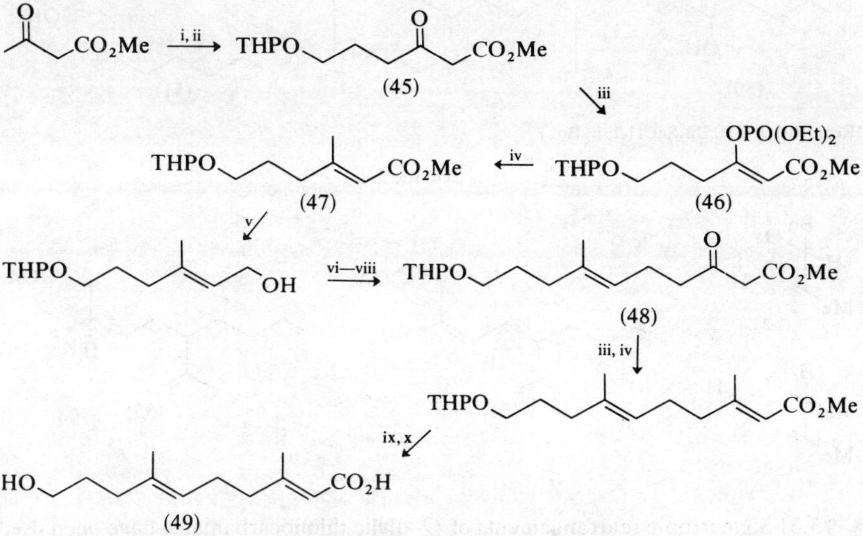

Reagents: i, NaH, Bu^nLi; ii, $THPOCH_2CH_2Br$; iii, NaH, $PO_2(OEt)_2Cl$; iv, $LiMe_2Cu$; v, $LiAlH_4$; vi, Bu^nLi, $MeSO_2Cl$; vii, LiBr; viii, $\bar{C}H_2CO\bar{C}HCO_2Et$; ix, NaOH; x, dil. HCl

Scheme 1

tetrahydropyranyl (THP) ether of 2-bromoethanol to yield the γ-alkylated product (45). Conversion into the enol diethyl phosphate (46) then followed, by treatment with sodium hydride and diethyl phosphorochloridate in ether. The key step is then reaction with lithium dimethylcuprate to produce the trisubstituted olefin (47). Invariably, in this reaction, the major product is formed with retention of the geometry of substitution of the phosphate. After formation of (48), repetition of this sequence leads to (49). A general method is, therefore, available to introduce isoprene units in a stereoselective fashion in a synthetic sequence.

[114] R. Rossi, *Synthesis*, 1978, **6**, 413.
[115] C. C. Leznoff, *Accounts Chem. Res.*, 1979, **11**, 327.
[116] J. M. Brant, J. C. Young, and R. M. Silverstein, *Fortschr. Chem. org. Naturstoffe*, 1979, **37**, 1.
[117] F. W. Sum and L. Weiler, *Canad. J. Chem.*, 1979, **57**, 1431.
[118] F. W. Sum and L. Weiler, *J.C.S. Chem. Comm.*, 1978, 985.

(*Z*)-Trisubstituted olefins have been formed *via* a [2,3]-sigmatropic rearrangement having a transition state with a preferential pseudo-axial substitution.[119] The allylic alcohol (50) was deprotonated and alkylated with iodomethyltributyltin. Direct treatment of this with butyl-lithium gave (51) (Scheme 2). It is clear that the transition state (52), having a pseudo-axial butyl substituent, is strongly preferred over (53). This rearrangement was used in the synthesis of (54), an active component in the sex attractant of the citrus pest the California red scale, *Aonidiella aurantii.* A second component (55) of the pheromone of this insect has been prepared, an aluminium-catalysed ene reaction of methyl propiolate with citronellyl acetate being a key step.[120]

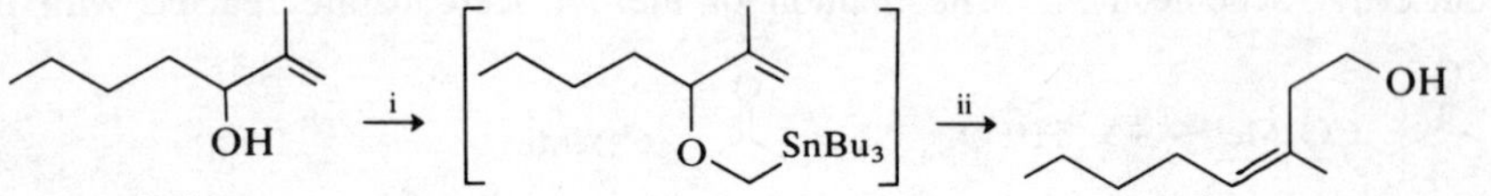

(50) (51)

Reagents: i, KH, Bu_3SnCH_2I; ii, BuLi

Scheme 2

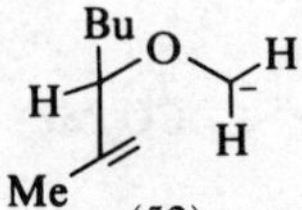

(52)

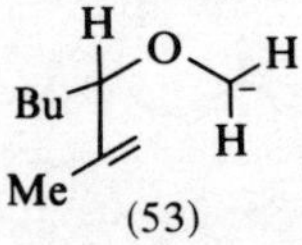

(53)

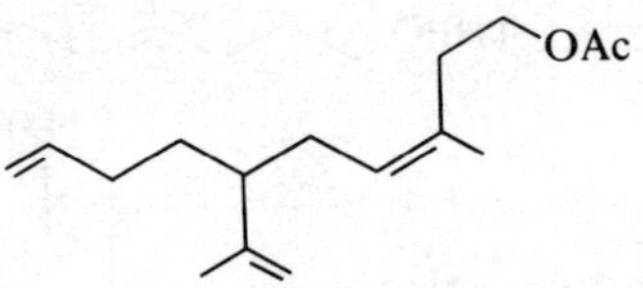

(54)

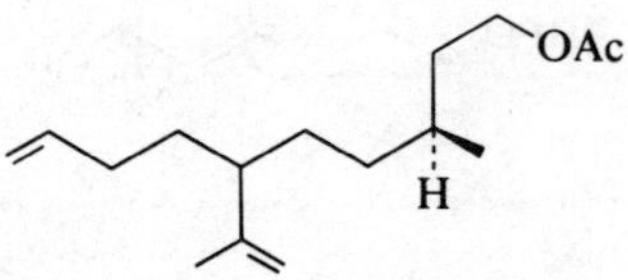

(55)

[3,3]-Sigmatropic rearrangements of *O*-allylic thionocarbamates have been used for stereospecific synthesis of $\alpha\beta$-unsaturated carboxylic esters and ketones, and applied to the formation of the alarm pheromones (*E*)-2,4-dimethylhex-2-enoic acid (57) and (*E*)-4,6-dimethyloct-4-en-3-one (58).[121] Treatment of the common precursor (56) with either one or two moles of dimethyl disulphide eventually leads to the formation of (57) and (58) (Scheme 3). In these cases the (*E*) stereochemistry was obtained exclusively but, in others, small amounts of (*Z*)-isomers were found.

The selective transformation of organoboranes into alkyl-magnesium compounds has been achieved by use of pentane-1,5-di(magnesium bromide).[122] The resulting Grignard reagents have been coupled with alkenyl halides in syntheses of a range of alkenes.

Considerable further effort has been expended on investigations of the value of Wittig reagents in pheromone synthesis. Reactions of Wittig reagents with undec-

[119] W. C. Still and A. Mitra, *J. Amer. Chem. Soc.*, 1978, **100**, 1927.
[120] B. B. Snider and D. Rodini, *Tetrahedon Letters*, 1978, 1399.
[121] T. Nakai, T. Mimura, and T. Kurokawa, *Tetrahedron Letters*, 1978, 2895.
[122] K. Kondo and S. Murahashi, *Tetrahedron Letters*, 1979, 1237.

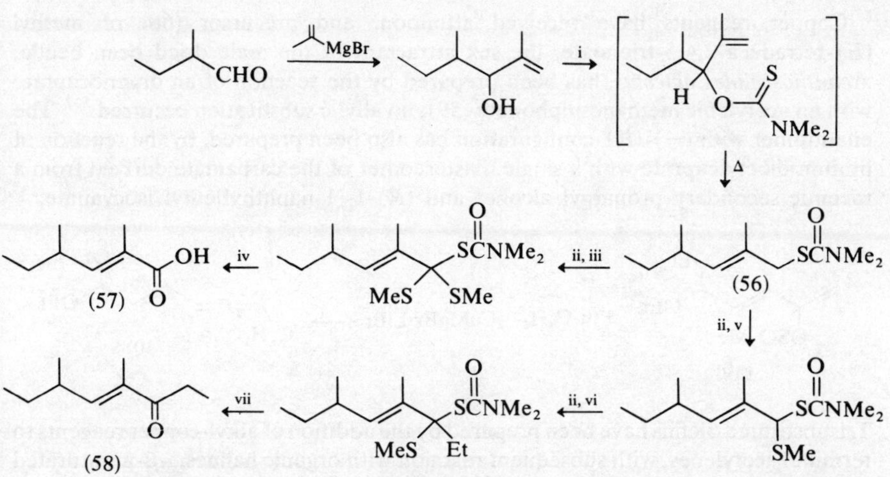

Reagents: i, Me_2NCSCl; ii, $LiNPr^i_2$; iii, $2 \times MeSSMe$; iv, $HgCl_2$, H_2O, v, MeSSMe (1 equiv.); vi, EtI; vii, HgO, BF_3

Scheme 3

10-enal and an ozonolysis product from vaccenyl acetate* have been used to form 1-substituted-(*Z*)-alk-11-enes.[123] The value of the base sodium bis(trimethylsilyl)amide in the controlled, specific formation of (*Z*)-olefins in Wittig reactions has been further demonstrated in preparations of (*Z*)-dec-5-enyl acetate (the sex attractant of the turnip moth, *Agrotis segetum*),[124] (*Z*)-octadec-13-enal (a component of the pheromone of the rice stem borer, *Chilo suppressalis*),[125] (*Z*)-alk-6-enyl acetates,[126] and (*Z*)-tetradec-7-enyl acetate and (*Z*)-dodec-7-enyl acetate.[127] Cycloheptanone has been converted into its methyl vinyl ether derivative and the ozonolysis product used in a synthesis of (*E,Z*)-dodeca-7,9-dienyl acetate.[128] A series of alkyl-branched analogues of lepidopteran pheromones have been prepared from alkyl(triphenyl)phosphonium salts.[129] Reduction of the appropriate diethylphosphonate with lithium aluminium hydride has led to formation of (*E,E*)-alka-$n,(n+3)$-dienes.[130] Selective ozonolysis of 1-methylocta-1,5-diene gave a keto-aldehyde, which has provided the starting point for stereospecific synthesis of (*Z*)-alkenyl acetates.[131]

* Octadec-11-en-1-yl acetate. *Senior Reporter.*

123 H. J. Bestmann, I. Kantardjiew, P. Rösel, W. Stransky, and O. Vostrowsky, *Chem. Ber.*, 1978, **111**, 248.
124 H. J. Bestmann, O. Vostrowsky, K. H. Koshatzky, H. Platz, T. Brosche, I. Kantardjiew, M. Rheinwald, and W. Knauf, *Angew. Chem. Internat. Edn.*, 1978, **17**, 769.
125 H. J. Bestmann, R. Wax, and O. Vostrowsky, *Chem. Ber.*, 1979, **112**, 3740.
126 H. Horicke, M. Tanouchi, and C. Hirano, *Agric. Biol. Chem.*, 1978, **42**, 1963.
127 H. J. Bestmann, O. Vostrowsky, H. Platz, Th. Brosche, and K. H. Koschatzky, *Tetrahedron Letters*, 1979, 497; H. J. Bestmann, K. H. Koschatzky, and O. Vostrowsky, *Chem. Ber.*, 1979, **112**, 1923.
128 H. J. Bestmann, J. Sub, and O. Vostrowsky, *Tetrahedron Letters*, 1979, 2467.
129 H. J. Bestmann, P. Rösel, and O. Vostrowsky, *Annalen*, 1979, 1189.
130 H. J. Bestmann, J. Sub, and O. Vostrowsky, *Tetrahedron Letters*, 1979, 245.
131 G. A. Tolstikov, V. N. Odinokov, R. I. Galeeva, R. S. Bakeeva, and V. R. Akhunova, *Tetrahedron Letters*, 1979, 4851.

Copper reagents have received attention, and precursor (60) of methyl (*E*)-tetradeca-2,4,5-trienoate, the sex attractant of the male dried bean beetle, *Acanthoscelides obtectus*, has been prepared by the reaction of an organocuprate with an acetylenic methanesulphonate (59); no allylic substitution occurred.[132] The enantiomer with (−)-(*R*)-configuration has also been prepared, by the reaction of lithium dioctylcuprate with a single diastereomer of the carbamate derived from a racemic secondary propargyl alcohol and (*R*)-1-(1-naphthyl)ethyl isocyanate.[133]

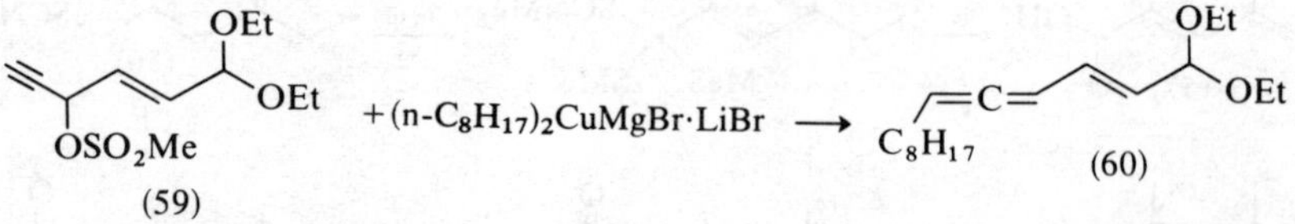

Trisubstituted olefins have been prepared by the addition of alkyl-copper reagents to terminal acetylenes, with subsequent reaction with organic halides, αβ-unsaturated carbonyl compounds, and epoxides.[134] This method was applied in the preparation of (*Z,Z*)-7-methyl-3-propyldeca-2,6-dien-1-ol, a component of the pheromone of the codling moth, *Laspeyresia pomonella*.[135]

General syntheses of conjugated dienes continue to generate interest. Hydrozirconation of acetylenes leads to (*E*)-alk-1-enylzirconium intermediates and addition to vinylic halides yields (*E,E*)-dienes in high yield.[136] In another approach, dienaminium salts (61), obtained from 2-alkyl-pyridines, are coupled with suitable Grignard reagents to yield (*Z,E*)-dienes (62) (Scheme 4).[137] A third approach

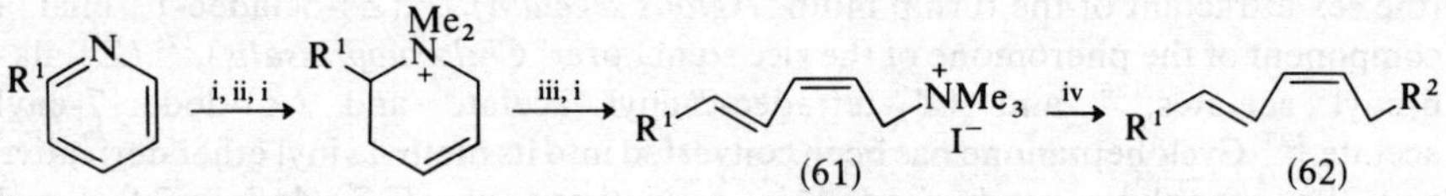

Reagents: i, MeI; ii, $NaBH_4$; iii, 9M-KOH; iv, Li_2CuCl_4, R^2MgBr

Scheme 4

involves the reaction of lithium dialkylcuprates with 3-acetoxypent-1-en-4-yne to give conjugated (*Z*)-enynes.[138] These derivatives are readily converted into the corresponding (*Z,Z*)- or (*E,Z*)-dienes. A similar scheme has been used for the synthesis of (*Z,E*)-hexa-7,11-dienyl acetate.[139]

A 1,3-isomerization of dicarbonyl(methylcyclopentadienyl)manganese complexes of electrophilic acetylenes (63) in the presence of basic aluminium oxide yields the corresponding allene complexes (64). Reaction with a phosphonate and removal of the metal complex with ferric chloride yields (65) (Scheme 5).[140]

132 H. Kleijn, H. Westmijze, K. Kruithof, and P. Vermeer, *Rec. Trav. chim.*, 1979, **98**, 27.
133 W. H. Pirkle and C. W. Boeder, *J. Org. Chem.*, 1978, **43**, 2091.
134 A. Marfat, P. R. McGuirk, and P. Helquist, *J. Org. Chem.*, 1979, **44**, 3888.
135 A. Marfat, P. R. McGuirk, and P. Helquist, *J. Org. Chem.*, 1979, **44**, 1345.
136 N. Okukado, D. E. Van Horn, and W. L. Klima, *Tetrahedron Letters*, 1978, 1027.
137 G. Decodts, G. Dressaire, and Y. Langlois, *Synthesis*, 1979, 510.
138 G. Cassani, P. Massardo, and P. Piccardi, *Tetrahedron Letters*, 1979, 633.
139 A. Hammoud and C. Descoins, *Bull. Soc. chim. France, Part II*, 1978, 299.
140 M. Francke-Neumann and F. Brion, *Angew. Chem. Internat. Edn.*, 1979, **18**, 688.

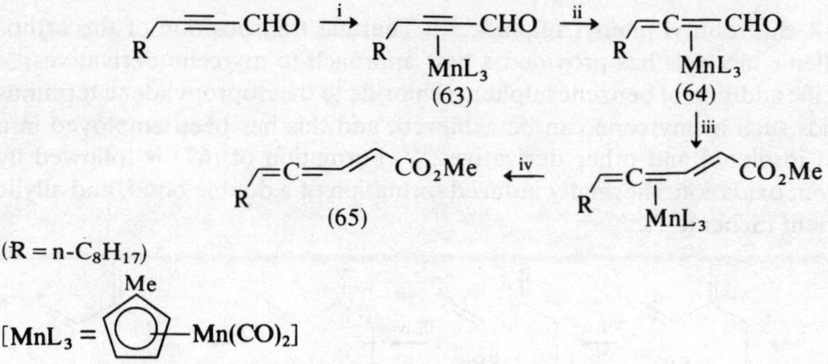

Reagents: i, $[MnL_3]$; ii, Al_2O_3; iii, $(MeO)_2PO\bar{C}HCO_2Me$; iv, $FeCl_3$

Scheme 5

Sorbyl acetate has been coupled with the appropriate Grignard reagent in the presence of lithium chlorocuprate to yield (*E,E*)-dodeca-8,10-dien-1-ol and other analogues.[141] Although not stereospecific, (*E*)-dodeca-9,11-dienyl acetate has been prepared from a butadiene telomer.[142] Acetylenic routes have been employed for preparations of (*Z,Z*)- and (*Z,E*)-octadeca-3,13-dienyl acetate, the attractants for the cherrytree borer, *Synanthedon hector*.[143] Greater than 98% selectivity has been found in the Claisen rearrangement of the t-butyldimethylsilyloxy vinyl ether of the acetates of secondary enynols, due presumably to the 1,3-interactions in the transition state.[144] This reaction has been used in the synthesis of (*E,Z*)-hexadeca-10,12-dienyl-1-ol.

A 1,4-dehydration of an allylic alcohol has been used in a synthesis of (*E*)-dodeca-9,11-dienyl acetate. Treatment of (66) with 2,4-dinitrobenzenesulphenyl chloride in the presence of triethylamine yields a sulphonate ester, which rapidly rearranges to the corresponding allylic sulphoxide, followed by elimination (Scheme 6).[145]

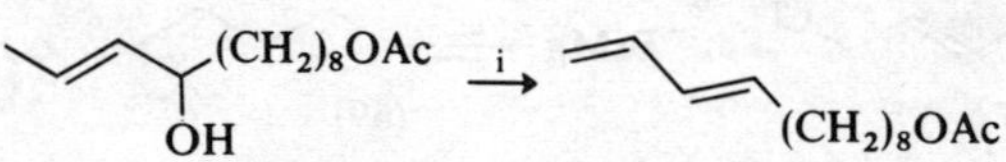

(66)

Reagents: i, $2,4\text{-}(NO_2)_2C_6H_3SCl$, Et_3N, CH_2Cl_2, Δ

Scheme 6

Ipsenol and ipsdienol, components of the aggregation pheromone of *Ips paraconfusus*, have again been prime targets. The reaction of 2-(trimethylsilylmethyl)buta-1,3-diene with isovaleraldehyde in the presence of titanium tetrachloride gave a good yield of ipsenol; this method clearly provides an isoprenylation technique.[146] A similar approach is found in the use of the Grignard reagent derived from 3-

[141] D. Samain and C. Descoins, *Synthesis*, 1978, 388; H. J. Bestmann, J. Sub, and O. Vostrowsky, *Tetrahedron Letters*, 1978, 3329.
[142] T. Mandai, H. Yasuda, M. Kaito, J. Tsuji, R. Yamaoka, and H. Fukami, *Tetrahedron*, 1979, **35**, 309.
[143] M. Uchida, K. Mori, and M. Matsui, *Agric. Biol. Chem.*, 1978, **42**, 1067.
[144] D. Samain and C. Descoins, *Bull. Soc. chim. France, Part II*, 1979, 71.
[145] J. H. Babler and B. J. Invergo, *J. Org. Chem.*, 1979, **44**, 3723.
[146] A. Hosomi, M. Saito, and H. Sakurai, *Tetrahedron Letters*, 1979, 429.

methylene-4-chlorobutyl phenyl sulphide.[147] Thermal transposition of the ortho-esters of allenic alcohols has provided a new approach to myrcene derivatives.[148]

Site-specific addition of benzenesulphenyl chloride to the isopropylidene terminus of terpenoids such as myrcene can be achieved, and this has been employed in a synthesis of ipsdienol and other derivatives.[149] Formation of (67) is followed by acetoxylation, oxidation, thermally induced formation of a double bond, and allylic rearrangement (Scheme 7).

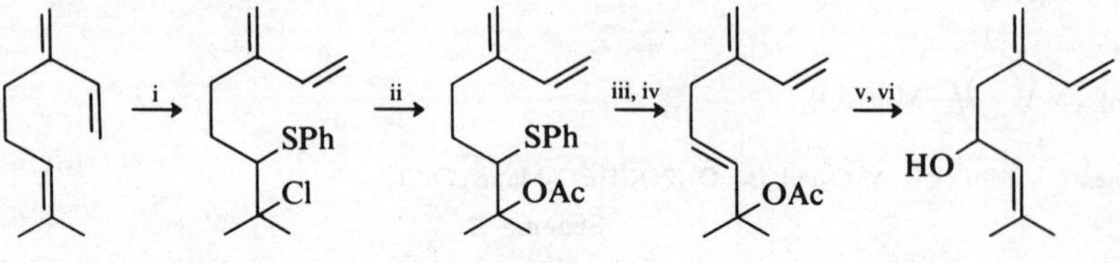

(67)

Reagents: i, PhSCl; ii, NaOAc, AcOH; iii, H_2O_2, AcOH; iv, 120 °C, toluene; v, *p*-TsOH, AcOH; vi, $LiAlH_4$

Scheme 7

Both antipodes of ipsdienol have been synthesized from (+)-(*R*)-glyceraldehyde acetonide and (+)-(*R*)-malic acid, and naturally occurring (+)-ipsdienol was established as having the (*S*) configuration.[150] The (*R*)-enantiomers of (*Z*)- and (*E*)-14-methylhexadec-8-en-1-ol and of (*Z*)- and (*E*)-14-methylhexadec-8-enal, components of the sex pheromone of the female dermestid beetles, have been prepared from (+)-(*R*)-citronellol as the starting point.[151] Conversion of the initially formed epoxide into the aldehyde was a key step.

Two components of the sex pheromone of *Phthorimaea operculla*, the potato tubeworm moth, have been prepared.[152] In both cases, (68) was used, and its reaction with (69) and (70) gave, after reduction, (*E,Z*)-trideca-4,7-dien-1-ol and

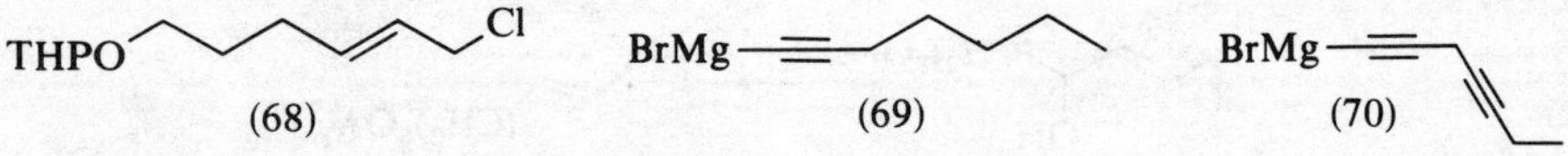

(68) (69) (70)

(*E,Z,Z*)-trideca-4,7,10-trien-1-ol. The former compound has also been prepared by a conjugate opening of an allylic epoxide by a vinylic organocopper reagent.[153] Another triene, (*Z,Z*)-pentacosa-1,7,13-triene (a major component of the cuticular lipid of the male stable fly, *Stomoxys calcitrans*), together with the three other isomers, has been prepared.[154] Hydrogenation of acetylenic linkages and metal reductions with metal and amine were used to prepare (*Z,Z*)- and (*Z,E*)-isomers, followed by inversion of the olefins to obtain the other two isomers. The (*E,E*)-isomer (72) was prepared from the (*Z,Z*)-isomer by the reaction of (71) with

[147] B. Cazes, E. Guittet, S. Julia and O. Ruel, *J. Organometallic Chem.*, 1979, **177**, 67.
[148] M. Bertrand and J. Viala, *Tetrahedron Letters*, 1978, 2575.
[149] Y. Masaki, K. Hashimoto, K. Sakuma, and K. Kaji, *J.C.S. Chem. Comm.*, 1979, 855.
[150] K. Mori, T. Takigawa, and T. Matsuo, *Tetrahedron*, 1979, **35**, 933.
[151] K. Mori, T. Suguro, and M. Uchida, *Tetrahedron*, 1978, **34**, 1; R. Rossi, P. A. Salvadori, and A. Carpita, *ibid.*, 1979, **35**, 2039; T. Suguro and K. Mori, *Agric. Biol. Chem.*, 1979 **43**, 409.
[152] S. Voerman and G. H. L. Rothschild, *J. Chem. Ecol.*, 1978, **4**, 531.
[153] A. Alexakis, G. Cahiez, and J. F. Normant, *Tetrahedron Letters*, 1978, 2027.
[154] P. E. Sonnett, *J. Chem. Ecol.*, 1979, **5**, 415.

m-chloroperbenzoic acid followed by treatment of the tris-epoxide with triphenylphosphine dibromide and then reductive dehalogenation with zinc. The (*Z,E*)-isomer was also converted into the tris-epoxide, and its reaction with excess trifluoroacetyl chloride followed by heating with sodium iodide gave the (*E,Z*)-isomer.

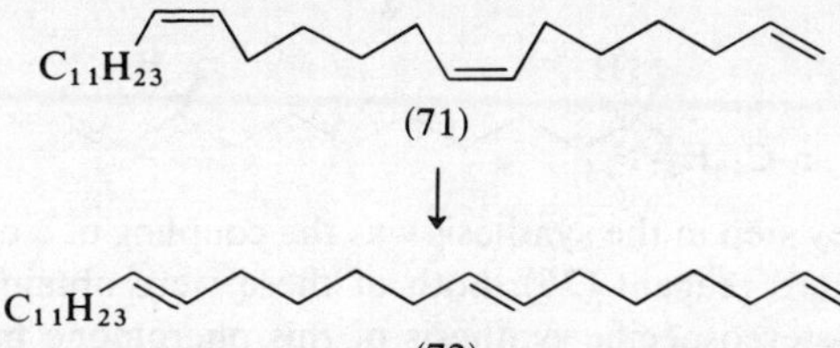

(71)

(72)

Syntheses of all the possible stereoisomers of *erythro*-3,7-dimethylpentadec-2-yl acetate and propionate have been reported.[155] The crucial steps were the attack of chiral organocopper reagents, to yield (+)-(2*R*,3*R*)-2,3-epoxybutane (73) and the (−)-(2*S*,3*S*)-epoxide, respectively. The second part of the molecule (−)-(*R*)-1-bromo-4-methyldodecane and its enantiomer were derived from (+)-(*R*)-citronellol. A coupling of (73) and (74) gave (2*R*,3*R*,7*R*)-*erythro*-3,7-dimethylpentadecan-2-ol, and the other three enantiomers were obtained similarly.

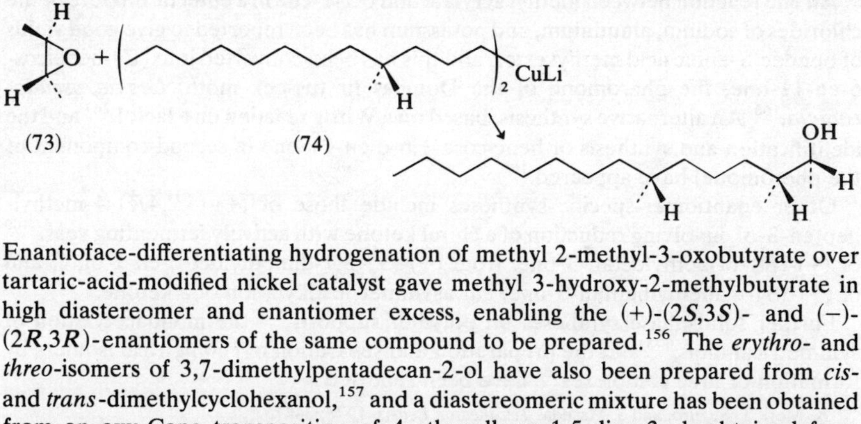

(73) (74)

Enantioface-differentiating hydrogenation of methyl 2-methyl-3-oxobutyrate over tartaric-acid-modified nickel catalyst gave methyl 3-hydroxy-2-methylbutyrate in high diastereomer and enantiomer excess, enabling the (+)-(2*S*,3*S*)- and (−)-(2*R*,3*R*)-enantiomers of the same compound to be prepared.[156] The *erythro*- and *threo*-isomers of 3,7-dimethylpentadecan-2-ol have also been prepared from *cis*- and *trans*-dimethylcyclohexanol,[157] and a diastereomeric mixture has been obtained from an oxy-Cope transposition of 4-ethynylhexa-1,5-dien-3-ol, obtained from reactions of a vinylallenic Grignard with acrolein.[158] Field studies on the effectiveness of racemic 3,7-dimethylpentadec-2-yl propionate with four species of pine sawflies have been made.[159]

The four stereoisomers of 3,11-dimethylnonacosan-2-one, the female sex pheromone of the German cockroach, *Blatella germanica*, have also been prepared and the natural pheromone has been shown to have the absolute configuration

[155] K. Mori, S. Tamada, and M. Matsui, *Tetrahedron Letters*, 1978, 901; K. Mori and S. Tamada, *Tetrahedron*, 1979, **35**, 1279.
[156] A. Tai, M. Imaida, T. Oda, and H. Watanabe, *Chem. Letters*, 1978, 61.
[157] G. Magnusson, *Tetrahedron*, 1978, **34**, 1385.
[158] P. Place, M. L. Roumestant, and G. Gore, *J. Org. Chem.*, 1978, **43**, 1001.
[159] D. M. Jewett, F. Matsumura, and H. C. Coppel, *J. Chem. Ecol.*, 1978, **4**, 277.

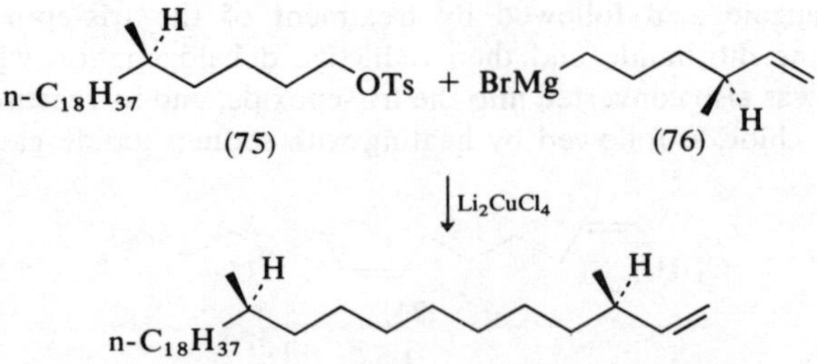

(3*S*,11*S*).[160] The key step in the synthesis was the coupling of a chiral tosylate (75) with a chiral Grignard reagent (76); both of these were obtained from (+)-(*R*)-citronellol. A non-stereospecific synthesis of this pheromone has been reported, involving an oxy-Cope reaction of 4-ethynylhexa-1,5-dien-3-ol.[161]

Both enantiomers of epoxy-2-methyloctadecane have been prepared from routes beginning with L-(+)-tartaric acid; the (+)-(7*R*,8*S*)-enantiomer was found to be biologically active.[162] A resolution technique involving liquid-chromatographic separation of diastereomeric (*R*)-1-(1-naphthyl)ethyl isocyanates has also been used for the same preparations.[163] Racemic disparlure was also synthesized, by selective ozonolysis of cyclo-octa-1,5-diene followed by Grignard coupling[164] or Kolbe electrolysis.[165]

An ene reaction between methyl acrylate and oct-1-ene in a eutectic mixture of the chlorides of sodium, aluminium, and potassium has been reported to give good yields of hendec-5-enoic acid methyl ester, and this has been converted into (*Z*)-heneicos-6-en-11-one, the pheromone of the Douglas fir tussock moth, *Orgyia pseudotsugata*.[166] An alternative synthesis, based on a Wittig reaction on a lactol,[167] and the identification and synthesis of heneicosa-1,6-dien-11-one (a second component of the pheromone) have appeared.[168]

Other enantiomer-specific syntheses include those of (+)-(3*R*,4*R*)-4-methylheptan-3-ol, involving reduction of a chiral ketone with actively fermenting yeast,[169] of (−)-(*R*)-6-methyloctan-3-one, from (−)-(*S*)-3,7-dimethyloct-6-en-1-ol,[170] and of (+)-(*S*)-4-methylheptan-3-one, *via* asymmetric alkylation of a ketone.[171]

Further solid-phase syntheses on polymer supports,[172] the monoacetylation of symmetrical diols,[173] and the preparation and separation of geometrical isomers by formation of urea complexes[174] have been reported.

160 K. Mori, T. Suguro, and S. Masuda, *Tetrahedron Letters*, 1978, 3447.
161 P. Place, M. L. Roumestant, and J. Gore, *Tetrahedron*, 1978, **34**, 1931.
162 K. Mori, T. Takigawa, and M. Matsui, *Tetrahedron*, 1979, **35**, 933.
163 W. H. Pirkle and P. L. Rinaldi, *J. Org. Chem.*, 1979, **44**, 1025.
164 G. A. Tolstikov, V. N. Odinokov, R. I. Galeeva, and R. S. Bakeeva, *Tetrahedron Letters*, 1978, 1857.
165 H. Kluneberg and H. J. Schafer, *Angew. Chem. Internat. Edn.*, 1978, **17**, 47.
166 B. Åkermark and A. Ljungqvist, *J. Org. Chem.*, 1978, **43**, 4387.
167 M. Fétizon and C. Lazare, *J.C.S. Perkin I*, 1978, 842.
168 L. M. Smith, R. G. Smith, T. M. Loehr, and G. D. Daves, jun., G. E. Daterman, and R. H. Wohleb, *J. Org. Chem.*, 1978, **43**, 2361.
169 G. Frater, *Helv. Chim. Acta*, 1979, **62**, 2829.
170 R. Rossi and P. A. Salvadori, *Synthesis*, 1979, 209.
171 D. Enders and H. Eichenauer, *Angew. Chem. Internat. Edn.*, 1979, **18**, 397.
172 C. C. Leznoff, *J.C.S. Chem. Comm.*, 1978, 327; T. M. Fyles, C. C. Leznoff, and J. Weatherston, *Canad. J. Chem.*, 1978, **56**, 1031; T. M. Fyles, C. C. Leznoff, and J. Weatherston, *J. Chem. Ecol.*, 1978, **4**, 109.
173 J. H. Babler and M. J. Coghlan, *Tetrahedron Letters*, 1979, 1971.
174 G. Leadbetter and J. R. Plimmer, *J. Chem. Ecol.*, 1979, **5**, 101.

Alicyclic Derivatives.—Both enantiomers of grandisol, one of the four pheromones of the male boll weevil, *Anthonomus grandis*, have been prepared.[175] Photocycloaddition of ethylene to (77) gave (78), which was then converted into the acid (79); this was resolved by treatment with optically active bases. Separation into the two enantiomers was followed by conversion into the two enantiomers of grandisol (Scheme 8), exemplified by the formation of (+)-(1*R*,2*S*)-grandisol (80).

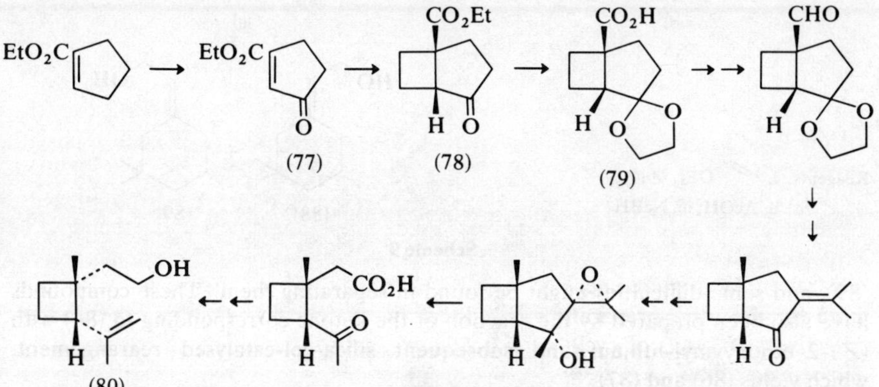

Scheme 8

An acid-catalysed rearrangement of an oxycyclopropanecarbinol into a cyclobutanone has also been used in a synthesis of (+)-grandisol.[176] Conversion of (81) into (82) with a Simmons–Smith reaction, followed by acid-induced hydrolytic rearrangement, gave (83), which was converted into (+)-(80). Treatment of racemic 6,7-epoxygeranyl t-butyl sulphide with butyl-lithium and 1,2-bis(dimethylamino)ethane has been shown to yield substantial amounts of a cyclobutyl derivative,[177] although other products were also formed. A further synthesis of grandisol involved the addition of an organocuprate to a cyclobutenyl ester.[178]

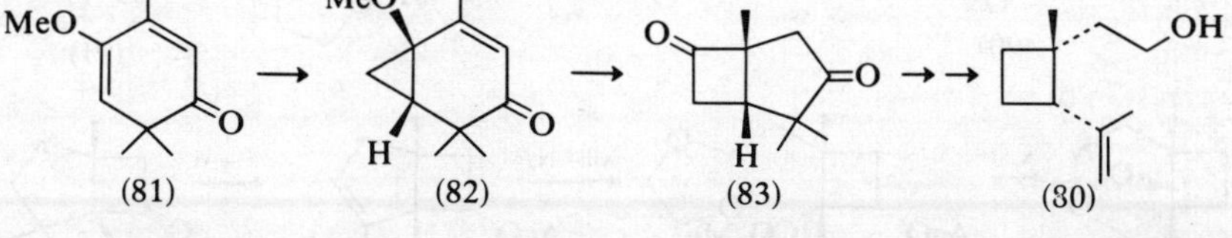

Syntheses of the three other components of the sex pheromone of the boll weevil have been reported. The addition of ethyl vinyl ether to (84) in the presence of $ZnCl_2$ gave (85), which on treatment with sodium acetate in acetic acid yielded (86) and (87).[179] Reduction with sodium borohydride gave (88) and (89), as shown in Scheme 9. The compounds required for the pheromone mixture are (86), (87), and

[175] K. Mori, *Tetrahedron*, 1978, **34**, 915.
[176] E. Wenkert, D. A. Bergers, and N. F. Golob, *J. Amer. Chem. Soc.*, 1978, **100**, 1263.
[177] V. Rautenstrauch, *J.C.S. Chem. Comm.*, 1978, 519.
[178] R. D. Clark, *Synth. Comm.*, 1979, **9**, 325.
[179] J. P. de Souza and A. M. R. Gonçalves, *J. Org. Chem.*, 1978, **43**, 2068.

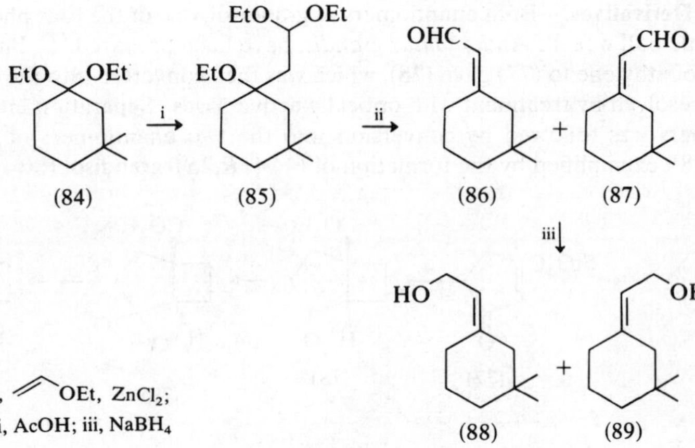

Reagents: i, $CH_2{=}CHOEt$, $ZnCl_2$; ii, AcOH; iii, $NaBH_4$

Scheme 9

(89), and some difficulties might be found in separating them. These compounds have also been prepared by the reaction of the ketone corresponding to (84) with (*Z*)-2-ethoxyvinyl-lithium and subsequent silica-gel-catalysed rearangement, which yields (86) and (87).[180]

Both enantiomers of seudenol, 3-methylcyclohex-2-en-1-ol, an aggregation pheromone of the Douglas fir beetle, *Dendroctonus pseudotsugae*, have been prepared by the reaction of 3-iodocyclohex-2-en-1-ol with lithium dimethylcuprate and their absolute configurations determined.[181] The structure of lineatin, an attractant compound isolated from frass produced by female beetles of *Trypodendron lineatum*, has been confirmed as (93), although the synthesis would not be appropriate on any scale (Scheme 10).[182] The two ketones (91) and (92) were prepared, beginning with

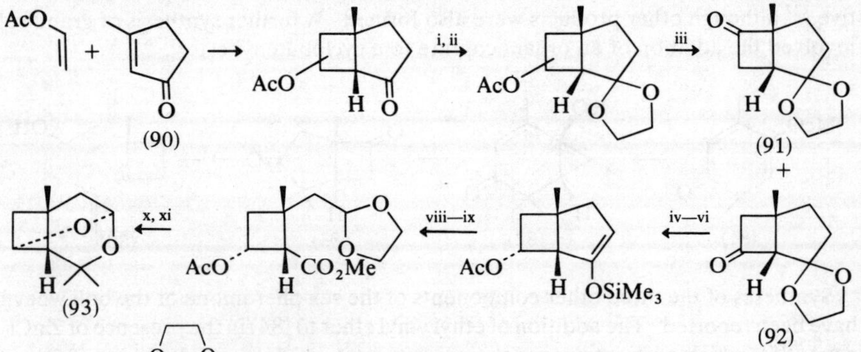

Reagents: i, (ethylene acetal of but-3-en-2-one), *p*-TsOH; ii, KCN, EtOH; iii, CrO_3, py, HCl; iv, $LiBu^s{}_3BH$; v, Ac_2O, py; vi, $LiNPr^i{}_2$, Me_3SiCl; vii, O_3, PPh_3; viii, CH_2N_2; ix, $CH_2(OH)CH_2OH$, *p*-TsOH; x, MeMgI; xi, dil. HCl

Scheme 10

[180] R. H. Wollenberg and R. Peries, *Tetrahedron Letters*, 1979, 297.

[181] K. Mori, S. Tamada, M. Uchida, N. Mizumachi, T. Tachibana, and M. Matsui, *Tetrahedron*, 1978, **34**, 1901.

[182] K. Mori and M. Sasaki, *Tetrahedron Letters*, 1979, 1329.

photocycloaddition of vinyl acetate and (90). Chromatography on silicic acid gave the two ketones, of which (91) was the major product.

A systematic study has been made on the synthesis of eleven 1,6-dioxaspiro[4.4]nonanes, and their mass spectra have been discussed.[183] The use of (+)-(*R*)-γ-caprolactone (94) has allowed the preparation of a mixture of the two diastereomers (97) and (98), which serve as the principal component of the aggregation pheromone of the beetle *Pityogenes chacographus*.[184] The reaction of (94) with the lithium salt of the propargyl alcohol derivative (95) gave (96), which on hydrogenation and treatment with hydrochloric acid led to a mixture of (97) and (98). The preparation of the other two enantiomers is possible, from (−)-(*S*)-γ-caprolactone. These two pairs of optically active diastereomers have also been prepared by a route in which the key step was the alkylation of the dianion of an α-acetyl-γ-butyrolactone (100) with (*R*)-1,2-epoxybutane (99).[185]

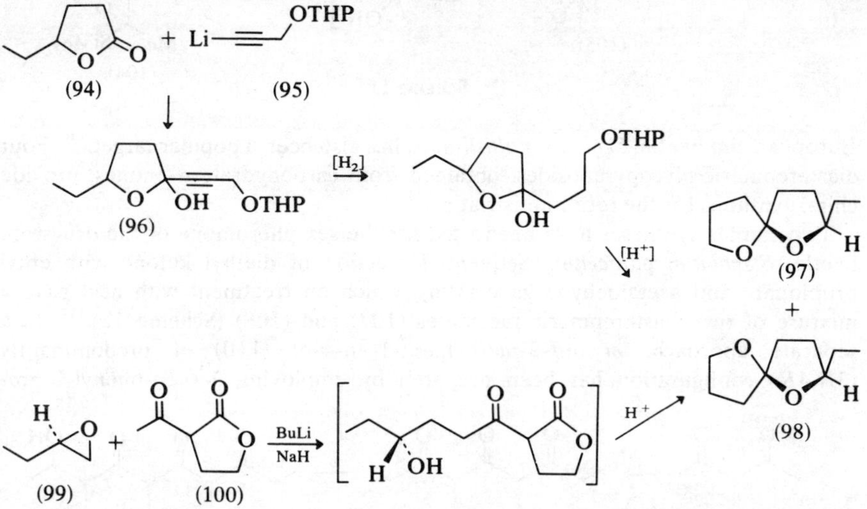

(+)-(*R*)-Frontalin (105), the aggregation pheromone of the southern pine beetle, *Dendroctonus frontalis*, has been prepared in a demonstration of the use of α-methyl-α-chlorotrimethylsilyl carbanion (102).[186] Beginning from (−)-(3*R*)-linalool, the αβ-epoxysilane (103) was prepared by the action of (102) on the aldehyde (101) (Scheme 11). The action of boron trifluoride etherate on (104) gave (105). The overall reaction represents the highest overall yield for a chiral synthesis of this derivative.

Frontalin has also been prepared in a route involving the reaction of the dianion of methyl acetoacetate[187] and by an impressive titanium-chloride-catalysed photoreaction of heptane-2,6-dione.[188] Multistriatin, one of the three components of the

[183] W. Francke and W. Reith, *Annalen*, 1979, 1.
[184] L. R. Smith, J. J. Williams, and R. M. Silverstein, *Tetrahedron Letters*, 1978, 3231.
[185] K. Mori, M. Sasaki, S. Tamada, T. Suguro, and S. Masuda, *Tetrahedron*, 1979, **35**, 1601; *Heterocycles*, 1978, **10**, 111.
[186] P. Magnus and G. Roy, *J.C.S. Chem. Comm.*, 1978, 297.
[187] P. E. Sum and L. Weiler, *Canad. J. Chem.*, 1979, **57**, 1475.
[188] T. Sato, S. Yamaguchi, and H. Kaneko, *Tetrahedron Letters*, 1979, 1863.

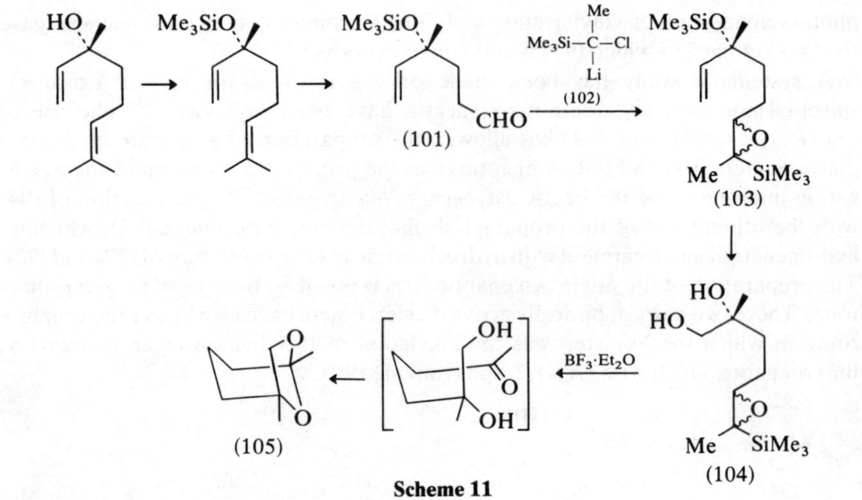

Scheme 11

European elm beetle, *Scolytus multistriatus*, has also been a popular target.[189] Four diastereomeric glycopyranosides, obtained from carbohydrate α-enones, provide chiral synthons for the four multistriatins.[190]

Two similar syntheses have been used for the sex pheromone of the drugstore beetle, *Stegobium paniceum*. Sequential reaction of diethyl ketone with ethyl propionate and acetaldehyde gave (106), which on treatment with acid gave a mixture of two diastereomeric racemates (107) and (108) (Scheme 12).[191] In a separate approach, *erythro*-3-methylpent-1-en-4-ol (110) of predominantly (3*R*,4*R*) configuration has been prepared by employing 3-*endo*-phenyl-2-*exo*-

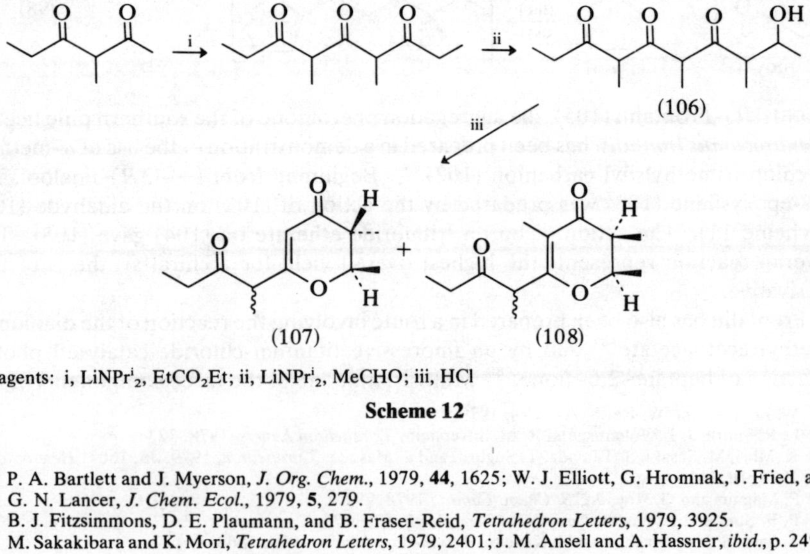

Reagents: i, $LiNPr^i_2$, $EtCO_2Et$; ii, $LiNPr^i_2$, MeCHO; iii, HCl

Scheme 12

[189] P. A. Bartlett and J. Myerson, *J. Org. Chem.*, 1979, **44**, 1625; W. J. Elliott, G. Hromnak, J. Fried, and G. N. Lancer, *J. Chem. Ecol.*, 1979, **5**, 279.
[190] B. J. Fitzsimmons, D. E. Plaumann, and B. Fraser-Reid, *Tetrahedron Letters*, 1979, 3925.
[191] M. Sakakibara and K. Mori, *Tetrahedron Letters*, 1979, 2401; J. M. Ansell and A. Hassner, *ibid.*, p. 2497.

dihydroxybornane (109).[192] This was then converted into (111), which on condensation with the dianion of 4-methylheptane-3,5-dione (112) gave (113) (Scheme 13). After cyclization, c.d. studies showed that the final product had (2*R*,3*S*) configuration and that the natural pheromone had (2*S*,3*R*) configuration.

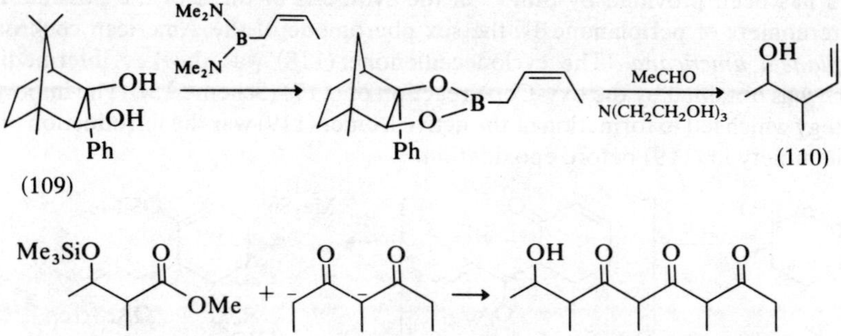

(109) (110) (111) (112) (113)

Scheme 13

Enantiomeric specific syntheses of a number of 4-alkyl(or alkenyl)-γ-lactones of known absolute configuration have been made from glutamic acid enantiomers.[193] The alkyl side-chain was introduced by conversion of the acid into a toluene-*p*-sulphonate followed by displacement with lithium dialkylcuprates. These enantiomerically pure lactones have also been prepared by the separation of diastereomeric carbamates, although large-scale syntheses by this method cannot be envisaged.[194]

The synthesis of pederamide, a key intermediate for a route to pederin, a vesicant compound of the beetle *Paederus fuscipes*, was accomplished from *trans*-2,3-epoxybutane.[195] Two syntheses of (+)-phoracantholid J (116), a compound found in the metasternal secretion of the eucalypt longicorn beetle, *Phoracantha synonyma*, have appeared. In the first the lactonization was achieved by silver-catalysed reaction of the 5-(2-pyridyl)thioester.[196] A second method involved the formation of the acetal (114) with the aid of benzeneselenyl bromide, followed by oxidation and thermolysis, which affords (116) after a Claisen rearrangement of (115) (Scheme 14).[197]

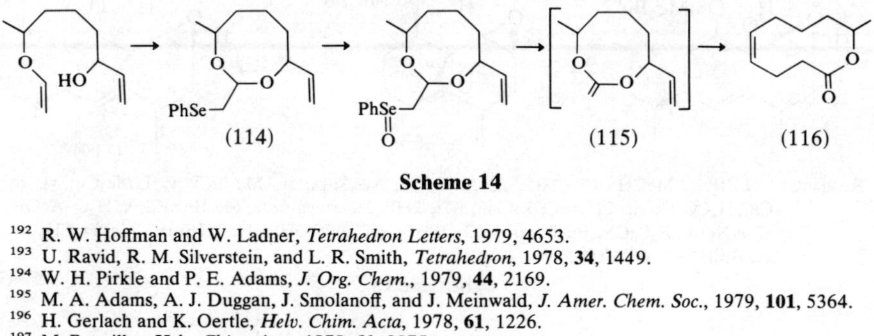

(114) (115) (116)

Scheme 14

[192] R. W. Hoffman and W. Ladner, *Tetrahedron Letters*, 1979, 4653.
[193] U. Ravid, R. M. Silverstein, and L. R. Smith, *Tetrahedron*, 1978, **34**, 1449.
[194] W. H. Pirkle and P. E. Adams, *J. Org. Chem.*, 1979, **44**, 2169.
[195] M. A. Adams, A. J. Duggan, J. Smolanoff, and J. Meinwald, *J. Amer. Chem. Soc.*, 1979, **101**, 5364.
[196] H. Gerlach and K. Oertle, *Helv. Chim. Acta*, 1978, **61**, 1226.
[197] M. Petrzilka, *Helv. Chim. Acta*, 1978, **61**, 3075.

This reaction is not completely stereospecific since some of the *trans*-isomer is also produced (*cis*: *trans* = 7:2), but could provide a new method for syntheses of other lactones.

Probably the most outstanding synthetic contribution to this field in the past two years has been provided by Still[198] in the synthesis of three of the possible four diastereomers of periplanone-B, the sex pheromone of the American cockroach, *Periplaneta americana*. The cyclodecadienone (118) was the key intermediate, which was obtained by the oxy-Cope reaction of (117) (Scheme 15). The important strategy which led to formation of the active isomer (119) was the introduction of the diene moiety in (119) before epoxidation.

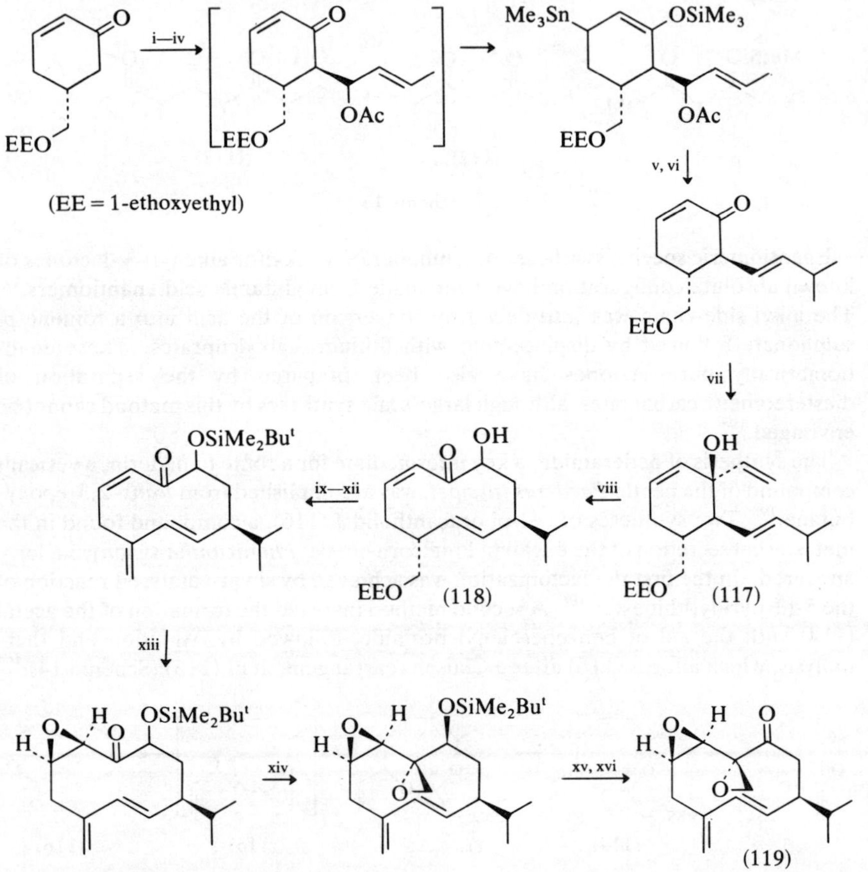

Reagents: i, $LiNPr^i_2$, MeCH=CHCHO; ii, Ac_2O; iii, Me_3SnLi; iv, Me_3SiCl; v, $LiMe_2Cu$; vi, *m*-$ClC_6H_4CO_3H$; vii, CH_2=CHLi; viii, KH, THF, 18-crown-6; ix, Me_2Bu^tSiCl; x, H_2O–AcOH; xi, *o*-$NO_2C_6H_4SeCN$, Bu_3P; xii, H_2O_2, THF; xiii, Bu^tOOH, KH, THF; xiv, $Me_2\overset{+}{S}$=CH_2 X^-; xv, Bu_4N^+ F^-; xvi, CrO_3, py

Scheme 15

[198] W. C. Still, *J. Amer. Chem. Soc.*, 1979, **101**, 2493.

The chemistry of antifeedants has gained considerable attention. A total synthesis of warburganal (120) involved sequential protection and oxidation of (121)[199] and, in a separate approach, has also been obtained from isodrimenine (122).[200] Ring B of this compound has also been prepared by using 2,2-dimethylcyclohexanone as a model.[201] A *cis*-decalin derivative (123) having similar functionalities to ajugarin was obtained by a route in which a key step was Lewis-acid-catalysed cyclization of (124) to (125).[202] This *cis*-decalin part of ajugarin has also been constructed by a Diels–Alder approach.[203]

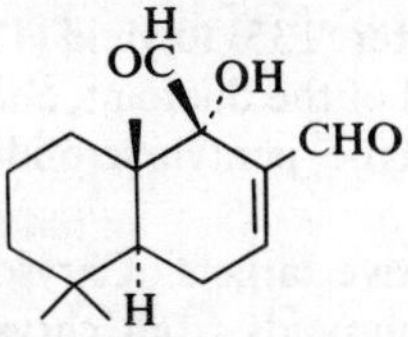

(120)

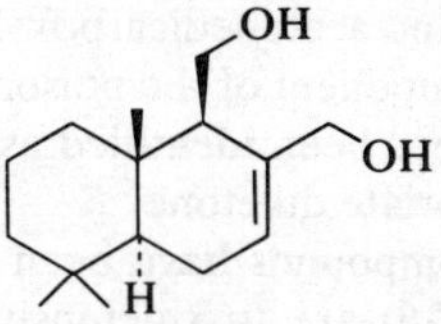

(121)

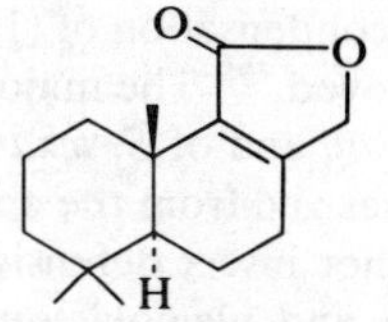

(122)

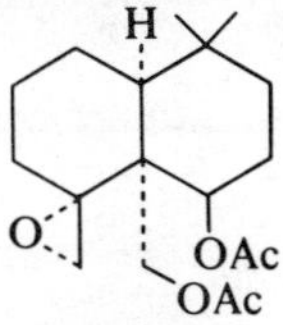

(123)

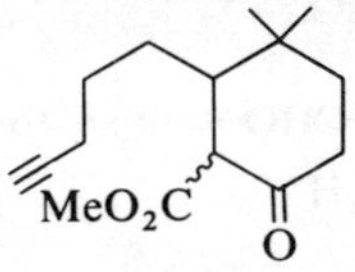

(124)

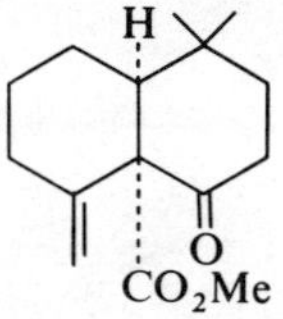

(125)

Precoccinellin (126) and coccinellin (127), defensive alkaloids of the ladybug, *Coccinella septempunctata*, have been prepared from perhydroboraphenalene (128), involving conversion into the enamine (129) and isomerization of the methylated derivative (130) to (131).[204] Alternatively, in another scheme, a Robinson–Schopf-

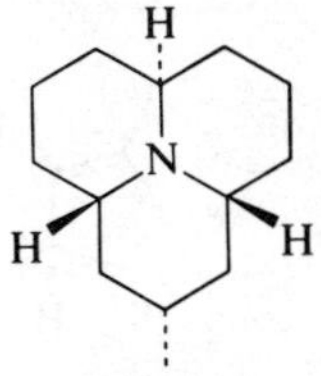

(126)

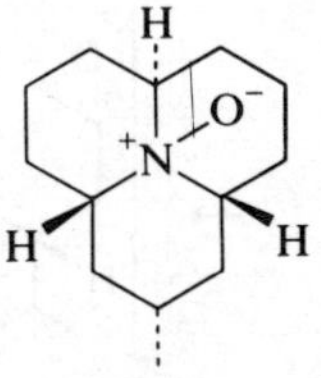

(127)

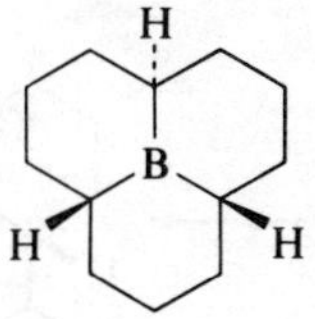

(128)

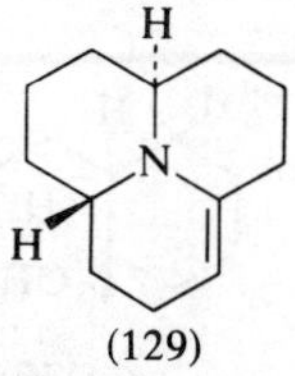

(129)

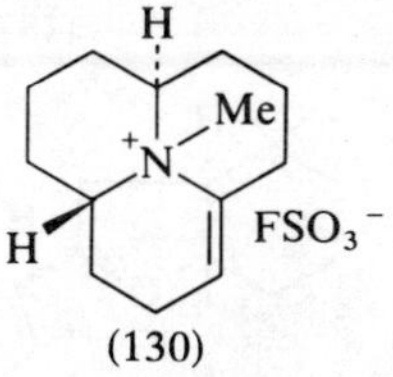

(130)

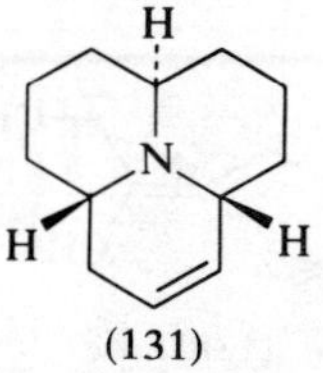

(131)

199 S. P. Tanis and K. Nakanishi, *J. Amer. Chem. Soc.*, 1979, **101**, 4398.
200 T. Nakata, H. Akita, T. Naito, and T. Oishi, *J. Amer. Chem. Soc.*, 1979, **101**, 4400.
201 A. J. G. M. Peters, J. M. Roskam, and A. de Groot, *Rec. Trav. Chim.*, 1978, **97**, 277.
202 W. P. Jackson and S. V. Ley, *J.C.S. Chem. Comm.*, 1979, 732.
203 D. J. Goldsmith, G. Srouji, and C. Kwong, *J. Org. Chem.*, 1978, **43**, 3182.
204 R. H. Mueller and M. E. Thompson, *Tetrahedron Letters*, 1979, 1991.

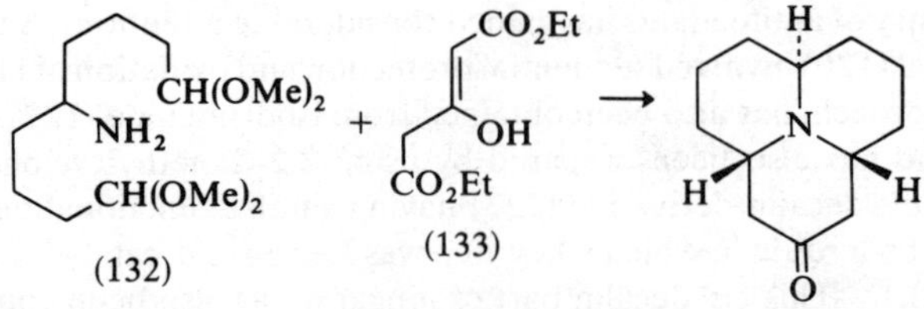

(132) (133) (134)

type condensation of (132) and acetonedicarboxylate ester (133) to yield (134) was employed.[205] The major component of the poison gland of the thief ant, *Solenopsis molesta*, and of *S. texanas* has been identified as 2-hexyl-5-pentylpyrrolidine and synthesized from the appropriate diketone.[206]

Other insect defensive compounds have been attractive targets. Chrysomedial (135) and plagiolactone (136) are two defensive compounds from chrysomelid

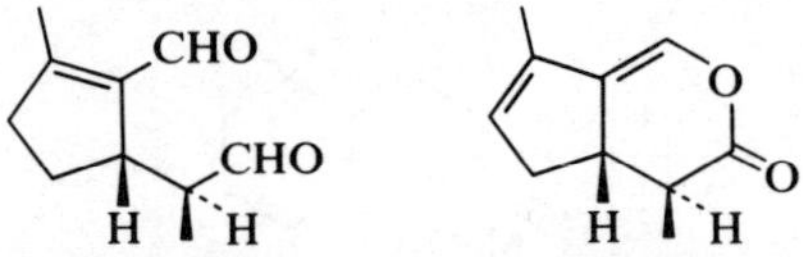

(135) (136)

beetles; both were prepared from limonene by initial functionalization of the terminal double bond, ozonolysis, and Knoevenagel condensation of (137).[207] Ancistrodial (138), from the termite *Ancistrotermes cavithorax*, has been prepared by a sequence involving Stobbe condensation of γ-cyclohomocitral and diethyl succinate.[208] A more difficult target was provided by ancistrofuran (139), found in the major soldiers of the same termite, and this was obtained by cyclization of (140).[209]

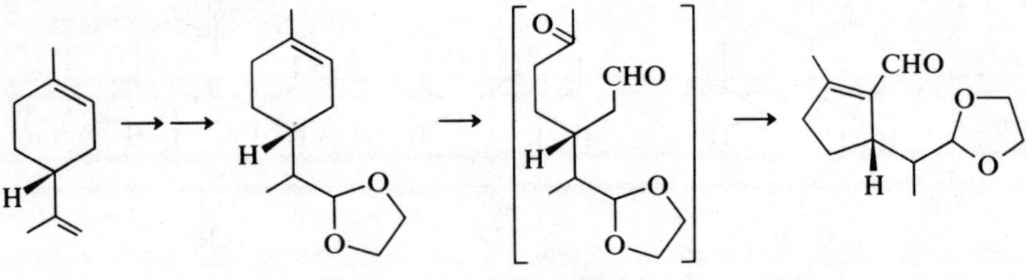

(137)

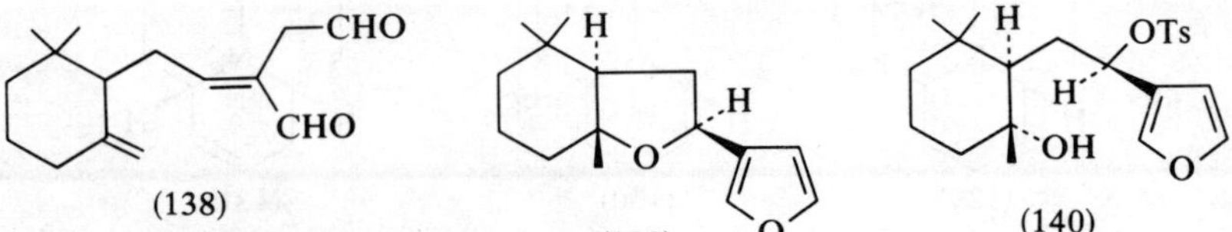

(138) (139) (140)

205 R. V. Stevens and A. W. M. Lee, *J. Amer. Chem. Soc.*, 1979, **101**, 7032.
206 T. H. Jones, M. S. Blum, and H. M. Fales, *Tetrahedron Letters*, 1979, 1031.
207 J. Meinwald and T. M. Jones, *J. Amer. Chem. Soc.*, 1978, **100**, 1883.
208 R. Baker, P. H. Briner, and D. A. Evans, *J.C.S. Chem. Comm.*, 1978, 410.
209 R. Baker, P. H. Briner, and D. A. Evans, *J.C.S. Chem. Comm.*, 1978, 981.

16 Biological Chemistry

Part (iv) Marine Natural Products

By C. CHRISTOPHERSEN
H. C. Ørsted Institute, University of Copenhagen, DK-2100 Copenhagen, Denmark

and

N. JACOBSEN
Cheminova Ltd., P.O. Box 9, Lemvig, DK-7620 Lemvig, Denmark

1 Introduction

This is the first report on marine natural products to appear in *Annual Reports*.

Although, basically, marine natural products could be accommodated in reviews on the various well-established classes of natural products, the last decades have seen such an increasing interest in the chemistry of marine organisms that a temporary segregation seems justified.

For quite a long time the main emphasis in marine natural products chemistry has been laid on the differences between marine and terrestrial organisms, spurred on by the discovery of a number of unique features in the marine metabolites. Apparently novel structural features now cause less stupefaction than earlier. Hence, there has been an increased desire to discover the biological significance of marine natural products.

In order to comply with this tendency, we have chosen to arrange the material phylogenetically, except for a section on sterols and carotenoids, devoting a relatively higher proportion of the space to the (alas) few papers dealing with biological aspects of marine natural products chemistry.

In 1979, comparatively few reviews have appeared, the most prominent and extensive being volume II of a series of reviews;[1] it includes chapters on marine carotenoids,[1a] sterols of marine invertebrates,[1b] diterpenoids,[1c] terpenoids from Coelenterates,[1d] and applications of ^{13}C n.m.r. spectroscopy to marine natural products.[1e] The translation of Hashimoto's book on biologically active marine compounds[2] is a most welcome and stimulating event. Reviews have appeared on compounds from microalgae; in particular, their influence on the field of marine natural products[3a] and molecular aspects of the halogen-based biosynthesis of

[1] 'Marine Natural Products. Chemical and Biological Perspectives', ed. P. J. Scheuer, Academic Press, New York, San Francisco, and London, 1978, Vol. II. (*a*) S. Liaaen-Jensen; (*b*) L. J. Goad; (*c*) W. Fenical; (*d*) B. Tursch, J. C. Braekman, D. Daloze, and M. Kaisin; (*e*) J. J. Sims, A. F. Rose, and R. R. Izac.

[2] Y. Hashimoto, 'Marine Toxins and Other Bioactive Marine Metabolites', Japan Scientific Societies Press, 1979.

[3] (*a*) Y. Shimizu, *Recent Adv. Phytochem.*, 1979, **13**, 199; (*b*) W. Fenical, *ibid.*, p. 219.

marine natural products.[3b] Pigments of marine invertebrates[4] and marine aliphatic natural products[5] have also been reviewed.

2 Marine Sterols and Carotenoids

Although there is much research on marine sterols, it will only be briefly mentioned because of the lack of criteria on which to base a selection of the most relevant investigations. The origin and fate of the diversity of sterol structures and their role in symbiotic relationships are mainly unaccounted for.[6]

A comprehensive review on the origin, distribution, and transformation of sterols in invertebrates has appeared.[1b]

Sterols having unusual side-chains are frequently encountered in marine organisms, particularly sponges, where they are sometimes major sterols, indicating that they serve an as yet unknown function.[7—9] Much evidence suggests the unicellular algae to be the prime producers of marine sterols, leaving most invertebrates in the role of accumulators or structural modifiers. The dinoflagellate *Cryptecodinium cohnii* introduces the 23- and 24-methyl groups of dinosterol from methionine.[10] This may have relevance to the biosynthesis of gorgosterol, a 23-methyl sterol originally isolated from gorgonians, which was recently found in the dinoflagellate *Peridinium foliaceum*;[11] gorgonians have zooxanthellae (symbiotic dinoflagellates).[1] Furthermore, 23-methyl-22-dehydrocholesterol has been characterized from zooxanthellae.[12]

Modification of the sterol ring system is exemplified by $\Delta^{5,7,22}$ sterols and $\Delta^{4,7}$-3,6-dioxo-sterols, isolated from the sponges *Biemna fortis*[13] and *Rhaphidostila incisa*[14] respectively.

The field of marine carotenoids shares several features with that of sterols, making it equally difficult to select the most relevant carotenoid papers for the present report. Most aspects of marine carotenoid research are dealt with in a review.[1a]

By analogy with the situation within the sterol field, it is suspected that only the algae are capable of carotenoid synthesis *de novo*. Many observations support this hypothesis; *e.g.*, the identification of the typical dinoflagellate carotenoid peridinin as the main carotenoid in several species of the Alcyonacea (soft corals) that have symbiotic dinoflagellates[15] and as the pigment (earlier known as sulcatoxanthin) in the sea anemone *Anemonia sulcata*.[16]

4 G. Y. Kennedy, *Adv. Marine Biol.*, 1979, **16**, 309.
5 R. E. Moore, in 'Aliphatic and Related Natural Products Chemistry', ed. F. D. Gunstone (Specialist Periodical Reports), The Chemical Society, London, 1979, Vol. 1, p. 20.
6 J. L. Boutry, A. Saliot, and M. Barbier, *Experientia*, 1979, **35**, 1541.
7 W. Hofheintz and G. Oesterhelt, *Helv. Chim. Acta*, 1979, **62**, 1307.
8 W. C. M. C. Kokke. C. Tarchini, D. B. Stierle, and C. Djerassi, *J. Org. Chem.*, 1979, **44**, 3385.
9 C. Delseth, L. Tolela, P. J. Scheuer, R. J. Wells, and C. Djerassi, *Helv. Chim. Acta*, 1979, **62**, 101.
10 N. W. Withers, R. C. Tuttle, L. J. Goad, and T. W. Goodwin, *Phytochemistry*, 1979, **18**, 71.
11 N. W. Withers, W. C. M. C. Kokke, M. Rohmer, W. H. Fenical, and C. Djerassi, *Tetrahedron Letters*, 1979, 3605.
12 W. C. M. C. Kokke, N. W. Withers, I. J. Massey, W. Fenical, and C. Djerassi, *Tetrahedron Letters*, 1979, 3601.
13 C. Delseth, Y. Kashman, and C. Djerassi, *Helv. Chim. Acta*, 1979, **62**, 2037.
14 A. Malorni, L. Minale, and R. Riccio, *Nouveau J. Chim.*, 1979, **2**, 351.
15 M. Hallenstvet and S. Liaaen-Jensen, *Biochem. Syst. Ecol.*, 1979, **7**, 171.
16 A. Fiksdahl, M. Hallenstvet, L. Beress, and S. Liaaen-Jensen, *Biochem. Syst. Ecol.*, 1979, **7**, 173.

Many recent studies have been concerned with the determination of absolute stereochemistry (*e.g.* tunaxanthin[17]) and of the structures of complexes between carotenoids and proteins (*e.g.* alloporin from the coral *Allopora californica*[18]).

3 Marine Plants

Like their terrestrial counterparts, marine plants (particularly algae) play a major role as producers of secondary metabolites. It is, however, characteristic of the marine environment that species from several phyla of invertebrates are known to contain copious amounts of substances strongly resembling plant metabolites. This has led several natural products chemists to believe that these compounds have their origin in the algae (or bacteria) which are symbionts, epiphytes, or simply the diet of these invertebrates. A few chemical relationships between algae and invertebrates have been demonstrated or indicated, such as the relationship between the sacoglossan mollusc *Placobranchus ocellatus* and its symbiotic algal chloroplasts.[19]

Although many of the substances reported from marine invertebrates are undoubtedly of algal origin, they will be mentioned in the appropriate section on invertebrates, but, where an algal origin has been indicated, this will be mentioned.

Cyanophyceae (Blue-green Algae).—The structures of a series of chlorine-containing amines, for example malyngamide A (1), from shallow-water strains of *Lyngbya majuscula*, were reported.[20,21] The authors point out the interesting structural similarity between the central amino-acid part of the malyngamides and dysidin, a metabolite from a sponge *Dysidia herbacea*, which is known to contain large amounts of symbiotic blue-green algae.

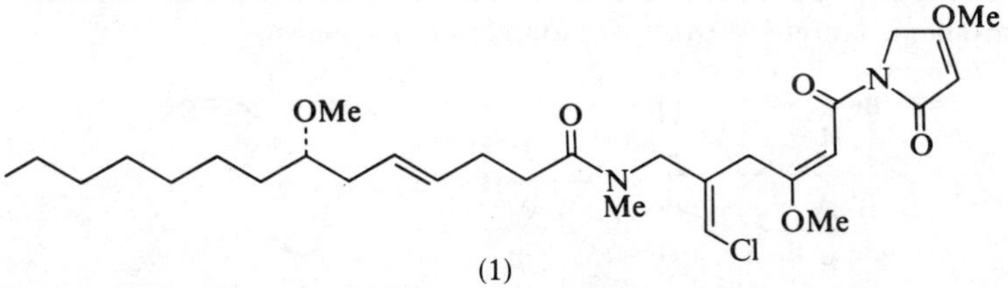

(1)

A series of amides that are structurally related to the malyngamides, for example pukelimide A (2), were also isolated from *L. majuscula*.[22,23] This alga is known to cause a dermatitis called 'swimmer's itch' in Hawaii. One of the substances responsible for this effect, *i.e.* lyngbyatoxin A, has been characterized as an indole alkaloid (3) that is closely related to a fungal metabolite, teleocidin B, which causes similar physiological effects.[24]

[17] A. Bingham, Jr., D. W. Wilkie, and H. S. Mosher, *Comp. Biochem. Physiol. (B)*, 1979, **62**, 489.
[18] H. Rønneberg, G. Borch, D. L. Fox, and S. Liaaen-Jensen, *Comp. Biochem. Physiol. (B)*, 1979, **62**, 309.
[19] C. Ireland and P. J. Scheuer, *Science*, 1979, **205**, 922.
[20] J. H. Cardellina, II., F.-J. Marner, and R. E. Moore, *J. Amer. Chem. Soc.*, 1979, **101**, 240.
[21] J. H. Cardellina, II., D. Dalietos, F.-J. Marner, J. S. Mynderse, and R. E. Moore, *Phytochemistry*, 1978, **17**, 2091.
[22] C. J. Simmons, F.-J. Marner, J. H. Cardellina, II., R. E. Moore, and K. Seff, *Tetrahedron Letters*, 1979, 2003.
[23] J. H. Cardellina, II. and R. E. Moore, *Tetrahedron Letters*, 1979, 2007.
[24] J. H. Cardellina, II., F.-J., Marner, and R. E. Moore, *Science*, 1979, **204**, 193.

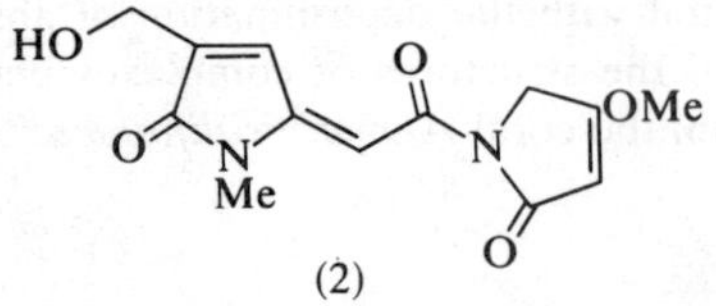

(2)

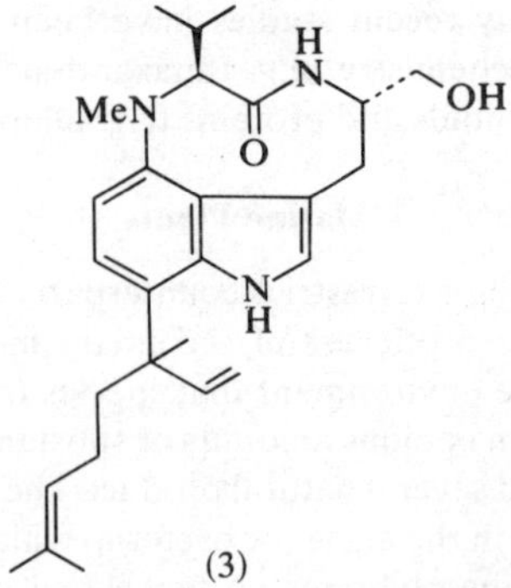

(3)

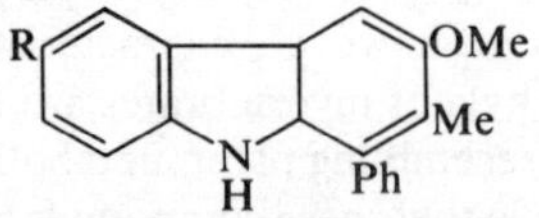

(4) R = H
(5) R = Cl

Alkaloids were also present in *Hyella caespitosa*, from which the two unusual carbazole alkaloids hyellazole (4) and chlorohyellazole (5) were isolated.[25]

Rhodophyceae (Red Algae).—A large portion of the pioneering work on metabolites of red algae has been done on members of the genus *Laurencia*; since then, a vast number of compounds, belonging mainly to two classes (*i.e.* halogenated terpenoids and halogenated compounds containing a C_{15} straight-chain skeleton) have been described from this genus. Most of the recent papers on *Laurencia* metabolites report only minor variations within already known structural types, or are concerned with details of known structures.

In the non-terpenoid class, a polycyclic acetylenic ketal, obtusin (6),[26] and an acetylenic chlorine- and bromine-containing nine-membered cyclic ether, obtusenyne (7),[27] have both been isolated from Mediterranean *L. obtusa*, whereas a bromo-allene, laurallene (8),[28] was found in *L. nipponica*.

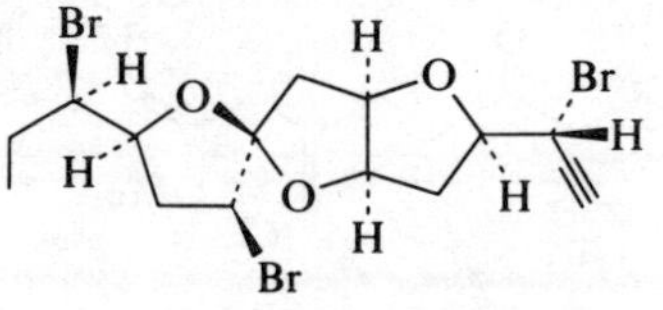

(6)

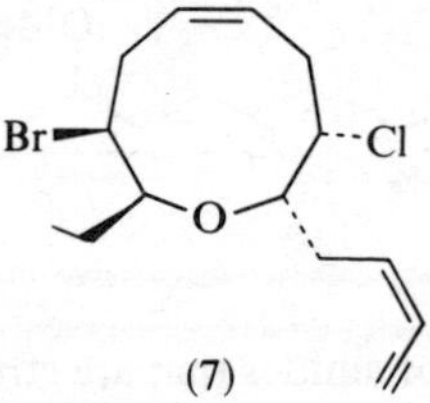

(7)

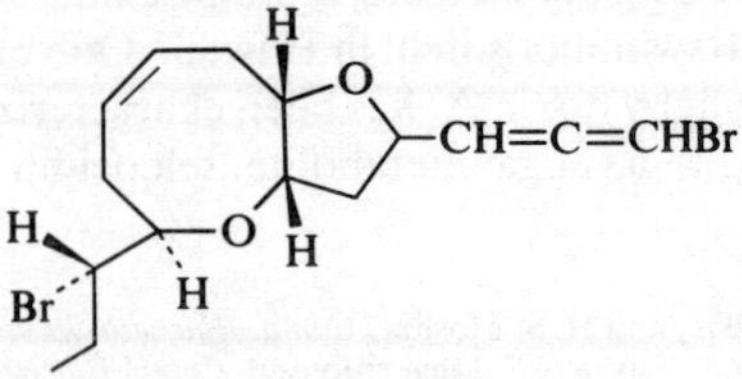

(8)

[25] J. H. Cardellina, II., M. P. Kirkup, R. E. Moore, J. S. Mynderse, K. Seff, and C. J. Simmons, *Tetrahedron Letters*, 1979, 4915.
[26] B. M. Howard, W. Fenical, E. V. Arnold, and J. Clardy, *Tetrahedron Letters*, 1979, 2841.
[27] T. J. King, S. Imre, A. Öztunc, and R. H. Thomson, *Tetrahedron Letters*, 1979, 1453.
[28] A. Fukuzawa and E. Kurosawa, *Tetrahedron Letters*, 1979, 2797.

The genus *Laurencia* has presented certain taxonomic problems because there is considerable morphological variation within the species. Preliminary attempts to employ gas chromatography–mass spectrometry of crude extracts as a chemotaxonomic aid have been reported.[29]

A number of *Laurencia* sesquiterpenoids, especially of the chamigrene types [*e.g.* obtusol (9)] contain a chloro- and a bromo-substituent in a vicinal arrangement. Determination of the exact positions of these substituents has, so far, been done with a high degree of uncertainty. Thus, after *X*-ray analysis of members of this group, some earlier assignments had to be revised,[30] and, using this information combined with a detailed ^{13}C n.m.r. analysis of a number of related structures, the structures of two new compounds, isofurocaespitan (10) and obtusan (11), from *L. caespitosa* and *L. obtusa*, respectively, could be determined with greater confidence.[31]

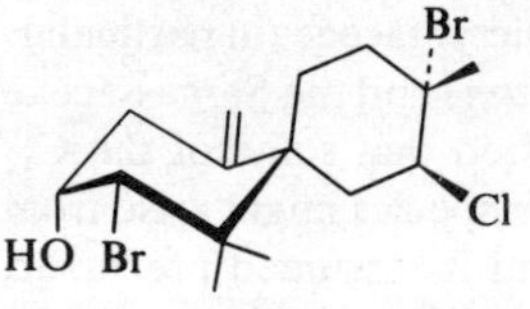

(9)

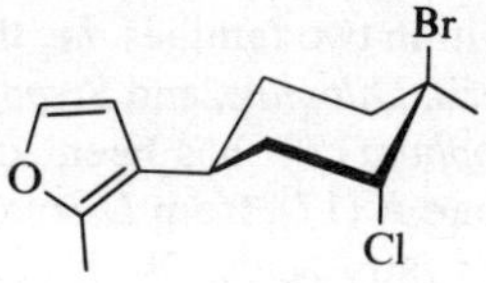

(10)

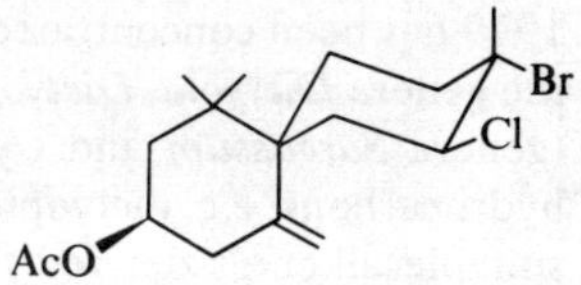

(11)

While bromo- and chloro-substituents have been found to be commonplace features of marine natural products, the two compounds (12) and (13) isolated from *L. nana* represent the first examples of iodinated sesquiterpenes.[32]

From a Californian *L. spectabilis*, an unusual collection of $\alpha\beta$-unsaturated α-branched aldehydes, exemplified by the major constituent, 2-dodecylhexadec-2-enal (14), was identified.[33]

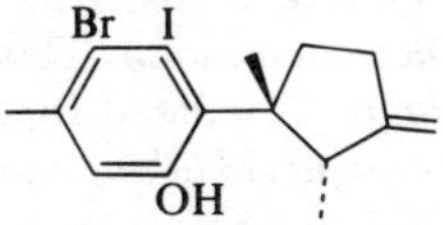

(12)

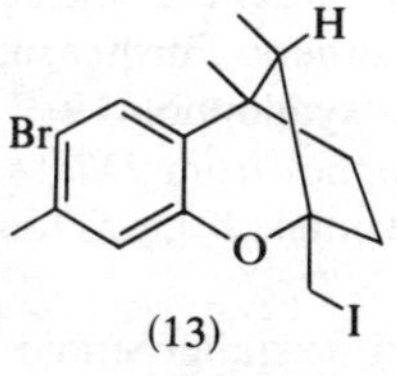

(13)

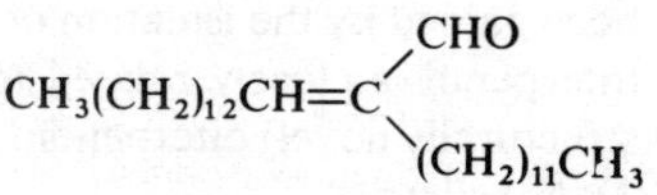

(14)

Prostaglandins PGE_2 (15) and $PGF_{2\alpha}$ (16), isolated from *Gracilaria lichenoides*, are the first examples of prostaglandins from a plant.[34] This finding suggests that the prostaglandins found in gorgonians may have their origin in symbiotic algae, since both (15) and (16) have been isolated from gorgonians.[35]

Fenical[36] has suggested that several of the simple halogenated compounds (acrylic acids, halogenomethanes, lactones, *etc.*) from members of the family Bonnemaisoniaceae arise by simple transformations of halogenated ketones. Recent

[29] S. Caccamese, L. P. Hager, K. L. Rinehart, Jr., and R. B. Setzer, *Botanica Marina*, 1979, **22**, 41.
[30] A. G. Gonzáles, J. D. Martin, V. S. Martin, M. Martinez-Ripoll, and J. Fayos, *Tetrahedron Letters*, 1979, 2717.
[31] A. G. Gonzáles, J. D. Martin, V. S. Martin, and M. Norte, *Tetrahedron Letters*, 1979, 2719.
[32] R. R. Izac and J. J. Sims, *J. Amer. Chem. Soc.*, 1979, **101**, 6136.
[33] R. P. Maiden and J. A. Pettus, Abstract 575, ACS/CSJ Chemical Congress, Honolulu, Hawaii, April 1–6, 1979.
[34] R. P. Gregson, J. F. Marwood, and R. J. Quinn, *Tetrahedron Letters*, 1979, 4505, and references therein.
[35] Y. Komoda, T. Kanayasu, and M. Ishikawa, *Chem. Pharm. Bull.*, 1979, **27**, 2491.

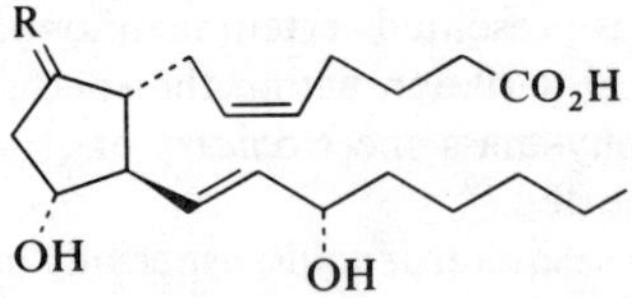

(15) R = O
(16) R = α-OH, β-H

work,[36,37] on the whole, supports this hypothesis, but further work is needed in order to determine which of these transformations occur *in vivo* and which during the isolation procedure.

Phaeophyceae (Brown Algae).—The work on metabolites of brown algae reported in 1979 has been concentrated within two families, *i.e.* the Dictyotaceae (in particular, the genera *Dictyota*, *Dictyopteris*, *Dilophus*, and *Stypopodium*) and the Sargassaceae (genera *Sargassum* and *Cystophora*). It has been suggested that some of the C_{11} hydrocarbons, *e.g.* dictyopterene A (17), from *Dictyopteris* species might arise from suitable alkenols derived from fatty acids. Both (17) and its assumed precursor, dictyoprolene (18), have been isolated from *Dictyopteris prolifera*, and it was shown that the absolute stereochemistry of (18) is consistent with this hypothesis.[38]

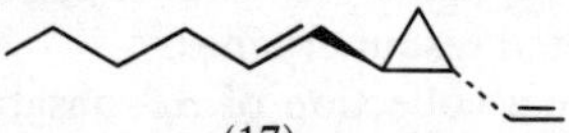

(17)

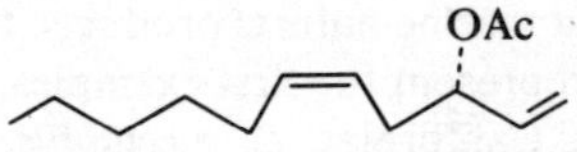

(18)

Within the Dictyotaceae, a group of genera (including *Dictyota* and *Dilophus*) appear more closely related than the others. Further support for this relationship has been gained by the isolation of hydroxydilophol (19)[39] from *Dictyota masonii*. This diterpenol is closely related to dilophol from *Dilophus ligulatus*. A related (but structurally novel) diterpenoid, dictyodial (20), was found in *Dictyota crenulata* and *D. flabellata*.[40]

Several compounds containing a sesquiterpenoid (drimane) substituent on a benzenoid moiety are known from *Dictyopteris undulata*, the latest being yahazunol

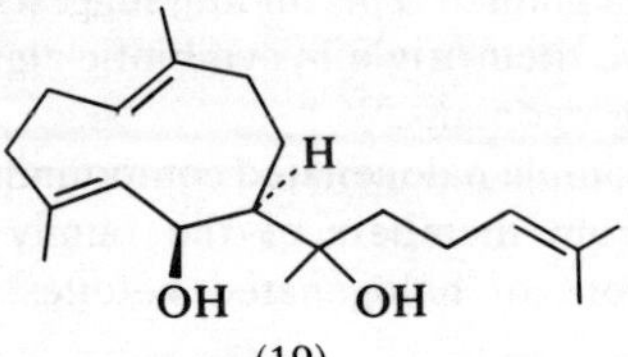

(19)

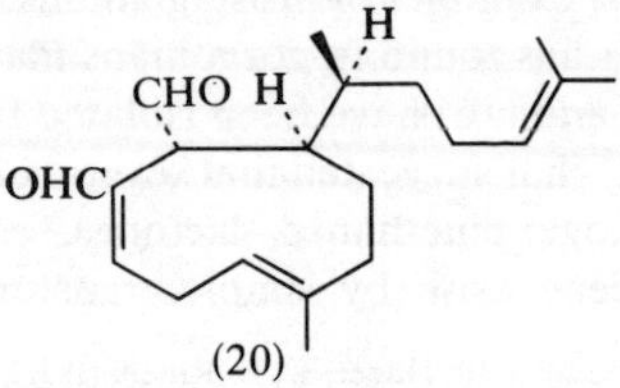

(20)

[36] F. X. Woolard, R. E. Moore, and P. P. Roller, *Phytochemistry*, 1979, **18**, 617.
[37] N. Jacobsen and J. Ø. Madsen, *Tetrahedron Letters*, 1978, 3065.
[38] K. Yamada, H. Tan, and H. Tatematsu, *J.C.S. Chem. Comm.*, 1979, 572.
[39] H. H. Sun and W. Fenical, *J. Org. Chem.*, 1979, **44**, 1354.
[40] J. Finer, J. Clardy, W. Fenical, L. Minale, R. Riccio, J. Battaile, M. Kirkup, and R. E. Moore, *J. Org. Chem.*, 1979, **44**, 2044.

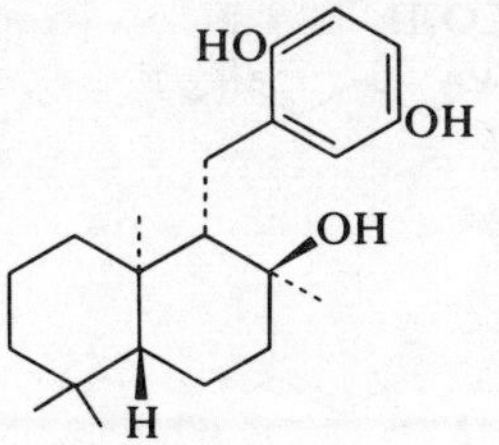

(21)

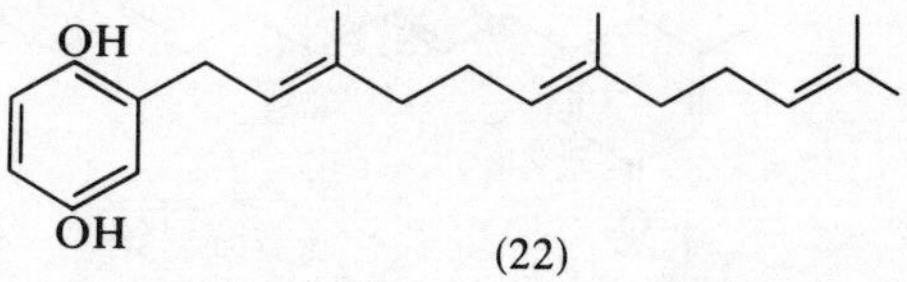

(22)

(21).[41] A likely precursor for these compounds, the polyprenylquinone (22), was found in the same alga.[42]

Stypopodium zonale is toxic to herbivorous fish. Compounds responsible for the toxicity have been isolated and shown to be a diterpenoid-substituted *o*-benzoquinone, stypoldione (23), and its corresponding hydroquinone, stypotriol (24). The latter is the most powerful toxin, but is rapidly aerially oxidized to (23) outside the alga.[43]

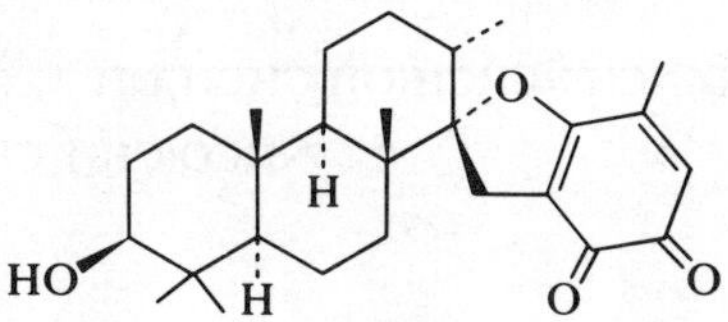

(23)

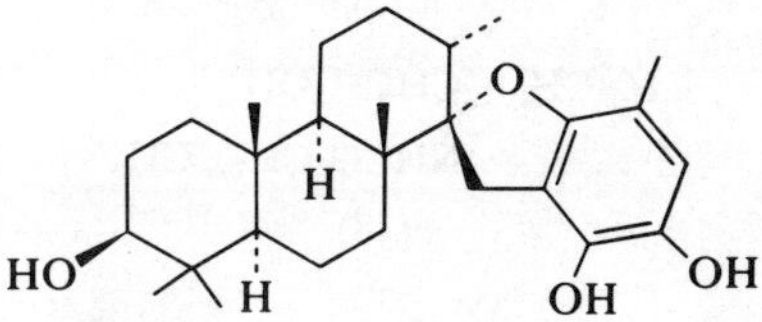

(24)

In the genus *Sargassum*, plastoquinones such as sargaquinoic acid (25) and sargaquinal (26) frequently occur.[44] It has been suggested that a series of derivatives of farnesylacetone (27) isolated from *S. micracanthum* are derived from plastoquinones by oxidative cleavage, as indicated in (25).[45]

Chlorophyceae (Green Algae).—The structures (28) were originally tentatively assigned to caulerpicin, a bioactive mixture of homologous compounds from a Philippine *Caulerpa racemosa*. Two of these, (28; $n = 24$) and (28; $n = 25$), have now been synthesized and shown to be different from the original caulerpicins.[46] A mixture of similar compounds, *N*-acyl-sphingosines (29), has been isolated from a Sri Lankan *C. racemosa*, and is claimed to be the same as caulerpicin.[47] A comparison of the physiological and physical properties of this mixture with those of caulerpicin has not, however, been reported.

Green algae from the families Codiaceae and Caulerpaceae are often avoided by herbivores. From *Rhipocephalus phoenix* (Codiaceae), two toxic feeding deterrents against fish were isolated, and identified as the sesquiterpenoids rhipocephalin (30) and rhipocephenal (31).[48]

41 M. Ochi, H. Kotsuki, K. Muraoka, and T. Tokoroyama, *Bull. Chem. Soc. Japan*, 1979, **52**, 629.
42 M. Ochi, H. Kotsuki, S. Inoue, M. Taniguchi, and T. Tokoroyama, *Chem. Letters*, 1979, 831.
43 W. H. Gerwick, W. Fenical, N. Fritsch, and J. Clardy, *Tetrahedron Letters*, 1979, 145.
44 T. Kusumi, Y. Shibata, M. Ishitsuka, T. Kinoshita, and H. Kakisawa, *Chem. Letters*, 1979, 277.
45 T. Kusumi, M. Ishitsuka, Y. Nomura, T. Konno, and H. Kakisawa, *Chem. Letters*, 1979, 1181.
46 U. Sørensen and C. Christophersen, unpublished results.
47 M. Mahendran, S. Somasundaram, and R. H. Thomson, *Phytochemistry*, 1979, **18**, 1885.
48 H. H. Sun and W. Fenical, *Tetrahedron Letters*, 1979, 685.

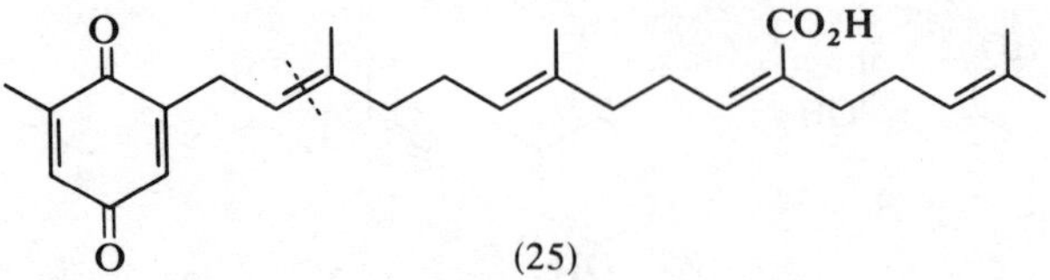

(25)

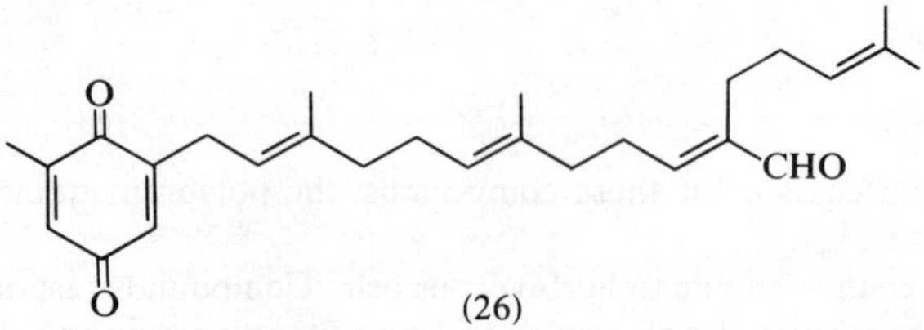

(26)

(27)

$CH_3(CH_2)_{13}CHCH_2OH$ with $NHCO(CH_2)_nCH_3$ on the CH

(28)

$CH_3(CH_2)_{12}CH\overset{E}{=}CHCH(OH)CHCH_2OH$ with $NHCO(CH_2)_nCH_3$ on the CH

(29)

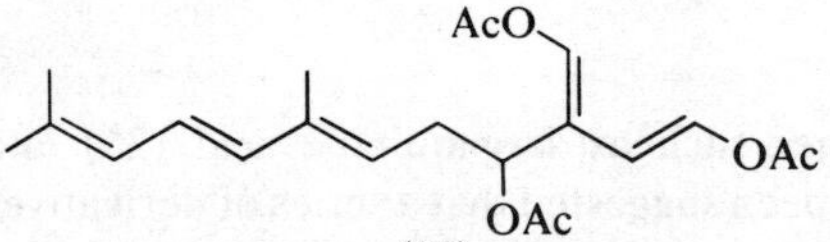

(30)

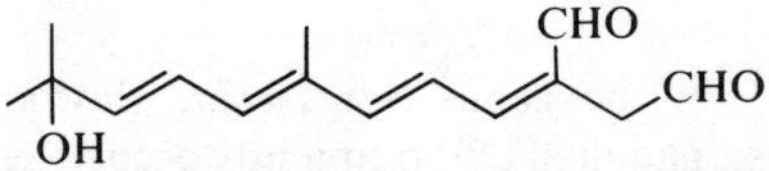

(31)

4 Marine Invertebrates

Porifera.—Marine sponges still remain popular sources of intriguing compounds. In spite of the wealth of structural information gathered from these species, study of biochemical, biosynthetic, and ecological information is virtually lacking. This sad state of affairs is mainly due to three inherited problems in sponge research. Sponge systematics is based on relatively few characters, and, since these animals display considerable natural variation, classification is often extremely difficult. Sponges are filter-feeders, and it is not known to what extent they may retain food-derived metabolites. Finally, many sponges are known to harbour a wealth of parasitic or symbiotic micro-organisms, especially microalgae and bacteria. Nothing is known of the role of this microflora; however, several compounds described from sponges have a striking resemblance to known algal and bacterial metabolites.

These points may be demonstrated by the recent isolation of alkylated scalarins from *Dysidea herbacea*, collected in the Gulf of Suez;[49] specimens of *D. herbacea* from other geographical regions, however, lack these compounds.[50] Furthermore, a

[49] Y. Kashman and M. Zviely, *Tetrahedron Letters*, 1979, 3879.

[50] C. Charles, J. C. Braekman, D. Daloze, B. Tursch, and R. Karlsson, *Tetrahedron Letters*, 1978, 1519; R. Kazlauskas, P. T. Murphy, and R. J. Wells, *ibid.*, 1978, 4945, 4949; R. J. Wells, *Pure Appl. Chem.*, 1979, **51**, 1829.

survey of the genus *Spongia* from Australian waters has revealed that species collected from tropical environments have never yielded C_{21} furano-terpenes, while diterpenoids have never been detected in non-tropical *Spongia*, in spite of the fact that the two classes of compounds are abundantly represented within this genus.[51]

The sesterterpenoid hydroxyquinols toxistylide A and B were only present in some collections of *Microciona toxistyla*, while all of them contained sesquiterpenoids.[52] These few examples also deliver the *coup de grace* to the hypothesis that sponge taxonomy may be based on simple analyses of secondary metabolites.

Several reports on peroxides isolated from sponge material have appeared. From *Chondrosia collectrix*, a mixture of (32) and its C-3 epimer has been characterized. The corresponding methyl esters were also found.[53] Apparently these peroxides are derived from fatty acids with methyl or ethyl side-chains, while muqubilin (33) is a C_{24} isoprenoid from an unidentified *Prianos* sp.[54]

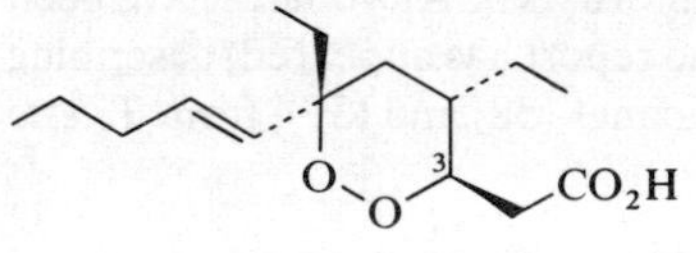

(32)

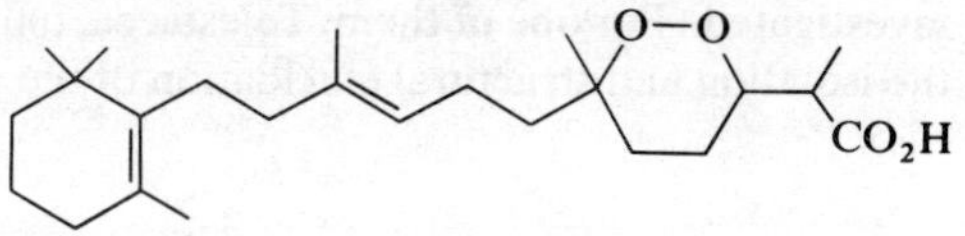

(33)

An ichthyotoxic peroxide, sigmosceptrellin-A (34), from *Sigmosceptrella laevis*, is a nor-sesterterpenoid[55] which, curiously enough, possesses the same sesquiterpenoid skeleton as the orange compound (35), poetically named ilimaquinone, from *Hippiospongia metachromia* (tentative classification).[56]

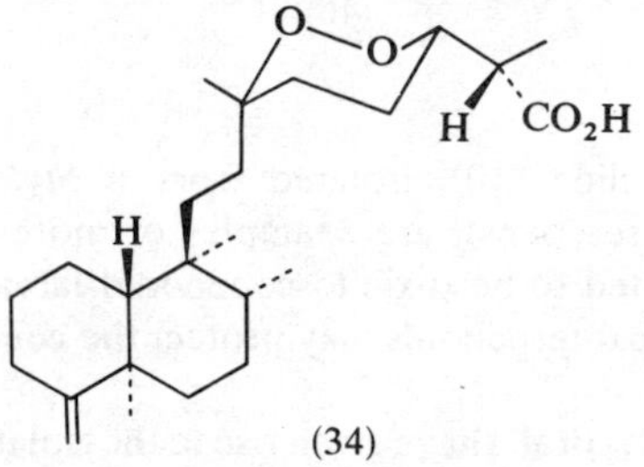

(34)

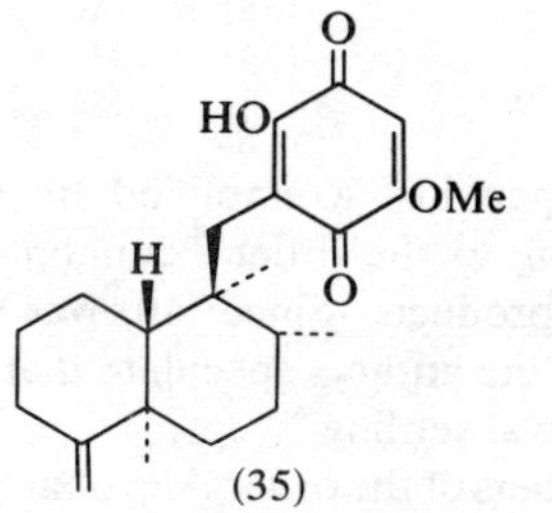

(35)

Quinones are well-known structural elements in marine natural products; however, renierone (36),[57] from an unidentified *Reniera* sp., is the first example of an isoquinoline quinone. Renierone displays strong antimicrobial activity. Other

[51] R. Kazlauskas, P. T. Murphy, R. J. Wells, K. Noack, W. E. Oberhänsli, and P. Schönholzer, *Austral. J. Chem.*, 1979, **32**, 867.
[52] G. Cimino, S. De Stefano, L. Minale, R. Riccio, K. Hirtosu, and J. Clardy, *Tetrahedron Letters*, 1979, 3619.
[53] D. B. Stierle and D. J. Faulkner, *J. Org. Chem.*, 1979, **44**, 964.
[54] Y. Kashman and M. Rotem, *Tetrahedron Letters*, 1979, 1707.
[55] M. Albericci, M. Collart-Lempereur, J. C. Braekman, D. Daloze, B. Tursch, J. P. Declercq, G. Germain, and M. Van Meerssche, *Tetrahedron Letters*, 1979, 2687.
[56] R. T. Luibrand, T. R. Erdman, J. J. Vollmer, P. J. Scheuer, J. Finer, and J. Clardy, *Tetrahedron*, 1979, **35**, 609.
[57] D. E. McIntyre, D. J. Faulkner, D. V. Engen, and J. Clardy, *Tetrahedron Letters*, 1979, 4163.

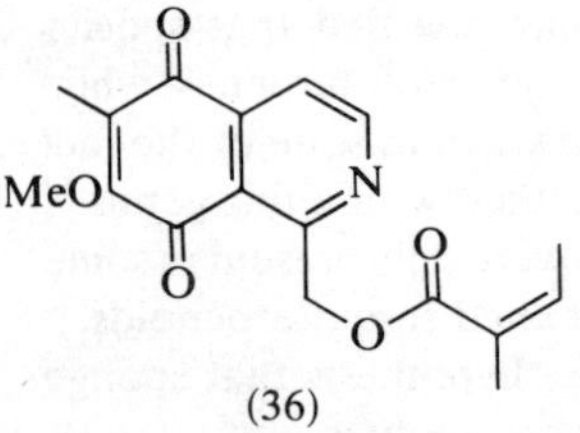
(36)

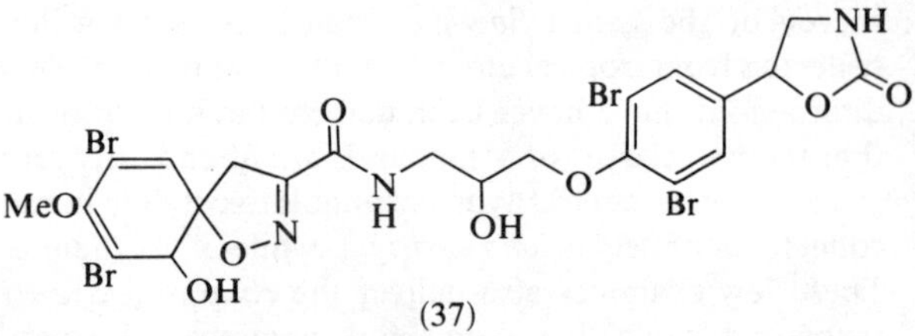
(37)

alkaloids, such as fistularine 1 (37), from *Aplysina fistularis* forma *fulva*, exhibit very interesting growth-inhibiting effects on cells *in vitro*.[58]

Coelenterata.—The two phyla Cnidaria and Ctenophora constitute the Coelenterata. Class Anthozoa in phylum Cnidaria is divided into the subclass Alcyonaria (Octocorallia) and Zoantharia (Hexacorallia). Four orders of Alcyonaria have been investigated. For one of them, Telestacea, only one report has appeared, describing the isolation and structural elucidation of the pregnanes (38) and (39), from *Telesto riisei*.[59]

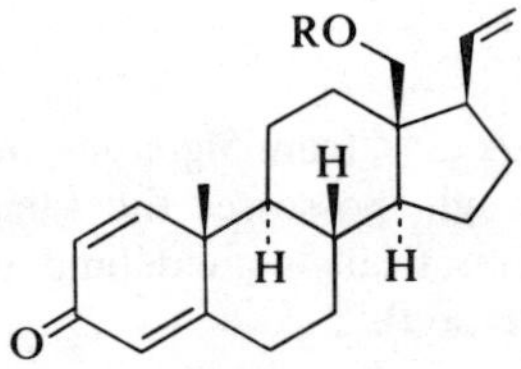
(38) R = Ac
(39) R = H

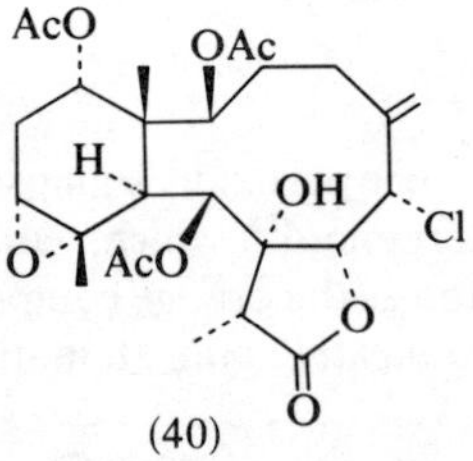
(40)

Diterpenoids, exemplified by stylatulide (40), isolated from a *Stylatula* sp. belonging to the order Pennatulacea (sea pens), are examples of more complex marine products. Since (40) was reported to be toxic to copepodid larvae (Crustaceae), the authors speculate that these diterpenoids may protect the coelenterate from larval settling.[60]

Members of the order Alcyonacea (soft corals) have given rise to the isolation of an amazing variety of terpenoids. The rich harvest of fourteen-membered-ring cembranolide diterpenes has continued, plus an occasional thirteen-membered ring, *e.g.* (41).[61] Diterpenoids from soft corals include nine-membered-ring representatives, *e.g.* isoxeniolide A (42) from *Xenia novae-britanniae*.[62]

Among the many sesquiterpenoids isolated from members of the Alcyonacea, those derived from the capnellane skeleton have been the subject of further investigations. The soft coral *Capnella imbricata* showed considerable geographical

[58] Y. Gopichand and F. J. Schmitz, *Tetrahedron Letters*, 1979, 3921.
[59] R. A. Ross and P. J. Scheuer, *Tetrahedron Letters*, 1979, 4701.
[60] S. J. Wratten and D. J. Faulkner, *Tetrahedron*, 1979, **35**, 1907.
[61] Y. Yamada, S. Suzuki, K. Iguchi, K. Hosaka, H. Kikuchi, Y. Tsukitani, H. Horiai, and F. Shibayama, *Chem. Pharm. Bull.*, 1979, **27**, 2394.
[62] J. C. Braekman, D. Daloze, B. Tursch, J. P. Declercq, G. Germain, and M. Van Meerssche, *Bull. Soc. chim. belges*, 1979, **88**, 71.

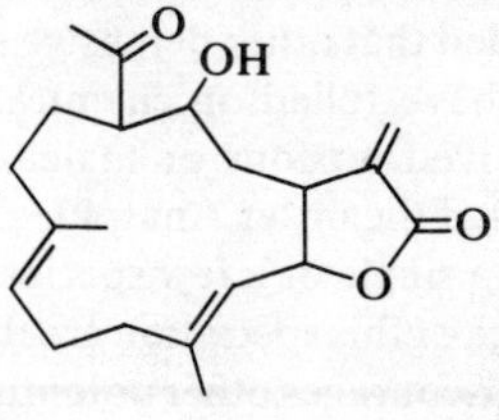

(41)

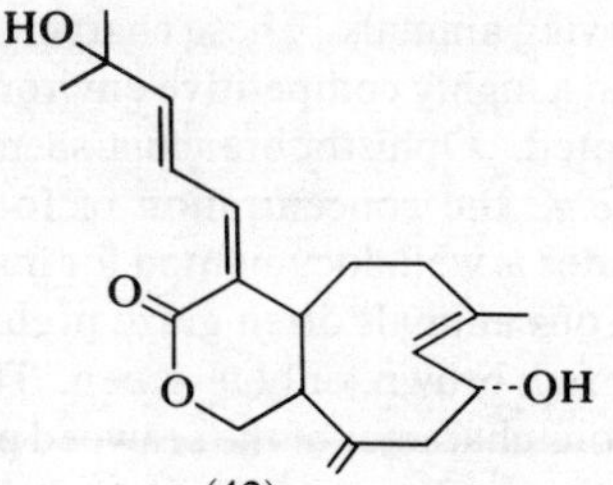

(42)

and individual variation in ratios of capnellane-derived metabolites.[63] Species from the northern coast of New Guinea, for example, were found to contain $\Delta^{9(12)}$-capnellene-2,5,8,10-tetraol (43). Interestingly enough, $\Delta^{9(12)}$-capnellene (44) was found to co-occur with precapnelladiene (45) in a hydrocarbon fraction. This observation has given rise to the suggestion that (45) is a biogenetic precursor of the capnellane skeleton.[64]

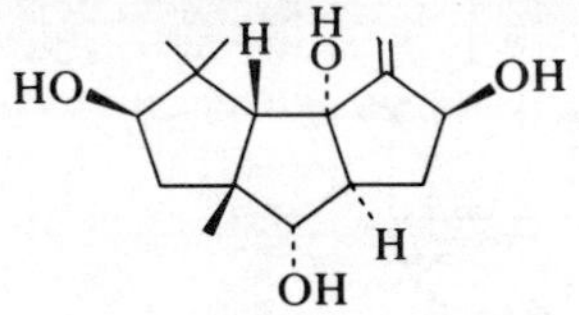

(43)

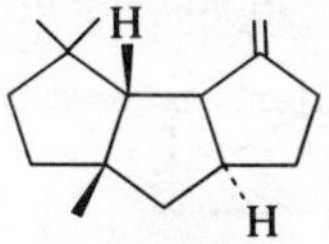

(44)

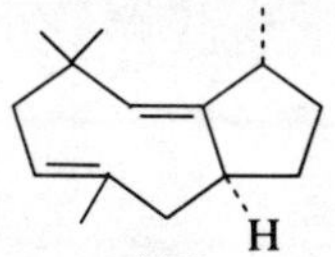

(45)

The first marine spermidine derivatives, (46) and dihydro-(46), have been reported from the soft coral *Sinularia brongersmai.*[65] These derivatives exhibit interesting levels of cytotoxicity, an effect which in other coelenterates has often been traced to the cembranolides.

The fourth order, the Gorgonacea, has received ample attention earlier; however, work published during 1979 has been extremely sparse.

A representative of the Hexacorallia, the gold coral, a Pacific zoanthid species of the genus *Gerardia*, has added two more variations, *e.g.* (47),[66] to the tetra-azacyclopentazulene theme.

$$CH_3(CH_2)_8\overset{Me}{C}\underset{E}{=}CHCONH(CH_2)_3\overset{Me}{N}(CH_2)_4NMe_2$$

(46)

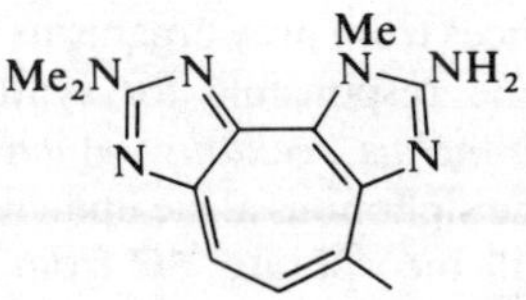

(47)

Mollusca.—Significant progress has been achieved in unravelling important questions in the chemical ecology of subclass Opisthobranchia (sea slugs). Members of this subclass lack external shells and are often conspicuously coloured, delicate,

[63] M. Kaisin, B. Tursch, J. P. Declercq, G. Germain, and M. Van Meerssche, *Bull. Soc. chim. belges.*, 1979, **88**, 253.
[64] E. Ayanoglu, T. Gebreyesus, C. M. Beechan, and C. Djerassi, *Tetrahedron*, 1979, **35**, 1035.
[65] F. J. Schmitz, K. H. Hollenbeak, and R. S. Prasad, *Tetrahedron Letters*, 1979, 3387.
[66] R. E. Schwartz, M. B. Yunker, P. J. Scheuer, and T. Ottersen, *Canad. J. Chem.*, 1979, **57**, 1707.

slow-moving animals. These characteristics are only compatible with survival of the species in a highly competitive environment provided that other defensive strategies are adopted. Ophisthobranchia seem mainly to have relied on chemical defence tactics; *e.g.*, the concentration of food-chain-derived noxious or toxic secondary metabolites is well documented for many sea hares of the order Anaspidea.[67] These herbivorous animals often graze preferentially on a single or a few species of algae, whether red, brown, or blue-green. The advantages of this speciation are dual, since the noxious character of the seaweed presumably discourages other potential grazers and the accumulation of the chemicals in the digestive glands of the sea slug deters would-be predators.

A beautiful illustration of the perfection evolved in this type of relationship is offered in the investigation of the nudibranch *Phyllidia varicosa*.[68] When provoked, this organism will secrete a mucous slime containing a mixture of 2-isocyanopupukeanane (48) and 9-isocyanopupukeanane (49), lethal to fish and crustaceans. The sesquiterpenoid isocyanides (48) and (49) originate from a sponge

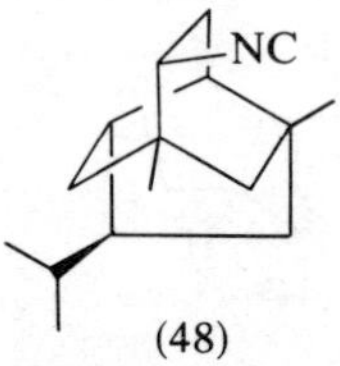

(48)

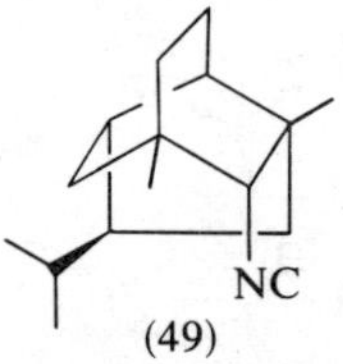

(49)

(*Hymeniacidon* sp.), the prey of *P. varicosa.* Undoubtedly the mixture of (48) and (49) functions also as an allomone in this predator–prey relationship. Total syntheses have been published for (48)[69] as well as for (49).[70] However, the biosynthesis of isocyanides in sponges is far from clear. Isocyanides are often accompanied by the corresponding isothiocyanates and formamides. Recent evidence seems to indicate that isocyanides are not formed *via* formamides, leaving the immediate precursor unknown.[71] Future research in this area may very well benefit from a closer examination of the role of symbiotic micro-organisms in the sponge tissue.

Opisthobranchia may not only benefit from the accumulation of chemical substances from prey organisms but may, in certain cases, secure the complete intact organelle responsible for synthesis, as exemplified by the sacoglossan mollusc *Placobranchus ocellatus*, which assimilates photosynthetically viable chloroplasts from the siphonous algae upon which they graze. The mollusc–chloroplast symbiotic pair will incorporate ^{14}C from radioactive hydrogen carbonate into 9,10-deoxytridachione (50). The activity appears slowly in photodeoxytridachione (51), as a result of an unprecedented photochemical isomerization *in vivo*. The authors speculate that this reaction may serve as a chemical sunscreen.[19]

[67] W. Fenical, H. L. Sleeper, V. J. Paul, M. O. Stallard, and H. H. Sun, *Pure Appl. Chem.*, 1979, **51**, 1865.
[68] M. R. Hagadone, B. J. Burreson, P. J. Scheuer, J. S. Finer, and J. Clardy, *Helv. Chim. Acta*, 1979, **62**, 2484.
[69] E. J. Corey and M. Ishiguro, *Tetrahedron Letters*, 1979, 2745.
[70] (*a*) E. J. Corey, M. Behforouz, and M. Ishiguro, *J. Amer. Chem. Soc.*, 1979, **101**, 1608; (*b*) H. Yamamoto and H. L. Sham, *ibid.*, p. 1609.
[71] A. Iengo, C. Santacrose, and G. Sodano, *Experientia*, 1979, **35**, 10.

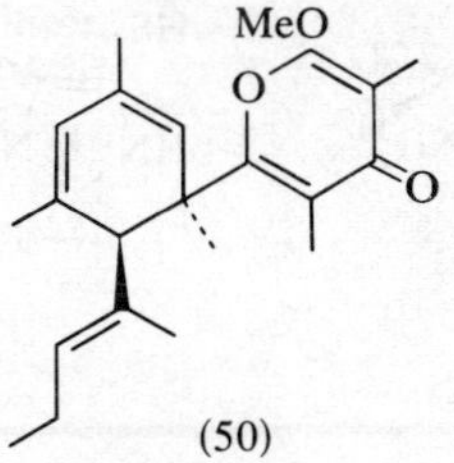

(50)

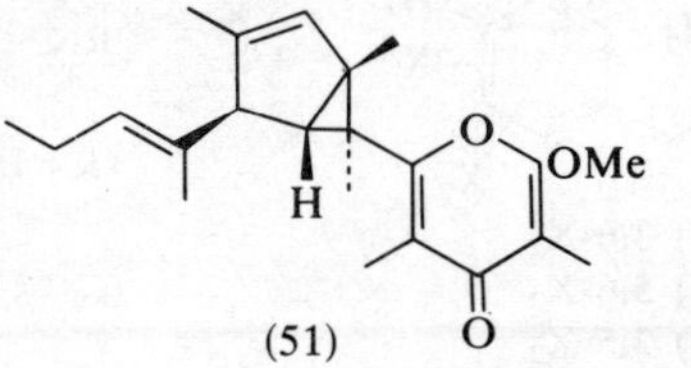

(51)

That some Opisthobranchia possess synthetic ability of their own is exemplified by an investigation of *Navanax inermis* (Cephalaspidea). The members of this species locate each other by following the slime trail laid down by a crawling animal. If heavily molested, the sea slug will deposit a bright yellow secretion containing three main components; navenones A (52), B (53), and C (54). The presence of this

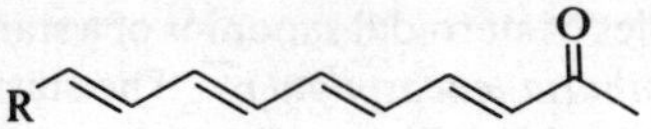

(52) R= 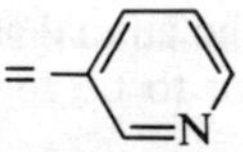 (53) R=Ph (54) R= 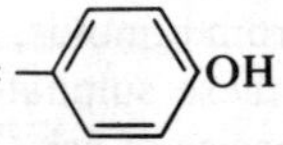

mixture elicits a specific avoidance behaviour in any *Navanax*, namely termination of trail-following and migration in a new direction that forms an angle of greater than 90° with that of the original direction. Labelled acetate (^{14}C), given with the food, was incorporated into the navenones, the highest activity being found in (54), followed by (53). Partly based on evidence from the labelling experiment, the authors have suggested a biogenetic relationship (54) → (53) → (52);[67] this hypothesis warrants further study.

Mycosporin analogues are known from fungi, algae, and coelenterates. Two such compounds, mytillins A (55) and B (56), have recently been isolated from the bivalve *Mytillus galloprovincialis*, a Mediterranean variant of *Mytillus edulis* (mussel), which is perhaps the most commonly eaten bivalve in Europe.[72] Compound (56) was also present in the red alga *Porphyra tenera*, which is a part of the common Japanese diet.[73]

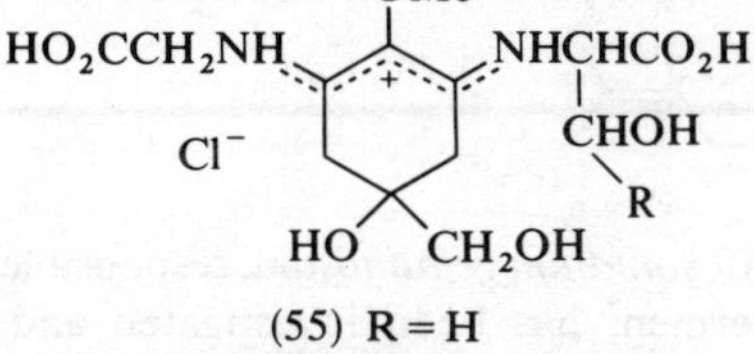

(55) R = H
(56) R = Me

The branchial heart of *Octopus vulgaris*, the common octopus, contains an iron-containing pigment, adenochrome. As a result of a considerable amount of work, desferriadenochrome has been shown to be a peptide composed of glycine and

[72] F. Chioccara, G. Misuraca, E. Novellino, and G. Prota, *Tetrahedron Letters*, 1979, 3181.
[73] S. Takano, A. Nakanishi, D. Uemura, and Y. Hirata, *Chem. Letters*, 1979, 419.

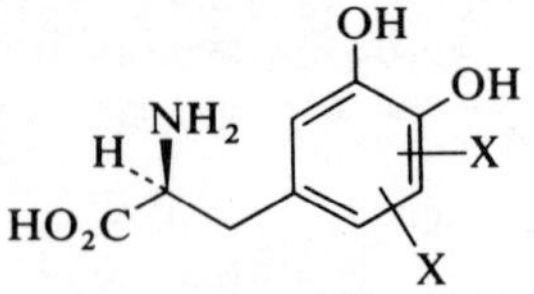

(57) 3,6-X_2
(58) 5,6-X_2
(59) 3,5-X_2

X=

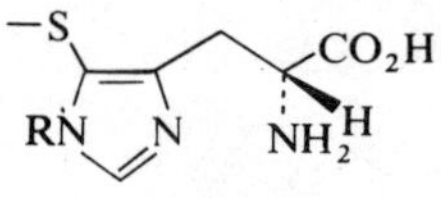

(R = H or Me)

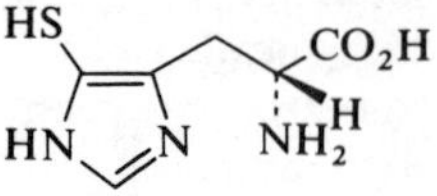

(60)

three iron-binding amino-acids (57), (58), and (59), named adenochromines A, B, and C, respectively.[74] The adenochromines were suggested to arise from addition of 5-mercapto-L-histidine (60) to L-dopaquinone, formed by enzymatic oxidation of L-dopa, and the suggestion was supported by the realization of the reaction *in vitro*.[75]

Other Phyla.—Work on echinoderms has included the elucidation of the structures of the carbohydrate moieties of steroidal saponins of a starfish (*Asterias amurensis*)[76] and a sea cucumber (*Holothuria leucospilota*).[77] The starfish saponin was suspected to play a regulatory role in the follicle cells. Of the many aromatic polyketides isolated from crinoids, most, if not all, are present in the animal as sulphate esters. Some of these sulphated derivatives have been shown to be food deterrents for several species of fish.[78]

The phylum Bryozoa (moss animals) has not, until now, attracted the attention of natural products chemists. Recently, the brominated alkaloids flustramines A and B [(61) and (62) respectively] were isolated from *Flustra foliacea*.[79]

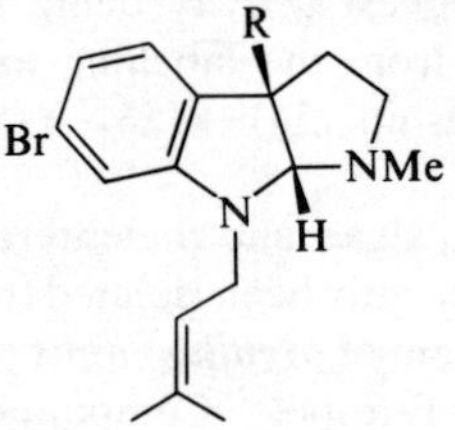

$Me_2\overset{+}{S}(\rightarrow O)CH_2CH_2OH$

(63)

(61) R=

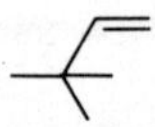

(62) R=

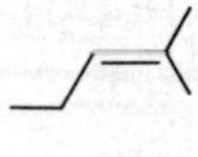

Another bryozoan, *Alcyonidium gelatinosum*, responsible for a severe eczematous dermatitis among fishermen, has been investigated and the allergen has been

[74] S. Ito, G. Nardi, A. Palumbo, and G. Prota, *J.C.S. Perkin I*, 1979, 2617.
[75] S. Ito, G. Nardi, A. Palumbo, and G. Prota, *Experientia*, 1979, **35**, 14.
[76] S. Ikegami, K. Okano, and H. Muragaki, *Tetrahedron Letters*, 1979, 1769.
[77] I. Kitagawa, T. Nishino, and Y. Kyogoku, *Tetrahedron Letters*, 1979, 1419.
[78] J. A. Rideout, N. B. Smith, and M. D. Sutherland, *Experientia*, 1979, **35**, 1273.
[79] J. S. Carlé and C. Christophersen, *J. Amer. Chem. Soc.*, 1979, **101**, 4012; *J. Org. Chem.*, 1979, **45**, 1586.

identified as a sulphoxonium compound (63).[80] Urochordates reportedly display a high incidence of compounds that have anti-cancer activity. In the case of the tunicate *Aplidium californicum*, this activity was accounted for by prenylhydroquinone, while two other metabolites, *i.e.* 6-hydroxy-2,2-dimethylchromene and prenylquinone, have anti-mutagenic activity.[81]

[80] J. S. Carlé and C. Christophersen, *J. Amer. Chem. Soc.*, 1980, **102**, 5107.
[81] B. M. Howard, K. Clarkson, and R. L. Bernstein, *Tetrahedron Letters*, 1979, 4449.

16 Biological Chemistry

Part (v) Nucleic Acids

(a) Nucleosides and Nucleotides

By G. SHAW

School of Chemistry, University of Bradford, Bradford BD7 1DP

1 Introduction

The previous report on nucleoside and nucleotide chemistry appeared in 1976, and in the intervening years there has been no abatement to the flood of publications in this field, stimulated not least by the continued isolation of new nucleosides from micro-organisms or as constituents of tRNAs. Several of these are *C*-nucleosides, and this has no doubt added to the enhanced interest in the chemistry of this type of substance over the period. Indeed, the close structural similarity of most of these compounds to components of nucleic acids or to intermediates in nucleotide biosynthesis *de novo* is remarkable, and adds weight to the search for synthetic chemotherapeutically active compounds in this area.

The first successors[1] to the invaluable series 'Synthetic Procedures in Nucleic Acid Chemistry' and the first volumes[2] in what promises to be a major review of the field have appeared. Several other reviews[3—11] covering a wide range of nucleic acid chemistry include specialist accounts[7] of modified nucleosides of tRNAs, nucleoside antibiotics,[8] cyclic nucleotides,[9,10] and the formation of purine nucleosides from imidazole precursors.[11]

[1] Nucleic Acid Chemistry' ed. L. Townsend and R. S. Tipson, Wiley–Interscience, New York, 1978, Vols. 1 and 2.
[2] 'The Chemistry of Nucleosides and Nucleotides', ed. R. K. Robins and L. B. Townsend, Plenum Press, New York, 1979.
[3] 'Chemistry and Biology of Nucleosides and Nucleotides', ed. R. E. Harmon, R. K. Robins, and L. B. Townsend, Academic Press, New York, 1978.
[4] *Nucleic Acids Res.*, 1978, Special Publication No. 4.
[5] 'Nucleoside Analogues', ed. R. T. Walker, E. De Clercq, and F. Eckstein, NATO Advanced Study Institute Series, Plenum Press, New York and London, 1979.
[6] K. Galankiewicz *Khim. geterotsikl. Soedinenii*, 1978, 723.
[7] J. A. McCloskey and S. Nishimura, *Accounts Chem. Res.*, 1977, **10**, 403.
[8] 'Nucleoside Antibiotics', A. Bloch, in Kirk–Othmer Encyclopaedia of Chemical Technology, ed. M. Grayson and D. Eckroth, Wiley, New York, 1978, Vol. 2, p. 962.
[9] Advances in Cyclic Nucleotide Research', Vol. 9, ed. W. J. George and L. J. Ignarro, Raven Press, New York, 1978.
[10] 'Advances in Cyclic Nucleotide Research, Vol. 10, ed. G. Brooker, P. Greengard, and G. A. Robison, Raven Press, New York, 1979.
[11] A. Yamazaki and M. Okutsu, *J. Heterocyclic Chem.*, 1978, **15**, 353.

2 Nucleoside Antibiotics and tRNA Constituents

Oxazinomycin (1) (which yields pseudouridine with ammonia) and its α-anomer have been synthesized[12] (Scheme 1) from the 2′,3′-*O*-isopropylidene-5′-*O*-trityl-D-ribofuranosylacetonitriles (2).

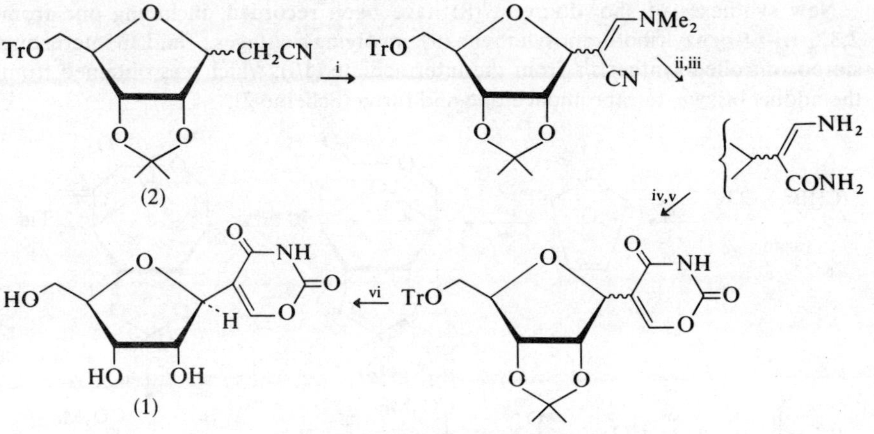

(Tr = Ph_3C; Im = 1-imidazolyl)

Reagents: i, $(Me_2N)_2CHOBu^t$; ii, NH_2OH; iii, H_2–Pt; iv, H_3O^+; v, Im_2CO; vi, CF_3CO_2H–H_2O (9 : 1)

Scheme 1

Formycin (3) and formycin B (4) have been synthesized[13] by cyclization of appropriate ribofuranosyl-pyrazoles (5), which were prepared by addition of N_2CHR^2 to $NCCR^3{=}CHCO_2R^1$ (R^3 = 2,3,5-tri-*O*-benzoyl-β-D-ribofuranosyl).

The total synthesis of wyosine (Nucleoside Yt) (6), which occurs in the next position to the 3′-end of the anticodon of yeast $tRNA^{Phe}$, has been reported,[14] by cyclization of 5-methylamino-4-cyano-1-(2,3,5-tri-*O*-acetyl-β-D-ribofuranosyl)-imidazole. Wyosine is strongly fluorescent, and its unusual extreme lability to hydrolysis is characteristic of 3-alkyl-guanosines.

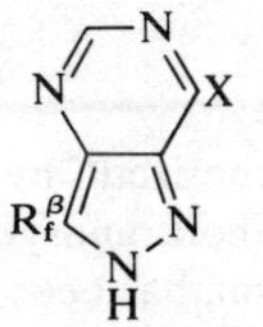

(3) X = NH_2
(4) X = OH

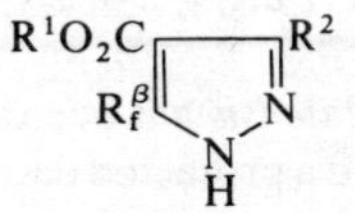

(5)

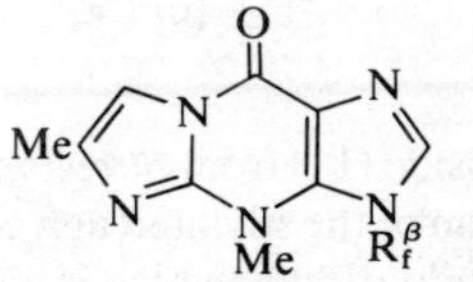

(6)

(R_f^β = β-D-ribofuranosyl)

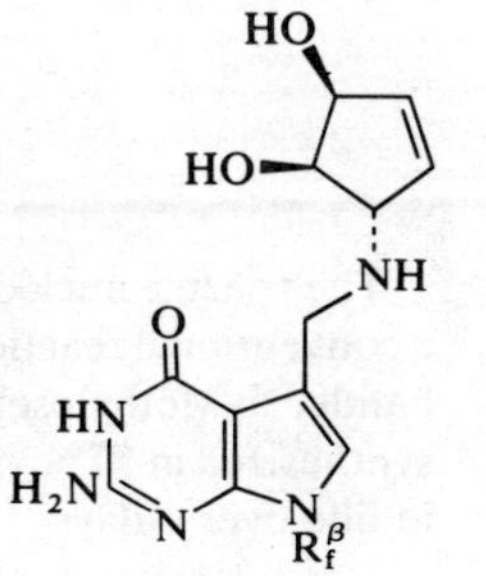

(7)

[12] S. DeBernardo and M. Weigele, *J. Org. Chem.*, 1977, **42**, 109.
[13] L. Kalvoda, *Coll. Czech. Chem. Comm.*, 1978, **43**, 1431.
[14] S. Nakatsuka, T. Ohgi, and T. Goto, *Tetrahedron Letters*, 1978, 2579.

A total synthesis of optically pure Nucleoside Q (7), the first position of the anticodon of the tRNA of *Escherichia coli* (Tyr, His, Asp, and Asn), has been recorded.[15] Two diastereoisomers were obtained, one of which [with a (3*S*,4*R*,5*S*) configuration in the cyclopentenyl side-chain] was identical to the natural nucleoside.

New syntheses of showdomycin (8) have been recorded, including one from 2,3,5-tri-*O*-benzylribofuranosylethyne (9), involving six stages,[16] and an interesting stereocontrolled synthesis[17] from the intermediate (10), which was obtained from the adduct of *sym*-tetrabromoacetone and furan (Scheme 2).

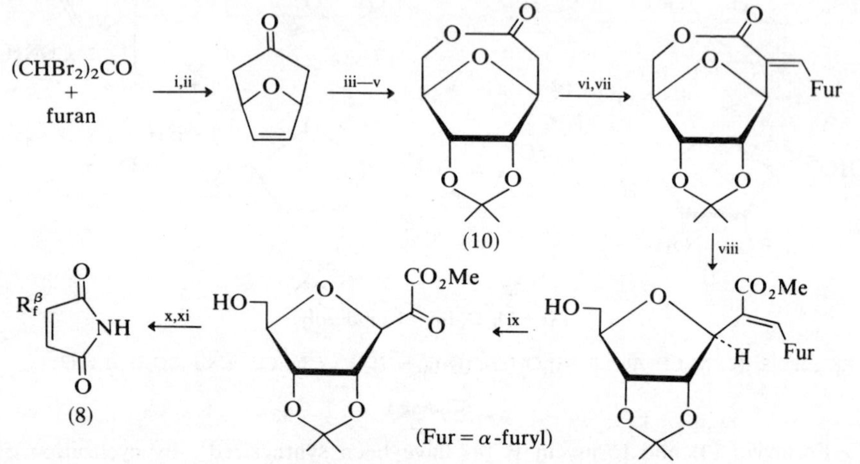

Reagents: i, $[Fe_2(CO)_9]$; ii, Zn–Cu couple; iii, OsO_4, H_2O_2; iv, $HClO_4$, Me_2CO; v, CF_3CO_3H; vi, furfural; vii, Bu^tCOCl; viii, NaOMe; ix, O_3; x, $Ph_3P{=}CHCONH_2$; xi, H_3O^+

Scheme 2

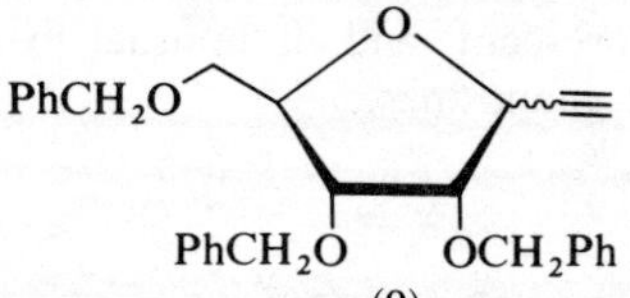

(9)

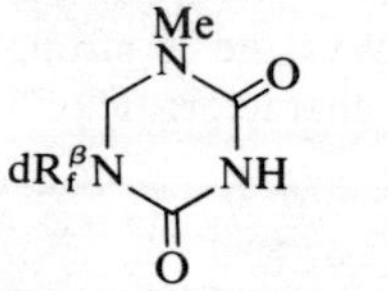

(11) dR_f^β = 2-deoxy-R_f^β

The triazine nucleoside (11) from *Streptomyces platensis* has been synthesized[18] by a conventional reaction of the silylated aglycone and a protected deoxyribofuranosyl halide. 1-Methylpseudouridine, which is also produced by this organism, has been synthesized in 57% yield by methylation of an acetylated pseudouridine with methyl iodide over 4 days.[19]

[15] T. Ohgi, T. Kondo, and T. Goto, *J. Amer. Chem. Soc.*, 1979, **101**, 3629.
[16] J. G. Buchanan, A. R. Edgar, M. J. Power, and C. T. Shanks, *J.C.S. Perkin I*, 1979, 225.
[17] T. Sato, R. Ito, Y. Hayakawa, and R. Noyori, *Tetrahedron Letters*, 1978, 1829.
[18] H. Skulnick, *J. Org. Chem.*, 1978, **43**, 3188.
[19] R. A. Earl and L. B. Townsend, *J. Heterocyclic Chem.*, 1977, **14**, 699.

New naturally occurring nucleosides continue to be discovered, both from micro-organisms and from tRNAs. They include the pyridine nucleoside clitidine (12), from *Clitocybe acromelalga*,[20] mildiomycin (13), from *Streptoverticillium rimofaciens*,[21] the somewhat analogous Nucleoside N (14), from the first letter position of the anticodon of $tRNA^{Gly}$ of *Bacillus subtilis*,[22] polyoxin N (15) (a *N*-nucleoside analogous to pyrazomycin), from *Streptomyces fiomogenes*,[23] anthelmycin (16)

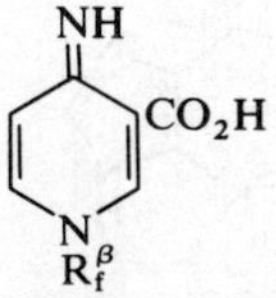

(12)

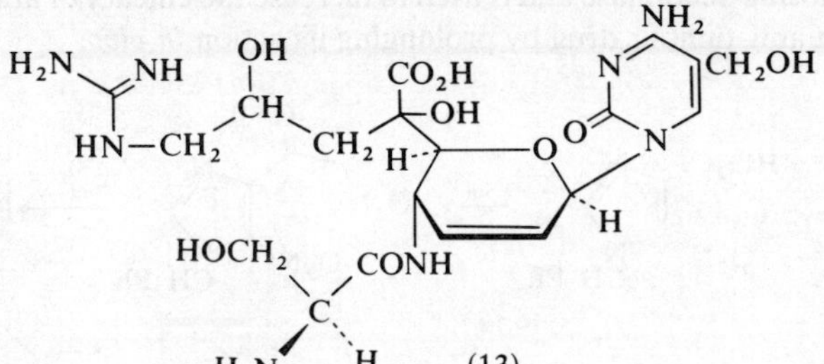

(13)

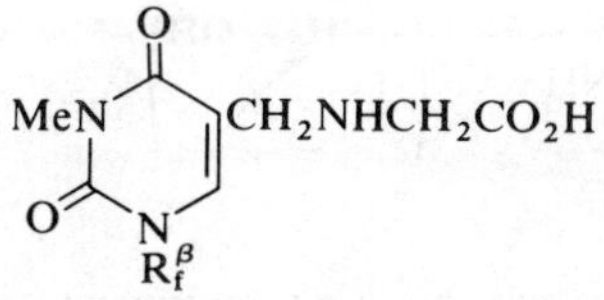

(14)

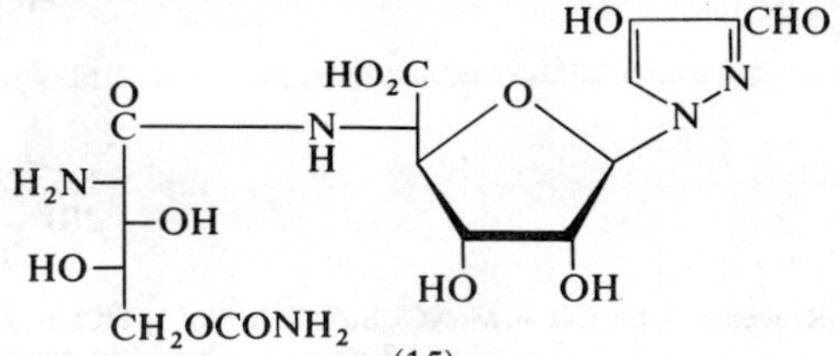

(15)

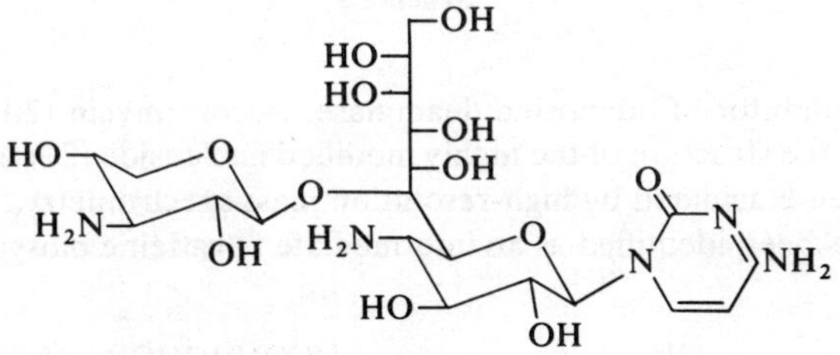

(16)

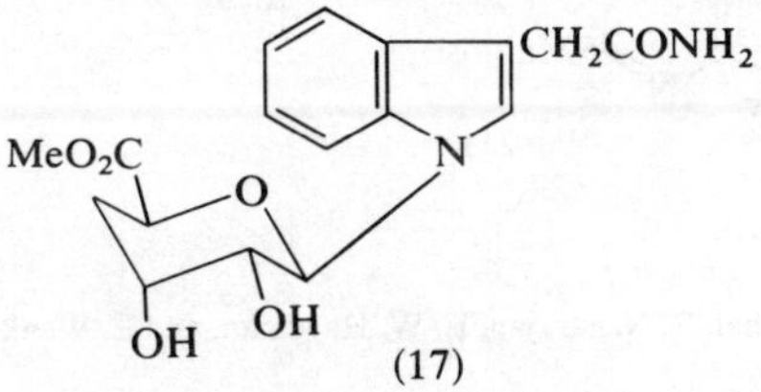

(17)

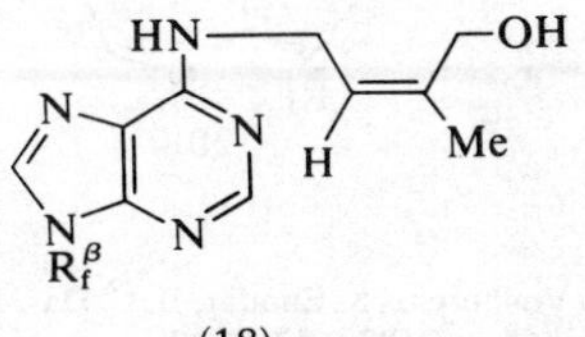

(18)

[20] K. Konno, K. Hayano, H. Shivahama, H. Saito, and T. Mastoumoto, *Tetrahedron Letters*, 1977, 481.
[21] S. Harada, E. Mizuta, and T. Kistii, *J. Amer. Chem. Soc.*, 1978, **100**, 4895.
[22] K. Murao and H. Ishikura, *Nucleic Acids Res.*, 1978, Special Publication No. 5, S333.
[23] M. Uramoto, J. Uzawa, S. Suzuki, K. Isono, J. G. Liehr and J. A. McCloskey, *Nucleic Acids Res.*, 1978, Special Publication No. 5, S327.

(identical with hikizimycin), from *S. longissimus*,[24] and the novel indole acetamide nucleoside neosidomycin (17), from *S. hygroscopicus*.[25]

The cytokinin-active ribonucleoside *cis*-zeatin-D-riboside (18) has also been isolated, from the tips of plants of *Nicotiana tabacum*.[26]

An interesting synthesis of pentostatin (2′-deoxycoformycin) (19) has been described[27] from imidazole precursors (Scheme 3). The compound is an inhibitor of adenosine deaminase and is used to increase the efficacy of arabinofuranosyladenine as an anti-tumour drug by prolonging its action *in vivo*.

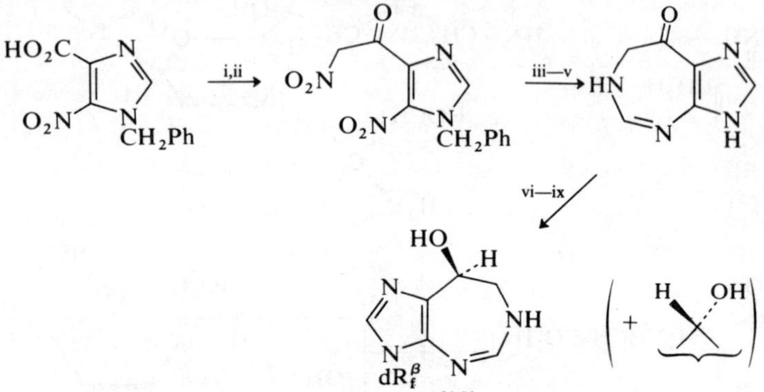

(19)

Reagents: i, Im_2CO; ii, $MeNO_2$ Bu^tOK; iii, $SnCl_2$, HCl; iv, H_2–Pd; v, $(ETO)_3CH$; vi, $(Me_3Si)_2NCOCF_3$; vii, 3,5-di-*O*-*p*-toluoyl-2-deoxy-D-ribose; viii, NaOMe: ix, $NaBH_4$

Scheme 3

A related inhibitor of adenosine deaminase, isocoformycin (20), has also been synthesized,[28] the structure of the highly modified nucleoside (21) from mammalian tRNA has been elucidated by high-resolution mass spectrometry,[29] and 7-methylxanthosine has been identified as an intermediate in caffeine biosynthesis.[30]

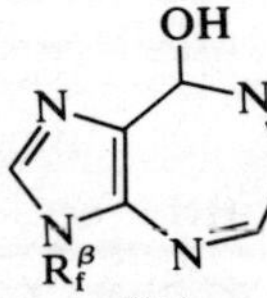

(20)

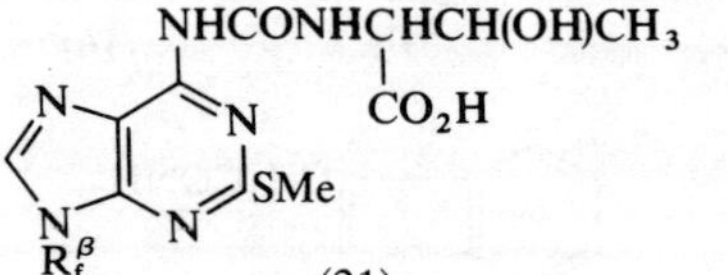

(21)

[24] M. Vuilhorgne, S. Ennifar, B. C. Das, J. W. Paschal, R. Nagarajan, E. W. Hagaman, and E. Wenkert, *J. Org. Chem.*, 1977, **42**, 3289.
[25] R. Furuta, S. Naruto, A. Tamura, and K. Yokogawa, *Tetrahedron Letters*, 1979, 1701.
[26] K. Kimura, T. Sugiyama, and T. Hashizume, *Nucleic Acids Res.*, 1978, Special Publication No. 5, S339.
[27] D. C. Baker and S. R. Putt, *J. Amer. Chem. Soc.*, 1979, **101**, 6127.
[28] M. Shimazaki, S. Kondo, K. Maeda, M. Ohno, and H. Umezawa, *J. Antibiotics*, 1979, **32**, 537.
[29] Z. Yamaizumi, S. Nishimura, K. Limburg, M. Raba, H. J. Gross, P. F. Crain, and J. A. McCloskey, *J. Amer. Chem. Soc.*, 1979, **101**, 2224.
[30] T. W. Baumann, E. Dupontlooser, and H. Wanner, *Phytochemistry*, 1978, **17**, 2075.

3 Nucleosides

Organopalladium derivatives[31,32] continue to be used for the preparation of 5-substituted uracil nucleosides, and an unusual alkylation of uridine at C-6 has been described[33] in which 2′,3′-*O*-isopropylideneuridine is lithiated in THF and treated with an alkyl halide at −78 °C to produce the 6-alkyl derivative in 46—60% yield, with no evidence for alkylation at the N-3 or 5′-*O* positions. Alkylation of purine and pyrimidine nucleosides or nucleotides with alkyl halides has been found to proceed in high yields in the presence of tetrabutylammonium fluoride in THF at room temperature.[34] A mass spectral study of eleven mono-benzylated nucleosides has enabled the site and extent of benzylation to be determined.[35] A mild nitrating agent, nitronium tetrafluoroborate, has been used[36] to produce 5-nitro-2′-deoxyuridine from 2′-deoxyuridine. New routes to pseudouridine and pseudocytidine continue to be reported. Pseudocytidine has been prepared in 60% yield by the reaction of 1,3-dimethylpseudouridine with guanidine[37] or directly[38] from the ribofuranosylacetonitrile (2). A general stereocontrolled synthesis of *C*-nucleosides, including pseudouridine, 2-thiopseudouridine, and pseudocytidine, from the non-carbohydrate intermediate (10) has been described[39] (Scheme 4). The acid derived from (10) has been resolved.[39]

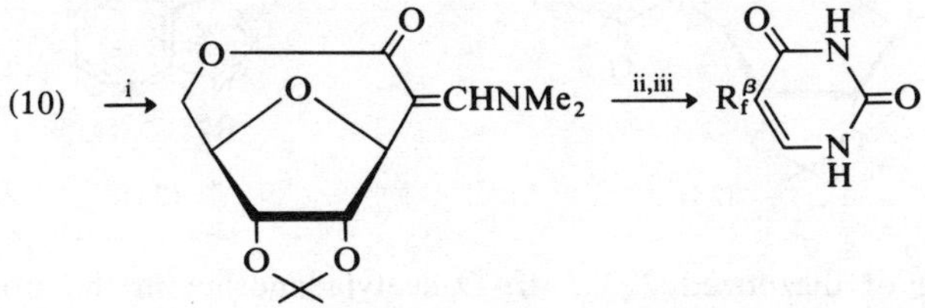

Reagents: i, Me_2NCHO; ii, NH_2CONH_2, ETONa; iii, H_3O^+

Scheme 4

Analogous reactions of pseudouridine that has been modified at C-5′ have been similarly achieved,[40] commencing with substituted derivatives of (10). 2′-Deoxypseudouridine and its α-anomer have also been prepared from 2,4-di-t-butoxy-5-lithiopyrimidine and 3,5-di-*O*-benzyl-2-deoxy-D-ribose, in yields of 22 and 25% respectively.[41]

[31] D. E. Bergstrom and M. K. Ogawa, *J. Amer. Chem. Soc.*, 1978, **100**, 8106.
[32] A. S. Jones, G. Verhelst, and R. T. Walker, *Tetrahedron Letters*, 1979, 4415.
[33] H. Tanaka, I. Nasu, and T. Miyasaka, *Tetrahedron Letters*, 1979, 4755.
[34] K. K. Ogilvie, S. L. Beaucage, M. F. Gillen, D. Entwistle, and M. Quillam, *Nucleic Acids Res.*, 1979, **6**, 1695, 2261.
[35] H. T. Cory, K. Yamaizumi, D. L. Smith, D. R. Knowles, A. D. Broom, and J. A. McCloskey, *J. Heterocyclic Chem.*, 1979, **16**, 585.
[36] G.-F. Huang and P. F. Torrence, *J. Org. Chem.*, 1977, **42**, 3821.
[37] K. Hirota, K. A. Watanabe, and J. J. Fox, *J. Org. Chem.*, 1978, **43**, 1193.
[38] C. K. Chu, U. Reichman, K. A. Watanabe, and J. J. Fox, *J. Org. Chem.*, 1977, **42**, 711.
[39] R. Noyori, T. Sato, and Y. Hayakawa, *J. Amer. Chem. Soc.*, 1978, **100**, 2561.
[40] T. Sato, M. Watanabe, and R. Noyori, *Tetrahedron Letters*, 1978, 4403.
[41] S. D. Bridges, D. M. Brown, and R. C. Ogden, *J.C.S. Chem Comm.*, 1977, 460.

5-β-D-Arabinofuranosylisocytosine (22), prepared from pseudoisocytidine by inversion at the 2′-position with acetoxyisobutyryl chloride, isomerizes at room temperature to the α-anomer.[42]

Photochemical reactions of nucleosides still continue to excite considerable interest. Irradiation of 5′-*O*-tritylthymidine in the presence of pyruvoyl chloride produced the 3′-oxo-derivative (23)[43] in 61% yield. 2′,3′-*O*-Isopropylideneuridine, when irradiated in methanol, gave the 6-hydroxymethyl-dihydrouridine (24),[44] and a *S–C* photorearrangement of a pyrimidine *S*-ribofuranoside has resulted in a novel route to pseudouridine.[45]

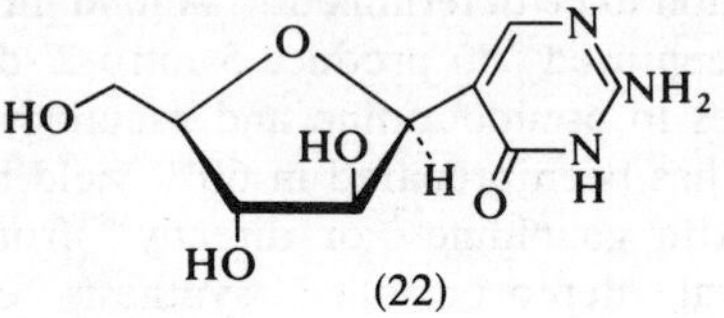

(22)

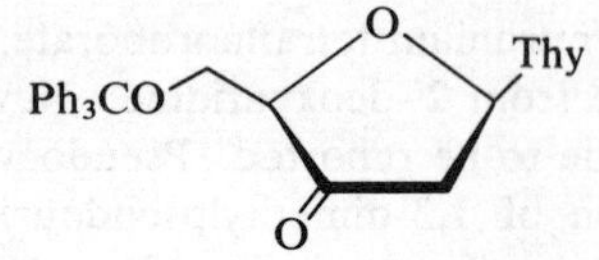

(23)

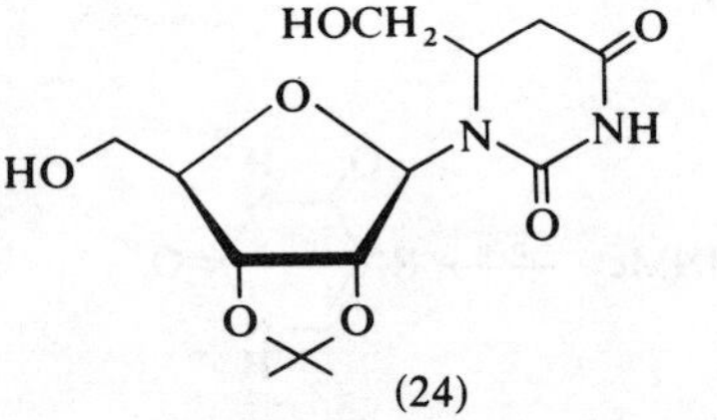

(24)

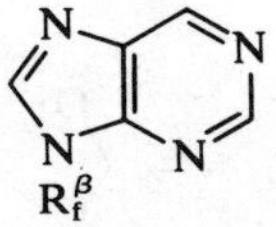

(25)

Irradiation of diazotized 2′,3′,5′-tri-*O*-acetyladenosine in the presence of a hydrogen abstractor has given an interesting synthesis of nebularine (25).[46]

Electrochemical reduction of the di-iododideoxythymidine derivative (26) gave the unusual cyclopropane derivative (27) in 36% yield.[47] A similar compound was obtained from the analogous uracil nucleoside. Electrolysis of the 3′-deoxy-3′-bromo-D-xylosyladenine (28) gave the 2′,3′-unsaturated nucleoside (29).[48]

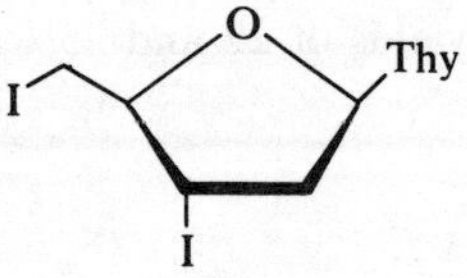

(26)

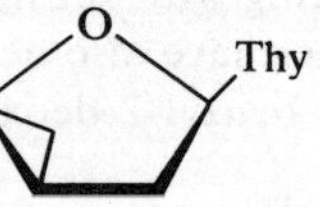

(27)

(Thy = thymine)

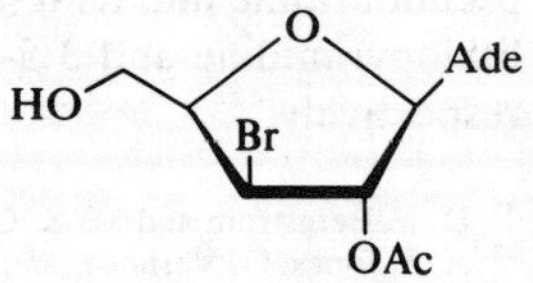

(28)

(Ade = adenine)

[42] C. K. Chu, U. Reichman, K. A. Watanabe, and J. J. Fox, *J. Medicin. Chem.*, 1978, **21**, 96.
[43] R. W. Binkley, D. G. Hehemann, and W. W. Binkley, *J. Org. Chem.*, 1978, **43**, 2573.
[44] J.-L. Fourrey, G. Henry, and P. Jouin, *Tetrahedron Letters*, 1979, 951.
[45] J.-L. Fourrey, G. Henry, and P. Jouin, *J. Amer. Chem. Soc.*, 1977, **99**, 6753.
[46] V. Nawi and S. G. Richardson, *Tetrahedron Letters*, 1979, 1181.
[47] T. Adachi, T. Iwasaki, M. Miyoshi, and I. Inoue, *J.C.S. Chem. Comm.*, 1977, 248.
[48] T. Adachi, T. Iwasaki, I. Inoue and M. Miyoshi, *J. Org. Chem.*, 1979, **44**, 1404.

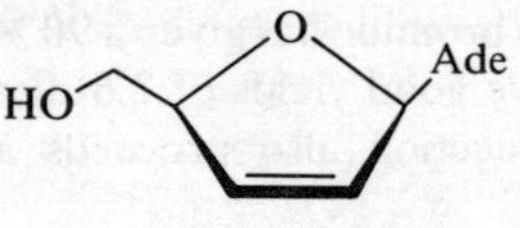

(29)

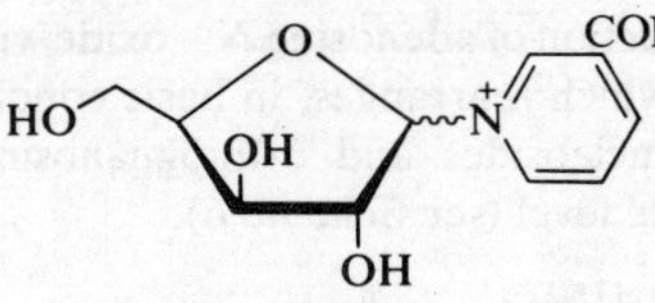

(30)

A novel synthesis of a nicotinamide D-xylofuranoside involves the reaction of 3′,5′-*O*-isopropylidene-D-xylosylamine with 1-(2,4-dinitrophenyl)-3-carbamoylpyridinium chloride in methanol at room temperature, when a good yield of a mixture of the anomers (30) was obtained.[49]

Stereospecific syntheses of xylofuranosyl- and of ribofuranosyl-uracil have also been reported, from 2,3-*O*-isopropylidene-D-ribofuranosylamine and 3,5-*O*-isopropylidene-D-xylofuranosylamine, which with ethyl *N*-(α-acetyl-β-ethoxyacryloyl)carbamate gave, after deblocking, 5-acetyluridine and 5-acetyl-1-α-D-xylofuranosyluracil respectively; these were the sole anomers formed.[50] The stereospecificity is explained by the formation of a cyclic intermediate (31).

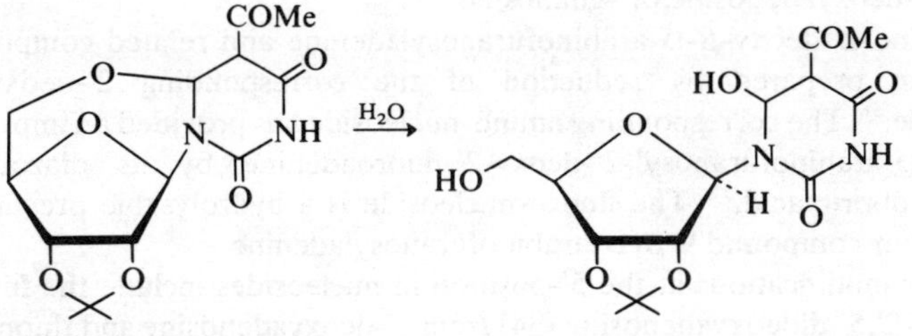

(31)

Other routes to 5-acetyluridine[51] and related compounds, including 5-ethynyluridine,[52] have also been recorded

N-Methyluridine may be converted into the diazepine nucleosides (32) in 28% yield by reaction with a carbene (Scheme 5).[53]

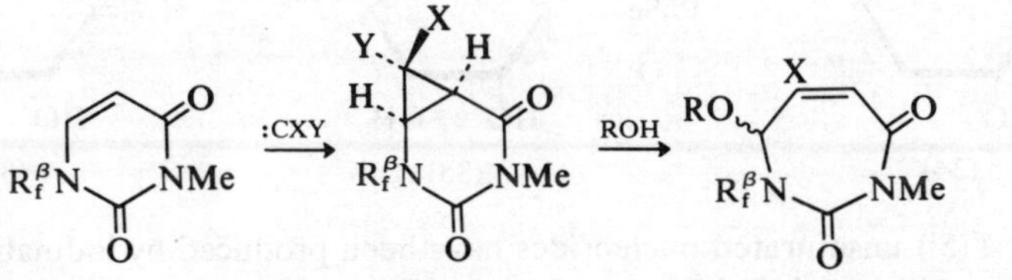

(32) X,Y = Cl or Br

Scheme 5

[49] N. I. Oppenheimer, *J. Carbohydr. Nucleosides, Nucleotides*, 1978, **5**, 251.
[50] R. Lofthouse, G. Shaw, P. S. Thomas, G. Mackenzie, D. H. Robinson, and P. W. Rugg, *J.C.S. Perkin I*, 1977, 997.
[51] A. S. Jones, G. P. Stephenson, and R. T. Walker, *Tetrahedron*, 1979, **35**, 1125.
[52] P. J. Barr, A. S. Jones, P. Serafinowski, and R. T. Walker, *J.C.S. Perkin I*, 1978, 1263.
[53] H. P. M. Thiellier, G. J. Koomen, and U. K. Pandit, *Tetrahedron*, 1977, **33**, 2609.

The reaction of adenosine N^1-oxide with cyanogen bromide has given a 90% yield of (33), which rearranges, in basic conditions, to give good yields of 2,6-diaminopurine nucleosides and 6-thioguanosine.[54] The reaction also proceeds at the nucleotide level (see Scheme 6).

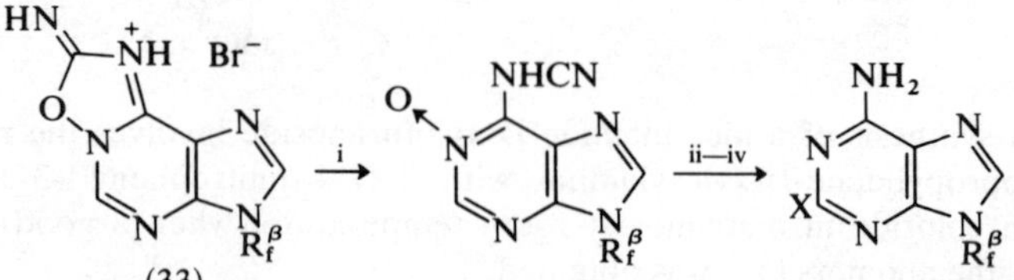

(33)

X = NH_2 or SH

Reagents: i, MeOH, NH_3; ii, MeI; iii, HO^-; iv, H_2–Ni, *or* H_2S

Scheme 6

A further application of the useful transglycosylation procedure involves condensation of 2′-amino-2′-deoxyuridine with N^6-octanoyladenine or N^2-palmitoylguanine in the presence of trimethylsilyl trifluoromethanesulphonate to give 2′-amino-2′-deoxyadenosine or -guanosine.[55]

2′-Amino-2′-deoxy-β-D-arabinofuranosyladenine and related compounds have also been prepared by reduction of the corresponding 2′-deoxy-2′-azido-nucleoside.[56] The corresponding amino-nucleoside has provided an improved route to 9-β-D-arabinofuranosyl-2′-deoxy-2′-fluoroadenine by its diazotization in tetrafluoroboric acid.[57] The fluoro-nucleoside is a hydrolysable precursor of the anti-tumour compound 9-β-D-arabinofuranosyladenine.

Further modifications at the 5′-position in nucleosides include the formation of 5′-chloro-2′,5′-dideoxyadenosine (34) from 2′-deoxyadenosine and thionyl chloride in hexamethylphosphoramide.[58] 5′-Chloro-5′-deoxyadenosine reacts with sodium phenylselenide and hydrogen peroxide to give 54% of the seleno-derivative (35), which is converted into the useful unsaturated derivative (36) in 94% yield by dimethyl sulphoxide.[59]

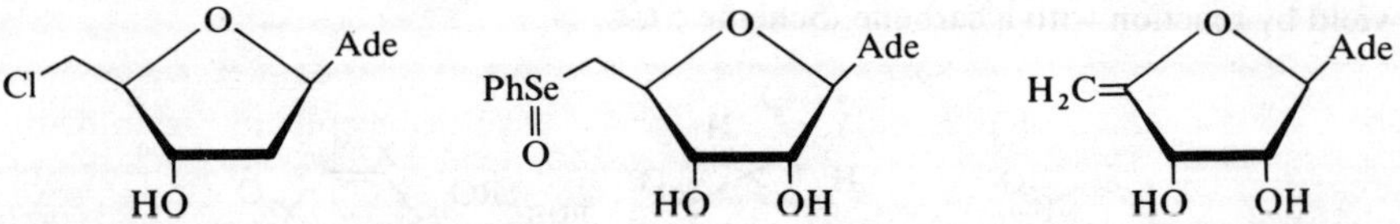

(34) (35) (36)

Similar 4′(5′)-unsaturated nucleosides have been produced by iodinating purine nucleosides with methyltriphenoxyphosphonium iodide, followed by dehydrohalogenation of the 5′-iodo-5′-deoxy-nucleosides that are formed.[60] The 5′(6′)-

[54] T. Ueda, K. Miura, and T. Kasai, *Chem. Pharm. Bull.* 1978, **26**, 2122.
[55] M. Imazawa and F. Eckstein, *J. Org. Chem.*, 1979, **44**, 2039.
[56] A. Sato, R. Imai, N. Nakamizo, and T. Hirata, *Chem. Pharm. Bull.* 1979, **27**, 821.
[57] J. A. Montgomery, S. D. Clayton, and A. T. Shortnacy, *J. Heterocyclic Chem.*, 1979, **16**, 157.
[58] L. M. Beacham, III, *J. Org. Chem.*, 1979, **44**, 3100.
[59] N. Zylber and J. Zylber, *J.C.S. Chem. Comm.*, 1978, 1084.
[60] S. D. Dimitrijevich, J. P. Verheyden, and J. G. Moffatt, *J. Org. Chem.*, 1979, **44**, 400.

acetylenic nucleosides (37) have also been obtained, *via* a Wittig reaction on the aldehydes (38).[61]

Crossed aldol reactions have been employed[62] to give 4′-substituted nucleosides, including the adenosine derivative (39), from the 5-aldehyde and formaldehyde.

Reduction of the deuteriated compound (40) with the adduct from (−)-pinene and 9-borabicyclo[3.3.1]nonane has given a sample of (5′*R*)-[5-$^{2}H_{1}$]adenosine with 60% optical purity at C-5.[63]

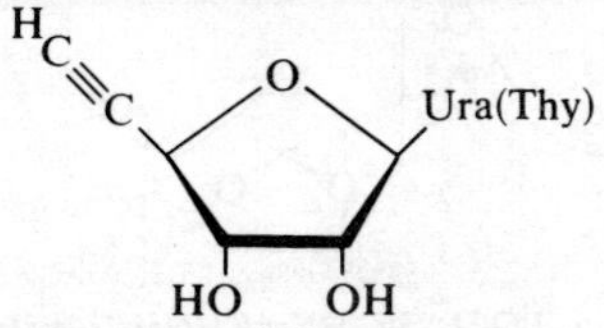

(37)

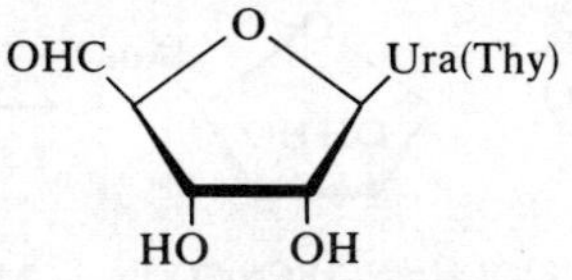

(38)

(Ura = uracil)

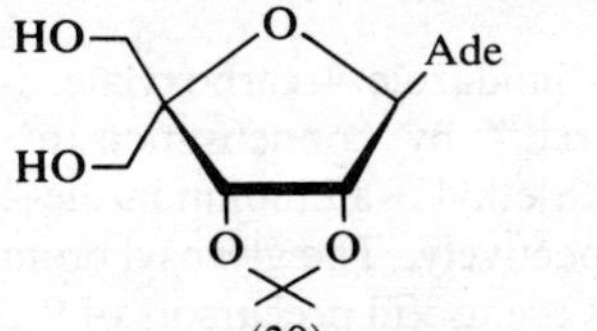

(39)

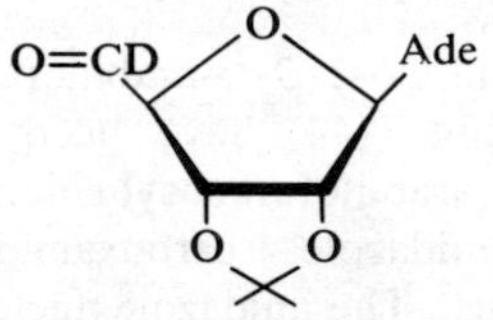

(40)

The 9-phenylxanthen-9-yl group, which is removable by acids, has been used to protect the 5′-OH of deoxyribonucleosides,[64] and a regioselective 2′-*O*-deacylation of fully acylated purine and pyrimidine nucleosides by hydrazine gives 39—70% of the intended products, although the 5′-*O*-acyl derivatives may be produced quantitatively.[65]

The tetraisopropyldisiloxane-1,3-diyl group has been used[66] to functionalize the 2′-hydroxy-group in nucleosides by simultaneous protection of the 3′- and 5′-hydroxy-groups, and 4-dimethylaminopyridine has been proposed as a selective catalyst for the silylation of secondary alcohols[67] and in a simplified procedure for the preparation of triphenylmethyl ethers.[68]

Imidazole and related nucleosides continue to attract attention. 5-Amino-1-α- and -β-D-mannofuranosylimidazole esters (41) have been prepared from di-isopropylidene-D-mannofuranosylamine (42) and converted into mannosyl-adenines and into lyxose nucleosides (Scheme 7). The ^{1}H n.m.r. spectra of comparable pairs of anomeric mannofuranosyl-imidazoles and 2′,3′-*O*-isopropylidene-D-mannofuranosyl-adenines showed that they did not accord with the widely used empirical rule relating anomer configuration with the field positions of H-1′.[69]

61 R. A. Sharma and M. Bobek, *J. Org. Chem.*, 1978, **43**, 367.
62 G. H. Jones, M. Taniguchi, D. Tegg, and J. G. Moffatt, *J. Org. Chem.*, 1979, **44**, 1309.
63 R. J. Parry, *J.C.S. Chem. Comm.*, 1978, 294.
64 J. B. Chattopadhyaya and C. B. Reese, *J.C.S. Chem. Comm.*, 1978, 639.
65 Y. Ishido, N. Nakazaki, and N. Sakain, *J.C.S. Perkin I*, 1979, 2088.
66 W. T. Markiewicz, *J. Chem. Res.*, 1979, (*S*)24, (*M*)181.
67 S. K. Chaudhary and O. Hernandez, *Tetrahedron Letters*, 1979, 99.
68 S. K. Chaudhary and O. Hernandez, *Tetrahedron Letters*, 1979, 95.
69 G. Mackenzie and G. Shaw, *J.C.S. Chem. Comm.*, 1977, 753.

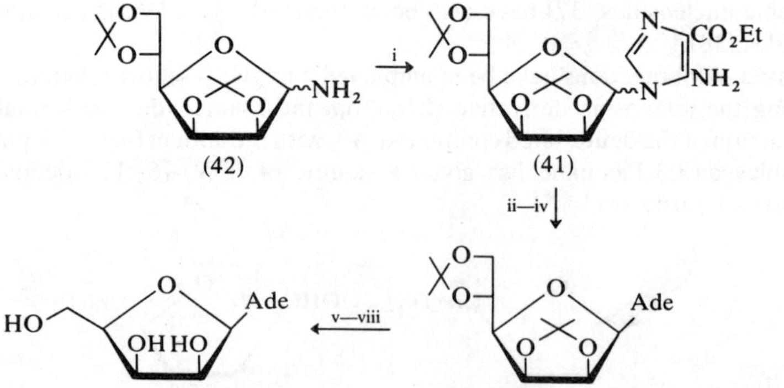

Reagents: i, EtOCH=NCH(CN)CO_2Et; ii, NH_3; iii, $POCl_3$; iv, HN=$CHNH_3^+$; v, H_3O^+; vi, KIO_4; vii, $NaBH_4$; viii, H_3O^+

Scheme 7

Ethyl 5-amino-1-*β*-D-arabinofuranosylimidazole-4-carboxylate (43) and the -4-carboxamide (44) have been prepared[70] by condensation of 2,3,5-tri-*O*-benzyl-*α*-D-arabinofuranosyl chloride with ethyl 5-aminoimidazole-4-carboxylate or 5-aminoimidazole-4-carboxamide, respectively. The glycosyl bromide or iodide failed to react. The imidazole nucleosides are useful precursors of 9-*β*-D-arabino-furanosyl-purines, including 9-*β*-D-arabinofuranosyladenine.

An acyclic D-arabinityl-aminoimidazole (45) has been isolated[71] from the reaction of *N*-D-arabinopyranosyl-*N'N'*-dimethylformamidine with *α*-amino-*α*-cyano-acetamide and methanolic acetic acid. The acyclic imidazole nucleoside was converted into the acyclic D-arabinityl-hypoxanthine (46) and the adenine (47).

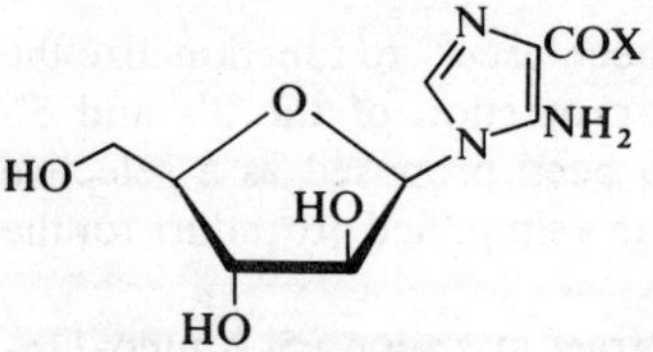

(43) X = OEt
(44) X = NH_2

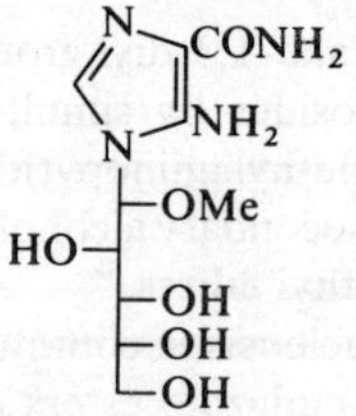

(45)

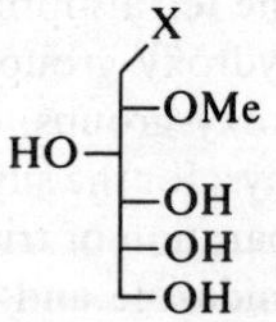

(46) X = Hyp
(47) X = Ade
(Hyp = hypoxanthine)

Methylation of ethyl 5-amino-1-(2,3-*O*-isopropylidene-*β*-D-ribofuranosyl)-imidazole-4-carboxylate has been found to produce the 3-methylquaternary salt (48).[72] An attempt to prepare compounds analogous to the antibiotic bredinin (49) by diazotization of the corresponding 5-amino-imidazole nucleoside gave an 18% yield of the 2-imidazolone nucleoside (50), the structure of which was confirmed by synthesis from the silylated aglycone and a ribose derivative.[73]

[70] G. Mackenzie and G. Shaw, *J.C.S. Chem. Comm.*, 1978, 882.
[71] G. Mackenzie, G. Shaw, and D. H. Robinson, *J.C.S. Perkin I*, 1977, 1094.
[72] T. Brown, K. Kadir, G. Mackenzie, and G. Shaw, *J.C.S. Perkin I*, 1979, 3107.
[73] P. C. Srivastava, R. J. Rousseau, and R. K. Robins, *J.C.S. Chem. Comm.*, 1977, 151.

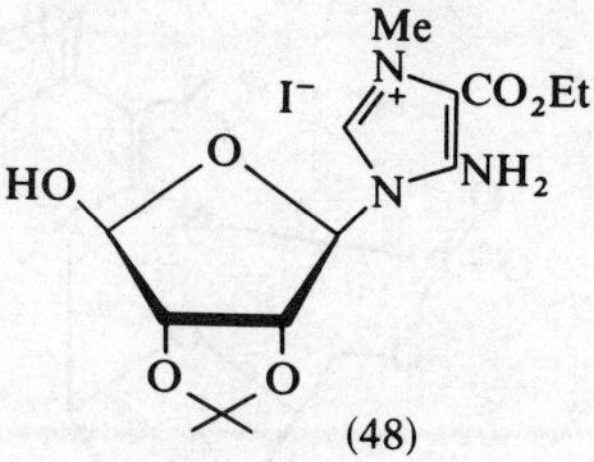

(48)

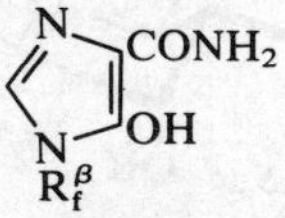

(49)

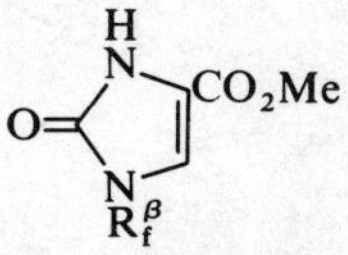

(50)

5-Amino-1-β-D-ribofuranosylimidazole-4-carboxylic acid (51) forms peptides, including the naturally occurring aspartate derivative (52), in aqueous solution with 1-cyclohexyl-3-(4-ethylmorpholin-2-yl)carbodi-imide in yields of up to 10%. The isopropylidene derivative of (51) reacts with acetic anhydride in pyridine to produce the oxazin-7-one nucleoside (53).[74]

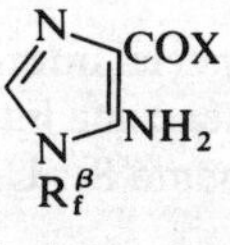

(51) X = OH
(52) X = L-Asp

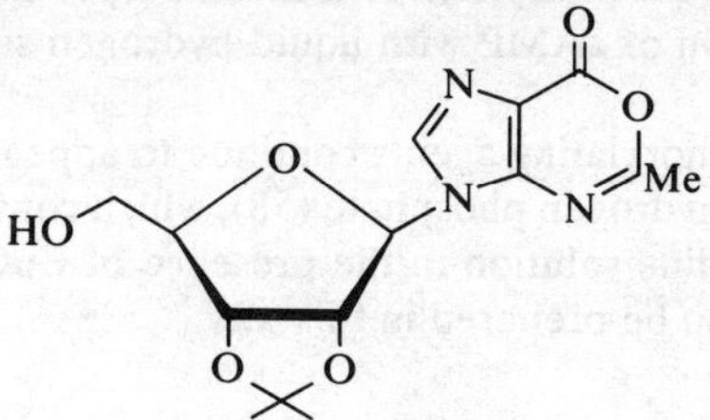

(53)

4 Nucleotides

The absolute configuration at the phosphorus atom in diastereoisomers of adenosine 5′-(α-thiotriphosphate) (54) has been determined,[75,76] and stereospecific syntheses of (S_P) and (R_P) adenosine 3′,5′-(cyclic)phosphorothioates (55) and (56) respectively have been recorded.[77] Compounds of these types, including compounds labelled with oxygen isotopes, have been used to obtain information on binding mechanisms and stereochemical interactions with various enzymes.[78,79]

In addition to the normal 5′-phosphorylation reaction, 3-β-D-ribofuranosyladenine was found[80] to react with phosphoryl chloride to produce the 9,5′-cyclic

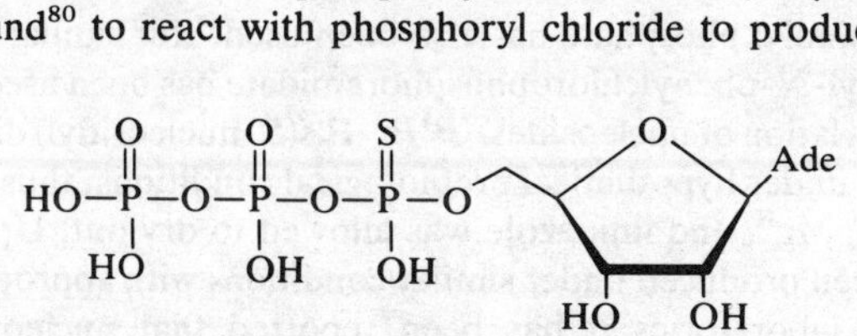

(54)

[74] G. Shaw, P. S. Thomas, C. A. H. Patey, and S. E. Thomas, *J.C.S. Perkin I*, 1979, 1415.
[75] R. L. Jarvest and G. Lowe, *J.C.S. Chem. Comm.*, 1979, 364.
[76] J. P. Richard, H. T. Ho, and P. A. Frey, *J. Amer. Chem. Soc.*, 1978, **100**, 7756.
[77] J. Baranak, R. W. Kinas, K. Lesiak, and W. J. Stec, *J.C.S. Chem. Comm.*, 1979, 940.
[78] W. A. Blattler and J. R. Knowles, *J. Amer. Chem. Soc.*, 1979, **101**, 510.
[79] J. P. Richard and P. A. Frey, *J. Amer. Chem. Soc.*, 1978, **100**, 7757.
[80] J. T. Uchic, *Tetrahedron Letters*, 1977, 3775.

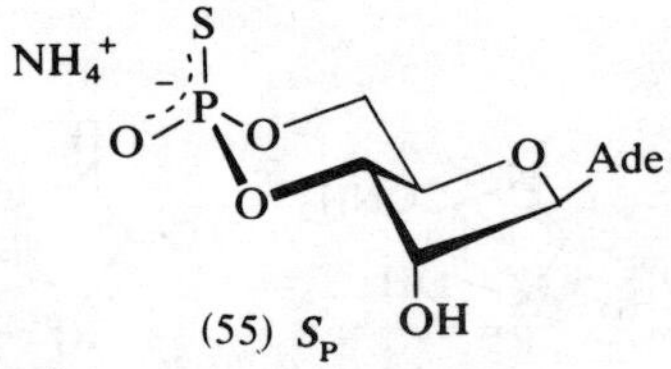

(55) S_P

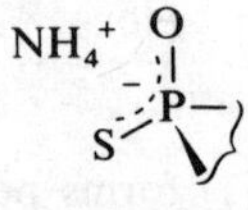

(56) R_P

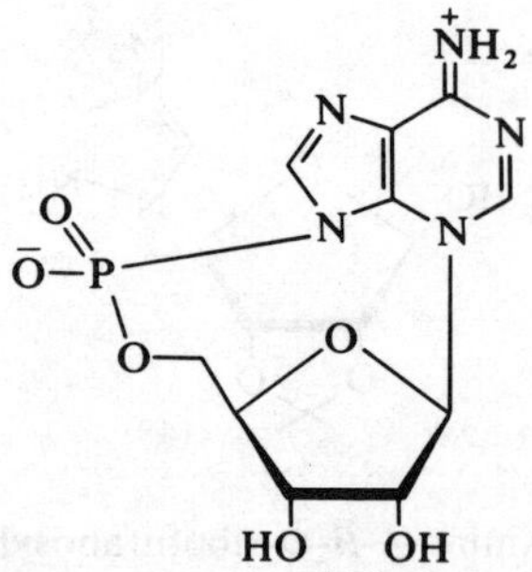

(57)

phosphonate (57), and 3′,5′-cyclic phosphoramidates and phosphites have been obtained from thymidine and hexamethylphosphoramidite.[81]

The 3′,5′-cyclic phosphate of 6-mercaptopurine has been prepared in 65% yield by the reaction of cAMP with liquid hydrogen sulphide in aqueous pyridine, over 7 days.[82]

New phosphorylating agents continue to appear. Of special interest is the use of 8-quinolyl dihydrogen phosphate (58), which converts adenosine into AMP in 75% yield, in pyridine solution in the presence of $CuCl_2$ (Scheme 8): UMP, CMP, and GMP may also be prepared in this way.[83]

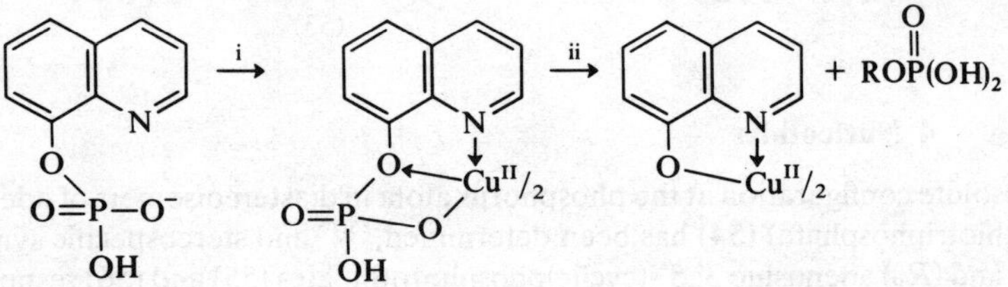

(58)

Reagents: i, $CuCl_2$; ii, ROH

Scheme 8

5-Chloro-8-quinolyl phosphate has also been used[84] for similar purposes.

p-Chlorophenyl-*N*-phenylchlorophosphoramidate has been used[85] to good effect for the phosphorylation of nucleosides. P^1P^2-Bis(5′-nucleosidyl) diphosphates have been produced[86] under hypothetical prebiological conditions; thus, when a solution of uridine, ADP, Mg^{II}, and imidazole was allowed to dry out, UppA was formed. NAD has also been produced under similar conditions with appropriate precursors. From the same laboratories it has been reported that nucleosides have been phosphorylated in aqueous alkaline solutions, using trimetaphosphates, to give good

81 G. S. Bajwa and W. G. Bentrude, *Tetrahedron Letters*, 1978, 421.
82 R. B. Meyer, T. E. Stone, and F. P. Heinzel, *J. Heterocyclic Chem.*, 1978, **15**, 1511.
83 H. Takaku, *Chem. Pharm. Bull.* 1977, **25**, 2121.
84 H. Takaku, R. Yamaguchi, and T. Nomoto, *Tetrahedron Letters*, 1979, 3857.
85 E. Ohtsuka, T. Tanaka, and M. Ikehara, *J. Amer. Chem. Soc.*, 1979, **101**, 6409.
86 R. Lohrmann and L. E. Orgel, *J. Mol. Evol.*, 1978, **11**, 17.

yields of nucleotides.[87] Deoxynucleosides gave mixtures of the 3′- and 5′-triphosphates in ≤44% yield, whereas adenosine in the presence of borate gave ATP in 9.9% yield, and 3′-deoxythymidine gave the 5′-triphosphate in 15% yield.

In a similar series of experiments, when solutions of adenosine 5′-pyrophosphates ($n \geq 3$) or P^1P^2-bis(5′-adenosyl) diphosphate were allowed to dry out in the presence of amines, appropriately substituted adenosine 5′-phosphoramidates were formed in yields of 10—50%.[88]

Dinucleoside phosphoramidates have been prepared[89] from silylated nucleoside phosphites and 5′-deoxy-5′-azido-nucleosides (Scheme 9).

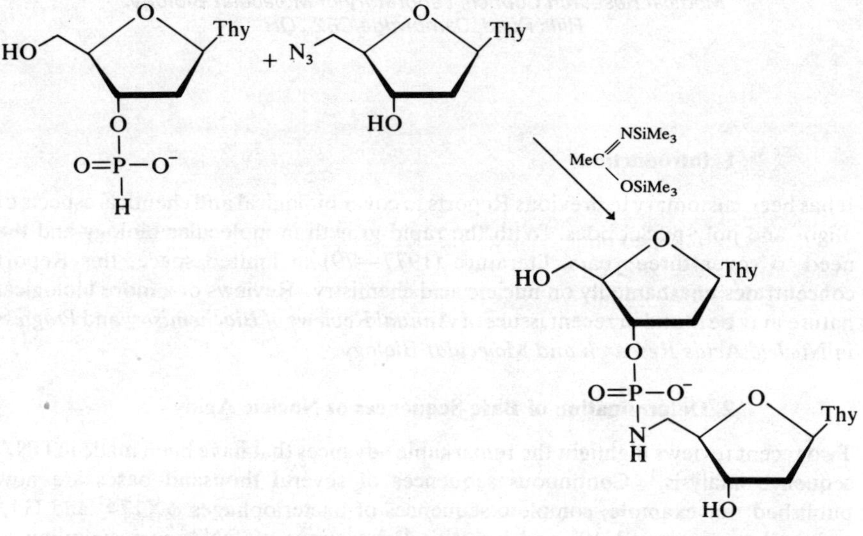

Scheme 9

X-Ray crystal structures that have been determined include 3′-UMP[90] and thymidine 3′,5′-cyclic *NN*-dimethylphosphoramidate,[91] and details of the binding of copper ions to IMP[92] and UMP have appeared.[93]

[87] E. Etaix and L. E. Orgel, *J. Carbohydr. Nucleosides, Nucleotides*, 1978, **5**, 91.
[88] R. Lohrmann, *J. Mol. Evol.*, 1977, **10**, 137.
[89] D. E. Gibbs, *Tetrahedron Letters*, 1977, 679.
[90] T. Srikrishnan, S. M. Fridey, and R. Parthasarathy, *J. Amer. Chem. Soc.*, 1978, **100**, 3739.
[91] M. G. Newton, N. S. Pantales, G. S. Bajwa, and W. G. Bentrude, *Tetrahedron Letters*, 1977, 4457.
[92] K. Aoki, *J.C.S. Chem. Comm.*, 1977, 600.
[93] B. E. Fischer and R. Bon., *J.C.S. Chem. Comm.*, 1977, 272.

16 Biological Chemistry

Part (v) Nucleic Acids

(b) Oligonucleotides and Polynucleotides

By M. J. GAIT

Medical Research Council, Laboratory of Molecular Biology, Hills Road, Cambridge CB2 2QH

1 Introduction

It has been customary in previous Reports to cover biological and chemical aspects of oligo- and poly-nucleotides. With the rapid growth in molecular biology and the need to cover three years' literature (1977—79) in limited space, this Report concentrates unashamedly on nucleic acid chemistry. Reviews of a more biological nature may be found in recent issues of *Annual Reviews of Biochemistry* and *Progress in Nucleic Acids Research and Molecular Biology*.

2 Determination of Base Sequences of Nucleic Acids

Two recent reviews highlight the remarkable advances that have been made in DNA sequence analysis.[1] Continuous sequences of several thousand bases are now published: for example, complete sequences of bacteriophages ϕX174[2] and G4,[3] mammalian viruses SV40[4] and hepatitis B,[5] sections of DNA corresponding to mammalian genes (insulin,[6] globin,[7] and ovalbumin[8]), and human mitochondrial DNA.[9] Such feats have been possible due to the emergence of two new sequencing methodologies. The first[10] relies on partial and base-selective chemical cleavage of ^{32}P-end-labelled DNA followed by separation of the resultant fragments by gel

[1] S. M. Weissman, *Analyt. Biochem.*, 1979, **98**, 243; R. Wu, *Ann. Rev. Biochem.*, 1978, **47**, 607.

[2] F. Sanger, A. R. Coulson, T. Friedmann, G. M. Air, B. G. Barrell, N. L. Brown, J. C. Fiddes, C. A. Hutchinson III, P. M. Slocombe, and M. Smith, *J. Mol. Biol.*, 1978, **125**, 225.

[3] G. N. Godson, B. G. Barrell, R. Staden, and J. C. Fiddes, *Nature*, 1978, **276**, 236.

[4] W. Fiers, R. Contreras, G. Haegeman, R. Rogiers, A. Van de Voorde, H. Van Heuverswyn, J. Van Herreweghe, G. Volckaert, and M. Ysebaert, *Nature*, 1978, **273**, 113; V. B. Reddy, B. Thimmappaya, R. Dhar, K. N. Subramanian, B. S. Zain, J. Pan, P. K. Ghosh, M. L. Celma, and S. M. Weissman, *Science*, 1978, **200**, 494.

[5] F. Galibert, E. Mandart, F. Fitoussi, P. Tiollais, and P. Charnay, *Nature*, 1979, **281**, 646.

[6] B. Cordell, G. Bell, E. Tischer, F. M. De Noto, A. Ullrich, R. Pictet, W. J. Rutter, and H. Goodman, *Cell*, 1979, **18**, 533; P. Lomedico, N. Rosenthal, A. Efstratiadis, W. Gilbert, R. Kolodner, and R. Tizard, *ibid.*, p. 545.

[7] A. Van Ooyen, J. Van den Berg, N. Mantel, and C. Weissmann, *Science*, 1979, **206**, 337.

[8] J. F. Catterall, B. W. O'Malley, M. A. Robertson, R. Staden, Y. Tanaka, and G. G. Brownlee, *Nature*, 1978, **275**, 510; K. O'Hare, R. Breathnach, C. Benoist, and P. Chambon, *Nucleic Acids Res.*, 1979, **7**, 321.

[9] B. G. Barrell, A. T. Bankier, and J. Drouin, *Nature*, 1979, **282**, 189.

[10] A. M. Maxam and W. Gilbert, *Proc. Nat. Acad. Sci. U.S.A.*, 1977, **74**, 560.

electrophoresis. Depurination is facilitated by prior methylation of guanines at N-7 and of adenines at N-3, using dimethyl sulphate, whereas depyrimidination is achieved by reaction with hydrazine (see Section 6). The DNA is then broken by treatment with piperidine. Judicious choice of overall reaction conditions further distinguishes bases. The second method[11] makes use of primed transcription of DNA with DNA polymerase in the presence of chain-terminating inhibitors, in particular 2′,3′-dideoxynucleoside 5′-triphosphates. Partial incorporation of just one inhibitor added to each of four separate reactions gives rise to a series of fragments, resolvable by gel electrophoresis. The method has been extended for use with double-stranded DNA.[12] The use of a computer considerably aids the storage and processing of DNA sequence information.[13]

Despite several elegant direct methods of RNA sequencing,[14] it is now in most cases quicker to copy mRNA (often unpurified) into its complementary DNA (cDNA), using reverse transcriptase. The crude DNA is engineered into a carrier DNA (plasmid) which is used to transform *Escherichia coli* cells. Particular colonies (clones) that exhibit characteristics assignable to the presence of the sequence of interest may be selected and the relevant DNA excised and sequenced (*e.g.* ACTH–β-lipotropin precursor mRNA[15] and ovalbumin mRNA[16]).

Cloning techniques are developing rapidly for analysis of genomic DNA. New vector classes capable of handling large DNA inserts have been constructed, some bearing peculiarly hybrid names (cosmid,[17] phasmid[18]). Of similar use is the single-stranded circular DNA from filamentous bacteriophages such as M13.[19] In 1977 the intriguing discovery was made that many eukaryotic structural genes are interrupted by pieces of non-coding DNA (intervening sequences, introns). A review and speculative discussion of this topic has recently appeared.[20] Functional RNA is seemingly derived by transcription of the genes into RNA precursors, which are then processed, including the removal of all introns. The non-contiguous nature of mammalian genes has enhanced the cDNA route to analysis and subsequent expression of hormone-specifying sequences (*e.g.* insulin[21] and growth hormone[22]).

[11] F. Sanger, S. Nicklen, and A. R. Coulson, *Proc. Nat. Acad. Sci. U.S.A.*, 1977, **74**, 5463.
[12] J. Maat and A. J. H. Smith, *Nucleic Acids Res.*, 1978, **5**, 4537.
[13] R. Staden, *Nucleic Acids Res.*, 1977, **4**, 4037.
[14] G. G. Brownlee and E. M. Cartwright, *J. Mol. Biol.*, 1977, **114**, 93; H. Donis-Keller, A. Maxam, and W. Gilbert, *Nucleic Acids Res.*, 1977, **4**, 2527; J. Stanley and S. Vassilenko, *Nature*, 1978, **274**, 87; A. Simoncsits, G. G. Brownlee, R. S. Brown, J. R. Rubin, and H. Guilley, *ibid.*, 1978, **269**, 833; D. A. Peattie, *Proc. Nat. Acad. Sci. U.S.A.*, 1979, **76**, 1760.
[15] S. Nakanishi, A. Inoue, T. Kita, M. Nakamura, A. C. Y. Chang, S. N. Cohen, and S. Numa, *Nature*, 1979, **278**, 423.
[16] L. McReynolds, B. W. O'Malley, A. D. Nisbet, J. E. Fothergill, D. Givol, S. Fields, M. Robertson, and G. G. Brownlee, *Nature*, 1978, **273**, 723.
[17] J. Collins and B. Hohn, *Proc. Nat. Acad. Sci. U.S.A.*, 1978, **75**, 4242.
[18] G. Cesarini, J. Karn, and S. Brenner, *Proc. Nat. Acad. Sci. U.S.A.*, 1980, **77**, 5217.
[19] P. Schreir and R. Cortese, *J. Mol. Biol.*, 1979, **129**, 169; J. Messing, B. Gronenborn, B. Muller-Hill, and P. H. Hofschneider, in 'Integration and Excision of DNA Molecules', ed. P. Hofschneider and P. Starlinger, Springer Verlag, Berlin, 1978, pp. 29—32.
[20] F. Crick, *Science*, 1979, **204**, 264.
[21] L. Villa-Komaroff, A. Efstratiadis, P. Broome, P. Lomedico, R. Tizard, S. P. Naber, W. L. Chick, and W. Gilbert, *Proc. Nat. Acad. Sci. U.S.A.*, 1978, **75**, 3727.
[22] P. H. Seeburg, J. Shine, J. A. Martial, J. D. Baxter, and H. M. Goodman, *Nature*, 1977, **270**, 486; D. V. Goeddel, H. L. Heyneker, T. Hozumi, R. Arentzen, K. Itakura, D. G. Yansura, M. J. Ross, G. Miozzari, R. Crea, and P. H. Seeburg, *ibid.*, 1979, **281**, 544.

3 Synthetic Oligonucleotides in Molecular Biology

The completion of the synthetic gene for the *E. coli* tyrosine suppressor tRNA and its control regions[23] required the work of some forty five scientists over ten years. In contrast, the inheritors of this technology can now carry out similar operations with considerably more ease. Despite the impressive nature of modern gene synthesis and subsequent expression in bacteria (*e.g.* somatostatin[24] and insulin[25]), there is a disappointing trend to omit the fine details of the chemical procedures. In the oligoribonucleotide field, the determination of the total synthesis of $tRNA^{fMet}$ is nearing completion.[26]

The synthesis of the *lac* operon and of a number of analogues with altered sequence or containing modified bases has been described.[27] Studies of repressor binding have identified particular functional groups on the DNA that are important to the interaction. Of great future potential is a new and generally applicable method of site-specific mutagenesis of DNA.[28] A synthetic oligonucleotide that is mismatched in a specific position on a DNA template is used to prime transcription by DNA polymerase. The progeny DNA contains a mutation at the site of the mismatch. Transition[28] and transversion[29] mutants of ϕX174 have been made.

Of general use in cloning are self-complementary oligonucleotides (linkers), which, after ligation to the end of a DNA, may be cut by a restriction enzyme to give a 'sticky end'.[30] Other adaptors will perform a similar function without the need for digestion with a restriction enzyme.[31] One type of sticky end can also be changed into another.[31] Oligonucleotides that are ^{32}P-labelled have been used as probes for the isolation of gastrin mRNA[32] and for the gene for yeast iso-1-cytochrome-*c*.[33]

The first crystal structure of a synthetic deoxytetranucleotide, d(A-T-A-T),[34] heralds a new era in the determination of structures of DNAs. Interestingly, the large difference in torsion angle of the O(3')–P bond observed between the sequences T-A and A-T hints at the importance of conformational effects in the recognition of specific DNA sequences by proteins. A useful review of the uses of synthetic oligodeoxyribonucleotides has been published.[35]

23 H. G. Khorana, *Science*, 1979, **203**, 614.
24 K. Itakura, T. Hirose, R. Crea, A. D. Riggs, H. Heyneker, F. Bolivar, and H. W. Boyer, *Science*, 1977, **198**, 1056.
25 R. Crea, A. Kraszewski, T. Hirose, and K. Itakura, *Proc. Nat. Acad. Sci. U.S.A.*, 1978, **75**, 5765.
26 E. Ohtsuka, T. Tanaka, S. Tanaka, K. Fujiyama, A. F. Markham, E. Nakagawa, T. Wakabayashi, Y. Taniyama, S. Nishikawa, R. Fukumoto, H. Uemura, T. Doi, and M. Ikehara, *Nucleic Acids Res.*, Special Publication No. 6, 1979, S195.
27 D. V. Geoddel, D. G. Yansura, C. Winston, and M. H. Caruthers, *J. Mol. Biol.*, 1978, **123**, 661; H. Sista, R. T. Loder, and M. H. Caruthers, *Nucleic Acids Res.*, 1979, **6**, 2583.
28 C. A. Hutchinson III, S. Phillips, M. H. Edgell, S. Gillam, P. Jahnke, and M. Smith, *J. Biol. Chem.*, 1978, **253**, 6551; H. Kössel, R. Buyer, S. Morioka, and H. Schott, *Nucleic Acids Res.*, Special Publication No. 4, 1978, S91.
29 S. Gillam, P. Jahnke, C. Astell, S. Phillips, C. A. Hutchinson III, and M. Smith, *Nucleic Acids Res.*, 1979, **6**, 2973.
30 R. H. Scheller, R. E. Dickerson, H. W. Boyer, A. D. Riggs, and K. Itakura, *Science*, 1977, **196**, 177.
31 C. P. Bahl, R. Wu, R. Brousseau, A. K. Sood, H. M. Hsiung, and S. A. Narang, *Biochem. Biophys. Res. Comm.*, 1978, **81**, 695.
32 M. Mevarech, B. E. Noyes, and K. L. Agarwal, *J. Biol. Chem.*, 1979, **254**, 7472.
33 D. L. Montgomery, B. D. Hall, S. Gillam, and M. Smith, *Cell*, 1978, **14**, 673.
34 M. A. Viswamitra, O. Kennard, Z. Shakked, P. G. Jones, G. M. Sheldrick, S. Salisbury, and L. Falvello, *Nature*, 1978, **273**, 687.
35 R. Wu, C. P. Bahl, and S. A. Narang, *Progr. Nucleic Acids Res. Mol. Biol.*, 1978, **21**, 101.

4 Synthesis of Oligonucleotides

Two general reviews have been published[36] and two more deal with phosphotriester methods of synthesis.[37] The advent of both DNA sequencing and h.p.l.c.[38] techniques has significantly aided the isolation and identification of oligonucleotides, and above all has led to an increased awareness of the need for improved synthetic methods.

In the synthesis of oligodeoxyribonucleotides in solution by the phosphodiester approach, the lipophilic t-butyldiphenylsilyl group has been used in place of acetyl for the 3′-protection of oligonucleotide chains.[39] The group is alkali-stable but cleaved by fluoride ion in buffered pyridine, and considerably aids h.p.l.c. separations of protected oligonucleotides. The triphenylphosphine–carbon tetrachloride complex has been proposed as a condensing agent for the rapid formation of phosphodiester bonds, and it is particularly useful for the preparation of 5′-dianilides from nucleoside 5′-phosphates.[40] In coupling reactions, the phosphorylpyridinium derivative (1) has been shown to be the most likely phosphorylating agent when

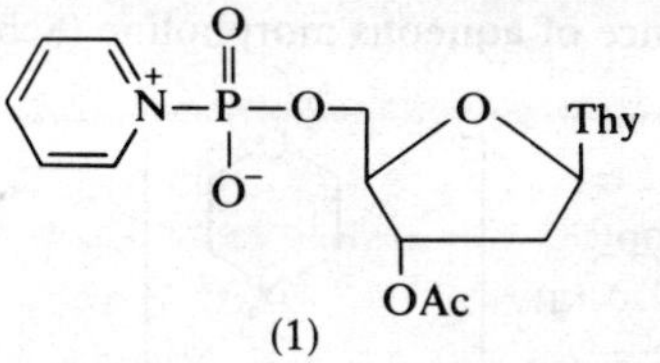

(1)

either $Ph_3P–CCl_4$ or arenesulphonyl chlorides are used to activate 3′-*O*-acetyl-2′-deoxythymidine 5′-phosphate.[41] Phosphodiester groups also become activated in coupling reactions, and comparative studies of rates suggest that the use of trinucleotide or longer blocks as 'phosphate components' leads to a higher probability of side-products being formed.[42] The use of short oligonucleotide blocks with protected phosphodiester bonds has been recommended.[43]

Phosphotriester methods of oligodeoxyribonucleotide synthesis are characterized by their greater selectivity in inter-nucleotide coupling reactions. However, until recently, inadequate phosphorylation and deprotection procedures have resulted in unreliable syntheses. The most popular phosphorylating agent for nucleoside 3′-hydroxyl-groups is the bis-triazolide of *p*-chlorophenyl phosphate (2).[44] Further reaction with 2-cyanoethanol results in the formation of a nucleoside 3′-*p*-chlorophenyl, 2-cyanoethyl phosphotriester. The cyanoethyl group may be removed

36 V. Amarnath and A. D. Broom, *Chem. Rev.*, 1977, **77**, 183; M. Ikehara, E. Ohtsuka and A. F. Markham, *Adv. Carbohydrate Chem. Biochem.*, 1979, **36**, 135.

37 C. B. Reese, *Tetrahedron*, 1978, **34**, 3143; J. H. van Boom, *Heterocycles*, 1977, **7**, 1197.

38 H.-J. Fritz, R. Belagaje, E. L. Brown, R. H. Fritz, R. A. Jones, R. G. Lees, and H. G. Khorana, *Biochemistry*, 1978, **17**, 1257.

39 R. A. Jones, H.-J. Fritz, and H. G. Khorana, *Biochemistry*, 1978, **17**, 1268.

40 G. F. Mishenina, V. V. Samukov, L. N. Semenova, and T. N. Shubina, *Bioorg. Khim.*, 1978, **4**, 735, 1137.

41 V. F. Zarytova, D. M. Graifer, E. I. Ivanova, D. G. Knorre, A. V. Lebedev, and A. I. Rezvukhin, *Nucleic Acids Res.*, Special Publication No. 4, 1978, S209.

42 D. G. Knorre, V. F. Zarytova, A. V. Lebedev, L. M. Khalimskaya, and E. A. Sheshegova, *Nucleic Acids Res.*, 1978, **5**, 1253.

43 D. G. Knorre, G. T. Mishenina, and T. N. Shubina, *Doklady Akad. Nauk. S.S.S.R.*, 1978, **243**, 662.

44 K. Itakura, N. Katagiri, and S. A. Narang, *Canad. J. Chem.*, 1974, **52**, 3689.

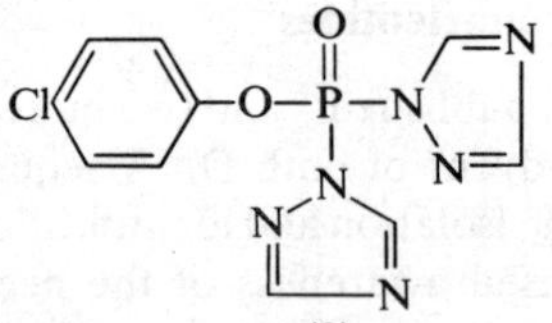
(2)

selectively, using triethylamine in pyridine.[45] Monofunctional 3′-phosphorylating agents have also been suggested.[46] The 4,4′-dimethoxytriphenylmethyl group (dimethoxytrityl) is usual for 5′-protection, and it has been shown to be rapidly cleaved by treatment with benzenesulphonic acid.[47] Loss of *N*-benzoyladenine is substantially reduced compared to treatment with acetic acid. The lability of the 9-phenylxanthen-9-yl protecting group to acid is similar to that of dimethoxytrityl,[48] but of greater future potential for 5′-protection is the 2-dibromomethylbenzoyl group (DBMB).[49] Treatment with silver perchlorate converts DBMB-protected deoxynucleosides into their 2-formylbenzoyl derivatives, which are then readily hydrolysed in the presence of aqueous morpholine (Scheme 1).

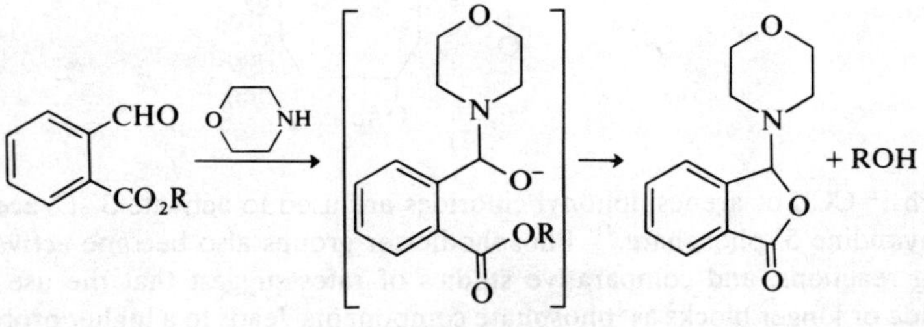
Scheme 1

Arylsulphonyltetrazoles are rapid new coupling agents that are used in phosphotriester synthesis,[47] and the use of 8-quinolylsulphonyl chloride has also been proposed.[50] However, 5′-*O*-sulphonylation by such coupling agents is a side-reaction that has yet to be fully eliminated.

Of much concern is the continued widespread use of ammonia and hydroxide ion for the removal of aryl protecting groups from phosphodiester bonds, despite further warnings of the resultant likelihood of inter-nucleotide cleavage[51] and other side-reactions.[52] The alternative use of 1,1,3,3-tetramethylguanidinium *p*-nitrobenzaldoximate or of pyridine-2-carboxaldoximate has been shown to effect deprotection smoothly, with minimal formation of side-products.[53]

45 A. K. Sood and S. A. Narang, *Nucleic Acids Res.*, 1977, **4**, 2753.
46 C. B. Reese and Y. T. Y. Kui, *J.C.S. Chem. Comm.*, 1977, 802; C. B. Reese and L. Yau, *ibid.*, 1978, 1050.
47 J. Stawinski, T. Hozumi, S. A. Narang, C. P. Bahl, and R. Wu, *Nucleic Acids Res.*, 1977, **4**, 353.
48 J. B. Chattopadhyaya, and C. B. Reese, *J.C.S. Chem. Comm.*, 1978, 639.
49 J. B. Chattopadhyaya, C. B. Reese, and A. H. Todd, *J.C.S. Chem. Comm.*, 1979, 987.
50 H. Takaku, M. Kato, M. Yoshida, and T. Hata, *Nucleic Acids Res.*, Special Publication No. 5, 1978, S345.
51 R. W. Adamiak, R. Arentzen, and C. B. Reese, *Tetrahedron Letters*, 1977, 1431.
52 J. F. M. de Rooij, G. Wille-Hazeleger, P. M. J. Burgers, and J. H. van Boom, *Nucleic Acids Res.*, 1979, **6**, 2237.
53 C. B. Reese, R. C. Titmas, and L. Yau, *Tetrahedron Letters*, 1978, 2727.

Phosphotriester routes now predominate in oligoribonucleotide synthesis, yet no consensus has been reached concerning the usage of protecting groups. Acid-labile 5′-protecting groups have been mostly chosen, in combination with acetyl,[54] *o*-nitrobenzyl (removable photolytically),[55] or t-butyldimethylsilyl[56] on the 2′-position. The last group is removable by fluoride ion, but serious danger of the formation of side-products has been shown in the deprotection of oligonucleotide phosphotriesters containing 2′-*O*-t-butyldimethylsilyl groups by fluoride ion.[52] Base-catalysed 2′–3′ migration of a silyl group is another potential hazard.[57] An alternative combination is the acid-labile methoxytetrahydropyranyl group for the 2′-position and the laevulinyl group for the 5′-position (removable with buffered hydrazine).[58] Useful 3′-phosphorylating agents are 2,2,2-trichloroethyl *o*-chlorophenyl phosphorochloridate[58] and *p*-chlorophenyl phosphoramidochloridate.[59] Corresponding nucleoside 3′-*O*-arylphosphates are obtained by treatment of the nucleoside 3′-phosphotriester with zinc and toluene-*p*-sulphonic acid[58] and isopentyl nitrite,[59] respectively. As a warning to those who may have been tempted to couple unprotected nucleosides chemically, the preferential formation of 2 → 5′ links has been shown to occur.[60]

The introduction of new resins has revitalized solid-phase synthesis of oligodeoxyribonucleotides. Copolymers based on polydimethylacrylamide[61] and polyacryloylmorpholide[62] are freely permeated by polar organic solvents. In contrast to polystyrene resins, good coupling yields were obtained using a phosphodiester approach. An impervious polymer formed by grafting of uncrosslinked polystyrene to Teflon beads has shown similarly good results in phosphodiester syntheses[63] (a polar polymer surface is obtained by high nucleoside loading). Improved yields were obtained by blocking the inter-nucleotide phosphate group after coupling with aniline, using the $Ph_3P–CCl_4$ reagent.[63] Phosphotriester methods of solid-phase synthesis have recently been shown to be both rapid and flexible.[64,65] Much higher yields were obtained in the synthesis of octa- and dodeca-deoxyribonucleotides on a poly(dimethylacrylamide) resin than could have been expected by a phosphodiester route.[65]

Enzymatic oligonucleotide synthesis has become increasingly important. Polynucleotide phosphorylase catalyses the reaction of a nucleoside 5′-diphosphate with the 3′-end of an oligonucleotide primer. Some improvements to addition reactions of ribonucleotides have recently been made,[66] but, owing to the preponderance of

54 R. W. Adamiak, E. Biala, K. Grzeskowiak, R. Kierzek, A. Kraszewski, W. T. Markiewicz, J. Okupniak, J. Stawinski, and M. Wiewiorowski, *Nucleic Acids Res.*, 1978, **5**, 1889.
55 E. Ohtsuka, T. Tanaka, S. Tanaka, and M. Ikehara, *J. Amer. Chem. Soc.*, 1978, **100**, 4580.
56 K. K. Ogilvie, S. L. Beaucage, A. L. Schifman, N. Y. Thieriault, and K. L. Sadana, *Canad. J. Chem.*, 1978, **56**, 2768.
57 S. J. Jones and C. B. Reese, *J.C.S. Perkin I*, 1979, 2762.
58 J. H. van Boom, P. M. J. Burgers, C. H. M. Verdegaal, and G. Wille, *Tetrahedron*, 1978, **34**, 1999.
59 E. Ohtsuka, T. Tanaka, T. Wakabayashi, Y. Taniyama, and M. Ikehara, *J.C.S. Chem. Comm.*, 1978, 824.
60 R. Lohrmann and L. E. Orgel, *Tetrahedron*, 1978, **34**, 853.
61 M. J. Gait and R. C. Sheppard, *Nucleic Acids Res.*, 1977, **4**, 1135, 4391; 1979, **6**, 1259.
62 C. K. Narang, K. Brunfeldt, and K. E. Norris, *Tetrahedron Letters*, 1977, 1819.
63 V. K. Potapov, V. P. Veiko, O. N. Koroleva, and Z. A. Shabarova, *Nucleic Acids Res.*, 1979, **6**, 2041.
64 K. Miyoshi and K. Itakura, *Tetrahedron Letters*, 1979, 3635.
65 M. J. Gait, M. Singh, R. C. Sheppard, M. D. Edge, A. R. Greene, G. R. Heathcliffe, T. C. Atkinson, C. R. Newton, and A. F. Markham, *Nucleic Acids Res.*, 1980, **8**, 1081.
66 B. W.-K. Shum and D. M. Crothers, *Nucleic Acids Res.*, 1978, **5**, 2297.

multi-addition products, only relatively simple syntheses are practical. Greater control is possible in the deoxyribonucleotide case, and oligonucleotides of chain length up to 13 have been prepared from a minimum primer length of 3.[67] T4-RNA ligase catalyses the joining of a 5′-phosphate of an oligonucleotide donor to a 3′-hydroxyl of an oligonucleotide acceptor, and, as distinct from DNA ligase, this does not require a template. 2′-*O*-*o*-nitrobenzyl-protected oligoribonucleotide donors will join to oligoribonucleotide acceptors without concomitant self-condensation or cyclization. The protecting group is photolytically cleavable.[68] Similarly, nucleoside 3′,5′-bisphosphates have been shown to be good donors.[69,70] After joining them (with RNA ligase), the 3′-phosphates are removed with phosphatase; stepwise oligoribonucleotide[70] and oligodeoxyribonucleotide[71] synthesis has been described. The joining reaction is also useful for the 3′-labelling of RNA with ^{32}P, to high specific activity.[72]

Mass spectral analysis of synthetic oligodeoxyribonucleotides is a useful tool that has yet to be fully exploited. It has been used to detect whether or not all blocking groups have been removed at the appropriate stage of a synthesis.[73]

5 Analogues of Oligo- and Poly-nucleotides

From studies of polypyrimidine nucleoside analogues, it had previously been thought that intact 2′-hydroxy-groups were essential for the induction of interferon by poly(I)·poly(C). It has now been shown that poly(2′-deoxy 2′-substituted purine nucleosides) have similar physical properties to their ribose counterparts, and that the 1 : 1 complexes of poly(C) with poly(2′-azido-2′-deoxy-I),[74] poly(2′-chloro-2′-deoxy-I),[75] and poly(2′-fluoro-2′-deoxy-I)[76] are potent inducers of interferon. Poly(2-methylthio-A) and poly(2-ethylthio-A) strongly inhibit viral reverse transcriptases and also form Hoogsteen-paired 1 : 1 complexes with poly(U).[77] The complexes of 2-aza-analogues of poly(A) and of poly(I) with complementary natural homopolymers are strongly destabilized, however, owing to repulsion of the lone pair of electrons on nitrogen.[78]

ApU analogues containing electron-withdrawing substituents in the 5-position of uridine showed decreased stacking interactions compared to ApU.[79] Similarly modified ApApU and ApA(pU)$_2$ analogues were less effective in stimulating the binding of aminoacyl-tRNA to ribosomes than their natural counterparts.[80] Whereas the template activity of ApAp5NH$_2$U was lower than that of ApApU, that of ApA(p5NH$_2$U)$_2$ was higher than that of ApA(pU)$_2$, suggesting the importance of

[67] S. Gillam, P. Jahnke, and M. Smith, *J. Biol. Chem.*, 1978, **253**, 2532.
[68] E. Ohtsuka, S. Nishikawa, A. F. Markham, S. Tanaka, T. Miyake, T. Wakabayashi, M. Ikehara, and M. Sugiura, *Biochemistry*, 1978, **17**, 4894.
[69] Y. Kikuchi, H. Hishinuma, and K. Sakaguchi, *Proc. Nat. Acad. Sci. U.S.A.*, 1978, **75**, 1270.
[70] T. E. England and O. C. Uhlenbeck, *Biochemistry*, 1978, **17**, 2069.
[71] D. M. Hinton and R. I. Gumport, *Nucleic Acids Res.*, 1979, **7**, 453.
[72] T. E. England and O. C. Uhlenbeck, *Nature*, 1978, **275**, 560.
[73] M. A. Armbruster and J. L. Wiebers, *Analyt. Biochem.*, 1977, **83**, 570.
[74] T. Fukui, N. Kakauchi, and M. Ikehara, *Nucleic Acids Res.*, 1977, **4**, 4249.
[75] N. Kakauchi, T. Fukui, and M. Ikehara, *Nucleic Acids Res.*, 1979, **6**, 2627.
[76] M. Ikehara, N. Kakauchi, and T. Fukui, *Nucleic Acids Res.*, 1978, **5**, 3315.
[77] T. Fukui and M. Ikehara, *Biochim. Biophys. Acta*, 1979, **562**, 527.
[78] T. Fukui, N. Kakauchi, and M. Ikehara, *Biochim. Biophys. Acta*, 1978, **520**, 441.
[79] W. Hillen and H. G. Gassen, *Biochim. Biophys. Acta*, 1978, **518**, 7.

correct base overlap in anticodon interactions.[80] 5-Methoxyuridine occurs in several tRNAs of *Bacillus subtilis*, but, although poly(5-methoxy-U) has been shown to form triplexes with poly(A), it did not direct the synthesis of polyPhe *in vitro*, unlike other 5-substituted homopolymers.[81]

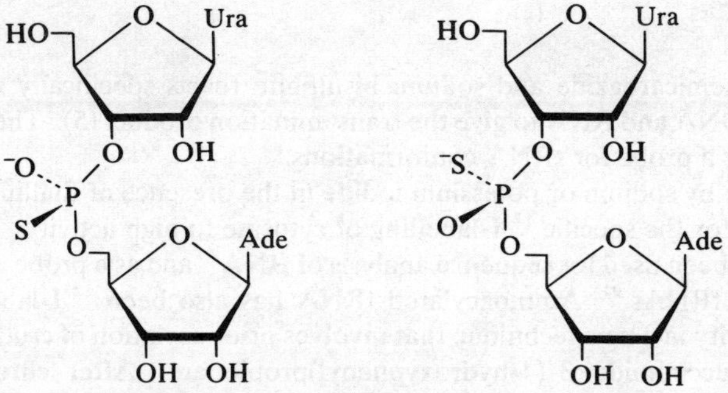

(R_P) (3) (S_P)

The two diastereomers of 5′-*O*-adenosyl 3′-*O*-uridylyl phosphorothioate (3) have been chemically synthesized *via* phosphotriester intermediates.[82] Only the (R_P) diastereomer is sensitive to digestion with snake venom exonuclease. Phosphorothioate linkages incorporated enzymatically into DNA do not significantly alter its properties, and should be useful for labelling DNAs with heavy metals[83] Chemically synthesized methylphosphonate analogues of oligodeoxyribonucleotides are resistant to digestion with an exonuclease and are useful model substrates for studies of the action of restriction enzymes.[84] Some non-phosphorus-containing carbamate[85] and acetamidate analogues[86] of oligonucleotides have also been prepared.

6 Chemical Reactions on Polynucleotides

Denaturation of DNA with formaldehyde is much used in electron microscopy. A detailed kinetic analysis of this reaction has now confirmed that, initially, hydrogen-bond rupture and base unstacking takes place at an A-T-rich region in the interior of DNA, due to local thermal fluctuations. This is followed by hydroxymethylation of the exocyclic amino-group of adenine and of the endocyclic imino-group of thymine.[87] The reaction of hydrazine with DNA is pyrimidine-specific, and is used in the Maxam–Gilbert sequencing method.[10] In addition to the expected 3-aminopyrazole, a major product of the reaction of cytosine has been identified as 3-ureidopyrazole (4),[88] and not *NN*′-di-(3-pyrazolyl)hydrazine, as first thought. A

[80] W. Hillen and H. G. Gassen, *Nucleic Acids Res.*, Special Publication No. 4, 1978, S149.
[81] W. Hillen and H. G. Gassen, *Biochim. Biophys. Acta*, 1979, **562**, 207.
[82] P. M. J. Burgers, and F. Eckstein, *Biochemistry*, 1979, **18**, 592.
[83] H.-P. Vosberg and F. Eckstein, *Biochemistry*, 1977, **16**, 3633.
[84] K. L. Agarwal and F. Riftina, *Nucleic Acids Res.*, 1979, **6**, 3009.
[85] W. S. Mungall and J. K. Kaiser, *J. Org. Chem.*, 1977, **42**, 703.
[86] M. J. Gait, A. S. Jones, M. D. Jones, M. J. Shepherd, and R. T. Walker, *J.C.S. Perkin I*, 1979, 1389.
[87] J. D. McGhee and P. H. von Hippel, *Biochemistry*, 1977, **16**, 3276.
[88] A. R. Cashmore and G. B. Peterson, *Nucleic Acids Res.*, 1978, **5**, 2485.

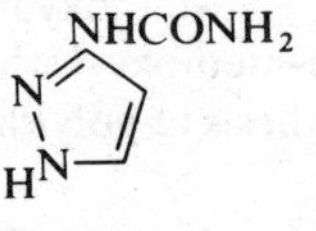

(4)

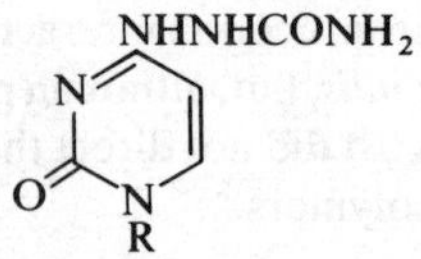

(5)

mixture of semicarbazide and sodium bisulphite reacts specifically with cytosine residues in DNA and RNA to give the transamination product (5). The reaction has been used as a probe for tRNA conformations.[89]

Iodination by sodium or potassium iodide in the presence of thallium(III) salts is well known for the specific ^{125}I-labelling of cytosine to high activity. This reaction has recently been used for sequence analysis of RNA[90] and as a probe of the tertiary structure of tRNAs.[91] Aminoacylated tRNA has also been ^{125}I-labelled to high specific activity in a new technique that involves prior acylation of crude tRNA with *N*-hydroxysuccinimido 3-(4-hydroxyphenyl)propionate. After chromatographic purification, the tRNA of interest is iodinated at the newly introduced benzene ring by reaction with [^{125}I]iodide ion and Chloramine T.[92] The 5′-termini of short oligoribonucleotides can be determined by cleavage with snake venom phosphodiesterase, followed by oxidation with periodate and then allowing the resultant nucleoside dialdehyde to react with *p*-hydrazinobenzene[^{35}S]sulphonic acid.[93] However, more generally useful in RNA sequencing is the new 3′-^{32}P-labelling procedure previously described (Section 4).[72] The 3′-^{32}P-labelling of DNA has been effected by the terminal-transferase-catalysed addition of 4-thiouridine, followed by thio-specific modification using α-halogeno-acetamide derivatives. Fluorescent, radioactive, and photoactivatable groups have been attached to DNA by this method.[94] Whereas the specific attachment of a fluorescent probe to the minor base Q[95] or to its under-modified replacement, 7-aminomethyl-7-deaza G,[96] in *E. coli* $tRNA^{Tyr}$ did not destroy the ability of the tRNA to be aminoacylated, fluorescent labelling of X base in *E. coli* $tRNA^{Phe}$ did.[97] The reaction of 1-fluoro-2,4-dinitrobenzene with $tRNA^{Phe}$ from yeast at pH 8 resulted in labelling of adenine residues only.[98] This reaction may prove useful for the introduction of antigenic sites into tRNA.

Osmium tetroxide, in the presence of pyridine or 2,2′-bipyridyl, adds across the 5,6-double-bond of pyrimidines in polynucleotides to give stable osmate esters that are useful in heavy-metal labelling.[99] The reaction with isopentenyladenosine is much faster, and permits single-site labelling of yeast $tRNA^{Tyr}$.[100] In the absence of

[89] K. Negishi, F. Harada, S. Nishimura, and H. Hayatsu, *Nucleic Acids Res.*, 1977, **7**, 2283.
[90] E. Dickson, L. K. Pape, and H. D. Robertson, *Nucleic Acids Res.*, 1979, **6**, 91.
[91] I. L. Batey and D. M. Brown, *Biochim. Biophys. Acta*, 1977, **474**, 378.
[92] G. M. Tener, A. D. Delaney, T. A. Grigliatti, G. J. Cowling, and I. C. Gillam, *Biochemistry*, 1978, **17**, 741.
[93] G. O. Osuji and M. W. Johnson, *F.E.B.S. Letters*, 1977, **83**, 85.
[94] H. Eshaghpour, D. Söll, and D. M. Crothers, *Nucleic Acids Res.*, 1979, **7**, 1485.
[95] A. Pingoud, R. Kownatzki, and G. Maass, *Nucleic Acids Res.*, 1977, **4**, 327.
[96] H. Kasai, N. Shindo-Okado, S. Noguchi, and S. Nishimura, *Nucleic Acids Res.*, 1979, **7**, 231.
[97] P. W. Schiller and A. N. Shechter, *Nucleic Acids Res.*, 1977, **4**, 2161.
[98] K. Watanabe and F. Cramer, *European J. Biochem.*, 1978, **89**, 425.
[99] C. H. Chang, M. Beer, and L. M. Marzilli, *Biochemistry*, 1977, **16**, 33.
[100] W. R. Midden and E. J. Behrman, *F.E.B.S. Letters*, 1979, **103**, 300.

tertiary nitrogen donors, osmium tetroxide reacts preferentially with thymidine in DNA to give 5,6-dihydroxythymidine. Subsequent glycosidic cleavage with piperidine forms the basis of a T-specific reaction that is useful in DNA sequence analysis.[101] A polymetallic agent, tetrakis(acetoxymercuri)methane, has been used for labelling of 4-thiouridine and 6-thioguanosine residues in tRNA,[102] and the use of platinum derivatives as probes of polynucleotide structure[103] and as anti-tumour agents[103,104] has been reviewed.

Of the known alkylating carcinogens, *N*-ethyl-*N*-nitrosourea has the highest reactivity with phosphodiester groups compared to other sites in DNA,[105] and has been used to help chart the points of interaction of RNA polymerase with DNA promoters.[106] The 2-halogenoethyl-nitrosoureas, on the other hand, have been shown to cause intra-strand cross-linking, probably between guanine and cytosine,[107] and this may be directly related to their antileukaemic properties.[108] The binding of ethidium bromide to DNA duplexes is stronger where there is a mismatch, and this suggests that this frameshifting agent can function by stabilizing transient mismatched intermediates in DNA.[109] The interaction of DNA with antibiotics and other drugs is an area of much current activity and has been extensively reviewed.[110]

101 T. Friedman and D. M. Brown, *Nucleic Acids Res.*, 1978, **5**, 615.
102 K. G. Strothkamp, J. Lehmann, and S. J. Lippard, *Proc. Nat. Acad. Sci. U.S.A.*, 1978, **75**, 1181.
103 S. J. Lippard, *Accounts Chem. Res.*, 1978, **11**, 211.
104 J. J. Roberts and A. J. Thomson, *Progr. Nucleic Acids Res. Mol. Biol.*, 1979, **22**, 71.
105 B. Singer, *J. Toxicol. Environ. Health*, 1977, **2**, 1279; D. H. Swenson, P. B. Farmer, and P. D. Lawley, *Biochem. J.*, 1978, **171**, 375.
106 U. Siebenlist and W. Gilbert, *Proc. Nat. Acad. Sci. U.S.A.*, 1980, **77**, 122.
107 K. W. Kohn, *Cancer Res.*, 1977, **37**, 1450.
108 J. W. Lown, L. W. McLaughlin and Y.-M. Chang, *Bioorg. Chem.*, 1978, **7**, 97.
109 D. C. Helfgott and N. R. Kallenbach, *Nucleic Acids Res.*, 1979, **7**, 1011; C. H. Lee and I. Tinoco, *Nature*, 1978, **274**, 609.
110 M. J. Waring in 'Drug Action at the Molecular Level', ed. G. C. K. Roberts, Macmillan, London, 1977; R. J. Suhadolnik, *Progr. Nucleic Acids Res. Mol. Biol.*, 1979, **22**, 193; W. A. Remers, in 'The Chemistry of Antitumour Antibiotics', Vol. I, Wiley–Interscience, New York, 1979.

16 Biological Chemistry
Part (vi) Enzyme Chemistry

By P. F. LEADLAY
Department of Biochemistry, University of Cambridge
Tennis Court Road, Cambridge CB2 1QW

1 Introduction

In the past year, useful reviews have been published on chiral methyl groups,[1] coenzyme B_{12}-dependent enzymes,[2] and positional isotope exchange studies of enzyme mechanisms.[3] Other good articles have appeared on two-dimensional n.m.r.,[4] the preparation and uses of immobilized enzymes,[5] cryoenzymology,[6] and the prediction of protein conformation from amino-acid sequence data.[7]

The proceedings of the Ciba Foundation Symposium on 'Molecular Interactions and Activity of Proteins' are now published,[8] as are the papers given at the Third European Symposium on Vitamin B_{12}.[9] Several new books on enzymes have also appeared.[10]

Recent *X*-ray crystal structure determinations include those of citrate synthase from pig heart,[11] phosphofructokinase from *Bacillus stearothermophilus*,[12] 6-phosphogluconate dehydrogenase from sheep liver,[13] and chicken heart mitochondrial aspartate aminotransferase.[14] Some aspects of natural-abundance ^{13}C n.m.r. as applied to proteins have been reviewed.[15] Curiously, ^{15}N and ^{13}C n.m.r. studies have given apparently quite different results concerning the mechanism of action of a typical serine proteinase.[16]

1. H. G. Floss and M. D. Tsai, *Adv. Enzymol.*, 1979, **50**, 243.
2. B. M. Babior and J. S. Krouwer, *C.R.C. Crit. Rev. Biochem.*, 1979, **6**, 35.
3. I. A. Rose, *Adv. Enzymol.*, 1979, **50**, 361.
4. K. Wüthrich, *Trends Biochem. Sci.*, 1979, **4**, N178.
5. G. G. Guilbault and M. H. Sadar, *Accounts Chem. Res.*, 1979, **6**, 204.
6. P. Douzou and C. Balny, *Adv. Protein Chem.*, 1978, **32**, 77.
7. P. Y. Chou and G. D. Fasman, *Adv. Enzymol.*, 1979, **49**, 45.
8. Ciba Foundation Symposium No. 60, Excerpta Medica, Amsterdam, 1978.
9. 'Vitamin B_{12}', ed. B. Zagalak and W. Friedrich, Walter de Gruyter, Berlin and New York, 1979.
10. (*a*) C. Walsh, 'Enzymatic Reaction Mechanisms', W. H. Freeman, New York, 1979; (*b*) 'Transition States of Biological Processes', ed. R. L. Schowen and R. D. Gandour, Plenum Press, New York, 1978; (*c*) R. L. Foster, 'The Nature of Enzymology', Croom Helm, London, 1980.
11. G. Wiegand, D. Kukla, H. Scholze, T. A. Jones, and R. Huber, *European J. Biochem.*, 1979, **93**, 41.
12. P. R. Evans and P. J. Hudson, *Nature*, 1979, **279**, 500.
13. M. A. Abdallah, M. J. Adams, I. G. Archibald, J.-F. Biellmann, J. R. Helliwell, and S. E. Jenkins, *European J. Biochem.*, 1979, **98**, 121.
14. G. Eichele, G. C. Ford, M. Glor, J. N. Jansonius, C. Mavrides, and P. Christen, *J. Mol. Biol.*, 1979, **133**, 161.
15. A. Allerhand, *Accounts Chem. Res.*, 1978, **11**, 469.
16. W. Bachovchin and J. D. Roberts, *J. Amer. Chem. Soc.*, 1978, **100**, 8041.

2 Enzyme-catalysed Methyl-transfer Reactions

In a large number of enzyme-catalysed reactions, methyl groups are either created or destroyed.[1] The methyl group can also be transferred intact either intramolecularly[17] or from a donor such as *S*-adenosylmethionine, to a wide range of acceptors.[18]

In all these reactions, the determination of the stereochemical consequences of reaction at the methyl group is of central importance in elucidating the mechanism. This was first made possible by the discovery of synthetic routes to acetyl-CoA (1) containing a chiral methyl group.[19] These molecules are chiral because the methyl group contains each of the three isotopes of hydrogen in a specific configuration. By exploiting the existence of an intramolecular kinetic isotope effect in the malate synthase reaction (Scheme 1), a method was developed[19] to provide both the absolute configuration and a quantitative measure of the enantiomeric excess at the methyl carbon.[20] The chirality of any methyl group can therefore be assayed by transforming it chemically into the methyl group of acetate.

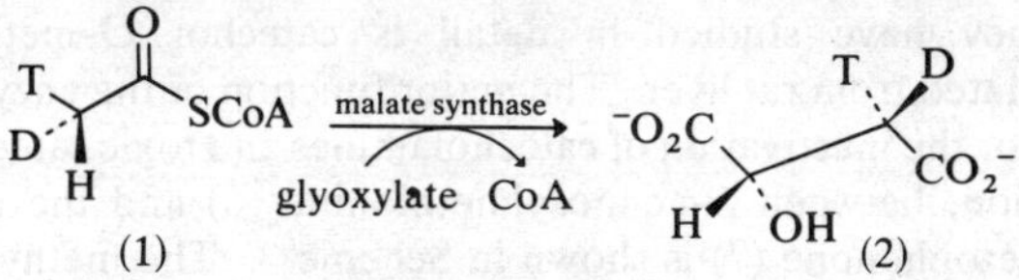

(1) (2)

($D = {}^2H$; $T = {}^3H$)

Scheme 1

In order to study methyl transfers in which *S*-adenosylmethionine (6) is the methyl donor, two almost identical synthetic routes to chiral [*Me*-$^3H_1$,$^2H_1$]methionine have been developed,[21] and in Arigoni's laboratory a way has also been devised to re-convert methionine into acetic acid, to check its chirality.[21a]

Transferases have now been studied that catalyse methyl transfer respectively to carbon, to oxygen, or to sulphur, but not as yet to nitrogen,[21a] and the only methyl donor examined so far is *S*-adenosylmethionine (6).[21a] A good example of a study on transfer to sulphur concerns the (essentially irreversible) methylation of homocysteine (4) that is catalysed by *S*-adenosylmethionine:homocysteine methyltransferase (EC 2.1.1.12).[21a] Here, *S*-methylmethionine (3) can deputize for *S*-adenosylmethionine (6) as the methyl donor, as outlined in Scheme 2, so that each transfer generates two molecules of methionine (5).

Since (3) was synthesized chemically from (2*S*)-[*Me*-$^3H_1$,$^2H_1$]methionine, the chiral methyl groups in (3) should be distributed equally between the two diastereotopic positions at sulphur.[22] Half the labelled methyl groups in the sample of methionine (5) that is produced will therefore be derived from (3) without transfer,

17 G. T. Phillips and K. H. Clifford, *European J. Biochem.*, 1976, **61**, 271.
18 G. L. Cantoni, *Ann. Rev. Biochem.*, 1975, **44**, 435.
19 (*a*) J. W. Cornforth, J. W. Redmond, H. Eggerer, W. Buckel, and C. Gutschow, *Nature*, 1969, **221**, 1212; (*b*) J. Lüthy, J. Rétey, and D. Arigoni, *Nature*, 1969, **221**, 1213.
20 H. Lenz and H. Eggerer, *European J. Biochem.*, 1976, **65**, 237.
21 (*a*) D. Arigoni, in ref. 8, p. 404; (*b*) L. Mascaro, Jr., R. Höhrhammer, S. Eisenstein, L. K. Sellers, K. Mascaro, and H. G. Floss, *J. Amer. Chem. Soc.*, 1977, **99**, 273.
22 G. Grue-Sorensen, E. Kelstrup, A. Kjaer, and J. O. Madsen, *J.C.S. Chem. Comm.*, 1980, 19.

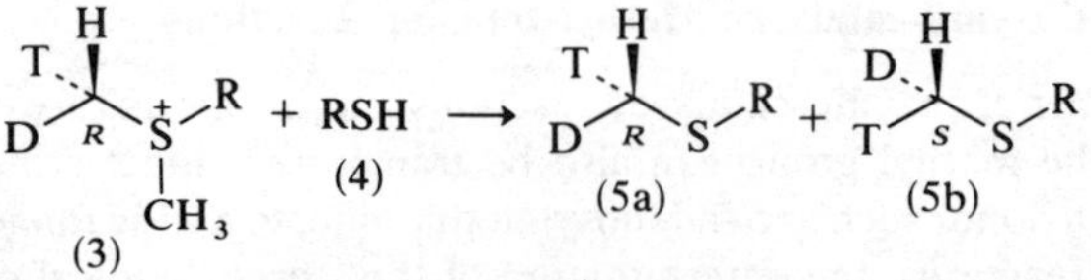

$[R = (CH_2)_2CH(\overset{+}{N}H_3)CO_2^-]$

Scheme 2

and will retain the original configuration. The other half will have been transferred to (4), and their configuration will depend on the mechanism. In fact, as indicated in Scheme 2, the methionine obtained proved, on degradation to acetic acid and chirality analysis, to be completely racemic. Therefore one can conclude that this methyl transfer involves a net inversion of configuration.[21a]

Schowen and his co-workers have used kinetic isotope effects to examine the nature of the transition state in enzyme-catalysed transfers of methyl groups.[23] An enzyme that they have studied in detail is catechol *O*-methyl transferase (EC 2.1.1.6), isolated from rat liver. The major function of the enzyme *in vivo* is to provide a route for the inactivation of catecholamines and toxic catechols in the diet. A typical reaction, between *S*-adenosylmethionine (6) and the model acceptor 3,4-dihydroxyacetophenone (7) is shown in Scheme 3. The methylation occurs at either hydroxy-group but not both.[23b]

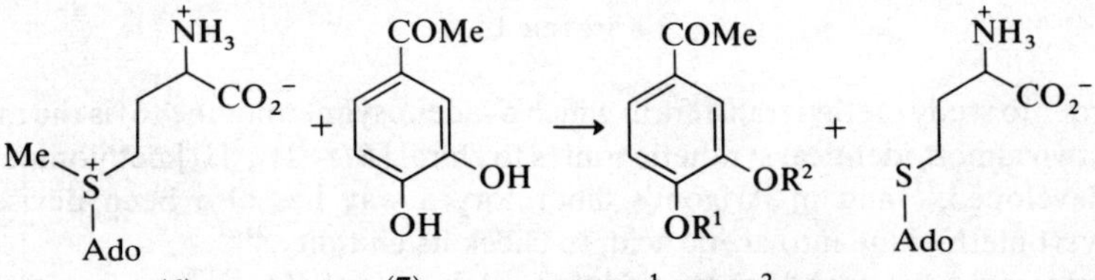

(6) (7) (8) $R^1 = H, R^2 = Me$

(9) $R^1 = Me, R^2 = H$

(Ado = adenosyl)

Scheme 3

A sample of (6), enriched with ^{13}C in the methyl group and with the correct chirality at sulphur, was prepared by the action of a yeast culture on the corresponding (2*S*)-[*Me*-^{13}C]methionine. The initial rate of methyl transfer from *S*-adenosyl[*Me*-^{13}C]methionine to (7), catalysed by catechol *O*-methyl transferase, was compared with the rate of transfer from unlabelled material (6) measured in a separate experiment. This non-competitive method has the advantage that isotope effects on V_{max}, the maximal velocity, can be separated from effects on K_m, the Michaelis constant.

An isotope effect of 1.09 was observed; this is consistent with the methyl transfer being clearly rate-limiting, and with other elementary steps (such as product release) being fast.[24] The size of the kinetic isotope effect indicates that the transition state

[23] (*a*) M. F. Hegazi, D. M. Quinn, and R. L. Schowen, in ref. 10*b*, p. 413; (*b*) M. F. Hegazi, R. T. Borchardt, and R. L. Schowen, *J. Amer. Chem. Soc.*, 1979, **101**, 4359.

[24] For a recent review of primary kinetic isotope effects see J. P. Klinman, *Adv. Enzymol.*, 1978, **46**, 415.

for the methyl transfer is symmetrical, with the methyl group 'in flight' between sulphur and oxygen.

These authors had previously[25] measured the secondary deuterium isotope effect on the methyl transfer to (7) from (2*S*)-[*Me*-$^2H_3$]methionine. Secondary deuterium isotope effects are generally considered to arise[26] from changes in bending force constant as the transition state is approached. The measured value, *i.e.* $V_{max}(CH_3)/V_{max}(CD_3) = 0.83$, is typical for a rather 'tight' transition state[23] (short oxygen–sulphur bond distance). Since the secondary isotope effect is even more strongly inverse than in model S_N2 reactions,[27] steric compression in the transition state has been proposed as a mechanism for the rate enhancement brought about by the enzyme.

An assumption in the above treatment is that the methyl transfer is direct, and does not involve prior methyl transfer from (6) to an enzyme nucleophile, with transfer to the acceptor (7) in a second step. Kinetic evidence against the existence of such a methyl–enzyme intermediate is equivocal.[28] However it has been shown that *S*-adenosylmethionine bearing a chiral methyl group transfers this group to adrenaline (epinephrine) with net inversion of its configuration,[29] and this is most easily explained by a direct methyl transfer. An exactly analogous problem, the vexed question of the existence of a phospho-enzyme intermediate in the acetate-kinase-catalysed reaction, has recently been illuminated by the use of ATP containing a chiral γ-phospho-group.[30]

3 Phosphorylase Kinase

Phosphorylase kinase (EC 2.7.1.38) is a central component of the enzyme system that controls glycogen breakdown in skeletal muscle. It catalyses the activation of glycogen phosphorylase *b* to glycogen phosphorylase *a* by the ATP-linked phosphorylation of a unique hydroxy-group of serine in each of the two subunits of phosphorylase *b*. This was the first recognized example of the control of enzyme activity by covalent modification.[31] As shown in Scheme 4, phosphorylase kinase is

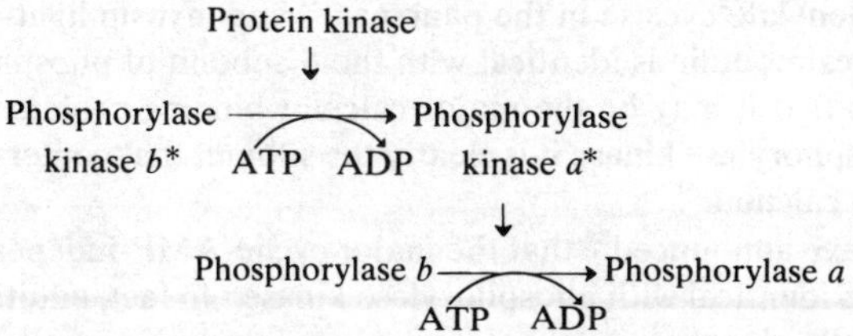

* *a* and *b* represent more-active and less-active forms respectively

Scheme 4

[25] M. F. Hegazi, R. T. Borchardt, and R. L. Schowen, *J. Amer. Chem. Soc.*, 1976, **98**, 3048.
[26] M. H. O'Leary, in ref. 10*b*, p. 298.
[27] I. Mihel, J. O. Knipe, J. K. Coward, and R. L. Schowen, *J. Amer. Chem. Soc.*, 1979, **101**, 4349.
[28] J. K. Coward, E. P. Spisz, and F. Y. H. Wu, *Biochemistry*, 1973, **12**, 2291.
[29] H. G. Floss, in ref. 1, p. 289.
[30] W. A. Blättler and J. R. Knowles, *Biochemistry*, 1979, **18**, 3927.
[31] For recent reviews, see (*a*) P. Cohen, *Biochem. Soc. Trans.*, 1979, **7**, 16; (*b*) G. M. Carlson, P. J. Bechtel, and D. J. Graves, *Adv. Enzymol.*, 1979, **50**, 41.

activated through ATP-linked phosphorylation by a 'general protein kinase', which in turn is activated by 3′,5′-cyclic adenosine monophosphate. The synthesis of cyclic AMP from ATP is triggered by the binding of hormones to specific receptors linked to adenylate cyclase. This 'cascade' system ensures a very rapid amplification of the initial hormonal stimulus. Phosphorylase kinase activity is also dependent on micromolar levels of calcium ions, which are released as a result of nervous stimuli.[31] Since this concentration of calcium ions also initiates muscle contraction, glycogenolysis and muscle contraction are synchronized.

Recent work[31] has greatly increased our understanding of this complex enzyme. The protein was at first found to contain four copies each of three different types of subunit (α, β, and γ, of M_r 145 000, 130 000, and 45 000 respectively; but see below). Rapid phosphorylation of the kinase by cyclic-AMP-dependent protein kinase occurs on both the α and the β subunits.[31a] This was the first observation of control by multi-site phosphorylation, and a similar pattern has since been found for several other enzymes, including glycogen synthase[31a] and mitochondrial pyruvate dehydrogenase.[32] The phosphorylation of β subunits occurs first, and correlates with the increase in catalytic activity. The slower phosphorylation on α subunits has no direct effect on the activity, but greatly increases the rate of dephosphorylation of the β subunits.

Recently, the important discovery was made[31] that rabbit skeletal muscle phosphorylase kinase actually has the subunit composition $(\alpha\beta\gamma\delta)_4$. The fourth subunit, δ, is small (of M_r 16 000) and highly acidic. It appears to be identical with a protein called calmodulin, or 'calcium-dependent regulator protein', which was first identified in 1970 as a heat-stable calcium-binding protein that is required as an activator of one form of brain cyclic nucleotide phosphodiesterase.[33] It was subsequently isolated and shown to be largely identical in its amino-acid sequence with troponin C, the calcium-binding protein of the complex that controls skeletal muscle contraction.[34] Calmodulin is present in high concentrations in a wide range of animal tissues, and has now been clearly implicated in the regulation of several other enzymes, including the calcium-dependent ATPase of erythrocyte membranes,[34] adenylate cyclase in the pancreas,[34] and myosin light-chain kinase.[34] The finding that calmodulin is identical with the δ subunit of phosphorylase kinase strongly suggests that it may be the major calcium-binding protein in mammalian cells. In the phosphorylase kinase it is clearly the subunit that confers the sensitivity of the enzyme to calcium.[31]

Four groups have announced[35] that the major cyclic-AMP-independent glycogen synthase kinase is identical with phosphorylase kinase. In fact, additional calmodulin stimulates both glycogen synthase and phosphorylase kinase. This provides evidence for direct involvement of calcium ions in the control of glycogen synthesis as well as in its breakdown.

[32] P. H. Sugden, A. L. Kerbey, P. J. Randle, C. A. Waller, and K. B. M. Reid, *Biochem. J.*, 1979, **181**, 419.

[33] (*a*) W. Y. Cheung, *Biochem. Biophys. Res. Comm.*, 1970, **38**, 533; (*b*) S. Kakiuchi, R. Yamazaki, and H. Nakajima, *Proc. Japan Acad.*, 1970, **46**, 589.

[34] For a review, see R. J. A. Grand and S. V. Perry, *Biochem. J.*, 1979, **183**, 285.

[35] (*a*) K. X. Walsh, D. M. Millikin, K. K. Schleuder, and E. M. Reimann, *J. Biol. Chem.*, 1979, **254**, 6611; (*b*) A. A. DePaoli-Roach, P. J. Roach, and J. Larner, *ibid.*, p. 4212; (*c*) N. Embi, D. B. Rylatt, and P. Cohen, *European J. Biochem.*, 1979, **100**, 339; (*d*) T. R. Soderling, A. K. Srivastava, M. A. Bass, and B. S. Khatra, *Proc. Nat. Acad. Sci. U.S.A.*, 1979, **76**, 2536.

4 Enzyme-activated Irreversible Inhibitors

An enzyme-activated inhibitor binds at the active site as an inert substrate analogue, and is converted by the target enzyme into a reactive species that inactivates the enzyme.[36] This type of specific inhibition, also referred to as k_{cat} or suicide inhibition,[37] continues to be used with great success against flavin-linked and pyridoxal-phosphate-linked enzymes for which the removal of a proton is thought to constitute an integral part of the normal catalytic pathway. For example, mono- and poly-fluoro-analogues are found to be efficient inhibitors of, *inter alia*, α-dialkyl-amino-acid transaminase,[38] ornithine decarboxylase,[39] and alanine racemase.[40]

A detailed study has been made[41] of the suicide inhibition of the flavin-linked (2*R*)-lactate dehydrogenase from the bacterium *Megasphaera elsdenii*. This enzyme catalyses the oxidation of (2*R*)-lactate to pyruvate, with an electron-transferring flavoprotein serving as the natural electron acceptor. The substrate analogue (2*R*)-2-hydroxybut-3-ynoic acid (10) is both a substrate and an efficient irreversible inhibitor, since the enzyme only undergoes an average of five catalytic turnovers, in air at pH 7.0, before being inactivated. The inactivation is accompanied by the conversion of the enzyme-bound flavin into a stable, pink flavin adduct (11), which can be released from the enzyme by acid treatment and then purified. The novel adduct (11), characterized by its spectroscopic and chemical properties and by comparison with model compounds, is the result of covalent linkage of the acetylenic inhibitor (10) to positions N-5 and C-6 of the flavin nucleus (Scheme 5). This contrasts with (2*S*)-lactate mono-oxygenase (EC 1.13.12.4) from *Mycobacterium*

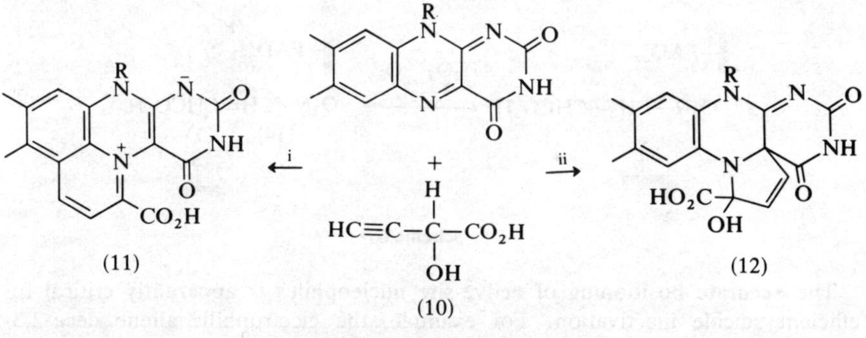

[R = $CH_2(CHOH)_3CH_2OPO_3^-H$]

Reagents: i, (2*R*)-lactate dehydrogenase; ii, (2*S*)-lactate mono-oxygenase

Scheme 5

[36] For recent reviews, see (*a*) C. Walsh, *Ann. Rev. Biochem.*, 1978, **47**, 881; (*b*) 'Enzyme-Activated Irreversible Inhibitors', ed. N. Seeler, M. Jung, and J. Koch-Weser, Elsevier/North-Holland, Amsterdam, 1979.

[37] M. C. Summers and S. C. Williams, *Ann. Reports (B)*, 1977, **74**, 432, and refs. cited therein.

[38] J. W. Keller and M. H. O'Leary, *Biochem. Biophys. Res. Comm.*, 1979, **90**, 1104.

[39] B. W. Metcalf, P. Bey, C. Danzin, M. J. Jung, P. Casara, and J. P. Vevert, *J. Amer. Chem. Soc.*, 1978, **100**, 2551.

[40] E. Wang and C. Walsh, *Biochemistry*, 1978, **17**, 1313.

[41] S. T. Olson, V. Massey, S. Ghisla, and C. D. Whitfield, *Biochemistry*, 1979, **18**, 4724, 4733.

smegmatis, where inactivation by the (2*S*)-enantiomer of (10) leads[42] to formation of the adduct (12), in which the inhibitor is covalently attached to the flavin nucleus through N-5 and C-4a. Massey and his co-workers[41] have interpreted this as evidence for the same face of the bound flavin being turned towards the substrate in both enzymes, and for a similar geometry of the active site in each case. From their results they favour a mechanism for inactivation in which an allenic carbanion, derived from (10) by proton abstraction, attacks reduced flavin. However, alternative mechanisms for both suicide inactivation and the processing of normal substrate, involving radical intermediates,[43] for example, cannot be excluded.

The need for thorough characterization of the inactivator–enzyme adduct is underlined by recent work[44] on the suicide inactivation of succinate dehydrogenase (EC 1.3.99.1), which catalyses the oxidation of succinate to fumarate. The enzyme is inhibited irreversibly by 3-nitropropionic acid (13), and it was originally suggested that the inactivator becomes covalently attached to the flavin nucleus through N-5.[45] However, re-investigation has now shown that the enzyme first oxidizes the 3-nitropropionate (13) to 3-nitroacrylate (14) [3-nitroprop-2-enoic acid], and this is then trapped by 1,4-nucleophilic addition of a thiol group of a cysteine residue at the active site (Scheme 6).

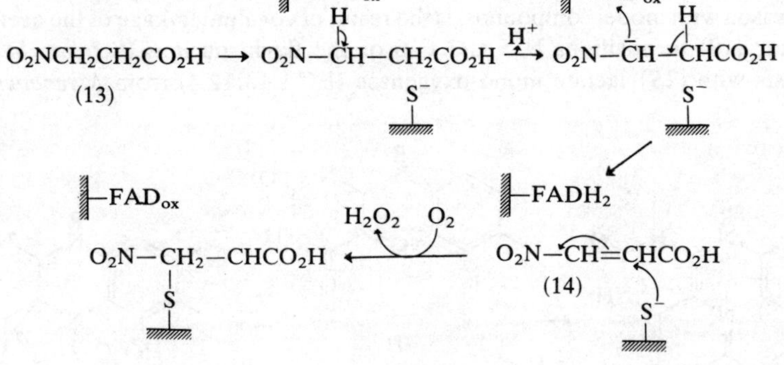

Scheme 6

The accurate positioning of active-site nucleophiles is apparently critical for efficient suicide inactivation. For example, the electrophilic allene deca-2,3-dienoyl-CoA, which inactivates 3-hydroxydecanoyl thioester dehydratase within a single turnover,[37,46] is a well-behaved substrate for dec-3-ynoyl thioester isomerase from pig liver.[46] A similar observation has recently been made[47] in Walsh's laboratory with vinylglycine (15) [2-aminobut-3-enoic acid], a potent suicide inhibitor of many transaminases that require pyridoxal phosphate. The accepted

[42] A. Schonbrunn, R. H. Abeles, C. T. Walsh, S. Ghisla, H. Ogata, and V. Massey, *Biochemistry*, 1976, **15**, 1798.
[43] S. Ghisla, V. Massey, and Y. S. Choong, *J. Biol. Chem.*, 1979, **254**, 10662 and refs. cited therein.
[44] C. J. Coles, D. E. Edmondson, and T. P. Singer, *J. Biol. Chem.*, 1979, **254**, 5161.
[45] T. A. Alston, L. Mela, and H. J. Bright, *Proc. Nat. Acad. Sci. U.S.A.*, 1977, **74**, 3767.
[46] F. M. Miesowicz and K. Bloch, *J. Biol. Chem.*, 1979, **254**, 5868, and refs. cited therein.
[47] M. Johnston, D. Jankowski, P. Marcotte, H. Tanaka, N. Esaki, K. Soda, and C. Walsh, *Biochemistry*, 1979, **18**, 4690.

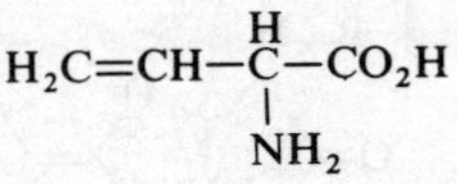

(15)

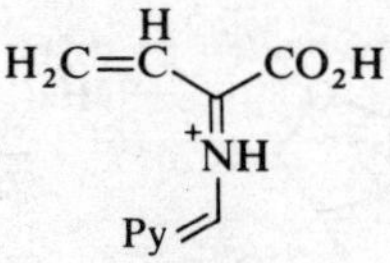

(16)

Py—CHO = Pyridoxal phosphate

mechanism for inactivation involves[37] the generation of the electrophilic adduct (16) at the active site. It now appears that this same adduct is a normal enzyme-bound intermediate in the reactions catalysed by cystathionine γ-synthase (EC 4.2.99.9) and related enzymes that catalyse pyridoxal-phosphate-dependent α,β-elimination and γ-replacement reactions. These enzymes may have evolved a geometry at the active site that prevents enzyme nucleophiles from causing inactivation.[47]

A recent explanation for the consistency of the stereochemical course of aldolase-catalysed reactions (a proton is replaced by another group, with retention of configuration[48]) has been advanced along the same lines. It is suggested that an aldolase with enzyme bases appropriately placed to bring about inversion of configuration would also be able to catalyse 'suicide' reactions, such as elimination.[49]

Most suicide inhibitors depend for their activation upon the initial removal of a proton, and this approach cannot be applied to a large number of enzymes such as hydrolases and group-transferring enzymes. There is great interest, therefore in other ways of generating the reactive species; for example by initial nucleophilic attack by a group in the active site of the enzyme.[50—52] Impetus for this search has been provided by the realization that several natural toxins and antibiotics appear to function as suicide inhibitors. Clavulanic acid (17), for example, a novel β-lactam isolated from *Streptomyces clavuligerus* and a potent inhibitor of β-lactamases from a variety of both Gram-positive and Gram-negative bacteria, both *in vivo* and *in vitro*,[53] appears to function as an enzyme-activated inhibitor.[54]

With the inducible β-lactamase from cell extracts of *Staphylococcus aureus*,[54a] the inhibition is active-site-directed, as judged by substrate-protection experiments. It is suggested[54a] that enzyme-catalysed opening of the β-lactam ring to (18) and subsequent cleavage of the oxazolidine ring might expose a reactive carbonyl function in (19) (Scheme 7). Formation of an imine between this carbonyl and an ε-amino-group on a lysine residue could then lead to decarboxylation and covalent modification of the active site. The proposed covalent linkage cannot be very stable, for excess clavulanate is gradually consumed and the original activity is then regained. Nevertheless, low concentrations of inhibitor are effective *in vivo* in abolishing the resistance of β-lactamase-producing staphylococci to synthetic penicillins.

48 K. R. Hanson and I. A. Rose, *Accounts Chem. Res.*, 1975, **8**, 1.
49 R. Motiu-DeGrood, W. Hunt, J. Wilde, and D. J. Hupe, *J. Amer. Chem. Soc.*, 1979, **101**, 2182.
50 D. V. Santi, T. Wataya, and A. Matsuda, in ref. 36*b*, p. 291.
51 M. Vilkas, in ref. 36*b*, p. 323.
52 C. M. Pickart and W. P. Jencks, *J. Biol. Chem.*, 1979, **254**, 9121.
53 T. T. Howarth, A. G. Brown, and T. J. King, *J.C.S. Chem. Comm.*, 1976, 266.
54 (*a*) C. Reading and P. Hepburn, *Biochem J.*, 1979, **179**, 67; (*b*) J. Fisher, R. L. Charnas, and J. R. Knowles, *Biochemistry*, 1978, **17**, 2180, 2185.

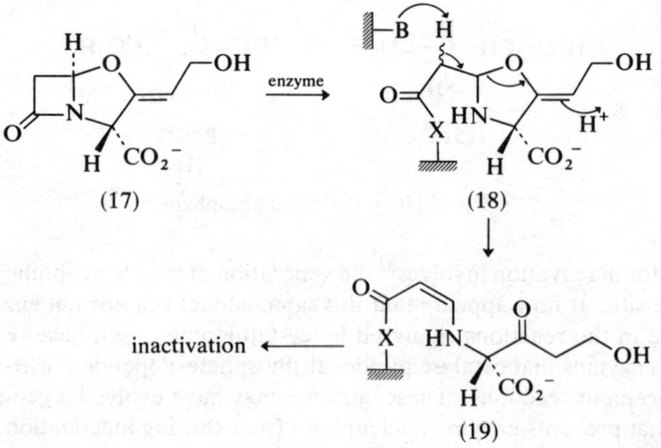

Scheme 7

A group at Harvard University[54b] has studied an *E. coli* plasmid-encoded β-lactamase, whose crystal structure at 0.4 nm resolution has recently been reported.[55] They found a rather complex pattern of inhibition by clavulanic acid; superimposed on a transient, reversible inactivation there was a slow and (in contrast to the *S. aureus* enzyme) irreversible inactivation. Clavulanate is hydrolysed more rapidly by this enzyme, and the average number of turnovers before each inactivation event is about 115. Evidently the formation of (19) is not the only process leading to covalent attachment of the inhibitor to the enzyme, since three types of irreversibly inactivated complex could be distinguished on the basis of their spectroscopic properties and chemical reactivity.[54b]

5 Dihydrofolate Reductase

Dihydrofolate reductase (EC 1.5.1.3) catalyses the NADPH-linked reduction of 7,8-dihydrofolate (20) to tetrahydrofolate (21).[56] Folate (22) is also slowly reduced to (21). Tetrahydrofolate (21) is an essential cofactor in the synthesis of thymidylate, inosinate, and methionine, and specific inhibitors of dihydrofolate reductase, such as methotrexate (23), block the synthesis of these metabolites. Methotrexate (23) has proved to be clinically useful in cancer chemotherapy, and related compounds are effective antimalarial and antibacterial drugs.[56]

Dihydrofolate reductase has been isolated from various bacterial sources.[57] The *E. coli* enzyme is typical in consisting of a single polypeptide chain of M_r 18 000, containing 159 amino-acid residues.[57] It shows extensive amino-acid sequence homology with the enzyme from both *Lactobacillus casei* and *Streptococcus faecium*.[57] Very recently, a Wellcome research group has discovered a new type of dihydrofolate reductase, encoded by an *E. coli* plasmid and containing four

[55] J. R. Knox, J. A. Kelly, P. C. Moews, and M. L. DeLucia, *Microb. Drug Resistance*, 1979, **2**, 313.
[56] (*a*) R. L. Blakley, in 'The Biochemistry of Folic Acid and Related Pteridines', ed. A. Neuberger and E. L. Tatum, Elsevier, New York, 1969, pp. 1—569; (*b*) 'Chemistry and Biology of Pteridines', ed. R. L. Kislink and G. M. Brown, Elsevier/North-Holland, Amsterdam, 1978.
[57] D. Stone, A. W. Phillips, and J. J. Burchall, *European J. Biochem.*, 1977, **72**, 613.

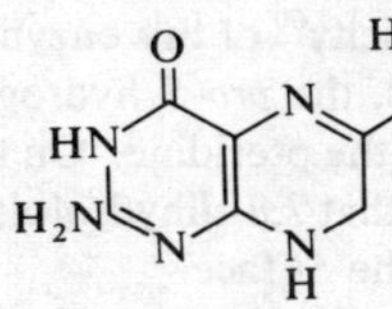

(20)

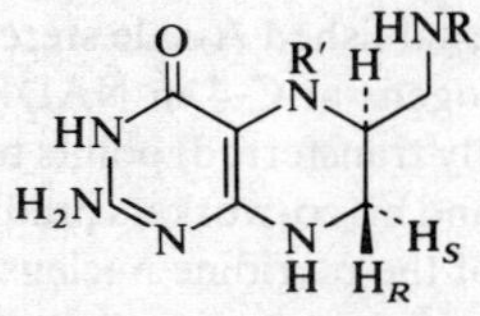

(21a) R′ = H, unlabelled
(21b) R′ = H, H_S = D
(25) R′ = CHO

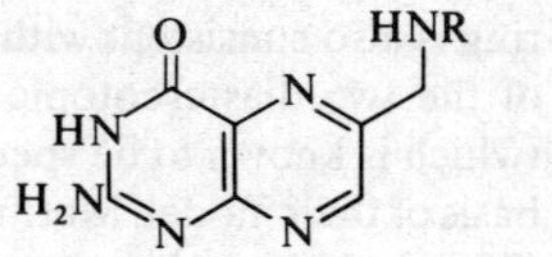

(22)

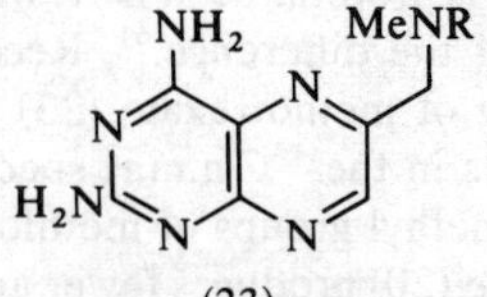

(23)

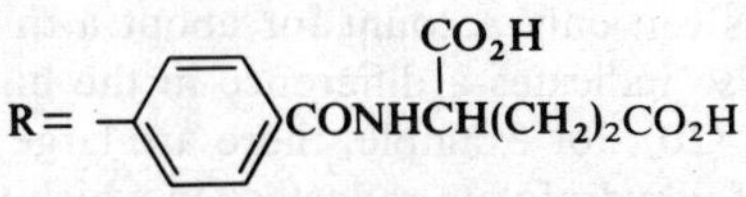

apparently identical subunits, each of M_r 8500. This shows no significant sequence homology with the chromosomal enzyme.[58]

Amino-acid sequences have also been determined for the enzymes from pig and chicken liver and a mouse cancer cell line.[59] The vertebrate enzymes typically are monomeric, of M_r *ca.* 21 000, and resemble each other closely, showing an overall sequence identity of 89%. They also show homology with bacterial enzymes, particularly in the *N*-terminal part of the chain that is thought to form the major active-site cleft.[60]

The crystal structure of *L. casei* dihydrofolate reductase, in a ternary complex with methotrexate (23) and NADPH (24), has been determined by *X*-ray diffraction

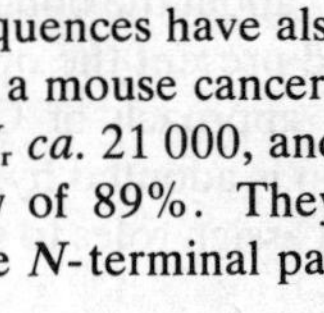

[R = (2′-*O*-phosphoryl) ADP-ribosyl]

(24a) unlabelled
(24b) H_R = D

methods at 0.25 nm resolution, and the same group has also determined the crystal structure of the binary complex between *E. coli* dihydrofolate reductase and methotrexate (23).[60] Since different enzymes were used, comparisons between these structures must be interpreted with caution. In the ternary complex, the folate analogue and the cofactor apparently bind to the same domain of the protein. The C-6 of the pteridine nucleus is directly above the C-4 position of the nicotinamide ring of NADPH (24), and only 0.45 nm away. This positioning would favour a direct transfer of hydride ion from the cofactor to C-6. The positioning of the nicotinamide

[58] D. Stone and S. L. Smith, *J. Biol. Chem.*, 1979, **254**, 10857.
[59] S. L. Smith, P. Patrick, D. Stone, A. W. Phillips, and J. J. Burchall, *J. Biol. Chem.*, 1979, **254**, 11475, and refs. cited therein.
[60] D. A. Matthews, R. A. Alden, S. T. Freer, N. Xuong, and J. Kraut, *J. Biol. Chem.*, 1978, **254**, 4144, and refs, cited therein.

ring is also consistent with the established A-side stereospecificity[60] of this enzyme: of the two diastereotopic hydrogens at C-4 in NADPH (24), the *pro-S* hydrogen (which is known to be specifically transferred) points towards the pteridine. On the basis of their model, Matthews and his co-workers predicted[60] that 7,8-dihydrofolate (20) would be reduced at C-6 of the pteridine nucleus from the *Si* face.

Despite the *X*-ray crystal structure evidence, other studies have indicated differences in the binding of methotrexate (23) and dihydrofolate (20) to the enzyme. Methotrexate (23) binds at least 10 000-fold more tightly to dihydrofolate reductase than does (20), and although bound methotrexate is protonated at N-1, while (20) is not,[61] this can only account for about a third of the difference.[61] Recent n.m.r. work[62] also indicates a difference in the binding of methotrexate (23) and dihydrofolate (20): for example, there are large shifts in the ^{13}C n.m.r. spectrum of a sample of dihydrofolate reductase in which the methyl groups of methionine have been enriched with ^{13}C when (23) binds to it, while (20) produces fewer and smaller shifts.[62]

Matthews and his co-workers[60] were careful to point out that their data do not exclude the possibility that methotrexate (23) and the natural cofactor bind differently within the same site. A rotation of 180° about the bond between C-6 and C-9, and slight movements in other angles, would present the opposite face of the pteridine to NADPH (24) and still allow close approach of C-6 to the C-4 of nicotinamide. Without knowing which orientation is adopted by the natural cofactor, it is impossible to interpret sequence data and assign roles to specific residues in the active site with any confidence.

A decision between these two possibilities can now be made because of the recent *X*-ray determination[63] of the absolute configuration at C-6 of both diastereomers of 5,10-methenyltetrahydrofolate. The biologically active diastereomer was established to have the (6*S*) configuration, and therefore the biologically active diastereomer of 5-formyltetrahydrofolate, from which it was derived, has the configuration shown in (25).[63] This knowledge has been used to establish the stereochemical course of the reductase-catalysed reaction.[64] Dihydrofolate (20) was reduced with NADPH and dihydrofolate reductase, and the product (21) was shown to possess the (6*S*) configuration by direct chemical correlation with (25).[64] This proves that attack by NADPH occurs at the *Re* face of dihydrofolate (20), and not at the *Si* face, as predicted by the *X*-ray results. Evidently methotrexate (23) does bind 'the wrong way up' compared to dihydrofolate (20), and this weakens the proposal that it functions as an analogue of the transition state of the enzyme.[65]

The stereochemical course of the much slower reduction of folate (22) was also established:[64] $NADP^+$ was reduced, using the B-side-specific enzyme glucose 6-phosphate dehydrogenase, and the resulting NADPH (24b) was used to reduce folate (22) in the presence of dihydrofolate reductase. A ^{1}H n.m.r. analysis of the

[61] K. Hood and G. C. K. Roberts, *Biochem. J.*, 1978, **171**, 357.
[62] R. L. Blakley, L. Cocco, R. E. London, T. E. Walker, and N. A. Matwiyoff, *Biochemistry*, 1978, **17**, 2284, and refs. cited therein.
[63] J. C. Fontecilla-Camps, C. E. Bugg, C. Temple, Jr., J. D. Rose, J. A. Montgomery, and R. L. Kislink, *J. Amer. Chem. Soc.*, 1979, **101**, 6114.
[64] P. A. Charlton, D. W. Young, B. Birdsall, J. Feeney, and G. C. K. Roberts, *J.C.S. Chem. Comm.*, 1979, 922.
[65] J. W. Williams, J. F. Morrison, and R. G. Duggleby, *Biochemistry*, 1979, **18**, 2567.

product established that the tetrahydrofolate has the configuration shown in (21b), with deuterium specifically in the (7*S*) position. Thus hydride transfer is from the 4-*pro-R* position of NADPH (24) to the *Si* face at C-7 of folic acid (22), and reduction at C-6 and C-7 occurs on the same face of the pteridine nucleus.[64]

6 Multi-enzyme Proteins

Enzymes in many metabolic pathways are physically associated in multienzyme proteins [*cf. Ann. Reports* (*B*), 1977, **74**, 443]. These proteins are of two basic types: multi-enzyme complexes (also termed multi-enzyme aggregates), in which some or all of the component enzymes are only held together by non-covalent interactions; and multi-enzyme polypeptides,[66a] (or multi-enzyme conjugates[66b]), in which all of the associated catalytic activities are carried on the same polypeptide chain. Multi-functional proteins are thought to have arisen as a result of fusion of the structural genes coding for the individual enzymes, and this arrangement provides a simple way of co-ordinating the control of their synthesis. The significance of multi-enzyme complexes in general is harder to assess, although possible advantages have frequently been discussed.[67]

The third, second, and tenth steps in the biosynthetic pathway to histidine in yeast are catalysed by a trifunctional protein, which was only isolable when proteinase inhibitors were present during the purification of the enzyme.[68] Similarly, tryptophan synthase from yeast has now been shown to be a multifunctional protein, rather than a non-covalently associated aggregate, by using a rapid purification procedure that is designed to minimize proteolytic damage.[69] Such difficulties in the isolation of intact complexes because of the action of endogenous proteinases can, however, be turned to good account. Brief exposure of a multi-enzyme complex to a specific proteolytic enzyme such as trypsin often leads to chain cleavage at a very limited number of particularly reactive sites. The resulting fragments often retain catalytic activity, and their characterization can provide very useful information about the organization of the native complex. This approach has been used, for example, with glutamine synthetase,[70] aspartokinase I–homoserine dehydrogenase I,[71] fatty acid synthase,[72] and tryptophan synthase.[73]

Brief treatment of the biotin-dependent transcarboxylase from *Propionibacterium shermanii* with trypsin has the interesting consequence that the part of the biotin carrier protein that bears the prosthetic group is specifically excised from the complex.[74] A very similar finding has been made for the pyruvate dehydrogenase

[66] (*a*) P. Karlson and H. B. F. Dixon, *Trends Biochem. Sci.*, 1979, **4**, N275; (*b*) F. H. Gaertner, *ibid.*, 1978, **3**, 63.
[67] (*a*) K. Kirschner and H. Bisswanger, *Ann. Rev. Biochem.*, 1976, **45**, 143; (*b*) G. R. Welch, *Prog. Biophys. Mol. Biol.*, 1977, **32**, 103.
[68] J. K. Keesey, Jr., R. Bigelis, and G. R. Fink, *J. Biol. Chem.*, 1979, **254**, 7427.
[69] M. Dettwiler and K. Kirschner, *European J. Biochem.*, 1979, **102**, 159.
[70] A. Dautry-Varsat, G. N. Cohen, and E. R. Stadtman, *J. Biol. Chem.*, 1979, **254**, 3124.
[71] P. A. Briley, L. Sibilli, M.-A. Chalvignac, P. Cossart, G. Le Bras, A. De Wolf, and G. N. Cohen, *J. Biol. Chem.*, 1978, **253**, 8867.
[72] J. K. Stoops, P. Ross, M. J. Arslanian, K. C. .'une, S. J. Wakil, and R. M. Oliver, *J. Biol. Chem.*, 1979, **254**, 7418, and refs. cited therein.
[73] (*a*) W. Higgins, T. Fairwell, and E. W. Miles, *Biochemistry*, 1979, **18**, 4827; (*b*) V. Rocha, E. F. Brennan, and S Plumb, *Arch. Biochem. Biophys.*, 1979, **193**, 34.
[74] H. G. Wood, *C.R.C. Crit. Rev. Biochem.*, 1979, **7**, 143, and refs cited therein.

complex from *E. coli*,[75] where limited proteolysis[76] appears to affect only those subunits (E2) that bear dihydrolipoyl residues. This is surprising, because the E2 subunits are regarded as forming the 'core' of this complex, and might have been expected to be screened from proteolytic attack.

Some elegant work has been reported[77] on mammalian multi-enzyme complexes in the pathway responsible for the production of pyrimidine nucleotides *de novo* for DNA biosynthesis. What follows mainly concerns the first three steps in the pathway to uridine 5′-monophosphate, in which carbamoyl phosphate is produced and converted (Scheme 8), by way of *N*-carbamoyl aspartate (26), into dihydro-orotate (27).

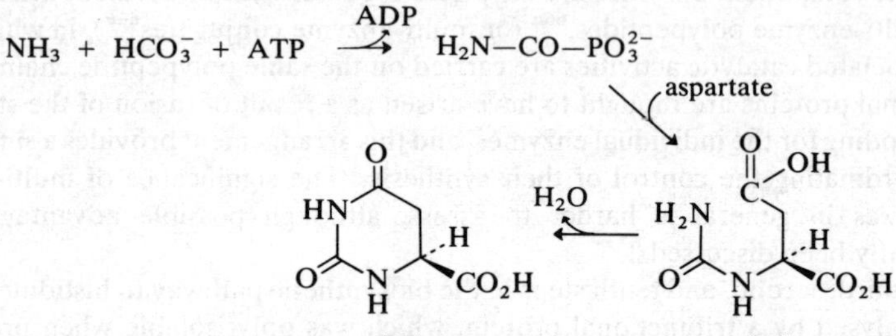

Scheme 8

The three enzymes involved, *i.e.* carbamoyl phosphate synthetase (EC 6.3.5.5), aspartate carbamoyltransferase (EC 2.1.3.2), and dihydro-orotase (EC 3.5.2.3), have been at least partially purified from a variety of animal tissues, and in every case all three activities are associated with a single multi-enzyme complex.[77a] The amounts of enzyme obtained were too small to permit extensive characterization but the enzyme from mouse ascites hepatoma cells[78] certainly appeared to be a multi-functional protein.

Large amounts of this complex have now become available through the use of a simple and effective technique: mammalian cells in tissue culture were exposed to gradually increasing concentrations of *N*-(phosphonoacetyl)-L-aspartate (28), which is a substrate analogue and also a potent inhibitor[79] of the second enzyme in

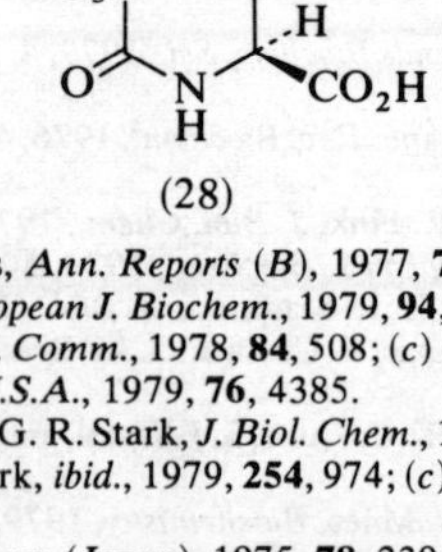

(28)

[75] M. C. Summers and D. C. Williams, *Ann. Reports (B)*, 1977, **74**, 445.

[76] (*a*) G. Hale and R. N. Perham, *European J. Biochem.*, 1979, **94**, 119; (*b*) C. Gebhardt, D. Mecke, and H. Bisswanger, *Biochem. Biophys. Res. Comm.*, 1978, **84**, 508; (*c*) D. M. Bleile, P. Munk, R. M. Oliver, and L. J. Reed, *Proc. Nat. Acad. Sci. U.S.A.*, 1979, **76**, 4385.

[77] (*a*) P. F. Coleman, D. P. Suttle, and G. R. Stark, *J. Biol. Chem.*, 1977, **252**, 6379; (*b*) R. A. Padgett, G. M. Wahl, P. F. Coleman, and G. R. Stark, *ibid.*, 1979, **254**, 974; (*c*) D. P. Suttle and G. R. Stark, *ibid.*, 1979, **254**, 4602.

[78] M. Mori and M. Tatibana, *J. Biochem. (Japan)*, 1975, **78**, 239.

[79] E. A. Swyryd, S. S. Seaver, and G. R. Stark, *J. Biol. Chem.*, 1974, **249**, 6945.

the pathway, *i.e.* aspartate carbamoyltransferase. Mutant cells were obtained which resist concentrations of this inhibitor 5000 times higher than required to cause 50% inhibition of the wild-type cell line. In these mutant cells all three activities of the complex were increased by nearly a hundred-fold, and the complex (constituting nearly 10% of the total cellular protein) could be readily purified and characterized. The pure protein contains only one type of polypeptide, of M_r 200 000, which carries all three enzyme activities of the complex. The isolated multifunctional protein is a mixture of trimers and hexamers. Over-accumulation of this multifunctional protein in the resistant cells has been shown[77b] to be due to an increased rate of enzyme synthesis, and a large RNA, purified from mutant cells, co-sediments in sucrose gradients with a fraction that has the capacity to direct the synthesis of the multifunctional protein *in vitro*.[77b] This approach to obtaining relatively large amounts of mammalian complexes has been extended to a second, bifunctional complex in the same biosynthetic pathway.[77c] By exposing cells to increasing amounts of specific inhibitors of orotidine 5′-phosphate decarboxylase (EC 4.1.1.23), which is the sixth and last enzyme in the pathway, resistant mutants have been obtained that over-accumulate both this enzyme and its neighbour, orotate phosphoribosyltransferase (EC 2.4.2.10), by nearly seventy-fold. In principle, this technique is applicable to any enzyme that is essential for cell growth for which a specific inhibitor is available.

Author Index